Mathematical Physics Studies

Series Editors
Giuseppe Dito, Dijon, France
Edward Frenkel, Berkeley, CA, USA
Sergei Gukov, Pasadena, CA, USA
Yasuyuki Kawahigashi, Tokyo, Japan
Maxim Kontsevich, Bures-sur-Yvette, France
Nicolaas P. Landsman, Nijmegen, The Netherlands
Bruno Nachtergaele, Davis, CA, USA

The series publishes original research monographs on contemporary mathematical physics. The focus is on important recent developments at the interface of Mathematics, and Mathematical and Theoretical Physics. These will include, but are not restricted to: application of algebraic geometry, D-modules and symplectic geometry, category theory, number theory, low-dimensional topology, mirror symmetry, string theory, quantum field theory, noncommutative geometry, operator algebras, functional analysis, spectral theory, and probability theory.

More information about this series at http://www.springer.com/series/6316

Asao Arai

Inequivalent Representations of Canonical Commutation and Anti-Commutation Relations

Representation-theoretical Viewpoint for Quantum Phenomena

Asao Arai
Department of Mathematics
Hokkaido University
Sapporo
Hokkaido, Japan

ISSN 0921-3767 ISSN 2352-3905 (electronic)
Mathematical Physics Studies
ISBN 978-981-15-2182-9 ISBN 978-981-15-2180-5 (eBook)
https://doi.org/10.1007/978-981-15-2180-5

© Springer Nature Singapore Pte Ltd. 2020
This work is subject to copyright. All rights are reserved by the Publisher, whether the whole or part of the material is concerned, specifically the rights of translation, reprinting, reuse of illustrations, recitation, broadcasting, reproduction on microfilms or in any other physical way, and transmission or information storage and retrieval, electronic adaptation, computer software, or by similar or dissimilar methodology now known or hereafter developed.
The use of general descriptive names, registered names, trademarks, service marks, etc. in this publication does not imply, even in the absence of a specific statement, that such names are exempt from the relevant protective laws and regulations and therefore free for general use.
The publisher, the authors and the editors are safe to assume that the advice and information in this book are believed to be true and accurate at the date of publication. Neither the publisher nor the authors or the editors give a warranty, expressed or implied, with respect to the material contained herein or for any errors or omissions that may have been made. The publisher remains neutral with regard to jurisdictional claims in published maps and institutional affiliations.

This Springer imprint is published by the registered company Springer Nature Singapore Pte Ltd.
The registered company address is: 152 Beach Road, #21-01/04 Gateway East, Singapore 189721, Singapore

Preface

As is well known, canonical commutation relations (CCR) and canonical anti-commutation relations (CAR) are fundamental algebraic relations in quantum physics. Models in quantum mechanics and quantum field theory are constructed based on Hilbert space representations of CCR and/or CAR. There are basically two categories for representations of CCR and CAR, respectively, i.e., reducible and irreducible, and each of them is divided into two classes: equivalent and inequivalent. Quantum theories based on equivalent representations of CCR and/or CAR are physically equivalent, being different only in the framework of the physical picture.[1] On the other hand, quantum theories based on inequivalent irreducible representations are essentially different from each other, describing non-comparable physical situations.

Roles of equivalent representations and inequivalent irreducible representations of CCR and/or CAR are different. Although equivalent representations are physically equivalent to each other as mentioned above, they may be mathematically important. For example, there may be mathematical problems which are not so easy to solve in a representation, but relatively are very easy to solve in other representations equivalent to the former.[2] Concerning inequivalent irreducible

[1]For example, the Schrödinger representation of the CCR with d degrees of freedom (see Sect. 2.6.1 for the definition) is within the framework of the physical picture in which the position of a quantum particle is measured in the d-dimensional position space, while the momentum representation of the CCR with the same degree, which is equivalent to the Schrödinger one, is within the framework of the physical picture in which the momentum of a quantum particle is measured in the d-dimensional momentum space. In this case, the d-dimensional Fourier transform is the unitary transformation which maps the Schrödinger representation to the momentum representation, giving the equivalence of them (see Sect. 2.6.1 for more details).

[2]A typical example is a one-dimensional quantum harmonic oscillator. The spectrum of the Hamiltonian is easily found in the Born–Heisenberg–Jordan representation of the CCR with one degree of freedom (see Sect. 2.6.2) rather than in the Schrödinger one with the same degree, which is equivalent to the former. In quantum field theory, the Q-space representation (functional Schrödinger representation), which is equivalent to the Fock representation of the CCR over a Hilbert space, is very useful (see Sect. 8.13).

representations of CCR and/or CAR, the following philosophical point of view is suggested by studies on models in quantum mechanics and quantum field theory:

> *The Universe uses inequivalent irreducible representations of CCR and/or CAR to create or produce "characteristic" quantum phenomena in which macroscopic objects may be involved.*

Examples of such phenomena include: Aharonov–Bohm effect, Casimir effect, Bose–Einstein condensation, superfluidity, superconductivity, boson masses, and fermion masses (see the text of this book for further details).

This book is concerned with mathematical analysis on representations of CCR and CAR, including both finite and infinite degrees of freedom. With a philosophy as stated above, the emphasis of the description is put on inequivalent irreducible representations of CCR and CAR and reading their roles in the context of quantum physics.

It may be asked why it still is necessary to consider or study representations of CCR with finite degrees of freedom; is not the von Neumann uniqueness theorem enough? The answer is no, because the von Neumann uniqueness theorem applies only to Weyl representations of CCR with finite degrees of freedom on a separable Hilbert space (unfortunately this point seems not to be understood precisely in general among non-experts of representation theory of CCR). For non-Weyl representations of CCR, even if their degrees of freedom are finite, the von Neumann uniqueness theorem is not valid anymore. But, interestingly enough, the Universe makes use of this invalidity to create or produce "characteristic" quantum phenomena, giving physically important roles to non-Weyl representations of CCR with finite degrees of freedom as well.

In the case of CCR with infinite degrees of freedom, even in the category of irreducible Weyl representations, inequivalent ones appear and some of them in quantum field theory correspond to physically interesting quantum phenomena as mentioned above.

There may be inequivalent representations of CCR and CAR which are related to spontaneous symmetry breaking.[3] In this book, however, we do not discuss this aspect. To do that, one needs approaches using $*$-algebras (C^*-algebras, von Neumann algebras, or O^*-algebras) and their representations. Representations of CCR and CAR treated in this book are basically the ones that are operator realizations of algebraic relations on Hilbert spaces.

This book is organized as follows. Chapter 1 is a summary of mathematical theories which may be used in the following chapters. It is mainly concerned with operator theories on Hilbert spaces. Those who are familiar with the subjects may skip this chapter.

In Chap. 2, we describe a general theory of representations of CCR with finite degrees of freedom. For each degree of freedom, three classes of representations of CCR exist basically: ordinary ones in each of which all the representatives are

[3] This is seen, e.g., in a Bose–Einstein condensation of an infinite system of free Bose gas at finite temperature.

not bounded, weak Weyl ones in each of which half of the representatives are unitary, and Weyl ones in each of which every representative is unitary. In addition to these classes, we consider also weak forms of CCR. We review Schrödinger representations and Born–Heisenberg–Jordan (BHJ) representations of CCR and give a direct proof of the equivalence of them with the same degrees of freedom, which corresponds to the equivalence of Schrödinger's "wave mechanics" as a quantum theory[4] and "matrix mechanics" by BHJ. Concerning the von Neumann uniqueness theorem mentioned above, we only state it without proof. A detailed analysis on weak Weyl representations is given. It is noteworthy that, in weak Weyl representations, each self-adjoint representative is absolutely continuous.

Chapter 3 is devoted to the analysis of a representation of the CCR with two degrees of freedom appearing in quantum mechanics of a two-dimensional system of a charged particle moving around a magnetic field concentrated on some points in the plane. It is well known that, in such a physical situation, a characteristic phenomenon, called the Aharonov–Bohm (AB) effect (a shift of the interference pattern of electron beams due to the existence of a solenoidal magnetic field), may occur. We show that the representation is irreducible and formulate a necessary and sufficient condition for the representation to be equivalent to the Schrödinger representation with the same degree. We show that the case where the representation is inequivalent to the Schrödinger one corresponds to the occurrence of the AB effect in the context under consideration. This is an interesting correspondence to note.

In Chap. 4, we present a general mathematical theory of time operators. In the first stage of cognition, a time operator T is defined to be a canonical conjugate of a Hamiltonian H, i.e., it is a representative (symmetric, but not necessarily self-adjoint) in a representation of the CCR with one degree of freedom ($[T, H] = i$ on a suitable subspace of the representation space under consideration). Concerning this definition, the following point may be remarked. There has been a misunderstanding that there exist no time operators as observables (self-adjoint operators). This may be due to Pauli's statement in the absence of time operators (see Sect. 4.2 for more details) or mathematically non-rigorous arguments which do not care about domains of unbounded operators. In fact, Pauli's statement is false. This is shown by an explicit construction of a self-adjoint time operator for a class of Hamiltonians (see Sect. 4.4) (but, every Hamiltonian does not have a self-adjoint time operator). We introduce five classes of time operators. A rigorous form of time–energy uncertainty relation is formulated in terms of time operator. If H is semi-bounded (bounded from below or above), then the representation (T, H) of the CCR with one degree of freedom is obviously inequivalent to the Schrödinger representation with the same degree. Hence, here we have classes of representations inequivalent to Schrödinger representations. We infer that each class of time operators plays different roles in physics. For example, as we shall show, a strong time operator controls the decay-in-time rate of transition probabilities of state vectors (see Sect. 4.5.14). There is

[4]In a one-body problem in three-dimensional space, Schrödinger's wave mechanics can be viewed also as a classical field theory.

a time operator which is related to time delay in a two-body scattering process in non-relativistic quantum mechanics (see Remark 4.15).

Chapter 5 is a brief description of part of representation theory of CAR with finite degrees of freedom. Representations of CAR with finite degrees of freedom are physically related to internal degrees of freedom like spin.

In Chap. 6 (resp. Chap. 7), we describe elements of the theory of boson (resp. fermion) Fock space, which is used to construct Bose (resp. Fermi) fields in quantum field theory. Only requisites for the following chapters are given.

In Chap. 8 (resp. Chap. 9), we analyze representations of CCR (resp. CAR) with infinite degrees of freedom in an abstract framework.

The last chapter is devoted to descriptions of various inequivalent representations of CCR or CAR with infinite degrees of freedom and physical correspondences of them, including scaling of quantum fields, infinite volume theories, Bose–Einstein condensation, temperatures, infrared and ultraviolet renormalizations, boson masses, and masses of Dirac particles.

For the reader's convenience, four appendices are added in this book. They are concerned with basic mathematical subjects.

This book is written as an introductory text to theories of representations of CCR and CAR in relation to quantum physics with special attention given to their inequivalent representations, being directed to general audiences including graduate students and researchers in mathematics and physics who are interested in mathematical structures of quantum mechanics and quantum field theory. For this reason, in the main text (Chaps. 2–10) of the book, the author tried to make mathematical descriptions as detailed as possible. No completeness is intended for the bibliography in this book, and references in the bibliography are restricted more or less to those which are directly related to the subjects discussed in the text. Articles and books written in Japanese also are included in the bibliography for convenience of readers who can read Japanese.

It would be the author's great pleasure if this book could demonstrate the above-mentioned philosophical point of view to a certain extent and suggest the unfathomable richness and depth of the world of representations of CCR and CAR as *fundamental concepts* underlying quantum phenomena.

Sapporo, Japan
October 2019

Asao Arai

Acknowledgments

Theory of representations of CCR has been one of my research subjects since I started my research career in 1978. During past years, I have benefited from conversations or discussions with many researchers on the subject. First of all, I would like to express my hearty thanks to Professor Hiroshi Ezawa (Emeritus Professor, Gakushuin University) who inspired me with the importance of the representation-theoretical point of view for quantum mechanics and quantum field theory, in particular, the importance of inequivalent representations of CCR. I have learned a lot from his excellent book *Structures of Quantum Mechanics* (Iwanami-Shoten, 1978, in Japanese) and from his seminars on mathematical physics at Gakushuin University. Professor Ezawa gave me also the opportunity to deliver a series of lectures on representations of CCR and physics in summer schools on mathematical physics of which he was one of the organizers. I would like to thank also Professor Shige-Toshi Kuroda (Emeritus Professor, Gakushuin University), who was my teacher when I was a graduate student at the University of Tokyo, and Professor Huzihiro Araki (Emeritus Professor, Kyoto University) for their kind interest in my work on representations of CCR. My thanks go also to the following researchers who gave me the opportunity to deliver a series of lectures or a seminar on representations of CCR at their institutions (honorific titles omitted): Daisuke Fujiwara, Harald Grosse, Masao Hirokawa, Fumio Hiroshima, Takashi Ichinose, Atsushi Inoue, Hiroshi Isozaki, Tadashi Kawanago, Hideki Kurose, Itaru Mitoma (deceased), Nobuaki Obata, Masanori Ohya (deceased), Izumi Ojima, Tomohiro Sasamoto, Herbert Spohn, Kenji Yajima.

Contents

List of Mathematical Symbols

$A \upharpoonright \mathscr{D}$	Restriction of an operator A to a subspace $\mathscr{D}$
$\mathrm{AC}[a, b]$	Absolutely continuous functions on closed interval $[a, b]$
A^*	Adjoint of a densely defined linear operator A on a Hilbert space
$\{\ ,\ \}$	Anti-commutator: $\{A, B\} := AB + BA$
$\mathfrak{B}(\mathscr{H}, \mathscr{K})$	Everywhere defined bounded linear operators from a Hilbert space $\mathscr{H}$ to a Hilbert space $\mathscr{K}$
$\mathfrak{B}(\mathscr{H})$	$\mathfrak{B}(\mathscr{H}, \mathscr{H})$
$\mathbb{C}$	Complex numbers
CAR	Canonical anti-commutation relations
CCR	Canonical commutation relations
$[\ ,\ ]$	Commutator: $[A, B] := AB - BA$
$\mathsf{CCR}(N)$	Representations of the CCR with N degrees of freedom
$\mathsf{CCR}_{\mathrm{sa}}(N)$	Self-adjoint representations of the CCR with N degrees of freedom
$\mathsf{CCR}_{\mathrm{W}}(N)$	Weyl representations of the CCR with N degrees of freedom
$\mathsf{CCR}_{\mathrm{WW}}(N)$	Weak Weyl representations of the CCR with N degrees of freedom
CONS	Complete orthonormal system
${}_nC_k$	Binomial coefficient with respect to (n, k): ${}_nC_k := \frac{n!}{(n-k)!k!}$
$C(\mathbb{R}^d)$	Continuous functions on $\mathbb{R}^d$
$C(D)$	Continuous functions on an open set D
$C^n(\mathbb{R}^d)$	n times continuously differentiable functions on $\mathbb{R}^d$
$C^n(D)$	n times continuously differentiable functions on an open set D
$C_0^n(D)$	Elements of $C^n(D)$ with compact support in D
$C_0^\infty(D)$	Elements of $C^\infty(D)$ with compact support in D
$C^\infty(A)$	C^∞-domain of a linear operator A
D^c	Complement of a subset D of a set X: $D^c := X \setminus D$
$D(A)$	Domain of a linear operator A
D_j	Generalized partial differential operator in variable x_j
Δ	Generalized Laplacian: $\Delta := \sum_{j=1}^d D_j^2$
$\mathscr{D}^\perp$	Orthogonal complement of a subset $\mathscr{D}$ of an inner product space
$\dim \mathscr{V}$	Dimension of a vector space $\mathscr{V}$

E_A	Spectral measure of a self-adjoint operator A
$\mathscr{F}_{\mathrm{b}}(\mathscr{H})$	Boson Fock space over a Hilbert space $\mathscr{H}$
$\mathscr{F}_d$	d-dimensional Fourier transform on $L^2(\mathbb{R}^d)$
$\mathscr{F}_{\mathrm{f}}(\mathscr{K})$	Fermion Fock space over a Hilbert space $\mathscr{K}$
F_L	Discrete Fourier transform
$F_{L,d}$	d-dimensional discrete Fourier transform
$\mathscr{H}_{\mathbb{C}}$	Complexification of a real Hilbert space $\mathscr{H}$
$\mathrm{Im}\, z$	The imaginary part of $z \in \mathbb{C}$
i	The imaginary unit
I	Identity
$I_{\mathscr{X}}$	Identity on a set $\mathscr{X}$
$\mathfrak{I}_2(\mathscr{H})$	Hilbert–Schmidt operators on a Hilbert space $\mathscr{H}$
$\langle\ ,\ \rangle_{\mathscr{H}}$	Inner product of an inner product space $\mathscr{H}$
$\langle\ ,\ \rangle$	Inner product
$\boldsymbol{k} \cdot \boldsymbol{x}$	Euclidean inner product of vectors $\boldsymbol{k}$ and $\boldsymbol{x}$
$\mathbb{K}$	$\mathbb{R}$ (real numbers) or $\mathbb{C}$ (complex numbers)
$\mathfrak{L}(\mathscr{H}, \mathscr{K})$	Linear operators (not necessarily bounded) from a Hilbert space $\mathscr{H}$ to a Hilbert space $\mathscr{K}$
$\mathfrak{L}(\mathscr{H})$	$\mathfrak{L}(\mathscr{H}, \mathscr{H})$
$L^2(M)$	Hilbert space of square (Lebesgue) integrable functions on $M \subset \mathbb{R}^d$
$L^1(\mathbb{R}^d)$	Lebesgue integrable functions on $\mathbb{R}^d$
$L^2(\mathbb{R}^d)$	Hilbert space of square (Lebesgue) integrable functions on $\mathbb{R}^d$
$L^\infty(\mathbb{R}^d)$	Essentially bounded functions on $\mathbb{R}^d$
$L^1_{\mathrm{loc}}(\Omega)$	Locally integrable functions on an open set $\Omega \subset \mathbb{R}^d$
M_F	Multiplication operator by a function F
$\lvert M \rvert$	Lebesgue measure of a set $M \subset \mathbb{R}^d$
$\mu_{\mathrm{L}}^{(d)}$	d-dimensional Lebesgue measure
$\mathbb{N}$	Natural numbers (positive integers)
$\\|\ \\|_{\mathscr{H}}$	Norm of an inner product space $\mathscr{H}$
$\\|\ \\|$	Norm
$\\|A\\|$	Operator norm of a bounded linear operator A
$\\|f\\|_\infty$	Supremum norm of an essentially bounded function f
ONS	Orthonormal system
$\oplus_{n=0}^{\infty} \mathscr{H}_n$	Infinite direct sum Hilbert space of Hilbert spaces $\{\mathscr{H}_n\}_{n=0}^{\infty}$
$\hat{\otimes}_{j=1}^{n} \mathscr{D}_j$	n-fold algebraic tensor product of vector spaces $\mathscr{D}_1, \ldots, \mathscr{D}_n$
$\hat{\otimes}_{\mathrm{s}}^{n} \mathscr{D}$	n-fold symmetric algebraic tensor product of a vector space $\mathscr{D}$
$\hat{\otimes}_{\mathrm{as}}^{n} \mathscr{D}$	n-fold anti-symmetric algebraic tensor product of a vector space $\mathscr{D}$
$\otimes_{j=1}^{n} \mathscr{H}_j$	n-fold tensor product of Hilbert spaces $\mathscr{H}_1, \ldots, \mathscr{H}_n$
$\otimes_{\mathrm{s}}^{n} \mathscr{H}$	n-fold symmetric tensor product of Hilbert space $\mathscr{H}$
$\otimes_{\mathrm{as}}^{n} \mathscr{H}$	n-fold anti-symmetric tensor product of Hilbert space $\mathscr{H}$
$\pi_{\mathrm{S}}^{(d)}$	Schrödinger representation of the CCR with degree d
Π_+	The open upper half plane in $\mathbb{C}$
Π_-	The open lower half plane in $\mathbb{C}$
$\mathrm{Re}\, z$	The real part of $z \in \mathbb{C}$

$\mathbb{R}$	Real numbers
$\mathbb{R}_+$	Positive real numbers
$\hat{\mathbb{R}}$	One-dimensional momentum (wave number) space
$\mathbb{R}^d$	d-dimensional vector space of d-tuples of real numbers: $\{(x_1, \ldots, x_d) \| x_i \in \mathbb{R}, i = 1, \ldots, d\}$
$\hat{\mathbb{R}}^d$	d-dimensional momentum (wave number vector) space
$\mathrm{Ran}\,(A)$	Range of a linear operator A
$\rho(A)$	Resolvent set of a linear operator A
$\mathscr{S}(\mathbb{R}^d)$	Rapidly decreasing C^∞-functions on $\mathbb{R}^d$
$\mathscr{S}_{\mathbb{R}}(\mathbb{R}^d)$	Real-valued rapidly decreasing C^∞-functions on $\mathbb{R}^d$
$\mathscr{S}'(\mathbb{R}^d)$	Tempered distributions on $\mathbb{R}^d$
$\mathscr{S}'_{\mathbb{R}}(\mathbb{R}^d)$	Real tempered distributions on $\mathbb{R}^d$
$\sigma(A)$	Spectrum of a linear operator A
$\sigma_{\mathrm{ac}}(A)$	Absolutely continuous spectrum of a self-adjoint operator A
$\sigma_{\mathrm{c}}(A)$	Continuous spectrum of a linear operator A
$\sigma_{\mathrm{d}}(A)$	Discrete spectrum of a linear operator A
$\sigma_{\mathrm{p}}(A)$	Point spectrum of a linear operator A
$\sigma_{\mathrm{sc}}(A)$	Singular continuous spectrum of a self-adjoint operator A
$\mathrm{span}\,\mathscr{D}$	Subspace generated by a subset $\mathscr{D}$ of a vector space
$\mathscr{V}_{\mathbb{C}}$	Complexification of a real vector space $\mathscr{V}$
$\mathbb{Z}$	Integers
$\mathbb{Z}_+$	Non-negative integers
z^*	The complex conjugate of $z \in \mathbb{C}$

Chapter 1
Mathematical Preliminaries

Abstract Operator theories on Hilbert spaces including those on tensor product Hilbert spaces are reviewed. We assume that the reader is familiar with elements of theories of Hilbert spaces and Banach spaces. (We will omit proofs for well-known facts.)

1.1 Linear Operators on Hilbert Spaces

Throughout this book, we use the term "Hilbert space" to mean "complex Hilbert space" unless otherwise stated. We denote the inner product and the norm of a Hilbert space $\mathscr{H}$ by $\langle\ ,\ \rangle_{\mathscr{H}}$ and $\|\cdot\|_{\mathscr{H}}$ respectively with the convention that the mapping $\mathscr{H} \times \mathscr{H} \ni (\Psi, \Phi) \mapsto \langle \Psi, \Phi \rangle_{\mathscr{H}} \in \mathbb{C}$ is linear in Φ and anti-linear in Ψ. We sometimes write $\langle\ ,\ \rangle$ and $\|\cdot\|$ instead of $\langle\ ,\ \rangle_{\mathscr{H}}$ and $\|\cdot\|_{\mathscr{H}}$ respectively, omitting the subscript $\mathscr{H}$, if there is no danger of confusion.

In this section, we collect basic definitions and facts in the theory of linear operators on Hilbert spaces.

1.1.1 Definitions and Some Basic Facts

Let $\mathscr{H}$ and $\mathscr{K}$ be Hilbert spaces and $\mathscr{D}$ be a subspace of $\mathscr{H}$. A mapping $A: \mathscr{D} \to \mathscr{K}; \mathscr{D} \ni \Psi \mapsto A(\Psi) \in \mathscr{K}$ is called a **linear operator** from $\mathscr{H}$ to $\mathscr{K}$ with **domain** $\mathscr{D}$ if, for all $\Psi, \Phi \in \mathscr{D}$ and $\alpha, \beta \in \mathbb{C}$,

$$A(\alpha\Psi + \beta\Phi) = \alpha A(\Psi) + \beta A(\Phi) \quad \text{(linearity)}.$$

We denote the domain $\mathscr{D}$ of A by $D(A)$. We sometimes write $A(\Psi) = A\Psi$ simply and use the word "operator" to mean "linear operator" unless otherwise stated. If $D(A) = \mathscr{H}$, then A is said to be **everywhere defined**. If $D(A)$ is dense, then A is said to be **densely defined**.

© Springer Nature Singapore Pte Ltd. 2020

A. Arai, *Inequivalent Representations of Canonical Commutation and Anti-Commutation Relations*, Mathematical Physics Studies,
https://doi.org/10.1007/978-981-15-2180-5_1

There is a notion which is a counterpart to the notion of a linear operator. A mapping $A: \mathscr{D} \to \mathscr{K}$ is called an **anti-linear operator** with domain $D(A) = \mathscr{D}$ if, for all $\Psi, \Phi \in \mathscr{D}$ and $\alpha, \beta \in \mathbb{C}$,

$$A(\alpha\Psi + \beta\Phi) = \alpha^* A(\Psi) + \beta^* A(\Phi) \quad \text{(anti-linearity)},$$

where α^* denotes the complex conjugate of α. An anti-linear operator is called also a **conjugate linear operator**.

The **identity** $I_{\mathscr{H}}$ on $\mathscr{H}$ defined by $D(I_{\mathscr{H}}) := \mathscr{H}$ and $I_{\mathscr{H}}(\Psi) = \Psi$, $\Psi \in \mathscr{H}$ is a simple example of a linear operator. We sometimes write $I_{\mathscr{H}} = I$ if there is no danger of confusion. The **zero operator** $0_{\mathscr{H},\mathscr{K}}$ from $\mathscr{H}$ to $\mathscr{K}$ is defined by $D(0_{\mathscr{H},\mathscr{K}}) := \mathscr{H}$ and $0_{\mathscr{H},\mathscr{K}}(\Psi) := 0_{\mathscr{K}}$, $\Psi \in \mathscr{H}$, where $0_{\mathscr{K}}$ is the zero vector in $\mathscr{K}$. Usually $0_{\mathscr{H},\mathscr{K}}$ (resp. $0_{\mathscr{K}}$) is simply written 0 (resp. 0) if there is no danger of confusion.

For each linear operator A from $\mathscr{H}$ to $\mathscr{K}$, the set

$$D_A^* := \{\Phi \in \mathscr{K} | \exists \Theta_\Phi \in \mathscr{H} \text{ such that } \langle \Phi, A\Psi \rangle_{\mathscr{K}} = \langle \Theta_\Phi, \Psi \rangle_{\mathscr{H}}, \forall \Psi \in D(A)\}$$

is a subspace of $\mathscr{K}$. If A is densely defined, then the vector Θ_Φ is uniquely determined for each $\Phi \in D_A^*$ and the correspondence $D_A^* \ni \Phi \mapsto \Theta_\Phi$ is linear. Therefore, *under the condition that A is densely defined*, one can define a linear operator A^* from $\mathscr{K}$ to $\mathscr{H}$ by

$$D(A^*) := D_A^*, \quad A^*\Phi := \Theta_\Phi, \quad \Phi \in D(A^*). \tag{1.1}$$

The operator A^* is called the **adjoint** of A.

We denote by $\mathfrak{L}(\mathscr{H}, \mathscr{K})$ the set of linear operators from $\mathscr{H}$ to $\mathscr{K}$ and set

$$\mathfrak{L}(\mathscr{H}) := \mathfrak{L}(\mathscr{H}, \mathscr{H}).$$

For convenience, we call an element of $\mathfrak{L}(\mathscr{H})$ a linear operator *on* $\mathscr{H}$, even if it is not everywhere defined on $\mathscr{H}$.

An operator $A \in \mathfrak{L}(\mathscr{H}, \mathscr{K})$ is said to be **equal** to $B \in \mathfrak{L}(\mathscr{H}, \mathscr{K})$ if $D(A) = D(B)$ and $A\Psi = B\Psi$, $\forall \Psi \in D(A)$. In this case we write $A = B$.

For an operator $A \in \mathfrak{L}(\mathscr{H}, \mathscr{K})$, an operator $B \in \mathfrak{L}(\mathscr{H}, \mathscr{K})$ is called an **extension** of A if $D(A) \subset D(B)$ and $A\Psi = B\Psi$, $\Psi \in D(A)$. In this case we write $A \subset B$.

It follows that $A = B$ if and only if $A \subset B$ and $B \subset A$.

It is easy to see that the relation "$\subset$" on $\mathfrak{L}(\mathscr{H}, \mathscr{K})$ is a partial ordering.

The adjoint operation makes the order of two densely defined linear operators converse:

Proposition 1.1 *Let $A \in \mathfrak{L}(\mathscr{H}, \mathscr{K})$ and $B \in \mathfrak{L}(\mathscr{H}, \mathscr{K})$ be densely defined. Suppose that $A \subset B$. Then $B^* \subset A^*$.*

Proof An easy exercise. ■

For A and B in $\mathfrak{L}(\mathscr{H}, \mathscr{K})$, the **sum** $A + B \in \mathfrak{L}(\mathscr{H}, \mathscr{K})$ of A and B is defined as follows:

$$D(A + B) := D(A) \cap D(B), \quad (A + B)\Psi := A\Psi + B\Psi, \quad \Psi \in D(A + B).$$

Proposition 1.2 *Let $A \in \mathfrak{L}(\mathscr{H}, \mathscr{K})$ and $B \in \mathfrak{L}(\mathscr{H}, \mathscr{K})$ be densely defined.*

(i) *If $A + B$ is densely defined, then $A^* + B^* \subset (A + B)^*$.*
(ii) *If $B \in \mathfrak{B}(\mathscr{H}, \mathscr{K})$, then $(A + B)^* = A^* + B^*$.*

For $\alpha \in \mathbb{C}$, $\alpha A \in \mathfrak{L}(\mathscr{H}, \mathscr{K})$ (the **scalar multiple** of A by α) is defined by

$$D(\alpha A) := D(A), \quad (\alpha A)\Psi = \alpha A\Psi, \quad \Psi \in D(\alpha A).$$

Let $A \in \mathfrak{L}(\mathscr{H}, \mathscr{K})$ and $B \in \mathfrak{L}(\mathscr{K}, \mathscr{L})$ ($\mathscr{L}$ is a Hilbert space). Then the **product** $BA \in \mathfrak{L}(\mathscr{H}, \mathscr{L})$ of A and B is defined by

$$D(BA) := \{\Psi \in D(A) | A\Psi \in D(B)\}, \quad (BA)\Psi := B(A\Psi), \quad \Psi \in D(BA).$$

The next proposition, which shows that the product operation for an operator from the left (resp. right) side preserves the partial ordering of operators, is often useful:

Proposition 1.3 *Let $A, B \in \mathfrak{L}(\mathscr{H}, \mathscr{K})$ and $A \subset B$. Then:*

(i) *For all $C \in \mathfrak{L}(\mathscr{K}, \mathscr{L})$ ($\mathscr{L}$ is a Hilbert space), $CA \subset CB$.*
(ii) *For all $D \in \mathfrak{L}(\mathscr{L}, \mathscr{H})$, $AD \subset BD$.*

Proof An easy exercise. ■

Concerning the adjoint operation for the product of two operators, the following facts should be kept in mind:

Proposition 1.4 *Let $A \in \mathfrak{L}(\mathscr{H}, \mathscr{K})$ and $B \in \mathfrak{L}(\mathscr{K}, \mathscr{L})$ be densely defined. Then:*

(i) *If BA is densely defined, then $A^* B^* \subset (BA)^*$.*
(ii) *If $B \in \mathfrak{B}(\mathscr{K}, \mathscr{L})$, then $D(BA) = D(A)$ and $(BA)^* = A^* B^*$.*

For a subspace $\mathscr{D} \subset D(A)$, an operator $A \restriction \mathscr{D} \in \mathfrak{L}(\mathscr{H}, \mathscr{K})$ is defined as follows:

$$D(A \restriction \mathscr{D}) := \mathscr{D}, \quad (A \restriction \mathscr{D})\Psi := A\Psi, \quad \Psi \in \mathscr{D}.$$

This operator is called the **restriction** of A to $\mathscr{D}$. It is obvious that $A \restriction \mathscr{D} \subset A$.

For $A, B \in \mathfrak{L}(\mathscr{H}, \mathscr{K})$ and a subspace $\mathscr{D} \subset D(A) \cap D(B)$, we say that "$A$ is equal to B on $\mathscr{D}$" if $A \restriction \mathscr{D} = B \restriction \mathscr{D}$, i.e., $A\Psi = B\Psi, \forall \Psi \in \mathscr{D}$. In this case we write

$$A = B \quad \text{on } \mathscr{D}.$$

For $A, B \in \mathfrak{L}(\mathscr{H})$, the **commutator** $[A, B] \in \mathfrak{L}(\mathscr{H})$ of A and B is defined by

$$[A, B] := AB - BA$$

with $D([A, B]) = D(AB) \cap D(BA)$. Two operators A and B in $\mathfrak{L}(\mathscr{H})$ are said to **commute** if $[A, B] \subset 0$ (i.e., $[A, B] = 0$ on $D([A, B])$).[1]

For any subspace $\mathscr{D}$ of $D([A, B])$, $[A, B]$ has the restriction to $\mathscr{D}$. But the domain of $[A, B]$ cannot be extended in the strict sense of commutator. In this sense, $D([A, B]) = D(AB) \cap D(BA)$ is called the **maximal domain** for the commutator $[A, B]$.

If there is a subspace $\mathscr{D} \subset D([A, B])$ such that $[A, B]\Psi = 0, \forall \Psi \in \mathscr{D}$, then we say that A and B commute on $\mathscr{D}$. In this case, we write "$[A, B] = 0$ on $\mathscr{D}$".

For each $A \in \mathfrak{L}(\mathscr{H}, \mathscr{K})$, the subset

$$\text{Ran}\,(A) := \{A\Psi | \Psi \in D(A)\}$$

is called the **range** of A, which is a subspace of $\mathscr{K}$. The subset

$$\ker A := \{\Psi \in D(A) | A\Psi = 0\} \tag{1.2}$$

is called the **kernel** of A, which is a subspace of $\mathscr{H}$.

Concepts concerning mapping properties of $A \in \mathfrak{L}(\mathscr{H}, \mathscr{K})$ are defined as follows: A is said to be **injective** or **one-to-one** if $\Psi_1 \neq \Psi_2$ $(\Psi_1, \Psi_2 \in D(A))$, then $A\Psi_1 \neq A\Psi_2$. In this case, the **inverse operator** A^{-1} from $\mathscr{K}$ to $\mathscr{H}$ is defined by

$$D(A^{-1}) := \text{Ran}\,(A), \quad A^{-1}\Phi := \Psi_\Phi, \quad \Phi \in D(A^{-1}),$$

where $\Psi_\Phi \in D(A)$ is the unique vector such that $A\Psi_\Phi = \Phi$.

It is easy to see that A is injective if and only if $\ker A = \{0\}$.

If $\text{Ran}\,(A) = \mathscr{K}$, then A is said to be **surjective**.

If A is injective and surjective, then A is said to be **bijective** and, in this case, A is called a **bijection**.

[1] Recall that the domain $D(0)$ of the zero operator 0 on $\mathscr{H}$ is $\mathscr{H}$.

If there exists a constant $C > 0$ such that $\|A\Psi\|_{\mathscr{K}} \le C\|\Psi\|_{\mathscr{H}}$, $\forall\Psi \in D(A)$, then A is said to be **bounded**. In this case, the quantity

$$\|A\| := \sup_{\Psi\in D(A),\Psi\neq 0} \frac{\|A\Psi\|_{\mathscr{K}}}{\|\Psi\|_{\mathscr{H}}}$$

is called the **operator norm** of A. It follows that $\|A\Psi\|_{\mathscr{K}} \le \|A\|\,\|\Psi\|$, $\forall\Psi \in D(A)$.

A densely defined bounded operator can be extended uniquely to an everywhere defined bounded operator without change of operator norm:

Theorem 1.1 (Extension Theorem) *Let A be a densely defined bounded operator on $\mathscr{H}$. Then there exists a unique bounded operator $\widetilde{A} \in \mathfrak{B}(\mathscr{H})$ such that $A \subset \widetilde{A}$. Moreover, $\|\widetilde{A}\| = \|A\|$.*

Let $\mathfrak{B}(\mathscr{H}, \mathscr{K})$ be the set of everywhere defined bounded linear operators from $\mathscr{H}$ to $\mathscr{K}$. Then $\mathfrak{B}(\mathscr{H}, \mathscr{K})$ is a Banach space with operator norm. We set

$$\mathfrak{B}(\mathscr{H}) := \mathfrak{B}(\mathscr{H}, \mathscr{H}).$$

It holds that, for all $A \in \mathfrak{B}(\mathscr{H}, \mathscr{K})$, the adjoint A^* of A is in $\mathfrak{B}(\mathscr{K}, \mathscr{H})$ and

$$\|A^*\| = \|A\|.$$

There is a weaker notion of boundedness of a linear operator. An operator $B \in \mathfrak{L}(\mathscr{H}, \mathscr{K})$ is said to be **relatively bounded** with respect to an operator $A \in \mathfrak{L}(\mathscr{H}, \mathscr{K})$ or **A-bounded** if $D(A) \subset D(B)$ and there exist constants $a, b \ge 0$ such that

$$\|B\Psi\| \le a\|A\Psi\| + b\|\Psi\|, \quad \forall\Psi \in D(A). \tag{1.3}$$

The infimum of the constants a satisfying (1.3) (b may depend on a) is called the **relative bound** of B with respect to A. We denote it by $\|B\|_A$:

$$\|B\|_A := \inf\{a \ge 0 | (1.3) \text{ holds}\}. \tag{1.4}$$

An operator $A \in \mathfrak{L}(\mathscr{H}, \mathscr{K})$ is said to be **closed** if it has the following property: for any sequence $\{\Psi_n\}_{n=1}^{\infty}$ in $D(A)$ (i.e., $\Psi_n \in D(A)$, $n \in \mathbb{N}$) such that $\lim_{n\to\infty} \Psi_n = \Psi \in \mathscr{H}$ and $\lim_{n\to\infty} A\Psi_n = \Phi \in \mathscr{K}$, it holds that $\Psi \in D(A)$ and $A\Psi = \Phi$.

An operator $A \in \mathfrak{L}(\mathcal{H}, \mathcal{K})$ is said to be **closable** if it has a closed extension (i.e. there exists a closed operator $B \in \mathfrak{L}(\mathcal{H}, \mathcal{K})$ such that $A \subset B$). In this case, an operator $\bar{A} \in \mathfrak{L}(\mathcal{H}, \mathcal{K})$, called the **closure** of A, is associated with A:

$$D(\bar{A}) = \{\Psi \in \mathcal{H} \mid \exists \{\Psi_n\}_{n=1}^{\infty} \text{ with } \Psi_n \in D(A) \text{ such that } \lim_{n\to\infty} \Psi_n = \Psi, \exists \lim_{n\to\infty} A\Psi_n\},$$

$$\bar{A}\Psi = \lim_{n\to\infty} A\Psi_n, \quad \Psi \in D(\bar{A}).$$

It follows that $\bar{A}$ is closed and it is the smallest closed extension of A (i.e., if B is a closed extension of A, then $\bar{A} \subset B$).

The closure operation $A \mapsto \bar{A}$ for a closable operator $A \in \mathfrak{L}(\mathcal{H}, \mathcal{K})$ preserves the partial ordering of operators:

Proposition 1.5 *Let A and B be closable operators from $\mathcal{H}$ to $\mathcal{K}$ satisfying $A \subset B$. Then $\bar{A} \subset \bar{B}$.*

A subspace $\mathcal{D} \subset D(A)$ is called a **core** for A if, for each $\Psi \in D(A)$, there exists a sequence $\{\Psi_n\}_{n=1}^{\infty}$ in $\mathcal{D}$ such that $\lim_{n\to\infty} \Psi_n = \Psi$ and $\lim_{n\to\infty} A\Psi_n = A\Psi$.

It is easy to see that, if A is closed and $\mathcal{D}$ is a core for A, then $A = \overline{A \restriction \mathcal{D}}$.

If A is closable, then a subspace $\mathcal{D} \subset D(A)$ is a core for A if and only if $\mathcal{D}$ is a core for $\bar{A}$. In that case, $\bar{A} = \overline{A \restriction \mathcal{D}}$.

The following proposition states basic facts on closability of a linear operator:

Proposition 1.6 *Let $A \in \mathfrak{L}(\mathcal{H}, \mathcal{K})$ be densely defined (hence A^* exists). Then (i) and (ii) below hold:*

(i) *A^* is closed.*
(ii) *A is closable if and only if $D(A^*)$ is dense in $\mathcal{K}$. In that case, it holds that $\overline{A} = A^{**}$. In particular, if A is closed, then $A^{**} = A$.*
(iii) *If A is closable, then $(\overline{A})^* = A^*$.*

Proof See, e.g., [130, Theorem VIII.1] or [23, Proposition 2.20]. ■

Proposition 1.6(ii) is often useful to prove the closability of a densely defined linear operator in applications.

The following fact also plays an important role:

Lemma 1.1 *Let A be a densely defined operator on a Hilbert space. Then* $\ker A^*A = \ker A$.

Proof It is obvious that $\ker A \subset \ker A^*A$. To prove the converse inclusion relation, let $\Psi \in \ker A^*A$. Then $\Psi \in D(A^*A)$ and $A^*A\Psi = 0$. Taking the inner product of both sides with the vector Ψ and using $\Psi \in D(A)$, we obtain $\|A\Psi\|^2 = 0$. Hence $A\Psi = 0$, i.e., $\Psi \in \ker A$. Thus $\ker A^*A \subset \ker A$. ■

The following decomposition theorem also is often useful:

Theorem 1.2 (Hilbert Space Decomposition Theorem) *Let $A \in \mathfrak{L}(\mathscr{H}, \mathscr{K})$ be a densely defined closed operator. Then*

$$\mathscr{H} = \ker A \oplus \overline{\mathrm{Ran}\,(A^*)}. \tag{1.5}$$

Proof See, e.g., [23, Lemma 2.28].[2] ∎

Theorem 1.2 immediately yields:

Corollary 1.1 *Let $A \in \mathfrak{L}(\mathscr{H}, \mathscr{K})$ be a densely defined closed operator. Then A is injective if and only if* $\mathrm{Ran}\,(A^*)$ *is dense in* $\mathscr{H}$.

Proof Suppose that A is injective. Then $\ker A = \{0\}$. Hence, by (1.5), $\mathscr{H} = \overline{\mathrm{Ran}\,(A^*)}$, implying that $\mathrm{Ran}\,(A^*)$ is dense in $\mathscr{H}$.

Conversely, suppose that $\mathrm{Ran}\,(A^*)$ is dense in $\mathscr{H}$. Then $\mathscr{H} = \overline{\mathrm{Ran}\,(A^*)}$. Hence, by (1.5), $\ker A = \{0\}$, implying that A is injective. ∎

For an operator $A \in \mathfrak{L}(\mathscr{H}, \mathscr{K})$ and a subset $\mathscr{D} \subset D(A)$, we define a subset $A\mathscr{D}$ of $\mathscr{K}$ by

$$A\mathscr{D} := \{A\Psi | \Psi \in \mathscr{D}\}.$$

If $\mathscr{K} = \mathscr{H}$, $\mathscr{D}$ is a subspace of $\mathscr{H}$ and $A\mathscr{D} \subset \mathscr{D}$, then we say that A leaves $\mathscr{D}$ invariant and $\mathscr{D}$ is called an **invariant subspace** of A.

The notions of sum and product for linear operators can be extended to the case of any finite number of linear operators. For operators $A_1, \ldots, A_n \in \mathfrak{L}(\mathscr{H}, \mathscr{K})$ $(n \geq 2)$, the **operator sum** $\sum_{j=1}^n A_j = A_1 + A_2 + \cdots + A_n \in \mathfrak{L}(\mathscr{H}, \mathscr{K})$ is defined by

$$D(\textstyle\sum_{j=1}^n A_j) := \cap_{j=1}^n D(A_j), \quad (\sum_{j=1}^n A_j)\Psi := \sum_{j=1}^n A_j\Psi, \quad \Psi \in D(\sum_{j=1}^n A_j).$$

Let $\mathscr{H}_1, \ldots, \mathscr{H}_n$ and $\mathscr{H}_{n+1}$ be Hilbert spaces and $B_j \in \mathfrak{L}(\mathscr{H}_j, \mathscr{H}_{j+1})$ $(j = 1, \ldots, n)$. Then the **operator product** $B_n \cdots B_1 \in \mathfrak{L}(\mathscr{H}_1, \mathscr{H}_{n+1})$ is defined by

$$D(B_n \cdots B_1) := \{\Psi \in D(B_1) |\, B_j \cdots B_1\Psi \in D(B_{j+1}),\ j = 1, \ldots n-1\},$$
$$(B_n \cdots B_1)\Psi := B_n(B_{n-1}(\cdots(B_2(B_1\Psi))\cdots)), \quad \Psi \in D(B_n \cdots B_1).$$

If $B_j \in \mathfrak{B}(\mathscr{H}_j, \mathscr{H}_{j+1})$ $(j = 1, \ldots, n)$, then $B_n \cdots B_1 \in \mathfrak{B}(\mathscr{H}_1, \mathscr{H}_{n+1})$ and

$$\|B_n \cdots B_1\| \leq \prod_{j=1}^n \|B_j\|.$$

[2] The proof of Theorem 1.2 is not so difficult.

In particular, for all $A \in \mathfrak{B}(\mathscr{H})$,

$$\|A^n\| \le \|A\|^n.$$

1.1.2 Convergences of Bounded Operators

There are three types of convergence for a sequence of bounded operators. Let $\mathscr{H}$ and $\mathscr{K}$ be Hilbert spaces and $A_n,\ A \in \mathfrak{B}(\mathscr{H}, \mathscr{K})$, $n \in \mathbb{N}$.

(1) A sequence $\{A_n\}_{n=1}^{\infty}$ is said to **uniformly converge** to A if $\lim_{n\to\infty} \|A_n - A\| = 0$. In this case we write $\text{u}-\lim_{n\to\infty} A_n = A$ and A is called the **uniform limit** of $\{A_n\}_{n=1}^{\infty}$.
(2) A sequence $\{A_n\}_{n=1}^{\infty}$ is said to **strongly converge** to A if, for all $\Psi \in \mathscr{H}$, $\lim_{n\to\infty} A_n\Psi = A\Psi$. In this case we write $\text{s}-\lim_{n\to\infty} A_n = A$ and A is called the **strong limit** of $\{A_n\}_{n=1}^{\infty}$.
(3) A sequence $\{A_n\}_{n=1}^{\infty}$ is said to **weakly converge** to A if, for all $\Psi \in \mathscr{H}$ and $\Phi \in \mathscr{K}$, $\lim_{n\to\infty} \langle \Phi, A_n\Psi\rangle = \langle \Phi, A\Psi\rangle$. In this case we write $\text{w}-\lim_{n\to\infty} A_n = A$ and A is called the **weak limit** of $\{A_n\}_{n=1}^{\infty}$.

1.1.3 Spectra

For each operator $A \in \mathfrak{L}(\mathscr{H})$, there exist mutually disjoint four subsets of $\mathbb{C}$ corresponding to mapping properties of $A - \lambda := A - \lambda I_{\mathscr{H}}$ ($\lambda \in \mathbb{C}$ and $I_{\mathscr{H}}$ is the identity on $\mathscr{H}$):

$$\begin{aligned}
\rho(A) &:= \{\lambda \in \mathbb{C} |\ A - \lambda \text{ is injective, Ran}\,(A-\lambda) \text{ is dense in } \mathscr{H} \text{ and} \\
&\qquad\qquad (A-\lambda)^{-1} \text{ is bounded}\}, \\
\sigma_{\text{c}}(A) &:= \{\lambda \in \mathbb{C} |\ A - \lambda \text{ is injective, Ran}\,(A-\lambda) \text{ is dense in } \mathscr{H} \text{ and} \\
&\qquad\qquad (A-\lambda)^{-1} \text{ is unbounded}\}, \\
\sigma_{\text{r}}(A) &:= \{\lambda \in \mathbb{C} |\ A - \lambda \text{ is injective and Ran}\,(A-\lambda) \text{ is not dense in } \mathscr{H}\}, \\
\sigma_{\text{p}}(A) &:= \{\lambda \in \mathbb{C} |\ A - \lambda \text{ is not injective}\}.
\end{aligned}$$

The subsets $\rho(A)$, $\sigma_{\text{c}}(A)$, $\sigma_{\text{r}}(A)$ and $\sigma_{\text{p}}(A)$ are called respectively the **resolvent set**, the **continuous spectrum**, the **residual spectrum** and the **point spectrum** of A. These subsets are mutually disjoint and satisfy

$$\rho(A) \cup \sigma_{\text{c}}(A) \cup \sigma_{\text{r}}(A) \cup \sigma_{\text{p}}(A) = \mathbb{C}. \tag{1.6}$$

The subset

$$\sigma(A) := \rho(A)^c \; (:= \mathbb{C} \setminus \rho(A)) \tag{1.7}$$

is called the **spectrum** of A. We have

$$\sigma(A) = \sigma_c(A) \cup \sigma_r(A) \cup \sigma_p(A) \tag{1.8}$$

and

$$\sigma(A) \cap \rho(A) = \emptyset, \quad \sigma(A) \cup \rho(A) = \mathbb{C}. \tag{1.9}$$

The point spectrum $\sigma_p(A)$ is nothing but the set of all eigenvalues[3] of A and, for each eigenvalue $\lambda \in \sigma_p(A)$, $\ker(A - \lambda)$ is the **eigenspace** of A with eigenvalue λ. The dimension of the eigenspace (i.e., $\dim \ker(A - \lambda)$) is called the **multiplicity** of the eigenvalue λ. If the multiplicity of λ is one (resp. more than one), then λ is said to be **simple** (resp. **degenerate**). If the multiplicity of λ is finite and λ is an isolated point of $\sigma(A)$ (i.e., there exists a constant $\delta > 0$ such that $\{z \in \mathbb{C} | 0 < |z - \lambda| < \delta\} \subset \rho(A)$), then λ is a **discrete eigenvalue** of A. We denote by $\sigma_d(A)$ the set of the discrete eigenvalues of A. It follows that

$$\sigma_d(A) \subset \sigma_p(A).$$

If $\sigma(A) = \sigma_d(A)$, then the spectrum of A is said to be **purely discrete** or A has a purely discrete spectrum.

The next proposition is important, which follows from the closed graph theorem:

Proposition 1.7 *Let A be a closed linear operator on $\mathscr{H}$. Then $\lambda \in \rho(A)$ if and only if $A - \lambda$ is bijective.*

The next proposition states a fundamental fact on the spectrum of a closable operator:

Proposition 1.8 *For any closable operator A on $\mathscr{H}$, $\rho(A) = \rho(\bar{A})$ and $\sigma(A) = \sigma(\bar{A})$.*

The next proposition is concerned with an elementary fact on a spectral mapping:

Proposition 1.9 *Let $A \in \mathfrak{L}(\mathscr{H})$. Then, for all $a \in \mathbb{R} \setminus \{0\}$ and $b \in \mathbb{R}$,*

$$\sigma(aA + b) = \{a\lambda + b | \lambda \in \sigma(A)\}. \tag{1.10}$$

In particular,

$$\sigma(-A) = \{-\lambda | \lambda \in \sigma(A)\}. \tag{1.11}$$

[3] Recall that a complex number $\lambda \in \mathbb{C}$ is called an **eigenvalue** of A if there exists a *non-zero* vector Ψ in $D(A)$ such that $A\Psi = \lambda\Psi$; such a vector Ψ is called an **eigenvector** of A with eigenvalue λ.

Proof We first show that

$$\rho(aA+b) = \{a\lambda + b | \lambda \in \rho(A)\}. \tag{1.12}$$

We denote the set on the right hand side by $R_{a,b}$. Let $z \in \rho(aA+b)$. Then $B := aA + b - z$ is injective with B^{-1} bounded and Ran B is dense in $\mathscr{H}$. One can rewrite $B = a(A - (z-b)/a)$, which implies that $(z-b)/a \in \rho(A)$. Hence there is an element $\lambda \in \rho(A)$ such that $(z-b)/a = \lambda$, i.e., $z = a\lambda + b$. Hence $\rho(aA+b) \subset R_{a,b}$.

By tracing conversely the arguments given in the preceding paragraph, one can easily show that $R_{a,b} \subset \rho(aA+b)$. Thus (1.12) holds. This fact implies that $\sigma(aA+b) = R_{a,b}^c$. Using (1.9), one can show that $R_{a,b}^c = \{a\mu + b | \mu \in \sigma(A)\}$. Thus (1.10) follows. ■

1.1.4 Unitary Operators and Related Facts

We recall that a unitary operator $U: \mathscr{H} \to \mathscr{K}$ is an operator such that Ran $(U) = \mathscr{K}$ (surjectivity) and $\langle U\Psi, U\Phi\rangle_{\mathscr{K}} = \langle \Psi, \Phi\rangle_{\mathscr{H}}$, $\Psi, \Phi \in \mathscr{H}$ (preservation of inner product), which is equivalent to that $U^*U = I_{\mathscr{H}}$ and $UU^* = I_{\mathscr{K}}$. Hence U is bijective with $U^{-1} = U^*$ which also is a unitary operator from $\mathscr{K}$ to $\mathscr{H}$. A unitary operator is sometimes called a unitary transformation.

For two Hilbert spaces $\mathscr{H}$ and $\mathscr{K}$, we write $\mathscr{H} \cong \mathscr{K}$ if there exists a unitary operator U from $\mathscr{H}$ to $\mathscr{K}$. It is easy to see that this relation $\cong$ is an equivalence relation in the set of Hilbert spaces. Based on this fact, one says that $\mathscr{H}$ and $\mathscr{K}$ are **isomorphic** if $\mathscr{H} \cong \mathscr{K}$. In the case where $\mathscr{H} \cong \mathscr{K}$ under a unitary operator U: $\mathscr{H} \to \mathscr{K}$, we write $\mathscr{H} \overset{U}{\cong} \mathscr{K}$ (but we sometimes omit U if there is no danger of confusion). Such a unitary operator U is called a **Hilbert space isomorphism** (or simply an isomorphism) .

If an isomorphism $U: \mathscr{H} \to \mathscr{K}$ is independent of the choice of complete orthonormal systems of $\mathscr{H}$ and $\mathscr{K}$, then U is called a **natural isomorphism**.

The next theorem is a basic one to find a Hilbert space isomorphism.

Theorem 1.3 (Isomorphism Theorem) *Let $\mathscr{H}$ and $\mathscr{K}$ be Hilbert spaces, and $\{e_\lambda\}_{\lambda\in\Lambda}$ and $\{f_\lambda\}_{\lambda\in\Lambda}$ (Λ is an index set not necessarily countable) be complete orthonormal systems of $\mathscr{H}$ and $\mathscr{K}$ respectively. Then there exists a unitary operator $U: \mathscr{H} \to \mathscr{K}$ such that*

$$Ue_\lambda = f_\lambda, \quad \lambda \in \Lambda.$$

For a proof of this theorem, see, e.g., [28, Theorem 2.4], [23, Theorem 2.5].

The spectra of a linear operator $A \in \mathfrak{L}(\mathscr{H})$ are unitarily invariant: for all unitary operators $U: \mathscr{H} \to \mathscr{K}$,

$$\sigma(UAU^{-1}) = \sigma(A), \quad \sigma_\#(UAU^{-1}) = \sigma_\#(A), \quad \# = \text{c, r, p}.$$

A linear operator A on $\mathscr{H}$ is said to be **unitarily equivalent** to a linear operator B on $\mathscr{K}$ if there exists a unitary operator $U: \mathscr{H} \to \mathscr{K}$ such that $UAU^{-1} = B$. In this case, we write

$$A \overset{U}{\cong} B \quad \text{or} \quad A \cong B \tag{1.13}$$

if U is understood. It follows from the above facts that

$$A \cong B \Longrightarrow \sigma_{\#}(A) = \sigma_{\#}(B), \quad \# = \text{ c, r, p}.$$

The next propositions are basic elementary facts often used.

Proposition 1.10 *Let U be a unitary operator from $\mathscr{H}$ to a Hilbert space $\mathscr{K}$ and $A, B \in \mathfrak{L}(\mathscr{H})$. Then*

$$U(A + B)U^{-1} = UAU^{-1} + UBU^{-1}. \tag{1.14}$$

Proof An easy exercise. ■

Remark 1.1 The point in Proposition 1.10 is that (1.14) is an *operator equality*, i.e., $D(U(A + B)U^{-1}) = D(UAU^{-1} + UBU^{-1})$ and, for all $\Psi \in D(U(A + B)U^{-1})$, $U(A + B)U^{-1}\Psi = UAU^{-1}\Psi + UBU^{-1}\Psi$.

Proposition 1.11 *Let U be a unitary operator on $\mathscr{H}$ and $\mathscr{D}$ be a subspace of $\mathscr{H}$ such that $U\mathscr{D} \subset \mathscr{D}$ and $U^{-1}\mathscr{D} \subset \mathscr{D}$. Then $U\mathscr{D} = \mathscr{D}$.*

Proof An easy exercise. ■

1.1.5 Symmetric and Self-adjoint Operators

An operator $A \in \mathfrak{L}(\mathscr{H})$ is said to be **symmetric** if $D(A)$ is dense and $A \subset A^*$.

It is easy to see that a densely defined operator $A \in \mathfrak{L}(\mathscr{H})$ is symmetric if and only if, for all $\Psi, \Phi \in D(A)$, $\langle \Psi, A\Phi \rangle_{\mathscr{H}} = \langle A\Psi, \Phi \rangle_{\mathscr{H}}$.

One can easily show that a densely defined operator $A \in \mathfrak{L}(\mathscr{H})$ is symmetric if and only if $\langle \Psi, A\Psi \rangle_{\mathscr{H}} \in \mathbb{R}$, $\forall \Psi \in D(A)$.

Proposition 1.12 *Every symmetric operator A on $\mathscr{H}$ is closable with $\bar{A} \subset A^*$.*

Proof By the definition of symmetric operator, $A \subset A^*$. By Proposition 1.6(i), A^* is closed. Hence A is closable and $\bar{A} \subset A^*$. ■

For an operator A on $\mathscr{H}$ and a dense subspace $\mathscr{D}$ of $\mathscr{H}$ such that $\mathscr{D} \subset D(A)$, A is said to be **symmetric on** $\mathscr{D}$ if $A \restriction \mathscr{D}$ is a symmetric operator.

If $A \in \mathfrak{L}(\mathscr{H})$ is densely defined and $A = A^*$, then A is called a **self-adjoint operator**.

It is obvious that a self-adjoint operator is a closed symmetric operator. But one should be careful, in that *the converse is not true*.

A densely defined linear operator A on $\mathscr{H}$ is said to be **maximally symmetric** if it is symmetric and has no non-trivial symmetric extensions (i.e., if B is a symmetric operator such that $A \subset B$, then $B = A$).

It follows that a maximally symmetric operator is closed.

It is easy to show that a self-adjoint operator is maximally symmetric. But the converse is not true.

A symmetric operator A is said to be **essentially self-adjoint** if its closure $\bar{A}$ is self-adjoint.

Let A be a symmetric operator on $\mathscr{H}$ and $\mathscr{D}$ be a dense subspace of $\mathscr{H}$ such that $\mathscr{D} \subset D(A)$. Then it is easy to see that $A \restriction \mathscr{D}$ is a symmetric operator. We say that A is essentially self-adjoint on $\mathscr{D}$ if $A \restriction \mathscr{D}$ is essentially self-adjoint. In that case, it follows that A is essentially self-adjoint and $\overline{A \restriction \mathscr{D}} = \bar{A}$ (hence $\mathscr{D}$ is a core for the self-adjoint operator $\bar{A}$). Conversely, if $\bar{A}$ is self-adjoint and $\mathscr{D}$ is a core for $\bar{A}$. then A is essentially self-adjoint on $\mathscr{D}$.

A symmetric operator A on $\mathscr{H}$ is said to be **bounded from below** if there exists a constant $\gamma \in \mathbb{R}$ such that $\langle \Psi, A\Psi \rangle \geq \gamma \|\Psi\|^2$, $\forall \Psi \in D(A)$. In this case we write $A \geq \gamma$. In particular, if $A \geq 0$ (the case $\gamma = 0$), then A is said to be **non-negative**.

A symmetric operator A satisfying $-A \geq -\gamma$ for some $\gamma \in \mathbb{R}$ is said to be **bounded from above**. In this case we write $A \leq \gamma$. In particular, if $A \leq 0$, then A is said to be **non-positive**.

A symmetric operator is said to be **semi-bounded** if it is bounded from below or bounded from above.

An operator $P \in \mathfrak{B}(\mathscr{H})$ is called an **orthogonal projection** if $P = P^*$ (self-adjoint) and $P^2 = P$ (idempotent).

1.1.6 *Lowest Energy and Ground State of a Self-adjoint Operator*

Let H be a self-adjoint operator on $\mathscr{H}$ which is bounded from below. Then the infimum of the spectrum of H

$$E_0(H) := \inf \sigma(H)$$

is finite. In view of quantum theory and by abuse of terminology, we call $E_0(H)$ the **lowest energy** of H. One can show that

$$E_0(H) = \inf_{\Psi \in D(H), \|\Psi\|=1} \langle \Psi, H\Psi \rangle \,.$$

This is called the **variational principle for the lowest energy**. If $E_0(H)$ is an eigenvalue of H, then a non-zero vector in the eigenspace $\ker(H - E_0(H))$ is called a **ground state** of H. In this case, H is said to have a ground state and $E_0(H)$ is called the **ground state energy** of H. The ground state of H is said to be unique (resp. degenerate) if $\dim\ker(H - E_0(H)) = 1$ (resp. $\dim\ker(H - E_0(H)) \geq 2$).

A ground state of H is called a **zero-energy ground state** if $E_0(H) = 0$.

1.1.7 Spectral Theorem and Some Facts in Functional Calculus

We recall that, with each self-adjoint operator A on a Hilbert space $\mathscr{H}$, a unique one-dimensional spectral measure[4] E_A is associated, called the spectral measure of A, and A has the spectral representation

$$A = \int_{\mathbb{R}} \lambda \, dE_A(\lambda),$$

which means that

$$D(A) = \left\{ \Psi \in \mathscr{H} \middle| \int_{\mathbb{R}} \lambda^2 d\|E_A(\lambda)\Psi\|^2 < \infty \right\},$$

$$\langle \Phi, A\Psi \rangle = \int_{\mathbb{R}} \lambda d \langle \Phi, E_A(\lambda)\Psi \rangle, \quad \Phi \in \mathscr{H}, \Psi \in D(A).$$

This fact is called the **spectral theorem** for self-adjoint operators (Theorem B.2 in Appendix B).

The spectral theorem allows one to develop the functional calculus based on E_A (see Appendix B). Namely, for each Borel measurable function $f : \mathbb{R} \to \mathbb{C} \cup \{\pm\infty\}$, there exists a linear operator $f(A)$ on $\mathscr{H}$ such that

$$f(A) = \int_{\mathbb{R}} f(\lambda) dE_A(\lambda), \tag{1.15}$$

which means that

$$D(f(A)) = \left\{ \Psi \in \mathscr{H} \middle| \int_{\mathbb{R}} |f(\lambda)|^2 d\|E_A(\lambda)\Psi\|^2 < \infty \right\}, \tag{1.16}$$

$$\langle \Phi, f(A)\Psi \rangle = \int_{\mathbb{R}} f(\lambda) d \langle \Phi, E_A(\lambda)\Psi \rangle, \quad \Phi \in \mathscr{H}, \Psi \in D(f(A)). \tag{1.17}$$

Example 1.1 For a complex number $z \in \mathbb{C}$, let $f_z(\lambda) := e^{iz\lambda}$, $\lambda \in \mathbb{R}$. Then we define e^{izA} by

$$e^{izA} := f_z(A).$$

---

[4]See Appendix B for the definition of a spectral measure and related facts.

It follows from Theorem B.1(v) that, for all $t \in \mathbb{R}$, e^{itA} is unitary.

Example 1.2 Let A be a non-negative self-adjoint operator on $\mathscr{H}$. Then, for all $\alpha > 0$, the function $g_\alpha(\lambda) := \chi_{[0,\infty)}(\lambda)|\lambda|^\alpha$ on $\mathbb{R}$ is Borel measurable. Hence one can define $g_\alpha(A)$. We denote it by A^α:

$$A^\alpha := \int_{\mathbb{R}} \chi_{[0,\infty)}(\lambda)|\lambda|^\alpha dE_A(\lambda).$$

Since the function g_α is a non-negative continuous function on $\mathbb{R}$, it follows from Theorem B.1(vi) that A^α is a non-negative self-adjoint operator. One can show that $(A^\alpha)^{1/\alpha} = A$.

Example 1.3 Let $T \in \mathfrak{B}(\mathscr{H}, \mathscr{K})$ ($\mathscr{K}$ is a Hilbert space). Then T^*T is a non-negative bounded self-adjoint operator on $\mathscr{H}$. Hence, by the preceding example,

$$|T| := (T^*T)^{1/2}$$

is a non-negative bounded self-adjoint operator. The operator $|T|$ is called the **absolute value** or **modulus** of T. It follows that $|T|^2 = T^*T$.

The following proposition is sometimes useful in applications.

Proposition 1.13 *Let A be a non-negative self-adjoint operator on $\mathscr{H}$ and $\mathscr{D}$ be a core for A. Then, for all $\alpha \in (0, 1)$, $D(A) \subset D(A^\alpha)$ and $\mathscr{D}$ is a core for A^α.*

Proof Let $\Psi \in D(A)$. Then (note that $\operatorname{supp} E_A \subset [0, \infty)$),

$$\begin{aligned}
\int_{[0,\infty)} \lambda^{2\alpha} d\|E_A(\lambda)\Psi\|^2 &= \int_{[0,1]} \lambda^{2\alpha} d\|E_A(\lambda)\Psi\|^2 + \int_{(1,\infty)} \lambda^{2\alpha} d\|E_A(\lambda)\Psi\|^2 \\
&\leq \int_{[0,1]} d\|E_A(\lambda)\Psi\|^2 + \int_{(1,\infty)} \lambda^2 d\|E_A(\lambda)\Psi\|^2 \\
&\leq \|E_A([0,1])\Psi\|^2 + \|A\Psi\|^2 < \infty.
\end{aligned}$$

Hence $\Psi \in D(A^\alpha)$. Therefore $D(A) \subset D(A^\alpha)$ and

$$\|A^\alpha \Psi\| \leq \|\Psi\| + \|A\Psi\|, \quad \Psi \in D(A). \tag{1.18}$$

By the assumption on $\mathscr{D}$, for each $\Psi \in D(A)$, there exists a sequence $\{\Psi_n\}_n$ in $\mathscr{D}$ such that $\Psi_n \to \Psi$ and $A\Psi_n \to A\Psi$ as $n \to \infty$. Hence, by (1.18), $A^\alpha \Psi_n \to A^\alpha \Psi$ as $n \to \infty$. This means that

$$A^\alpha \restriction D(A) \subset \overline{A^\alpha \restriction \mathscr{D}}.$$

Hence

$$\overline{A^\alpha \restriction D(A)} \subset \overline{A^\alpha \restriction \mathscr{D}}. \tag{1.19}$$

Next, let $\Psi \in D(A^\alpha)$ and $\Psi_n := E_A([0,n])\Psi$, $n \in \mathbb{N}$. Then $\Psi_n \in D(A)$ and $\Psi_n \to \Psi$ as $n \to \infty$. We have

$$\|A^\alpha \Psi_n - A^\alpha \Psi\|^2 = \int_{\lambda > n} \lambda^{2\alpha} d\|E_A(\lambda)\Psi\|^2.$$

Since $\int_{[0,\infty)} \lambda^{2\alpha} d\|E_A(\lambda)\Psi\|^2 < \infty$, it follows that the right hand side tends to 0 as $n \to \infty$. Hence $A^\alpha \Psi_n \to A^\alpha \Psi$ as $n \to \infty$. Therefore $A^\alpha \subset \overline{A^\alpha \restriction D(A)}$. By this and (1.19), we obtain $A^\alpha \subset \overline{A^\alpha \restriction \mathscr{D}}$. It is obvious that $A^\alpha \restriction \mathscr{D} \subset A^\alpha$ and hence $\overline{A^\alpha \restriction \mathscr{D}} \subset A^\alpha$. Thus $\overline{A^\alpha \restriction \mathscr{D}} = A^\alpha$, implying that $\mathscr{D}$ is a core for A^α. ■

Proposition 1.14 *Let A be a self-adjoint operator on $\mathscr{H}$. Then, for all $\Psi \in D(A)$,*

$$\|e^{iA}\Psi - \Psi\| \le \|A\Psi\|. \tag{1.20}$$

Proof By the functional calculus, we have

$$\|e^{iA}\Psi - \Psi\|^2 = \int_{\mathbb{R}} |e^{i\lambda} - 1|^2 d\|E_A(\lambda)\Psi\|^2.$$

It is easy to see that

$$|e^{i\lambda} - 1| \le |\lambda|, \quad \lambda \in \mathbb{R}.$$

Hence

$$\|e^{iA}\Psi - \Psi\|^2 \le \int_{\mathbb{R}} \lambda^2 d\|E_A(\lambda)\Psi\|^2 = \|A\Psi\|^2.$$

Thus (1.20) follows. ■

1.1.8 Trace Class and Hilbert–Schmidt Operators

Let $\mathscr{H}$ be a separable Hilbert space and T be a non-negative bounded self-adjoint operator on $\mathscr{H}$. Then, for a complete orthonormal system (CONS) $\{e_n\}_{n=1}^\infty$ of $\mathscr{H}$, $\sum_{n=1}^\infty \langle e_n, Te_n\rangle$ is a non-negative finite number or infinite, independently of the choice of $\{e_n\}_{n=1}^\infty$. Hence one can define

$$\operatorname{Tr} T := \sum_{n=1}^\infty \langle e_n, Te_n\rangle. \tag{1.21}$$

It is called the trace of T.

A bounded linear operator $T \in \mathfrak{B}(\mathscr{H})$ is called a **trace class operator** if $\operatorname{Tr}|T| < \infty$. In this case, T is said to be trace class and the **trace** $\operatorname{Tr} T \in \mathbb{C}$ of T can be defined by (1.21), independently of the choice of $\{e_n\}_{n=1}^{\infty}$.

Remark 1.2 If $\mathscr{H}$ is finite dimensional with $\dim \mathscr{H} = N$, then each everywhere defined linear operator T on $\mathscr{H}$ is bounded and, for any CONS $\{e_n\}_{n=1}^{N}$ of $\mathscr{H}$, $\operatorname{Tr} T := \sum_{n=1}^{N} \langle e_n, Te_n \rangle$ is in $\mathbb{C}$, independently of the choice of $\{e_n\}_{n=1}^{N}$.

Example 1.4 Let $f, g \in \mathscr{H}$ and define an operator $P_{f,g}$ on $\mathscr{H}$ by

$$P_{f,g}\Psi := \langle f, \Psi \rangle\, g, \quad \Psi \in \mathscr{H}. \tag{1.22}$$

Then $P_{f,g} \in \mathfrak{B}(\mathscr{H})$ with $\|P_{f,g}\| = \|f\|\,\|g\|$ and

$$P_{f,g}^{*}\Psi = \langle g, \Psi \rangle\, f, \quad \Psi \in \mathscr{H}.$$

Hence $P_{f,f}$ is self-adjoint and

$$|P_{f,g}|^2 = P_{f,g}^{*} P_{f,g} = \|g\|^2 P_{f,f}.$$

In particular, $P_{f,f}^2 = \|f\|^2 P_{f,f}$ and $P_{f,f} \geq 0$.

Let $f \neq 0$. Then it follows that $P_{f,f}^{1/2} = P_{f,f}/\|f\|$. Hence

$$|P_{f,g}| = \frac{\|g\|}{\|f\|} P_{f,f}.$$

Let $\{e_n\}_{n=1}^{\infty}$ be a CONS of $\mathscr{H}$. Then

$$\operatorname{Tr}|P_{f,g}| = \frac{\|g\|}{\|f\|} \sum_{n=1}^{\infty} |\langle e_n, f \rangle|^2 = \|g\|\,\|f\| < \infty.$$

Hence $P_{f,g}$ is a trace class operator on $\mathscr{H}$ and

$$\operatorname{Tr} P_{f,g} = \langle f, g \rangle.$$

If $f \neq 0$ and $g \neq 0$, then $P_{f,g}$ is an example of one-rank operators.

In what follows, we sometimes write

$$P_{f,g} = \langle f, \cdot \rangle\, g. \tag{1.23}$$

A bounded linear operator $T \in \mathfrak{B}(\mathscr{H}, \mathscr{K})$ is said to be **Hilbert–Schmidt** if T^*T is a trace class operator on $\mathscr{H}$. It follows that $T \in \mathfrak{B}(\mathscr{H}, \mathscr{K})$ is Hilbert–Schmidt if and only if, for a CONS $\{e_n\}_n$ of $\mathscr{H}$, $\sum_n \|Te_n\|^2 < \infty$. The non-negative number

$$\|T\|_2 := \sqrt{\operatorname{Tr} T^*T} = \sqrt{\sum_n \|Te_n\|^2}.$$

is called the **Hilbert–Schmidt norm** of T. It follows that

$$\|T\| \leq \|T\|_2$$

Moreover, T^* also is Hilbert–Schmidt and

$$\|T^*\|_2 = \|T\|_2.$$

Example 1.5 The operator $P_{f,g}$ in Example 1.4 is Hilbert–Schmidt with

$$\|P_{f,g}\|_2 = \|f\| \, \|g\|.$$

The following representation theorem for Hilbert–Schmidt operators is useful:

Theorem 1.4 *Let T be a Hilbert–Schmidt operator on a Hilbert space $\mathscr{H}$. Then there exist (not necessarily complete) ONS's $\{e_n\}_{n=1}^N$ and $\{f_n\}_{n=1}^N$ of $\mathscr{H}$ ($N < \infty$ or $N = \infty$) and positive real numbers $\mu_n > 0$ ($n = 1, \ldots, N$) such that $\sum_{n=1}^{\infty} \mu_n^2 < \infty$ in the case $N = \infty$ and*

$$T = \sum_{n=1}^{N} \mu_n \langle f_n, \cdot \rangle \, e_n. \tag{1.24}$$

Proof It is well known that a Hilbert–Schmidt operator is a compact operator and that a compact operator has a canonical form which, applied to T, takes form (1.24) (see, e.g., [130, Theorems VI.17]). Hence $Tf_n = \mu_n e_n$. Consider the case $N = \infty$. Since T is Hilbert–Schmidt, $\sum_{n=1}^{\infty} \|Tf_n\|^2 < \infty$. Hence $\sum_{n=1}^{\infty} \mu_n^2 < \infty$. ∎

Remark 1.3 The numbers μ_n ($n = 1, 2, \ldots, N$) are positive eigenvalues of $|T|$ (it is easy to see that $T^*Tf_n = \mu_n^2 f_n$). They are called the singular values of T.

Remark 1.4 In the case $N = \infty$, the convergence in (1.24) can be taken in operator norm topology.

If T in Theorem 1.4 is self-adjoint (i.e., T is a self-adjoint Hilbert–Schmidt operator), then one has a stronger result:

Theorem 1.5 *Let T be a self-adjoint Hilbert–Schmidt operator on an infinite dimensional Hilbert space $\mathscr{H}$. Then there exist a CONS $\{e_n\}_{n=1}^{\infty}$ of $\mathscr{H}$ and real numbers $\{\lambda_n\}_{n=1}^{\infty}$ such that $Te_n = \lambda_n e_n$, $n \geq 1$ and*

$$\|T\|_2^2 = \sum_{n=1}^{\infty} \lambda_n^2.$$

Proof By the Hilbert–Schmidt theorem on a self-adjoint compact operator (e.g., [130, Theorem VI.16]), there are a CONS $\{e_n\}_{n=1}^{\infty}$ of $\mathscr{H}$ and real numbers $\{\lambda_n\}_{n=1}^{\infty}$

such that $Te_n = \lambda_n e_n$, $n \geq 1$ and $\lambda_n \to 0$ as $n \to \infty$. Since T is Hilbert–Schmidt, $\|T\|_2^2 = \sum_{n=1}^\infty \|Te_n\|^2 = \sum_{n=1}^\infty \lambda_n^2$. ■

The following proposition formulates a sufficient condition for a densely defined linear operator on $\mathscr{H}$ to have an extension to a Hilbert–Schmidt operator on $\mathscr{H}$.

Proposition 1.15 *Let $T \in \mathfrak{L}(\mathscr{H})$ be densely defined. Suppose that there exists a CONS $\{e_n\}_{n=1}^\infty$ of $\mathscr{H}$ such that $e_n \in D(T^*)$, $n \geq 1$ and $\sum_{n=1}^\infty \|T^*e_n\|^2 < \infty$. Then T is closable and $\overline{T}$ is Hilbert–Schmidt.*

Proof Let $\Psi \in D(T)$ and, for each $N \in \mathbb{N}$, define $\Phi_N \in \mathscr{H}$ by

$$\Phi_N := \sum_{n=1}^{N} \langle e_n, T\Psi\rangle\, e_n.$$

Then $\lim_{N\to\infty} \Phi_N = T\Psi$. In particular, $\lim_{N\to\infty} \|\Phi_N\|^2 = \|T\Psi\|^2$. We have

$$\begin{aligned}\|\Phi_N\|^2 &= \sum_{n=1}^{N} |\langle e_n, T\Psi\rangle|^2 = \sum_{n=1}^{N} |\langle T^*e_n, \Psi\rangle|^2 \\ &\leq \left(\sum_{n=1}^{N} \|T^*e_n\|^2\right) \|\Psi\|^2 \leq C^2 \|\Psi\|^2\end{aligned}$$

with $C := \sqrt{\sum_{n=1}^\infty \|T^*e_n\|^2} < \infty$, where we have used the Schwarz inequality. Hence, taking the limit $N \to \infty$, we obtain $\|T\Psi\|^2 \leq C^2\|\Psi\|^2$. Hence T is bounded with $\|T\| \leq C$. It is well known that a densely defined bounded operator is closable and its closure is an everywhere defined bounded operator. Hence T is closable and $\overline{T} \in \mathfrak{B}(\mathscr{H})$. This implies also that $(\overline{T})^* \in \mathfrak{B}(\mathscr{H})$. On the other hand, $(\overline{T})^* = T^*$. Hence $T^* \in \mathfrak{B}(\mathscr{H})$. It follows from this result and the assumption $\sum_{n=1}^\infty \|T^*e_n\|^2 < \infty$ that T^* is Hilbert–Schmidt. Hence $(T^*)^* = \overline{T}$ also is Hilbert–Schmidt. ■

Let $\mathfrak{I}_2(\mathscr{H})$ be the set of Hilbert–Schmidt operators on $\mathscr{H}$. Then $\mathfrak{I}_2(\mathscr{H})$ is a Hilbert space with inner product

$$\langle S, T\rangle_2 := \mathrm{Tr}\, S^*T, \quad S, T \in \mathfrak{I}_2(\mathscr{H})$$

(see, e.g., [130, Theorem VI.22]). We set

$$\|T\|_2 := \sqrt{\langle T, T\rangle_2}, \quad T \in \mathfrak{I}_2(\mathscr{H}).$$

It is not so difficult to show that, for all $S \in \mathfrak{B}(\mathscr{H})$ and $T \in \mathfrak{I}_2(\mathscr{H})$, TS and ST are in $\mathfrak{I}_2(\mathscr{H})$ and

$$\|TS\|_2 \leq \|T\|_2\|S\|, \quad \|ST\|_2 \leq \|S\|\,\|T\|_2. \tag{1.25}$$

The following lemma is a basic fact on $\mathfrak{I}_2(\mathscr{H})$:

Lemma 1.2 *Let* $\{e_n\}_{n=1}^{\infty}$ *be a CONS of* $\mathscr{H}$. *Then* $\{P_{e_m,e_n}\}_{m,n\in\mathscr{H}}$ *is a CONS of* $\mathfrak{I}_2(\mathscr{H})$.

Proof It is easy to see that $\{P_{e_m,e_n}\}_{m,n\in\mathscr{H}}$ is an ONS of $\mathfrak{I}_2(\mathscr{H})$. To prove the completeness of it, let $T \in \mathfrak{I}_2(\mathscr{H})$ be such that $\langle T, P_{e_m,e_n}\rangle_2 = 0$, $m, n \in \mathbb{N}$. It is easy to see that $\langle T, P_{e_m,e_n}\rangle_2 = \langle Te_m, e_n\rangle$. Hence $\langle Te_m, e_n\rangle = 0$, $m, n \in \mathbb{N}$, which implies that $T = 0$. Thus $\{P_{e_m,e_n}\}_{m,n\in\mathscr{H}}$ is a CONS of $\mathfrak{I}_2(\mathscr{H})$. ■

1.1.9 Reduction of Operators

For a closed subspace $\mathscr{M} \subset \mathscr{H}$, there exists a unique orthogonal projection $P_{\mathscr{M}}$ such that $\mathrm{Ran}\,(P_{\mathscr{M}}) = \mathscr{M}$. The operator $P_{\mathscr{M}}$ is called the orthogonal projection onto $\mathscr{M}$.

A linear operator A on $\mathscr{H}$ is said to be **reduced** by $\mathscr{M}$ if $P_{\mathscr{M}}A \subset AP_{\mathscr{M}}$ (i.e., for all $\Psi \in D(A)$, $P_{\mathscr{M}}\Psi \in D(A)$ and $AP_{\mathscr{M}}\Psi = P_{\mathscr{M}}A\Psi$). In this case, the operator $A_{\mathscr{M}}$ on $\mathscr{M}$ defined by

$$D(A_{\mathscr{M}}) := D(A) \cap \mathscr{M}, \quad A_{\mathscr{M}}\Psi = A\Psi, \quad \Psi \in D(A_{\mathscr{M}})$$

is called the **reduced part** of A to $\mathscr{M}$.

We denote by $\mathscr{M}^{\perp}$ the orthogonal complement of $\mathscr{M}$:

$$\mathscr{M}^{\perp} := \{\Phi \in \mathscr{H} \mid \langle \Phi, \Psi\rangle_{\mathscr{H}} = 0,\ \forall \Psi \in \mathscr{M}\},$$

which is a closed subspace of $\mathscr{H}$. Note that $Q_{\mathscr{M}} := 1 - P_{\mathscr{M}}$ is the orthogonal projection onto $\mathscr{M}^{\perp}$ (by the projection theorem) and $P_{\mathscr{M}}A \subset AP_{\mathscr{M}}$ is equivalent to $Q_{\mathscr{M}}A \subset AQ_{\mathscr{M}}$. Hence, if A is reduced by $\mathscr{M}$, then $\mathscr{M}^{\perp}$ also reduces A.

Many properties of a linear operator reduced by a closed subspace are inherited by the reduced part of it:

Proposition 1.16 *Let* $\mathscr{M}$ *be a closed subspace of* $\mathscr{H}$ *and* A *be a linear operator on* $\mathscr{H}$ *which is reduced by* $\mathscr{M}$.

(i) *If* A *is densely defined, then so is* $A_{\mathscr{M}}$.
(ii) *If* A *is closed, then so is* $A_{\mathscr{M}}$.
(iii) *If* A *is a densely defined closed operator, then so is* $A_{\mathscr{M}}$.
(iv) *If* A *is symmetric, then so is* $A_{\mathscr{M}}$.

(v) *If A is self-adjoint, then so is $A_{\mathcal{M}}$. Moreover, for any Borel measurable function f on $\mathbb{R}$, $f(A)$ (see (1.15)) is reduced by $\mathcal{M}$ and its reduced part $f(A)_{\mathcal{M}}$ is given by $f(A)_{\mathcal{M}} = f(A_{\mathcal{M}})$.*
(vi) *If A is unitary, then so is $A_{\mathcal{M}}$.*

1.1.10 Strongly Continuous One-Parameter Unitary Groups

Let $\mathcal{X}$ be a Banach space with norm $\|\cdot\|_{\mathcal{X}}$ and $\mathbb{I}$ be an interval of $\mathbb{R}$. A mapping $F: \mathbb{I} \to \mathcal{X}$ ($\mathbb{I} \ni t \mapsto F(t) \in \mathcal{X}$) is called an $\mathcal{X}$-valued function on $\mathbb{I}$. Let $a \in \mathbb{I}$. If

$$\lim_{\varepsilon\to 0} \|F(a+\varepsilon) - F(a)\|_{\mathcal{X}} = 0,$$

then F is said to be **strongly continuous** at point a. If F is strongly continuous at all points in $\mathbb{I}$, then it is said to be strongly continuous on $\mathbb{I}$.

If there exists a vector $F'(a) \in \mathcal{X}$ such that

$$\lim_{\varepsilon\to 0} \left\| \frac{1}{\varepsilon}(F(a+\varepsilon) - F(a)) - F'(a) \right\|_{\mathcal{X}} = 0,$$

then F is said to be **strongly differentiable** at point a and $F'(a)$ is called the strong differential vector of F at point a. If F is strongly differentiable at all points in $\mathbb{I}$, then it is said to be strongly differentiable on $\mathbb{I}$. In this case the mapping F': $\mathbb{I} \to \mathcal{X}; \mathbb{I} \ni t \mapsto F'(t)$ is called the **strong derivative** of F. The strong derivative $F'(t)$ is written also as

$$F'(t) = \frac{dF(t)}{dt} = \frac{d}{dt}F(t).$$

The notion of boundedness of a linear operator on $\mathcal{X}$ is defined in the same way as in the case of a linear operator on a Hilbert space.[5] We denote by $\mathfrak{B}(\mathcal{X})$ the set of everywhere defined bounded linear operators on $\mathcal{X}$.

Let $\mathbb{I}$ be an interval of $\mathbb{R}$ as above and $T(\cdot): \mathbb{I} \to \mathfrak{B}(\mathcal{X}); \mathbb{I} \ni t \mapsto T(t) \in \mathfrak{B}(\mathcal{X})$ be a mapping from $\mathbb{I}$ to $\mathfrak{B}(\mathcal{X})$ (i.e., $\mathfrak{B}(\mathcal{X})$-valued function on $\mathbb{I}$). For each $\Psi \in \mathcal{X}$, we define an $\mathcal{X}$-valued function F_Ψ on $\mathbb{I}$ by

$$F_\Psi(t) := T(t)\Psi.$$

Let $a \in \mathbb{I}$. The mapping $T(\cdot)$ is said to be **strongly continuous** at point a if, for all $\Psi \in \mathcal{X}$, F_Ψ is strongly continuous at point a. If F_Ψ is strongly continuous on $\mathbb{I}$, then $T(\cdot)$ is said to be strongly continuous on $\mathbb{I}$.

---

[5] A linear operator A on $\mathcal{X}$ is said to be bounded if there exists a constant $C > 0$ such that $\|A\Psi\|_{\mathcal{X}} \leq C\|\Psi\|_{\mathcal{X}}$, $\Psi \in D(A)$. In this case, the operator norm $\|A\|$ is defined by $\|A\| := \sup_{\Psi\in D(A),\Psi\neq 0}(\|A\Psi\|_{\mathcal{X}}/\|\Psi\|_{\mathcal{X}})$.

Let $\mathscr{D}$ be a subspace of $\mathscr{X}$ and $\Psi \in \mathscr{D}$. Then the $\mathscr{X}$-valued function $T(\cdot)\Psi$: $\mathbb{I} \to \mathscr{X}; \mathbb{I} \ni t \mapsto T(t)\Psi \in \mathscr{X}$ is said to be **strongly differentiable** at point a if F_Ψ is strongly differentiable at point a. In this case, we write

$$F'_\Psi(a) = \left.\frac{dT(t)}{dt}\right|_{t=a} \Psi \quad \text{or} \quad T'(a)\Psi.$$

If, for all $\Psi \in \mathscr{D}$, F_Ψ is strongly differentiable on $\mathbb{I}$, then $T(\cdot)$ is said to be strongly differentiable on $\mathbb{I}$ with domain $\mathscr{D}$. In this case, we write

$$F'_\Psi(t) = \frac{dT(t)}{dt}\Psi \quad \text{or} \quad T'(t)\Psi.$$

The operator $dT(t)/dt$ or $T'(t)$ is called the **strong derivative** of $T(t)$ on $\mathscr{D}$.

Let $\mathscr{H}$ be a Hilbert space. A family $\{U(t)\}_{t\in\mathbb{R}}$ of unitary operators on $\mathscr{H}$ is called a **strongly continuous one-parameter unitary group** if the following hold:

(i) The mapping $\mathbb{R} \ni t \mapsto U(t)$ is strongly continuous.
(ii) (group property) For all $s, t \in \mathbb{R}$, $U(t+s) = U(t)U(s)$.

It follows from (ii) and the unitarity of $U(0)$ that $U(0) = I$ (identity).

Example 1.6 For each self-adjoint operator H on a Hilbert space $\mathscr{H}$, $\{e^{-itH}\}_{t\in\mathbb{R}}$ is a strongly continuous one-parameter unitary group. Moreover, for all $\Psi \in D(H)$, $\Psi_H(t) := e^{-itH}\Psi$ is in $D(H)$ and strongly differentiable in t, satisfying the differential equation

$$i\frac{d}{dt}\Psi_H(t) = H\Psi_H(t). \tag{1.26}$$

In general, for an operator A on a Hilbert space $\mathscr{H}$, the differential equation

$$i\frac{d}{dt}\Psi(t) = A\Psi(t) \quad (\Psi(t) \in D(A), \forall t \in \mathbb{R}) \tag{1.27}$$

is called the **abstract Schrödinger equation** with respect to A, where $d\Psi(t)/dt$ is taken in the sense of strong derivative.

Equation (1.26) shows that, for each $\Psi \in D(H)$, $\Psi_H(t)$ is a solution of the abstract Schrödinger equation with respect to H such that $\Psi_H(0) = \Psi$.

Conversely, the initial value problem for the abstract Schrödinger equation with respect to a self-adjoint operator H has a unique solution which is of the form $\Psi_H(t)$.

A fundamental fact on a strongly continuous one-parameter unitary group is stated in the next theorem.

Theorem 1.6 (Stone's Theorem) *For each strongly continuous one-parameter unitary group $\{U(t)\}_{t\in\mathbb{R}}$ on a Hilbert space $\mathscr{H}$, there exists a unique self-adjoint operator S on $\mathscr{H}$ such that $U(t) = e^{itS}$, $t \in \mathbb{R}$.*

For a proof of this theorem, see, e.g., [130, Theorem VIII.8] or [23, Theorem 4.5].

The self-adjoint operator S in Theorem 1.6 is called the **generator** of the unitary group $U(t)$.[6]

The next theorem may be useful in proving essential self-adjointness for a class of symmetric operators.

Theorem 1.7 *Let $\{U(t)\}_{t\in\mathbb{R}}$ be a strongly continuous one-parameter unitary group on $\mathscr{H}$ and suppose that there exists a dense subspace $\mathscr{D}$ in $\mathscr{H}$ such that, for all $t \in \mathbb{R}$, $U(t)\mathscr{D} \subset \mathscr{D}$ and, for all $\Psi \in \mathscr{D}$, $U(t)\Psi$ is strongly differentiable in $t \in \mathbb{R}$. Define an operator S_0 on $\mathscr{H}$ by $D(S_0) := \mathscr{D}$ and*

$$S_0\Psi := \frac{1}{i}\lim_{t\to 0}\frac{1}{t}(U(t)-1)\Psi, \quad \Psi \in \mathscr{D}.$$

Then S_0 is essentially self-adjoint and $U(t) = e^{it\overline{S_0}}$, $t \in \mathbb{R}$.

This theorem immediately implies the following fact:

Corollary 1.2 *Let S be a self-adjoint operator on $\mathscr{H}$. Suppose that there exists a dense subspace $\mathscr{D}$ in $\mathscr{H}$ such that $e^{itS}\mathscr{D} \subset \mathscr{D}$, $t \in \mathbb{R}$. Then S is essentially self-adjoint on $\mathscr{D}$.*

Example 1.7 In quantum theory, the total energy of a quantum system is represented by a self-adjoint operator H on a Hilbert space $\mathscr{H}$, called the **Hamiltonian** of the system, where non-zero vectors in $\mathscr{H}$ describe state vectors of the system. In the case where H is independent of time $t \in \mathbb{R}$, the time development of the system is described by the strongly continuous one-parameter unitary group $\{e^{-itH/\hbar}\}_{t\in\mathbb{R}}$, where $\hbar := h/2\pi$ is the reduced Planck constant (h is the Planck constant). If $\Psi \in \mathscr{H}$ is the initial state (the state vector at $t = 0$), then the state vector at time t is given by $e^{-itH/\hbar}\Psi$.

1.1.11 Strongly Continuous One-Parameter Semi-Groups

Let $\mathscr{X}$ be a Banach space and $T(t) \in \mathfrak{B}(\mathscr{X})$ with $t \geq 0$ be given. A family $\{T(t)\}_{t\geq 0}$ is called a **strongly continuous one-parameter semi-group** or (C_0)-**semi-group** if the following hold:

(i) The mapping $[0,\infty) \ni t \mapsto T(t)$ is strongly continuous:
(ii) (semi-group property) For all $s, t \in [0,\infty)$, $T(t+s) = T(t)T(s)$.
(iii) $T(0) = I$.

[6] There is a case where iS is called the generator of $U(t)$.

Example 1.8 Let H be a self-adjoint operator on a Hilbert space $\mathscr{H}$ and bounded below. Then $\{e^{-\beta H}\}_{\beta\geq 0}$ is a strongly continuous one-parameter semi-group. Moreover, for all $\Psi \in D(H)$, $\Psi(\beta) := e^{-\beta H}\Psi$ is in $D(H)$ and strongly differentiable in β, satisfying the differential equation

$$\frac{d}{d\beta}\Psi(\beta) = -H\Psi(\beta). \tag{1.28}$$

For a (C_0)-semi-group $\{T(t)\}_{t\geq 0}$, one can define an operator A as follows:

$$D(A) := \{\Psi \in \mathscr{X} | \text{the limit s}-\lim_{\varepsilon\downarrow 0}\varepsilon^{-1}(T(\varepsilon)\Psi - \Psi) \text{ exists}\}, \tag{1.29}$$

$$A\Psi := \lim_{\varepsilon\downarrow 0}\varepsilon^{-1}(T(\varepsilon)\Psi - \Psi), \quad \Psi \in D(A). \tag{1.30}$$

The operator A is called the **generator** of the (C_0)-semi-group $\{T(t)\}_{t\geq 0}$ and *symbolically* written

$$T(t) = e^{tA}, \quad t \geq 0.$$

The following theorem is fundamental:

Theorem 1.8 (Hille–Yosida Theorem) *A linear operator A on a Banach space $\mathscr{X}$ is the generator of a (C_0)-semi-group $\{T(t)\}_{t\geq 0}$ on $\mathscr{X}$ if and only if the following hold:*

(i) *The operator A is a densely defined closed operator.*
(ii) *There exist real constants M and ω such that the semi-infinite interval (ω, ∞) is included in the resolvent set $\rho(A)$ and*

$$\|(A - \lambda)^{-n}\| \leq \frac{M}{(\lambda - \omega)^n}, \quad \lambda > \omega, \ n \in \mathbb{N}.$$

In that case, $\{T(t)\}_{t\geq 0}$ is uniquely determined by A and $\|T(t)\| \leq Me^{\omega t}$, $t \geq 0$. For all $\Psi \in D(A)$, $T(t)\Psi \in D(A)$, $t > 0$ and $T(t)\Psi$ is strongly differentiable in $t \in [0, \infty)$ with

$$\frac{d}{dt}T(t)\Psi = AT(t)\Psi = T(t)A\Psi.$$

For a proof of this theorem, see textbooks of functional analysis (e.g., [101, 104, 131, 165]).

As a special case of the Hille-Yosida theorem, one has the following theorem (the (C_0)-semi-group version of Stone's theorem):

Theorem 1.9 *Let $\{T(t)\}_{t\geq 0}$ be a strongly continuous one-parameter semi-group on a Hilbert space $\mathscr{H}$ such that, for all $t \geq 0$, $T(t)$ is self-adjoint. Then there exists a unique self-adjoint operator H on $\mathscr{H}$ bounded from below such that*

$$T(t) = e^{-tH}, \quad t \geq 0.$$

1.2 Strong Commutativity of Self-adjoint Operators

For self-adjoint operators on a Hilbert space, there is a stronger concept of commutativity.

Two self-adjoint operators A an B on $\mathscr{H}$ are said to **strongly commute** if the spectral measure E_A of A and the spectral measure E_B of B commute, i.e., for all Borel sets $J, K \subset \mathbb{R}$, $E_A(J)E_B(K) = E_B(K)E_A(J)$.

An n-tuple $A = (A_1, \ldots, A_n)$ of self-adjoint operators on $\mathscr{H}$ is said to be **strongly commuting** if, for all $j, k = 1, \ldots, n$ with $j \neq k$, A_j and A_k strongly commute.

Conditions equivalent to the strong commutativity of two self-adjoint operators on a Hilbert space are summarized in the following proposition[7]:

Proposition 1.17 *Let A and B be self-adjoint operators on a Hilbert space $\mathscr{H}$. The following are equivalent:*

(i) *A and B strongly commute.*
(ii) *For all $s, t \in \mathbb{R}$, e^{isA} and e^{itB} commute: $e^{isA}e^{itB} = e^{itB}e^{isA}$.*
(iii) *For all $z, w \in \mathbb{C} \setminus \mathbb{R}$, the resolvents $(A - z)^{-1}$ and $(B - w)^{-1}$ commute:*

$$(A - z)^{-1}(B - w)^{-1} = (B - w)^{-1}(A - z)^{-1}.$$

There is a simpler criterion for the strong commutativity of two self-adjoint operators.

Proposition 1.18 *Let A and B be self-adjoint operators on a Hilbert space $\mathscr{H}$. Then A and B strongly commute if and only if, for all $t \in \mathbb{R}$, $e^{itB}A \subset Ae^{itB}$. In that case the operator equality $e^{itB}A = Ae^{itB}$, $t \in \mathbb{R}$ holds.*

Proof Suppose that A and B strongly commute. Then part (ii) in Proposition 1.17 holds. This implies that

$$e^{isA} = e^{-itB}e^{isA}e^{itB} = e^{isA(t)} \tag{1.31}$$

[7] For a proof, see, e.g., [130, Theorem VIII.13], [23, Theorem 4.9] or [143, Propositions 5.27 and 6.15].

with $A(t) = e^{-itB}Ae^{itB}$, where, in the second equality, we have used the unitary covariance of functional calculus (see Theorem B.3 in Appendix B). Hence, for all $t \in \mathbb{R}$, $A = A(t)$, i.e., $A = e^{-itB}Ae^{itB}$. This is equivalent to

$$e^{itB}A = Ae^{itB}, \quad t \in \mathbb{R}. \tag{1.32}$$

In particular, $e^{itB}A \subset Ae^{itB}$ for all $t \in \mathbb{R}$.

Conversely, suppose that $e^{itB}A \subset Ae^{itB}$ for all $t \in \mathbb{R}$. Then, by applying Proposition 1.3(i), we obtain $A \subset e^{-itB}Ae^{itB}$ (multiply e^{-itB} from the left). Since the both sides are self-adjoint, they coincide: $A = e^{-itB}Ae^{itB}$. By the unitary covariance of functional calculus, we obtain the first equation in (1.31), which implies that part (ii) in Proposition 1.17 holds. Thus A and B strongly commute. ■

The next proposition shows that the concept of strong commutativity of self-adjoint operators is stronger than that of the ordinary commutativity:

Proposition 1.19 *Let A and B be strongly commuting self-adjoint operators on a Hilbert space $\mathscr{H}$. Then, for all $\Psi \in D(BA) \cap D(B)$, $\Psi \in D(AB)$ and $[A, B]\Psi = 0$. In particular, $[A, B] = 0$ on $D([A, B])$, i.e., A and B commute.*

Proof By Proposition 1.18, we have (1.32). Let $\Psi \in D(BA) \cap D(B)$ and $f(t) := e^{itB}A\Psi$ and $g(t) := Ae^{itB}\Psi$. Then $f(t) = g(t)$. Since $A\Psi \in D(B)$, it follows that $f(t)$ is strongly differentiable in t with strong derivative $f'(t) = ie^{itB}BA\Psi$. Hence $g(t)$ also is strongly differentiable in t. Since $\Psi \in D(B)$, $h(t) := e^{itB}\Psi$ is strongly differentiable in t with strong derivative $h'(t) = ie^{itB}B\Psi$. Since A is closed, it follows that $h'(t) \in D(A)$ (hence $\Psi \in D(Ae^{itB}B), t \in \mathbb{R}$) and $g'(t) = Ah'(t)$. Therefore $e^{itB}BA\Psi = Ae^{itB}B\Psi$. Considering the case $t = 0$, we see that $\Psi \in D(AB)$ and $AB\Psi = BA\Psi$. Thus the first half of the present proposition is proved. The second half is a simple consequence of it. ■

Proposition 1.18 suggests a concept of strong commutativity of a (not necessarily self-adjoint) linear operator with a self-adjoint operator.

Definition 1.1 A linear operator A on $\mathscr{H}$ (i.e., $A \in \mathfrak{L}(\mathscr{H})$) is said to **strongly commute** with a self-adjoint operator B on $\mathscr{H}$ if, for all $t \in \mathbb{R}$, $e^{itB}A \subset Ae^{itB}$.

Proposition 1.20 *Let $A \in \mathfrak{L}(\mathscr{H})$ and B be a self-adjoint operator. Then A strongly commutes with B if and only if*

$$e^{itB}Ae^{-itB} = A, \ \forall t \in \mathbb{R}. \tag{1.33}$$

Proof Suppose that A strongly commutes with B. Then, for all $t \in \mathbb{R}$, $e^{itB}A \subset Ae^{itB}$. Hence, by Proposition 1.3, $A \subset e^{-itB}Ae^{itB}$ and $e^{itB}Ae^{-itB} \subset A$. Since $t \in \mathbb{R}$ is arbitrary, we can take $-t$ as t in the former relation to obtain $A \subset e^{itB}Ae^{-itB}$. Hence (1.33) holds.

Conversely, if (1.33) holds, then $e^{itB}A = Ae^{itB}$, $t \in \mathbb{R}$. Thus A strongly commutes with B. ∎

Proposition 1.21 *Let $A \in \mathfrak{L}(\mathscr{H})$ and B be a self-adjoint operator. Suppose that A strongly commutes with B. Then A and B commute: $[A, B] = 0$ on $D(AB) \cap D(BA)$.*

Proof The proof of this proposition is similar to that of Proposition 1.19. ∎

Remark 1.5 The converse of Proposition 1.21 does not hold.

1.3 Criteria of Self-adjointness of Symmetric Operators

In quantum mechanics, physical quantities (observables) are represented by self-adjoint operators. But, in concrete models of quantum mechanics, it is usual that, for unbounded observables, only their symmetry is known a priori. Therefore it is important to have criteria for a symmetric operator to be self-adjoint or essentially self-adjoint. Here, for the reader's convenience, we recall some theorems on criteria for self-adjointness and essential self-adjointness of symmetric operators.

Theorem 1.10 (von Neumann's Theorem) *For any densely defined closed operator A on $\mathscr{H}$, A^*A is a non-negative self-adjoint operator.*

For a proof of this theorem, see, e.g., [16, Theorem 2.57] or [131, Theorem X.25].

Theorem 1.11 *Let A be a symmetric operator on $\mathscr{H}$. Then the following are equivalent:*

(i) *A is self-adjoint.*
(ii) *A is closed and* $\ker(A^* \pm i) = \{0\}$.
(iii) $\mathrm{Ran}\,(A \pm i) = \mathscr{H}$.

For a proof of this theorem, see, e.g., [130, Theorem VIII.3]. The next theorem immediately follows from Theorem 1.11.

Theorem 1.12 *Let A be a symmetric operator on $\mathscr{H}$. Then the following are equivalent:*

(i) *A is essentially self-adjoint.*
(ii) $\ker(A^* \pm i) = \{0\}$.
(iii) $\mathrm{Ran}\,(A \pm i)$ *are dense.*

Proof

(i)$\Longrightarrow$(ii) Suppose that A is essentially self-adjoint. Then $\bar{A}$ is self-adjoint. Hence, by applying Theorem 1.11(ii), $\ker(\bar{A}^* \pm i) = \{0\}$. But $\bar{A}^* = A^*$. Hence (ii) holds.

(ii)$\Longrightarrow$(iii) Condition (ii) implies that $\ker(\overline{A}^* \pm i) = \{0\}$. The operator $\overline{A}$ is a closed symmetric operator. Hence condition (ii) in Theorem 1.11 with A replaced by $\overline{A}$ is satisfied. Therefore $\mathrm{Ran}\,(\overline{A} \pm i) = \mathscr{H}$. It is easy to see that $\mathrm{Ran}\,(\overline{A} \pm i) = \overline{\mathrm{Ran}\,(A \pm i)}$. Hence $\mathrm{Ran}\,(A \pm i)$ are dense.

(iii)$\Longrightarrow$(i) As just shown, condition (iii) is equivalent to $\mathrm{Ran}\,(\overline{A} \pm i) = \mathscr{H}$. Hence condition (iii) in Theorem 1.11 is satisfied with A replaced by $\overline{A}$. Hence $\overline{A}$ is self-adjoint, i.e., A is essentially self-adjoint. ■

Theorems 1.11 and 1.12 can be used to find theorems on (essential) self-adjointness of symmetric operators.

Theorem 1.13 (Commutator Theorem) *Let S be a self-adjoint operator on $\mathscr{H}$ such that $S \geq 1$ and A be a symmetric operator on $\mathscr{H}$. Suppose that there exists a dense subspace $\mathscr{D}$ of $\mathscr{H}$ satisfying the following conditions:*

(i) *$\mathscr{D} \subset D(A) \cap D(S)$ and $\mathscr{D}$ is a core for S.*
(ii) *There exists a constant $a > 0$ such that, for all $\Psi \in \mathscr{D}$, $\|A\Psi\| \leq a\|S\Psi\|$.*
(iii) *There exists a constant $b > 0$ such that, for all $\Psi \in \mathscr{D}$,*
$|\langle A\Psi, S\Psi\rangle - \langle S\Psi, A\Psi\rangle| \leq b\|S^{1/2}\Psi\|^2$.

Then A is essentially self-adjoint on $\mathscr{D}$ and every core for S is a core for the self-adjoint operator $\overline{A \restriction \mathscr{D}}$.

For a proof of this theorem, see, e.g., [16, Theorem 2.32] or [131, Theorem X.37].
For a linear operator A on $\mathscr{H}$, we define

$$C^\infty(A) := \cap_{n=1}^\infty D(A^n), \tag{1.34}$$

called the C^∞-**domain** of A. A vector Ψ in $C^\infty(A)$ is said to be **analytic** if there exists a constant $t > 0$ such that $\sum_{n=0}^\infty \|A^n\Psi\| t^n/n! < \infty$.

Theorem 1.14 (Analytic Vector Theorem) *Let A be a symmetric operator on $\mathscr{H}$. Suppose that there exists a dense subspace $\mathscr{D} \subset D(A)$ such that each $\Psi \in \mathscr{D}$ is an analytic vector of A and $A\mathscr{D} \subset \mathscr{D}$. Then A is essentially self-adjoint on $\mathscr{D}$.*

For a proof, see, e.g., [131, p.203, Corollary 2] or [120].
There is a stronger notion of an analytic vector. A vector Ψ in $C^\infty(A)$ is said to be **entire analytic** if, for all $t > 0$, $\sum_{n=0}^\infty \|A^n\Psi\| t^n/n! < \infty$.
A basic theorem on self-adjointness of perturbations of a self-adjoint operator is given as follows:

Theorem 1.15 (The Kato–Rellich Theorem) *Let A be a self-adjoint operator on a Hilbert space $\mathscr{H}$ and B be a symmetric operator on $\mathscr{H}$ such that $D(A) \subset D(B)$ and*

$$\|B\Psi\| \leq a\|A\Psi\| + b\|\Psi\|, \quad \Psi \in D(A)$$

for constants a and $b \geq 0$ with $0 \leq a < 1$. Then $A + B$ is self-adjoint with $D(A+B) = D(A)$ and essentially self-adjoint on any core for A. Moreover, if A is bounded from below, then so is $A + B$.

For proofs of this theorem, we refer the reader to [98, Chapter V, §4] or [131, Theorem X.12].

We want to add a theorem on self-adjointness which is often useful. To prove the theorem, we recall a basic fact in Hilbert space theory:

Lemma 1.3 *Let $\{e_n\}_{n=1}^{\infty}$ be an ONS of a Hilbert space $\mathscr{H}$ and $\alpha := \{\alpha_n\}_{n=1}^{\infty}$ be a complex sequence such that $\sum_{n=1}^{\infty} |\alpha_n|^2 < \infty$. Then the infinite series*

$$\Psi(\alpha) := \sum_{n=1}^{\infty} \alpha_n e_n$$

converges in $\mathscr{H}$ and

$$\|\Psi(\alpha)\|^2 = \sum_{n=1}^{\infty} |\alpha_n|^2.$$

Proof An easy exercise. ■

For a subset $\mathscr{E} \neq \emptyset$ of a complex vector space, we denote by span($\mathscr{E}$) or span $\mathscr{E}$ the subspace algebraically spanned by $\mathscr{E}$:

$$\text{span}(\mathscr{E}) := \left\{ \sum_{j=1}^{n} \alpha_j \Psi_j | n \in \mathbb{N},\ \alpha_j \in \mathbb{C},\ \Psi_j \in \mathscr{E},\ j = 1, \ldots, n \right\}. \tag{1.35}$$

The following theorem also is often useful.

Theorem 1.16 *Let S be a closed symmetric operator on a Hilbert space. Suppose that there exists a CONS $\{e_n\}_{n=1}^{\infty}$ of $\mathscr{H}$ such that, for each $n \in \mathbb{N}$, $e_n \in D(S)$ and $Se_n = \lambda_n e_n$ with a real constant λ_n (i.e., λ_n is an eigenvalue of S). Then S is essentially self-adjoint on* span$\{e_n | n \in \mathbb{N}\}$ *and hence self-adjoint. Moreover,*

$$\sigma(S) = \overline{\{\lambda_n | n \in \mathbb{N}\}}. \tag{1.36}$$

Proof Let $\mathscr{D} := \text{span}\{e_n | n \in \mathbb{N}\}$. We prove that $(S \pm i)\mathscr{D}$ are dense in $\mathscr{H}$. Then, by Theorem 1.12, S is essentially self-adjoint on $\mathscr{D}$ and hence $\overline{S \restriction \mathscr{D}} = S$ is self-adjoint. Let Ψ be in $((S+i)\mathscr{D})^{\perp}$. Then, for all $n \in \mathbb{N}$, $\langle (S+i)e_n, \Psi \rangle = 0$. The left hand side is equal to $(\lambda_n - i)\langle e_n, \Psi \rangle$. Hence $\langle e_n, \Psi \rangle = 0$. Therefore $\Psi = 0$. Thus $(S+i)\mathscr{D}$ is dense in $\mathscr{H}$. Similarly, one can show that $(S-i)\mathscr{D}$ is dense in $\mathscr{H}$.

Let us prove (1.36). It is obvious that $\{\lambda_n | n \in \mathbb{N}\} \subset \sigma(S)$ and hence $\overline{\{\lambda_n | n \in \mathbb{N}\}} \subset \sigma(S)$. To prove the converse inclusion relation, let $z \in (\overline{\{\lambda_n | n \in \mathbb{N}\}})^c$. Then there exists a constant $\delta > 0$ such that $|\lambda_n - z| \geq \delta$, $n \in \mathbb{N}$.

Let $\Psi \in \ker(S - z)$. Then $(S - z)\Psi = 0$. Hence $\langle e_n, (S - z)\Psi \rangle = 0$. The left hand side is equal to $\langle (S - z^*)e_n, \Psi \rangle = (\lambda_n - z) \langle e_n, \Psi \rangle$. Hence $\langle e_n, \Psi \rangle = 0$. Therefore $\Psi = 0$. Thus $S - z$ is injective.

To show that $S - z$ is surjective, let $\Phi \in \mathscr{H}$. Then

$$\sum_{n=1}^{\infty} \left| \frac{\langle e_n, \Phi \rangle}{\lambda_n - z} \right|^2 \leq \frac{1}{\delta^2} \sum_{n=1}^{\infty} |\langle e_n, \Phi \rangle|^2 = \frac{1}{\delta^2} \|\Phi\|^2 < \infty.$$

Hence, by Lemma 1.3, the infinite series

$$\Psi := \sum_{n=1}^{\infty} \frac{\langle e_n, \Phi \rangle}{\lambda_n - z} e_n$$

converges in $\mathscr{H}$. Let $\Psi_N := \sum_{n=1}^{N} \frac{\langle e_n, \Phi \rangle}{\lambda_n - z} e_n$. Then $\Psi_N \to \Psi$ as $N \to \infty$. It is easy to see that $\Psi_N \in D(S)$ and $(S - z)\Psi_N = \sum_{n=1}^{N} \langle e_n, \Phi \rangle e_n \to \Phi$ $(N \to \infty)$. Hence, by the closedness of $S - z$, $\Psi \in D(S - z) = D(S)$ and $(S - z)\Psi = \Phi$. Hence S is surjective. Thus $z \in \rho(S)$, implying that $(\overline{\{\lambda_n | n \in \mathbb{N}\}})^c \subset \rho(S)$. Hence $\sigma(S) \subset \overline{\{\lambda_n | n \in \mathbb{N}\}}$. ■

Remark 1.6 If $\{\lambda_n\}_n$ in Theorem 1.16 has no accumulation points, then $\sigma(S) = \sigma_{\mathrm{p}}(S) = \{\lambda_n | n \in \mathbb{N}\}$.

1.4 Basic Properties of Symmetric Operators

Non-essential self-adjoint symmetric operators also play important roles in quantum mechanics. For later use, here we state basic properties of a closed symmetric operator. Let

$$\Pi_+ := \{z \in \mathbb{C} | \operatorname{Im} z > 0\} \text{ (the open upper half-plane in } \mathbb{C}), \tag{1.37}$$

$$\Pi_- := \{z \in \mathbb{C} | \operatorname{Im} z < 0\} \text{ (the open lower half-plane in } \mathbb{C}), \tag{1.38}$$

where $\operatorname{Im} z$ denotes the imaginary part of z.

Proposition 1.22 *Let A be a symmetric operator on $\mathscr{H}$. Then:*

(i) *$\overline{A}$ is a closed symmetric operator and*

$$\sigma(A) = \sigma(\overline{A}), \quad \rho(A) = \rho(\overline{A}). \tag{1.39}$$

(ii) *For all $z \in \mathbb{C} \setminus \mathbb{R}$ and $\Psi \in D(A)$,*

$$\|(A - z)\Psi\| \geq |\operatorname{Im} z| \, \|\Psi\|. \tag{1.40}$$

In particular, $A - z$ is injective and $(A - z)^{-1}$ is bounded with $D((A - z)^{-1}) = \operatorname{Ran}(A - z)$ and $\|(A - z)^{-1}\| \leq 1/|\operatorname{Im} z|$.

Proof

(i) It is an elementary fact that A is closable and $\overline{A}$ is a closed symmetric operator. Therefore, by Proposition 1.8, (1.39) holds.
(ii) This is a well-known fact. ∎

Proposition 1.23 *Let A be a symmetric operator on $\mathscr{H}$. Then:*

(i) $\dim\ker(A^* - z)$ *is constant throughout* Π_+ ($z \in \Pi_+$).
(ii) $\dim\ker(A^* - w)$ *is constant throughout* Π_- ($w \in \Pi_-$).

For a proof of this proposition, see, e.g., [131, Theorem X.1] (note that $A^* = (\overline{A})^*$).

For a symmetric operator A, the non-negative integers

$$n_+(A) := \dim\ker(A^* - i) \tag{1.41}$$

and

$$n_-(A) := \dim\ker(A^* + i) \tag{1.42}$$

are called the **deficiency indices** of A. Since $A^* = (\overline{A})^*$, we have

$$n_\pm(A) = n_\pm(\overline{A}).$$

By Proposition 1.23,

$$n_+(A) = \dim\ker(A^* - z), \quad z \in \Pi_+ \tag{1.43}$$

and

$$n_-(A) = \dim\ker(A^* - w), \quad w \in \Pi_-. \tag{1.44}$$

Theorem 1.17 *Let A be a symmetric operator on $\mathscr{H}$. Then:*

(i) *If* $\ker(A^* - z_0) \neq \{0\}$ *for some* $z_0 \in \Pi_+$, *then* $\overline{\Pi_-} \subset \sigma(A)$.
(ii) *If* $\ker(A^* - w_0) \neq \{0\}$ *for some* $w_0 \in \Pi_-$, *then* $\overline{\Pi_+} \subset \sigma(A)$.
(iii) *If* $\ker(A^* - z_0) \neq \{0\}$ *and* $\ker(A^* - w_0) \neq \{0\}$ *for some* $z_0 \in \Pi_+$ *and some* $w_0 \in \Pi_-$, *then* $\sigma(A) = \mathbb{C}$.
(iv) *On the spectrum* $\sigma(A)$ *of A, one of the following holds:*
    (a) $\sigma(A) = \overline{\Pi_+}$.
    (b) $\sigma(A) = \overline{\Pi_-}$.
    (c) $\sigma(A) = \mathbb{C}$.
    (d) $\sigma(A) \subset \mathbb{R}$.
(v) *A is self-adjoint if and only if case (d) in (iv) holds. In that case,* $\sigma_r(A) = \emptyset$.

For a proof of the theorem, see, e.g., [131, Theorem X.1].

As for self-adjoint extensions of a symmetric operator, the following theorem holds:

Theorem 1.18 *Let A be a symmetric operator on $\mathscr{H}$. Then:*

(i) *A is self-adjoint if and only if A is closed and $n_\pm(A) = 0$.*
(ii) *A has self-adjoint extensions if and only if $n_+(A) = n_-(A)$. There exists a one-to-one correspondence between self-adjoint extensions of A and unitary operators from $\ker(A^* - i)$ to $\ker(A^* + i)$.*
(iii) *Let $n_+(A) = n_-(A) \geq 1$. Then A has uncountably many different self-adjoint extensions.*
(iv) *If $n_+(A) = 0$ and $n_-(A) \neq 0$ or $n_+(A) \neq 0$ and $n_-(A) = 0$, then $\overline{A}$ is maximally symmetric.*

For a proof of this theorem, see, e.g., [131, p. 141, Corollary].

Theorem 1.19 *Let A be a symmetric operator on $\mathscr{H}$.*

(i) *If $n_+(A) = 0$ and $n_-(A) \neq 0$, then $\sigma(A) = \overline{\Pi}_+$.*
(ii) *If $n_+(A) \neq 0$ and $n_-(A) = 0$, then $\sigma(A) = \overline{\Pi}_-$.*
(iii) *If $n_+(A) \neq 0$ and $n_-(A) \neq 0$, then $\sigma(A) = \mathbb{C}$.*

Proof

(i) The condition $n_-(A) \neq 0$ and Theorem 1.17(ii) imply that $\overline{\Pi}_+ \subset \sigma(A)$. By condition $n_+(A) = 0$ and (1.43), $\ker(A^* - z) = \{0\}$ for all $z \in \Pi_+$. Hence, by applying Theorem 1.2, we have $\mathscr{H} = \overline{\operatorname{Ran}(A - z^*)} = \operatorname{Ran}(\bar{A} - z^*)$. Hence, for all $z \in \Pi_+, z^* \in \rho(\bar{A}) = \rho(A)$. Therefore $\Pi_- \subset \rho(A)$. Thus $\sigma(A) = \overline{\Pi}_+$.
(ii) The present assumption implies that $n_+(-A) = 0$ and $n_-(-A) \neq 0$. Hence, by applying (i) with A replaced by $-A$, $\sigma(-A) = \overline{\Pi}_+$. Hence, by (1.11), we obtain $\sigma(A) = \overline{\Pi}_-$.
(iii) The present assumption and Theorem 1.2 imply that, for all $z \in \Pi_+$ and $w \in \Pi_-$, $\overline{\operatorname{Ran}(A - z^*)}$ and $\overline{\operatorname{Ran}(z - w^*)}$ are not equal to $\mathscr{H}$. Hence $z^*, w^* \in \sigma(A)$, which implies that $\sigma(A) = \mathbb{C}$. ■

1.5 Tensor Product of Hilbert Spaces

For complex vector spaces $\mathscr{V}_1, \ldots, \mathscr{V}_n$ with $n \in \mathbb{N}$, a mapping $f: \mathscr{V}_1 \times \cdots \times \mathscr{V}_n \to \mathbb{C}$ is called a **conjugate n-linear functional** if it is separately anti (conjugate)-linear, i.e., for each $j = 1, \ldots, n$ and all $u_j \in \mathscr{V}_j$, $(v_1, \ldots, v_n) \in \mathscr{V}_1 \times \cdots \times \mathscr{V}_n$, $\alpha \in \mathbb{C}$,

$$f(v_1, \ldots, \underbrace{v_j + u_j}_{j\text{-th}}, \ldots, v_n) = f(v_1, \ldots, \underbrace{v_j}_{j\text{-th}}, \ldots, v_n) + f(v_1, \ldots, \underbrace{u_j}_{j\text{-th}}, \ldots, v_n),$$

$$f(v_1, \ldots, \underbrace{\alpha v_j}_{j\text{-th}}, \ldots, v_n) = \alpha^* f(v_1, \ldots, \underbrace{v_j}_{j\text{-th}}, \ldots, v_n).$$

The set $\mathscr{L}_*(\mathscr{V}_1 \times \cdots \times \mathscr{V}_n)$ of conjugate n-linear functionals on $\mathscr{V}_1 \times \cdots \times \mathscr{V}_n$ becomes a complex vector space with the usual operation of addition of functionals and that of scalar multiplication.

Let $\mathscr{H}_1, \ldots, \mathscr{H}_n$ be Hilbert spaces ($n \geq 2$). Then, for each n-tuple $(\Psi_1, \ldots, \Psi_n) \in \mathscr{H}_1 \times \cdots \times \mathscr{H}_n$, a conjugate n-linear functional is defined by

$$(\otimes_{j=1}^n \Psi_j)(\Phi_1, \ldots, \Phi_n) := \prod_{j=1}^n \langle \Phi_j, \Psi_j \rangle, \quad (\Phi_1, \ldots, \Phi_n) \in \mathscr{H}_1 \times \cdots \times \mathscr{H}_n.$$

The conjugate n-linear functional $\otimes_{j=1}^n \Psi_j$ is called the **tensor product** of $\Psi_1, \ldots, \Psi_n$. It is also written $\Psi_1 \otimes \cdots \otimes \Psi_n$.

Let $\mathscr{D}_j$ be a subspace of $\mathscr{H}_j$ ($j = 1, \ldots, n$). Then the subspace

$$\hat{\otimes}_{j=1}^n \mathscr{D}_j := \text{span}\{\otimes_{j=1}^n \Psi_j | \Psi_j \in \mathscr{D}_j,\ j = 1, \ldots, n\}$$

of $\mathscr{L}_*(\mathscr{H}_1 \times \cdots \times \mathscr{H}_n)$ is called the **algebraic tensor product** of $\mathscr{D}_1, \ldots, \mathscr{D}_n$. If $\mathscr{D}_j = \mathscr{D}$, $j = 1, \ldots, n$, then we write $\hat{\otimes}_{j=1}^n \mathscr{D}_j = \hat{\otimes}^n \mathscr{D}$.

It is shown[8] that there is an inner product $\langle\ ,\ \rangle$ of $\hat{\otimes}_{j=1}^n \mathscr{H}_j$ such that

$$\left\langle \otimes_{j=1}^n \Phi_j, \otimes_{j=1}^n \Psi_j \right\rangle = \prod_{j=1}^n \langle \Phi_j, \Psi_j \rangle, \quad \Phi_j, \Psi_j \in \mathscr{H}_j,\ j = 1, \ldots, n.$$

The completion of $\hat{\otimes}_{j=1}^n \mathscr{H}_j$ with respect to this inner product is called the **tensor product Hilbert space** (or simply the tensor product) of $\mathscr{H}_1, \ldots, \mathscr{H}_n$ and written $\otimes_{j=1}^n \mathscr{H}_j$. By the definition of completion of an inner product space, the algebraic tensor product $\hat{\otimes}_{j=1}^n \mathscr{H}_j$ is identified with a dense subspace of $\otimes_{j=1}^n \mathscr{H}_j$. In the case where $\mathscr{H}_j = \mathscr{H}$, $j = 1, \ldots, n$, we write

$$\otimes_{j=1}^n \mathscr{H}_j = \otimes^n \mathscr{H}.$$

It is called the n-**fold tensor product** of $\mathscr{H}$.

The following theorem is a basic theorem in the theory of tensor product Hilbert spaces:

Theorem 1.20 *Let $\{e_k^{(j)}\}_{k=1}^\infty$ be a CONS of $\mathscr{H}_j$. Then*

$$\{e_{k_1}^{(1)} \otimes \cdots \otimes e_{k_n}^{(n)} | k_j \in \mathbb{N}, j = 1, \ldots, n\}$$

is a CONS of $\otimes_{j=1}^n \mathscr{H}_j$.

[8] See, e.g., [130, §II.4], [25, Lemma 1.2], [28, Lemma 2.3] or [143, Lemma 7.19].

Proof See, e.g., [28, Proposition 2.2(iv)]. ■

In concluding this section, we describe basic facts on continuity and differentiability of a mapping from an interval of $\mathbb{R}$ to $\otimes_{j=1}^n \mathscr{H}_j$. Let $\mathbb{I}$ be an interval of $\mathbb{R}$ and, for each $j = 1, \ldots, n$, $\Psi_j(\cdot)\colon \mathbb{I} \to \mathscr{H}_j; \mathbb{I} \ni t \mapsto \Psi_j(t) \in \mathscr{H}_j$ be a mapping from $\mathbb{I}$ to $\mathscr{H}_j$. Then a mapping $F\colon \mathbb{I} \to \otimes_{j=1}^n \mathscr{H}_j$ is defined by

$$F(t) := \otimes_{j=1}^n \Psi_j(t), \quad t \in \mathbb{I}. \tag{1.45}$$

Proposition 1.24

(i) *If each $\Psi_j(\cdot)$ is strongly continuous at point $a \in \mathbb{I}$, then so is F.*
(ii) *If each $\Psi_j(\cdot)$ is strongly continuous on $\mathbb{I}$, then so is F.*
(iii) *If each $\Psi_j(\cdot)$ is strongly differentiable at point $a \in \mathbb{I}$, then so is F and*

$$F'(a) = \sum_{j=1}^n \Psi_1(a) \otimes \cdots \otimes \Psi_{j-1}(a) \otimes \Psi_j'(a) \otimes \Psi_{j+1}(a) \otimes \cdots \otimes \Psi_n(a), \tag{1.46}$$

where $\Psi_j'(a)$ is the strong differential vector of $\Psi_j(\cdot)$ at point a.
(iv) *If each $\Psi_j(\cdot)$ is strongly differentiable on $\mathbb{I}$, then so is F and $F'(t)$ is given by the right hand side of (1.46) with a replaced by t.*

Proof

(i) We have for all $\varepsilon \in \mathbb{R}$ such that $a + \varepsilon \in \mathbb{I}$

$$\begin{aligned} F(a+\varepsilon) - F(a) = &\sum_{j=1}^n \Psi_1(a) \otimes \cdots \otimes \Psi_{j-1}(a) \\ &\otimes (\Psi_j(a+\varepsilon) - \Psi_j(a)) \otimes \Psi_{j+1}(a+\varepsilon) \otimes \cdots \otimes \Psi_n(a+\varepsilon). \end{aligned} \tag{1.47}$$

Hence

$$\begin{aligned} \|F(a+\varepsilon) - F(a)\| \le &\sum_{j=1}^n \|\Psi_1(a)\| \cdots \|\Psi_{j-1}(a)\| \\ &\times \|\Psi_j(a+\varepsilon) - \Psi_j(a)\| \, \|\Psi_{j+1}(a+\varepsilon)\| \cdots \|\Psi_n(a+\varepsilon)\| \\ \to &\, 0 \quad (\varepsilon \to 0). \end{aligned}$$

Hence F is strongly continuous at point a.
(ii) This follows from (i).

(iii) By (1.47) with $\varepsilon \neq 0$, we have

$$\frac{1}{\varepsilon}(F(a+\varepsilon) - F(a)) = \sum_{j=1}^{n} \Psi_1(a) \otimes \cdots \otimes \Psi_{j-1}(a) \otimes \frac{1}{\varepsilon}(\Psi_j(a+\varepsilon) - \Psi_j(a)) \otimes \Psi_{j+1}(a+\varepsilon) \otimes \cdots \otimes \Psi_n(a+\varepsilon).$$

By the present assumption, $\lim_{\varepsilon \to 0} \varepsilon^{-1}(\Psi_j(a+\varepsilon) - \Psi_j(a)) = \Psi_j'(a)$. Hence, by an application of (i), we obtain

$$\lim_{\varepsilon \to 0} \frac{1}{\varepsilon}(F(a+\varepsilon) - F(a)) = \sum_{j=1}^{n} \Psi_1(a) \otimes \cdots \otimes \Psi_{j-1}(a) \otimes \Psi_j'(a) \otimes \Psi_{j+1}(a) \otimes \cdots \otimes \Psi_n(a).$$

Thus F is strongly differentiable at point a and (1.46) holds.

(iv) This follows from (iii). ■

1.6 Symmetric and Anti-symmetric Tensor Product Hilbert Spaces

There are special classes of tensor product Hilbert spaces. We denote by $\mathfrak{S}_n$ the symmetry group of order n (i.e., the set of permutations of $1, \ldots, n$). Let $\mathscr{H}$ be a Hilbert space. Then, for each $\sigma \in \mathfrak{S}_n$, there is a unique unitary operator U_σ on $\otimes^n \mathscr{H}$ such that

$$U_\sigma(\otimes_{j=1}^n \Psi_j) = \otimes_{j=1}^n \Psi_{\sigma(j)}, \quad \Psi_j \in \mathscr{H}, \ j = 1, \ldots, n. \tag{1.48}$$

Using the unitary operators U_σ, $\sigma \in \mathfrak{S}_n$, one can define the following operators:

$$S_n := \frac{1}{n!} \sum_{\sigma \in \mathfrak{S}_n} U_\sigma, \tag{1.49}$$

$$A_n := \frac{1}{n!} \sum_{\sigma \in \mathfrak{S}_n} \operatorname{sgn}(\sigma) U_\sigma, \tag{1.50}$$

where $\operatorname{sgn}(\sigma)$ means the signature of σ. One can prove that S_n and A_n are orthogonal projections:

$$S_n^* = S_n, \quad S_n^2 = S_n, \tag{1.51}$$

$$A_n^* = A_n, \quad A_n^2 = A_n. \tag{1.52}$$

The operator S_n (resp. A_n) is called the **symmetrizer** or the **symmetrization operator** (resp. the **anti-symmetrizer** or the **anti-symmetrization operator**) on $\otimes^n \mathscr{H}$. Therefore the ranges of S_n and A_n are closed and hence Hilbert spaces. We write

$$\otimes_{\mathrm{s}}^n \mathscr{H} := \operatorname{Ran}(S_n), \quad \otimes_{\mathrm{as}}^n \mathscr{H} := \operatorname{Ran}(A_n). \tag{1.53}$$

The Hilbert space $\otimes_{\mathrm{s}}^n \mathscr{H}$ (resp. $\otimes_{\mathrm{as}}^n \mathscr{H}$) is called the n-**fold symmetric** (resp. **anti-symmetric**) **tensor product** of $\mathscr{H}$. The anti-symmetric tensor product $\otimes_{\mathrm{as}}^n \mathscr{H}$ is sometimes written as $\wedge^n \mathscr{H}$ or $\wedge^n(\mathscr{H})$, being called the n-**fold wedge product** of $\mathscr{H}$.

For a subspace $\mathscr{D}$ of $\mathscr{H}$, the subspaces

$$\hat{\otimes}_{\mathrm{s}}^n \mathscr{D} := S_n(\hat{\otimes}^n \mathscr{D})$$

and

$$\hat{\otimes}_{\mathrm{as}}^n \mathscr{D} := A_n(\hat{\otimes}^n \mathscr{D})$$

are called respectively the n-**fold algebraic symmetric tensor product** of $\mathscr{D}$ and the n-**fold algebraic anti-symmetric tensor product** of $\mathscr{D}$.

Using (1.51), we have for all $\Psi_j, \Phi_j \in \mathscr{H}$, $j = 1, \ldots, n$

$$\begin{aligned}\langle S_n(\Psi_1 \otimes \cdots \otimes \Psi_n), S_n(\Phi_1 \otimes \cdots \otimes \Phi_n)\rangle &= \langle \Psi_1 \otimes \cdots \otimes \Psi_n, S_n(\Phi_1 \otimes \cdots \otimes \Phi_n)\rangle \\ &= \frac{1}{n!} \sum_{\sigma \in \mathfrak{S}_n} \prod_{j=1}^n \left\langle \Psi_j, \Phi_{\sigma(j)} \right\rangle. \end{aligned} \tag{1.54}$$

Let $\{e_j\}_{j=1}^\infty$ be a CONS of $\mathscr{H}$ and

$$\Psi_{j_1 \ldots j_n} := S_n(e_{j_1} \otimes \cdots \otimes e_{j_n}), \quad j_1, \ldots, j_n \in \mathbb{N}, \quad j_1 \le j_2 \le \cdots \le j_n.$$

Then, by (1.54), one sees that $\{\Psi_{j_1 \ldots j_n} | j_1, \ldots, k_n \in \mathbb{N}, j_1 \le j_2 \le \cdots \le j_n\}$ is an orthogonal system of $\otimes_{\mathrm{s}}^n \mathscr{H}$. Note that

$$\Psi_{j_1 \cdots j_n} = S_n((\otimes^{n_1} e_{i_1}) \otimes (\otimes^{n_2} e_{i_2}) \otimes \cdots \otimes (\otimes^{n_r} e_{i_r})$$

for some $(e_{i_1}, \ldots, e_{i_r})$ with $i_1 < i_2 < \cdots i_r$ $(i_1, \ldots, i_r \in \{j_1, j_2, \ldots, j_n\}$, $r \in \mathbb{N})$. Hence

$$\|\Psi_{j_1 \ldots j_n}\|^2 = \frac{n_1! \cdots n_r!}{n!}.$$

Using these facts, one obtains the following theorem[9]:

Theorem 1.21 *Let*

$$C_{j_1\dots j_n} := \sqrt{\frac{n!}{n_1!\cdots n_r!}}. \tag{1.55}$$

Then $\{C_{j_1\dots j_n}\Psi_{j_1\dots j_n}|j_1,\dots,j_n \in \mathbb{N},\ j_1 \le j_2 \le \cdots \le j_n\}$ *is a CONS of* $\otimes_{\mathrm{s}}^n \mathscr{H}$.

For $u_1,\dots,u_n \in \mathscr{H}$, we define the **wedge product** (or **exterior product**) of them $u_1 \wedge \cdots \wedge u_n \in \wedge^n(\mathscr{H})$ by

$$u_1 \wedge \cdots \wedge u_n := \sqrt{n!}A_n(\otimes_{j=1}^n u_j). \tag{1.56}$$

It is easy to see that, for all $u_j,\ v_j \in \mathscr{H}$ $(j = 1,\dots,n)$,

$$\langle u_1 \wedge \cdots \wedge u_n, v_1 \wedge \cdots \wedge v_n\rangle = \sum_{\sigma\in\mathfrak{S}_n} \mathrm{sgn}(\sigma)\left\langle u_1, v_{\sigma(1)}\right\rangle\cdots\left\langle u_n, v_{\sigma(n)}\right\rangle. \tag{1.57}$$

Using this formula, one can prove, in a similar way to above, the following theorem:

Theorem 1.22 *Let* $\{e_j\}_{j=1}^\infty$ *be a CONS of* $\mathscr{H}$. *Then*

$$\{e_{i_1} \wedge \cdots \wedge e_{i_n}|i_1,\dots,i_n \in \mathbb{N}, i_1 < i_2 < \dots < i_n\}$$

is a CONS of $\wedge^n(\mathscr{H})$.

Proof See [28, Proposition 2.11]. ■

1.7 Tensor Products of L^2-Spaces

Let (X, Σ, μ) be a σ-finite measure space and $L^2(X, d\mu)$ be the Hilbert space of equivalence classes of Σ-measurable functions on X (see Appendix A).

Let $n \in \mathbb{N}$, $n \ge 2$ and, for $j = 1,\dots,n$, (X_j, Σ_j, μ_j) be a σ-finite measure space. For each function f_j on X_j $(j = 1,\dots,n)$, one can define a function $f_1 \times \cdots \times f_n$ on the product space $X_1 \times \cdots \times X_n$ by

$$(f_1 \times \cdots \times f_n)(x_1,\dots,x_n) := \prod_{j=1}^n f_j(x_j), \quad (x_1,\dots,x_n) \in X_1 \times \cdots \times X_n.$$

We denote the product measure of $\mu_1,\ \dots,\ \mu_n$ by $\mu_1 \otimes \cdots \otimes \mu_n$.

[9]For more details, see [28, Proposition 2.9].

Theorem 1.23 *Suppose that each $L^2(X_j, d\mu_j)$ $(j = 1, \ldots, n)$ is separable. Then there exists a unitary operator $U: \otimes_{j=1}^n L^2(X_j, d\mu_j) \to L^2(X_1 \times \cdots \times X_n, d(\mu_1 \otimes \cdots \times \mu_n))$ such that*

$$U(\otimes_{j=1}^n f_j) = f_1 \times \cdots \times f_n, \quad f_j \in L^2(X_j, d\mu_j),\ j = 1, \ldots, n.$$

Proof See, e.g., [28, Theorem 2.5]. ■

The unitary operator U in Theorem 1.23 is independent of the choice of complete orthonormal systems of $L^2(X_j, d\mu_j)$, $j = 1, \ldots, n$. For this reason, we call it the **natural isomorphism** between $\otimes_{j=1}^n L^2(X_j, d\mu_j)$ and $L^2(X_1 \times \cdots \times X_n, d(\mu_1 \otimes \cdots \otimes \mu_n))$ and write

$$\otimes_{j=1}^n L^2(X_j, d\mu_j) \cong L^2(X_1 \times \cdots \times X_n, d(\mu_1 \otimes \cdots \otimes \mu_n)). \tag{1.58}$$

We now restrict ourselves to the case where

$$X := X_1 = X_2 = \cdots = X_n, \quad \mu := \mu_1 = \mu_2 = \cdots = \mu_n$$

and write

$$\mu^n := \underbrace{\mu \otimes \cdots \otimes \mu}_{n \text{ factors}}.$$

For an element f of $L^2(X^n, d\mu^n)$ and a permutation $\sigma \in \mathfrak{S}_n$, we define a function f_σ on X^n by

$$f_\sigma(x_1, \ldots, x_n) := f(x_{\sigma(1)}, \ldots, x_{\sigma(n)}), \quad \mu\text{-a.e.} x = (x_1, \ldots, x_n) \in X^n.$$

It follows that $f_\sigma \in L^2(X^n, d\mu^n)$.

An element f of $L^2(X^n, d\mu^n)$ is said to be **symmetric** (resp. **anti-symmetric**) if $f_\sigma = f$ (resp. $f_\sigma = \text{sgn}(\sigma) f$) for all $\sigma \in \mathfrak{S}_n$.

Let

$$L^2_{\text{sym}}(X^n, d\mu^n) := \{f \in L^2(X^n, d\mu^n) | \forall \sigma \in \mathfrak{S}_n,\ f_\sigma = f\}, \tag{1.59}$$

the set of symmetric elements of $L^2(X^n, d\mu^n)$. It is easy to see that $L^2_{\text{sym}}(X^n, d\mu^n)$ is a closed subspace of $L^2(X^n, d\mu^n)$ and hence it is a Hilbert space. The Hilbert space $L^2_{\text{sym}}(X^n, d\mu^n)$ is called the **Hilbert space of square integrable symmetric functions** on X^n with respect to μ^n. Contrary to this Hilbert space, one has

$$L^2_{\text{as}}(X^n, d\mu^n) := \{f \in L^2(X^n, d\mu^n) | \forall \sigma \in \mathfrak{S}_n,\ f_\sigma = \text{sgn}(\sigma) f\}, \tag{1.60}$$

which is called the **Hilbert space of square integrable anti-symmetric functions** on X^n with respect to μ^n.

Theorem 1.24 *Assume that $L^2(X, d\mu)$ is separable.*

(i) *There exists a unitary operator U_s: $\otimes_s^n L^2(X, d\mu) \to L^2_{sym}(X^n, d\mu^n)$ such that, for all $f_j \in L^2(X, d\mu)$ $(j = 1, \ldots, n)$,*

$$U_s(S_n(\otimes_{j=1}^n f_j)) = \frac{1}{n!} \sum_{\sigma \in \mathfrak{S}_n} f_{\sigma(1)} \times \cdots \times f_{\sigma(n)}.$$

(ii) *There exists a unitary operator U_{as}: $\wedge^n(L^2(X, d\mu)) \to L^2_{as}(X^n, d\mu^n)$ such that, for all $f_j \in L^2(X, d\mu)$ $(j = 1, \ldots, n)$,*

$$U_{as}(A_n(\otimes_{j=1}^n f_j)) = \frac{1}{n!} \sum_{\sigma \in \mathfrak{S}_n} \mathrm{sgn}(\sigma)(f_{\sigma(1)} \times \cdots \times f_{\sigma(n)}).$$

Proof See, e.g., [28, Theorems 2.10 and 2.11]. ∎

We call the unitary operator U_s the **natural isomorphism** between $\otimes_s^n L^2(X, d\mu)$ and $L^2_{sym}(X^n, d\mu^n)$, and the unitary operator U_{as} the **natural isomorphism** between $\wedge^n(L^2(X, d\mu))$ and $L^2_{as}(X^n, d\mu^n)$. These natural isomorphisms are symbolically written

$$\otimes_s^n L^2(X, d\mu) \cong L^2_{sym}(X^n, d\mu^n), \quad \wedge^n(L^2(X, d\mu)) \cong L^2_{as}(X^n, d\mu^n). \tag{1.61}$$

1.8 Tensor Products of Operators

We next define basic operators on tensor product Hilbert spaces. Let A_j be a densely defined closable operator from $\mathcal{H}_j$ $(j = 1, \ldots, n)$ to a Hilbert space $\mathcal{K}_j$. Then it is shown[10] that there exists a unique densely defined closed operator $\otimes_{j=1}^n A_j$ from $\otimes_{j=1}^n \mathcal{H}_j$ to $\otimes_{j=1}^n \mathcal{K}_j$ such that $\hat{\otimes}_{j=1}^n D(A_j)$ is a core for it and

$$(\otimes_{j=1}^n A_j)(\otimes_{j=1}^n \Psi_j) = \otimes_{j=1}^n A_j \Psi_j, \quad \Psi_k \in D(A_k), \quad k = 1, \ldots, n.$$

It follows that

$$\otimes_{j=1}^n A_j = \otimes_{j=1}^n \overline{A}_j.$$

[10] See the above-mentioned literature.

The operator $\otimes_{j=1}^n A_j$ is called the **tensor product operator** (or simply the tensor product) of $A_1, \ldots, A_n$.

In the case where $\mathscr{H}_j = \mathscr{H}, \mathscr{K}_j = \mathscr{K}$ $(j = 1, \ldots, n)$ and $A_j = A$ $(j = 1, \ldots)$, we write $\otimes_{j=1}^n A_j = \otimes^n A \in \mathfrak{L}(\otimes^n \mathscr{H}, \otimes^n \mathscr{K})$, which is called the n-fold tensor product of $A \in \mathfrak{L}(\mathscr{H}, \mathscr{K})$.

The next lemma will be used later:

Lemma 1.4 *Let A be a densely defined closable operator on a Hilbert space $\mathscr{H}$ and B be a densely defined closable operator on a Hilbert space $\mathscr{K}$. Suppose that $A \otimes I = I \otimes B$ on $D(A)\hat{\otimes}D(B)$. Then there is a constant $c \in \mathbb{C}$ such that $A = cI$ on $D(A)$ and $B = cI$ on $D(B)$.*

Proof By the present assumption, for all $\Psi \in D(A)$ and $\Phi \in D(B)$, $A\Psi \otimes \Phi = \Psi \otimes B\Phi$. Hence, for all $\eta \in \mathscr{H}$ and $\chi \in \mathscr{K}$, $\langle \eta, A\Psi \rangle \langle \chi, \Phi \rangle = \langle \eta, \Psi \rangle \langle \chi, B\Phi \rangle$. Therefore

$$\langle \chi, \langle \eta, A\Psi \rangle \Phi \rangle = \langle \chi, \langle \eta, \Psi \rangle B\Phi \rangle, \quad \langle \eta, \langle \chi, \Phi \rangle A\Psi \rangle = \langle \eta, \langle \chi, B\Phi \rangle \Psi \rangle.$$

Thus $\langle \eta, A\Psi \rangle \Phi = \langle \eta, \Psi \rangle B\Phi$, $\langle \chi, \Phi \rangle A\Psi = \langle \chi, B\Phi \rangle \Psi$. Taking $\Psi = \eta$ with $\|\eta\| = 1$ in the first equation, we obtain $B\Phi = c\Phi$ with $c = \langle \eta, A\eta \rangle$. Putting this into the second equation, we obtain $A\Psi = c\Psi$. ∎

We collect some basic facts on tensor product operators.[11]

Theorem 1.25 *Let $A_j \in \mathfrak{B}(\mathscr{H}_j, \mathscr{K}_j)$, $j = 1, \ldots, n$.*

(i) *$\otimes_{j=1}^n A_j$ is bounded with $\| \otimes_{j=1}^n A_j \| = \prod_{j=1}^n \|A_j\|$.*
(ii) *If each A_j is unitary, then $\otimes_{j=1}^n A_j$ is unitary.*
(iii) *Let $A_1 \in \mathfrak{B}(\mathscr{H}_1)$ and $A_2 \in \mathfrak{B}(\mathscr{H}_2)$. Then*

$$A_1 \otimes A_2 = (A_1 \otimes I)(I \otimes A_2) = (I \otimes A_2)(A_1 \otimes I).$$

Remark 1.7 Theorem 1.25(iii) can be extended to $\otimes_{j=1}^n A_j$ with $A_j \in \mathfrak{B}(\mathscr{H}_j)$.

Theorem 1.26 *Consider the case where $\mathscr{K}_j = \mathscr{H}_j$ $(j = 1, \ldots, n)$.*

(i) *If each A_j $(j = 1, \ldots, n)$ is symmetric, then $\otimes_{j=1}^n A_j$ is a closed symmetric operator.*
(ii) *If each A_j $(j = 1, \ldots, n)$ is self-adjoint, then $\otimes_{j=1}^n A_j$ is self-adjoint.*
(iii) *For a densely defined closable operator A_j on $\mathscr{H}_j$, we define*

$$\tilde{A}_j := I \otimes \cdots \otimes I \otimes \underbrace{A_j}_{j\text{-th}} \otimes I \otimes \cdots \otimes I, \quad j = 1, \ldots, n.$$

[11] For details of proof, see, e.g., [28, Theorem 3.8], [130, Theorem VIII.33 (a)], [143, Theorem 7.23].

Let A_j be a symmetric operator. Then $\tilde{A}_j$ is a closed symmetric operator. Moreover, if A_j is self-adjoint, then $\tilde{A}_j$ is self-adjoint and $(\tilde{A}_1, \ldots, \tilde{A}_n)$ is strongly commuting.

(iv) *Let A_j be self-adjoint and $\mathscr{D}_j$ be a core for A_j. Then*

$$S(A) := \sum_{j=1}^{n} \tilde{A}_j \tag{1.62}$$

is essentially self-adjoint on $\hat{\otimes}_{j=1}^{n} \mathscr{D}_j$. Moreover, for all $t \in \mathbb{R}$,

$$e^{it\overline{S(A)}} = \otimes_{j=1}^{n} e^{itA_j}. \tag{1.63}$$

(v) *If A_j is self-adjoint and bounded from below with $A_j \geq \gamma_j$ (γ_j is a real constant), then $S(A)$ is self-adjoint and bounded from below with $S(A) \geq \sum_{j=1}^{n} \gamma_j$.*

Remark 1.8 For a proof of (1.63), Proposition 1.24 can be applied.

Concerning the spectrum of $S(A)$, we state only a basic property:

Theorem 1.27 *Let A_j be a self-adjoint operator on $\mathscr{H}_j$ ($j = 1, \ldots, n$) and $S(A)$ be defined by (1.62). Then*

$$\sigma(S(A)) = \overline{\left\{ \sum_{j=1}^{n} \lambda_j | \lambda_j \in \sigma(A_j),\ j = 1, \ldots, n \right\}}.$$

For a proof of this theorem, see, e.g., [28, Theorem 3.8], [130, Theorem VIII.33 (b)], [143, Corollary 7.25].

The following proposition is about a continuity property of tensor product operator.

Proposition 1.25 *Let $\mathscr{H}$ and $\mathscr{K}$ be Hilbert spaces and $T_k \in \mathfrak{B}(\mathscr{H})$ ($k \in \mathbb{N}$). Suppose that* $\mathrm{s}-\lim_{k\to\infty} T_k = T \in \mathfrak{B}(\mathscr{H})$ *and* $C := \sup_{k\geq 1} \|T_k\| < \infty$. *Then*

$$\mathrm{s}-\lim_{k\to\infty} T_k \otimes I = T \otimes I. \tag{1.64}$$

Proof It is easy to see that, for all $\Psi \in \mathscr{H} \hat{\otimes} \mathscr{K}$, $\lim_{k\to\infty}(T_k \otimes I)\Psi = (T \otimes I)\Psi$. Let $\Psi \in \mathscr{H} \otimes \mathscr{K}$. Then, by the density of $\mathscr{H} \hat{\otimes} \mathscr{K}$ in $\mathscr{H} \otimes \mathscr{K}$, for any $\varepsilon > 0$, there exists a vector $\Psi_\varepsilon \in \mathscr{H} \hat{\otimes} \mathscr{K}$ such that $\|\Psi_\varepsilon - \Psi\| < \varepsilon$. By the triangle inequality, we have

$$\begin{aligned}\|(T_k \otimes I)\Psi - (T \otimes I)\Psi\| &\leq \|(T_k \otimes I)(\Psi - \Psi_\varepsilon)\| + \|(T_k \otimes I)\Psi_\varepsilon - (T \otimes I)\Psi_\varepsilon\| \\ &\quad + \|(T \otimes I)(\Psi_\varepsilon - \Psi)\|\end{aligned}$$

$$\begin{aligned} &\leq \|T_k \otimes I\| \, \|\Psi - \Psi_\varepsilon\| + \|(T_k \otimes I)\Psi_\varepsilon - (T \otimes I)\Psi_\varepsilon\| \\ &\quad + \|T \otimes I\| \, \|\Psi - \Psi_\varepsilon\| \\ &\leq \|T_k\|\varepsilon + \|(T_k \otimes I)\Psi_\varepsilon - (T \otimes I)\Psi_\varepsilon\| + \|T\|\varepsilon \\ &\leq (C + \|T\|)\varepsilon + \|(T_k \otimes I)\Psi_\varepsilon - (T \otimes I)\Psi_\varepsilon\|. \end{aligned}$$

Hence $\limsup_{k\to\infty} \|(T_k \otimes I)\Psi - (T \otimes I)\Psi\| \leq (C + \|T\|)\varepsilon$. Taking $\varepsilon \to 0$, we obtain $\lim_{k\to\infty} \|(T_k \otimes I)\Psi - (T \otimes I)\Psi\| = 0$. Thus (1.64) follows. ■

Corollary 1.3 *Let $\mathscr{H}$ and $\mathscr{K}$ be Hilbert spaces and $T_k \in \mathfrak{B}(\mathscr{H})$, $S_k \in \mathfrak{B}(\mathscr{K})$ $(k \in \mathbb{N})$. Suppose that* $\text{s-}\lim_{k\to\infty} T_k = T \in \mathfrak{B}(\mathscr{H})$, $\text{s-}\lim_{k\to\infty} S_k = S \in \mathfrak{B}(\mathscr{K})$ *with* $\sup_{k\geq 1} \|T_k\| < \infty$ *and* $\sup_{k\geq 1} \|S_k\| < \infty$. *Then*

$$\text{s-}\lim_{k\to\infty} T_k \otimes S_k = T \otimes S. \tag{1.65}$$

Proof We have $T_k \otimes S_k = (T_k \otimes I)(I \otimes S_k)$. By Proposition 1.25,

$$\text{s-}\lim_{k\to\infty} (T_k \otimes I)(I \otimes S_k) = (T \otimes I)(I \otimes S).$$

Thus (1.65) follows. ■

1.9 The Natural Isomorphism Between $\mathfrak{I}_2(\mathscr{H})$ and $\mathscr{H} \otimes \mathscr{H}$

In Sect. 1.1.8, we have seen that the set $\mathfrak{I}_2(\mathscr{H})$ of Hilbert–Schmidt operators on a Hilbert space $\mathscr{H}$ is a Hilbert space. In this section we prove a theorem on a natural isomorphism of $\mathfrak{I}_2(\mathscr{H})$.

We recall a basic notion:

Definition 1.2 Let $\mathscr{V}$ be an inner product space with inner product $\langle\ ,\ \rangle_{\mathscr{V}}$ and norm $\|\cdot\|_{\mathscr{V}}$. A mapping $C\colon \mathscr{V} \to \mathscr{V}$ is called a **conjugation** on $\mathscr{V}$ if it is an anti-linear mapping satisfying that $C^2 = I$ and

$$\|Cf\|_{\mathscr{V}} = \|f\|_{\mathscr{V}}, \quad f \in \mathscr{V}. \tag{1.66}$$

Let C be a conjugation on $\mathscr{V}$. Then, using the polarization identity, norm-preserving property (1.66) of C and the anti-linearity of C, one can prove the following formula:

$$\langle Cf, Cg\rangle_{\mathscr{V}} = \langle g, f\rangle_{\mathscr{V}}, \quad f, g \in \mathscr{V}. \tag{1.67}$$

It is easy to see that the subset

$$\mathscr{V}_C := \{f \in \mathscr{V} | Cf = f\} \tag{1.68}$$

of $\mathscr{V}$ is a real inner product space. For a vector $f \in \mathscr{V}$, the vectors

$$\operatorname{Re} f := \frac{1}{2}(f + Cf) \tag{1.69}$$

and

$$\operatorname{Im} f := \frac{1}{2i}(f - Cf) \tag{1.70}$$

are in $\mathscr{V}_C$. One has

$$f = \operatorname{Re} f + i \operatorname{Im} f.$$

The vector $\operatorname{Re} f$ (resp. $\operatorname{Im} f$) is called the **real** (resp. **imaginary**) **part** of f with respect to C. We call $\mathscr{V}_C$ the real part of $\mathscr{V}$ with respect to C.

The following isomorphism theorem is often useful:

Theorem 1.28 *Let $\mathscr{H}$ be a separable Hilbert space and C be a conjugation on $\mathscr{H}$. Then there exists a unitary operator $U_C : \mathfrak{I}_2(\mathscr{H}) \to \mathscr{H} \otimes \mathscr{H}$ such that*

$$U_C(\langle f, \cdot \rangle\, g) = Cf \otimes g, \quad f, g \in \mathscr{H}. \tag{1.71}$$

Proof Let $\{e_n\}_{n=1}^{\infty}$ be a CONS of $\mathscr{H}$. Then it is easy to see that $\{Ce_n\}_{n=1}^{\infty}$ is a CONS of $\mathscr{H}$. Hence, by Theorem 1.20, $\{Ce_n \otimes e_m | n, m \in \mathbb{N}\}$ is a CONS of $\mathscr{H} \otimes \mathscr{H}$. Therefore, by Lemma 1.2 and the isomorphism theorem for Hilbert spaces, there exists a unitary operator $U_C : \mathfrak{I}_2(\mathscr{H}) \to \mathscr{H} \otimes \mathscr{H}$ such that

$$U_C P_{e_m, e_n} = Ce_m \otimes e_n, \quad m, n \in \mathbb{N}.$$

By expanding $f, g \in \mathscr{H}$ as $f = \sum_{m=1}^{\infty} \langle e_m, f \rangle\, e_m$ and $g = \sum_{n=1}^{\infty} \langle e_n, g \rangle\, e_n$, we obtain

$$P_{f,g} = \sum_{m,n=1}^{\infty} \langle f, e_m \rangle \langle e_n, g \rangle\, P_{e_m, e_n}$$

in the Hilbert–Schmidt norm $\| \cdot \|_2$. Hence

$$U_C P_{f,g} = \sum_{m,n=1}^{\infty} \langle f, e_m \rangle \langle e_n, g \rangle\, Ce_m \otimes e_n = Cf \otimes g.$$

Thus (1.71) holds. ■

Theorems 1.4 and 1.28 immediately yield the following corollary:

Corollary 1.4 *Let $T \in \mathfrak{I}_2(\mathscr{H})$. Then, with the notation in Theorem 1.4,*

$$U_C T = \sum_{n=1}^{\infty} \mu_n C f_n \otimes e_n. \tag{1.72}$$

1.10 Convergence of Self-adjoint Operators in Strong Resolvent Sense

Let H_n ($n \in \mathbb{N}$) and H be self-adjoint operators on a Hilbert space $\mathscr{H}$. Then $\mathbb{C}\setminus\mathbb{R} \subset \rho(H_n) \cap \rho(H)$. Hence, for all $z \in \mathbb{C} \setminus \mathbb{R}$, $(H_n - z)^{-1}$ and $(H - z)^{-1}$ are in $\mathfrak{B}(\mathscr{H})$.

Definition 1.3 A sequence $\{H_n\}_n$ of self-adjoint operators is said to **converge to** H **in the strong resolvent sense** if, for all $z \in \mathbb{C} \setminus \mathbb{R}$,

$$\text{s}-\lim_{n\to\infty} (H_n - z)^{-1} = (H - z)^{-1},$$

where s$-$ lim means strong limit (i.e. $\lim_{n\to\infty}(H_n - z)^{-1}\Psi = (H - z)^{-1}\Psi$, $\Psi \in \mathscr{H}$).

A basic criterion on strong resolvent convergence of a sequence of self-adjoint operators is given in the following theorem.

Theorem 1.29 *Suppose that there exists a core $\mathscr{D}$ for H such that $\mathscr{D} \subset \cap_{n=1}^{\infty} D(H_n)$ and $\lim_{n\to\infty} H_n\Psi = H\Psi$ for all $\Psi \in \mathscr{D}$. Then $\{H_n\}_n$ converges to H in the strong resolvent sense.*

Proof Since $\mathscr{D}$ is a core for H, for all $z \in \mathbb{C}\setminus\mathbb{R}$, $(H - z)\mathscr{D} := \{(H - z)\Psi | \Psi \in \mathscr{D}\}$ is dense in $\mathscr{H}$. Let $\Phi \in (H - z)\mathscr{D}$. Then $\Psi := (H - z)^{-1}\Phi$ is in $\mathscr{D}$. We have

$$(H_n - z)^{-1}\Phi - (H - z)^{-1}\Phi = (H_n - z)^{-1}(H - H_n)\Psi.$$

Hence

$$\|(H_n - z)^{-1}\Phi - (H - z)^{-1}\Phi\| \le \frac{1}{|\text{Im}\, z|}\|(H - H_n)\Psi\|.$$

Hence $\lim_{n\to\infty}(H_n - z)^{-1}\Phi = (H - z)^{-1}\Phi$. Since $\sup_{n\in\mathbb{N}} \|(H_n - z)^{-1}\| \le 1/|\text{Im}\, z|$ and $(H - z)\mathscr{D}$ is dense, it follows from a limiting argument that s$-\lim_{n\to\infty}(H_n - z)^{-1} = (H - z)^{-1}$. ■

There is an important theorem:

Theorem 1.30 *A sequence* $\{H_n\}_n$ *of self-adjoint operators on* $\mathscr{H}$ *converges to* H *in the strong resolvent sense if and only if* $\mathrm{s}\!-\!\lim_{n\to\infty} e^{-itH_n} = e^{-itH}$ *for all* $t \in \mathbb{R}$.

For a proof of this theorem, we refer the reader to [130, Theorem VIII.21].

1.11 Sesquilinear Form and Representation Theorem

Let $\mathscr{H}_1$ and $\mathscr{H}_2$ be Hilbert spaces, and $\mathscr{D}_1$ and $\mathscr{D}_2$ be subspaces of $\mathscr{H}_1$ and $\mathscr{H}_2$ respectively. A mapping $\mathfrak{s}$: $\mathscr{D}_1 \times \mathscr{D}_2 \to \mathbb{C}$ is called a **sesquilinear form** or a **quadratic form** with domain $\mathscr{D}_1 \times \mathscr{D}_2$ if, for all $\Phi, \Phi' \in \mathscr{D}_1, \Psi, \Psi' \in \mathscr{D}_2$ and $\alpha, \beta \in \mathbb{C}$,

$$\mathfrak{s}(\Phi, \alpha\Psi + \beta\Psi') = \alpha\, \mathfrak{s}(\Phi, \Psi) + \beta\, \mathfrak{s}(\Phi, \Psi'),$$

$$\mathfrak{s}(\alpha\Phi + \beta\Phi', \Psi) = \alpha^* \mathfrak{s}(\Phi, \Psi) + \beta^* \mathfrak{s}(\Phi', \Psi).$$

The domain $\mathscr{D}_1 \times \mathscr{D}_2$ is called the **form domain** of $\mathfrak{s}$ and written $\mathscr{D}_1 \times \mathscr{D}_2 = \mathscr{Q}(\mathfrak{s})$.

In the case where $\mathscr{H} = \mathscr{H}_1 = \mathscr{H}_2$ and $\mathscr{D} = \mathscr{D}_1 = \mathscr{D}_2$, we call a sesquilinear form $\mathfrak{s}$ with domain $\mathscr{D} \times \mathscr{D}$ a sesquilinear form on $\mathscr{H}$ with form domain $\mathscr{D}$ and write $\mathscr{D} = \mathscr{Q}(\mathfrak{s})$.

Let $\mathscr{H}$ be a Hilbert space. A sesquilinear form $\mathfrak{s}$ on $\mathscr{H}$ is said to be **Hermitian** if $\mathfrak{s}(\Psi, \Phi)^* = \mathfrak{s}(\Phi, \Psi)$ for all $\Psi, \Phi \in \mathscr{Q}(\mathfrak{s})$. It is easy to see that a sesquilinear form $\mathfrak{s}$ is Hermitian if and only if $\mathfrak{s}(\Psi, \Psi) \in \mathbb{R}$ for all $\Psi \in \mathscr{Q}(\mathfrak{s})$.

A Hermitian sesquilinear form $\mathfrak{s}$ on $\mathscr{H}$ is said to be **bounded from below** if there exists a constant $\gamma \in \mathbb{R}$ such that $\mathfrak{s}(\Psi, \Psi) \geq \gamma\|\Psi\|^2$, $\Psi \in \mathscr{Q}(\mathfrak{s})$. If γ can be taken to be zero (i.e., $\mathfrak{s}(\Psi, \Psi) \geq 0$, $\Psi \in \mathscr{Q}(\mathfrak{s})$), then $\mathfrak{s}$ is said to be **non-negative**.

A sesquilinear form $\mathfrak{s}$ on $\mathscr{H}$ is called a **symmetric sesquilinear form** if it is Hermitian and $\mathscr{Q}(\mathfrak{s})$ is dense in $\mathscr{H}$.

A sesquilinear form $\mathfrak{s}$ on $\mathscr{H}$ is said to be **closed** if, for any sequence $\{\Psi_n\}_{n=1}^{\infty}$ in $\mathscr{Q}(\mathfrak{s})$ satisfying the condition that $\lim_{n\to\infty} \Psi_n = \Psi \in \mathscr{H}$ and $\mathfrak{s}(\Psi_n - \Psi_m, \Psi_n - \Psi_m) \to 0$ ($n, m \to \infty$), it holds that $\Psi \in \mathscr{Q}(\mathfrak{s})$ and $\lim_{n\to\infty} \mathfrak{s}(\Psi_n - \Psi, \Psi_n - \Psi) = 0$.

Example 1.9 Let A be a self-adjoint operator on $\mathscr{H}$. Then, by functional calculus, one can define a sesquilinear form $\mathfrak{s}_A$ with form domain $\mathscr{Q}(\mathfrak{s}_A) := D(|A|^{1/2})$ by

$$\mathfrak{s}_A(\Psi, \Phi) := \int_{\mathbb{R}} \lambda d\, \langle \Psi, E_A(\lambda)\Phi \rangle\,, \quad \Psi, \Phi \in \mathscr{Q}(\mathfrak{s}_A).$$

This sesquilinear form $\mathfrak{s}_A$ on $\mathscr{H}$ is called the sesquilinear form associated with A. It is easy to see that $\mathfrak{s}_A$ is a symmetric form. Moreover, if A is bounded from below with $A \geq \gamma$ for a constant $\gamma \in \mathbb{R}$, then $\mathfrak{s}_A$ is closed and one has

$$\mathfrak{s}_A(\Psi, \Phi) = \left\langle (A-\gamma)^{1/2}\Psi, (A-\gamma)^{1/2}\Phi \right\rangle + \gamma \langle \Psi, \Phi \rangle, \quad \Psi, \Phi \in \mathscr{Q}(\mathfrak{s}_A). \tag{1.73}$$

A non-negative closed symmetric form is the sesquilinear form of a non-negative self-adjoint operator:

Theorem 1.31 (Representation Theorem) *Let $\mathfrak{s}$ be a non-negative closed symmetric form on $\mathscr{H}$. Then there exists a unique non-negative self-adjoint operator A on $\mathscr{H}$ such that $\mathscr{Q}(\mathfrak{s}) = D(A^{1/2})$ and*

$$\mathfrak{s}(\Psi, \Phi) = \left\langle A^{1/2}\Psi, A^{1/2}\Phi \right\rangle, \quad \Psi, \Phi \in \mathscr{Q}(\mathfrak{s}).$$

For a proof of this theorem, see, e.g., [98, Chapter VI, Theorem 2.23], [130, Theorem VIII.15] or [16, Theorem 2.56]).

In application to quantum mechanics, Theorem 1.31 can be used to define a Hamiltonian.

1.12 Fourier Analysis

In this section we recall basic facts in Fourier analysis without proof. For more details, we refer the reader to [131, Chapter IX].

For each natural number d, we denote by

$$\mathbb{R}^d := \{\boldsymbol{x} = (x_1, \ldots, x_d) | x_j \in \mathbb{R}, j = 1, \ldots, d\}$$

the d-dimensional Euclidean space. The dual space of $\mathbb{R}^d$ (i.e. the space of linear functionals on $\mathbb{R}^d$) is naturally identified with the Euclidean space

$$\hat{\mathbb{R}}^d := \{\boldsymbol{k} = (k_1, \ldots, k_d) | k_j \in \mathbb{R}, j = 1, \ldots, d\}$$

through the Euclidean inner product

$$\boldsymbol{k} \cdot \boldsymbol{x} := \sum_{j=1}^{d} k_j x_j,$$

which is called also the canonical bilinear form in the variables $\boldsymbol{k}$ and $\boldsymbol{x}$. In applications to quantum physics below, each point in $\mathbb{R}^d$ represents a position in the space where quantum particles exist, while an element $\boldsymbol{k}$ in $\hat{\mathbb{R}}^d$ represents a wave number vector of a quantum particle.

We denote by $L^2(\mathbb{R}^d)$ the Hilbert space of equivalence classes of square integrable functions on $\mathbb{R}^d$ with respect to the d-dimensional Lebesgue measure:

$$L^2(\mathbb{R}^d) := \left\{ f : \mathbb{R}^d \to \mathbb{C} \cup \{\pm\infty\}, \text{ Borel measurable} \middle| \int_{\mathbb{R}^d} |f(\boldsymbol{x})|^2 d\boldsymbol{x} < \infty \right\}$$

with inner product

$$\langle f, g \rangle_{L^2(\mathbb{R}^d)} := \int_{\mathbb{R}^d} f(\boldsymbol{x})^* g(\boldsymbol{x}) d\boldsymbol{x}, \quad f, g \in L^2(\mathbb{R}^d).$$

The Fourier transform $\mathscr{F}_d : L^2(\mathbb{R}^d) \to L^2(\hat{\mathbb{R}}^d)$ is defined by

$$(\mathscr{F}_d f)(\boldsymbol{k}) := \frac{1}{(2\pi)^{d/2}} \int_{\mathbb{R}^d} e^{-i\boldsymbol{k}\cdot\boldsymbol{x}} f(\boldsymbol{x}) d\boldsymbol{x}, \quad \text{a.e.}\boldsymbol{k} \in \hat{\mathbb{R}}^d,\ f \in L^2(\mathbb{R}^d) \tag{1.74}$$

in the L^2-sense,[12] where "a.e." means "almost everywhere" with respect to the d-dimensional Lebesgue measure. One can show that $\mathscr{F}_d$ is a unitary transformation with the inverse $\mathscr{F}_d^{-1} : L^2(\hat{\mathbb{R}}^d) \to L^2(\mathbb{R}^d)$ given by

$$(\mathscr{F}_d^{-1} g)(\boldsymbol{x}) = \frac{1}{(2\pi)^{d/2}} \int_{\hat{\mathbb{R}}^d} e^{i\boldsymbol{k}\cdot\boldsymbol{x}} g(\boldsymbol{k}) d\boldsymbol{k}, \quad \text{a.e.}\boldsymbol{x},\ g \in L^2(\hat{\mathbb{R}}^d) \tag{1.75}$$

in the L^2-sense. We use also the following symbols:

$$\hat{f} = \mathscr{F}_d f, \quad \check{g} = \mathscr{F}_d^{-1} g. \tag{1.76}$$

For each $j = 1, \ldots, d$, a linear operator ∂_j on $L^2(\mathbb{R}^d)$ is defined as follows:

$$D(\partial_j) := C_0^\infty(\mathbb{R}^d), \quad \partial_j f := \frac{\partial f}{\partial x_j}, \quad f \in D(\partial_j), \tag{1.77}$$

where $C_0^\infty(\mathbb{R}^d)$ is the space of infinitely differentiable functions on $\mathbb{R}^d$ with compact support in $\mathbb{R}^d$ and $\partial f/\partial x_j$ is the partial derivative of f in $x_j \in \mathbb{R}$, the j-th

[12] This means that

$$\lim_{R_j\to\infty, j=1,\ldots,d} \int_{\hat{\mathbb{R}}^d} \left| (\mathscr{F}_d f)(\boldsymbol{k}) - (2\pi)^{-d/2} \int_{|x_j|\le R_j, j=1,\ldots,d} e^{-i\boldsymbol{k}\cdot\boldsymbol{x}} f(\boldsymbol{x}) d\boldsymbol{x} \right|^2 d\boldsymbol{k} = 0.$$

But, if $f \in L^2(\mathbb{R}^d)$ is integrable, then $(\mathscr{F}_d f)(\boldsymbol{k}) = (2\pi)^{-d/2} \int_{\mathbb{R}^d} e^{-i\boldsymbol{k}\cdot\boldsymbol{x}} f(\boldsymbol{x}) d\boldsymbol{x}$ (the usual Lebesgue integral) and $\mathscr{F}_d f$ is a bounded continuous function on $\hat{\mathbb{R}}^d$.

component of point $\boldsymbol{x} = (x_1, \ldots, x_d) \in \mathbb{R}^d$. The operator ∂_j is called the **partial differential operator** in the j-th variable x_j. It is shown that ∂_j is closable and $\partial_j^* = -\bar{\partial}_j$. The closure

$$D_j := \bar{\partial}_j \tag{1.78}$$

of ∂_j is called the **generalized partial differential operator** in x_j, acting in $L^2(\mathbb{R}^d)$.

Let

$$\hat{p}_j := -iD_j, \quad j = 1, \ldots, d. \tag{1.79}$$

Then it is shown that

$$\mathscr{F}_d \hat{p}_j \mathscr{F}_d^{-1} = M_{k_j}, \quad j = 1, \ldots, d, \tag{1.80}$$

where M_{k_j} is the multiplication operator by the function k_j on $\hat{\mathbb{R}}^d$ (see Appendix A for the definition and basic properties of a multiplication operator). Since M_{k_j} is self-adjoint, it follows that $\hat{p}_j$ is self-adjoint with

$$D(\hat{p}_j) = \left\{ f \in L^2(\mathbb{R}^d) | \int_{\hat{\mathbb{R}}^d} |k_j \hat{f}(\boldsymbol{k})|^2 d\boldsymbol{k} < \infty \right\}. \tag{1.81}$$

In the context of quantum mechanics, $\hat{p}_j$ is called the j-th **momentum operator.**[13]

By (1.80) and the unitary covariance of functional calculus (Theorem B.3 in Appendix B), one has for all $t \in \mathbb{R}$

$$\mathscr{F}_d e^{-it\hat{p}_j} \mathscr{F}_d^{-1} = e^{-itM_{k_j}}.$$

Hence, for all $f \in L^2(\mathbb{R}^d)$,

$$(e^{-it\hat{p}_j} f)(\boldsymbol{x}) = f(\boldsymbol{x} - t\boldsymbol{e}_j), \quad \text{a.e.} \boldsymbol{x} \in \mathbb{R}^d, \tag{1.82}$$

where $(\boldsymbol{e}_1, \ldots, \boldsymbol{e}_d) \in (\mathbb{R}^d)^d$ is the standard basis of $\mathbb{R}^d$:

$$\boldsymbol{e}_j = (0, \ldots, 0, \underset{\smile}{\overset{j\text{-th}}{}} 1, 0, \ldots, 0) \in \mathbb{R}^d, \quad j = 1, \ldots, d. \tag{1.83}$$

[13]In applications to quantum physics, we will use the physical unit system where the reduced Planck constant $\hbar$ is equal to 1. In the original physical unit system (SI), $\hat{p}_j = -i\hbar D_j$.

Formula (1.82) shows that $e^{-it\hat{p}_j} f$ means the translation of f by the vector $t\boldsymbol{e}_j$. It implies also that, for all $j, k = 1, \ldots, d$ and $t, s \in \mathbb{R}$, $e^{-it\hat{p}_j}$ and $e^{-is\hat{p}_k}$ commute with

$$(e^{-ia_1\hat{p}_1} \cdots e^{-ia_d\hat{p}_d} f)(\boldsymbol{x}) = f(\boldsymbol{x} - \boldsymbol{a}), \quad \forall \boldsymbol{a} = (a_1, \ldots, a_d) \in \mathbb{R}^d, \text{ a.e.} \boldsymbol{x} \in \mathbb{R}^d. \tag{1.84}$$

Hence the d-tuple

$$\hat{\boldsymbol{p}} := (\hat{p}_1, \ldots, \hat{p}_d) \tag{1.85}$$

of self-adjoint operators is strongly commuting (see Proposition 1.17).

1.13 Discrete Fourier Transform

In this section, we describe a finite box version of Fourier analysis on $\mathbb{R}^d$. Let us take the bounded closed interval $[a, b] \subset \mathbb{R}$ and consider the L^2-space

$$L^2([a,b]) := \left\{ f : [a,b] \to \mathbb{C} \cup \{\pm\infty\}, \text{Borel measurable} \middle| \int_{[a.b]} |f(x)|^2 dx < \infty \right\},$$

which is a Hilbert space with inner product $\langle f, g \rangle := \int_{[a,b]} f(x)^* g(x) dx$ ($f, g \in L^2([a,b])$) and with equality $f = g$ in the sense that $f(x) = g(x)$ a.e.$x \in [a, b]$ (see Appendix A). We first show that there is a natural isomorphism between $L^2([a,b])$ and $L^2([c,d])$ ($[c,d]$ is another interval). For this purpose, we define a mapping τ: $[a,b] \to [c,d]$ by

$$\tau(x) := \frac{d-c}{b-a}(x-a) + c, \quad x \in [a,b].$$

Then it is easy to see that τ is bijective. Hence, for all $f \in L^2([c,d])$, we have

$$\int_{[a,b]} |f(\tau(x))|^2 dx = \frac{b-a}{d-c} \int_{[c,d]} |f(y)|^2 dy < \infty.$$

Therefore, defining a mapping U_τ: $L^2([c,d]) \to L^2([a,b])$ by

$$(U_\tau f)(x) := \sqrt{\frac{d-c}{b-a}} f(\tau(x)), \quad f \in L^2([c,d]), \text{ a.e.} x \in [a,b], \tag{1.86}$$

we see that U_τ is a unitary operator. We call the unitary operator U_τ the **natural isomorphism** from $L^2([c,d])$ to $L^2([a,b])$. Thus, with regard to L^2-spaces on bounded closed intervals, it is sufficient to consider one of them.

Now we consider a closed interval

$$\mathbb{I}_L := [-L/2, L/2] \tag{1.87}$$

with length $L > 0$ and introduce

$$\Gamma_L := \frac{2\pi}{L}\mathbb{Z} := \left\{ k = \frac{2\pi}{L} n | n \in \mathbb{Z} \right\}, \tag{1.88}$$

where $\mathbb{Z}$ is the set of integers. For each $k \in \Gamma_L$, we define a function ϕ_k on $\mathbb{I}_L$ by

$$\phi_k(x) := \frac{1}{\sqrt{L}} e^{ikx}, \quad x \in \mathbb{I}_L. \tag{1.89}$$

The following theorem is well known:

Theorem 1.32 *The set $\{\phi_k\}_{k\in\Gamma_L}$ is a CONS of $L^2(\mathbb{I}_L)$.*

By Theorem 1.32, every $f \in L^2(\mathbb{I}_L)$ has the expansion

$$f = \sum_{k\in\Gamma_L} \langle \phi_k, f \rangle \phi_k$$

in $L^2(\mathbb{I}_L)$ and

$$\|f\|^2 = \sum_{k\in\Gamma_L} |\langle \phi_k, f \rangle|^2, \tag{1.90}$$

The set

$$\ell^2(\Gamma_L) := \left\{ u = \{u_k\}_{k\in\Gamma_L} | u_k \in \mathbb{C}, k \in \Gamma_L, \sum_{k\in\Gamma_L} |u_k|^2 < \infty \right\} \tag{1.91}$$

of absolutely square summable sequences on Γ_L is an infinite-dimensional complex vector space with the usual addition of two sequences and the usual scalar multiplication for sequences. Moreover, $\ell^2(\Gamma_L)$ becomes a Hilbert space with inner product

$$\langle u, v \rangle := \sum_{k\in\Gamma_L} u_k^* v_k, \quad u, v \in \ell^2(\Gamma_L).$$

Theorem 1.33 *There exists a unique unitary operator F_L: $L^2(\mathbb{I}_L) \to \ell^2(\Gamma_L)$ such that*

$$(F_L f)_k = \langle \phi_k, f \rangle = \frac{1}{\sqrt{L}} \int_{-L/2}^{L/2} e^{-ikx} f(x) dx, \quad f \in L^2(\mathbb{I}_L), k \in \Gamma_L. \tag{1.92}$$

Proof By (1.90), one can define a mapping $F_L\colon L^2(\mathbb{I}_L) \to \ell^2(\Gamma_L)$ by (1.92). Then (1.90) implies that F_L is isometric[14] and hence preserves the inner product. To show that F_L is surjective, let $u \in \ell^2(\Gamma_L)$. Then $\sum_{k\in\Gamma_L} |u_k|^2 < \infty$. Hence, by Lemma 1.3,

$$f_u := \sum_{k\in\Gamma_L} u_k\phi_k$$

exists in $L^2(\mathbb{I}_L)$. Then $\langle\phi_k, f_u\rangle = u_k$. Hence $F_L f_u = u$. Therefore F_L is surjective. Thus F_L is unitary. The uniqueness of F_L is easily shown. ■

We call the unitary operator $F_L\colon L^2(\mathbb{I}_L) \to \ell^2(\Gamma_L)$ the **one-dimensional discrete Fourier transform**. It follows from (1.92) that

$$(F_L\phi_k)_{k'} = \delta_{kk'}, \quad k, k' \in \Gamma_L. \tag{1.93}$$

By abuse of notation, we write

$$\hat{f} = F_L f, \quad f \in L^2(\mathbb{I}_L).$$

There will be no danger of confusion, provided that one is aware of the space to which f belongs.

The inverse discrete Fourier transform $F_L{}^{-1}\colon \ell^2(\Gamma_L) \to L^2(\mathbb{I}_L)$ is given by

$$(F_L{}^{-1}u)(x) = \sum_{k\in\Gamma_L} u_k\phi_k(x) = \frac{1}{\sqrt{L}}\sum_{k\in\Gamma_L} u_k e^{ikx}, \quad u \in \ell^2(\Gamma_L), \tag{1.94}$$

in the sense of L^2-convergence.

1.14 Momentum Operator with Periodic Boundary Condition

We define a linear operator $\hat{p}_L$ on $L^2(\mathbb{I}_L)$ as follows:

$$D(\hat{p}_L) := \left\{ f \in L^2(\mathbb{I}_L) | \sum_{k\in\Gamma_L} |k\hat{f}_k|^2 < \infty \right\}, \tag{1.95}$$

$$\hat{p}_L f := F_L^{-1}(k\hat{f}) = \sum_{k\in\Gamma_L} k\hat{f}_k\phi_k, \quad f \in D(\hat{p}_L), \tag{1.96}$$

[14] A linear operator T from a Hilbert space $\mathscr{H}$ to a Hilbert space $\mathscr{K}$ is called an **isometry** if $D(T) = \mathscr{H}$ and $\|T\Psi\|_{\mathscr{K}} = \|\Psi\|_{\mathscr{H}}$, $\Psi \in \mathscr{H}$. An isometry $T\colon \mathscr{H} \to \mathscr{K}$ preserves the inner product: $\langle T\Psi, T\Phi\rangle_{\mathscr{K}} = \langle\Psi, \Phi\rangle_{\mathscr{H}}$, $\Psi, \Phi \in \mathscr{H}$. This follows from the polarization identity.

where $\hat{f}_k := (F_L f)_k$. By (1.93), we have

$$\hat{p}_L \phi_k = k\phi_k, \quad k \in \Gamma_L. \tag{1.97}$$

Namely, ϕ_k is an eigenvector of $\hat{p}_L$ with eigenvalue k.

Theorem 1.34 *The operator $\hat{p}_L$ is self-adjoint and*

$$\sigma(\hat{p}_L) = \sigma_{\mathrm{p}}(\hat{p}_L) = \Gamma_L, \tag{1.98}$$

where the multiplicity of eigenvalue k of $\hat{p}_L$ is one.

Proof By the definition of $\hat{p}_L$, we have

$$F_L \hat{p}_L F_L^{-1} = M_k, \tag{1.99}$$

where the right hand side is the multiplication operator on $\ell^2(\Gamma_L)$ by k. The operator M_k is self-adjoint. Hence $\hat{p}_L$ is self-adjoint. As we have seen above, $\{\phi_k\}_{k\in\Gamma_L}$ is a set of eigenvectors of $\hat{p}_L$ and is a CONS of $L^2(\mathbb{I}_L)$. Thus, by Theorem 1.16, (1.98) follows. ■

One can identify $\hat{p}_L$ with an differential operator. To show this, we recall a notion. In general, a function f on the interval $[a, b]$ is said to be **absolutely continuous** if there exists an integrable function g on $[a, b]$ and a constant $c \in \mathbb{R}$ such that $f(x) = \int_a^x g(y)dy + c$. In this case, it is easy to see that f is continuous on $[a, b]$ and differentiable at a.e. $x \in [a, b]$ and $f'(x) = g(x)$, a.e.x. We denote by AC$([a, b])$ (or AC$[a, b]$) the set of absolutely continuous Borel measurable functions on $[a, b]$.

Theorem 1.35

$$D(\hat{p}_L) = \left\{ f \in L^2(\mathbb{I}_L) \cap \mathrm{AC}(\mathbb{I}_L) | \, f(-L/2) = f(L/2), \ f' \in L^2(\mathbb{I}_L) \right\} \tag{1.100}$$

$$\hat{p}_L f = -if', \quad f \in D(\hat{p}_L). \tag{1.101}$$

Proof Let the set on the right hand side of (1.100) be D. Let $f \in D$ first. Then, by integration by parts, we have for all $k \in \Gamma_L$

$$\langle \phi_k, f' \rangle = ik \langle \phi_k, f \rangle = ik\hat{f}_k. \tag{1.102}$$

Hence

$$\sum_{k\in\Gamma_L} |k\hat{f}_k|^2 = \sum_{k\in\Gamma_L} |\langle \phi_k, f' \rangle|^2 = \|f'\|^2 < \infty.$$

Therefore $f \in D(\hat{p}_L)$. Moreover, by using (1.102), we have

$$\langle \phi_k, \hat{p}_L f \rangle = \langle F_L \phi_k, F_L \hat{p}_L f \rangle = k \hat{f}_k = \langle \phi_k, -if' \rangle.$$

Since this holds for all $k \in \Gamma_L$ and $\{\phi_k\}_{k \in \Gamma_L}$ is a CONS of $L^2(\mathbb{I}_L)$, it follows that $\hat{p}_L f = -if'$. Thus $D \subset D(\hat{p}_L)$ and $\hat{p}_L f = -if'$ for all $f \in D$.

Next, let $f \in D(\hat{p}_L)$ and $v = \{v_k\}_{k \in \Gamma_L}$ with $v_k := ik \hat{f}_k$. Then $v \in \ell^2(\Gamma_L)$. Hence, by the Cauchy–Schwarz inequality,

$$\sum_{k \in \Gamma_L \setminus \{0\}} |\hat{f}_k| = \sum_{k \in \Gamma_L \setminus \{0\}} \frac{1}{|k|} \cdot |v_k| \le \sqrt{\sum_{k \in \Gamma_L \setminus \{0\}} \frac{1}{k^2}} \sqrt{\sum_{k \in \Gamma_L} |v_k|^2} < \infty.$$

Hence

$$f(x) = \sum_{k \in \Gamma_L} \hat{f}_k \phi_k(x)$$

converges uniformly in $x \in \mathbb{I}_L$ and hence f is a continuous function on $\mathbb{I}_L$. It is easy to see that $f(-L/2) = f(L/2)$.

It follows from the condition $v \in \ell^2(\Gamma_L)$ and Lemma 1.3 that $g := \sum_{k \in \Gamma_L} v_k \phi_k$ is in $L^2(\mathbb{I}_L)$. In particular, by the Schwarz inequality, g is integrable on $\mathbb{I}_L$. Hence one can define an absolutely continuous function

$$u(x) := \int_{-L/2}^{x} g(y) dy = \langle \chi_{[-L/2,x]}, g \rangle_{L^2(\mathbb{I}_L)}, \quad x \in \mathbb{I}_L,$$

where $\chi_{[a,b]}$ is the characteristic function of $[a, b]$. By the unitarity of F_L,

$$u(x) = \langle F_L \chi_{[-L/2,x]}, F_L g \rangle_{\ell^2(\Gamma_L)} = \langle F_L \chi_{[-L/2,x]}, v \rangle_{\ell^2(\Gamma_L)}.$$

We have for all $k \in \Gamma_L$

$$\begin{aligned} (F_L \chi_{[-L/2,x]})_k^* v_k &= ik \hat{f}_k \frac{1}{\sqrt{L}} \int_{-L/2}^{x} e^{iky} dy \\ &= \hat{f}_k \frac{1}{\sqrt{L}} \int_{-L/2}^{x} \frac{d}{dy} e^{iky} dy \\ &= \hat{f}_k (\phi_k(x) - \phi_k(-L/2)). \end{aligned}$$

Hence $u(x) = f(x) - f(-L/2)$. Therefore $f(x) = \int_{-L/2}^{x} g(y) dy + f(-L/2)$, implying that $f \in \mathrm{AC}(\mathbb{I}_L)$, $f'(x) = g(x)$, a.e.$x \in \mathbb{I}_L$ and $f' \in L^2(\mathbb{I}_L)$. Hence $f \in D$. Thus $D(\hat{p}_L) \subset D$. ■

For a continuous function f on $\mathbb{I}_L$, condition $f(L/2) = f(-L/2)$ is called the **periodic boundary condition** (PBC) on f. As is seen from (1.100), each element of $D(\hat{p}_L)$ obeys the PBC. Based on this property, we call the self-adjoint operator $\hat{p}_L$ the **momentum operator with PBC** on $L^2(\mathbb{I}_L)$.

For later use, we compute the unitary group $\{e^{-it\hat{p}_L}\}_{t\in\mathbb{R}}$. We first note that, for each $x \in \mathbb{R}$, there exists a unique integer $n_x \in \mathbb{Z}$ such that $x - n_x L \in \mathbb{I}_L$. Concretely, n_x is given as follows:

$$n_x = \begin{cases} -n & \text{if } -\frac{(2n+1)L}{2} \leq x < -\frac{(2n-1)L}{2} (n \in \mathbb{N}) \\ 0 & \text{if } x \in \mathbb{I}_L \\ n & \text{if } \frac{(2n-1)L}{2} < x \leq \frac{(2n+1)L}{2} (n \in \mathbb{N}) \end{cases} . \tag{1.103}$$

For each $x \in \mathbb{I}_L$ and $t \in \mathbb{R}$, we define

$$[x - t]_L := x - t - n_{x-t} L \in \mathbb{I}_L. \tag{1.104}$$

Lemma 1.5 *For each $t \in \mathbb{R}$ and $f \in L^2(\mathbb{I}_L)$, the function $\mathbb{I}_L \ni x \mapsto f([x - t]_L)$ is Borel measurable and*

$$\int_{\mathbb{I}_L} |f([x - t]_L)|^2 dx = \|f\|^2_{L^2(\mathbb{I}_L)}. \tag{1.105}$$

Proof It follows from (1.103) that the function $\mathbb{R} \ni x \mapsto x - n_x L$ is Borel measurable and so is $[x - t]_L$. Hence the function $\mathbb{I}_L \ni x \mapsto f([x - t]_L)$ is Borel measurable as a composite function. Let $J_n := ((2n - 1)L/2, (2n + 1)L/2]$, $J_{-n} := [(2(-n)-1)L/2, (2(-n)+1)L/2), n \in \mathbb{N}$ and $J_0 := \mathbb{I}_L$. Then $\mathbb{R} = \cup_{n\in\mathbb{Z}} J_n$. Hence

$$\begin{aligned} \int_{\mathbb{I}_L} |f([x - t]_L)|^2 dx &= \int_{\mathbb{R}} |f([x - t]_L)|^2 \chi_{(-L/2,L/2)}(x) dx \\ &= \sum_{n=-\infty}^{\infty} \int_{x-t\in J_n} |f(x - t - nL)|^2 \chi_{(-L/2,L/2)}(x) dx \\ &= \lim_{N\to\infty} \int_{\mathbb{I}_L} |f(y)|^2 \theta_N(y) dy \end{aligned}$$

with

$$\theta_N(y) := \sum_{n=-N}^{N} \chi_{(-L/2,L/2)}(y + t + nL).$$

It is easy to see that, for each $y \in \mathbb{I}_L$, there exists a unique integer n_y such that $y + t + n_y L \in (-L/2, L/2)$ and $y + t + mL \notin (-L/2, L/2)$ if $m \neq n_y$. Hence $\chi_{(-L/2,L/2)}(y+t+n_y L) = 1$ and $\chi_{(-L/2,L/2)}(y+t+mL) = 0$ for all $m \neq n_y$. This implies that $0 \leq \theta_N(y) \leq 1$ and $\lim_{N\to\infty} \theta_N(y) = 1$. Therefore, by the Lebesgue dominated convergence theorem, $\lim_{N\to\infty} \int_{\mathbb{I}_L} |f(y)|^2 \theta_N(y) dy = \int_{\mathbb{I}_L} |f(y)|^2 dy$. Thus (1.105) holds. ■

By Lemma 1.105, one can define a linear operator $T_L(t)$ on $L^2(\mathbb{I}_L)$ by

$$(T_L(t)f)(x) := f([x - t]_L), \quad \text{a.e.}x, \quad f \in L^2(\mathbb{I}_L). \tag{1.106}$$

By (1.105), $T_L(t)$ is bounded with operator norm $\|T_L(t)\| = 1$.

The operator $T_L(t)f$ is interpreted as the mod L translation of f by t.

Theorem 1.36 *For all $t \in \mathbb{R}$,*

$$e^{-it\hat{p}_L} = T_L(t). \tag{1.107}$$

In particular, $T_L(t)$ is unitary.

Proof By (1.99) and the unitary covariance of functional calculus (Theorem B.3 in Appendix B), we have

$$F_L e^{-it\hat{p}_L} F_L^{-1} = e^{-itM_k}.$$

Hence, for all $f \in C_0^\infty(-L/2, L/2)$ (the set of C^∞-functions on $(-L/2, L/2)$ with support in $(-L/2, L/2)$),

$$(e^{-it\hat{p}_L} f)(x) = \frac{1}{L} \sum_{k\in\Gamma_L} e^{ik(x-t)} \hat{f}_k, \quad x \in \mathbb{I}_L,$$

where the convergence is uniform in x. Since $e^{ik(x-t)} = e^{ik[x-t]_L}$ for all $k \in \Gamma_L$, the right hand side of the above equation is equal to $f([x-t]_L) = (T_L(t)f)(x)$. Hence $e^{-it\hat{p}_L} f = T_L(t)f$. Since $C_0^\infty(-L/2, L/2)$ is dense in $L^2(\mathbb{I}_L)$, operator equality (1.107) holds. ■

1.15 Momentum Operators with Other Boundary Conditions

One can consider also differential operators on $L^2(\mathbb{I}_L)$ with boundary conditions different from PBC. For each $\alpha \in \mathbb{C}$ with $|\alpha| = 1$, we define an operator $p(\alpha)$ on $L^2(\mathbb{I}_L)$ as follows:

$$D(p(\alpha)) := \left\{ f \in L^2(\mathbb{I}_L) \cap \mathrm{AC}(\mathbb{I}_L) | f(-L/2) = \alpha f(L/2),\ f' \in L^2(\mathbb{I}_L) \right\}, \tag{1.108}$$

$$p(\alpha)f := -if', \quad f \in D(p(\alpha)). \tag{1.109}$$

As is seen, the case $\alpha = 1$ corresponds to the case of PBC.

Theorem 1.37 *The operator $p(\alpha)$ is self-adjoint. Moreover, the spectrum of $p(\alpha)$ is purely discrete with*

$$\sigma(p(\alpha)) = \sigma_{\mathrm{d}}(p(\alpha)) = \left\{k - \frac{\theta_\alpha}{L} | k \in \Gamma_L\right\}, \tag{1.110}$$

where $\theta_\alpha \in [0, 2\pi)$ is the argument of α:

$$\alpha = e^{i\theta_\alpha}. \tag{1.111}$$

The multiplicity of each eigenvalue $k - (\theta_\alpha/L)$ is one.

Proof We first recall that, for all $f, g \in \mathrm{AC}(\mathbb{I}_L)$, the following integration by parts formula holds[15]:

$$\int_{-L/2}^{L/2} f(x)^* g'(x)dx = f(L/2)^* g(L/2) - f(-L/2)^* g(-L/2) - \int_{-L/2}^{L/2} f'(x)^* g(x)dx.$$

Let $f, g \in D(p(\alpha))$. Then $f(L/2)^* g(L/2) - f(-L/2)^* g(-L/2) = 0$ by $|\alpha|^2 = 1$. Hence it follows that $\langle f, p(\alpha)g\rangle = \langle p(\alpha)f, g\rangle$. Therefore $p(\alpha)$ is symmetric.

We next prove the closedness of $p(\alpha)$. Hence let $f_n \in D(p(\alpha))$, $f_n \to f \in L^2(\mathbb{I}_L)$ and $g_n := p(\alpha)f_n \to g \in L^2(\mathbb{I}_L)$ as $n \to \infty$. We have

$$f_n(x) = i\int_{-L/2}^{x} g_n(y)dy + f_n(-L/2), \quad x \in \mathbb{I}_L.$$

Let $F_n(x) := f_n(x) - f_n(-L/2)$. Then $F_n(x) = \langle \chi_{[-L/2,x]}, g_n\rangle$. Hence $\lim_{n\to\infty} F_n(x) = F(x) := \langle \chi_{[-L/2,x]}, g\rangle$. Moreover

$$\begin{aligned}
\|F_n - F\|^2 &= \int_{-L/2}^{L/2} |\langle \chi_{[-L/2,x]}, g_n\rangle - \langle \chi_{[-L/2,x]}, g\rangle|^2 dx \\
&\le \int_{-L/2}^{L/2} \|\chi_{[-L/2,x]}\|^2 \|g_n - g\|^2 dx \\
&\le L^2 \|g_n - g\|^2 \to 0 \ (n \to \infty).
\end{aligned}$$

[15]For a proof, see textbooks on Lebesgue integration.

Hence $\lim_{n\to\infty} F_n = F$ in the L^2-convergence. Note that

$$f_n(-L/2) = f_n(x) - F_n(x), \quad x \in \mathbb{I}_L. \tag{1.112}$$

Hence $\lim_{n\to\infty} \|f_n(-L/2) - (f - F)\| = 0$. Therefore there exists a subsequence $\{n_j\}_{j=1}^{\infty}$ such that $\lim_{j\to\infty} f_{n_j}(-L/2) = f(x) - F(x)$ a.e.$x \in \mathbb{I}_L$. The left hand side is a constant. Hence $f(x) = F(x) + c$ a.e.$x \in \mathbb{I}_L$ with $c := \lim_{j\to\infty} f_{n_j}(-L/2)$. Since F is absolutely continuous with $F'(x) = g(x)$ a.e.$x \in \mathbb{I}_L$, it follows that f is identified with an absolutely continuous function on $\mathbb{I}_L$ and $f'(x) = g(x)$ a.e.$x \in \mathbb{I}_L$, implying that $f' \in L^2(\mathbb{I}_L)$ and

$$f(x) = F(x) + c, \quad x \in \mathbb{I}_L. \tag{1.113}$$

In particular, $f(-L/2) = c$. By (1.112)

$$\lim_{j\to\infty} f_{n_j}(L/2) = c + F(L/2).$$

Using the boundary condition $f_{n_j}(-L/2) = \alpha f_{n_j}(L/2)$, we obtain that $c = \alpha(c + F(L/2))$. Hence $f(-L/2) = \alpha f(L/2)$. Therefore $f \in D(p(\alpha))$ and (1.113) implies that $p(\alpha)f = g$. Thus $p(\alpha)$ is closed.

Finally, we prove the self-adjointness of $p(\alpha)$ and (1.110). For each $k \in \Gamma_L$, we define a function

$$\phi_{\alpha,k}(x) := e^{-i\theta_\alpha x/L}\phi_k(x), \quad x \in \mathbb{I}_L.$$

Then it is easy to see that $\phi_{\alpha,k} \in D(p(\alpha))$ and

$$p(\alpha)\phi_{\alpha,k} = (k - L^{-1}\theta_\alpha)\phi_{\alpha,k}$$

and $\{\phi_{\alpha,k}\}_{k\in\Gamma_L}$ is a CONS of $L^2(\mathbb{I}_L)$ (this can be proved by using the fact that $\{\phi_k\}_{k\in\Gamma_L}$ is a CONS of $L^2(\mathbb{I}_L)$). Therefore, by Theorem 1.16, $p(\alpha)$ is self-adjoint and

$$\sigma(p(\alpha)) = \sigma_{\mathrm{p}}(p(\alpha)) = \left\{k - \frac{\theta_\alpha}{L} | k \in \Gamma_L\right\}$$

(note that the set $\left\{k - L^{-1}\theta_\alpha | k \in \Gamma_K\right\}$ has no accumulation points). It is obvious that $k - L^{-1}\theta_\alpha = k' - L^{-1}\theta_\alpha$ $(k, k' \in \Gamma_L)$ if and only if $k = k'$. Hence the completeness of $\{\phi_{\alpha,k}\}_{k\in\Gamma_L}$ implies that the multiplicity of each eigenvalue $k - L^{-1}\theta_\alpha$ is one. Hence $\sigma_{\mathrm{p}}(p(\alpha)) = \sigma_{\mathrm{d}}(p(\alpha))$ in fact. ■

We call the operator $p(\alpha)$ the **momentum operator with α-boundary condition** on $L^2(\mathbb{I}_L)$.

1.16 Higher-Dimensional Discrete Fourier Transform

A d-dimensional extension of the one-dimensional discrete Fourier transform with $d \in \mathbb{N}$ is straightforward. The interval $\mathbb{I}_L$ has the d-dimensional extension

$$\mathbb{I}_L^d = \underbrace{\mathbb{I}_L \times \cdots \times \mathbb{I}_L}_{d \text{ factors}} = \{x = (x_1, \ldots, x_d) | x_j \in \mathbb{I}_L, j = 1, \ldots, d\}, \tag{1.114}$$

a cubic box in $\mathbb{R}^d$. For each

$$\boldsymbol{k} \in \Gamma_L^d = \underbrace{\Gamma_L \times \cdots \times \Gamma_L}_{d} = \{\boldsymbol{k} = (k_1, \ldots, k_d) | k_j \in \Gamma_L, j = 1, \ldots, d\}, \tag{1.115}$$

we define a function $\phi_{\boldsymbol{k}}^{(d)}$ on $\mathbb{I}_L^d$ by

$$\phi_{\boldsymbol{k}}^{(d)}(\boldsymbol{x}) := \phi_{k_1}(x_1) \cdots \phi_{k_d}(x_d), \quad \boldsymbol{x} \in \mathbb{I}_L^d, \tag{1.116}$$

where ϕ_k is defined by (1.89). By a general theorem on CONS,[16] the set $\{\phi_{\boldsymbol{k}}^{(d)} | \boldsymbol{k} \in \Gamma_L^d\}$ is a CONS of $L^2(\mathbb{I}_L^d)$.

We denote by $\ell^2(\Gamma_L^d)$ the Hilbert space of absolutely square summable sequences on Γ_L^d:

$$\ell^2(\Gamma_L^d) := \left\{ u = \{u_{\boldsymbol{k}}\}_{\boldsymbol{k} \in \Gamma_L^d} | u_{\boldsymbol{k}} \in \mathbb{C}, \sum_{\boldsymbol{k} \in \Gamma_L^d} |u_{\boldsymbol{k}}|^2 < \infty \right\} \tag{1.117}$$

with inner product

$$\langle u, v \rangle := \sum_{\boldsymbol{k} \in \Gamma_L^d} u_{\boldsymbol{k}}^* v_{\boldsymbol{k}}, \quad u, v \in \ell^2(\Gamma_L^d).$$

[16]Let (X_j, Σ_j, μ_j) $(j = 1, \ldots, n)$ be a σ-finite measure space and $\{f_j^{(n)}\}_{j=1}^\infty$ be a CONS of $L^2(X_j, d\mu_j)$. Let $f_{j_1,\ldots,j_n}(x) := f_{j_1}(x_1) \cdots f_{j_n}(x_n)$, a.e.$x = (x_1, \ldots, x_n) \in X := X_1 \times \cdots \times X_n$ $(j_1, \ldots, j_n \in \mathbb{N})$. Then $\{f_{j_1,\ldots,j_n} | j_1, \ldots, j_n \in \mathbb{N}\}$ is a CONS of $L^2(X, \otimes^n d\mu_j)$. For a proof, see, e.g., [32, Theorem 4.2].

A d-dimensional version of Theorem 1.33 is as follows:

Theorem 1.38 *There exists a unique unitary operator* $F_{L,d}\colon L^2(\mathbb{I}_L^d) \to \ell^2(\Gamma_L^d)$ *such that*

$$(F_{L,d}f)_{\boldsymbol{k}} = \left\langle \phi_{\boldsymbol{k}}^{(d)}, f \right\rangle = \frac{1}{\sqrt{L^d}} \int_{\mathbb{I}_L^d} e^{-i\boldsymbol{k}\cdot\boldsymbol{x}} f(\boldsymbol{x}) d\boldsymbol{x}, \quad f \in L^2(\mathbb{I}_L^d), \boldsymbol{k} \in \Gamma_L^d, \tag{1.118}$$

where $\boldsymbol{k} \cdot \boldsymbol{x} = \sum_{j=1}^d k_j x_j$.

The unitary operator $F_{L,d}$ is called the d-**dimensional discrete Fourier transform** from $L^2(\mathbb{I}_L^d)$ to $\ell^2(\Gamma_L^d)$.

Chapter 2
Representations of Canonical Commutation Relations with Finite Degrees of Freedom

Abstract A general theory of representations of canonical commutation relations (CCR) with finite degrees of freedom is presented. For each $N = 1, 2, \ldots,$ the representations of the CCR with N degrees of freedom have some subclasses. Fundamental properties of each of them are derived. Some examples of representations are given, including Schrödinger and Born–Heisenberg–Jordan representations. Physical correspondences of representations of CCR are discussed.

2.1 Definitions and Fundamental Properties

2.1.1 Definition of Representation of Canonical Commutation Relations and Remarks

Definition 2.1

(i) A triple $\pi_N := (\mathscr{H}, \mathscr{D}, (Q_1, \ldots, Q_N, P_1, \ldots, P_N))$ consisting of a Hilbert space $\mathscr{H}$, a subspace $\mathscr{D} \neq \{0\}$ of $\mathscr{H}$ (not necessarily dense in $\mathscr{H}$) and a $2N$-tuple $(Q_1, \ldots, Q_N, P_1, \ldots, P_N)$ of symmetric operators Q_j and P_j $(j = 1, \ldots, N)$ on $\mathscr{H}$ is called a **representation of the canonical commutation relations** (CCR) with N degrees of freedom (or with degree N) if

$$\mathscr{D} \subset \cap_{j,k=1}^{N} D([Q_j, P_k]) \cap D([Q_j, Q_k]) \cap D([P_j, P_k])$$

and the CCR with N degrees of freedom

$$[Q_j, P_k] = i\delta_{jk}, \tag{2.1}$$

$$[Q_j, Q_k] = 0, \quad [P_j, P_k] = 0 \quad (j, k = 1, \ldots, N) \tag{2.2}$$

hold on $\mathscr{D}$, where i is the imaginary unit and δ_{jk} is the Kronecker delta: $\delta_{jj} = 1$, $\delta_{jk} = 0$ for $j \neq k$ $(j, k = 1, \ldots, N)$. The Hilbert space $\mathscr{H}$

© Springer Nature Singapore Pte Ltd. 2020

A. Arai, *Inequivalent Representations of Canonical Commutation and Anti-Commutation Relations*, Mathematical Physics Studies,
https://doi.org/10.1007/978-981-15-2180-5_2

is called the **representation space** of π_N. The operators Q_j and P_j ($j = 1, \ldots, N$) are called the **representatives** in π_N. We call $\mathscr{D}$ a **CCR-domain** for $(Q_1, \ldots, Q_N, P_1, \ldots, P_N)$.

(ii) If Q_j and P_j are essentially self-adjoint for all $j = 1, \ldots, N$ (i.e., the closures $\overline{Q}_j$ and $\overline{P}_j$ are self-adjoint for $j = 1, \ldots, N$), then we call π_N a **self-adjoint representation** of the CCR with N degrees of freedom.

Some remarks on Definition 2.1 may be necessary.

1. The concept of representation of CCR in Definition 2.1 is not that of representation of an algebra; it is rather a concept of *operator realizations* of CCR as algebraic relations (but, see Remarks 4 and 5 below). In this book, we are mainly interested in operator theoretical properties of representatives $Q_1, \ldots, Q_N, P_1, \ldots, P_N$ in connection with quantum phenomena.
2. Since we want to consider also the case where representatives in π_N are not necessarily essentially self-adjoint, we introduce the term "self-adjoint representation of CCR". But it should be kept in mind that this is different from the concept of *self-adjointness of representations of algebras* (see, e.g., [6, 94, 97, 128]). The term "self-adjoint representation of CCR" is just for indicating the essential self-adjointness of the representatives in a representation of CCR.
3. There is another type of CCR which will be discussed in Sect. 2.7 and Chap. 8. To distinguish the CCR mentioned above from them, commutation relations (2.1) and (2.2) are called the **Heisenberg CCR** or the **Heisenberg commutation relations** with N degrees of freedom. But, in this section, we mainly use the term "CCR" in the sense defined above.
4. A representation of CCR can be viewed as a representation of a Lie algebra. Recall that a Lie algebra $\mathfrak{g}$ over $\mathbb{K}$ ($\mathbb{K} = \mathbb{R}$ or $\mathbb{C}$) is a vector space over $\mathbb{K}$ which has a mapping $[\ ,\ ]: \mathfrak{g} \times \mathfrak{g} \to \mathfrak{g}$ ($\mathfrak{g} \times \mathfrak{g} \ni (X, Y) \mapsto [X, Y] \in \mathfrak{g}$), called a Lie bracket, such that the following hold:

   (i) (linearity) $[aX + bY, Z] = a[X, Z] + b[Y, Z], \quad X, Y, Z \in \mathfrak{g},\ a, b \in \mathbb{K}$.
   (ii) (anti-symmetry) $[X, Y] = -[Y, X], \quad X, Y \in \mathfrak{g}$.
   (iii) (Jacobi identity) $[X, [Y, Z]] + [Y, [Z, X]] + [Z, [X, Y]] = 0, \quad X, Y, Z \in \mathfrak{g}$.

   A Lie algebra over $\mathbb{R}$ (resp. $\mathbb{C}$) is called a real (resp. complex) Lie algebra.

   A subspace $\mathfrak{h}$ of a Lie algebra $\mathfrak{g}$ over $\mathbb{K}$ is called a Lie sub-algebra of $\mathfrak{g}$ if $\mathfrak{h}$ is a Lie algebra with the Lie bracket of $\mathfrak{g}$.

   Two Lie algebras $\mathfrak{g}$ and $\mathfrak{g}'$ over $\mathbb{K}$ are said to be isomorphic if there is a linear bijection $f: \mathfrak{g} \to \mathfrak{g}'$ such that $[f(X), f(Y)] = f([X, Y])$ for all $X, Y \in \mathfrak{g}$. In this case, f is called a Lie algebra isomorphism.

   For a vector space V over $\mathbb{K}$, we denote by $L(V)$ the space of all everywhere defined linear operators on V. It is easy to see that $L(V)$ is a Lie algebra over $\mathbb{K}$ whose Lie bracket is the commutator $[X, Y] := XY - YX,\ X, Y \in L(V)$. This Lie algebra is denoted by $\mathfrak{gl}(V)$.

   A mapping $\eta: \mathfrak{g} \to \mathfrak{gl}(V)$ is called a representation of $\mathfrak{g}$ if it is linear (i.e., $\eta(aX + bY) = a\eta(X) + b\eta(Y),\ X, Y \in \mathfrak{g},\ a, b \in \mathbb{K}$) and Lie bracket-preserving (i.e., $\eta([X, Y]) = [\eta(X), \eta(Y)],\ X, Y \in \mathfrak{g}$). In this case, V is called the representation space of η. If the dimension $\dim V$ of V is finite (resp. infinite), then η is called a finite (resp. an infinite)-dimensional representation of $\mathfrak{g}$.

A $(2N + 1)$-dimensional real Lie algebra $\mathfrak{h}_{2N+1}$ with $N \in \mathbb{N}$ is called a $(2N + 1)$-**dimensional Heisenberg Lie algebra** if it has a basis $(X_1, \ldots, X_N, Y_1, \ldots, Y_N, Z)$ (an ordered basis) such that, for $j, k = 1, \ldots, N$,

$$[X_j, Y_k] = \delta_{jk} Z, \quad [X_j, X_k] = 0, \quad [Y_j, Y_k] = 0, \quad [X_j, Z] = 0, \quad [Y_j, Z] = 0.$$

It is easily shown that $\mathfrak{h}_{2N+1}$ is unique up to Lie algebra isomorphisms.

Example 2.1 A simple example of a three-dimensional Heisenberg Lie algebra (the case $N = 1$) is given by the set of 3×3 upper triangular real matrices with zero in the diagonal:

$$\mathfrak{h}_3 = \left\{ \begin{pmatrix} 0 & a & b \\ 0 & 0 & c \\ 0 & 0 & 0 \end{pmatrix} \middle| a, b, c \in \mathbb{R} \right\}.$$

In this case, one can take X_1 and Y_1 and Z as follows:

$$X_1 = \begin{pmatrix} 0 & 1 & 1 \\ 0 & 0 & 0 \\ 0 & 0 & 0 \end{pmatrix}, \quad Y_1 = \begin{pmatrix} 0 & 0 & 0 \\ 0 & 0 & 1 \\ 0 & 0 & 0 \end{pmatrix}, \quad Z = \begin{pmatrix} 0 & 0 & 1 \\ 0 & 0 & 0 \\ 0 & 0 & 0 \end{pmatrix}.$$

We denote the complexification[1] of $\mathfrak{h}_{2N+1}$ by $\mathfrak{h}_{2N+1,\mathbb{C}}$.

Let $(\mathscr{H}, \mathscr{D}, (Q_1, \ldots, Q_N, P_1, \ldots, P_N))$ be a representation of the CCR with N degrees of freedom and

$$Q_{j,\mathscr{D}} := Q_j \restriction \mathscr{D}, \quad P_{j,\mathscr{D}} := P_j \restriction \mathscr{D}, \quad j = 1, \ldots, N,$$

the restrictions of Q_j and P_j to $\mathscr{D}$ respectively. Suppose that *$\mathscr{D}$ is an invariant dense subspace for Q_j and P_j ($j = 1, \ldots, N$)* (i.e., $\mathscr{D}$ is dense in $\mathscr{H}$ and $Q_j \mathscr{D} \subset \mathscr{D}$, $P_j \mathscr{D} \subset \mathscr{D}$). We denote by $I_{\mathscr{D}}$ the identity on $\mathscr{D}$. Let

$$\mathfrak{A}_N := \{Q_{1,\mathscr{D}}, \ldots, Q_{N,\mathscr{D}}, P_{1,\mathscr{D}}, \ldots, P_{N,\mathscr{D}}, iI_{\mathscr{D}}\}.$$

[1]The complexification of a real Lie algebra $\mathfrak{g}$ is defined to be the complex vector space $\mathfrak{g}_{\mathbb{C}}$ (:=the complexification of $\mathfrak{g}$ as a real vector space) with Lie bracket $[\ ,\ ]: \mathfrak{g}_{\mathbb{C}} \times \mathfrak{g}_{\mathbb{C}} \to \mathfrak{g}_{\mathbb{C}}$ given by

$$[X + iY, X' + iY'] := [X, X'] + i[X, Y'] + i[Y, X'] - [Y, Y'], \quad X, Y, X', Y' \in \mathfrak{g}.$$

Then $\mathfrak{g}_{\mathbb{C}}$ is a complex Lie algebra.

Then the subspace

$$l(\mathfrak{A}_N) := \left\{ \sum_{j=1}^{N} \alpha_j Q_{j,\mathscr{D}} + \sum_{j=1}^{N} b_j P_{j,\mathscr{D}} + ciI_{\mathscr{D}} | \alpha_j, b_j, c \in \mathbb{C}, \ j = 1, \ldots, N \right\} \subset L(\mathscr{D})$$

generated by $\mathfrak{A}_N$ is a $(2N+1)$-dimensional Lie sub-algebra over $\mathbb{C}$ with $\mathfrak{A}_N$ being a basis. Hence there is a unique vector space isomorphism φ: $\mathfrak{h}_{2N+1,\mathbb{C}} \to l(\mathfrak{A}_N)$ such that

$$\varphi(X_j) = Q_{j,\mathscr{D}}, \quad \varphi(Y_j) = P_{j,\mathscr{D}}, \quad \varphi(Z) = iI_{\mathscr{D}}, \quad j = 1, \ldots, N.$$

The mapping φ can be viewed as a linear operator from $\mathfrak{h}_{2N+1,\mathbb{C}}$ into $\mathfrak{gl}(\mathscr{D})$. Then it follows that φ is a representation of $\mathfrak{h}_{2N+1,\mathbb{C}}$. Thus

$$\rho_{\mathscr{D}} := (\mathscr{H}, \mathscr{D}, (Q_{1,\mathscr{D}}, \ldots, Q_{N,\mathscr{D}}, P_{1,\mathscr{D}}, \ldots, P_{N,\mathscr{D}})), \tag{2.3}$$

which is a representation of the CCR with N degrees of freedom, can be viewed as a representation of the complex Heisenberg Lie algebra $\mathfrak{h}_{2N+1,\mathbb{C}}$. From this viewpoint, $(\mathscr{H}, \mathscr{D}, (Q_1, \ldots, Q_N, P_1, \ldots, P_N))$ can be regarded as an extension of $\rho_{\mathscr{D}}$ in the sense that the representatives $Q_{j,\mathscr{D}}$ and $P_{j,\mathscr{D}}$ $(j = 1, \ldots, N)$ are replaced by extensions of them.
5. Suppose that $\mathscr{D}$ is an invariant dense subspace for Q_j and P_j $(j = 1, \ldots, N)$ as in Remark 4. The complex vector space $L(\mathscr{D})$ of everywhere defined linear operators on $\mathscr{D}$ becomes a complex algebra with product operation $L(\mathscr{D}) \times L(\mathscr{D}) \ni (A, B) \mapsto AB$ (the operator product of A and B). Then $\mathfrak{A}_N$ generates a sub-algebra of $L(\mathscr{D})$, which is denoted by $\mathfrak{A}_{\pi_N}$. On the other hand, there is a concept of algebra. An algebra $\mathfrak{A}$ is called a **CCR-algebra** if $\mathfrak{A}$ is generated by elements $X_1, \ldots, X_N, Y_1, \ldots, Y_N$ and unit element I satisfying

$$[X_j, Y_k] = i\delta_{jk}I, \quad [X_j, X_k] = 0, \quad [Y_j, Y_k] = 0 \quad (j, k = 1, \ldots, N).$$

Then there is an algebraic isomorphism τ: $\mathfrak{A} \to \mathfrak{A}_{\pi_N}$ such that $\tau(X_j) = Q_{j,\mathscr{D}}$, $\tau(Y_j) = P_{j,\mathscr{D}}$ $(j = 1, \ldots, N)$ and $\tau(I) = I_{\mathscr{D}}$. Hence $(\mathscr{H}, \mathscr{D}, (Q_{1,\mathscr{D}}, \ldots, Q_{N,\mathscr{D}}, P_{1,\mathscr{D}}, \ldots, P_{N,\mathscr{D}}))$ can be viewed as a representation of a CCR-algebra $\mathfrak{A}$. For analysis from this point of view, see, e.g., [6, 94, 97, 142]. Although this approach is interesting, we do not take it in the present book.
6. In Definition 2.1, $\mathscr{D}$ is not assumed to be a common core for Q_j and P_j $(j = 1, \ldots, N)$, unlike the usual treatment of representations of CCR-algebra. This makes it possible to treat wider classes of representations of CCR.
7. There may be other CCR-domains for $(Q_1, \ldots, Q_N, P_1, \ldots, P_N)$.
8. For each N-dimensional vector $\boldsymbol{x} = (x_1, \ldots, x_N) \in \mathbb{R}^N$, one can define

$$Q(\boldsymbol{x}) := \sum_{j=1}^{N} x_j Q_j, \quad P(\boldsymbol{x}) := \sum_{j=1}^{N} x_j P_j.$$

Suppose that $\cap_{j=1}^{N} Q_j$ and $\cap_{j=1}^{N} D(P_j)$ are dense in $\mathscr{H}$. Then it is easy to see that $Q(\boldsymbol{x})$ and $P(\boldsymbol{x})$ are symmetric operators satisfying

$$[Q(\boldsymbol{x}),\ P(\boldsymbol{y})] = i\boldsymbol{x}\cdot\boldsymbol{y}, \tag{2.4}$$

$$[Q(\boldsymbol{x}),\ Q(\boldsymbol{y})] = 0, \quad [P(\boldsymbol{x}),\ P(\boldsymbol{y})] = 0, \quad \boldsymbol{x},\ \boldsymbol{y} \in \mathbb{R}^N \tag{2.5}$$

on $\mathscr{D}$, where $\boldsymbol{x}\cdot\boldsymbol{y} := \sum_{j=1}^{N} x_j y_j$. Let $(\boldsymbol{e}_1,\ldots,\boldsymbol{e}_N) \in (\mathbb{R}^N)^N$ is the standard basis of $\mathbb{R}^N$: $(\boldsymbol{e}_j)_k = \delta_{jk},\ j,k = 1,\ldots,N$. Then

$$Q_j = Q(\boldsymbol{e}_j), \quad P_j = P(\boldsymbol{e}_j), \quad j = 1,\ldots,N.$$

Therefore, instead of π_N, one can use the triple $(\mathscr{H},\ \mathscr{D},\ \{Q(\boldsymbol{x}),\ P(\boldsymbol{x}) | \boldsymbol{x} \in \mathbb{R}^N\})$ as a representation of the CCR with N degrees of freedom. The latter may be simpler in notation. Moreover, it has a direct extension to a representation of CCR with infinite degrees of freedom (see Chap. 8). But, in this book, as far as representations of the CCR with N degrees of freedom are concerned, we use the symbol π_N to be more concrete.

We denote by $\mathsf{CCR}(N)$ (resp. $\mathsf{CCR}_{\mathrm{sa}}(N)$) the set of all (resp. self-adjoint) representations of the CCR with N degrees of freedom. Obviously

$$\mathsf{CCR}_{\mathrm{sa}}(N) \subset \mathsf{CCR}(N).$$

The elements in $\mathsf{CCR}_{\mathrm{sa}}(N)$ with $N \geq 2$ are classified into two classes:

A self-adjoint representation $\pi_N := (\mathscr{H}, \mathscr{D}, (Q_1,\ldots,Q_N, P_1,\ldots,P_N))$ is said to be **strong** if $\overline{Q} := (\overline{Q}_1,\ldots,\overline{Q}_N)$ and $\overline{P} := (\overline{P}_1,\ldots,\overline{P}_N)$ are respectively strongly commuting. We denote by $\mathsf{CCR}_{\mathrm{ssa}}(N)$ the set of strong self-adjoint representations in $\mathsf{CCR}_{\mathrm{sa}}(N)$.

A self-adjoint representation $\pi_N := (\mathscr{H}, \mathscr{D}, (Q_1,\ldots,Q_N, P_1,\ldots,P_N))$ is said to be **semi-strong** if at least one of $\overline{Q}$ and $\overline{P}$ is strongly commuting. We denote by $\mathsf{CCR}_{\mathrm{ssa}}^{\mathrm{semi}}(N)$ the set of semi-strong self-adjoint representations in $\mathsf{CCR}_{\mathrm{sa}}(N)$.

Thus we have the following inclusion relations ($N \geq 2$):

$$\mathsf{CCR}_{\mathrm{ssa}}(N) \subset \mathsf{CCR}_{\mathrm{ssa}}^{\mathrm{semi}}(N) \subset \mathsf{CCR}_{\mathrm{sa}}(N) \subset \mathsf{CCR}(N).$$

2.1.2 Some General Properties

We now study some general aspects of representations of CCR. We first note the infinite dimensionality of representation spaces of CCR, which, in the context of quantum mechanics, has a significance:

Proposition 2.1 *Let* $(\mathscr{H}, \mathscr{D}, (Q_1,\ldots,Q_N, P_1,\ldots,P_N)) \in \mathsf{CCR}(N)$ *be such that* $\mathscr{D}$ *is dense in* $\mathscr{H}$. *Then* $\mathscr{H}$ *is infinite dimensional.*

Proof Suppose that $\dim \mathscr{H} = n < \infty$. Then $\mathscr{D} = \mathscr{H}$ and $[Q_j, P_j] = i$ on $\mathscr{H}$. Taking the traces of both sides and using the symmetry of trace, one obtains $0 = i\,n$. But this is a contradiction. ■

This proposition tells us in particular that *any representation of CCR with finite degrees of freedom with a dense CCR-domain is infinite dimensional in the sense that the representation space is infinite dimensional*

The next lemma is a basis for studies on representations of CCR.

Lemma 2.1 *Let A and B be symmetric operators on a Hilbert space $\mathscr{H}$ and suppose that there exist a dense subspace $\mathscr{D} \subset D([A, B])$ and a closed operator C on $\mathscr{H}$ such that $\mathscr{D} \subset D(C)$ and*

$$[A, B] = C \quad \text{on } \mathscr{D}. \tag{2.6}$$

Then:

(i) *If $\mathscr{D}$ is a core for C, then $D([A, B]) \subset D(C^*)$ and*

$$[A, B] = -C^* \quad \text{on } D([A, B]). \tag{2.7}$$

Moreover, $C \subset -C^*$.

(ii) *If $C \in \mathfrak{B}(\mathscr{H})$, then C is anti-self-adjoint (i.e., $C^* = -C$) and*

$$[A, B] = C \quad \text{on } D([A, B]). \tag{2.8}$$

(iii) *Let $C \in \mathfrak{B}(\mathscr{H})$ and injective with C^{-1} bounded, satisfying that $C\mathscr{D} \subset D(A)$ and $CA = AC$ on $\mathscr{D}$. Then at least one of A and B is unbounded.*

Proof

(i) By (2.6) and the symmetry of A and B, we have for all $\Psi \in D([A, B])$ and $\Phi \in \mathscr{D}$,

$$\langle \Phi, [A, B]\Psi \rangle = \langle [B, A]\Phi, \Psi \rangle = \langle C\Phi, -\Psi \rangle .$$

Since $\mathscr{D}$ is a core for C, it follows that $-\Psi \in D(C^*)$ and $-C^*\Psi = [A, B]\Psi$. Hence $D([A, B]) \subset D(C^*)$ and (2.7) holds.

By (2.6) and (2.7), we have $C \restriction \mathscr{D} \subset -C^*$. Taking the closure of both sides, we obtain $C \subset -C^*$.

(ii) In this case, C and $-C^*$ are bounded. Hence, by (i), $C = -C^*$ and (2.7) implies (2.8).

(iii) We prove this part by reduction to absurdity. Hence suppose that both A and B were bounded. Then $\bar{A}$ and $\bar{B}$ are bounded with $D(\bar{A}) = D(\bar{B}) = \mathscr{H}$. By a simple limiting argument, we obtain $\bar{A}\bar{B} - \bar{B}\bar{A} = C$ and $\bar{A}C = C\bar{A}$. By the injectivity of C, $C \neq 0$. Hence it follows that $\bar{A} \neq 0$ and $\bar{B} \neq 0$. By induction, we can show that, for all $n \in \mathbb{N}$,

$$\bar{A}^n \bar{B} - \bar{B}\bar{A}^n = nC\bar{A}^{n-1}. \tag{2.9}$$

This implies that, for all $n \in \mathbb{N}$, $\bar{A}^n \neq 0$. Indeed, if $\bar{A}^n = 0$ for some $n \geq 2$, then $nC\bar{A}^{n-1} = 0$ and hence, by the injectivity of C, $\bar{A}^{n-1} = 0$. By iteration, we obtain $\bar{A} = 0$. But this is a contradiction. Commutation relation (2.9) yields

$$n\|C\bar{A}^{n-1}\| \leq 2\|\bar{A}^n\|\,\|\bar{B}\| \leq 2\|\bar{A}^{n-1}\|\,\|\bar{A}\|\,\|\bar{B}\| \leq 2\|C^{-1}\|\,\|C\bar{A}^{n-1}\|\,\|\bar{A}\|\,\|\bar{B}\|.$$

If $C\bar{A}^{n-1} = 0$, then, by the injectivity of C, $\bar{A}^{n-1} = 0$. But this is a contradiction. Hence $\|C\bar{A}^{n-1}\| \neq 0$. Therefore $n \leq 2\|C^{-1}\|\,\|\bar{A}\|\,\|\bar{B}\|$. Since $n \in \mathbb{N}$ is arbitrary, this is a contradiction. ■

Lemma 2.1(ii) shows that, under a suitable condition, a commutation relation on a dense subspace may extend to the same commutation relation on the maximal domain of the commutator. As a remark on (iii) in Lemma 2.1, we mention that there are cases where one of A and B is bounded (see Sect. 2.6.4).

A simple application of Lemma 2.1 to representations of CCR yields the next proposition:

Proposition 2.2 *Let* $(\mathscr{H}, \mathscr{D}, (Q_1, \ldots, Q_N, P_1, \ldots, P_N)) \in \mathsf{CCR}(N)$ *such that* $\mathscr{D}$ *is dense. Then:*

(i)

$$[Q_j, P_k] = i\delta_{jk} \quad \text{on } D(Q_j P_k) \cap D(P_k Q_j), \tag{2.10}$$

$$[Q_j, Q_k] = 0 \quad \text{on } D(Q_j Q_k) \cap D(Q_k Q_j), \tag{2.11}$$

$$[P_j, P_k] = 0 \quad \text{on } D(P_j P_k) \cap D(P_k P_j). \tag{2.12}$$

(ii) *For each* $j = 1, \ldots, N$, *at least one of* Q_j *and* P_j *is unbounded.*

Proof

(i) Applying Lemma 2.1 to the case where $A = Q_j$, $B = P_k$ and $C = i\delta_{jk}$, we obtain (2.10). Similarly, one can prove (2.11) and (2.12).
(ii) This follows from Lemma 2.1(iii) with $A_j = Q_j$, $B = P_j$ and $C = i$. ■

Proposition 2.2(i) shows that, *in any representation of the CCR with N degrees of freedom with a dense CCR-domain, the representatives satisfy the CCR with N degrees of freedom on the relevant maximal domains*. Proposition 2.2(ii) also is an important fact which should be kept in mind.

In general, a representation of the CCR with N degrees of freedom is said to be **unbounded** if at least one of its representatives is unbounded.

Proposition 2.2(ii) shows that any representation of the CCR with N degrees of freedom with a dense CCR-domain is unbounded.

Canonical commutation relations give a restriction on spectral properties of the representatives in π_N:

Proposition 2.3 *Let* $(\mathcal{H}, \mathcal{D}, (Q_1, \ldots, Q_N, P_1, \ldots, P_N)) \in \mathsf{CCR}(N)$. *Then, for each* $j = 1, \ldots, N$, $Q_j \restriction \mathcal{D}$ *and* $P_j \restriction \mathcal{D}$ *have no eigenvalues.*

Proof Let $A := Q_j \restriction \mathcal{D}$ and $B := P_j \restriction \mathcal{D}$. Suppose that A had an eigenvalue λ with an eigenvector $\Psi \in \mathcal{D} \setminus \{0\}$: $A\Psi = \lambda\Psi$. By the symmetry of Q_j, λ is real. We have $Q_j B\Psi - P_j A\Psi = i\Psi$. By the symmetry of Q_j and P_j, we obtain $\langle A\Psi, B\Psi\rangle - \langle B\Psi, A\Psi\rangle = i\|\Psi\|^2$. The left hand side $= \lambda\langle\Psi, B\Psi\rangle - \lambda\langle B\Psi, \Psi\rangle = 0$. Hence $\|\Psi\|^2 = 0$, i.e., $\Psi = 0$. But this is a contradiction. Thus A has no eigenvalues. Similarly, one can show that B has no eigenvalues. ■

Remark 2.1 Proposition 2.3 does not mean that Q_j or P_j has no eigenvalues, but tells us that, if Q_j or P_j has an eigenvalue, then any eigenvector belonging to it is not in $\mathcal{D}$. A simple example of this kind of representation of CCR is given in Sect. 2.6.4 below.

2.1.3 *Unitary Equivalence*

It is easy to see that, for any representation $\pi_N := (\mathcal{H}, \mathcal{D}, (Q_1, \ldots, Q_N, P_1, \ldots, P_N))$ in $\mathsf{CCR}(N)$ and any unitary transformation $U: \mathcal{H} \to \mathcal{K}$ ($\mathcal{K}$ is a Hilbert space), $\pi_N' := (\mathcal{K}, U\mathcal{D}, (UQ_1U^{-1}, \ldots, UQ_NU^{-1}, UP_1U^{-1}, \ldots, UP_NU^{-1}))$ is in $\mathsf{CCR}(N)$. In this case, if the former is a self-adjoint representation, then so is the latter. We call π_N' a unitary transformation of π_N.

We say that two representations $\rho = (\mathcal{H}, \mathcal{D}, (Q_1, \ldots, Q_N, P_1, \ldots, P_N))$ and $\rho' = (\mathcal{H}', \mathcal{D}', (Q_1', \ldots, Q_N', P_1', \ldots, P_N'))$ in $\mathsf{CCR}(N)$ are said to be **equivalent** (or ρ is equivalent to ρ') if there exists a unitary transformation $U: \mathcal{H} \to \mathcal{H}'$ such that $U\overline{Q}_jU^* = \overline{Q_j'}$ and $U\overline{P}_jU^* = \overline{P_j'}$, $j = 1, \ldots, N$. In this case, we write $\rho \cong \rho'$.

The reason for using the closures of representatives in defining the equivalence of ρ and ρ' is to exclude a trivial inequivalence.[2] It is easy to see that this relation $\cong$ is an equivalence relation in $\mathsf{CCR}(N)$. Thus $\mathsf{CCR}(N)$ is classified with equivalence classes.

Example 2.2 Let $\rho = (\mathcal{H}, \mathcal{D}, (Q_1, \ldots, Q_N, P_1, \ldots, P_N)) \in \mathsf{CCR}(N)$ such that $\mathcal{D}$ is an invariant dense subspace for Q_j and P_j $(j = 1, \ldots, N)$. Then, as already remarked, $\rho_{\mathcal{D}}$ given by (2.3) is in $\mathsf{CCR}(N)$. Suppose that $\mathcal{D}$ is a common core for Q_j and P_j, $j = 1, \ldots, N$. Then $\overline{Q}_{j,\mathcal{D}} = \overline{Q}_j$ and $\overline{P}_{j,\mathcal{D}} = \overline{P}_j$. Hence ρ and $\rho_{\mathcal{D}}$

---

[2] Recall that a unitary transformation of a closed operator is a closed operator and hence a non-closed operator cannot be unitarily equivalent to a closed operator (this is a trivial inequivalence).

are trivially equivalent. But, if $\mathscr{D}$ is not a core for some Q_j or P_j, then ρ is not necessarily equivalent to $\rho_{\mathscr{D}}$. For example, suppose that Q_j is self-adjoint for some $j \in \{1, \ldots, N\}$ and $\mathscr{D}$ is not a core for Q_j. Then $\overline{Q}_{j,\mathscr{D}}$ is not self-adjoint. Hence Q_j cannot be unitarily equivalent to $\overline{Q}_{j,\mathscr{D}}$. Therefore ρ is not equivalent to $\rho_{\mathscr{D}}$.

Example 2.3 Let $N \geq 2$. In general, for every N-tuple $(A_1, \ldots, A_N)$ of strongly commuting self-adjoint operators on a Hilbert space $\mathscr{H}$ and for any unitary transformation U from $\mathscr{H}$ to a Hilbert space $\mathscr{K}$, $(UA_1U^{-1}, \ldots, UA_NU^{-1})$ is strongly commuting. Hence, each representation $\rho \in \mathsf{CCR}_{\mathrm{ssa}}(N)$ is inequivalent to all elements $\rho' \in \mathsf{CCR}^{\mathrm{semi}}_{\mathrm{ssa}}(N) \setminus \mathsf{CCR}_{\mathrm{ssa}}(N)$.

Remark 2.2 We shall see in Chap. 3 that, for all $N \geq 2$, $\mathsf{CCR}^{\mathrm{semi}}_{\mathrm{ssa}}(N) \setminus \mathsf{CCR}_{\mathrm{ssa}}(N) \neq \emptyset$ (see Theorem 3.7 and Remark 3.3). Hence inequivalent self-adjoint representations of the CCR with degree $N \geq 2$ do exist indeed. In addition, there do exist inequivalent self-adjoint representations in $\mathsf{CCR}(1)$ too (see Sect. 2.6.4 or [75]). We want to emphasize that it is only for restricted classes of representations of CCR that a uniqueness theorem may hold.[3] Since there seems to be misunderstandings in this respect among researchers, we dare to remark this point.

Inequivalence of two representations of CCR can be read off from spectral properties of the representatives:

Proposition 2.4 *Let* $\rho = (\mathscr{H}, \mathscr{D}, (Q_1, \ldots, Q_N, P_1, \ldots, P_N))$ *and* $\rho' = (\mathscr{H}', \mathscr{D}', (Q'_1, \ldots, Q'_N, P'_1, \ldots, P'_N))$ *be in* $\mathsf{CCR}(N)$*. Suppose that, for some* j*,* $\sigma(Q_j) \neq \sigma(Q'_j)$ *or* $\sigma(P_j) \neq \sigma(P'_j)$*. Then* ρ *and* ρ' *are inequivalent.*

Proof Suppose that $\rho \cong \rho'$. Then, for all $j = 1, \ldots, N$, $\overline{Q}_j \cong \overline{Q'_j}$ and $\overline{P}_j \cong \overline{P'_j}$ (see (1.13)). Hence $\sigma(\overline{Q}_j) = \sigma(\overline{Q'_j})$ and $\sigma(\overline{P}_j) = \sigma(\overline{P'_j})$ for $j = 1, \ldots, N$. For any closable operator A, $\sigma(A) = \sigma(\overline{A})$ (Proposition 1.8). Hence we arrive at a contradiction. ∎

2.2 Heisenberg Uncertainty Relation

For any representation $(\mathscr{H}, \mathscr{D}, (Q_1, \ldots, Q_N, P_1, \ldots, P_N)) \in \mathsf{CCR}(N)$, one has

$$[Q_j, P_j] = i \quad \text{on } \mathscr{D}. \tag{2.13}$$

This may be regarded as expressing one of the simplest non-commutative relations of operators. To derive an implication of (2.13), we introduce a quantity.

[3] In Theorem 2.19, we state a uniqueness theorem for a class of strong self-adjoint representations of CCR, called Weyl representations.

Let A be a symmetric operator on $\mathscr{H}$ and, for each unit vector $\Psi \in D(A)$ ($\|\Psi\| = 1$), define

$$(\Delta A)_\Psi := \|(A - \langle \Psi, A\Psi \rangle)\Psi\|. \tag{2.14}$$

We call it the **uncertainty** of A in the vector Ψ.

Remark 2.3 Let A be self-adjoint and E_A be the spectral measure of it. Then, for all unit vectors $\Psi \in \mathscr{H}$, the measure μ_Ψ on the measurable space $(\mathbb{R}, \boldsymbol{B}^1)$ defined by

$$\mu_\Psi(B) := \|E_A(B)\Psi\|^2, \quad B \in \boldsymbol{B}^1 \tag{2.15}$$

is a probability measure (i.e., $\mu_\Psi(\mathbb{R}) = 1$). We have

$$(\Delta A)_\Psi^2 = \int_{\mathbb{R}} (\lambda - m)^2 d\mu_\Psi(\lambda)$$

with $m := \langle \Psi, A\Psi \rangle = \int_{\mathbb{R}} \lambda d\mu_\Psi(\lambda)$. The function λ on $\mathbb{R}$ can be regarded as a random variable on the probability space $(\mathbb{R}, \boldsymbol{B}^1, \mu_\Psi)$.[4] In this interpretation, $(\Delta A)_\Psi^2$ is the variance of λ and hence $(\Delta A)_\Psi$ is the standard deviation of λ. The above definition comes from a generalization of this structure.

It is easy to see that

$$(\Delta A)_\Psi^2 = \|A\Psi\|^2 - |\langle \Psi, A\Psi \rangle|^2. \tag{2.16}$$

Proposition 2.5 *Let A be a symmetric operator on $\mathscr{H}$. Then, for all $a \in \mathbb{R}$ and each unit vector $\Psi \in D(A)$,*

$$\|(A - a)\Psi\| \geq (\Delta A)_\Psi. \tag{2.17}$$

The equality holds if and only if $a = \langle \Psi, A\Psi \rangle$.

Proof We have

$$\begin{aligned} \|(A - a)\Psi\|^2 &= \|A\Psi\|^2 - 2a \langle \Psi, A\Psi \rangle + a^2 \\ &= (a - \langle \Psi, A\Psi \rangle)^2 + (\Delta A)_\Psi^2 \quad \text{(by (2.16))} \\ &\geq (\Delta A)_\Psi^2. \end{aligned}$$

Hence (2.17) holds and $\|(A - a)\Psi\| = (\Delta A)_\Psi$ if and only if $a = \langle \Psi, A\Psi \rangle$. ■

[4]A measure space (X, Σ, μ) is called a **probability space** if $\mu(X) = 1$. A Σ-measurable function f on X is called a **random variable** on (X, Σ, μ) if it is real-valued and finite μ-a.e.

This proposition means that the minimum of the function: $a \mapsto \|(A-a)\Psi\|$ is attained at $a = \langle \Psi, A\Psi \rangle$, the expectation of value of A in the vector Ψ.

Lemma 2.2 *Let A and B be operators on $\mathscr{H}$. Suppose that there exist a constant $c \in \mathbb{C}$ and a non-zero subspace $\mathscr{D}$ (not necessarily dense) such that $\mathscr{D} \subset D(A) \cap D(B)$ and*

$$\langle A\Psi, B\Psi \rangle - \langle B\Psi, A\Psi \rangle = c \langle \Psi, \Psi \rangle, \quad \Psi \in \mathscr{D}. \tag{2.18}$$

Then, for all $\Psi \in \mathscr{D}$,

$$\|A\Psi\| \, \|B\Psi\| \geq \frac{|c|}{2} \|\Psi\|^2. \tag{2.19}$$

Proof By the Schwarz inequality, we have for all $\Psi \in \mathscr{D}$

$$|\mathrm{Im}\, \langle A\Psi, B\Psi \rangle| \leq |\langle A\Psi, B\Psi \rangle| \leq \|A\Psi\| \, \|B\Psi\|, \tag{2.20}$$

where, for a complex number $z \in \mathbb{C}$, $\mathrm{Im}\, z$ denotes the imaginary part of z. On the other hand, we have

$$\mathrm{Im}\, \langle A\Psi, B\Psi \rangle = \frac{1}{2i}(\langle A\Psi, B\Psi \rangle - \langle B\Psi, A\Psi \rangle) = \frac{c}{2i} \|\Psi\|^2. \tag{2.21}$$

Hence (2.19) follows. ∎

Remark 2.4 It follows from (2.18) or (2.21) that the constant c in (2.18) is in fact a pure imaginary number.

Lemma 2.3 *Let A and B be symmetric operators on $\mathscr{H}$. Suppose that there exist a constant $c \in \mathbb{C}$ and a subspace $\mathscr{D}$ (not necessarily dense) such that $\mathscr{D} \subset D(A) \cap D(B)$ and (2.18) holds. Then, for all $\Psi \in \mathscr{D}$ and $a, b \in \mathbb{R}$,*

$$\|(A-a)\Psi\| \, \|(B-b)\Psi\| \geq \frac{|c|}{2} \|\Psi\|^2. \tag{2.22}$$

Proof Let $\tilde{A} := A - a$ and $\tilde{B} := B - b$. Then it is easy to see that (2.18) with (A, B) replaced by $(\tilde{A}, \tilde{B})$ holds. Hence, by Lemma 2.2, (2.22) follows. ∎

We now derive important inequalities associated with representations of CCR:

Theorem 2.1 *Let $(\mathscr{H}, \mathscr{D}, (Q_1, \ldots, Q_N, P_1, \ldots, P_N)) \in \mathsf{CCR}(N)$ and $j = 1, \ldots, N$. Then:*

(i) *For all unit vectors $\Psi \in \mathscr{D}$,*

$$(\Delta Q_j)_\Psi (\Delta P_j)_\Psi \geq \frac{1}{2}. \tag{2.23}$$

(ii) *If $\mathscr{D}$ is dense in $\mathscr{H}$, then, for all unit vectors $\Psi \in D(Q_j P_j) \cap D(P_j Q_j)$, (2.23) holds.*

Proof

(i) Equation (2.1) on $\mathscr{D}$ and the symmetry of Q_j and P_j imply (2.18) with $A = Q_j$, $B = P_j$ and $c = i$. Hence a simple application of Lemma 2.3 with $A = Q_j$, $B = P_j$, $c = i$, $a = \langle \Psi, Q_j \Psi \rangle$ and $b = \langle \Psi, P_j \Psi \rangle$ yields (2.23).
(ii) In the present case, by Lemma 2.1, we have $[Q_j, P_j] = i$ on $D([Q_j, P_j])$. Hence, by a simple application of Lemma 2.3, we obtain (2.23) for all unit vectors $\Psi \in D(Q_j P_j) \cap D(P_j Q_j)$. ■

Inequality (2.23) is called the **Heisenberg uncertainty relation** with respect to (Q_j, P_j). In the case where both Q_j and P_j represent physical quantities in the context of quantum mechanics (and hence they are self-adjoint), the Heisenberg uncertainty relation means that the uncertainties (standard deviations) of Q_j and P_j in the state vector Ψ has a correlation, implying that there are no state vectors Ψ in $\mathscr{D}$ (or in $D(Q_j P_j) \cap D(P_j Q_j)$ in the case where $\mathscr{D}$ is dense) such that $(\Delta Q_j)_\Psi = 0$ and $(\Delta P_j)_\Psi = 0$. In this sense, Q_j and P_j are not independent. This is one of the interesting aspects of CCR, which shows a big difference of quantum mechanics from classical mechanics.

It would be interesting to investigate for what vectors Ψ the equality in (2.23) holds. For this problem, we recall a notion:

A unit vector $\Psi \in D(Q_j P_j) \cap D(P_j Q_j)$ which attains the equality

$$(\Delta Q_j)_\Psi (\Delta P_j)_\Psi = \frac{1}{2}$$

in (2.23) is called a **minimal uncertainty state** with respect to (Q_j, P_j) if it exists.

We remark, however, that a minimal uncertainty state with respect to (Q_j, P_j) does not always exist; it depends on the pair (Q_j, P_j). Later, we shall give examples of (Q_j, P_j) which have a minimal uncertainty state (see Proposition 2.11). For a general sufficient condition under which no minimal uncertainty states exist, see [15, Theorem 7.2] (also [110]).

There is another form of uncertainty relations which is a general form for (2.23).

Lemma 2.4 *Let A and B be symmetric operators on a Hilbert space $\mathscr{H}$. Then, for all $\Psi \in D([A, B])$ and $a, b \in \mathbb{R}$,*

$$\|(A - a)\Psi\| \, \|(B - b)\Psi\| \geq \frac{1}{2} |\langle \Psi, [A, B]\Psi \rangle|. \tag{2.24}$$

Proof Using the symmetry of A and B, one sees easily that

$$\langle \Psi, [A, B]\Psi \rangle = \langle \Psi, [A - a, B - b]\Psi \rangle = 2i \mathrm{Im} \, \langle (A - a)\Psi, (B - a)\Psi \rangle .$$

Then, applying by (2.20) with (A, B) replaced by $(A - a, B - b)$, we obtain (2.24). ■

Theorem 2.2 *Let A and B be symmetric operators on a Hilbert space $\mathscr{H}$. Then, for all unit vectors $\Psi \in D([A, B])$,*

$$(\Delta A)_\Psi (\Delta B)_\Psi \geq \frac{1}{2} |\langle \Psi, [A, B]\Psi \rangle|. \tag{2.25}$$

Proof We need only to take $a = \langle \Psi, A\Psi \rangle$ and $b = \langle \Psi, B\Psi \rangle$ in (2.24). ■

Inequality (2.25) is called the **Heisenberg–Robertson inequality** with respect to (A, B). This inequality can be non-trivial only if $[A, B] \neq 0$, i.e., A and B are non-commutative. In that case, it means that the non-commutativity of A and B may give a restriction on the product of the uncertainties of A and B depending on the vector Ψ, implying non-independence of A and B.

2.3 Direct Sum Representations

Let $M \in \mathbb{N}$ with $M \geq 2$ and suppose that M representations

$$\pi_m = (\mathscr{H}_m, \mathscr{D}_m, (Q_{m1}, \ldots, Q_{mN}, P_{m1}, \ldots, P_{mN})) \in \mathsf{CCR}(N) \quad (m = 1, \ldots, M)$$

are given. Then one can construct a representation of the CCR with N degrees of freedom on the direct sum Hilbert space (see Appendix C)

$$\mathscr{H}^{(M)} := \oplus_{m=1}^{M} \mathscr{H}_m = \{\Psi = (\Psi_1, \ldots, \Psi_M) | \Psi_m \in \mathscr{H}_m,\ m = 1, \ldots, M\}$$

as follows. The *algebraic* direct sum

$$\mathscr{D}^{(M)} := \hat{\oplus}_{m=1}^{M} \mathscr{D}_m$$

of $\{\mathscr{D}_m\}_{m=1}^{M}$ is a subspace of $\mathscr{H}^{(M)}$ (we use the symbol $\hat{\oplus}$ to indicate algebraic direct sum). Let

$$Q_j^{(M)} := \oplus_{m=1}^{M} Q_{mj}, \quad P_j^{(M)} := \oplus_{m=1}^{M} P_{mj},$$

the direct sum operators of $\{Q_{mj}\}_{m=1}^{M}$ and $\{P_{mj}\}_{m=1}^{M}$ respectively (see Appendix C for the definition of a direct sum operator). Then it follows that

$$\oplus_{m=1}^{M} \pi_m := (\mathscr{H}^{(M)}, \mathscr{D}^{(M)}, (Q_1^{(M)}, \ldots, Q_N^{(M)}, P_1^{(M)}, \ldots, P_N^{(M)}))$$

is a representation of the CCR with N degrees of freedom. If each $\mathscr{D}_m$ is dense in $\mathscr{H}_m$, then $\mathscr{D}^{(M)}$ is dense in $\mathscr{H}^{(M)}$. The representation $\oplus_{m=1}^M \pi_m$ is called the **direct sum** of $\{\pi_m\}_{m=1}^M$. In this case each π_m is called the ***m*th component** of $\oplus_{m=1}^M \pi_m$.

In general, a representation $\rho = (\mathscr{H}, \mathscr{D}, (Q_1, \ldots, Q_N, P_1, \ldots, P_N)) \in \mathsf{CCR}(N)$ is called a **finite direct sum representation** if there exist an $M \in \mathbb{N}$ and M representations $\pi_m \in \mathsf{CCR}(N)$, $m = 1, \ldots, M$ such that $\rho = \oplus_{m=1}^M \pi_m$. In this case, ρ is said to be decomposed into an M direct sum of representations in $\mathsf{CCR}(N)$. This decomposition is said to be **non-trivial** if $M \geq 2$.

Proposition 2.6 *Let π_m be as above and*

$$\pi'_m := (\mathscr{H}'_m, \mathscr{D}'_m, (Q'_{m1}, \ldots, Q'_{mN}, P'_{m1}, \ldots, P'_{mN})) \in \mathsf{CCR}(N)$$

be another representation of CCR. Suppose that, for each $m = 1, \ldots, M$, $\pi_m \cong \pi'_m$. Then $\oplus_{m=1}^M \pi_m \cong \oplus_{m=1}^M \pi'_m$.

Proof By the present assumption, there is a unitary transformation $U_m : \mathscr{H}_m \to \mathscr{H}'_m$ such that $U_m \overline{Q}_{mj} U_m^{-1} = \overline{Q'}_{mj}$, $U_m \overline{P}_{mj} U_m^{-1} = \overline{P'}_{mj}$, $j = 1, \ldots, N$. It is easy to see that the operator $U := \oplus_{m=1}^M U_m$ is a unitary transformation from $\oplus_{m=1}^M \mathscr{H}_m$ onto $\oplus_{m=1}^M \mathscr{H}'_m$ and

$$U \overline{\oplus_{m=1}^M Q_{mj}} U^{-1} = \overline{\oplus_{m=1}^M Q'_{mj}}, \quad U \overline{\oplus_{m=1}^M P_{mj}} U^{-1} = \overline{\oplus_{m=1}^M P'_{mj}},$$

where the following elementary fact is used: for each closable operator A_m on $\mathscr{H}_m$, $\overline{\oplus_{m=1}^M A_m} = \oplus_{m=1}^M \overline{A}_m$. Thus $\oplus_{m=1}^M \pi_m \cong \oplus_{m=1}^M \pi'_m$. ■

The concept of finite direct sum representation of CCR can be extended to a concept of infinite direct sum representation. Let $\{\pi_m\}_{m=1}^\infty$ with $\pi_m \in \mathsf{CCR}(N)$ be given and $Q_j^{(\infty)} := \oplus_{m=1}^\infty Q_{mj}$, $P_j^{(\infty)} := \oplus_{m=1}^\infty P_{mj}$. Then

$$\oplus_{m=1}^\infty \pi_m := (\oplus_{m=1}^\infty \mathscr{H}_m, \hat{\oplus}_{m=1}^\infty \mathscr{D}_m, (Q_1^{(\infty)}, \ldots, Q_N^{(\infty)}, P_1^{(\infty)}, \ldots, P_N^{(\infty)}))$$

is a representation of the CCR with N degrees of freedom (see Appendix C for notation). This representation is called the **infinite direct sum representation** of $\{\pi_m\}_{m=1}^\infty$.

2.4 Tensor Product Representations

In addition to direct sum representations of CCR, there is another type of construction of representations of CCR. Let

$$\rho_1(q, p) := (\mathscr{H}, \mathscr{D}, (q, p)) \tag{2.26}$$

be a representation of the CCR with one degree of freedom, i.e., $\rho_1(q,p) \in \mathsf{CCR}(1)$ such that $\mathscr{D}$ is dense in $\mathscr{H}$. Then one can construct tensor product operators

$$Q_j := I \otimes \cdots \otimes I \otimes \underbrace{q}_{j\text{-th}} \otimes I \otimes \cdots \otimes I \tag{2.27}$$

and

$$P_j := I \otimes \cdots \otimes I \otimes \underbrace{p}_{j\text{-th}} \otimes I \otimes \cdots \otimes I \quad (j = 1, \ldots, N) \tag{2.28}$$

on the N-fold tensor product Hilbert space $\otimes^N \mathscr{H}$ of $\mathscr{H}$. It follows that, for $j = 1, \ldots, N$, Q_j and P_j are closed symmetric operators on $\otimes^N \mathscr{H}$ (see Theorem 1.26(iii)). It is straightforward to see that

$$\rho_N(q,p) := (\otimes^N \mathscr{H}, \hat{\otimes}^N \mathscr{D}, (Q_1, \ldots, Q_N, P_1, \ldots, P_N)) \tag{2.29}$$

is in $\mathsf{CCR}(N)$. We call $\rho_N(q,p)$ the N-**tensor product representation** of CCR based on $\rho_1(q,p)$. In this context, $\rho_1(q,p)$ is called the original representation of $\rho_N(q,p)$.

If q and p are self-adjoint, then, by Theorem 1.26(iii), $\rho_N(q,p)$ is a self-adjoint representation of CCR, i.e., $\rho_N(q,p) \in \mathsf{CCR}_{\mathrm{sa}}(N)$. In this case, $(Q_1, \ldots, Q_N)$ and $(P_1, \ldots, P_N)$ are strongly commuting respectively (Theorem 1.26(iii)). Hence, for any $\rho_1(q,p) \in \mathsf{CCR}_{\mathrm{sa}}(1)$,

$$\rho_N(q,p) \in \mathsf{CCR}_{\mathrm{ssa}}(N).$$

This is noteworthy.

To prove an equivalence theorem on $\rho_N(q,p)$, we first prove a lemma.

Lemma 2.5 *Let A be a densely defined closed operator on a Hilbert space $\mathscr{H}_1$ and $\mathscr{H}_2$ be a Hilbert space. Let U_j $(j = 1, 2)$ be a unitary transformation from $\mathscr{H}_j$ to a Hilbert space $\mathscr{K}_j$. Let $U := U_1 \otimes U_2$ (this is a unitary operator from $\mathscr{H}_1 \otimes \mathscr{H}_2$ to $\mathscr{K}_1 \otimes \mathscr{K}_2$ by Theorem 1.25(ii)). Then operator equality*

$$U(A \otimes I)U^{-1} = (U_1 A U_1^{-1}) \otimes I \tag{2.30}$$

holds.

Proof Let $A' := U_1 A U_1^{-1}$. Then A' is a densely defined closed operator on $\mathscr{K}_1$. It is easy to see that $U(A \otimes I)U^{-1} = A' \otimes I$ on $D(A')\hat{\otimes}\mathscr{K}_2$. Since $D(A')\hat{\otimes}\mathscr{K}_2$ is a core for $A' \otimes I$, it follows that $A' \otimes I \subset U(A \otimes I)U^{-1}$. Interchanging the role of A and A', one has $A \otimes I \subset U^{-1}(A' \otimes I)U$, which implies that $U(A \otimes I)U^{-1} \subset A' \otimes I$. Thus (2.30) follows. ■

The next theorem shows that, if the original representation of an N-tensor product representation is equivalent to that of another N-tensor product representation, then the N-tensor product representations are equivalent, a harmonious structure:

Theorem 2.3 *Suppose that* $\rho_1(q,p) \cong \rho_1(q',p') := (\mathscr{H}', \mathscr{D}', (q',p')) \in$ $\mathsf{CCR}(1)$. *Then* $\rho_N(q,p) \cong \rho_N(q',p')$.

Proof The present assumption implies that there is a unitary transformation U from $\mathscr{H}$ to $\mathscr{H}'$ such that $U\bar{q}U^{-1} = \overline{q'}$ and $U\bar{p}U^{-1} = \overline{p'}$. By Theorem 1.25(ii), the N-fold tensor product $U_N := \otimes^N U$ of U is a unitary transformation from $\otimes^N \mathscr{H}$ to $\otimes^N \mathscr{H}'$. Then, by a straightforward extension of Lemma 2.5, we have

$$U_N Q_j U_N^{-1} = Q'_j, \quad U_N P_j U_N^{-1} = P'_j,$$

where Q'_j (resp. P'_j) is defined by (2.27) (resp. (2.28) with q (resp. p) replaced by q' (resp. p'). Thus $\rho_N(q,p) \cong \rho_N(q',p')$. ∎

2.5 Irreducibility

To exclude trivial inequivalences among elements of $\mathsf{CCR}(N)$, one needs a concept of *irreducibility*. We next describe this aspect.

Let $\mathscr{H}$ be a Hilbert space and $\mathfrak{A}$ be a set of (not necessarily bounded) linear operators on $\mathscr{H}$: $\mathfrak{A} \subset \mathfrak{L}(\mathscr{H})$.

A closed subspace $\mathscr{M}$ of $\mathscr{H}$ is said to **reduce** $\mathfrak{A}$ if each $A \in \mathfrak{A}$ is reduced by $\mathscr{M}$. In this case, $\mathscr{M}$ is called a **reducing subspace** of $\mathfrak{A}$.

Obviously, $\{0\}$ and $\mathscr{H}$ reduce $\mathfrak{A}$. These spaces are called the **trivial reducing subspaces**.

The set $\mathfrak{A}$ is said to be **reducible** if it has a non-trivial reducing subspace (i.e. there exists a closed subspace $\mathscr{M} \neq \{0\}, \mathscr{H}$ which reduces $\mathfrak{A}$).

If $\mathfrak{A}$ is not reducible (i.e. $\mathfrak{A}$ has no non-trivial reducing subspaces), then $\mathfrak{A}$ is said to be **irreducible**.

In the case where $\mathfrak{A}$ consists of everywhere defined bounded operators only, the reducibility of $\mathfrak{A}$ is characterized as follows:

Proposition 2.7 *Let* $\mathfrak{A} \subset \mathfrak{B}(\mathscr{H})$. *Then a closed subspace* $\mathscr{M}$ *of* $\mathscr{H}$ *reduces* $\mathfrak{A}$ *if and only if* $\mathfrak{A}$ *leaves* $\mathscr{M}$ *and* $\mathscr{M}^\perp$ *invariant (i.e. for all* $A \in \mathfrak{A}$, $A\mathscr{M} \subset \mathscr{M}$ *and* $A\mathscr{M}^\perp \subset \mathscr{M}^\perp$*).*

Proof Suppose that $\mathscr{M}$ reduces $\mathfrak{A}$ and let P be the orthogonal projection onto $\mathscr{M}$. Then, for all $A \in \mathfrak{A}$ and $\Psi \in \mathscr{M}$, $PA\Psi = AP\Psi = A\Psi$. Hence $A\Psi \in \mathscr{M}$. Therefore $A\mathscr{M} \subset \mathscr{M}$. Similarly, condition $(I-P)A = A(I-P)$ implies that $A\mathscr{M}^\perp \subset \mathscr{M}^\perp$.

Conversely, suppose that, for all $A \in \mathfrak{A}$, $A\mathcal{M} \subset \mathcal{M}$ and $A\mathcal{M}^\perp \subset \mathcal{M}^\perp$. Then, for all $\Psi \in \mathcal{M}$, $AP\Psi = A\Psi = PA\Psi$ and, for all $\Phi \in \mathcal{M}^\perp$, $AP\Phi = 0 = PA\Phi$. Hence $AP = PA$. Thus $\mathcal{M}$ reduces $\mathfrak{A}$. ■

We next show that reducibility and irreducibility are unitarily invariant concepts. Let $\mathcal{H}$ and $\mathcal{K}$ be Hilbert spaces. An operator set $\mathfrak{A} \subset \mathfrak{L}(\mathcal{H})$ is said to be **unitarily equivalent** to an operator set $\mathfrak{B} \subset \mathfrak{L}(\mathcal{K})$ if there exists a unitary operator U: $\mathcal{H} \to \mathcal{K}$ such that $\mathfrak{B} = \{UAU^{-1} | A \in \mathfrak{A}\}$.

It follows that $\mathfrak{A}$ is unitarily equivalent to $\mathfrak{B}$ if and only if $\mathfrak{B}$ is unitarily equivalent to $\mathfrak{A}$. Based on this fact, we say that $\mathfrak{A}$ and $\mathfrak{B}$ are equivalent or $\mathfrak{A}$ is equivalent to $\mathfrak{B}$ if $\mathfrak{A}$ is unitarily equivalent to $\mathfrak{B}$.

Proposition 2.8 *Let $\mathfrak{A} \subset \mathfrak{L}(\mathcal{H})$ and $\mathfrak{B} \subset \mathfrak{L}(\mathcal{K})$ and suppose that $\mathfrak{A}$ is equivalent to $\mathfrak{B}$. Then $\mathfrak{A}$ is reducible (resp. irreducible) if and only if $\mathfrak{B}$ is reducible (resp. irreducible).*

Proof By the present assumption, there exists a unitary operator $U: \mathcal{H} \to \mathcal{K}$ such that $\mathfrak{B} = \{UAU^{-1} | A \in \mathfrak{A}\}$. Suppose that $\mathfrak{A}$ is reducible. Then there exists a closed subspace $\mathcal{M} \subset \mathcal{H}$ such that $\mathcal{M} \neq \{0\}, \mathcal{H}$ and each $A \in \mathfrak{A}$ is reduced by $\mathcal{M}$. Let $\mathcal{N} := U\mathcal{M}$. Then $\mathcal{N}$ is a closed subspace of $\mathcal{K}$ and $\mathcal{N} \neq \{0\}, \mathcal{K}$. Each $B \in \mathfrak{B}$ is written $B = UAU^{-1}$ with some $A \in \mathfrak{A}$. It is easy to see B is reduced by $\mathcal{N}$. Thus $\mathfrak{B}$ is reducible.

Conversely, if $\mathfrak{B}$ is reducible, then, by symmetry (one needs only to replace $(\mathcal{H}, U, A)$ by $(\mathcal{K}, U^{-1}, B)$ in the above argument), one can conclude that $\mathfrak{A}$ is reducible. ■

To characterize the irreducibility of an operator set $\mathfrak{A} \subset \mathfrak{L}(\mathcal{H})$, we introduce a subset of $\mathfrak{B}(\mathcal{H})$ by

$$\mathfrak{A}' := \{T \in \mathfrak{B}(\mathcal{H}) \,|\, TA \subset AT, \;\; A \in \mathfrak{A}\}, \tag{2.31}$$

which is called the **strong commutant** of $\mathfrak{A}$. It is easy to see that $\mathfrak{A}'$ is a sub-algebra of $\mathfrak{B}(\mathcal{H})$.

Lemma 2.6 *Let $\mathfrak{A}, \mathfrak{B} \subset \mathfrak{L}(\mathcal{H})$ and $\mathfrak{A} \subset \mathfrak{B}$. Then $\mathfrak{B}' \subset \mathfrak{A}'$.*

Proof An easy exercise. ■

An operator set $\mathfrak{A} \subset \mathfrak{L}(\mathcal{H})$ is said to be **$*$-invariant** if each element A of $\mathfrak{A}$ is densely defined closed and $A^* \in \mathfrak{A}$. In this case, A^* is densely defined and $A^{**} = A \in \mathfrak{A}$. Hence the concept "$*$-invariant" is well-defined.

We denote by $\mathbb{C}I$ the set of scalar operators on $\mathcal{H}$:

$$\mathbb{C}I := \{cI \,|\, c \in \mathbb{C}\}.$$

Proposition 2.9 *Let $\mathfrak{A} \subset \mathfrak{L}(\mathcal{H})$.*

(i) *If $\mathfrak{A}' = \mathbb{C}I$, then $\mathfrak{A}$ is irreducible.*
(ii) *If $\mathfrak{A}$ is $*$-invariant and irreducible, then $\mathfrak{A}' = \mathbb{C}I$.*

Proof

(i) Let $\mathscr{M}$ be a reducing subspace of $\mathfrak{A}$ and P be the orthogonal projection onto $\mathscr{M}$. Then, for each $A \in \mathfrak{A}$, $PA \subset AP$. Hence $P \in \mathfrak{A}'$. Hence, by the present assumption, $P = cI$ for some $c \in \mathbb{C}$. Since $P = P^*$ and $P^2 = P$, it follows that $c = 0$ or 1, i.e., $P = 0$ or I. This means that $\mathscr{M} = \{0\}$ or $\mathscr{H}$. Thus $\mathfrak{A}$ is irreducible.

(ii) Let $T \in \mathfrak{A}'$. Then, for all $A \in \mathfrak{A}$, $TA \subset AT$. Hence $(AT)^* \subset (TA)^*$. We have $(AT)^* \supset T^*A^*$ and $(TA)^* = A^*T^*$ (by the boundedness of T). Hence $T^*A^* \subset A^*T^*$. Since $A^* \in \mathfrak{A}$, taking A^* instead of A and using the fact $(A^*)^* = A$, we obtain $T^*A \subset AT^*$. Hence $T^* \in \mathfrak{A}'$. Therefore the bounded self-adjoint operators $S_+ := (T + T^*)/2$ and $S_- := (T - T^*)/2i$ are in $\mathfrak{A}'$. Let $E(\cdot)$ be the spectral measure of S_+. Then, by Theorem B.4 in Appendix B, $E(B)A \subset AE(B)$ for all Borel sets $B \subset \mathbb{R}$ and $A \in \mathfrak{A}$. Since $E(B)$ is the orthogonal projection onto Ran $E(B)$, it follows that Ran $E(B)$ reduces $\mathfrak{A}$. By the irreducibility of $\mathfrak{A}$, Ran $E(B) = \{0\}$ or $\mathscr{H}$. Let $a = \inf \sigma(S_+)$. Then $a \in \sigma(S_+)$. Since supp E, the support of E, is equal to $\sigma(S_+)$, it follows that, for all $\varepsilon > 0$, $E([a, a + \varepsilon)) \neq 0$. Hence Ran $E([a, a + \varepsilon)) = \mathscr{H}$, which implies that, for all $\Psi \in \mathscr{H}$, $E([a, a + \varepsilon))\Psi = \Psi$. On the other hand, one has $\lim_{\varepsilon \downarrow 0} E([a, a + \varepsilon))\Psi = E(\{a\})\Psi$. Hence $E(\{a\})\Psi = \Psi$. Therefore $E(\{a\}) = I$. This implies that $S_+ = aI$. Similarly, one can show that $S_- = bI$ for some $b \in \mathbb{R}$. Therefore $T = S_+ + iS_- = (a + ib)I$. Thus $\mathfrak{A}' = \mathbb{C}I$. ∎

We now define reducibility and irreducibility of a representation of CCR.

Definition 2.2 Let $\rho = (\mathscr{H}, \mathscr{D}, (Q_1, \ldots, Q_N, P_1, \ldots, P_N)) \in \mathsf{CCR}(N)$. We say that ρ is **reducible** (resp. **irreducible**) if the operator set $\{Q_j, P_j | j = 1, \ldots, N\}$ is reducible (resp. irreducible).

It is easy to see that a non-trivial direct sum representation $\rho = \oplus_{m=1}^M \pi_m$ of $\pi_m \in \mathsf{CCR}(N)$, $m = 1, \ldots, M$ $(2 \le M \le \infty)$ is reducible. The converse of this also holds:

Proposition 2.10 *Suppose that* $\rho = (\mathscr{H}, \mathscr{D}, (Q_1, \ldots, Q_N, P_1, \ldots, P_N)) \in \mathsf{CCR}(N)$ *is reducible. Then* ρ *is decomposed into a non-trivial direct sum representation.*

Proof The reducibility of ρ implies that there exists a non-trivial closed subspace $\mathscr{M}$ of $\mathscr{H}$ which reduces Q_j and P_j, $j = 1, \ldots, N$. We denote by $Q_{j,\mathscr{M}}$ and $P_{j,\mathscr{M}}$ the reduced part of Q_j and P_j to $\mathscr{M}$ respectively. Let $P_{\mathscr{M}}$ be the orthogonal projection onto $\mathscr{M}$ and $j, k = 1, \ldots, N$. Since $\mathscr{D} \subset D(Q_j) \cap D(P_k)$, it follows that, for all $\Psi \in \mathscr{D}$, $P_{\mathscr{M}}\Psi \in D(Q_j) \cap D(P_k)$ and $Q_{j,\mathscr{M}} P_{\mathscr{M}}\Psi = P_{\mathscr{M}} Q_j \Psi$ and $P_{k,\mathscr{M}} P_{\mathscr{M}}\Psi = P_{\mathscr{M}} P_k \Psi$. One has $\mathscr{D} \subset D(Q_j P_k) \cap D(P_k Q_j)$. Hence, for all $\Psi \in \mathscr{D}$, $P_k \Psi \in D(Q_j)$ and $Q_j \Psi \in D(P_k)$. Hence $P_{\mathscr{M}} P_k \Psi \in D(Q_{j,\mathscr{M}})$ and $Q_{j,\mathscr{M}} P_{\mathscr{M}} P_k \Psi = P_{\mathscr{M}} Q_j P_k \Psi$. Similarly, $P_{\mathscr{M}} Q_j \Psi \in D(P_{k,\mathscr{M}})$

and $P_{k,\mathcal{M}}P_{\mathcal{M}}Q_j\Psi = P_{\mathcal{M}}P_kQ_j\Psi$. Therefore $Q_{j,\mathcal{M}}P_{\mathcal{M}}\Psi \in D(P_{k,\mathcal{M}})$ and $P_{k,\mathcal{M}}P_{\mathcal{M}}\Psi \in D(Q_{j,\mathcal{M}})$, i.e., $P_{\mathcal{M}}\Psi \in D(Q_{j,\mathcal{M}}P_{k,\mathcal{M}}) \cap D(P_{k,\mathcal{M}}Q_{j,\mathcal{M}})$ and

$$Q_{j,\mathcal{M}}P_{k,\mathcal{M}}P_{\mathcal{M}}\Psi = P_{\mathcal{M}}Q_jP_k\Psi, \quad P_{k,\mathcal{M}}Q_{j,\mathcal{M}}P_{\mathcal{M}}\Psi = P_{\mathcal{M}}P_kQ_j\Psi.$$

Hence $[Q_{j,\mathcal{M}}, P_{k,\mathcal{M}}]P_{\mathcal{M}}\Psi = i\delta_{jk}P_{\mathcal{M}}\Psi$. Similarly, one can show that, for $j, k = 1, \ldots, N$, $P_{\mathcal{M}}\mathcal{D} \subset D([Q_{j,\mathcal{M}}, Q_{k,\mathcal{M}}]) \cap D([P_{j,\mathcal{M}}, P_{k,\mathcal{M}}])$ and

$$[Q_{j,\mathcal{M}}, Q_{k,\mathcal{M}}] = 0, \quad [P_{j,\mathcal{M}}, P_{k,\mathcal{M}}] = 0 \quad \text{on } P_{\mathcal{M}}\mathcal{D}.$$

Hence

$$\pi_1 := (\mathcal{M}, P_{\mathcal{M}}\mathcal{D}, (Q_{1,\mathcal{M}}, \ldots, Q_{N,\mathcal{M}}, P_{1,\mathcal{M}}, \ldots, P_{N,\mathcal{M}}))$$

is in $\mathsf{CCR}(N)$. Similarly,

$$\pi_2 := (\mathcal{M}^\perp, P_{\mathcal{M}^\perp}\mathcal{D}, (Q_{1,\mathcal{M}^\perp}, \ldots, Q_{N,\mathcal{M}^\perp}, P_{1,\mathcal{M}^\perp}, \ldots, P_{N,\mathcal{M}^\perp}))$$

is in $\mathsf{CCR}(N)$. It is easy to see that $Q_j = Q_{j,\mathcal{M}} \oplus Q_{j,\mathcal{M}^\perp}$, $P_j = P_{j,\mathcal{M}} \oplus P_{j,\mathcal{M}^\perp}$ and $\mathcal{D} = P_{\mathcal{M}}\mathcal{D} \oplus P_{\mathcal{M}^\perp}\mathcal{D}$. Thus $\rho = \pi_1 \oplus \pi_2$. ■

To prove irreducibility of a given representation of CCR, the following theorem may be useful:

Theorem 2.4 *Let* $\rho = (\mathcal{H}, \mathcal{D}, (Q_1, \ldots, Q_N, P_1, \ldots, P_N)) \in \mathsf{CCR}(N)$. *Suppose that* $\{Q_j, P_j | j = 1, \ldots, N\}' = \mathbb{C}I$. *Then* ρ *is irreducible.*

Proof We need only to apply Proposition 2.9(i) to the case $\mathfrak{A} = \{Q_j, P_j | j = 1, \ldots, N\}$. ■

In the case of self-adjoint representations, the converse of Theorem 2.4 also is true:

Theorem 2.5 *Let* $\rho = (\mathcal{H}, \mathcal{D}, (Q_1, \ldots, Q_N, P_1, \ldots, P_N)) \in \mathsf{CCR}_{\mathrm{sa}}(N)$. *Then* ρ *is irreducible if and only if* $\{Q_j, P_j | j = 1, \ldots, N\}' = \mathbb{C}I$.

Proof We need only to prove "only if" part. Suppose that ρ is irreducible. It is obvious that $\{Q_j, P_j | j = 1, \ldots, N\}$ is $*$-invariant. Hence Proposition 2.9(ii), $\{Q_j, P_j | j = 1, \ldots, N\}' = \mathbb{C}I$. ■

For a self-adjoint representation $\rho = (\mathcal{H}, \mathcal{D}, (Q_1, \ldots, Q_N, P_1, \ldots, P_N))$ (i.e., $\rho \in \mathsf{CCR}_{\mathrm{sa}}(N)$), one has unitary operators e^{itQ_j} and e^{itP_j} ($t \in \mathbb{R}$). Hence there may be a possibility of characterizing the irreducibility of ρ in terms of the set $\{e^{itQ_j}, e^{itP_j} | t \in \mathbb{R}, j = 1, \ldots, N\}$ of unitary operators. To prove a theorem on this aspect, we prepare some lemmas.

Lemma 2.7 *Let A be a self-adjoint operator on $\mathscr{H}$ and $T \in \mathfrak{B}(\mathscr{H})$. Then the following are equivalent:*

(i) $TA \subset AT$.
(ii) *For all* $t \in \mathbb{R}$, $Te^{itA} = e^{itA}T$.

Proof (i) $\Longrightarrow$ (ii) By induction in $n \in \mathbb{N}$, one can show that, for all $n \in \mathbb{N}$, $TA^n \subset A^nT \cdots (*)$. For each $N \in \mathbb{N}$ and $t \in \mathbb{R}$, let $S_N(t) := \sum_{n=0}^{N}(it)^n A^n/n!$. Then, by $(*)$, for all $\Psi, \Phi \in C^\infty(A)$,

$$\langle T^*\Phi, S_N(t)\Psi\rangle = \langle S_N(-t)\Phi, T\Psi\rangle .$$

Let $\mathscr{D} := \{E_A([-R, R])\Psi | \Psi \in \mathscr{H}, R > 0\}$, where E_A is the spectral measure of A. By the functional calculus, one can show that, for all $\Theta \in \mathscr{D}$, $\lim_{N\to\infty} S_N(t)\Theta = e^{itA}\Theta$. Hence

$$\left\langle T^*\Phi, e^{itA}\Psi\right\rangle = \left\langle e^{-itA}\Phi, T\Psi\right\rangle, \Psi, \Phi \in \mathscr{D}.$$

Since $\mathscr{D}$ is dense in $\mathscr{H}$, it follows that $Te^{itA} = e^{itA}T$.

(ii) $\Longrightarrow$ (i) By the assumed operator equality, we have for all $\Psi \in D(A)$ and $t \in \mathbb{R} \setminus \{0\}$

$$\frac{1}{t}(e^{itA} - 1)T\Psi = T\frac{1}{t}(e^{itA} - 1)\Psi.$$

The right hand side converges to $iTA\Psi$ as $t \to 0$. Hence the left hand side also converges to $iTA\Psi$ as $t \to 0$. This implies that $T\Psi \in D(A)$ and $iAT\Psi = iTA\Psi$, i.e., $TA \subset AT$. ■

Lemma 2.8 *Let A be a self-adjoint operator on a Hilbert space $\mathscr{H}$ and $\mathscr{M}$ be a closed subspace of $\mathscr{H}$. Then the following are equivalent:*

(i) *For all* $t \in \mathbb{R}$, $e^{itA}\mathscr{M} \subset \mathscr{M}$.
(ii) *A is reduced by $\mathscr{M}$. In this case,*

$$e^{itA}\Psi = e^{itA_{\mathscr{M}}}\Psi, \quad \Psi \in \mathscr{M}, t \in \mathbb{R}, \tag{2.32}$$

where $A_{\mathscr{M}}$ is the reduced part of A to $\mathscr{M}$.

Proof (i) $\Longrightarrow$ (ii) Let P be the orthogonal projection onto $\mathscr{M}$. Then the present assumption implies that, for all $t \in \mathbb{R}$, $Pe^{itA}P = e^{itA}P$. Taking the adjoint of this equation and replacing t by $-t$, one has $Pe^{itA}P = Pe^{itA}$. Hence $e^{itA}P = Pe^{itA}$. Hence, by Lemma 2.7, $PA \subset AP$. Hence A is reduced by $\mathscr{M}$.

To prove (2.32), for each $\Psi \in D(A_{\mathcal{M}}) = D(A) \cap \mathcal{M}$, let $\Psi(t) := e^{itA}\Psi$, $t \in \mathbb{R}$. Then $\Psi(t)$ is in $D(A) \cap \mathcal{M} = D(A_{\mathcal{M}})$ and strongly differentiable in t, satisfying

$$\frac{d\Psi(t)}{dt} = iA\Psi(t) = iA_{\mathcal{M}}\Psi(t).$$

Since $A_{\mathcal{M}}$ is self-adjoint, it follows that $\Psi(t) = e^{itA_{\mathcal{M}}}\Psi$. Since $D(A_{\mathcal{M}})$ is dense in $\mathcal{M}$, (2.32) follows.

(ii) $\Longrightarrow$ (i) By the present assumption, $PA \subset AP$. Hence, by Lemma 2.7, $e^{itA}P = Pe^{itA}$, $t \in \mathbb{R}$. This implies that $e^{itA}\mathcal{M} \subset \mathcal{M}$. ∎

Theorem 2.6 *Let $\rho = (\mathcal{H}, \mathcal{D}, (Q_1, \ldots, Q_N, P_1, \ldots, P_N))$ be in $\mathsf{CCR}_{\mathrm{sa}}(N)$. Then ρ is irreducible if and only if $\{e^{itQ_j}, e^{itP_j} | t \in \mathbb{R}, j = 1, \ldots, N\}$ is irreducible.*

Proof *"only if" part.* Suppose that $\{e^{itQ_j}, e^{itP_j} | t \in \mathbb{R}, j = 1, \ldots, N\}$ is reducible. Hence there exists a non-trivial closed subspace $\mathcal{M}$ such that, for $j = 1, \ldots, N$, $e^{itQ_j}\mathcal{M} \subset \mathcal{M}$ and $e^{itP_j}\mathcal{M} \subset \mathcal{M}$. Therefore, by Lemma 2.8, Q_j and P_j are reduced by $\mathcal{M}$. Thus $\{Q_j, P_j | j = 1, \ldots, N\}$ is reducible.

"if" part. Suppose that $\{Q_j, P_j | j = 1, \ldots, N\}$ is reducible. Then there exists a non-trivial closed subspace $\mathcal{M}$ which reduces Q_j and P_j. Therefore, by Lemma 2.8, $e^{itQ_j}\mathcal{M} \subset \mathcal{M}$ and $e^{itP_j}\mathcal{M} \subset \mathcal{M}$ for all $t \in \mathbb{R}$. Thus $\{e^{itQ_j}, e^{itP_j} | j = 1, \ldots, N\}$ is reducible. ∎

Corollary 2.1 *Let $\rho = (\mathcal{H}, \mathcal{D}, (Q_1, \ldots, Q_N, P_1, \ldots, P_N)) \in \mathsf{CCR}_{\mathrm{sa}}(N)$. Then ρ is irreducible if and only if $\{e^{itQ_j}, e^{itP_j} | t \in \mathbb{R}, j = 1, \ldots, N\}' = \mathbb{C}I$.*

Proof Let ρ be irreducible. Then, by Theorem 2.6, $\mathfrak{W} := \{e^{itQ_j}, e^{itP_j} | t \in \mathbb{R}, j = 1, \ldots, N\}$ is irreducible. Since $(e^{itQ_j})^* = e^{-itQ_j}$ and $(e^{itP_j})^* = e^{-itP_j}$, $\mathfrak{W}$ is $*$-invariant. Hence, by Proposition 2.9(ii), $\mathfrak{W}' = \mathbb{C}I$.

Conversely, let $\mathfrak{W}' = \mathbb{C}I$. Then, by Proposition 2.9(i), $\mathfrak{W}$ is irreducible. Hence, by Theorem 2.6, ρ is irreducible. ∎

In concluding this section, we prove a theorem on irreducibility of tensor product representations of CCR.

Lemma 2.9 *Let $\mathfrak{A}$ be a set of densely defined closed operators on a Hilbert space $\mathcal{H}$ such that $\mathfrak{A}' = \mathbb{C}I$ and $\mathcal{K}$ be a Hilbert space. Suppose that $T \in \mathfrak{B}(\mathcal{H} \otimes \mathcal{K})$ is in $\{A \otimes I | A \in \mathfrak{A}\}'$. Then there exists a unique bounded operator S on $\mathcal{K}$ such that $T = I \otimes S$.*

Proof Let $\Phi, \Phi' \in \mathcal{K}$ be arbitrarily fixed and define $F: \mathcal{H} \times \mathcal{H} \to \mathbb{C}$ by

$$F(\Psi, \Psi') := \langle \Psi \otimes \Phi, T(\Psi' \otimes \Phi') \rangle, \quad \Psi, \Psi' \in \mathcal{H}.$$

Then it is easy to see that F is a bounded sesquilinear form with $|F(\Psi, \Psi')| \le (\|T\| \, \|\Phi\| \, \|\Phi'\|)\|\Psi\| \, \|\Psi'\|$. Hence, by using the Riesz representation theorem, one can show that there exists a unique bounded operator $T(\Phi, \Phi')$ on $\mathscr{H}$ such that

$$F(\Psi, \Psi') = \langle \Psi, T(\Phi, \Phi')\Psi' \rangle, \quad \Psi, \Psi' \in \mathscr{H}$$

with $\|T(\Phi, \Phi')\| \le \|T\| \, \|\Phi\| \, \|\Phi'\|$. Hence

$$\langle \Psi \otimes \Phi, T(\Psi' \otimes \Phi') \rangle = \langle \Psi, T(\Phi, \Phi')\Psi' \rangle, \quad \Psi, \Psi' \in \mathscr{H}. \tag{2.33}$$

Now let $A \in \mathfrak{A}$, $\Psi' \in D(A)$ and $\Psi \in D(A^*)$. By the present assumption, $T(A \otimes I) \subset (A \otimes I)T$. Hence $\langle \Psi \otimes \Phi, T(A\Psi' \otimes \Phi') \rangle = \langle A^*\Psi \otimes \Phi, T(\Psi' \otimes \Phi') \rangle$. By this relation and the above equation, we obtain

$$\langle \Psi, T(\Phi, \Phi')A\Psi' \rangle = \langle A^*\Psi, T(\Phi, \Phi')\Psi' \rangle.$$

Hence $T(\Phi, \Phi')\Psi' \in D(A^{**}) = D(A)$ and $AT(\Phi, \Phi')\Psi' = T(\Phi, \Phi')A\Psi'$, implying that $T(\Phi, \Phi')A \subset AT(\Phi, \Phi')$. Hence $T(\Phi, \Phi') \in \mathfrak{A}' = \mathbb{C}I$. Therefore there exists a scalar $c(\Phi, \Phi') \in \mathbb{C}$ such that $T(\Phi, \Phi') = c(\Phi, \Phi')I$. Hence, by (2.33),

$$\langle \Psi \otimes \Phi, T(\Psi' \otimes \Phi') \rangle = c(\Phi, \Phi') \langle \Psi, \Psi' \rangle.$$

Taking $\Psi = \Psi'$ with $\|\Psi\| = 1$, we have

$$c(\Phi, \Phi') = \langle \Psi \otimes \Phi, T(\Psi \otimes \Phi') \rangle.$$

This implies that $c(\ ,\ ) \colon \mathscr{K} \times \mathscr{K} \to \mathbb{C}$ is a bounded sesquilinear form with $|c(\Phi, \Phi')| \le \|T\| \, \|\Phi\| \, \|\Phi'\|$. Hence, by the Riesz representation theorem again, there exists a unique bounded operator S on $\mathscr{K}$ such that $c(\Phi, \Phi') = \langle \Phi, S\Phi' \rangle$. Hence we obtain

$$\langle \Psi \otimes \Phi, T(\Psi' \otimes \Phi') \rangle = \langle \Psi \otimes \Phi, \Psi' \otimes S\Phi' \rangle = \langle \Psi \otimes \Phi, (I \otimes S)\Psi' \otimes \Phi' \rangle.$$

Since $\mathscr{H} \hat{\otimes} \mathscr{K}$ is dense in $\mathscr{H} \otimes \mathscr{K}$, it follows that $T(\Psi' \otimes \Phi') = (I \otimes S)(\Psi' \otimes \Phi')$, which implies that $T\Theta = (I \otimes S)\Theta$, $\forall \Theta \in \mathscr{H} \hat{\otimes} \mathscr{K}$. The boundedness of T and $I \otimes S$ and the density of $\mathscr{H} \hat{\otimes} \mathscr{K}$ imply the operator equality $T = I \otimes S$. The uniqueness of S can be easily proved. ■

Lemma 2.10 *Let $\mathfrak{A}$ be a set of densely defined closed operators on a Hilbert space $\mathscr{H}$ such that $\mathfrak{A}' = \mathbb{C}I$ and $\mathfrak{B}$ be a set of densely defined closed operators on a Hilbert space $\mathscr{K}$ such that $\mathfrak{B}' = \mathbb{C}I$. Then $\{A \otimes I, I \otimes B \,|\, A \in \mathfrak{A},\ B \in \mathfrak{B}\}' = \mathbb{C}I$. In particular, $\{A \otimes I, I \otimes B \,|\, A \in \mathfrak{A},\ B \in \mathfrak{B}\}$ is irreducible.*[5]

[5]Apply Proposition 2.9(i).

Proof Let $T \in \{A \otimes I, I \otimes B | A \in \mathfrak{A}, B \in \mathfrak{B}\}'$. Then $T \in \{A \otimes I | A \in \mathfrak{A}\}'$. Hence, by Lemma 2.9, there exists a unique bounded operator S on $\mathscr{H}$ such that $T = I \otimes S$. The operator T is also in $\{I \otimes B | B \in \mathfrak{B}\}'$. Hence, by Lemma 2.9 again, there exists a unique bounded operator R on $\mathscr{H}$ such that $T = R \otimes I$. Hence $R \otimes I = I \otimes S$. By Lemma 1.4, this implies that $S = R = cI$ for some $c \in \mathbb{C}$. Thus $T = c(I \otimes I)$. ■

Theorem 2.7 *Let $\rho_1(q, p)$ be given by (2.26) such that q and p are closed symmetric. Suppose that $\{q, p\}' = \mathbb{C}I$. Then the tensor product representation $\rho_N(q, p)$ given by (2.29) is irreducible.*

Proof Let $\mathfrak{A} := \{q, p\}$. Then, by the present assumption, $\mathfrak{A}' = \mathbb{C}I$. Hence, by Lemma 2.10,

$$\{Q_j, P_j | j = 1, \ldots, N\}' = \{I \otimes \cdots I \otimes \underbrace{A}_{j\text{-th}} \otimes I \cdots \otimes I | A \in \mathfrak{A}, \ j = 1, \ldots, N\}' = \mathbb{C}I$$

Hence, by Proposition 2.9(i), $\rho_N(q, p)$ is irreducible. ■

Corollary 2.2 *If $\rho_1(q, p)$ is an irreducible self-adjoint representation, then $\rho_N(q, p)$ is an irreducible strong self-adjoint representation.*

Proof Under the present assumption, $\{q, p\}$ is a $*$-invariant irreducible set. Hence, by Proposition 2.9(ii), $\{q, p\}' = \mathbb{C}I$. Hence, by Theorem 2.7, $\rho_N(q, p)$ is irreducible. Since Q_j and P_j are self-adjoint in the present case, $\rho_N(q, p)$ is a strong self-adjoint representation. ■

2.6 Quantum Mechanics as Representations of CCR

We are now ready to discuss basic examples of representations of CCR appearing in quantum mechanics. But, before going into the details, we first present a structural understanding for quantum mechanics in terms of the concept of representations of CCR [161].

As is shown in the examples below as well as other examples in quantum mechanics, we see that, as far as *external degrees of freedom* are concerned, the fundamental structure of quantum mechanics lies in representations of CCR.[6] Considerations based on this structure lead us to the following symbolic statement: *quantum mechanics with N external degrees of freedom is* $\mathsf{CCR}(N)$ *in the following sense*:

(QM1) For each quantum system S with N external degrees of freedom, there corresponds to an equivalence class $\mathscr{E}_{\mathsf{S}}$ in the set of irreducible repre-

[6]In regard to *internal degrees of freedom* like spin, representations of other algebraic relations describe them.

sentations in $\mathsf{CCR}_{\mathrm{ssa}}(N)$, the set of strongly self-adjoint representations in $\mathsf{CCR}(N)$. Differences of representation spaces and representatives of elements of $\mathscr{E}_{\mathsf{S}}$ correspond to those of the kinds of observables and physical pictures that S can have according to measurement frameworks for S.[7]

(QM2) For each element of $\mathscr{E}_{\mathsf{S}}$, the representatives of it form the basic physical quantities from which the other quantum mechanical quantities are made. Here we mean by "a quantum mechanical quantity" an object which is not necessarily represented by a self-adjoint operator, in other words, not necessarily an observable ("physical" quantity). The form of each quantum mechanical quantity is independent of individual elements of $\mathscr{E}_{\mathsf{S}}$ in the following sense. Let $\pi_N = (\mathscr{H}, \mathscr{D}, (Q, P))$ and $\pi'_N = (\mathscr{H}', \mathscr{D}', (Q', P'))$ be in $\mathscr{E}_{\mathsf{S}}$, where we write

$$Q := (Q_1, \ldots, Q_N), \quad P := (P_1, \ldots, P_N),$$
$$Q' := (Q'_1, \ldots, Q'_N), \quad P' := (P'_1, \ldots, P'_N)$$

for notational simplicity. Then a quantum mechanical quantity F of S may be given by $f(Q, P)$ in the representation π_N, where $f \colon \mathbb{R}^{2N} \to \mathbb{C} \cup \{\pm\infty\}$ is Borel measurable and $f(Q, P)$ is supposed to be "suitably" defined (note that the definition of $f(Q, P)$ is non-trivial, because, for $j = 1, \ldots, N$, Q_j does not commute with P_j). Then F is represented by $f(Q', P')$ in the representation π'_N and $f(Q, P)$ is unitarily equivalent to $f(Q', P')$. In relation to this, we remark that, in the case where $f(x_1, \ldots, x_N, p_1, \ldots, p_N) = f_1(x_1, \ldots, x_N) + f_2(p_1, \ldots, p_N)$ with Borel measurable functions f_1 and f_2 on $\mathbb{R}^N$ ($x_j, p_j \in \mathbb{R}$, $j = 1, \ldots, N$), $f(Q, P)$ is defined by $f(Q, P) := f_1(Q) + f_2(P)$ via functional calculus, since Q (resp. P) is strongly commuting.

(QM3) An element in $\mathsf{CCR}(N)$ may appear as a secondary one in the theory described by an irreducible element π_N in $\mathsf{CCR}_{\mathrm{ssa}}(N)$ (i.e., it is formed out of the representatives of π_N).[8]

(QM4) Inequivalent representations of CCR may appear as those associated with "characteristic" quantum phenomena[9] or different physical conditions.[10]

[7]Recall that a quantum system may have different physical pictures according to measurement frameworks. For example, a single electron system has at least two physical pictures; one is based on measuring the position of the electron and the other is based on measuring the momentum of the electron. Note, however, that these measurements are not compatible due the *Heisenberg uncertainty principle*, so that one cannot have the two pictures simultaneously. This is an example of Bohr's complementary principle.

[8]A basic example of such a representation of CCR is the pair of a Hamiltonian and a time operator. See Chap. 4.

[9]An example in this category is given by the Aharonov–Bohm effect. See Chap. 3.

[10]For example, boundary conditions for a quantum system in a finite region. See Sect. 2.6.4 below.

2.6.1 Schrödinger Representation

Let $d \in \mathbb{N}$. We denote by $\hat{q}_j := M_{x_j}$ the multiplication operator by the j-th coordinate function x_j:

$$D(\hat{q}_j) = \left\{ f \in L^2(\mathbb{R}^d) | \int_{\mathbb{R}^d} |x_j f(\boldsymbol{x})|^2 d\boldsymbol{x} < \infty \right\}, \tag{2.34}$$

$$(\hat{q}_j f)(\boldsymbol{x}) = x_j f(\boldsymbol{x}), \quad \text{a.e.} \boldsymbol{x} \in \mathbb{R}^d, \ f \in D(\hat{q}_j). \tag{2.35}$$

The operator $\hat{q}_j$ is called the j-th **position operator**. It is easy to see that $\hat{q}_j$ is self-adjoint (see Theorem A.1(iv) in Appendix A). By Example B.1 in Appendix B, the spectral measure $E_{\hat{q}_j}$ of $\hat{q}_j$ is given by

$$(E_{\hat{q}_j}(B) f)(\boldsymbol{x}) = \chi_{\{\boldsymbol{x} \in \mathbb{R}^d | x_j \in B\}}(\boldsymbol{x}) f(\boldsymbol{x}), \quad \text{a.e.} \boldsymbol{x} \in \mathbb{R}^d, \ B \in \boldsymbol{B}^1, \ f \in L^2(\mathbb{R}^d). \tag{2.36}$$

Hence, for all $B, C \in \boldsymbol{B}^1$ and $j, k = 1, \ldots, d$, $E_{\hat{q}_j}(B)$ commutes with $E_{\hat{q}_k}(C)$. Therefore $\hat{q}_j$ strongly commutes with $\hat{q}_k$. Thus

$$\hat{\boldsymbol{q}} := (\hat{q}_1, \ldots, \hat{q}_d) \tag{2.37}$$

is strongly commuting. Moreover, $\hat{p}_j$ and $\hat{q}_j$ leave $C_0^\infty(\mathbb{R}^d)$ invariant and, for $j, k = 1, \ldots, d$,

$$[\hat{q}_j, \hat{p}_k] = i\delta_{jk}, \quad [\hat{q}_j, \hat{q}_k] = 0, \quad [\hat{p}_j, \hat{p}_k] = 0 \tag{2.38}$$

on $C_0^\infty(\mathbb{R}^d)$. We already know (see Sect. 1.12) that $\hat{p} = (\hat{p}_1, \ldots, \hat{p}_d)$ is strongly commuting. Thus

$$\pi_S^{(d)} := (L^2(\mathbb{R}^d), C_0^\infty(\mathbb{R}^d), (\hat{q}_1, \ldots, \hat{q}_d, \hat{p}_1, \ldots, \hat{p}_d)) \tag{2.39}$$

is a strong self-adjoint representation of the CCR with d degrees of freedom, i.e., $\pi_S^{(d)} \in \mathsf{CCR}_{\text{ssa}}(d)$. The representation $\pi_S^{(d)}$ is called the **Schrödinger representation** of the CCR with d degrees of freedom. This mathematical form gives a strict meaning to the so-called "coordinate representation" or "q-representation" used in the physics literature.

Remark 2.5 A characteristic of the Schrödinger representation $\pi_S^{(d)}$ is that, for each $j = 1, \ldots, d$, $\hat{q}_j$ and $\hat{p}_j$ are unitarily equivalent, i.e.,

$$\hat{q}_j = \mathscr{F}_d \hat{p}_j \mathscr{F}_d^{-1},$$

where $\mathscr{F}_d$ is the d-dimensional Fourier transform as a unitary operator on $L^2(\mathbb{R}^d)$.

Remark 2.6 The space $C_0^\infty(\mathbb{R}^d)$ is an invariant CCR-domain for $\hat{q}_j$ and $\hat{p}_j$ ($j = 1, \ldots, d$). As another invariant CCR-domain for $\hat{q}_j$ and $\hat{p}_j$, one can take $\mathscr{S}(\mathbb{R}^d)$, the set of rapidly decreasing C^∞-functions on $\mathbb{R}^d$.

By (1.80) and Example D.1 in Appendix D, $\hat{q}_j$ and $\hat{p}_j$ are absolutely continuous and

$$\sigma(\hat{q}_j) = \sigma_{\mathrm{ac}}(\hat{q}_j) = \mathbb{R}, \quad \sigma(\hat{p}_j) = \sigma_{\mathrm{ac}}(\hat{p}_j) = \mathbb{R}, \tag{2.40}$$

$$\sigma_{\mathrm{sc}}(\hat{q}_j) = \emptyset, \quad \sigma_{\mathrm{p}}(\hat{q}_j) = \emptyset, \tag{2.41}$$

$$\sigma_{\mathrm{sc}}(\hat{p}_j) = \emptyset, \quad \sigma_{\mathrm{p}}(\hat{p}_j) = \emptyset. \tag{2.42}$$

Theorem 2.1 immediately yields the next theorem:

Theorem 2.8 (Position–Momentum Uncertainty Relation) *For all unit vectors* $f \in D(\hat{q}_j\hat{p}_j) \cap D(\hat{p}_j\hat{q}_j)$ ($j = 1, \ldots, d$),

$$(\Delta\hat{q}_j)_f(\Delta\hat{p}_j)_f \geq \frac{1}{2}. \tag{2.43}$$

Remark 2.7 The position–momentum uncertainty relation (2.43) can be extended to all unit vectors $f \in D(\hat{q}_j) \cap D(\hat{p}_j)$. See Example 2.14 given later.

As already mentioned, in the context of quantum mechanics, $\mathbb{R}^d$ denotes the position space of a quantum particle. According to the standard interpretation of quantum mechanics, for a state vector $f \in L^2(\mathbb{R}^d)$ ($\|f\| = 1$) and a Borel set $B \subset \mathbb{R}^d$, $\int_B |f(\boldsymbol{x})|^2 d\boldsymbol{x}$ gives the probability that the position of the quantum particle under consideration is found in B when a measurement of it is made. Hence the Schrödinger representation $\pi_{\mathrm{S}}^{(d)}$ gives a theoretical framework based on the physical picture that the position of a quantum particle is observed.

Each $(\hat{q}_j, \hat{p}_j)$ has minimal uncertainty states:

Proposition 2.11 *Let* $\alpha > 0$, $\boldsymbol{c} = (c_1, \ldots, c_d) \in \mathbb{R}^d$ *be arbitrarily fixed and define*

$$f_{\alpha,\boldsymbol{c}}(\boldsymbol{x}) := \left(\frac{\alpha}{\pi}\right)^{d/4} e^{-\alpha(\boldsymbol{x}-\boldsymbol{c})^2/2}, \quad \boldsymbol{x} \in \mathbb{R}^d. \tag{2.44}$$

Then, for each $j = 1, \ldots, d$, $f_{\alpha,\boldsymbol{c}}$ *is a minimal uncertainty state with respect to* $(\hat{q}_j, \hat{p}_j)$.

Proof It is easy to see that $f_{\alpha,\boldsymbol{c}}$ is in $\mathscr{S}(\mathbb{R}^d)$ (hence $f_{\alpha,\boldsymbol{c}} \in D(\hat{q}_j\hat{p}_j) \cap D(\hat{p}_j\hat{q}_j)$) and $\|f_{\alpha,\boldsymbol{c}}\| = 1$. We have

$$\langle f_{\alpha,\boldsymbol{c}}, \hat{q}_j f_{\alpha,\boldsymbol{c}}\rangle = c_j, \quad \langle f_{\alpha,\boldsymbol{c}}, \hat{p}_j f_{\alpha,\boldsymbol{c}}\rangle = 0.$$

Hence

$$(\Delta\hat{q}_j)_{f_{\alpha,\boldsymbol{c}}} = \frac{1}{\sqrt{2\alpha}}, \quad (\Delta\hat{p}_j)_{f_{\alpha,\boldsymbol{c}}} = \sqrt{\frac{\alpha}{2}}.$$

Thus $(\Delta\hat{q}_j)_{f_{\alpha,c}}(\Delta\hat{p}_j)_{f_{\alpha,c}} = 1/2$. ∎

It follows from (1.82) that

$$f_{\alpha,\boldsymbol{c}} = e^{-ic_d\hat{p}_d}\cdots e^{-ic_1\hat{p}_1} f_\alpha \tag{2.45}$$

with $f_\alpha = f_{\alpha,\mathbf{0}}$.

A state of the form $f_{\alpha,\boldsymbol{c}}$ is said to be **Gaussian**.

Remark 2.8 One can define a measure $\mu_{\alpha,\boldsymbol{c}}$ on $(\mathbb{R}^d, \boldsymbol{B}^d)$ by

$$\mu_{\alpha,\boldsymbol{c}}(B) := \int_B f_{\alpha,\boldsymbol{c}}(\boldsymbol{x})^2 d\boldsymbol{x}, \quad B \in \boldsymbol{B}^d.$$

This type of measure is called a d-dimensional **Gaussian measure**. One has $\mu_{\alpha,\boldsymbol{c}}(\mathbb{R}^d) = 1$. Hence $(\mathbb{R}^d, \boldsymbol{B}^d, \mu_{\alpha,\boldsymbol{c}})$ is a probability space. It follows from the proof given above that the function x_j on $\mathbb{R}^d$ is a Gaussian random variable on this probability space with mean c_j and variance $1/(2\alpha)$.[11]

In the physical picture associated with the Schrödinger representation $\pi_{\mathrm{S}}^{(d)}$, there may be a potential $V: \mathbb{R}^d \to \mathbb{R} \cup \{\pm\infty\}$ (Borel measurable and a.e. finite) which yields a force acting on the particle. Then, in this case, the Hamiltonian for the particle of mass $m > 0$ is defined by

$$H_{\mathrm{S}}(V) := \frac{1}{2m}\sum_{j=1}^d \hat{p}_j^2 + V(\hat{\boldsymbol{q}}) = -\frac{1}{2m}\Delta + V(\hat{\boldsymbol{q}}), \tag{2.46}$$

where

$$\Delta := \sum_{j=1}^d D_j^2 \tag{2.47}$$

is the d-dimensional generalized Laplacian acting in $L^2(\mathbb{R}^d)$ and $V(\hat{\boldsymbol{q}})$ is the self-adjoint operator defined via the functional calculus for strongly commuting

[11] A random variable f on a probability space (X, Σ, μ) is called a **Gaussian random variable** with mean $m \in \mathbb{R}$ and variance $a > 0$ if the probability distribution $P^f(B) := \mu(f^{-1}(B))$ $(B \in \boldsymbol{B}^1)$ is of the form

$$P^f(B) = \int_B \frac{1}{\sqrt{2\pi a}} e^{-\frac{(x-m)^2}{2a}} dx.$$

self-adjoint operators $\hat{q}_1, \ldots, \hat{q}_d$. It is shown that $V(\hat{\boldsymbol{q}})$ is in fact the multiplication operator by the function V,[12] so that we have

$$H_{\mathrm{S}}(V) = -\frac{1}{2m}\Delta + V, \tag{2.48}$$

where V denotes the multiplication operator by V. An operator of the type $H_{\mathrm{S}}(V)$ is called a **Schrödinger operator**.

Remark 2.9 In any representation $\pi_d := (\mathscr{H}, \mathscr{D}, (Q, P)) \in \mathsf{CCR}_{\mathrm{ssa}}(d)$ with $Q := (Q_1, \ldots, Q_N)$, $P := (P_1, \ldots, P_N)$, which is different from $\pi_{\mathrm{S}}^{(d)}$, the Hamiltonian corresponding to $H_{\mathrm{S}}(V)$ is given by

$$H := \frac{1}{2m}\sum_{j=1}^{d} P_j^2 + V(Q).$$

Then the physical interpretation for each term on the right hand side is different from the case of the Schrödinger representation depending on the physical picture associated with π_d.

The *wave mechanics* presented by Schrödinger (1926) [144], if it is interpreted as a quantum theory,[13] can be understood as a quantum mechanics using the representation $\pi_{\mathrm{S}}^{(d)}$ of CCR. The name "Schödinger representation" for $\pi_{\mathrm{S}}^{(d)}$ comes from this interpretation.

As already mentioned, the d-dimensional Fourier transform $\mathscr{F}_d$ (see (1.74)) is a unitary transformation from $L^2(\mathbb{R}^d)$ to $L^2(\hat{\mathbb{R}}^d)$. Hence the Fourier transforms

$$\hat{q}_j' := \mathscr{F}_d \hat{q}_j \mathscr{F}_d^{-1}, \quad \hat{p}_j' := \mathscr{F}_d \hat{p}_j \mathscr{F}_d^{-1} \tag{2.49}$$

of $\hat{q}_j$ and $\hat{p}_j$ yield the representation

$$\pi_{\mathrm{S}}^{(d)\prime} := (L^2(\hat{\mathbb{R}}^d), \mathscr{F}_d C_0^\infty(\mathbb{R}^d), (\hat{q}_1', \ldots, \hat{q}_d', \hat{p}_1', \ldots, \hat{p}_d')) \tag{2.50}$$

[12] See, e.g., [23, Theorem 6.6]).

[13] The Schrödinger equation

$$i\frac{\partial\psi(\boldsymbol{x}, t)}{\partial t} = -\frac{1}{2m}\Delta\psi(\boldsymbol{x}, t) + V(\boldsymbol{x})\psi(\boldsymbol{x}, t) \quad ((\boldsymbol{x}, t) \in \mathbb{R}^3 \times \mathbb{R})$$

with potential V in the four-dimensional space–time $\mathbb{R}^3 \times \mathbb{R}$ ($\psi : \mathbb{R}^3 \times \mathbb{R} \to \mathbb{C}$) has also an interpretation as an equation for a non-relativistic matter wave advocated by de Broglie. But, the Schrödinger equation of the same type in the $(3n+1)$-dimensional space–time $\mathbb{R}^{3n} \times \mathbb{R}$ with $n \geq 2$ cannot be interpreted as an equation for a matter wave, since a matter wave as a physical reality should be described by a function on the four-dimensional space–time $\mathbb{R} \times \mathbb{R}^3$.

in $\mathsf{CCR}_{\mathrm{ssa}}(N)$, which is equivalent to the Schrödinger representation $\pi_{\mathrm{S}}^{(d)}$. By the Fourier analysis, we have

$$\hat{q}'_j = i D_{k_j}, \quad \hat{p}'_j = M_{k_j}, \quad j = 1, \ldots, d. \tag{2.51}$$

Since $\hat{\mathbb{R}}^d$ denotes the momentum space of a quantum particle, the representation $\pi_{\mathrm{S}}^{(d)\prime}$ gives a theoretical framework based on the physical picture that the momentum of the quantum particle is observed.

The representation $\pi_{\mathrm{S}}^{(d)\prime}$ is called the **momentum representation** or the ***p*-representation** with d degrees of freedom.

The equivalence of $\pi_{\mathrm{S}}^{(d)}$ and $\pi_{\mathrm{S}}^{(d)\prime}$ is interpreted as an expression of wave–particle duality (note that, in the physical unit system $\hbar = 1$, a momentum is also a wave number vector).

Let us show that $\pi_{\mathrm{S}}^{(d)}$ is *irreducible*. For this purpose, we prove a lemma first. We denote by $L^\infty(\mathbb{R}^d)$ the set of essentially bounded Borel measurable functions on $\mathbb{R}^d$. For each $F \in L^\infty(\mathbb{R}^d)$, we denote by M_F the multiplication operator by F (see Appendix A), which is a bounded operator.

Lemma 2.11 $\{e^{it\hat{q}_j} | t \in \mathbb{R},\ j = 1, \ldots, d\}' = \{M_F | F \in L^\infty(\mathbb{R}^d)\}$.

Proof It is obvious that $\{M_F | F \in L^\infty(\mathbb{R}^d)\} \subset \{e^{it\hat{q}_j} | t \in \mathbb{R},\ j = 1, \ldots, d\}'$. Hence it is sufficient to prove the converse inclusion relation. So let $T \in \{e^{it\hat{q}_j} | t \in \mathbb{R},\ j = 1, \ldots, d\}'$. Then $T e^{ik_1\hat{q}_1} \cdots e^{ik_d\hat{q}_d} = e^{ik_1\hat{q}_1} \cdots e^{ik_d\hat{q}_d} T$ for all $\boldsymbol{k} = (k_1, \ldots, k_d) \in \hat{\mathbb{R}}^d$. There exists an element $f_0 \in L^2(\mathbb{R}^d) \cap L^\infty(\mathbb{R}^d)$ such that $f_0(\boldsymbol{x}) > 0$ for all $\boldsymbol{x} \in \mathbb{R}^d$ (e.g., $f_0(\boldsymbol{x}) = e^{-\boldsymbol{x}^2}$, $\boldsymbol{x} \in \mathbb{R}^d$). Let $F_0 := f_0^{-1} T f_0$. Then, by the above commutativity,

$$T e^{ik_1\hat{q}_1} \cdots e^{ik_d\hat{q}_d} f_0 = M_{F_0} e^{ik_1\hat{q}_1} \cdots e^{ik_d\hat{q}_d} f_0.$$

Hence, introducing

$$\mathscr{D} := \mathrm{span}\{e^{ik_1\hat{q}_1} \cdots e^{ik_d\hat{q}_d} f_0 | k_j \in \hat{\mathbb{R}},\ j = 1, \ldots, d\},$$

we have $Tf = M_{F_0} f$ for all $f \in \mathscr{D}$.

Let us show that $\mathscr{D}$ is dense in $L^2(\mathbb{R}^d)$. So let $g \in \mathscr{D}^\perp$. Then, for all $\boldsymbol{k} \in \hat{\mathbb{R}}^d$, $\int_{\mathbb{R}^d} g(\boldsymbol{x})^* e^{i\boldsymbol{k}\boldsymbol{x}} f_0(\boldsymbol{x}) d\boldsymbol{x} = 0$. Since $g^* f_0$ is in $L^2(\mathbb{R}^d) \cap L^1(\mathbb{R}^d)$, it follows from the injectivity of the Fourier transform that $g(\boldsymbol{x})^* f_0(\boldsymbol{x}) = 0$, a.e.$\boldsymbol{x}$. Hence $g = 0$. Thus $\mathscr{D}$ is dense. Since T is bounded and M_{F_0} is closed, it follows that $D(M_{F_0}) = L^2(\mathbb{R}^d)$ and $T = M_{F_0}$. Hence, by Theorem A.1(ii), F_0 is essentially bounded, i.e., $F_0 \in L^\infty(\mathbb{R}^d)$. Thus $T \in \{M_F | F \in L^\infty(\mathbb{R}^d)\}$. ■

Lemma 2.12 *Let $F \in L^\infty(\mathbb{R}^d)$ such that $M_F \in \{e^{it\hat{p}_j} | t \in \mathbb{R},\ j = 1, \ldots, d\}'$. Then F is a constant function (i.e., for some constant $c \in \mathbb{C}$, $F(\boldsymbol{x}) = c$ for a.e.$\boldsymbol{x} \in \mathbb{R}^d$).*

Proof By the present assumption, $e^{ia_1\hat{p}_1}\cdots e^{ia_d\hat{p}_d}M_F f = M_F e^{ia_1\hat{p}_1}\cdots e^{ia_d\hat{p}_d} f$ for all $f \in L^2(\mathbb{R}^d)$ and $\boldsymbol{a} = (a_1, \ldots, a_d) \in \mathbb{R}^d$. By (1.84), this implies that, for some null set $N_{\boldsymbol{a}} \subset \mathbb{R}^d$,

$$F(\boldsymbol{x}+\boldsymbol{a})f(\boldsymbol{x}+\boldsymbol{a}) = F(\boldsymbol{x})f(\boldsymbol{x}+\boldsymbol{a}), \quad \boldsymbol{x} \in \mathbb{R}^d \setminus N_{\boldsymbol{a}}.$$

Hence $F(\boldsymbol{x}+\boldsymbol{a}) = F(\boldsymbol{x})$ for $\boldsymbol{x} \in \mathbb{R}^d \setminus N_{\boldsymbol{a}}$. Since $L^\infty(\mathbb{R}^d) \subset \mathscr{S}'(\mathbb{R}^d)$ (the space of tempered distributions on $\mathbb{R}^d$), it follows that F is a translation invariant tempered distribution. Hence there is a constant $c \in \mathbb{C}$ such that $F(\boldsymbol{x}) = c$ for a.e.$\boldsymbol{x} \in \mathbb{R}^d$. ■

Lemmas 2.11 and 2.12 immediately yield the following theorem:

Theorem 2.9 *The operator set* $\{e^{it\hat{q}_j}, e^{it\hat{p}_j} | t \in \mathbb{R}, j = 1, \ldots, d\}$ *is irreducible.*

Proof By Lemmas 2.11 and 2.12, $\{e^{it\hat{q}_j}, e^{it\hat{p}_j} | t \in \mathbb{R}, j = 1, \ldots, d\}' = \mathbb{C}I$. Hence, by Proposition 2.9(i), $\{e^{it\hat{q}_j}, e^{it\hat{p}_j} | t \in \mathbb{R}, j = 1, \ldots, d\}$ is irreducible. ■

By this theorem and Theorem 2.6, we obtain the following corollary:

Corollary 2.3 *The Schrödinger representation* $\pi_{\mathrm{S}}^{(d)}$ *is irreducible.*

We next show that $\pi_{\mathrm{S}}^{(d)}$ is equivalent to the d-tensor product representation

$$\rho_d(\hat{q}, \hat{p}) = (\otimes^d L^2(\mathbb{R}), \hat{\otimes}^d C_0^\infty(\mathbb{R}), (Q_1^{(\mathrm{S})}, \ldots, Q_d^{(\mathrm{S})}, P_1^{(\mathrm{S})}, \ldots, P_d^{(\mathrm{S})})), \tag{2.52}$$

of CCR constructed from the Schrödinger representation

$$\pi_{\mathrm{S}}^{(1)}(\hat{q}, \hat{p}) = (L^2(\mathbb{R}), C_0^\infty(\mathbb{R}), (\hat{q}, \hat{p})) \in \mathsf{CCR}_{\mathrm{sa}}(1) \tag{2.53}$$

of the CCR with one degree of freedom, where

$$\hat{q} := \hat{q}_1 = M_x, \quad \hat{p} := \hat{p}_1 = -iD_x \tag{2.54}$$

with D_x being the generalized differential operator in the variable $x \in \mathbb{R}$ acting in $L^2(\mathbb{R})$, and

$$Q_j^{(\mathrm{S})} := I \otimes \cdots \otimes I \otimes \underbrace{\hat{q}}_{j\text{-th}} \otimes I \otimes \cdots \otimes I, \tag{2.55}$$

$$P_j^{(\mathrm{S})} := I \otimes \cdots \otimes I \otimes \underbrace{\hat{p}}_{j\text{-th}} \otimes I \otimes \cdots \otimes I \quad (j = 1, \ldots, d). \tag{2.56}$$

For this purpose, we recall a basic fact.

Lemma 2.13 *There exists a unitary transformation $U_d \colon \otimes^d L^2(\mathbb{R}) \to L^2(\mathbb{R}^d)$ such that*

$$U_d(f_1 \otimes \cdots \otimes f_d) = f_1 \times \cdots \times f_d, \quad f_j \in L^2(\mathbb{R}),\ j = 1, \ldots, d,$$

where

$$(f_1 \times \cdots \times f_d)(\boldsymbol{x}) := f_1(x_1) \cdots f_d(x_d), \quad \text{a.e.}\boldsymbol{x} = (x_1, \ldots, x_d) \in \mathbb{R}^d,$$

Proof A simple application of Theorem 1.23. ■

Theorem 2.10 $\rho_d(\hat{q}, \hat{p}) \cong \pi_{\mathrm{S}}^{(d)}$.

Proof Let U_d be as in Lemma 2.13. Then one can show that

$$U_d Q_j^{(\mathrm{S})} U_d^{-1} = \hat{q}_j, \quad U_d P_j^{(\mathrm{S})} U_d^{-1} = \hat{p}_j, \quad j = 1, \ldots, d. \tag{2.57}$$

For proofs of these facts, see, e.g., [25, Examples 2.2 and 2.3], [32, Theorem 4.8, §4.2.3] or [28, Example 3.1]. Hence $\rho_d(\hat{q}, \hat{p})$ is equivalent to $\pi_{\mathrm{S}}^{(d)}$. ■

Remark 2.10 Theorem 2.10 gives another way to prove the irreducibility of the Schrödinger representation $\pi_{\mathrm{S}}^{(d)}$. Namely, one first proves that $\{\hat{q}, \hat{p}\}' = \mathbb{C}I$. Then, by Theorem 2.7, $\rho_d(\hat{q}, \hat{p})$ is irreducible. Hence, by Proposition 2.8 and Theorem 2.10, $\pi_{\mathrm{S}}^{(d)}$ is irreducible.

2.6.2 Born–Heisenberg–Jordan Representation

As is well known, there is another form of quantum theory discovered by Heisenberg [86], Born and Jordan [46] and Born et al. [47]. Their quantum theory was called "matrix mechanics", since it uses infinite matrices. The quantum theory by Born–Heisenberg–Jordan is interpreted as a quantum mechanics using a representation of CCR different from a Schrödinger one. In this subsection, we describe the representation of CCR on which the quantum theory of Born–Heisenberg–Jordan is based.

Let $\mathbb{Z}_+ := \{0\} \cup \mathbb{N}$ be the set of non-negative integers and

$$\mathbb{Z}_+^d := \{\boldsymbol{n} = (n_1, \ldots, n_d) \mid n_j \in \mathbb{Z}_+, j = 1, \ldots, d\}, \tag{2.58}$$

the d-fold direct product of $\mathbb{Z}_+$. We denote by $\ell^2(\mathbb{Z}_+^d)$ the set of absolutely square summable sequences on $\mathbb{Z}_+^d$:

$$\ell^2(\mathbb{Z}_+^d) := \left\{ \psi = \{\psi^{(\boldsymbol{n})}\}_{\boldsymbol{n} \in \mathbb{Z}_+^d} \,\middle|\, \psi^{(\boldsymbol{n})} \in \mathbb{C}, \boldsymbol{n} \in \mathbb{Z}_+^d, \sum_{\boldsymbol{n} \in \mathbb{Z}_+^d} |\psi^{(\boldsymbol{n})}|^2 < \infty \right\}. \tag{2.59}$$

This set becomes a Hilbert space with inner product

$$\langle \psi, \phi \rangle_{\ell^2(\mathbb{Z}_+^d)} := \sum_{\boldsymbol{n} \in \mathbb{Z}_+^d} \psi^{(\boldsymbol{n})*} \phi^{(\boldsymbol{n})}, \quad \psi, \phi \in \ell^2(\mathbb{Z}_+^d),$$

where the infinite series on the right hand side is absolutely convergent. A natural dense subspace of $\ell^2(\mathbb{Z}_+^d)$ is given by

$$\ell_0(\mathbb{Z}_+^d) := \left\{ \psi = \{\psi^{(\boldsymbol{n})}\}_{\boldsymbol{n} \in \mathbb{Z}_+^d} | \exists R > 0 \text{ such that, for all } |\boldsymbol{n}| > R,\ \psi^{(\boldsymbol{n})} = 0 \right\}. \tag{2.60}$$

For each $j = 1, \ldots, d$, a linear operator a_j on $\ell^2(\mathbb{Z}_+^d)$ is defined as follows:

$$D(a_j) := \left\{ \psi \in \ell^2(\mathbb{Z}_+^d) | \sum_{\boldsymbol{n} \in \mathbb{Z}_+^d} |\sqrt{n_j} \psi^{(\boldsymbol{n})}|^2 < \infty \right\}, \tag{2.61}$$

$$(a_j \psi)^{(\boldsymbol{n})} := \sqrt{n_j + 1}\, \psi^{(n_1, \ldots, n_j + 1, \ldots, n_d)}, \quad \psi \in D(a_j),\ \boldsymbol{n} \in \mathbb{Z}_+^d, \tag{2.62}$$

where $(n_1, \ldots, n_j + 1, \ldots, n_d)$ means $\boldsymbol{n}$ with n_j replaced by $n_j + 1$. It is obvious that $\ell_0(\mathbb{Z}_+^d) \subset D(a_j)$. Hence a_j is densely defined. Hence its adjoint a_j^* exists. Moreover, it is easy to show that a_j is closed.

To identify the domain and the operation of a_j^*, let $\psi \in D(a_j^*)$. Then, for all $\phi \in D(a_j)$,

$$\left\langle a_j^* \psi, \phi \right\rangle = \langle \psi, a_j \phi \rangle = \sum_{\boldsymbol{n} \in \mathbb{Z}_+^d, n_j \geq 1} (\sqrt{n_j} \psi^{(n_1, \ldots, n_j - 1, \ldots, n_d)})^* \phi^{(\boldsymbol{n})},$$

where $(n_1, \ldots, n_j - 1, \ldots, n_d)$ means $\boldsymbol{n}$ with n_j replaced by $n_j - 1$. Hence

$$(a_j^* \psi)^{(n_1, \ldots, 0, \ldots, n_d)} = 0, \quad (n_1, \ldots, n_{j-1}, n_{j+1}, \ldots, n_d) \in \mathbb{Z}_+^{d-1}, \tag{2.63}$$

$$(a_j^* \psi)^{(\boldsymbol{n})} = \sqrt{n_j}\, \psi^{(n_1, \ldots, n_j - 1, \ldots, n_d)}, \quad \boldsymbol{n} \in \mathbb{Z}_+^d,\ n_j \geq 1. \tag{2.64}$$

In particular,

$$\sum_{\boldsymbol{n} \in \mathbb{Z}_+^d, n_j \geq 1} n_j |\psi^{(n_1, \ldots, n_j - 1, \ldots, n_d)}|^2 < \infty, \tag{2.65}$$

since $a_j^*\psi \in \ell^2(\mathbb{Z}_+^d)$. Conversely, if $\psi \in \ell^2(\mathbb{Z}_+^d)$ satisfies (2.65), then one has $\langle \psi, a_j\phi \rangle = \langle \chi, \phi \rangle$, $\phi \in D(a_j)$ with $\chi \in \ell^2(\mathbb{Z}_+^d)$ satisfying

$$\chi^{(n_1,\ldots,0,\ldots,n_d)} = 0, \quad (n_1, \ldots, n_{j-1}, n_{j+1}, \ldots, n_d) \in \mathbb{Z}_+^{d-1},$$
$$\chi^{(\boldsymbol{n})} = \sqrt{n_j}\psi^{(n_1,\ldots,n_j-1,\ldots,n_d)}, \quad \boldsymbol{n} \in \mathbb{Z}_+^d,\ n_j \geq 1.$$

Hence $\psi \in D(a_j^*)$ and $a_j^*\psi = \chi$. Thus $D(a_j^*)$ is identified with

$$D(a_j^*) = \left\{ \psi \in \ell^2(\mathbb{Z}_+^d) | \sum_{\boldsymbol{n}\in\mathbb{Z}_+^d, n_j \geq 1} n_j |\psi^{(n_1,\ldots,n_j-1,\ldots,n_d)}|^2 < \infty \right\} \tag{2.66}$$

and the operation of a_j^* is given by (2.63) and (2.64).

It is easy to see that a_j and a_j^* leave $\ell_0(\mathbb{Z}_+^d)$ invariant. Moreover, for $j = 1, \ldots, d$,

$$(a_j^* a_j \psi)^{(\boldsymbol{n})} = n_j \psi^{(\boldsymbol{n})}, \quad \boldsymbol{n} \in \mathbb{Z}_+^d,\ \psi \in D(a_j^* a_j), \tag{2.67}$$
$$(a_j a_j^* \psi)^{(\boldsymbol{n})} = (n_j + 1)\psi^{(\boldsymbol{n})}, \quad \boldsymbol{n} \in \mathbb{Z}_+^d,\ \psi \in D(a_j a_j^*). \tag{2.68}$$

Hence

$$D(a_j^* a_j) = D(a_j a_j^*) = \left\{ \psi \in \ell^2(\mathbb{Z}_+^d) | \sum_{\boldsymbol{n}\in\mathbb{Z}_+^d} n_j^2 |\psi^{(\boldsymbol{n})}|^2 < \infty \right\} \tag{2.69}$$

and

$$[a_j, a_j^*] = 1 \quad \text{on } D(a_j^* a_j). \tag{2.70}$$

Similarly, one can easily prove the following commutation relations:

$$[a_j, a_k^*] = 0 \quad (j \neq k), \tag{2.71}$$
$$[a_j, a_k] = 0, \quad [a_j^*, a_k^*] = 0 \quad (j, k = 1, \ldots, d) \tag{2.72}$$

on $\ell_0(\mathbb{Z}_+^d)$.

It is obvious that the operators $a_j + a_j^*$ and $i(a_j^* - a_j)$ are symmetric. Hence, for any non-zero real constant $\lambda \in \mathbb{R} \setminus \{0\}$,

$$q_j^{(\mathrm{H})} := \frac{1}{\sqrt{2}\,\lambda}\overline{(a_j^* + a_j)}, \quad p_j^{(\mathrm{H})} := i\frac{\lambda}{\sqrt{2}}\overline{(a_j^* - a_j)} \tag{2.73}$$

are closed symmetric operators on $\ell^2(\mathbb{Z}_+^d)$ with $D(q_j^{(\mathrm{H})}) \cap D(p_j^{(\mathrm{H})}) \supset \ell_0(\mathbb{Z}_+^d)$. It follows that $q_j^{(\mathrm{H})}\ell_0(\mathbb{Z}_+^d) \subset \ell_0(\mathbb{Z}_+^d)$. $p_j^{(\mathrm{H})}\ell_0(\mathbb{Z}_+^d) \subset \ell_0(\mathbb{Z}_+^d)$ and, for $j, k = 1, \ldots, d$,

$$[q_j^{(\mathrm{H})}, p_k^{(\mathrm{H})}] = i\delta_{jk}, \quad [q_j^{(\mathrm{H})}, q_k^{(\mathrm{H})}] = 0, \quad [p_j^{(\mathrm{H})}, p_k^{(\mathrm{H})}] = 0 \tag{2.74}$$

on $\ell_0(\mathbb{Z}_+^d)$. Therefore

$$\pi_{\mathrm{BHJ}}^{(d)} := (\ell^2(\mathbb{Z}_+^d), \ell_0(\mathbb{Z}_+^d), \{q_j^{(\mathrm{H})}, p_j^{(\mathrm{H})} | j = 1, \ldots, d\}) \tag{2.75}$$

is in $\mathsf{CCR}(d)$. We call the representation $\pi_{\mathrm{BHJ}}^{(d)}$ the **Born–Heisenberg–Jordan (BHJ) representation** of the CCR with d degrees of freedom. The quantum theory by Born–Heisenberg–Jordan is interpreted as a quantum mechanics using this representation.

Remark 2.11 Since $\ell^2(\mathbb{Z}_+^d)$ can be identified with a boson Fock space discussed later (see Chap. 6), $\pi_{\mathrm{BHJ}}^{(d)}$ is called also the **Fock representation** of the CCR with d degrees of freedom.

The representation $\pi_{\mathrm{BHJ}}^{(d)}$ is in fact a self-adjoint representation, i.e., $\pi_{\mathrm{BHJ}}^{(d)} \in \mathsf{CCR}_{\mathrm{sa}}(d)$. Let us prove this fact. There are some different methods for that. Here we use the commutator theorem (Theorem 1.13). By von Neumann's theorem (Theorem 1.10),

$$\hat{N}_j := a_j^* a_j \quad (j = 1, \ldots, d) \tag{2.76}$$

is a non-negative self-adjoint operator on $\ell^2(\mathbb{Z}_+^d)$. We first show that the subspace

$$\mathscr{D}_0 := \ell_0(\mathbb{Z}_+^d)$$

is a core for $\hat{N}_j$. For all $\psi \in D(\hat{N}_j)$ and $R > 0$, we define $\psi^{(R)} \in \mathscr{D}_0$ by $\psi_{\boldsymbol{n}}^{(R)} := \psi^{(\boldsymbol{n})}$ for $|\boldsymbol{n}| < R$ and $\psi_{\boldsymbol{n}}^{(R)} = 0$ for $|\boldsymbol{n}| \geq R$. Then it is easy to see that $\lim_{R\to\infty} \psi^{(R)} = \psi$. Since $(\hat{N}_j\psi^{(R)})_{\boldsymbol{n}} = n_j\psi_{\boldsymbol{n}}^{(R)}$ by (2.67) and $\sum_{\boldsymbol{n}\in\mathbb{Z}_+^d} |n_j|^2\|\psi^{(\boldsymbol{n})}\|^2 < \infty$, it follows that $\lim_{R\to\infty} \hat{N}_j\psi^{(R)} = \hat{N}_j\psi$. Hence $\mathscr{D}_0$ is a core for $\hat{N}_j$. Let $T := \hat{N}_j + 1$. Then T is a self-adjoint operator with $T \geq 1$ and $\mathscr{D}_0$ is a core for T.

Let $\psi \in \mathscr{D}_0$. Then

$$\|q_j^{(H)}\psi\| \leq \frac{1}{\sqrt{2}|\lambda|}(\|a_j\psi\| + \|a_j^*\psi\|).$$

We have

$$\|a_j\psi\|^2 = \left\langle \psi, \hat{N}_j\psi \right\rangle \leq \langle \psi, T\psi \rangle = \|T^{1/2}\psi\|^2 \leq \|T\psi\|^2.$$

By (2.70), we have

$$\|a_j^*\psi\|^2 = \left\langle\psi, a_j a_j^*\psi\right\rangle = \langle\psi, T\psi\rangle \le \|T^{1/2}\psi\|^2 \le \|T\psi\|^2.$$

Hence $\|q_j^{(\mathrm{H})}\psi\| \le (\sqrt{2}/|\lambda|)\|T\psi\|$. Moreover

$$\left\langle q_j^{(\mathrm{H})}\psi, T\psi\right\rangle - \left\langle T\psi, q_j^{(\mathrm{H})}\psi\right\rangle = \frac{1}{\sqrt{2}\,\lambda}(\langle\psi, a_j\psi\rangle - \langle a_j\psi, \psi\rangle).$$

Hence

$$\left|\left\langle q_j^{(\mathrm{H})}\psi, T\psi\right\rangle - \left\langle T\psi, q_j^{(\mathrm{H})}\psi\right\rangle\right| \le (\sqrt{2}/|\lambda|)\|\psi\|\,\|a_j\psi\| \le (\sqrt{2}/|\lambda|)\|T^{1/2}\psi\|^2,$$

where we have used the inequality $\|\psi\| \le \|T^{1/2}\psi\|$. Therefore we can apply Theorem 1.13 to conclude that $q_j^{(\mathrm{H})}$ is essentially self-adjoint on $\mathscr{D}_0$. Hence $q_j^{(\mathrm{H})}$ is self-adjoint. Similarly, one can prove the self-adjointness of $p_j^{(\mathrm{H})}$. Thus $\pi_{\mathrm{BHJ}}^{(d)}$ is a self-adjoint representation.

We remark that one can prove the essential self-adjointness of $q_j^{(\mathrm{H})}$ on $\mathscr{D}_0$ by showing directly that $(q_j^{(\mathrm{H})} \pm i)\mathscr{D}_0$ are dense (see Theorem 1.12) or by using the analytic vector theorem (Theorem 1.14). The same applies to $p_j^{(\mathrm{H})}$.

The BHJ representation has interesting properties as shown below. For each $\boldsymbol{n} \in \mathbb{Z}_+^d$, we define $e_{\boldsymbol{n}}^{(d)} \in \ell_0(\mathbb{Z}_+^d)$ by

$$(e_{\boldsymbol{n}}^{(d)})(\boldsymbol{m}) := \delta_{\boldsymbol{nm}}, \quad \boldsymbol{m} \in \mathbb{Z}_+^d.$$

It is easy to see that $\{e_{\boldsymbol{n}}^{(d)}\}_{\boldsymbol{n}\in\mathbb{Z}_+^d}$ is a CONS of $\ell^2(\mathbb{Z}_+^d)$ and

$$\ell_0(\mathbb{Z}_+^d) = \mathrm{span}\{e_{\boldsymbol{n}}^{(d)} | \boldsymbol{n} \in \mathbb{Z}_+^d\}.$$

Hence, introducing the one-dimensional subspace

$$\mathscr{M}_{\boldsymbol{n}} := \{\alpha e_{\boldsymbol{n}}^{(d)} \,|\, \alpha \in \mathbb{C}\}, \tag{2.77}$$

one has the orthogonal decomposition

$$\ell^2(\mathbb{Z}_+^d) = \oplus_{\boldsymbol{n}\in\mathbb{Z}_+^d} \mathscr{M}_{\boldsymbol{n}}. \tag{2.78}$$

Moreover, for all $\boldsymbol{n} \in \mathbb{Z}_+^d$,

$$e_{\boldsymbol{n}}^{(d)} = \frac{1}{\sqrt{n_1! \cdots n_d!}}(a_1^*)^{n_1} \cdots (a_d^*)^{n_d} e_{\boldsymbol{0}}^{(d)}. \tag{2.79}$$

It follows that, for each $j = 1, \ldots, d$,

$$a_j e_{\boldsymbol{n}}^{(d)} = 0 \quad \text{for all } \boldsymbol{n} \in \mathbb{Z}_+^d \text{ with } n_j = 0. \tag{2.80}$$

Using this property, (2.71) and (2.72), one can show that

$$a_j e_{\boldsymbol{n}}^{(d)} = \sqrt{n_j} e_{(n_1,\ldots,n_j-1,\ldots,n_d)}^{(d)}, \quad n_j \geq 1, \tag{2.81}$$

$$a_j^* e_{\boldsymbol{n}}^{(d)} = \sqrt{n_j+1} e_{(n_1,\ldots,n_j+1,\ldots,n_d)}^{(d)}, \quad \boldsymbol{n} \in \mathbb{Z}_+^d. \tag{2.82}$$

Hence

$$\hat{N}_j e_{\boldsymbol{n}}^{(d)} = n_j e_{\boldsymbol{n}}^{(d)}. \tag{2.83}$$

Therefore the eigenvalue problem of $\hat{N}_j$ is completely solved with

$$\sigma(\hat{N}_j) = \sigma_{\mathrm{p}}(\hat{N}_j) = \mathbb{Z}_+$$

and, for each $\ell \in \sigma_{\mathrm{p}}(\hat{N}_j)$,

$$\ker(\hat{N}_j - \ell) = \mathrm{span}\{e_{\boldsymbol{n}}^{(d)} | \boldsymbol{n} \in \mathbb{Z}_+^d \text{ with } n_j = \ell\}.$$

By (2.83), the operator

$$\hat{N} := \sum_{j=1}^{d} \hat{N}_j \tag{2.84}$$

satisfies

$$\hat{N} e_{\boldsymbol{n}}^{(d)} = \left(\sum_{j=1}^{d} n_j \right) e_{\boldsymbol{n}}^{(d)}.$$

This means that the non-negative symmetric operator $\hat{N}$ has a CONS consisting of its eigenvectors. Hence $\hat{N}$ is a non-negative self-adjoint operator with

$$\sigma(\hat{N}) = \sigma_{\mathrm{p}}(\hat{N}) = \left\{ \sum_{j=1}^{d} n_j | n_j \in \mathbb{Z}_+,\ j = 1, \ldots, d \right\} = \mathbb{Z}_+. \tag{2.85}$$

and, for each $n \in \mathbb{Z}_+$,

$$\ker(\hat{N} - n) = \bigoplus_{n_1,\ldots,n_d \in \mathbb{Z}_+,\, n_1+\cdots+n_d=n} \mathscr{M}_{(n_1,\ldots,n_d)}. \tag{2.86}$$

Thus we obtain

$$\ell^2(\mathbb{Z}_+^d) = \bigoplus_{n=0}^{\infty} \ker(\hat{N} - n). \tag{2.87}$$

A standard physical interpretation for $(\ell^2(\mathbb{Z}_+^d), \{a_j, a_j^* | j = 1, \ldots, d\}, \hat{N}_j, \hat{N})$ in the context of quantum mechanics is as follows. The Hilbert space $\ell^2(\mathbb{Z}_+^d)$ is a Hilbert space of state vectors of a quantum system. By (2.83), the operator $\hat{N}_j$ may be interpreted as counting the number of quantum particles labeled by j and $e_{\boldsymbol{n}}^{(d)}$ represents a state in which there are n_j quantum particles with label j, $j = 1, \ldots, d$. Equation (2.81) shows that a_j annihilates one quantum particle with label j and (2.82) shows that a_j^* creates one quantum particle with label j. Each label j may physically denote a physical quantity such as momentum, energy and spin. Since $n_j \in \mathbb{Z}_+$ is arbitrary, it follows that a quantum particle with label j is a boson.[14] We call a quantum particle with label j a j-mode boson. Thus the quantum system under consideration is a system of d mode bosons. The operator $\hat{N}_j$ counts the number of j-mode bosons and $\hat{N}$ describes the total number of d mode bosons. With this interpretation, a_j (resp. a_j^*) is called the **annihilation** (resp. **creation**) **operator** for the j-mode boson. The operator $\hat{N}_j$ is called the **number operator** for the j-mode boson and $\hat{N}$ is called the **total number operator** of the quantum system under consideration.

Let $\alpha = (\alpha_1, \ldots, \alpha_d) \in \mathbb{C}^d$. Since $\{e_{\boldsymbol{n}}^{(d)}\}_{\boldsymbol{n}\in\mathbb{Z}_+^d}$ is a CONS of $\ell^2(\mathbb{Z}_+^d)$ and

$$\sum_{\boldsymbol{n}\in\mathbb{Z}_+^d} \frac{|\alpha_1^{n_1} \cdots \alpha_d^{n_d}|^2}{n_1! \cdots n_d!} = e^{|\alpha|^2} < \infty,$$

the infinite series

$$\psi_\alpha := \sum_{\boldsymbol{n}\in\mathbb{Z}_+^d} \frac{\alpha_1^{n_1} \cdots \alpha_d^{n_d}}{\sqrt{n_1! \cdots n_d!}} e_{\boldsymbol{n}}^{(d)} \tag{2.88}$$

converges with $\|\psi_\alpha\|^2 = e^{|\alpha|^2}$. This type of vector is called a **coherent state**.

[14]Here a "boson" means a quantum particle with a mode whose n particle state exists for all $n \in \mathbb{Z}_+$.

It is easy to see that $\psi_\alpha \in \cap_{j=1}^d D(a_j)$ and

$$a_j\psi_\alpha = \alpha_j\psi_\alpha. \tag{2.89}$$

Namely, ψ_α is an eigenvector of a_j with eigenvalue α_j. Since $\alpha_j \in \mathbb{C}$ is arbitrary, we obtain

$$\sigma(a_j) = \sigma_{\mathrm{p}}(a_j) = \mathbb{C}, \tag{2.90}$$

an interesting property.

In regard to the spectra of a_j^*, it is easy to see that, for all $z \in \mathbb{C}$, $\ker(a_j^* - z) = \{0\}$. Hence $\sigma_{\mathrm{p}}(a_j^*) = \emptyset$. Namely, for all $z \in \mathbb{C}$, $a_j^* - z$ is injective. But, by (2.89), for all $\psi \in D(a_j^*)$ and $\alpha \in \mathbb{C}$,

$$\left\langle (a_j^* - \alpha_j^*)\psi, \psi_\alpha \right\rangle = 0.$$

Hence Ran $(a_j^* - \alpha_j^*)$ is not dense. Hence $\alpha_j^* \in \sigma_{\mathrm{r}}(a_j^*)$ (see Sect. 1.1.3). Thus

$$\sigma(a_j^*) = \sigma_{\mathrm{r}}(a_j^*) = \mathbb{C}, \tag{2.91}$$

which is also interesting.

2.6.3 Equivalence of Schrödinger Representation—"Wave Mechanics"—and BHJ Representation—"Matrix Mechanics"

We show that, for each $d \in \mathbb{N}$, $\pi_{\mathrm{S}}^{(d)}$ is equivalent to $\pi_{\mathrm{BHJ}}^{(d)}$. A direct proof for this fact is possible. But, here, we employ facts on tensor product representations. So we first prove the following theorem:

Theorem 2.11 $\pi_{\mathrm{S}}^{(1)} \cong \pi_{\mathrm{BHJ}}^{(1)}$.

Proof Let $\hat{q}$ and $\hat{p}$ be given by (2.54) and define

$$a := \frac{1}{\sqrt{2}}\left(\lambda\hat{q} + \frac{i}{\lambda}\hat{p}\right), \tag{2.92}$$

where $\lambda \in \mathbb{R} \setminus \{0\}$ is a constant. Then it is easy to see that a is densely defined closable operator with $D(a) \supset \mathscr{S}(\mathbb{R})$ (the space of rapidly decreasing C^∞-functions on $\mathbb{R}$) and

$$a^* \supset \frac{1}{\sqrt{2}}\left(\lambda\hat{q} - \frac{i}{\lambda}\hat{p}\right), \tag{2.93}$$

Both a and a^* leave $\mathscr{S}(\mathbb{R})$ invariant (i.e. $a\mathscr{S}(\mathbb{R}) \subset \mathscr{S}(\mathbb{R})$ and $a^*\mathscr{S}(\mathbb{R}) \subset \mathscr{S}(\mathbb{R})$) satisfying another canonical commutation relation

$$[a, a^*] = I \quad \text{on } \mathscr{S}(\mathbb{R}). \tag{2.94}$$

By (2.92) and (2.93),

$$\hat{q} = \frac{1}{\sqrt{2}\,\lambda}(a^* + a), \quad \hat{p} = \frac{i\lambda}{\sqrt{2}}(a^* - a) \quad \text{on } D(\hat{q}) \cap D(\hat{p}). \tag{2.95}$$

Let

$$\psi_0(x) = \left(\frac{|\lambda|}{\sqrt{\pi}}\right)^{1/2} e^{-\lambda^2 x^2/2}, \quad x \in \mathbb{R}. \tag{2.96}$$

Then $\psi_0 \in \mathscr{S}(\mathbb{R})$ with $\|\psi_0\| = 1$ and

$$a\psi_0 = 0. \tag{2.97}$$

Let

$$\psi_n := \frac{1}{\sqrt{n!}}(a^*)^n \psi_0, \quad n \in \mathbb{Z}_+. \tag{2.98}$$

Then, using (2.94) and (2.97), one can show that $\{\psi_n\}_{n=0}^{\infty}$ is an ONS (orthonormal system) of $L^2(\mathbb{R})$:

$$\langle \psi_n, \psi_m \rangle = \delta_{nm}, \quad n, m \in \mathbb{Z}_+.$$

Moreover, using the Fourier analysis, one can prove that $\{\psi_n\}_{n=0}^{\infty}$ is complete.[15] Hence, by Theorem 1.3, there exists a unitary transformation $U_1 \colon L^2(\mathbb{R}) \to \ell^2(\mathbb{Z}_+)$ such that

$$U_1\psi_n = e_n^{(1)}, \quad n \in \mathbb{Z}+. \tag{2.99}$$

It is easy to see that

$$U_1 a U_1^{-1} = a_1, \quad U_1 a^* U_1^{-1} = a_1^* \quad \text{on } \ell_0(\mathbb{Z}_+). \tag{2.100}$$

[15] As is well known, ψ_n's can be written in terms of the Hermite functions Hence the completeness of $\{\psi_n\}_{n=0}^{\infty}$ comes from that of the Hermite functions. See, e.g., [131, Problem 7] or [23, Problem 4].

It follows from (2.100) that that

$$U_1 \hat{q} U_1^{-1} = q_1^{(\mathrm{H})}, \quad U_1 \hat{p} U_1^{-1} = p_1^{(\mathrm{H})} \quad \text{on } \ell_0(\mathbb{Z}_+). \tag{2.101}$$

Hence

$$q_1^{(\mathrm{H})} \restriction \ell_0(\mathbb{Z}_+) \subset U_1 \hat{q} U_1^{-1}, \quad p_1^{(\mathrm{H})} \restriction \ell_0(\mathbb{Z}_+) \subset U_1 \hat{p} U_1^{-1}.$$

As already shown that $q_1^{(\mathrm{H})}$ and $p_1^{(\mathrm{H})}$ are essentially self-adjoint on $\ell_0(\mathbb{Z}_+)$, it follows that

$$q_1^{(\mathrm{H})} \subset U_1 \hat{q} U_1^{-1}, \quad p_1^{(\mathrm{H})} \subset U_1 \hat{p} U_1^{-1}.$$

Since a self-adjoint operator has no non-trivial symmetric extension, we obtain

$$q_1^{(\mathrm{H})} = U_1 \hat{q} U_1^{-1}, \quad p_1^{(\mathrm{H})} = U_1 \hat{p} U_1^{-1}. \tag{2.102}$$

Thus $\pi_{\mathrm{S}}^{(1)}$ is equivalent to $\pi_{\mathrm{BHJ}}^{(1)}$. ■

Remark 2.12 The above proof shows also that $\hat{q}$ and $\hat{p}$ are essentially self-adjoint on $\mathrm{span}\{\psi_n | n \in \mathbb{Z}_+\}$.

Based on (2.100), we call the operator a (resp. a^*) the **annihilation** (resp. **creation**) **operator** in the Schrödinger representation $\pi_{\mathrm{S}}^{(1)}$.

Theorem 2.12 *For all $d \in \mathbb{N}$, $\pi_{\mathrm{S}}^{(d)} \cong \pi_{\mathrm{BHJ}}^{(d)}$.*

Proof By Theorems 2.11 and 2.3, we have $\rho_d(\hat{q}, \hat{p}) \cong \rho_d(q_1^{(\mathrm{H})}, p_1^{(\mathrm{H})})$. In quite the same way as in the proof of Theorem 2.10, one can show that $\rho_d(q_1^{(\mathrm{H})}, p_1^{(\mathrm{H})}) \cong \pi_{\mathrm{BHJ}}^{(d)}$. Hence $\rho_d(\hat{q}, \hat{p}) \cong \pi_{\mathrm{BHJ}}^{(d)}$. This fact and Theorem 2.10 imply the desired result. ■

Remark 2.13 Theorem 2.12 implies that the BHJ representation $\pi_{\mathrm{BHJ}}^{(d)}$ is irreducible.

Remark 2.14 The BHJ representation $\pi_{\mathrm{BHJ}}^{(d)}$ contains a non-zero real parameter λ, which may be interpreted as a parameter of a scale transformation. To make explicit the dependence of $\pi_{\mathrm{BHJ}}^{(d)}$ on λ, we write it $\pi_{\mathrm{BHJ}}^{(d)}(\lambda)$. Then Theorem 2.12 implies that, for all $\lambda, \lambda' \in \mathbb{R} \setminus \{0\}$,

$$\pi_{\mathrm{BHJ}}^{(d)}(\lambda) \cong \pi_{\mathrm{BHJ}}^{(d)}(\lambda'). \tag{2.103}$$

Therefore $\{\pi_{\mathrm{BHJ}}^{(d)}(\lambda)\}_{\lambda \in \mathbb{R} \setminus \{0\}}$ is a family consisting of equivalent irreducible representations in $\mathsf{CCR}_{\mathrm{sa}}(d)$. But, as we shall see later (Sect. 10.2), *this structure breaks down in the case of infinite degrees of freedom.*

In the context of quantum mechanics, Theorem 2.12 gives a mathematically rigorous form to the statement that "wave mechanics" by Schrödinger as a quantum theory is equivalent to "matrix mechanics" by Born–Heisenberg–Jordan.

In general, the equivalence of representations of CCR may play important roles in solving eigenvalue problems of observables (physical quantities) in quantum mechanics. Namely, there appears the case where, in a representation of CCR, it is not so easy to solve the eigenvalue problem of an observable (e.g., a Hamiltonian), but, in another representation of CCR, it is very easy to solve it. The next example is a typical one.

Example 2.4 (Quantum Harmonic Oscillator) For simplicity, we consider a one-dimensional quantum harmonic oscillator. In any representation $(\mathscr{H}, \mathscr{D}, (Q, P)) \in \mathsf{CCR}_{\rm sa}(1)$, the form H of its Hamiltonian is given by

$$H = \frac{1}{2m}P^2 + \frac{m\omega^2}{2}Q^2,$$

where $m > 0$ and $\omega > 0$ are the mass and the circular frequency of the oscillator respectively. We remark that, as is easily seen, H itself can be defined in any representation in $\mathsf{CCR}(1)$ which is not necessarily self-adjoint. In the Schrödinger representation $\pi_{\rm S}^{(1)}$, H takes the form

$$H_{\rm os} := -\frac{1}{2m}\Delta + \frac{m\omega^2}{2}x^2, \tag{2.104}$$

where Δ is the one-dimensional generalized Laplacian and x^2 denotes the multiplication operator by the function:$\mathbb{R} \ni x \mapsto x^2$. On the other hand, in the BHJ representation $\pi_{\rm BHJ}^{(1)}$, H takes the form

$$H_{\rm BHJ} := \frac{1}{2m}(p^{(\rm H)})^2 + \frac{m\omega^2}{2}(q^{(\rm H)})^2.$$

We take $\lambda = \sqrt{m\omega}$ in (2.73). Then, by direct computations using (2.73), we have

$$H_{\rm BHJ} = \omega\hat{N} + \frac{\omega}{2} \quad \text{on } \ell_0(\mathbb{Z}_+),$$

where $\hat{N} := \hat{N}_1 = a_1^* a_1$. As we have already seen, $\ell_0(\mathbb{Z}_+)$ is a core for the self-adjoint operator $\hat{N}$. Hence $H_{\rm BHJ}$ is essentially self-adjoint on $\ell_0(\mathbb{Z}_+)$ and

$$\overline{H}_{\rm BHJ} = \omega\hat{N} + \frac{1}{2}\omega.$$

Hence $\overline{H}_{\rm BHJ}$ has a purely discrete spectrum with

$$\sigma(\overline{H}_{\rm BHJ}) = \sigma_{\rm d}(\overline{H}_{\rm BHJ}) = \{E_n | n \in \mathbb{Z}_+\},$$

where

$$E_n := \omega \left(n + \frac{1}{2} \right),$$

which is a simple eigenvalue with a normalized eigenvector $e_n^{(1)}$. Thus the eigenvalue problem for H_{BHJ} is completely solved easily.

We have

$$U_1 H_{\mathrm{os}} U_1^{-1} = H_{\mathrm{BHJ}}.$$

and $U^{-1}\ell_0(\mathbb{Z}_+) \subset \mathscr{S}(\mathbb{R})$. Hence H_{os} is essentially self-adjoint on $\mathscr{S}(\mathbb{R})$ and

$$U_1 \overline{H}_{\mathrm{os}} U_1^{-1} = \overline{H}_{\mathrm{BHJ}}.$$

Hence, by the unitary invariance of spectra, we obtain

$$\sigma(\overline{H}_{\mathrm{os}}) = \sigma_{\mathrm{d}}(\overline{H}_{\mathrm{os}}) = \{E_n | n \in \mathbb{Z}_+\}. \tag{2.105}$$

A normalized eigenvector of $\overline{H}_{\mathrm{os}}$ belonging to E_n is given by

$$U_1^{-1} e_n^{(1)} = \psi_n.$$

We remark that H_{os} and H_{BHJ} are in fact self-adjoint, i.e., $\overline{H}_{\mathrm{os}} = H_{\mathrm{os}}$ and $\overline{H}_{\mathrm{BHJ}} = H_{\mathrm{BHJ}}$.[16]

2.6.4 *Quantum System in a Finite Box*

(A) A System with a Periodic Boundary Condition

Let us consider a quantum particle in the closed interval $\mathbb{I}_L$ given by (1.87). A Hilbert space of state vectors for this quantum system can be taken to be $L^2(\mathbb{I}_L) := L^2(\mathbb{I}_L, dx)$. We denote by $\hat{q}_L$ the multiplication operator on $L^2(\mathbb{I}_L)$ by x:

$$D(\hat{q}_L) := \left\{ f \in L^2(\mathbb{I}_L) \middle| \int_{\mathbb{I}_L} |xf(x)|^2 dx < \infty \right\},$$

$$(\hat{q}_L f)(x) := xf(x), \quad f \in D(\hat{q}_L), \text{ a.e.} x \in \mathbb{I}_L.$$

[16]One needs only to apply the following proposition: *Let A and B be closed operators on a Hilbert space such that $A+B$ is closable and suppose that there exist a core $\mathscr{D}$ for $\overline{A+B}$ and a constant $c>0$ such that $\mathscr{D} \subset D(A) \cap D(B)$ and $\|A\Psi\| + \|B\Psi\| \le c(\|(A+B)\Psi\| + \|\Psi\|)$, $\Psi \in \mathscr{D}$. Then $D(\overline{A+B}) = D(A) \cap D(B)$ and $A+B$ is closed.*

It is obvious that the function x is bounded on $\mathbb{I}_L$ with $\|x\|_\infty \leq L/2$. Hence, by Theorem A.1(ii) and (iv), $\hat{q}_L$ is a bounded self-adjoint operator with $D(\hat{q}_L) = L^2(\mathbb{I}_L)$ and $\|\hat{q}_L\| \leq L/2$. In fact, one can show that $\|\hat{q}_L\| = L/2$.

For all $B \subset \mathbb{I}_L$, $B \in \boldsymbol{B}^1$ with $|B| = 0$ and $f \in L^2(\mathbb{I}_L)$,

$$\int_{\{x|x\in B\}} |f(x)|^2 dx = \int_B |f(x)|^2 dx = 0.$$

Hence, by Proposition D.3 in Appendix D, $\hat{q}_L$ is absolutely continuous. It is easy to see that the essential range of the function x on $\mathbb{I}_L$ is $\mathbb{I}_L$. Thus

$$\sigma(\hat{q}_L) = \sigma_{\mathrm{ac}}(\hat{q}_L) = [-L/2, L/2], \quad \sigma_{\mathrm{sc}}(\hat{q}_L) = \emptyset, \quad \sigma_{\mathrm{p}}(\hat{q}_L) = \emptyset.$$

We denote by $C_0^\infty(-L/2, L/2)$ the set of infinitely differentiable functions on the open interval $(-L/2, L/2)$ with support in $(-L/2, L/2)$. This set is a dense subspace of $L^2(\mathbb{I}_L)$. Let $\hat{p}_L$ be the momentum operator acting in $L^2(\mathbb{I}_L)$ with PBC introduced in Sect. 1.14. Then it is easy to see that $C_0^\infty(-L/2, L/2) \subset D(\hat{q}_L) \cap D(\hat{p}_L)$, $\hat{q}_L C_0^\infty(-L/2, L/2) \subset C_0^\infty(-L/2, L/2)$, $\hat{p}_L C_0^\infty(-L/2, L/2) \subset C_0^\infty(-L/2, L/2)$ and

$$[\hat{q}_L, \hat{p}_L] = i$$

on $C_0^\infty(-L/2, L/2)$. Thus

$$\pi_L^{(1)} := (L^2(\mathbb{I}_L), C_0^\infty(-L/2, L/2), (\hat{q}_L, \hat{p}_L)) \tag{2.106}$$

is a self-adjoint representation of the CCR with one degree of freedom. Note that, by Theorem 1.34,

$$\sigma(p_L) = \sigma_{\mathrm{p}}(p_L) = \sigma_{\mathrm{d}}(p_L) = \Gamma_L. \tag{2.107}$$

One can show that $\pi_L^{(1)}$ is irreducible. To prove this fact, we prepare two lemmas.

Lemma 2.14 *If $T \in \mathfrak{B}(L^2(\mathbb{I}_L))$ satisfies $T\hat{q}_L = \hat{q}_L T$, then $T = M_F$ with F being an essentially bounded function on $\mathbb{I}_L$.*

Proof The present assumption implies that, for all $f \in L^2(\mathbb{I}_L)$, $T\hat{q}_L f = \hat{q}_L T f$, which implies that, for all $n \in \mathbb{Z}_+$, $T\hat{q}_L^n f = \hat{q}_L^n T f$. Let $F := T1$ (note that the constant function 1 is in $L^2(\mathbb{I}_L)$). Then $Tp_n = Fp_n$, where $p_n(x) := x^n$, $x \in \mathbb{I}_L$. Hence, for all f in the subspace

$$\mathscr{D} := \mathrm{span}\{p_n | n \geq 0\},$$

$Tf = Ff$. On the other hand, it follows from the Weierstrass theorem and the density of $C(\mathbb{I}_L)$ (the space of continuous functions on $\mathbb{I}_L$) in $L^2(\mathbb{I}_L)$ that $\mathscr{D}$ is dense in $L^2(\mathbb{I}_L)$. Hence, by a limiting argument, one can show that, for all $f \in L^2(\mathbb{I}_L)$, $(Tf)(x) = F(x)f(x)$, a.e.$x \in \mathbb{I}_L$. Hence $T = M_F$. Since $D(T) = L^2(\mathbb{I}_L)$, it follows from Theorem A.1(iii) in Appendix A that F is essentially bounded. ■

Lemma 2.15 *Let F be an essentially bounded function on $\mathbb{I}_L$. Suppose that $M_F \hat{p}_L \subset \hat{p}_L M_F$. Then F is a constant function on $\mathbb{I}_L$.*

Proof The present assumption implies that, for all $f \in D(\hat{p}_L)$, $Ff \in D(\hat{p}_L)$ and $F\hat{p}_L f = p_L(Ff)$. Taking $f = 1 \in D(\hat{p}_L)$, we see that $F \in D(\hat{p}_L)$ and $\hat{p}_L F = 0$. Hence F is a constant. ■

Theorem 2.13 *The representation $\pi_L^{(1)}$ is irreducible.*

Proof Let $T \in \{\hat{q}_L, \hat{p}_L\}'$. Then $T\hat{q}_L = \hat{q}_L T$ and $T\hat{p}_L \subset \hat{p}_L T$. The former and Lemma 2.14 imply that $T = M_F$ with F being an essentially bounded function on $\mathbb{I}_L$. Then it follows from the latter and Lemma 2.15 that F is a constant function on $\mathbb{I}_L$. Hence $T = cI$ for some $c \in \mathbb{C}$. Thus $\{\hat{q}_L, \hat{p}_L\}$ is irreducible. ■

Let $\hat{q}$ and $\hat{p}$ be the representatives of the Schrödinger representation $\pi_S^{(1)}$ of the CCR with one degree of freedom (see (2.53)). Since $\sigma(\hat{q}_L) \neq \sigma(\hat{q})$ or $\sigma(\hat{p}_L) \neq \sigma(\hat{p})$, it follows from Proposition 2.4 that $\pi_S^{(1)}$ and $\pi_L^{(1)}$ are inequivalent:

$$\boxed{\pi_S^{(1)} \not\cong \pi_L^{(1)}.}$$

Physically, this may be interpreted as expressing the difference of the finiteness and the infiniteness of configuration space of a quantum system.

We note also that, for all $L > 0$ and $L' > 0$ with $L \neq L'$, $\sigma(\hat{q}_L) \neq \sigma(\hat{q}_{L'})$. Hence

$$\boxed{\pi_L^{(1)} \not\cong \pi_{L'}^{(1)}.}$$

Therefore $\{\pi_L^{(1)}\}_{L>0}$ is a family of mutually inequivalent irreducible self-adjoint representations of the CCR with one degree of freedom. Thus, in the context of quantum mechanics, the set $\{L|L > 0\}$, which consists of macroscopic quantities indexing the sizes of a one-dimensional quantum system in a finite interval, can be viewed as the index set of a family of mutually inequivalent irreducible representations in $\mathsf{CCR}_{\mathrm{sa}}(1)$. This is a simple example which shows a correspondence between inequivalent irreducible representations of CCR and macroscopic quantities.

The d-dimensional version of $\pi_L^{(1)}$ is easily constructed. Its representation space is taken to be the Hilbert space $L^2(\mathbb{I}_L^d)$. On this Hilbert space, we define operators $\hat{q}_{Lj}, \hat{p}_{Lj}$ $(j = 1, \ldots, d)$ as follows:

$$\hat{q}_{Lj} := M_{x_j}, \tag{2.108}$$

$$\hat{p}_{Lj} := F_{L,d}^{-1} M_{k_j} F_{L,d}, \tag{2.109}$$

where M_{x_j} is the multiplication operator on $L^2(\mathbb{I}_L)$ by the j-th coordinate function x_j on $\mathbb{I}_L^d$ ($\boldsymbol{x} = (x_1, \ldots, x_d) \in \mathbb{I}_L^d$) and M_{k_j} is the multiplication operator on $\ell^2(\Gamma_L^d)$ by the j-th coordinate function k_j on Γ_L^d ($\boldsymbol{k} = (k_1, \ldots, k_d) \in \Gamma_L^d$). Let $C_0^\infty((-L/2, L/2)^d)$ be the set of infinitely differentiable functions on $(-L/2, L/2)^d$ with support in $(-L/2, L/2)^d$. Then

$$\pi_L^{(d)} := (L^2(\mathbb{I}_L^d), C_0^\infty((-L/2, L/2)^d), (\hat{q}_{L1}, \ldots, \hat{q}_{Ld}, \hat{p}_{L1}, \ldots, \hat{p}_{Ld}))$$

is a self-adjoint representation of the CCR with d degrees of freedom, i.e., $\pi_L^{(d)} \in \mathsf{CCR}_{\mathrm{sa}}(d)$. In the same way as in the case of $\pi_L^{(1)}$, one can show that $\pi_L^{(d)}$ is irreducible.

Since $\hat{q}_{Lj}$ is a bounded self-adjoint operator, it follows that $\pi_L^{(d)}$ is inequivalent to the Schrödinger representation $\pi_S^{(d)}$:

$$\boxed{\pi_L^{(d)} \not\cong \pi_S^{(d)}.}$$

As in the case $d = 1$, one has the following inequivalence:

$$\boxed{\pi_L^{(d)} \not\cong \pi_{L'}^{(d)}} \quad (L \neq L').$$

The free Hamiltonian of a non-relativistic quantum particle of mass $m > 0$ moving in $\mathbb{I}_L^d$ is defined by

$$H_L := \frac{1}{2m} \sum_{j=1}^{d} \hat{p}_{Lj}^2.$$

It follows from (2.109) that

$$F_{L,d} H_L F_{L,d}^{-1} = \frac{\boldsymbol{k}^2}{2m},$$

where the right hand side denotes the multiplication operator by the function $\boldsymbol{k}^2/(2m)$ on Γ_L^d. Hence H_L is a non-negative self-adjoint operator and has a purely discrete spectrum:

$$\sigma(H_L) = \sigma_{\mathrm{d}}(H_L) = \left\{ \frac{\boldsymbol{k}^2}{2m} \middle| \boldsymbol{k} \in \Gamma_L^d \right\}.$$

(B) A System with a Dirichlet Boundary Condition

As is easily seen, the differential operator d/dx with domain $C_0^\infty(-L/2, L/2)$ acting in $L^2(\mathbb{I}_L)$ is closable. We denote its closure by $D_{L,0}$:

$$D_{L,0} := \overline{d/dx \restriction C_0^\infty(-L/2, L/2)}. \tag{2.110}$$

Since $-id/dx$ is symmetric on $C_0^\infty(-L/2, L/2)$, it follows that $-iD_{L,0}$ is a closed symmetric operator on $L^2(\mathbb{I}_L)$. Based on this fact, one can consider, as another candidate for a momentum operator on $L^2(\mathbb{I}_L)$, the following operator:

$$\hat{p}_{L,0} := -iD_{L,0}. \tag{2.111}$$

It is easy to see that

$$\pi_{L,0}^{(1)} := (L^2(\mathbb{I}_L), C_0^\infty(-L/2, L/2), (\hat{q}_L, \hat{p}_{L,0}))$$

is in $\mathsf{CCR}(1)$. But we have:

Lemma 2.16 *The operator $\hat{p}_{L,0}$ is not self-adjoint and has uncountably many different self-adjoint extensions.*

Proof Let $f_0(x) := e^x$, $x \in \mathbb{I}_L$. Then $f_0 \in L^2(\mathbb{I}_L) \cap C^\infty(\mathbb{I}_L)$ with $f_0' = f_0$. Hence, for all $u \in C_0^\infty(-L/2, L/2)$, we have, by integration by parts, $\langle f_0, \hat{p}_{L,0}u \rangle = \langle -if_0, u \rangle$. Hence $f_0 \in D(\hat{p}_{L,0}^*)$ and $\hat{p}_{L,0}^* f_0 = -if_0$. Hence $f_0 \in \ker(\hat{p}_{L,0}^* + i)$. Conversely, let $f \in \ker(\hat{p}_{L,0}^* + i)$. Then, for all $u \in C_0^\infty(-L/2, L/2)$,

$$0 = \langle (\hat{p}_{L,0}^* + i)f, u \rangle = \langle f, (\hat{p}_{L,0} - i)u \rangle,$$

which implies that $\langle f, u' \rangle = \langle -f, u \rangle$. Hence it follows that

$$Df = f$$

in $\mathscr{D}'(-L/2, L/2)$, the space of distributions on $(-L/2, L/2)$, where Df means the generalized derivative of $f \in \mathscr{D}'(-L/2, L/2)$. Let $g(x) := e^{-x} f(x)$, $x \in \mathbb{I}_L$. Then

$g \in \mathscr{D}'(-L/2, L/2)$ and $Dg = -e^{-x}f + e^{-x}Df = 0$. Hence, for some constant $c \in \mathbb{C}$, $g(x) = c$, a.e.$x \in \mathbb{I}_L$. Therefore $f = ce^x = cf_0$. Thus we obtain

$$\ker(\hat{p}_{L,0}^* + i) = \{\alpha e^x | \alpha \in \mathbb{C}\}. \tag{2.112}$$

Similarly, one can show that

$$\ker(\hat{p}_{L,0}^* - i) = \{\alpha e^{-x} | \alpha \in \mathbb{C}\}. \tag{2.113}$$

Hence

$$n_+(\hat{p}_{L,0}) = 1 = n_-(\hat{p}_{L,0}), \tag{2.114}$$

where, for a closed symmetric operator A, $n_\pm(A)$ denote the deficiency indices of A (see Sect. 1.4). In particular, $\hat{p}_{L,0}$ is not self-adjoint.

Property (2.114) and Theorem 1.18(iii) imply also that $\hat{p}_{L,0}$ has uncountably many different self-adjoint extensions. ■

Lemma 2.16 immediately yields:

Proposition 2.12 *$\pi_{L,0}^{(1)}$ is not a self-adjoint representation. In particular, $\pi_{L,0}^{(1)}$ is inequivalent to $\pi_L^{(1)}$:*

$$\boxed{\pi_{L,0}^{(1)} \not\cong \pi_L^{(1)}.} \tag{2.115}$$

We also note that, by (2.114) and Theorem 1.17,

$$\sigma(\hat{p}_{L,0}) = \mathbb{C}. \tag{2.116}$$

Inequivalence (2.115) corresponds to the difference of boundary conditions in $\mathbb{I}_L$ of the differential operator $-id/dx$.

The operator $\hat{p}_{L,0}$ is called the **momentum operator with Dirichlet boundary condition**, although it is not self-adjoint.

The uncountably many different self-adjoint extensions mentioned in Lemma 2.16 correspond to different boundary conditions for $-id/dx$ in $\mathbb{I}_L$. It is obvious that $\hat{p}_L$ is a self-adjoint extension of $\hat{p}_{L,0}$.

Let $\widetilde{p}_{L,0}$ be one of the self-adjoint extensions of $\hat{p}_{L,0}$. Then

$$\widetilde{\pi}_{L,0}^{(1)} := (L^2(\mathbb{I}_L), C_0^\infty(-L/2, L/2), (\hat{q}_L, \widetilde{p}_{L,0}))$$

is a self-adjoint representation of the CCR with one degree of freedom. Since $\sigma(\widetilde{p}_{L,0}) \neq \sigma(\hat{p}_{L,0})$, this representation is inequivalent to $\pi_{L,0}^{(1)}$. Different self-adjoint extensions of $\hat{p}_{L,0}$ correspond to different boundary conditions in $\mathbb{I}_L$ for

$-id/dx$. Therefore the inequivalence between $\pi_{L,0}^{(1)}$ and $\tilde{\pi}_{L,0}^{(1)}$ corresponds to the difference of boundary conditions.

(C) A System with α-Boundary Condition

The momentum operator $p(\alpha)$ with α-boundary condition introduced in Sect. 1.15 is a self-adjoint extension of $\hat{p}_{L,0}$ and

$$\pi_L(\alpha) := (L^2(\mathbb{I}_L), C_0^\infty(-L/2, L/2), (\hat{q}_L, p(\alpha))) \tag{2.117}$$

is a self-adjoint representation of the CCR with one degree of freedom. In the same way as in the case of $\pi_L^{(1)}$, one can show that $\pi_L(\alpha)$ is irreducible. The following fact is interesting:

Theorem 2.14 *If $\alpha \neq \beta$ ($\alpha, \beta \in \{z \in \mathbb{C}||z| = 1\}$), then $\pi_L(\alpha)$ is inequivalent to $\pi_L(\beta)$.*

Proof Let $\alpha \neq \beta$. Then $\theta_\alpha \neq \theta_\beta$ (see (1.111)). Hence, by Theorem 1.37, $\sigma(p(\alpha)) \neq \sigma(p(\beta))$. Hence $p(\alpha)$ is not unitarily equivalent to $p(\beta)$. Thus $\pi(\alpha)$ is inequivalent to $\pi(\beta)$. ∎

Theorem 2.14 shows that a family of boundary conditions in $\mathbb{I}_L$ forms an index set of a family of mutually inequivalent irreducible representations of the CCR with one degree of freedom. The boundary conditions are of macroscopic nature and distinguishable from each other experimentally. Thus we see here an example of significant relations between macroscopic objects and inequivalent representations of CCR.

2.6.5 *Quantum System in a Half-Line*

We consider a system of a quantum particle of mass $m > 0$ in a half-line $[a, \infty)$ ($a \in \mathbb{R}$). We denote an element of $[a, \infty)$ by r. A natural Hilbert space of state vectors of the system is given by $L^2([a, \infty))$. The position operator $\hat{q}_a$ acting in $L^2([a, \infty))$ is defined by

$$\hat{q}_a := M_r,$$

the multiplication operator by r. It follows from Theorem A.1(v) in Appendix A that $\hat{q}_a$ is self-adjoint and

$$\sigma(\hat{q}_a) = [a, \infty). \tag{2.118}$$

A momentum operator $\hat{p}_a$ acting in $L^2([a,\infty))$ may be defined as follows:

$$D(\hat{p}_a) := C_0^1(a,\infty), \quad \hat{p}_a := -i\frac{d}{dr}.$$

One can easily show that $\hat{p}_a$ is symmetric and the triple

$$\pi_a := (L^2([a,\infty)), C_0^\infty(a,\infty), (\hat{q}_a, \hat{p}_a))$$

is a representation of the CCR with degree one: $\pi_a \in \mathsf{CCR}(1)$.

Lemma 2.17 *The representation π_a is irreducible.*

Proof Let $T \in \{\hat{q}_a, \hat{p}_a\}'$. Then $T \in \mathfrak{B}(L^2([a,\infty)))$ and

$$T\hat{q}_a \subset \hat{q}_a T, \quad T\hat{p}_a \subset \hat{p}_a T. \tag{2.119}$$

By the first relation and Lemma 2.7, we have $Te^{it\hat{q}_a} = e^{it\hat{q}_a}T$ for all $t \in \mathbb{R}$. Then, in the same way as in the proof of Lemma 2.11, $T = M_F$ for some $F \in L^\infty([a,\infty))$. The second relation in (2.119) implies that, for all $f \in C_0^1(a,\infty)$, $Ff \in C_0^1(a,\infty)$ and $Ff' = (Ff)'$. Since f is an arbitrary element of $C_0^1(a,\infty)$, it follows that $F \in C^1(a,\infty)$ and $F' = 0$ on (a,∞). Hence $F(r) = \alpha$, $r > a$ for some $\alpha \in \mathbb{C}$. Therefore $T = \alpha I$. Thus π_a is irreducible. ■

It follows from (2.118) and Lemma 2.17 that π_a is inequivalent to the Schrödinger representation $\pi_{\mathrm{S}}^{(1)}$.

Formula (2.118) implies that, if $a \neq b$ $(a, b \in \mathbb{R})$, the spectrum of $\hat{q}_a$ is different from that of $\hat{q}_b$. Hence π_a is inequivalent to π_b for all $b \neq a$:

$$\pi_a \not\cong \pi_b, \quad a \neq b. \tag{2.120}$$

Thus the real number a, which is physically a macroscopic quantity defining the half-line $[a,\infty)$, may be regarded as the parameter indexing a family of mutually inequivalent representations of the CCR with degree one.

In the same way as in the proof of Lemma 2.16, one has

$$\ker(\hat{p}_a^* - i) = \{\alpha e^{-r} | \alpha \in \mathbb{C}\}, \quad \ker(\hat{p}_a^* + i) = \{0\}, \tag{2.121}$$

where we have used the fact that the function e^r is not in $L^2([a,\infty))$. Hence

$$n_+(\hat{p}_a) = 1, \quad n_-(\hat{p}_a) = 0.$$

Therefore, by Theorem 1.18, one has the following:

(i) $\hat{p}_a$ is not essentially self-adjoint.
(ii) $\hat{p}_a$ has no self-adjoint extensions.
(iii) The closure $\overline{\hat{p}_a}$ is maximally symmetric.

We note also that (2.121) and Theorem 1.17 imply

$$\sigma(\hat{p}_a) = \overline{\Pi}_-. \tag{2.122}$$

It is easy to see that $\hat{p}_a^2$ is symmetric. Hence a free Hamiltonian of the system may be given by the closure of $\hat{p}_a^2/(2m)$. But it is not self-adjoint, since $\hat{p}_a^2/(2m)$ is not essentially self-adjoint as shown in the next proposition.

Proposition 2.13 *One has*

$$n_+(\hat{p}_a^2) = n_-(\hat{p}_a^2) = 1. \tag{2.123}$$

Moreover, the following hold:

(i) $\hat{p}_a^2$ *is not essentially self-adjoint.*
(ii) $\hat{p}_a^2$ *has uncountably many different self-adjoint extensions.*

Proof In the same way as in the proof of Lemma 2.16 (or see [131, pp. 144–145]), one can prove that

$$\ker((\hat{p}_a^2)^* + i) = \{\alpha e^{-r/\sqrt{2}} e^{-ir/\sqrt{2}} | \alpha \in \mathbb{C}\},$$

$$\ker((\hat{p}_a^2)^* - i) = \{\alpha e^{-r/\sqrt{2}} e^{ir/\sqrt{2}} | \alpha \in \mathbb{C}\}.$$

Hence (2.123) holds.

Part (i) and (ii) follow from Theorem 1.18. ■

It is known (see, e.g., [131, pp. 144–145]) that the self-adjoint extensions of $\hat{p}_a^2/(2m)$ are parameterized by $\mathbb{R} \cup \{\infty\}$. We denote them by $\{H_\lambda | \lambda \in \mathbb{R} \cup \{\infty\}\}$. One can take any H_λ as a free Hamiltonian of the system under consideration. If $\lambda \neq \mu$ ($\lambda, \mu \in \mathbb{R} \cup \{\infty\}$), then $D(H_\lambda) \neq D(H_\mu)$ corresponding to different boundary conditions at $r = a$. It is shown (see [131, p. 144–145]) that the different H_λ's correspond to different physics.

2.7 Cyclic Representations

The BHJ representation $\pi_{\text{BHJ}}^{(d)}$ is constructed from the annihilation operators a_j and the creation operators a_j^* which obey characteristic commutation relations (2.70)–(2.72). This structure can be generalized.

Commutation relations

$$[X_j, X_k^*] = \delta_{jk} I, \quad [X_j, X_k] = 0, \quad [X_j^*, X_k^*] = 0, \quad j, k = 1, \dots, N \quad (N \in \mathbb{N}) \tag{2.124}$$

for elements $X_1, \ldots, X_N$ in a $*$-algebra $\mathfrak{A}$ (I denotes the identity in $\mathfrak{A}$) are called the CCR with N degrees of freedom for $\{X_j\}_{j=1}^N$. This is another type of CCR. It may be called the **Dirac CCR** with N degrees of freedom [64]. In contrast to this, CCR (2.1)–(2.2) may be called the **Heisenberg CCR** with N degrees of freedom. But, in the case where there will be no danger of confusion, we will simply call each of them the CCR with N degrees of freedom.

Definition 2.3 Let $N \in \mathbb{N}$ and $\mathscr{D}$ be a non-zero subspace of a Hilbert space $\mathscr{H}$. Let $\{A_j\}_{j=1}^N$ be a set of densely defined closed operators on $\mathscr{H}$. The triple $\pi_A := (\mathscr{H}, \mathscr{D}, \{A_j, A_j^*\}_{j=1}^N)$ is called a **representation of the Dirac CCR** with N degrees of freedom if $\mathscr{D} \subset \cap_{j,k=1}^N D(A_j^* A_k) \cap D(A_k A_j^*) \cap D(A_j A_k) \cap D(A_j^* A_k^*)$ and, for all $j, k = 1, \ldots, N$,

$$[A_j, A_k^*] = \delta_{jk}, \quad [A_j, A_k] = 0, \quad [A_j^*, A_k^*] = 0 \quad \text{on } \mathscr{D}. \tag{2.125}$$

The subspace $\mathscr{D}$ is called a **CCR-domain** for $\{A_j, A_j^* | j = 1, \ldots, N\}$. If $\{A_j, A_j^* | j = 1, \ldots, N\}$ is irreducible, then π_A is said to be **irreducible**.

Let $(\mathscr{H}, \mathscr{D}, \{A_j, A_j^*\}_{j=1}^N)$ be a representation of the Dirac CCR with N degrees of freedom with $\mathscr{D}$ dense. Then one can define symmetric operators Q_j and P_j ($j = 1, \ldots, N$) by

$$Q_j := \frac{1}{\sqrt{2}}(A_j^* + A_j), \quad P_j := \frac{i}{\sqrt{2}}(A_j^* - A_j). \tag{2.126}$$

It is easy to see that $\mathscr{D} \subset \cap_{j,k=1}^N \big(D(Q_j P_k) \cap D(P_k Q_j) \cap D(Q_j Q_k) \cap D(P_j P_k)\big)$ and $\{Q_j, P_j | j = 1, \ldots, N\}$ obeys the Heisenberg CCR with N degrees of freedom. Thus $(\mathscr{H}, \mathscr{D}, (Q_1, \ldots, Q_N, P_1, \ldots, P_N))$ is a representation of the Heisenberg CCR with N degrees of freedom with a dense CCR-domain. We call this representation the **representation of the Heisenberg CCR associated with** $(\mathscr{H}, \mathscr{D}, \{A_j, A_j^*\}_{j=1}^N)$.

Conversely, suppose that a representation $(\mathscr{H}, \mathscr{D}, (Q_1, \ldots, Q_N, P_1, \ldots, P_N))$ of the Heisenberg CCR with N degrees of freedom with $\mathscr{D}$ dense is given. Then it is easy to see that

$$A_{j,0} := \frac{1}{\sqrt{2}}(Q_j + iP_j)$$

is densely defined with $D(A_{j,0}) \supset \mathscr{D}$ and $A_{j,0}^* \supset (Q_j - iP_j)/\sqrt{2}$. Hence $A_{j,0}^*$ is densely defined. Therefore, by Proposition 1.6(ii), $A_{j,0}$ is closable. We define

$$A_j := \overline{A}_{j,0} = \frac{1}{\sqrt{2}}\overline{(Q_j + iP_j)}. \tag{2.127}$$

Then A_j is densely defined closed operator. Moreover, one can show that $(\mathscr{H}, \mathscr{D}, \{A_j, A_j^*\}_{j=1}^N)$ is a representation of the Dirac CCR with N degrees of freedom. We call this representation the **representation of the Dirac CCR associated with** $(\mathscr{H}, \mathscr{D}, (Q_1, \ldots, Q_N, P_1, \ldots, P_N))$.

Thus we see that one can construct a representation of the Heisenberg CCR with N degrees of freedom from a representation of the Dirac CCR with N degrees of freedom with a dense CCR-domain and vice versa.

On irreducibility of a representation of Dirac CCR and the associated representation of Heisenberg CCR, we have the following.

Proposition 2.14 *Let $(\mathscr{H}, \mathscr{D}, \{A_j, A_j^*\}_{j=1}^N)$ be an irreducible representation of the Dirac CCR with N degrees of freedom such that, for each $j = 1, \ldots, N$, $D(A_j) \cap D(A_j^*)$ is a common core for A_j and A_j^*. Then the representation $(\mathscr{H}, \mathscr{D}, (Q_1, \ldots, Q_N, P_1, \ldots, P_N))$ of the Heisenberg CCR associated with it is irreducible.*

Proof Let $T \in \{Q_j, P_j | j = 1, \ldots, N\}'$. Then, for all $j = 1, \ldots, N$, $TQ_j \subset Q_jT$ and $TP_j \subset P_jT$. It follows from (2.126) that, for all $\Psi \in D(A_j) \cap D(A_j^*)$, $T_j\Psi \in D(A_j) \cap D(A_j^*)$ and

$$T(A_j^* + A_j)\Psi = (A_j^* + A_j)T\Psi, \quad T(A_j^* - A_j)\Psi = (A_j^* - A_j)T\Psi.$$

Hence $TA_j^*\Psi = A_j^*T\Psi$ and $TA_j\Psi = A_jT\Psi$. The former equation implies that, for all $\Phi \in D(A_j)$,

$$\langle T^*A_j\Phi, \Psi\rangle = \left\langle T^*\Phi, A_j^*\Psi\right\rangle.$$

Since $D(A_j) \cap D(A_j^*)$ is a core for A_j^* by the assumption, it follows that $T_j^*\Phi \in D(A_j^{**}) = D(A_j)$ and $A_jT^*\Phi = T^*A_j\Phi$. Hence $T^*A_j \subset A_jT^*$. Similarly, the equation $TA_j\Psi = A_jT\Psi$ implies that $T^*A_j^* \subset A_j^*T^*$. By the irreducibility and the $*$-invariance of $\{A_j, A_j^* | j = 1, \ldots, N\}$, $T^* = \alpha I$ with a constant $\alpha \in \mathbb{C}$. Hence $T = \alpha^* I$. Thus $\{Q_j, P_j | j = 1, \ldots, N\}$ is irreducible. ■

Proposition 2.15 *Let $(\mathscr{H}, \mathscr{D}, (Q_1, \ldots, Q_N, P_1, \ldots, P_N))$ be an irreducible self-ad-joint representation of the Heisenberg CCR with N degrees of freedom such that, for each $j = 1, \ldots, N$, $D(Q_j) \cap D(P_j)$ is a common core for Q_j and P_j. Then the representation $(\mathscr{H}, \mathscr{D}, \{A_j, A_j^* | j = 1, \ldots, N\})$ of the Dirac CCR associated with it is irreducible.*

Proof Let $T \in \{A_j, A_j^* | j = 1, \ldots, N\}'$. Then $TA_j \subset A_jT$ and $TA_j^* \subset A_j^*T$. It follows from these relations and (2.127) that, for all $\Phi, \Psi \in D(Q_j) \cap D(P_j)$,

$$\langle T^*\Phi, (Q_j + iP_j)\Psi\rangle = \langle T^*(Q_j - iP_j)\Phi, \Psi\rangle, \tag{2.128}$$

$$\langle T^*\Phi, (Q_j - iP_j)\Psi\rangle = \langle T^*(Q_j + iP_j)\Phi, \Psi\rangle. \tag{2.129}$$

Hence

$$\langle T^*\Phi, Q_j\Psi\rangle = \langle T^*Q_j\Phi, \Psi\rangle, \quad \langle T^*\Phi, P_j\Psi\rangle = \langle T^*P_j\Phi, \Psi\rangle.$$

By the present assumption, $D(Q_j) \cap D(P_j)$ is a common core for Q_j and P_j, and Q_j and P_j are self-adjoint. Hence it follows that $T^*\Phi \in D(Q_j) \cap D(P_j)$ and

$$Q_jT^*\Phi = T^*Q_j\Phi, \quad P_jT^*\Phi = T^*P_j\Phi.$$

By a limiting argument using the assumption that $D(Q_j) \cap D(P_j)$ is a core for Q_j and P_j again, one can show that these equations imply that $T^*Q_j \subset Q_jT^*$ and $T^*P_j \subset P_jT^*$. Hence, by the irreducibility and the $*$-invariance of the operator set $\{Q_j, P_j | j = 1, \ldots, N\}$, $T^* = \alpha I$ for a constant $\alpha \in \mathbb{C}$, i.e., $T = \alpha^* I$. Thus $\{A_j, A_j^* | j = 1, \ldots, N\}$ is irreducible. ∎

To discuss representations of Dirac CCR further, we introduce a general notion:

Definition 2.4 Let $\mathfrak{A}$ be a set of linear operators on a Hilbert space $\mathscr{H}$. A non-zero vector Ψ is called a **cyclic vector** for $\mathfrak{A}$ if $\Psi \in D(A)$ for all $A \in \mathfrak{A}$ and $\mathrm{span}\{\Psi, A\Psi | A \in \mathfrak{A}\}$ is dense in $\mathscr{H}$.

Example 2.5 Consider the BHJ representation $\pi_{\mathrm{BHJ}}^{(d)}$. Let $e_0^{(d)} \in \ell_0(\mathbb{Z}_+^d)$ be as in Sect. 2.6.2. Then $e_0^{(d)}$ is a cyclic vector for $\{(a_1^*)^{n_1} \cdots (a_d^*)^{n_d} | n_1, \ldots, n_d \in \mathbb{Z}_+\}$.

Definition 2.5 Let $\pi_A := (\mathscr{H}, \mathscr{D}, \{A_j, A_j^*\}_{j=1}^N)$ be a representation of the Dirac CCR with N degrees of freedom such that $\mathscr{D} \subset \cap_{j=1}^N D(A_j) \cap D(A_j^*)$ and $A_j\mathscr{D} \subset \mathscr{D}$ and $A_j^*\mathscr{D} \subset \mathscr{D}$.

(i) If there is a non-zero vector $\Psi_0 \in \cap_{j=1}^N D(A_j)$ satisfying

$$A_j\Psi_0 = 0, \quad j = 1, \ldots, N, \tag{2.130}$$

then Ψ_0 is called a **vacuum** for π_A.

(ii) If there is a vacuum $\Psi_0 \in \mathscr{D}$ for π_A such that Ψ_0 is a cyclic vector for $\{(A_1^*)^{n_1} \cdots (A_N^*)^{n_N} | n_1, \ldots, n_N \in \mathbb{Z}_+\}$, then π_A is called a **cyclic representation** of the Dirac CCR with N degrees of freedom with cyclic vector Ψ_0. In this case we write $\pi_A = (\mathscr{H}, \mathscr{D}, \{A_j, A_j^*\}_{j=1}^N, \Psi_0)$.

Example 2.6 Let a_j be the annihilation operator in Sect. 2.6.2. Then

$$(\ell^2(\mathbb{Z}_+^d), \ell_0(\mathbb{Z}_+^d), \{a_j, a_j^* | j = 1, \ldots, d\})$$

is a cyclic representation of the Dirac CCR with d degrees of freedom with a cyclic vector $e_0^{(d)}$.

Lemma 2.18 *Let $(\mathcal{H}, \mathcal{D}, \{A_j, A_j^*\}_{j=1}^N, \Psi_0)$ be a cyclic representation of Dirac CCR. For each $\boldsymbol{n} = (n_1, \ldots, n_N) \in \mathbb{Z}_+^N$, define a vector $e_{\boldsymbol{n}} \in \mathcal{H}$ by*

$$e_{\boldsymbol{0}} := \Psi_0, \quad e_{\boldsymbol{n}} := \frac{1}{\sqrt{n_1! \cdots n_N!}} (A_1^*)^{n_1} \cdots (A_N^*)^{n_N} \Psi_0 \quad (\boldsymbol{n} \neq \boldsymbol{0}).$$

Then $\{e_{\boldsymbol{n}}\}_{\boldsymbol{n} \in \mathbb{Z}_+^N}$ is a CONS of $\mathcal{H}$.

Proof Let $I_{\boldsymbol{n}} := \|(A_1^*)^{n_1} \cdots (A_N^*)^{n_N} \Psi_0\|^2$ and $n_1 \neq 0$. Then, using the Dirac CCR (2.125), we have

$$\begin{aligned} I_{\boldsymbol{n}} &= \left\langle (A_1^*)^{n_1-1} \cdots (A_N^*)^{n_N} \Psi_0, A_1 (A_1^*)^{n_1} \cdots (A_N^*)^{n_N} \Psi_0 \right\rangle \\ &= \left\langle (A_1^*)^{n_1-1} \cdots (A_N^*)^{n_N} \Psi_0, n_1 (A_1^*)^{n_1-1} \cdots (A_N^*)^{n_N} \Psi_0 \right\rangle \\ &+ \left\langle (A_1^*)^{n_1-1} \cdots (A_N^*)^{n_N} \Psi_0, (A_1^*)^{n_1} A_1 (A_2^*)^{n_2} \cdots (A_N^*)^{n_N} \Psi_0 \right\rangle. \end{aligned}$$

By the commutativity of A_1 with $A_2^*, \ldots, A_N^*$ on $\mathcal{D}$ and (2.130), we obtain

$$A_1 (A_2^*)^{n_2} \cdots (A_N^*)^{n_N} \Psi_0 = 0.$$

Hence

$$I_{\boldsymbol{n}} == n_1 I_{(n_1-1, n_2, \ldots, n_N)}.$$

Iterating this recursion relation, we obtain

$$I_{\boldsymbol{n}} = n_1! I_{(0, n_2, \ldots, n_N)}.$$

Repeating this process, we obtain

$$I_{\boldsymbol{n}} = n_1! n_2! \cdots n_N! I_{\boldsymbol{0}} = n_1! n_2! \cdots n_N!.$$

Hence $\|e_{\boldsymbol{n}}\| = 1$. Similarly, one can show that $\langle e_{\boldsymbol{n}}, e_{\boldsymbol{m}} \rangle = 0$ for $\boldsymbol{n} \neq \boldsymbol{m}$. Thus $\{e_{\boldsymbol{n}}\}_{\boldsymbol{n} \in \mathbb{Z}_+^N}$ is an ONS of $\mathcal{H}$. By the cyclicity of Ψ_0, $\text{span}\{e_{\boldsymbol{n}} | \boldsymbol{n} \in \mathbb{Z}_+^N\}$ is dense in $\mathcal{H}$. Hence $\{e_{\boldsymbol{n}}\}_{\boldsymbol{n} \in \mathbb{Z}_+^N}$ is complete. ∎

Proposition 2.16 *Let $\pi_A := (\mathcal{H}, \mathcal{D}, \{A_j, A_j^*\}_{j=1}^N, \Psi_0)$ be a cyclic representation of Dirac CCR. Then π_A is irreducible.*

Proof Let $T \in \{A_j, A_j^* | j = 1, \ldots, N\}'$. Then $TA_j \subset A_j T$ and $TA_j^* \subset A_j^* T$ for all $j = 1, \ldots, N$. For each $\boldsymbol{n} \in \mathbb{Z}_+^N$, we have

$$\langle e_{\boldsymbol{n}}, T e_{\boldsymbol{n}} \rangle = \frac{1}{n_1! \cdots n_N!} \left\langle A_N^{n_N} \cdots A_1^{n_1} (A_1^*)^{n_1} \cdots (A_N^*)^{n_N} \Psi_0, T \Psi_0 \right\rangle.$$

In the same way as in the proof of Lemma 2.18, we have

$$A_1^{n_1}(A_1^*)^{n_1}\cdots(A_N^*)^{n_N}\Psi_0 = n_1!(A_2^*)^{n_2}\cdots(A_N^*)^{n_N}\Psi_0.$$

Hence

$$A_N^{n_N}\cdots A_1^{n_1}(A_1^*)^{n_1}\cdots(A_N^*)^{n_N}\Psi_0 = n_1!\cdots n_N!\Psi_0.$$

Therefore, putting $\alpha := \langle\Psi_0, T\Psi_0\rangle$, we obtain

$$\langle e_{\boldsymbol{n}}, Te_{\boldsymbol{n}}\rangle = \alpha.$$

We next consider the case $\boldsymbol{m} \neq \boldsymbol{n}$. Then there is a $j \in \{1, \ldots, N\}$ such that $n_j \neq m_j$. We first consider the case where $n_j > m_j$. Then, in the same manner as above, we have

$$A_N^{n_N}\cdots A_1^{n_1}(A_1^*)^{m_1}\cdots(A_N^*)^{m_N}\Psi_0 = 0,$$

since $A_j^{n_j}(A_j^*)^{m_j}\Psi_0 = 0$. Hence

$$\langle e_{\boldsymbol{m}}, Te_{\boldsymbol{n}}\rangle = 0.$$

In the case $n_j < m_j$, we have

$$\langle e_{\boldsymbol{m}}, Te_{\boldsymbol{n}}\rangle = \frac{1}{\sqrt{m_1!\cdots m_N!n_1!\cdots n_N!}}\left\langle\Psi_0, TA_1^{m_1}\cdots A_N^{m_N}(A_1^*)^{n_1}\cdots(A_N^*)^{n_N}\Psi_0\right\rangle = 0.$$

Thus, for all $\boldsymbol{n}, \boldsymbol{n} \in \mathbb{Z}_+^N$,

$$\langle e_{\boldsymbol{m}}, Te_{\boldsymbol{n}}\rangle = \alpha\,\langle e_{\boldsymbol{m}}, e_{\boldsymbol{n}}\rangle\,.$$

Since $\{e_{\boldsymbol{n}}\}_{n\in\mathbb{Z}_+^N}$ is a CONS of $\mathscr{H}$ and T is bounded, it follows that $T = \alpha I$. Thus π_A is irreducible. ■

To establish a uniqueness theorem on representations of Dirac CCR, one needs a notion of regularity.

Definition 2.6 We say that a cyclic representation $(\mathscr{H}, \mathscr{D}, \{A_j, A_j^*\}_{j=1}^N, \Psi_0)$ of Dirac CCR is **regular** if

$$\mathrm{span}\{(A_1^*)^{n_1}\cdots(A_N^*)^{n_N}\Psi_0 | n_1, \ldots, n_N \in \mathbb{Z}_+\}$$

is a common core for A_j and A_j^*.

Example 2.7 Let a_j be the annihilation operator in the construction of the BHJ representation $\pi_{\mathrm{BHJ}}^{(d)}$ (see Sect. 2.6.2). Then the cyclic representation $(\ell^2(\mathbb{Z}_+^d), \ell_0(\mathbb{Z}_+^d), \{a_j, a_j^*\}_{j=1}^N, e_0^{(d)})$ is regular.

Theorem 2.15 (Uniqueness Theorem) *Let* $\pi_A := (\mathscr{H}, \mathscr{D}, \{A_j, A_j^*\}_{j=1}^N)$ *and* $\pi_B := (\mathscr{K}, \mathscr{F}, \{B_j, B_j^*\}_{j=1}^N)$ *be regular cyclic representations of the Dirac CCR with N degrees of freedom. Then* $\pi_A \cong \pi_B$.

Proof Let Ψ_0 and Φ_0 be cyclic vectors of π_A and π_B respectively and

$$\mathscr{D}_A := \mathrm{span}\{(A_1^*)^{n_1} \cdots (A_N^*)^{n_N} \Psi_0 | n_1, \ldots, n_N \in \mathbb{Z}_+\}, \tag{2.131}$$

$$\mathscr{D}_B := \mathrm{span}\{(B_1^*)^{n_1} \cdots (B_N^*)^{n_N} \Phi_0 | n_1, \ldots, n_N \in \mathbb{Z}_+\}. \tag{2.132}$$

By Lemma 2.18 and the isomorphism theorem on Hilbert spaces (Theorem 1.3), there is a unique unitary operator U from $\mathscr{H}$ to $\mathscr{K}$ such that

$$U(A_1^*)^{n_1} \cdots (A_N^*)^{n_N} \Psi_0 = (B_1^*)^{n_1} \cdots (B_N^*)^{n_N} \Phi_0, \quad \boldsymbol{n} \in \mathbb{Z}_+^N.$$

This implies that $U\mathscr{D}_A = \mathscr{D}_B$. Let $\Psi = (A_1^*)^{n_1} \cdots (A_N^*)^{n_N} \Psi_0$. Then, for each $j = 1, \ldots, N$, we have

$$\begin{aligned} UA_j^*\Psi &= U(A_1^*)^{n_1} \cdots (A_j^*)^{n_j+1} \cdots (A_N^*)^{n_N} \Psi_0 \\ &= (B_1^*)^{n_1} \cdots (B_j^*)^{n_j+1} \cdots (B_N^*)^{n_N} \Phi_0 \\ &= B_j^* U\Psi. \end{aligned}$$

Hence $UA_j^*U^{-1} = B_j^*$ on $\mathscr{D}_B$. By regularity of π_A and π_B, $\mathscr{D}_A$ $(= U^{-1}\mathscr{D}_B)$ and $\mathscr{D}_B$ are cores for A_j^* and B_j^* respectively. Therefore operator equality $UA_j^*U^{-1} = B_j^*$ holds, which implies also that $UA_jU^{-1} = B_j$. Thus $\pi_A \cong \pi_B$. ■

Example 2.8 The proof of Theorem 2.11 (the equivalence of the Schrödinger representation $\pi_{\mathrm{S}}^{(1)}$ and the BHJ representation $\pi_{\mathrm{BHJ}}^{(1)}$) is essentially an application of Theorem 2.15.

2.8 Weyl Representations

2.8.1 Heuristics

As we have seen in Proposition 2.2(ii), any representation of the CCR with N degrees of freedom with a dense CCR-domain is unbounded.

Generally speaking, unbounded operators are more difficult to treat than bounded ones. Therefore it may be of some worth to seek forms formally equivalent to CCR in terms of bounded operators constructed from operators satisfying CCR. An idea for that is to recall that a self-adjoint operator A on a Hilbert space generates the strongly continuous one-parameter unitary group $\{e^{itA}\}_{t\in\mathbb{R}}$. Hence, in a self-adjoint representation $\pi_N = (\mathscr{H}, \mathscr{D}, (Q_1, \ldots, Q_N, P_1, \ldots, P_N)) \in \mathsf{CCR}_{\mathrm{sa}}(N)$, the CCR may be transformed to relations among the unitary operators e^{itQ_j} and e^{isP_k} $(j, k = 1, \ldots, n,\ s, t \in \mathbb{R})$. To discover such relations, we first proceed heuristically.

For simplicity, suppose that $\mathscr{D}$ is a dense subspace and left invariant by Q_j and P_j, $j = 1, \ldots, N$. Then, in the same manner as in the proof of (2.9), one can show that, for all $n \in \mathbb{N}$,

$$Q_j P_k^n - P_k^n Q_j = i\delta_{jk} n P_k^{n-1} \quad \text{on } \mathscr{D}.$$

Hence, for all $s \in \mathbb{R}$ and $M \in \mathbb{N}$,

$$Q_j S_M - S_M Q_j = -s\delta_{jk} S_{M-1} \quad \text{on } \mathscr{D} \tag{2.133}$$

with $S_M := \sum_{n=0}^{M} (is)^n P_k^n / n!$. By the functional calculus for P_k, one can show that, for all entire analytic vectors Ψ of P_k, $\lim_{M\to\infty} S_M \Psi = e^{isP_k}\Psi$. Hence, by taking the limit $M \to \infty$ in (2.133), one obtains *formally* (*heuristically*)

$$Q_j e^{isP_k} - e^{isP_k} Q_j = -s\delta_{jk} e^{isP_k} \quad \text{(formal)}, \tag{2.134}$$

and hence

$$e^{isP_k} Q_j e^{-isP_k} = Q_j + s\delta_{jk} \quad \text{(formal)}.$$

If this holds rigorously as an operator equality, then, by the unitary covariance of functional calculus (Theorem B.3 in Appendix B), one obtains

$$e^{isP_k} e^{itQ_j} e^{-isP_k} = e^{it(Q_j + s\delta_{jk})} = e^{ist\delta_{jk}} e^{itQ_j},$$

which is equivalent to

$$e^{itQ_j} e^{isP_k} = e^{-ist\delta_{jk}} e^{isP_k} e^{itQ_j}, \quad s, t \in \mathbb{R},\ j, k = 1, \ldots, N. \tag{2.135}$$

Similarly, one infers that

$$e^{itQ_j} e^{isQ_k} = e^{isQ_k} e^{itQ_j}, \quad e^{itP_j} e^{isP_k} = e^{isP_k} e^{itP_j}, \quad s, t \in \mathbb{R},\ j, k = 1, \ldots, N. \tag{2.136}$$

Note that (2.136) implies that $Q = (Q_1, \ldots, Q_N)$ and $P = (P_1, \ldots, P_N)$ are respectively strongly commuting. But, in general, Q or P is not necessarily strongly commuting. Indeed, we shall see later examples of self-adjoint representations of CCR in which Q or P is not strongly commuting (see Remark 3.3). Therefore CCR

(2.1) and (2.2) are *not equivalent to (2.135) and (2.136)* even if all Q_j and P_j $(j = 1, \ldots, N)$ are self-adjoint. We emphasize that, even in the case $N = 1$, CCR (2.1) does not imply (2.135) as is shown in the next example.

Example 2.9 Let us consider the representation $\pi_L^{(1)} \in \mathsf{CCR}_{\mathrm{sa}}(1)$ defined by (2.106). By (1.107), we have for all $t, s \in \mathbb{R}$ and $f \in L^2(\mathbb{I}_L)$

$$(e^{it\hat{q}_L} e^{is\hat{p}_L} f)(x) = e^{itx} f([x+s]_L), \quad (e^{is\hat{p}_L} e^{it\hat{q}_L} f)(x) = e^{it[x+s]_L} f([x+s]_L) \quad \text{a.e.} x \in \mathbb{I}_L.$$

Hence

$$(e^{it\hat{q}_L} e^{is\hat{p}_L} f)(x) = e^{-its+itLn_s(x)} (e^{is\hat{p}_L} e^{it\hat{q}_L} f)(x), \tag{2.137}$$

where $n_s \colon \mathbb{I}_L \to \mathbb{Z}$ is defined by

$$n_s(x) := \frac{s + x - [x+s]_L}{L}, \quad x \in \mathbb{I}_L.$$

It is easy to see that, for all $t \notin \{2\pi q/L | q \in \mathbb{Q}\}$ ($\mathbb{Q}$ is the set of rational numbers) and $|s| > L$, $e^{itLn_s} \neq I$ and hence, for such t and s,

$$e^{it\hat{q}_L} e^{is\hat{p}_L} \neq e^{-its} e^{is\hat{p}_L} e^{it\hat{q}_L}.$$

Thus (2.135) with $N = 1$, $Q_1 = \hat{q}_L$ and $P_1 = \hat{p}_L$ does not hold.

Hence the above deduction is not always true. But, anyway, we have heuristically arrived at relations (2.135) and (2.136), which may be a candidate for a bounded operator form for representations of CCR.

2.8.2 Weyl Relations and CCR

Relations (2.135) and (2.136) are called the **Weyl relations** for a $2N$-tuple $(Q_1, \ldots, Q_N, P_1, \ldots, P_N)$ of self-adjoint operators. As already mentioned, in this case, $(Q_1, \ldots, Q_N)$ and $(P_1, \ldots, P_N)$ are respectively strongly commuting.

We now take (2.135) and (2.136) as a starting point of a theory. We first show that $(Q_1, \ldots, Q_N, P_1, \ldots, P_N)$ obeying the Weyl relations (2.135) and (2.136) gives a self-adjoint representation of the CCR with N degrees of freedom:

Theorem 2.16 *Let $Q = (Q_1, \ldots, Q_N)$ and $P = (P_1, \ldots, P_N)$ be two N-tuples of self-adjoint operators on a Hilbert space $\mathscr{H}$ satisfying (2.135) and (2.136). Then there exists a dense subspace $\mathscr{D}_{\mathrm{G}}$ of $\mathscr{H}$ such that, for all $t \in \mathbb{R}$ and $j = 1, \ldots, N$,*

$$e^{itQ_j} \mathscr{D}_{\mathrm{G}} \subset \mathscr{D}_{\mathrm{G}}, \quad e^{itP_j} \mathscr{D}_{\mathrm{G}} \subset \mathscr{D}_{\mathrm{G}}, \tag{2.138}$$

$$Q_j \mathscr{D}_{\mathrm{G}} \subset \mathscr{D}_{\mathrm{G}}, \quad P_j \mathscr{D}_{\mathrm{G}} \subset \mathscr{D}_{\mathrm{G}} \tag{2.139}$$

and $(\mathscr{H}, \mathscr{D}_{\mathrm{G}}, (Q_1, \ldots, Q_N, P_1, \ldots, P_N)) \in \mathsf{CCR}_{\mathrm{sa}}(N)$.

Proof For $\boldsymbol{x} = (x_1, \ldots, x_N) \in \mathbb{R}^N$ and $\boldsymbol{k} = (k_1, \ldots, k_N) \in \mathbb{R}^N$, we define a unitary operator $U(\boldsymbol{x}, \boldsymbol{k})$ by

$$U(\boldsymbol{x}, \boldsymbol{k}) := e^{ix_1 P_1} \cdots e^{ix_N P_N} e^{ik_1 Q_1} \cdots e^{ik_N Q_N}.$$

For each $f \in C_0^\infty(\mathbb{R}^{2N})$ and $\Psi \in \mathscr{H}$, the mapping:

$$\mathbb{R}^N \times \mathbb{R}^N \ni (\boldsymbol{x}, \boldsymbol{k}) \mapsto f(\boldsymbol{x}, \boldsymbol{k}) U(\boldsymbol{x}, \boldsymbol{k}) \Psi \in \mathscr{H}$$

is strongly continuous and

$$\int_{\mathbb{R}^{2N}} \| f(\boldsymbol{x}, \boldsymbol{k}) U(\boldsymbol{x}, \boldsymbol{k}) \Psi \| d\boldsymbol{x} d\boldsymbol{k} = \int_{\mathbb{R}^{2N}} |f(\boldsymbol{x}, \boldsymbol{k})| \, \|\Psi\| d\boldsymbol{x} d\boldsymbol{k} < \infty.$$

Hence

$$\Psi_f := \int_{\mathbb{R}^{2N}} f(\boldsymbol{x}, \boldsymbol{k}) U(\boldsymbol{x}, \boldsymbol{k}) \Psi d\boldsymbol{x} d\boldsymbol{k} \in \mathscr{H}$$

is defined, where the integral on the right hand side is taken in the sense of strong Riemann integral. Let

$$\mathscr{D}_{\mathrm{G}} := \mathrm{span}\{\Psi_f \,|\, f \in C_0^\infty(\mathbb{R}^{2N}),\ \Psi \in \mathscr{H}\}. \tag{2.140}$$

We first show that $\mathscr{D}_{\mathrm{G}}$ is dense in $\mathscr{H}$.

Let $\Psi \in \mathscr{H}$ and $g \in C_0^\infty(\mathbb{R}^{2N})$ be such that $g \geq 0$, $\int_{\mathbb{R}^{2N}} g(\boldsymbol{x}, \boldsymbol{k}) d\boldsymbol{x} d\boldsymbol{k} = 1$ and define f_n ($n \in \mathbb{N}$) by

$$f_n(\boldsymbol{x}, \boldsymbol{k}) := n^{2N} g(n\boldsymbol{x}, n\boldsymbol{k}), \quad (\boldsymbol{x}, \boldsymbol{k}) \in \mathbb{R}^{2N}.$$

Then $f_n \in C_0^\infty(\mathbb{R}^{2N})$ and $\int_{\mathbb{R}^{2N}} f_n(\boldsymbol{x}, \boldsymbol{k}) d\boldsymbol{x} d\boldsymbol{k} = 1$. Hence

$$\Psi_{f_n} - \Psi = \int_{\mathbb{R}^{2N}} f_n(\boldsymbol{x}, \boldsymbol{k})(U(\boldsymbol{x}, \boldsymbol{k}) - 1)\Psi d\boldsymbol{x} d\boldsymbol{k},$$

which implies that

$$\begin{aligned} \|\Psi_{f_n} - \Psi\| &\leq \int_{\mathbb{R}^{2N}} f_n(\boldsymbol{x}, \boldsymbol{k}) \|U(\boldsymbol{x}, \boldsymbol{k})\Psi - \Psi\| d\boldsymbol{x} d\boldsymbol{k} \\ &= \int_{\mathbb{R}^{2N}} g(\boldsymbol{x}, \boldsymbol{k}) \|U(\boldsymbol{x}/n, \boldsymbol{k}/n)\Psi - \Psi\| d\boldsymbol{x} d\boldsymbol{k}. \end{aligned}$$

We have

$$\lim_{n\to\infty} g(\boldsymbol{x},\boldsymbol{k})\|U(\boldsymbol{x}/n,\boldsymbol{k}/n)\Psi-\Psi\|=0,$$

$$|g(\boldsymbol{x},\boldsymbol{k})\|U(\boldsymbol{x}/n,\boldsymbol{k}/n)\Psi-\Psi\|\le 2g(\boldsymbol{x},\boldsymbol{k})\|\Psi\|$$

and $\int_{\mathbb{R}^{2N}} g(\boldsymbol{x},\boldsymbol{k})d\boldsymbol{x}d\boldsymbol{k} < \infty$. Hence, by the Lebesgue dominated convergence theorem, we obtain $\lim_{n\to\infty}\|\Psi_{f_n}-\Psi\|=0$, i.e., $\lim_{n\to\infty}\Psi_{f_n}=\Psi$. Since Ψ_{f_n} is in $\mathscr{D}_{\mathrm{G}}$, this result means that $\mathscr{D}_{\mathrm{G}}$ is dense in $\mathscr{H}$.

Let $\Psi_f\in\mathscr{D}_{\mathrm{G}}$. Then, by the boundedness of e^{itP_j}, we have for all $t\in\mathbb{R}$

$$\begin{aligned} e^{itP_j}\Psi_f &= \int_{\mathbb{R}^{2N}} f(\boldsymbol{x},\boldsymbol{k})e^{itP_j}U(\boldsymbol{x},\boldsymbol{k})\Psi d\boldsymbol{x}d\boldsymbol{k} \\ &= \int_{\mathbb{R}^{2N}} f(\boldsymbol{x},\boldsymbol{k})e^{ix_1P_1}\cdots e^{i(x_j+t)P_j}\cdots e^{ix_NP_N}e^{ik_1Q_1}\cdots e^{ik_NQ_N}\Psi d\boldsymbol{x}d\boldsymbol{k} \\ &= \int_{\mathbb{R}^{2N}} f(\boldsymbol{x}-t\boldsymbol{e}_j,\boldsymbol{k})U(\boldsymbol{x},\boldsymbol{k})\Psi d\boldsymbol{x}d\boldsymbol{k}, \end{aligned} \tag{2.141}$$

where $(\boldsymbol{e}_j)_{j=1}^N$ is the standard basis of $\mathbb{R}^N$ (see (1.83)). Hence $e^{itP_j}\Psi_f\in\mathscr{D}_{\mathrm{G}}$. Therefore $e^{itP_j}\mathscr{D}_{\mathrm{G}}\subset\mathscr{D}_{\mathrm{G}}$.

With regard to Q_j, using (2.135) and (2.136), we have for all $t\in\mathbb{R}$

$$e^{itQ_j}U(\boldsymbol{x},\boldsymbol{k})=e^{-itx_j}U(\boldsymbol{x},\boldsymbol{k}+t\boldsymbol{e}_j).$$

Hence, in the same way as above, one can show that $e^{itQ_j}\mathscr{D}_{\mathrm{G}}\subset\mathscr{D}_{\mathrm{G}}$.

We next show that $\mathscr{D}_{\mathrm{G}}\subset D(Q_j)\cap D(P_j)$, $j=1,\ldots,N$, $Q_j\mathscr{D}_{\mathrm{G}}\subset\mathscr{D}_{\mathrm{G}}$ and $P_j\mathscr{D}_{\mathrm{G}}\subset\mathscr{D}_{\mathrm{G}}$. Let $\Psi_f\in\mathscr{D}_{\mathrm{G}}$. Then, by (2.141), we have for $t\neq 0$

$$\frac{1}{t}(e^{itP_j}-1)\Psi_f=\int_{\mathbb{R}^{2N}} F_t(\boldsymbol{x},\boldsymbol{k})U(\boldsymbol{x},\boldsymbol{k})\Psi d\boldsymbol{x}d\boldsymbol{k}$$

with

$$F_t(\boldsymbol{x},\boldsymbol{k}):=\frac{f(\boldsymbol{x}-t\boldsymbol{e}_j,\boldsymbol{k})-f(\boldsymbol{x},\boldsymbol{k})}{t}.$$

We have

$$\lim_{t\to 0}F_t(\boldsymbol{x},\boldsymbol{k})=-\partial_{x_j}f(\boldsymbol{x},\boldsymbol{k}),$$

where $\partial_{x_j} := \partial/\partial x_j$. There is a positive constant R such that $\text{supp}\, f \subset \{(\boldsymbol{x}, \boldsymbol{k}) \in \mathbb{R}^{2N} \mid |\boldsymbol{x}| \le R,\ |\boldsymbol{k}| \le R\}$. Then $\text{supp}\, F_t \subset K := \{(\boldsymbol{x}, \boldsymbol{k}) \in \mathbb{R}^{2N} \mid |\boldsymbol{x}| \le R + \delta,\ |\boldsymbol{k}| \le R\}$ for all $|t| < \delta$ (a positive constant). Hence

$$\left\| \frac{1}{t}(e^{itP_j} - 1)\Psi_f + \Psi_{\partial_{x_j} f} \right\| \le \left(\int_K |F_t(\boldsymbol{x}, \boldsymbol{k}) + \partial_{x_j} f(\boldsymbol{x}, \boldsymbol{k})| d\boldsymbol{x} d\boldsymbol{k} \right) \|\Psi\|$$

It is easy to see that $|F_t(\boldsymbol{x}, \boldsymbol{k})| \le \|\partial_{x_j} f\|_\infty := \sup_{(\boldsymbol{x}, \boldsymbol{k}) \in \mathbb{R}^{2N}} |\partial_{x_j} f(\boldsymbol{x}, \boldsymbol{k})|$ and hence $|F_t(\boldsymbol{x}, \boldsymbol{k}) + \partial_{x_j} f(\boldsymbol{x}, \boldsymbol{k})| \le 2\|\partial_{x_j} f\|_\infty$, which is integrable on K. Therefore, by the Lebesgue dominated convergence theorem, we obtain

$$\lim_{t \to 0} \left\| \frac{1}{t}(e^{itP_j} - 1)\Psi_f + \Psi_{\partial_{x_j} f} \right\| = 0,$$

i.e.

$$\lim_{t \to 0} \frac{1}{t}(e^{itP_j} - 1)\Psi_f = -\Psi_{\partial_{x_j} f}.$$

This means that $\Psi_f \in D(P_j)$ and $iP_j\Psi_f = -\Psi_{\partial_{x_j} f}$, i.e.

$$P_j\Psi_f = i\Psi_{\partial_{x_j} f} \in \mathscr{D}_\mathrm{G}. \tag{2.142}$$

Thus $\mathscr{D}_\mathrm{G} \subset D(P_j)$ and $P_j\mathscr{D}_\mathrm{G} \subset \mathscr{D}_\mathrm{G}$.

In the same way as in the case of P_j, we have

$$\frac{1}{t}(e^{itQ_j} - 1)\Psi_f = \int_{\mathbb{R}^{2N}} G_t(\boldsymbol{x}, \boldsymbol{k}) U(\boldsymbol{x}, t) \Psi d\boldsymbol{x} d\boldsymbol{k} \quad (t \ne 0)$$

with

$$G_t(\boldsymbol{x}, \boldsymbol{k}) := \frac{1}{t}(e^{-itx_j} f(\boldsymbol{x}, \boldsymbol{k} - t\boldsymbol{e}_j) - f(\boldsymbol{x}, \boldsymbol{k})).$$

It is easy to see that

$$\lim_{t \to 0} G_t(\boldsymbol{x}, \boldsymbol{k}) = -\partial_{k_j} f(\boldsymbol{x}, \boldsymbol{k}) - ix_j f(\boldsymbol{x}, \boldsymbol{k}).$$

Hence, in the same way as in the case of P_j, one can show that

$$\lim_{t \to 0} \frac{1}{t}(e^{itQ_j} - 1)\Psi_f = -\Psi_{\partial_{k_j} f} - i\Psi_{x_j f},$$

implying that $\Psi_f \in D(Q_j)$ and

$$Q_j \Psi_f == i\Psi_{\partial_{k_j} f} - \Psi_{x_j f} \in \mathscr{D}_{\mathrm{G}}. \tag{2.143}$$

Thus $\mathscr{D}_{\mathrm{G}} \subset D(Q_j)$ and $Q_j \mathscr{D}_{\mathrm{G}} \subset \mathscr{D}_{\mathrm{G}}$.

It follows from (2.142) and (2.143) that, for $j, k = 1, \ldots, N$,

$$[Q_j, P_k] = i\delta_{jk}, \quad [Q_j, Q_k] = 0, \quad [P_j, P_k] = 0 \quad \text{on} \mathscr{D}_{\mathrm{G}}.$$

Thus $(\mathscr{H}, \mathscr{D}_{\mathrm{G}}, (Q_1, \ldots, Q_N, P_1, \ldots, P_N)) \in \mathsf{CCR}_{\mathrm{sa}}(N)$. ∎

The dense subspace $\mathscr{D}_{\mathrm{G}}$ defined by (2.140) is called the **Gårding domain** associated with the self-adjoint operators $(Q_1, \ldots, Q_N, P_1, \ldots, P_N)$.

Corollary 2.4 *Under the same assumption as in Theorem 2.16, $\mathscr{D}_G$ is a common core for Q_j and P_j, $j = 1, \ldots, N$.*

Proof This follows from (2.138) and Theorem 1.7. ∎

Corollary 2.5 *Let $Q = (Q_1, \ldots, Q_N)$ and $P = (P_1, \ldots, P_N)$ be two N-tuples of self-adjoint operators on a Hilbert space $\mathscr{H}$ satisfying (2.135) and (2.136). Then, for $j, k = 1, \ldots, N$,*

$$[Q_j, P_k] = i\delta_{jk} \quad \text{on } D(Q_j P_k) \cap D(P_k Q_j), \tag{2.144}$$

$$[Q_j, Q_k] = 0 \quad \text{on } D(Q_j Q_k) \cap D(Q_k Q_j), \tag{2.145}$$

$$[P_j, P_k] = 0 \quad \text{on } D(P_j P_k) \cap D(P_k P_j). \tag{2.146}$$

Proof This follows from Theorem 2.16 and Proposition 2.2. ∎

2.8.3 Weyl Representation of CCR

Based on Theorem 2.16, we define a subclass of $\mathsf{CCR}_{\mathrm{sa}}(N)$:

Definition 2.7 Let $(Q_1, \ldots, Q_N)$ and $(P_1, \ldots, P_N)$ be respectively N-tuples of self-adjoint operators on $\mathscr{H}$.

(i) $(\mathscr{H}, (Q_1, \ldots, Q_N, P_1, \ldots, P_N))$ is called a **Weyl representation** of the CCR with N degrees of freedom if $(Q_1, \ldots, Q_N, P_1, \ldots, P_N)$ obeys the Weyl relations (2.135) and (2.136) with N degrees of freedom. We denote by $\mathsf{CCR}_{\mathrm{W}}(N)$ the set of all Weyl representations of the CCR with N degrees of freedom.

(ii) A Weyl representation $(\mathscr{H}, (Q_1, \ldots, Q_N, P_1, \ldots, P_N))$ is said to be **irre-duci-ble** if $\{e^{itQ_j}, e^{itP_j} | t \in \mathbb{R}, j = 1, \ldots, N\}$ is irreducible.

(iii) A self-adjoint representation $\pi_N = (\mathscr{H}, \mathscr{D}, (Q_1, \dots, Q_N, P_1, \dots, P_N))$ of CCR is said to be a Weyl representation of the CCR with N degrees of freedom and written $\pi_N \in \mathsf{CCR}_{\mathrm{W}}(N)$ if $(\mathscr{H}, (Q_1, \dots, Q_N, P_1, \dots, P_N))$ is a Weyl representation of the CCR with N degrees of freedom.

It follows from this definition and Theorem 2.16 that

$$\boxed{\mathsf{CCR}_{\mathrm{W}}(N) \subset \mathsf{CCR}_{\mathrm{sa}}(N) \subset \mathsf{CCR}(N).} \tag{2.147}$$

2.8.4 Spectral Properties

Concerning Weyl representations, we first discuss spectral properties of their representatives. There is a general fact:

Lemma 2.19 *Let A and B be self-adjoint operators on a Hilbert space $\mathscr{H}$ satisfying*

$$e^{itA}e^{isB} = e^{-icts}e^{isB}e^{itA}, \quad s, t \in \mathbb{R} \tag{2.148}$$

with some non-zero constant $c \in \mathbb{R} \setminus \{0\}$. Then, for all $t \in \mathbb{R}$,

$$e^{itA}Be^{-itA} = B - ct. \tag{2.149}$$

Moreover, the following hold:

(i) $\sigma(A) = \sigma(B) = \mathbb{R}$.
(ii) *If $\mathscr{H}$ is separable, then $\sigma_{\mathrm{p}}(A) = \sigma_{\mathrm{p}}(B) = \emptyset$.*

Proof

(i) Relation (2.148) is equivalent to

$$e^{itA}e^{isB}e^{-itA} = e^{is(B-ct)}.$$

By the unitary covariance of functional calculus (Theorem B.3 in Appendix B), the left hand side is equal to $\exp(ise^{itA}Be^{-itA})$. Hence (2.149) follows.

Since e^{itA} is unitary and the spectrum of a linear operator is unitarily invariant, we obtain

$$\sigma(B) = \sigma(B - ct) = \{\lambda - ct | \lambda \in \sigma(B)\},$$

where we have used (1.10) to obtain the second equality. Since $t \in \mathbb{R}$ is arbitrary and $\sigma(B) \neq \emptyset$, it follows that

$$\sigma(B) = \cup_{t \in \mathbb{R}}\{\lambda - ct | \lambda \in \sigma(B)\} = \mathbb{R}.$$

Relation (2.148) is equivalent to

$$e^{isB}e^{it(-A)} = e^{-icts}e^{it(-A)}e^{isB}, \quad s, t \in \mathbb{R}.$$

Hence, by the preceding result, we have $\sigma(-A) = \mathbb{R}$. Therefore, by (1.11), $\mathbb{R} = \sigma(-(-A)) = \sigma(A)$.

(ii) Suppose that B has an eigenvalue $b \in \mathbb{R}$ and Ψ_b is a unit eigenvector of B with eigenvalue b: $B\Psi_b = b\Psi_b$, $\|\Psi_b\| = 1$. Then, by (2.149), we have

$$B\Phi_t = (b - ct)\Phi_t, \quad \Phi_t := e^{-itA}\Psi_b.$$

Hence $b - ct$ is an eigenvalue of B with Φ_t being an eigenvector belonging to $b - ct$ and, if $t \neq s$ ($s \in \mathbb{R}$), then Φ_t is orthogonal to Φ_s. Hence $\{\Phi_t\}_{t\in\mathbb{R}}$ is an ONS of $\mathscr{H}$. But every ONS of a separable Hilbert space is at most countable. Hence we arrive at a contradiction. Thus B has no eigenvalues. Similarly, it follows that A has no eigenvalues. ■

Theorem 2.17 *Let* $(\mathscr{H}, (Q_1, \ldots, Q_N, P_1, \ldots, P_N)) \in \mathsf{CCR}_{\mathrm{W}}(N)$. *Then:*

(i)

$$\sigma(Q_j) = \sigma(P_j) = \mathbb{R}, \quad j = 1, \ldots, N. \tag{2.150}$$

(ii) *If $\mathscr{H}$ is separable, then* $\sigma_{\mathrm{p}}(Q_j) = \sigma_{\mathrm{p}}(P_j) = \emptyset$, $j = 1, \ldots, N$.

Proof One needs only to apply Lemma 2.19 to the case $A = Q_j$, $B = P_j$, $c = 1$. ■

2.8.5 Irreducibility

We next consider irreducibility of a Weyl representation.

Proposition 2.17 *A Weyl representation* $(\mathscr{H}, (Q_1, \ldots, Q_N, P_1, \ldots, P_N))$ *is irreduci-ble if and only if* $\{e^{itQ_j}, e^{itP_j} | j = 1, \ldots, N\}' = \mathbb{C}I$.

Proof Suppose that the set $\mathfrak{W} := \{e^{itQ_j}, e^{itP_j} | j = 1, \ldots, N\}$ of unitary operators is irreducible. Since $(e^{itQ_j})^* = e^{-itQ_j}$ and $(e^{itP_j})^* = e^{-itP_j}$, it follows that $\mathfrak{W}$ is $*$-invariant. Hence, by Proposition 2.9(ii), $\mathfrak{W}' = \mathbb{C}I$. Conversely, suppose that $\mathfrak{W}' = \mathbb{C}I$. Then, by Proposition 2.9(i), $\mathfrak{W}$ is irreducible. ■

The following proposition is harmonious:

Proposition 2.18 *A Weyl representation* $(\mathscr{H}, (Q_1, \ldots, Q_N, P_1, \ldots, P_N))$ *is irreduci-ble if and only if* $\{Q_j, P_j | j = 1, \ldots, N\}' = \mathbb{C}I$.

Proof This follows from Theorem 2.5 and Corollary 2.1. ■

2.8.6 Unitary Invariance

The concept of Weyl representation of CCR is an unitarily invariant one in the sense that a unitary transformation of each Weyl representation of CCR is a Weyl representation of CCR:

Proposition 2.19 *Let* $(\mathscr{H}, (Q_1, \ldots, Q_N, P_1, \ldots, P_N)) \in \mathsf{CCR}_{\mathrm{W}}(N)$ *and* U *be a unitary transformation from* $\mathscr{H}$ *to a Hilbert space* $\mathscr{K}$. *Then* $(\mathscr{K}, (UQU^{-1}, UPU^{-1})) \in \mathsf{CCR}_{\mathrm{W}}(N)$, *where, for an* N*-tuple* $A = (A_1, \ldots, A_N)$ *of operators on* $\mathscr{H}$,

$$UAU^{-1} := (UA_1U^{-1}, \ldots, UA_NU^{-1}).$$

Proof Let $Q'_j := UQ_jU^{-1}$ and $P'_j := UP_jU^{-1}$ $(j = 1, \ldots, N)$. Then Q'_j and P'_j are self-adjoint and, by the unitary covariance of functional calculus, we have

$$Ue^{itQ_j}U^{-1} = e^{itQ'_j}, \quad Ue^{isP_j}U^{-1} = e^{isP'_j}, \quad s, t \in \mathbb{R}.$$

Hence, for $j, k = 1, \ldots, N$,

$$e^{itQ'_j}e^{isP'_k} = Ue^{itQ_j}e^{isP_j}U^{-1} = Ue^{-ist\delta_{jk}}e^{isP_j}e^{itQ_j}U^{-1} = e^{-ist\delta_{jk}}e^{isP'_k}e^{itQ'_j}.$$

Hence (2.135) with (Q_j, P_k) replaced by (Q'_j, P'_k) holds. Similarly, one can prove (2.136) with (Q_j, P_k) replaced by (Q'_j, P'_k). ■

Example 2.10 Let us show that the Schrödinger representation $\pi_{\mathrm{S}}^{(d)}$ is an irreducible Weyl representation. The irreducibility of $\{e^{it\hat{q}_j}, e^{it\hat{p}_j} | t \in \mathbb{R}, j = 1, \ldots, d\}$ follows from Theorem 2.9. To prove the Weyl relations with d degrees of freedom, we recall that, for all $s \in \mathbb{R}$ and $k = 1, \ldots, d$,

$$(e^{is\hat{p}_k}f)(\boldsymbol{x}) = f(\boldsymbol{x} + s\boldsymbol{e}_k), \quad f \in L^2(\mathbb{R}^d), \text{ a.e.} \boldsymbol{x} \in \mathbb{R}^d.$$

Hence

$$(e^{it\hat{q}_j}e^{is\hat{p}_k}f)(\boldsymbol{x}) = e^{itx_j}f(\boldsymbol{x} + s\boldsymbol{e}_k), \quad (e^{is\hat{p}_k}e^{it\hat{q}_j}f)(\boldsymbol{x}) = e^{it(\boldsymbol{x}+s\boldsymbol{e}_k)_j}f(\boldsymbol{x} + s\boldsymbol{e}_k).$$

This implies that $(e^{it\hat{q}_j}e^{is\hat{p}_k}f)(\boldsymbol{x}) = e^{-its\delta_{jk}}(e^{is\hat{p}_k}e^{it\hat{q}_j}f)(\boldsymbol{x})$. Hence operator equality

$$e^{it\hat{q}_j}e^{is\hat{p}_k} = e^{-its\delta_{jk}}e^{is\hat{p}_k}e^{it\hat{q}_j}$$

follows. Similarly, one can easily show that

$$e^{it\hat{q}_j}e^{is\hat{q}_k} = e^{is\hat{q}_k}e^{it\hat{q}_j}, \quad e^{it\hat{p}_j}e^{is\hat{p}_k} = e^{is\hat{p}_k}e^{it\hat{p}_j}.$$

Thus

$$\pi_{\mathrm{S}}^{(d)} \in \mathsf{CCR}_{\mathrm{W}}(d).$$

Then, by Theorem 2.12 and Proposition 2.19, the BHJ representation $\pi_{\mathrm{BHJ}}^{(d)}$ is an irreducible Weyl representation of CCR:

$$\pi_{\mathrm{BHJ}}^{(d)} \in \mathsf{CCR}_{\mathrm{W}}(d).$$

There exists a non-Weyl representation of CCR:

Example 2.11 By Example 2.9, $\pi_L^{(1)} \in \mathsf{CCR}_{\mathrm{sa}}(1)$ is not a Weyl representation: $\pi_L^{(1)} \notin \mathsf{CCR}_{\mathrm{W}}(1)$. A physical interpretation of this fact will be discussed in Example 2.12.

2.8.7 Real Linear Combinations of Representatives

It may be interesting to consider real linear combinations of representatives in a Weyl representation.

Theorem 2.18 *Let* $(\mathscr{H}, (Q_1, \ldots, Q_N, P_1, \ldots, P_N)) \in \mathsf{CCR}_{\mathrm{W}}(N)$ *and, for* $\boldsymbol{a} = (a_1, \ldots, a_N)$, $\boldsymbol{b} = (b_1, \ldots, b_N) \in \mathbb{R}^N$, *define*

$$Q(\boldsymbol{a}) := \sum_{j=1}^{N} a_j Q_j, \quad P(\boldsymbol{b}) := \sum_{j=1}^{N} b_j P_j.$$

Then $Q(\boldsymbol{a}) + P(\boldsymbol{b})$ *is essentially self-adjoint on* $\mathscr{D}_{\mathrm{G}}$ *and*

$$e^{it\overline{(Q(\boldsymbol{a})+P(\boldsymbol{b}))}} = e^{it^2\boldsymbol{a}\cdot\boldsymbol{b}/2} e^{ita_1Q_1} \cdots e^{ita_NQ_N} e^{itb_1P_1} \cdots e^{itb_NP_N}, \tag{2.151}$$

where $\boldsymbol{a} \cdot \boldsymbol{b} := \sum_{j=1}^{N} a_j b_j$.

Proof The symmetry of $Q(\boldsymbol{a})$ and $P(\boldsymbol{b})$ is obvious. For each $t \in \mathbb{R}$, let

$$U(t) := e^{it^2\boldsymbol{a}\cdot\boldsymbol{b}/2} e^{ita_1Q_1} \cdots e^{ita_NQ_N} e^{itb_1P_1} \cdots e^{itb_NP_N}.$$

Then it is easy to see that $U(t))$ is a unitary operator and strongly continuous in $t \in \mathbb{R}$. Moreover, using the Weyl relations, one can show that

$$U(t+s) = U(t)U(s), \quad t, s \in \mathbb{R}.$$

Hence $\{U(t)\}_{t\in\mathbb{R}}$ is a strongly continuous one-parameter unitary group. Therefore, by Stone's theorem (Theorem 1.6), there exists a unique self-adjoint operator S on $\mathscr{H}$ such that $U(t) = e^{itS}$. Since $U(t)\mathscr{D}_G \subset \mathscr{D}_G$, it follows from Corollary 1.2 that S is essentially self-adjoint on $\mathscr{D}_G$. On the other hand, it is easy to see that, for all $\Psi \in \mathscr{D}_G$,

$$S\Psi = (-i)\lim_{t\to 0}\frac{1}{t}(U(t)\Psi - \Psi) = (Q(\boldsymbol{a}) + P(\boldsymbol{b}))\Psi.$$

Thus $Q(\boldsymbol{a}) + P(\boldsymbol{b})$ is essentially self-adjoint on $\mathscr{D}_G$ and (2.151) holds. ■

Remark 2.15 One can consider a set of generalized Weyl relations:

$$e^{itA_j}e^{isA_k} = e^{-itsa_{jk}}e^{isA_k}e^{itA_j}, \quad s, t \in \mathbb{R}, \quad j = 1, \ldots, N,$$

where each A_j is a self-adjoint operator on a Hilbert space $\mathscr{H}$ and $a = (a_{jk})$ is an anti-symmetric real matrix. For these operators, one can obtain generalized results of those of Weyl representations. See [22] for further details.

Let $(\mathscr{H}, (Q_1, \ldots, Q_N, P_1, \ldots, P_N)) \in \mathsf{CCR}_W(N)$ and $\boldsymbol{a}, \boldsymbol{b} \in \mathbb{R}^N$. Then, by Theorem 2.18,

$$X(\boldsymbol{a}, \boldsymbol{b}) := \overline{Q(\boldsymbol{a}) + P(\boldsymbol{b})} \tag{2.152}$$

is self-adjoint.

Corollary 2.6 *For all $\boldsymbol{a}, \boldsymbol{b}, \boldsymbol{c}, \boldsymbol{d} \in \mathbb{R}^N$ and $s, t \in \mathbb{R}$,*

$$e^{itX(\boldsymbol{a},\boldsymbol{b})}e^{isX(\boldsymbol{c},\boldsymbol{d})} = e^{-its(\boldsymbol{a}\cdot\boldsymbol{d}-\boldsymbol{b}\cdot\boldsymbol{c})}e^{isX(\boldsymbol{c},\boldsymbol{d})}e^{itX(\boldsymbol{a},\boldsymbol{b})}. \tag{2.153}$$

In particular, if

$$\boldsymbol{a}\cdot\boldsymbol{d} - \boldsymbol{b}\cdot\boldsymbol{c} = 1, \tag{2.154}$$

then $(\mathscr{H}, (X(\boldsymbol{a}, \boldsymbol{b}), X(\boldsymbol{c}, \boldsymbol{d})))$ is in $\mathsf{CCR}_W(1)$.

Proof By (2.151), we have

$$\begin{aligned} e^{itX(\boldsymbol{a},\boldsymbol{b})}e^{isX(\boldsymbol{c},\boldsymbol{d})} &= e^{it^2\boldsymbol{a}\cdot\boldsymbol{b}/2}e^{is^2\boldsymbol{c}\cdot\boldsymbol{d}/2}e^{ita_1Q_1}\cdots e^{ita_NQ_N}e^{itb_1P_1}\cdots e^{itb_NP_N} \\ &\quad \times e^{isc_1Q_1}\cdots e^{isc_NQ_N}e^{isd_1P_1}\cdots e^{isd_NP_N}. \end{aligned}$$

By using the Weyl relations for $e^{i\alpha Q_j}$ and $e^{i\beta P_j}$ $(\alpha, \beta \in \mathbb{R})$, one can show that

$$\begin{aligned} &e^{itb_1P_1}\cdots e^{itb_NP_N}e^{isc_1Q_1}\cdots e^{isc_NQ_N}e^{isd_1P_1}\cdots e^{isd_NP_N} \\ &= e^{its\boldsymbol{b}\cdot\boldsymbol{c}}e^{isc_1Q_1}\cdots e^{isc_NQ_N}e^{isd_1P_1}\cdots e^{isd_NP_N}e^{itb_1P_1}\cdots e^{itb_NP_N}. \end{aligned}$$

Similarly,

$$
\begin{aligned}
&e^{ita_1Q_1}\cdots e^{ita_NQ_N}e^{isc_1Q_1}\cdots e^{isc_NQ_N}e^{isd_1P_1}\cdots e^{isd_NP_N}\\
&=e^{-its\boldsymbol{a}\cdot\boldsymbol{d}}e^{itc_1Q_1}\cdots e^{itc_NQ_N}e^{itd_1P_1}\cdots e^{itd_NP_N}e^{ita_1Q_1}\cdots e^{ita_NQ_N}.
\end{aligned}
$$

Combining these equations, we obtain (2.153). ■

By Corollary 2.6, we have a Weyl representation $(\mathscr{H},(X(\boldsymbol{a},\boldsymbol{b}),X(\boldsymbol{c},\boldsymbol{d})))$ under condition (2.154). Let us examine irreducibility of this Weyl representation.

Proposition 2.20 *Assume that a Weyl representation $(\mathscr{H},(Q_1,\ldots,Q_N,P_1,\ldots,P_N))\in \mathsf{CCR}_{\mathrm{W}}(N)$ is irreducible.*

(i) *Let $N=1$ and $ad-bc=1$ $(a,b,c,d\in\mathbb{R})$. Then $(\mathscr{H},(X(a,b),X(c,d)))$ is irreducible.*
(ii) *Let $N\geq 2$ and (2.154) be satisfied. Then $(\mathscr{H},(X(\boldsymbol{a},\boldsymbol{b}),X(\boldsymbol{c},\boldsymbol{d})))$ is not irreducible.*

Proof

(i) We have

$$X(a,b)=\overline{aQ_1+bP_1},\quad X(c,d)=\overline{cQ_1+dP_1}.$$

Let $T\in\{X(a,b),X(c,d)\}'$ (the strong commutant of the set $\{X(a,b),X(c,d)\}$), i.e. $T\in\mathfrak{B}(\mathscr{H})$, $TX(a,b)\subset X(a,b)T$, $TX(c,d)\subset X(c,d)T$. Let $\mathscr{D}_G$ be the Gårding domain associated with (Q_1,P_1). Then, for all $\Psi,\Phi\in\mathscr{F}_G$,

$$
\begin{aligned}
\langle\Phi,T(aQ_1+bP_1)\Psi\rangle&=\langle(aQ_1+bP_1)\Phi,T\Psi\rangle,\\
\langle\Phi,T(cQ_1+dP_1)\Psi\rangle&=\langle(cQ_1+dP_1)\Phi,T\Psi\rangle.
\end{aligned}
$$

Hence

$$
\begin{aligned}
a\langle\Phi,TQ_1\Psi\rangle+b\langle\Phi,TP_1\Psi\rangle&=a\langle Q_1\Phi,T\Psi\rangle+b\langle P_1\Phi,T\Psi\rangle,\\
c\langle\Phi,TQ_1\Psi\rangle+d\langle\Phi,TP_1\Psi\rangle&=c\langle Q_1\Phi,T\Psi\rangle+d\langle P_1\Phi,T\Psi\rangle.
\end{aligned}
$$

By condition $ad-bc=1$, we obtain

$$\langle\Phi,TQ_1\Psi\rangle=\langle Q_1\Phi,T\Psi\rangle,\quad \langle\Phi,TP_1\Psi\rangle=\langle P_1\Phi,T\Psi\rangle.$$

Since $\mathscr{D}_G$ is a common core for Q_1 and P_1 by Corollary 2.4, it follows that $T\Psi\in D(Q_1)\cap D(P_1)$ and $TQ_1\Psi=Q_1T\Psi$, $TP_1\Psi=P_1T\Psi$. Using the fact just mentioned again, we obtain $TQ_1\subset Q_1T$ and $TP_1\subset P_1T$. Since $\{Q_1,P_1\}'=\mathbb{C}I$, it follows that $T=\alpha I$ with some $\alpha\in\mathbb{C}$. Hence

$\{X(a,b), X(c,d)\}' = \mathbb{C}I$. Thus, by Proposition 2.18, the Weyl representation $(\mathscr{H}, (X(a,b), X(c,d))$ is irreducible.

(ii) Let $N \geq 2$. Under the condition (2.154), the two vectors $\boldsymbol{u} := (\boldsymbol{a}, \boldsymbol{b})$ and $\boldsymbol{v} := (\boldsymbol{c}, \boldsymbol{d})$ in $\mathbb{R}^{2N}$ are linearly independent. Since $2N \geq 4$ for $N \geq 2$, there exists a non-zero vector $\boldsymbol{w} = (\boldsymbol{w}_1, \boldsymbol{w}_2) \in \mathbb{R}^{2N}$ which is orthogonal to $\boldsymbol{u}$ and $\boldsymbol{v}$. Then, for all $s, t \in \mathbb{R}$, $e^{isX(\boldsymbol{w}_2, -\boldsymbol{w}_1)}$ commutes with $e^{itX(\boldsymbol{a},\boldsymbol{b})}$ and $e^{itX(\boldsymbol{c},\boldsymbol{d})}$ (use (2.153)). By an application of Lemma 2.19, we have $\sigma(X(\boldsymbol{w}, -\boldsymbol{w}_1)) = \mathbb{R}$. Hence it follows from the spectral mapping theorem that, if $s \neq 0$, then $\sigma(e^{isX(\boldsymbol{w}_2, -\boldsymbol{w}_1)}) = \{e^{ist} | t \in \mathbb{R}\} = \{z \in \mathbb{C} | |z| = 1\}$. Hence, for $s \neq 0$, $e^{isX(\boldsymbol{w}_2, -\boldsymbol{w}_1)}$ is not a scalar operator. Therefore $\{e^{itX(\boldsymbol{a},\boldsymbol{b})}, e^{itX(\boldsymbol{c},\boldsymbol{d})}\}' \neq \mathbb{C}I$. Since $\{e^{itX(\boldsymbol{a},\boldsymbol{b})}, e^{itX(\boldsymbol{c},\boldsymbol{d})}\}$ is $*$-invariant, it follows from Proposition 2.9(ii) that $\{e^{itX(\boldsymbol{a},\boldsymbol{b})}, e^{itX(\boldsymbol{c},\boldsymbol{d})}\}$ is not irreducible. ■

A simple application of Theorem 2.18 to Schrödinger representations leads one to the following fact:

Corollary 2.7 *Let $\hat{q}_j$ and $\hat{p}_j$ be the j-th position and the j-th momentum operator on $L^2(\mathbb{R}^d)$ defined by (2.35) and (1.79) respectively. Then, for all $\boldsymbol{a} = (a_1, \ldots, a_d) \in \mathbb{R}^d$ and $\boldsymbol{b} = (b_1, \ldots, b_d) \in \mathbb{R}^d$, the operator*

$$S(\boldsymbol{a}, \boldsymbol{b}) := \sum_{j=1}^{d} (a_j \hat{q}_j + b_j \hat{p}_j)$$

is essentially self-adjoint on $C_0^\infty(\mathbb{R}^d)$ and, for all $t \in \mathbb{R}$,

$$e^{it\overline{S(\boldsymbol{a},\boldsymbol{b})}} = e^{it^2 \boldsymbol{a}\cdot\boldsymbol{b}/2} e^{ita_1\hat{q}_1} \cdots e^{ita_d\hat{q}_d} e^{itb_1\hat{p}_1} \cdots e^{itb_d\hat{p}_d}. \tag{2.155}$$

Proof By Example 2.10, $\pi_{\mathrm{S}}^{(d)}$ is a Weyl representation. Hence, by Theorem 2.18, $\overline{S(\boldsymbol{a}, \boldsymbol{b})}$ is self-adjoint and (2.155) holds. Note that $e^{itb_j\hat{p}_j}$ and $e^{ita_j\hat{q}_j}$ leave $C_0^\infty(\mathbb{R}^d)$ invariant. Hence $e^{it\overline{S(\boldsymbol{a},\boldsymbol{b})}} C_0^\infty(\mathbb{R}^d) \subset C_0^\infty(\mathbb{R}^d)$. Therefore, by Corollary 1.2, $\overline{S(\boldsymbol{a}, \boldsymbol{b})}$ is essentially self-adjoint on $C_0^\infty(\mathbb{R}^d)$. It is obvious that $\overline{S(\boldsymbol{a}, \boldsymbol{b})} = S(\boldsymbol{a}, \boldsymbol{b})$ on $C_0^\infty(\mathbb{R}^d)$. Thus $S(\boldsymbol{a}, \boldsymbol{b})$ is essentially self-adjoint on $C_0^\infty(\mathbb{R}^d)$. ■

2.8.8 Von Neumann's Uniqueness Theorem and Remarks

For Weyl representations of CCR on *separable* Hilbert spaces, there exists a uniqueness theorem:

Theorem 2.19 (von Nemann's Uniqueness Theorem) *Let $\mathscr{H}$ be a separable Hilbert space and $(\mathscr{H}, (Q_1, \ldots, Q_N, P_1, \ldots, P_N)) \in \mathsf{CCR}_{\mathrm{W}}(N)$. Then there exist* *mutually orthogonal closed subspaces* $\mathscr{H}_m$ *of* $\mathscr{H}$ *(*$m = 1, \ldots, M$ *with* $M < \infty$ *or* $M = \infty$*) such that the following hold:*

(i) $\mathscr{H} = \oplus_{m=1}^{M} \mathscr{H}_m$.
(ii) *For each* $j = 1, \ldots, N$, Q_j *and* P_j *are reduced by* $\mathscr{H}_m$.
(iii) *Let* $Q_j^{(m)}$ *(resp.* $P_j^{(m)}$*) be the reduced part of* Q_j *(resp.* P_j*) to* $\mathscr{H}_m$*. Then, for each* m*, there exists a unitary transformation* $U_m \colon \mathscr{H}_m \to L^2(\mathbb{R}^N)$ *such that*

$$U_m Q_j^{(m)} U_m^{-1} = \hat{q}_j, \quad U_m P_j^{(m)} U_m^{-1} = \hat{p}_j, \quad j = 1, \ldots, N,$$

where $(\hat{q}_1, \ldots, \hat{q}_N)$ *and* $(\hat{p}_1, \ldots, \hat{p}_N)$ *are the representatives in the Schrödinger representation* $\pi_{\mathrm{S}}^{(N)}$.

We do not give a proof to the theorem here, since it takes much space and there are textbooks in which the theorem is proved (e.g., [129, Chapter IV], [126, Theorem 1.2], [16, §3.5] or [151, §3.2]; also see the original paper [160]).

Theorem 2.19 means that every Weyl representation with N degrees of freedom on a separable Hilbert space is equivalent to a direct sum of the Schrödinger representation $\pi_{\mathrm{S}}^{(N)}$:

$$\mathsf{CCR}_{\mathrm{W}}(N) \ni (\mathscr{H}, (Q_1, \ldots, Q_N, P_1, \ldots, P_N)) \cong \oplus_{m=1}^{M} \pi_{\mathrm{S}}^{(N)}.$$

Theorem 2.19 and Example 2.10 imply the following result:

Corollary 2.8 *Let* $\pi_{\mathrm{W}} \in \mathsf{CCR}_{\mathrm{W}}(N)$ *be an irreducible Weyl representation on a separable Hilbert space. Then* $\pi_{\mathrm{W}} \cong \pi_{\mathrm{S}}^{(N)}$*, i.e., every irreducible Weyl representation of the CCR with* N *degrees of freedom on a separable Hilbert space is equivalent to the Schrödinger representation* $\pi_{\mathrm{S}}^{(N)}$.

Thus, as far as Weyl representations of CCR on separable Hilbert spaces are concerned, the classification problem is completely solved. But we emphasize that, with regard to *non-Weyl representations* of CCR (i.e. elements in $\cup_{N=1}^{\infty}(\mathsf{CCR}(N) \setminus \mathsf{CCR}_{\mathrm{W}}(N))$), this is not the case.

Example 2.12 As we have already seen, for each $L > 0$, $\pi_L^{(1)}$ is an irreducible self-adjoint representation of the CCR with one degree of freedom which is *inequivalent* to the Schrödinger one $\pi_{\mathrm{S}}^{(1)}$. Note that (2.137) shows that it is not a Weyl representation. It is obvious that the breakdown of the Weyl relation with degree one comes from the multiplication unitary operator e^{itLn_s} $(t, s \in \mathbb{R})$ which is not always the identity. On the other hand, the integer-valued function n_s has a physical meaning. In the present system, the periodic boundary condition is imposed on the domain of the momentum operator $\hat{p}_L$. Hence the configuration space $\mathbb{I}_L$ may be identified with a circle C_L with the length L in circumference. Then, for each point $x \in C_L$ and $s > 0$ (resp. $s < 0$), $n_s(x)$ means the rotation number of a point which starts at x and moves on C_L by length $|s|$ anti-clockwise (resp.

clockwise). Therefore, to the breakdown of the Weyl relation in $\pi_L^{(1)}$, a physical situation corresponds. This is an interesting structure.

For all $k \in \Gamma_L$, we have $e^{ikLn_s} = 1$. Hence

$$e^{ik\hat{q}_L}e^{is\hat{p}_L} = e^{-iks}e^{is\hat{p}_L}e^{ik\hat{q}_L},$$

which means that the subset $\{e^{ik\hat{q}_L}, e^{is\hat{p}_L}|k \in \Gamma_L,\ s \in \mathbb{R}\}$ $(\subset \{e^{it\hat{q}_L}, e^{is\hat{p}_L}|s, t \in \mathbb{R}\})$ obeys a restricted Weyl relation. This also may be interesting to note.

Remark 2.16 It is known that there is a class of representations of Weyl-CCR algebras which are not regular, but may be useful in analyzing structural properties of "nonregular" quantum systems [1, 2]. In [2], a nonregular representation of an extended algebra of $\{e^{ik\hat{q}_L}, e^{is\hat{p}_L}|k \in \Gamma_L, s \in \mathbb{R}\}$ is given. It may be interesting to ask how such a nonregular representation is related to the non-Weylness of $\pi_L^{(1)}$.

To study non-Weyl representations, one should keep the following fact in mind:

Proposition 2.21 *Let $\pi_N \in \mathsf{CCR}(N) \setminus \mathsf{CCR}_{\mathrm{W}}(N)$. Then π_N is not equivalent to any direct sum representation of the Schrödinger representation $\pi_{\mathrm{S}}^{(N)}$.*

Proof If $\pi_N \in \mathsf{CCR}(N) \setminus \mathsf{CCR}_{\mathrm{sa}}(N)$ (i.e., π_N is a non-self-adjoint representation of CCR), then it is obvious that π_N cannot be equivalent to any direct sum of $\pi_{\mathrm{S}}^{(N)}$. So let $\pi_N \in \mathsf{CCR}_{\mathrm{sa}}(N) \setminus \mathsf{CCR}_{\mathrm{W}}(N)$. If π_N were equivalent to a direct sum of $\pi_{\mathrm{S}}^{(N)}$, then, by Proposition 2.19 and Example 2.10, the representatives in π_N satisfy (2.135) and (2.136). But this is a contradiction. ■

Remark 2.17 Other conditions than Weyl relations for a representation of CCR $\in \mathsf{CCR}(N)$ to be equivalent to a direct sum of the Schrödinger representation $\pi_{\mathrm{S}}^{(N)}$ are known. See, e.g., [129, Chapter IV] and references therein.

We give a simple application of Corollary 2.7:

Example 2.13 Consider $(L^2(\mathbb{R}), (\hat{q}, \hat{p})) \in \pi_{\mathrm{S}}^{(1)}$, the Schrödinger representation of the CCR with one degree of freedom. Let $a, b, c, d \in \mathbb{R}$ such that $ad - bc = 1$ and

$$Q := a\hat{q} + b\hat{p}, \quad P := c\hat{q} + d\hat{p}.$$

Then, by Proposition 2.20(i), $\pi_{a,b,c,d} := (L^2(\mathbb{R}), C_0^\infty(\mathbb{R}), (\overline{Q}, \overline{P}))$ is an irreducible Weyl representation with one degree of freedom. Hence, by the von Neumann uniqueness theorem, $\pi_{a,b,c,d} \cong \pi_{\mathrm{S}}^{(1)}$, i.e., there exists a unitary operator U on $L^2(\mathbb{R})$ such that

$$U\hat{q}U^{-1} = \overline{a\hat{q} + b\hat{p}}, \quad U\hat{p}U^{-1} = \overline{c\hat{q} + d\hat{p}}. \tag{2.156}$$

Hence the operator

$$H(a, b, c, d) := \frac{1}{2m}\left(\overline{c\hat{q} + d\hat{p}}\right)^2 + V\left(\overline{a\hat{q} + b\hat{p}}\right)$$

with a constant $m > 0$ and $V: \mathbb{R} \to \mathbb{R} \cup \{\pm\infty\}$ (Borel measurable) is unitarily equivalent to the Schrödinger operator H_S defined by (2.48) with $d = 1$. In particular, if $V(x) = m\omega^2 x^2/2$, $x \in \mathbb{R}$ ($\omega > 0$ is a constant), then $H(a, b, c, d)$ is unitarily equivalent to the one-dimensional quantum harmonic oscillator Hamiltonian H_{os} given by (2.104). Hence, in this case, the spectrum of $H(a, b, c, d)$ is the same as that of H_{os}.

A typical example of (a, b, c, d) is given as follows:

$$a = d = \cosh\lambda, \quad b = c = \sinh\lambda, \quad \lambda \in \mathbb{R}.$$

Remark 2.18 The correspondence $(\hat{q}, \hat{p}) \mapsto (a\hat{q} + b\hat{p}, c\hat{q} + d\hat{p})$ is called a Bogoliubov transformation with degree one. Equation (2.156) shows that this transformation is unitarily implementable. In Sect. 8.6, we shall discuss Bogoliubov transformations in an abstract framework.

2.9 Weak Form of CCR

In this section, we derive a weak form of CCR and apply it to derive an uncertainty relation strengthened in a sense. We first prove a lemma, which will be used later too:

Lemma 2.20 *Let A be a self-adjoint operator on a Hilbert space and B be a symmetric operator on $\mathscr{H}$. Suppose that there exists a $\mathfrak{B}(\mathscr{H})$-valued function K: $\mathbb{R} \to \mathfrak{B}(\mathscr{H})$; $\mathbb{R} \ni t \mapsto K(t) \in \mathfrak{B}(\mathscr{H})$ such that K is strongly differentiable on $\mathbb{R}$ and*

$$e^{itA}B \subset (B + K(t))e^{itA}, \quad t \in \mathbb{R}. \tag{2.157}$$

Then, $K(0) = 0$ and, for all $\Psi, \Phi \in D(A) \cap D(B)$,

$$\langle A\Psi, B\Phi\rangle - \langle B\Psi, A\Phi\rangle = -i\left\langle \Psi, K'(0)\Phi\right\rangle, \tag{2.158}$$

where $K'(t) := dK(t)/dt$ denotes the strong derivative of $K(t)$ in t.

Proof Putting $t = 0$ in (2.157), one obtains $K(0) = 0$. Let $\Psi, \Phi \in D(A) \cap D(B)$. Then, by (2.157),

$$\left\langle e^{-itA}\Psi, B\phi\right\rangle = \left\langle B\Psi, e^{itA}\Phi\right\rangle + \left\langle \Psi, K(t)e^{itA}\Phi\right\rangle, \quad t \in \mathbb{R}. \tag{2.159}$$

The $\mathscr{H}$-valued functions $e^{-itA}\Psi$, $e^{itA}\Phi$ and $K(t)e^{itA}\Phi$ of t are strongly differentiable in t with the strong derivatives

$$\frac{d}{dt}e^{-itA}\Psi = -iAe^{-itA}\Psi, \quad \frac{d}{dt}e^{itA}\Phi = iAe^{itA}\Phi,$$

$$\frac{d}{dt}K(t)e^{itA}\Phi = K'(t)e^{itA}\Phi + iK(t)Ae^{itA}\Phi.$$

Hence, differentiating in t both sides of (2.159) at $t = 0$, we obtain (2.158). ■

Lemma 2.21 *Let A be a self-adjoint operator on a Hilbert space $\mathscr{H}$ and B be a symmetric operator on $\mathscr{H}$ such that*

$$e^{itA}B \subset (B - ct)e^{itA}, \quad t \in \mathbb{R}, \tag{2.160}$$

where $c \in \mathbb{R}$ is a constant. Then, for all $\Psi, \Phi \in D(A) \cap D(B)$,

$$\langle A\Psi, B\Phi\rangle - \langle B\Psi, A\Phi\rangle = ic\,\langle \Psi, \Phi\rangle\,, \tag{2.161}$$

Proof One needs only to apply Lemma 2.20 with $K(t) = -ct$. ■

Now we can apply Lemma 2.21 to Weyl representations:

Theorem 2.20 *Let $(\mathscr{H}, (Q_1, \ldots, Q_N, P_1, \ldots, P_N)) \in \mathsf{CCR}_\mathrm{W}(N)$. Then:*

(i) *For all $t \in \mathbb{R}$ and $j, k = 1, \ldots, N$,*

$$e^{itQ_j}P_k e^{-itQ_j} = P_k - \delta_{jk}t, \tag{2.162}$$

$$e^{itQ_j}Q_k e^{-itQ_j} = Q_k, \tag{2.163}$$

$$e^{itP_j}P_k e^{-itP_j} = P_k. \tag{2.164}$$

(ii) *For each (j, k) $(j, k = 1, \ldots, N)$,*

$$\left\langle Q_j\Psi, P_k\Phi\right\rangle - \left\langle P_k\Psi, Q_j\Phi\right\rangle = i\delta_{jk}\,\langle \Psi, \Phi\rangle\,,\ \ \Psi, \Phi \in D(Q_j) \cap D(P_k), \tag{2.165}$$

$$\left\langle Q_j\Psi, Q_k\Phi\right\rangle - \left\langle Q_k\Psi, Q_j\Phi\right\rangle = 0, \quad \Psi, \Phi \in D(Q_j) \cap D(Q_k), \tag{2.166}$$

$$\left\langle P_j\Psi, P_k\Phi\right\rangle - \left\langle P_k\Psi, P_j\Phi\right\rangle = 0, \quad \Psi, \Phi \in D(P_j) \cap D(P_k). \tag{2.167}$$

Proof

(i) Formulas (2.162)–(2.164) follow from Lemma 2.19.
(ii) This part follows from (i) and Lemma 2.21. ■

Corollary 2.9 *Let $(\mathscr{H}, (Q_1, \ldots, Q_N, P_1, \ldots, P_N)) \in \mathsf{CCR}_\mathrm{W}(N)$. Then, for each $j = 1, \ldots, N$ and all unit vectors $\Psi \in D(Q_j) \cap D(P_j)$,*

$$(\Delta Q_j)_\Psi(\Delta P_j)_\Psi \geq \frac{1}{2}. \tag{2.168}$$

Proof This follows from (2.165) with $k = j$ and Lemma 2.3. ■

The uncertainty relation (2.168) may be regarded as a stronger form of (2.23) in the sense that it holds for state vectors in $D(Q_j) \cap D(P_j)$ which is larger than $\mathscr{D}$ in (2.23) or $D(Q_j P_j) \cap D(P_j Q_j)$.

For symmetric operators Q_j and P_j $(j = 1, \ldots, N)$, we call relations (2.165)–(2.167) the **weak form of the CCR** with N degrees of freedom.

Example 2.14 In Example 2.10, we have seen that the Schrödinger representation $\pi_{\mathrm{S}}^{(d)}$ is a Weyl representation. Hence it follows from Theorem 2.20 and Corollary 2.9 that, for all $j, k = 1, \ldots, d$,

$$\langle \hat{q}_j f, \hat{p}_k g \rangle - \langle \hat{p}_k f, \hat{q}_j g \rangle = i\delta_{jk} \langle f, g \rangle, \quad f, g \in D(\hat{q}_j) \cap D(\hat{p}_k),$$

$$\langle \hat{q}_j f, \hat{q}_k g \rangle - \langle \hat{q}_k f, \hat{q}_j g \rangle = 0, \quad f, g \in D(\hat{q}_j) \cap D(\hat{q}_k),$$

$$\langle \hat{p}_j f, \hat{p}_k g \rangle - \langle \hat{p}_k f, \hat{p}_j g \rangle = 0, \quad f, g \in D(\hat{p}_j) \cap D(\hat{p}_k)$$

and, for all unit vectors $f \in D(\hat{q}_j) \cap D(\hat{p}_j)$,

$$(\Delta \hat{q}_j)_f (\Delta \hat{p}_j)_f \geq \frac{1}{2}.$$

2.10 Weak Weyl Representations

There is a concept of representation of CCR weaker than the Weyl representation and stronger than the ordinary one. This is obtained by noting Theorem 2.20(i), which is derived from Weyl relations.

Definition 2.8 Let H_j be a self-adjoint operator and T_j be a symmetric operator $(j = 1, \ldots, N)$ on a Hilbert space on $\mathscr{H}$ and $\mathscr{D}$ be a *dense* subspace of $\mathscr{H}$ such that, for $j = 1, \ldots, N$, $\mathscr{D} \subset \cap_{j,k=1}^{N} D(T_j T_k) \cap D(T_k T_j)$. Then the following operator relations (2.169)–(2.171) are called the **weak Weyl relations** with N degrees of freedom for $(T_1, \ldots, T_N, H_1, \ldots, H_N)$:

$$e^{itH_k} T_j e^{-itH_k} = T_j + t\delta_{jk}, \tag{2.169}$$

$$e^{itH_k} e^{isH_j} = e^{isH_j} e^{itH_k}, \tag{2.170}$$

$$[T_j, T_k] = 0 \quad \text{on } \mathscr{D} \quad (t, s \in \mathbb{R},\ j, k = 1, \ldots, N). \tag{2.171}$$

Remark 2.19

(1) In the case $N = 1$, there is no need to refer to the subspace $\mathscr{D}$. Namely, we say that a self-adjoint operator H and a symmetric operator T on $\mathscr{H}$ obey the weak Weyl relation with one degree of freedom if

$$e^{itH} T e^{-itH} = T + t, \quad \forall t \in \mathbb{R}. \tag{2.172}$$

(2) Equation (2.170) is equivalent to that $(H_1, \dots, H_N)$ is strongly commuting. Hence, by Proposition 1.19, for $j, k = 1, \dots, N$, H_j and H_k commute:

$$[H_j, H_k] = 0 \quad \text{on } D(H_j H_k) \cap D(H_k H_j). \tag{2.173}$$

Remark 2.20 If a symmetric operator T_j satisfies (2.169), then so does the closure $\overline{T}_j$ of T_j:

$$e^{itH_k}\overline{T}_j e^{-itH_k} = \overline{T}_j + t\delta_{jk}.$$

Lemma 2.22 *Let H_j be a self-adjoint operator and T_j be a symmetric operator $(j = 1, \dots, N)$ on a Hilbert space $\mathscr{H}$. Suppose that*

$$e^{itH_k} T_j \subset (T_j + t\delta_{jk})e^{itH_k} \quad (t \in \mathbb{R},\ j, k = 1, \dots, N). \tag{2.174}$$

Then (2.169) holds.

Proof Condition (2.174) implies that, for all $t \in \mathbb{R}$, $e^{itH_k} D(T_j) \subset D(T_j)$ and

$$e^{itH_k} T_j \Psi = (T_j + t\delta_{jk})e^{itH_k}\Psi, \quad \Psi \in D(T) \in \mathbb{R},\ j, k = 1, \dots, N.$$

By an application of Proposition 1.11 with $U = e^{itH_k}$, one obtains $e^{itH_k} D(T_j) = D(T_j)$. Hence operator equality

$$e^{itH_k} T_j = (T_j + t\delta_{jk})e^{itH_k} \tag{2.175}$$

follows. This gives (2.169). ∎

In applications, it may be easier to check (2.174) than to show (2.169) directly.

Proposition 2.22 *Suppose that H_k and T_j $(j, k = 1, \dots, N)$ satisfy (2.169). Then, for all $\Psi \in D(H_k T_j) \cap D(H_k)$, $H_k\Psi \in D(\overline{T}_j)$ and*

$$\overline{T}_j H_k \Psi - H_k T_j \Psi = i\delta_{jk}\Psi. \tag{2.176}$$

In particular,

$$[T_j, H_k] = i\delta_{jk} \quad \textit{on } D([T_j, H_k]). \tag{2.177}$$

Proof Let $\Psi \in D(H_k T_j) \cap D(H_k)$. Then, for all $t \in \mathbb{R}$, $T_j e^{itH_k} = \Phi(t)$ with $\Phi(t) := e^{itH_k} T_j \Psi - t\delta_{jk} e^{itH_k}\Psi$. The vector-valued function $\Phi(t)$ is strongly differentiable in t with strong derivative $\Phi'(t) = ie^{itH_k} H_k T_j \Psi - \delta_{jk} e^{itH_k}\Psi - it\delta_{jk} e^{itH_k} H_k \Psi$. In particular, $\Phi'(0) = iH_k T_j \Psi - \delta_{jk}\Psi$. Hence $T_j e^{itH_k}\Psi$ is strongly differentiable in t. It follows that that $H_k\Psi \in D(\overline{T}_j)$ and $i\overline{T}_j H_k \Psi = \Phi'(0)$. This gives (2.176). ∎

Formulae (2.171), (2.173) and (2.177) show that $(T_1, \ldots, T_N, H_1, \ldots, H_N)$ obeys the CCR with N degrees of freedom on the subspace $\mathscr{D}\cap(\cap_{j,k=1}^N D([H_j, H_k])$ $\cap D([T_j, H_k]))$. Based on this property, we call $(\mathscr{H}, \mathscr{D}, (T_1, \ldots, T_N, H_1, \ldots, H_N))$ in Definition 2.8 a **weak Weyl representation** of the CCR with degree $N \geq 2$.

A weak Weyl representation of the CCR with degree one is defined to be $(\mathscr{H}, (T, H))$ with T a symmetric operator on $\mathscr{H}$ and H a self-adjoint operator on $\mathscr{H}$ such that

$$e^{itH}Te^{-itH} = T + t, \quad t \in \mathbb{R}. \tag{2.178}$$

We denote by $\mathsf{CCR}_{\mathrm{WW}}(N)$ the set of all weak Weyl representations of the CCR with degree N.

A simple example of a weak Weyl representation which is not a Weyl representation is given as follows.

Example 2.15 We consider the case $\mathscr{H} = L^2([a, \infty))$ with $a \in \mathbb{R}$. Let $\hat{q}_a$ and $\hat{p}_a$ be as in Sect. 2.6.5. Then, for all $f \in D(\hat{p}_a) = C_0^1(R, \infty)$ and $t \in \mathbb{R}$, $(e^{-it\hat{q}_a} f)(r) = e^{-itr} f(r),\ r \in [a, \infty)$. Hence $e^{-it\hat{q}_a} f \in D(\hat{p}_a)$ and

$$\hat{p}_a e^{-it\hat{q}_a} f = -te^{-itr} f - ie^{-itr} f' = -te^{-it\hat{q}_a} f + e^{-it\hat{q}_a}\hat{p}_a f.$$

Therefore

$$e^{it\hat{q}_a}(-\hat{p}_a)e^{-it\hat{q}_a} f = (-\hat{p}_a + t) f.$$

Hence it follows that

$$e^{it\hat{q}_a}(-\hat{p}_a)e^{-it\hat{q}_a} = (-\hat{p}_a) + t.$$

Taking the closure of both sides, we obtain

$$e^{it\hat{q}_a} T_a e^{-it\hat{q}_a} = T_a + t, \quad T_a := -\overline{\hat{p}_a}.$$

Thus

$$\pi_{\mathrm{W},a} := (L^2([a, \infty)), (T_a, \hat{q}_a)) \tag{2.179}$$

is a weak Weyl representation of the CCR with degree one. As we already know, the closed symmetric operator T_a is not self-adjoint. Hence $\pi_{\mathrm{W},a}$ is not a Weyl representation.

It should be remarked that it is not clear if the subspace $\mathscr{D}\cap(\cap_{j,k=1}^N D([H_j, H_k])\cap$ $D([T_j, H_k]))$, which is a CCR-domain for $(T_1, \ldots, T_N, H_1, \ldots, H_N)$ as we have seen above, is a non-zero subspace. On this aspect, we want to show that there exists a dense CCR-domain for $(\overline{T}_1, \ldots, \overline{T}_N, H_1, \ldots, H_N)$. Such a CCR-domain is constructed in analogy with the Gårding domain $\mathscr{D}_{\mathrm{G}}$ defined by (2.140) (see below).

Let $(\mathscr{H}, \mathscr{D}, (T_1, \ldots, T_N, H_1, \ldots, H_N))$ be a weak Weyl representation of the CCR with degree $N \geq 2$. For each $f \in C_0^\infty(\mathbb{R}^N)$ and $\Psi \in \mathscr{H}$, one can define a vector $\Psi_f \in \mathscr{H}$ by

$$\Psi_f := \int_{\mathbb{R}^N} f(t_1, \ldots, t_N) e^{it_1 H_1} \cdots e^{it_N H_N} \Psi dt_1 \cdots dt_N,$$

where the integral on the right hand side is taken in the sense of strong Riemann integral. Let

$$\mathscr{D}_{\boldsymbol{H}} := \text{span}\{\Psi_f | f \in C_0^\infty(\mathbb{R}^N), \Psi \in \mathscr{H}\} \tag{2.180}$$

with $\boldsymbol{H} := (H_1, \ldots, H_N)$. Then, in the same way as in the case of the Gårding domain $\mathscr{D}_{\mathrm{G}}$, one can prove the following lemma:

Lemma 2.23 *The subspace $\mathscr{D}_{\boldsymbol{H}}$ is dense in $\mathscr{H}$. Moreover, the following hold:*

(i) *For all $t \in \mathbb{R}$ and $j = 1, \ldots, N$, $e^{itH_j} \mathscr{D}_{\boldsymbol{H}} = \mathscr{D}_{\boldsymbol{H}}$.*
(ii) *For all $f \in C_0^\infty(\mathbb{R}^N)$, $\Psi \in \mathscr{H}$ and $j = 1, \ldots, N$, $\Psi_f \in D(H_j)$ and*

$$H_j \Psi_f = i \Psi_{\partial_j f}. \tag{2.181}$$

In particular, $H_j \mathscr{D}_{\boldsymbol{H}} \subset \mathscr{D}_{\boldsymbol{H}}$.

Proof One needs only to mimic the arguments leading to (2.142). ∎

We introduce a subspace

$$\mathscr{E}_{\boldsymbol{T},\boldsymbol{H}} := \text{span}\{\Psi_f | f \in C_0^\infty(\mathbb{R}^N), \Psi \in \mathscr{D}\}. \tag{2.182}$$

It is obvious that

$$\mathscr{E}_{\boldsymbol{T},\boldsymbol{H}} \subset \mathscr{D}_{\boldsymbol{H}}.$$

In the same way as in the case of the Gårding domain $\mathscr{D}_{\mathrm{G}}$, one can prove the following:

(TH.1) $\mathscr{E}_{\boldsymbol{T},\boldsymbol{H}}$ is dense in $\mathscr{H}$.[17]
(TH.2) For all $t \in \mathbb{R}$ and $j = 1, \ldots, N$, $e^{itH_j} \mathscr{E}_{\boldsymbol{T},\boldsymbol{H}} = \mathscr{E}_{\boldsymbol{T},\boldsymbol{H}}$. Moreover $H_j \mathscr{E}_{\boldsymbol{T},\boldsymbol{H}} \subset \mathscr{E}_{\boldsymbol{T},\boldsymbol{H}}$.
(TH.3) For $j = 1, \ldots, N$ and all $\Psi_f \in \mathscr{E}_{\boldsymbol{T},\boldsymbol{H}}$, $\Psi_f \in D(\overline{T}_j)$ and

$$\overline{T}_j \Psi_f = (T_j \Psi)_{\partial_j f} - \Psi_{t_j f}, \tag{2.183}$$

where formula (2.169) is used.[18]

---

[17] The density of $\mathscr{D}$ also is used.

[18] Since one needs a limiting argument to prove this fact, the closure $\overline{T}_j$ of T_j appears on the left hand side of (2.183).

Property (TH.2) implies that $\mathscr{E}_{\boldsymbol{T},\boldsymbol{H}} \subset \cap_{j,k=1}^N D([H_j, H_k])$. It follows from (TH.2), (TH.3) and Lemma 2.23 that

$$\mathscr{E}_{\boldsymbol{T},\boldsymbol{H}} \subset \cap_{j,k=1}^N (D([\overline{T}_j, \overline{T}_k]) \cap D([\overline{T}_j, H_k])).$$

Therefore

$$\mathscr{E}_{\boldsymbol{T},\boldsymbol{H}} \subset \cap_{j,k=1}^N (D([H_j, H_k]) \cap (D([\overline{T}_j, \overline{T}_k]) \cap D([\overline{T}_j, H_k]))).$$

Thus $(\mathscr{H}, \mathscr{E}_{\boldsymbol{T},\boldsymbol{H}}, (\overline{T}_1, \ldots, \overline{T}_N, H_1, \ldots, H_N))$ with $N \geq 2$ is a representation of the CCR with N degrees of freedom with a dense CCR-domain.

As for a weak Weyl representation of the CCR with degree one, we have the following:

Proposition 2.23 *Let $(\mathscr{H}, (T, H))$ be a weak Weyl representation of the CCR with degree one and*

$$\mathscr{D}_{T,H} := \text{span}\{\Psi_f | f \in f \in C_0^\infty(\mathbb{R}), \Psi \in D(T)\},$$

where $\Psi_f := \int_{\mathbb{R}} f(t) e^{itH} \Psi dt$. Then $\mathscr{D}_{T,H}$ is dense in $\mathscr{H}$ and $\mathscr{D}_{T,H}$ is a CCR-domain for $(\overline{T}, H)$, i.e., $\mathscr{D}_{T,H} \subset D(\overline{T}H) \cap D(H\overline{T})$ and $[\overline{T}, H] = i$ on $\mathscr{D}_{T,H}$. In particular, $D(\overline{T}H) \cap D(H\overline{T})$ is dense in $\mathscr{H}$ and

$$[\overline{T}, H] = i \quad \text{on } D(\overline{T}H) \cap D(H\overline{T}). \tag{2.184}$$

Proof In the case $N = 1$, we need only to take $\mathscr{D} = D(T)$ in the definition $\mathscr{E}_{\boldsymbol{T},\boldsymbol{H}}$, because only one T is involved. Then the arguments in the case $N \geq 2$ work without essential changes. Commutation relation (2.184) follows from the density of $\mathscr{D}_{T,H}$ and Proposition 2.2. ■

By Theorem 2.20 and (2.147), every Weyl representation of the CCR with degree N is a weak Weyl representation of the CCR with degree N (but the converse is not true). Thus we have the following inclusion relations:

$$\boxed{\mathsf{CCR}_{\mathrm{W}}(N) \subset \mathsf{CCR}_{\mathrm{WW}}(N) \subset \mathsf{CCR}(N).}$$

Remark 2.21 Needless to say, a non-self-adjoint weak Weyl representation $(\mathscr{H}, (T_1, \ldots, T_N, H_1, \ldots, H_N))$ of the CCR with degree N (i.e., at least one T_j is not self-adjoint) cannot be a Weyl representation of the CCR with degree N.

The next proposition answers the question regarding when a weak Weyl representation is a Weyl representation:

Proposition 2.24 *Let* $(\mathscr{H}, \mathscr{D}, (T_1, \ldots, T_N, H_1, \ldots, H_N)) \in \mathsf{CCR}_{\mathrm{WW}}(N)$ *and suppose that* $(T_1, \ldots, T_N)$ *is an* N*-tuple of strongly commuting self-adjoint operators on* $\mathscr{H}$*. Then* $(\mathscr{H}, (T_1, \ldots, T_N, H_1, \ldots, H_N)) \in \mathsf{CCR}_{\mathrm{W}}(N)$.

Proof By (2.169) and the unitary covariance of functional calculus (Theorem B.3 in Appendix B), we have

$$e^{itH_k}e^{isT_j}e^{-itH_k} = e^{is(T_j+t\delta_{jk})} = e^{isT_j}e^{ist\delta_{jk}}, \quad s, t \in \mathbb{R}.$$

Hence

$$e^{isT_j}e^{itH_k} = e^{-ist\delta_{jk}}e^{itH_k}e^{isT_j}.$$

By the strong commutativity of T_j and T_k, we have $e^{itT_j}e^{isT_k} = e^{isT_k}e^{itT_j}$, $s, t \in \mathbb{R}$. By these facts and (2.170), we see that $(T_1, \ldots, T_N, H_1, \ldots, H_N)$ obeys the Weyl relations with N degrees of freedom. Thus $(\mathscr{H}, (T_1, \ldots, T_N, H_1, \ldots, H_N))$ is a Weyl representation. ■

The next proposition formulates a necessary and sufficient condition for a self-adjoint weak Weyl representation to be equivalent to a Schrödinger representation:

Proposition 2.25 *Let* $\mathscr{H}$ *be a separable Hilbert space and* $(\mathscr{H}, \mathscr{D}, (T_1, \ldots, T_N, H_1, \ldots, H_N)) \in \mathsf{CCR}_{\mathrm{WW}}(N)$ *with* T_j *being self-adjoint* $(j = 1, \ldots, N)$*. Then* $(\mathscr{H}, (T_1, \ldots, T_N, H_1, \ldots, H_N))$ *is equivalent to a direct sum of the Schrödinger representation* $\pi_{\mathrm{S}}^{(N)}$ *if and only if* $(T_1, \ldots, T_N)$ *is strongly commuting.*

Proof Suppose that $(\mathscr{H}, (T_1, \ldots, T_N, H_1, \ldots, H_N))$ is equivalent to a direct sum of the Schrödinger representation $\pi_{\mathrm{S}}^{(N)}$. Then there exists an $M \in \mathbb{N} \cup \{\infty\}$ such that, for each $j = 1, \ldots, N$, $\oplus^M \hat{q}_j \cong T_j$. Since $(\hat{q}_1, \ldots, \hat{q}_N)$ is strongly commuting, it follows that $(T_1, \ldots, T_N)$ is strongly commuting.

Conversely, suppose that $(T_1, \ldots, T_N)$ is strongly commuting. Then, by Proposition 2.24, $(\mathscr{H}, (T_1, \ldots, T_N, H_1, \ldots, H_N))$ is a Weyl representation. Hence, by von Neumann's uniqueness theorem (Theorem 2.19), it is equivalent to a direct sum of $\pi_{\mathrm{S}}^{(N)}$. ■

Example 2.16 In Example 2.12, we have seen that $\pi_L^{(1)} \in \mathsf{CCR}_{\mathrm{sa}}(1)$ is not a Weyl representation. Hence, by Proposition 2.24, it is not a weak Weyl representation either. It may be interesting to see the "difference" from a weak Weyl representation. It follows from (2.137) that, for all $t \in \mathbb{R}$,

$$e^{it\hat{p}_L}\hat{q}_L e^{-it\hat{p}_L} = \hat{q}_L + t + Ln_{-t}.$$

This shows that the breakdown of the weak Weyl relation comes from the term Ln_{-t}, which has a physical meaning as explained in Example 2.12.

2.11 Absolute Continuity of Self-adjoint Representatives in Weak Weyl Representations

Let $(\mathscr{H}, (T_1, \ldots, T_N, H_1, \ldots, H_N)) \in \mathsf{CCR}_{\mathrm{WW}}(N)$. Then, by (2.169), for all $j, k = 1, \ldots, N$ and $t \in \mathbb{R}$,

$$e^{itH_j} T_j e^{-itH_j} = T_j + t, \tag{2.185}$$

$$e^{itH_k} T_j e^{-itH_k} = T_j, \quad j \neq k. \tag{2.186}$$

Relation (2.186) means that, for $j \neq k$, T_j strongly commutes with H_k (see Definition 1.1 and Proposition 1.20). But (2.185) shows that T_j does not strongly commute with H_j. Hence it may be interesting to develop a functional calculus for (T_j, H_j).

In what follows, for mathematical generality, we consider a slightly generalized weak Weyl relation

$$e^{itH} T e^{-itH} = T + tC, \quad t \in \mathbb{R}, \tag{2.187}$$

where T is a symmetric operator on $\mathscr{H}$, H is a self-adjoint operator on $\mathscr{H}$ and C is a bounded self-adjoint operator on $\mathscr{H}$ such that

$$CT \subset TC. \tag{2.188}$$

We first note the following fact:

Lemma 2.24

$$e^{itH} C e^{-itH} = C, \quad t \in \mathbb{R}. \tag{2.189}$$

Hence C strongly commutes with H.[19]

Proof Operator equality (2.187) is equivalent to

$$T = e^{-itH} T e^{itH} + t e^{-itH} C e^{itH}.$$

By (2.187), the first term on the right hand side is equal to $T - tC$. Hence $te^{-itH} C e^{itH} = tC$, which implies that $e^{-itH} C e^{itH} = C$ for $t \neq 0$. It is obvious that this equation holds for $t = 0$ too. Hence (2.189) follows. ■

Lemma 2.25 *The operator H is reduced by $\overline{\mathrm{Ran}\,(C)}$.*

[19] See Proposition 1.20.

Proof We denote by P the orthogonal projection onto $\overline{\mathrm{Ran}\,(C)}$. By (2.189), for all $\Psi \in \mathscr{H}$, $e^{itH}C\Psi = Ce^{itH}\Psi \in \mathrm{Ran}\,(C)$. Hence $e^{itH}\mathrm{Ran}\,(C) \subset \mathrm{Ran}\,(C)$. Therefore

$$e^{itH}\overline{\mathrm{Ran}\,(C)} \subset \overline{\mathrm{Ran}\,(C)}.$$

This implies that $Pe^{itH}P = e^{itH}P$. Taking the adjoint of this equation and replacing t with $-t$, we obtain $Pe^{itH}P = Pe^{itH}$. Hence $e^{itH}P = Pe^{itH}$. This implies that $PH \subset HP$ (see the proof of Proposition 1.19). Hence H is reduced by $\overline{\mathrm{Ran}\,(C)}$. ■

We denote by H_C the reduced part of H to $\overline{\mathrm{Ran}\,(C)}$:

$$H_C := H_{\overline{\mathrm{Ran}\,(C)}}. \tag{2.190}$$

For $u \in L^1(\mathbb{R})$ (the set of Lebesgue integrable functions on $\mathbb{R}$), we denote its Fourier transform by $\hat{u}$:

$$\hat{u}(E) := (\mathscr{F}_1 u)(E) := \frac{1}{\sqrt{2\pi}} \int_{\mathbb{R}} u(t)e^{-itE}dt, \quad E \in \mathbb{R}. \tag{2.191}$$

It is easy to see that $\hat{u}$ is bounded and

$$\|u\|_\infty \leq \frac{1}{\sqrt{2\pi}} \|u\|_{L^1(\mathbb{R})}, \tag{2.192}$$

where $\|u\|_\infty$ is the essential supremum of u (see (A.1) in Appendix A) and

$$\|u\|_{L^1(\mathbb{R})} := \int_{\mathbb{R}} |u(t)|dt,$$

the L^1-norm of u.

We introduce a subspace of $L^1(\mathbb{R})$:

$$L^1_1(\mathbb{R}) := \left\{ u \in L^1(\mathbb{R}) \cap C(\mathbb{R}) | \int_{\mathbb{R}} |t\,u(t)|dt < \infty \right\}, \tag{2.193}$$

where $C(\mathbb{R})$ is the set of continuous functions on $\mathbb{R}$. We set

$$\mathscr{M}_1 := \mathscr{F}_1(L^1_1(\mathbb{R})). \tag{2.194}$$

Since $\mathscr{S}(\mathbb{R}) \subset L^1_1(\mathbb{R})$, it follows that $\mathscr{F}_1(\mathscr{S}(\mathbb{R})) \subset \mathscr{M}_1$. It is well known that $\mathscr{F}_1(\mathscr{S}(\mathbb{R})) = \mathscr{S}(\mathbb{R})$. Hence

$$\mathscr{S}(\mathbb{R}) \subset \mathscr{M}_1. \tag{2.195}$$

The next lemma is easily proved:

Lemma 2.26 *For each $u \in L^1_1(\mathbb{R})$, $\hat{u}$ is a bounded continuously differentiable function on $\mathbb{R}$ with*

$$\hat{u}'(E) = \frac{(-i)}{\sqrt{2\pi}} \int_{\mathbb{R}} tu(t)e^{-iEt}dt$$

and $\hat{u}'$ is bounded on $\mathbb{R}$ with

$$\|\hat{u}'\|_\infty \leq \frac{1}{\sqrt{2\pi}}\|tu\|_{L^1(\mathbb{R})}.$$

By this lemma and the functional calculus for a self-adjoint operator, for each $f \in \mathscr{M}_1$, operators $f(H)$ and $f'(H)$ are bounded, i.e., $f(H), f'(H) \in \mathfrak{B}(\mathscr{H})$. The next theorem is concerned with a commutation relation of $f(H)$ and $\overline{T}$:

Theorem 2.21 *For all $f \in \mathscr{M}_1$, $f(H)D(T) \subset D(\overline{T})$ and, for all $\Psi \in D(T)$,*

$$\overline{T}f(H)\Psi - f(H)T\Psi = if'(H)C\Psi. \tag{2.196}$$

Proof By using the spectral theorem for H, we have

$$f(H)\Psi = \frac{1}{\sqrt{2\pi}} \int_{\mathbb{R}} \check{f}(t)e^{-itH}\Psi dt,$$

where $\check{f} = \mathscr{F}_1^{-1}f$ (the inverse Fourier transform of f) and the integral on the right hand side is taken in the sense of strong Riemann integral. Note that, for all $t \in \mathbb{R}$, $\check{f}(t)e^{-itH}\Psi \in D(T)$ and, by (2.172),

$$T(\check{f}(t)e^{-itH}\Psi) = \check{f}(t)e^{-itH}T\Psi + t\check{f}(t)e^{-itH}C\Psi.$$

Each term on the right hand is integrable in the sense of strong Riemann integral. Hence, by the closability of T, $f(H)\Psi \in D(\overline{T})$ and

$$\overline{T}f(H)\Psi = f(H)T\Psi + \frac{1}{\sqrt{2\pi}} \int_{\mathbb{R}} t\check{f}(t)e^{-itH}C\Psi dt.$$

It is easy to see that

$$f'(E) = -\frac{i}{\sqrt{2\pi}} \int_{\mathbb{R}} t\check{f}(t)e^{-itE}dt.$$

Hence the second term on the right hand side of the above equation is equal to $if'(H)C\Psi$. Thus (2.196) is derived. ■

Now we proceed to proving the absolute continuity of H_C.

Theorem 2.22 *The self-adjoint operator H_C is absolutely continuous.*

Proof Let $f \in \mathscr{S}(\mathbb{R})$ and $\Psi \in D(T)$. Then, by (2.188), $C\Psi \in D(T)$. By (2.195) and (2.196), we have

$$\langle TC\Psi, f(H)\Psi\rangle - \langle C\Psi, f(H)T\Psi\rangle = i\left\langle C\Psi, f'(H)C\Psi\right\rangle. \tag{2.197}$$

Let

$$\mu(\lambda) := \langle C\Psi, E_H((-\infty, \lambda])C\Psi\rangle, \quad \lambda \in \mathbb{R},$$

where E_H is the spectral measure of H, and

$$\sigma(\lambda) := \langle TC\Psi, E_H((-\infty, \lambda])\Psi\rangle - \langle C\Psi, E_H((-\infty, \lambda])T\Psi\rangle.$$

Then, it is easy to see that μ and σ are functions of bounded variation with

$$V[\mu] \le \|C\Psi\|^2, \quad V[\sigma] \le (\|TC\Psi\|\,\|\Psi\| + \|C\Psi\|\,\|T\Psi\|),$$

where $V[\mu]$ denotes the total variation of μ. By (2.197) and the spectral theorem, we have

$$\int_{\mathbb{R}} f(\lambda)d\sigma(\lambda) = i\int_{\mathbb{R}} f'(\lambda)d\mu(\lambda), \tag{2.198}$$

where the integrals are taken in the sense of Stieltjes integral. By the fact that $\text{s-}\lim_{\lambda\to\infty} E((-\infty, \lambda]) = I$ (s- lim means strong limit) and (2.188), we have

$$\lim_{\lambda\to\infty} \sigma(\lambda) = 0.$$

Since $\text{s-}\lim_{\lambda\to-\infty} E_H((-\infty, \lambda]) = 0$, it follows that $\lim_{\lambda\to-\infty}\sigma(\lambda) = 0$. Hence, applying the integration by parts formula on the Stieltjes integral, we have

$$\int_{\mathbb{R}} f(\lambda)d\sigma(\lambda) = -\int_{\mathbb{R}} f'(\lambda)\sigma(\lambda)d\lambda.$$

Hence (2.198) yields

$$-\int_{\mathbb{R}} f'(\lambda)\sigma(\lambda)d\lambda = i\int_{\mathbb{R}} f'(\lambda)d\mu(\lambda). \tag{2.199}$$

It is well known (or easy to see) that

$$\left|\int_{\mathbb{R}} f(\lambda)d\mu(\lambda)\right| \le \|f\|_\infty V[\mu], \quad \left|\int_{\mathbb{R}} f(\lambda)\sigma(\lambda)d\lambda\right| \le \|f\|_\infty V[\sigma].$$

Hence the functional $\varphi : \mathscr{S}(\mathbb{R}) \to \mathbb{C}$ defined by

$$\varphi(f) := i \int_{\mathbb{R}} f(\lambda) d\mu(\lambda) + \int_{\mathbb{R}} f(\lambda)\sigma(\lambda) d\lambda, \quad f \in \mathscr{S}(\mathbb{R}) \tag{2.200}$$

is a tempered distribution on $\mathbb{R}$. Equation (2.199) implies that $\varphi(f') = 0$ for all $f \in \mathscr{S}(\mathbb{R})$. Hence the distribution φ is identified with a constant $\alpha \in \mathbb{C}$ so that $\varphi(f) = \alpha \int_{\mathbb{R}} f(\lambda) d\lambda$ for all $f \in \mathscr{S}(\mathbb{R})$. This equation and (2.200) imply that the measure μ is absolutely continuous with its Radon–Nikodým derivative $d\mu/d\lambda$ being equal to $-i\alpha + i\sigma(\lambda)$, a.e.$\lambda$. Hence it follows that $\langle C\Psi, E_H(\cdot)C\Psi\rangle$ is absolutely continuous. Since $D(T)$ is dense in $\mathscr{H}$ and C is bounded, it follows from a limiting argument that, for all $\Phi \in \operatorname{Ran}(C)$, $\langle \Phi, E_H(\cdot)\Phi\rangle$ is absolutely continuous. By a limiting argument again, the absolute continuity of $\langle \Phi, E_H(\cdot)\Phi\rangle$ can be extended to all $\Phi \in \overline{\operatorname{Ran}(C)}$ (see Proposition D.1 in Appendix D). Thus H_C is absolutely continuous. ■

Remark 2.22 Theorem 2.22 with $C = I$ is due to Dorfmeisters [65] in the case where H is bounded and Schmüdgen [140] in the case where H is not necessarily bounded (see also [141]). Theorem 2.22 with $C \neq I$ is proved in [15]. Theorem 2.22 suggests that a (generalized) weak Weyl relation may be a relatively strong relation. We note that, in Theorem 2.22, separability of $\mathscr{H}$ is not assumed.

A simple application of Theorem 2.22 is:

Corollary 2.10 *If H is not absolutely continuous, then there are no symmetric operators T which satisfy (2.172).*

Theorem 2.22 gives an interesting result on a weak Weyl representation of CCR:

Corollary 2.11 *Let $(\mathscr{H}, \mathscr{D}, (T_1, \ldots, T_N, H_1, \ldots, H_N)) \in \mathsf{CCR}_{\mathrm{WW}}(N)$. Then, for $j = 1, \ldots, N$, H_j is absolutely continuous.*

As a corollary to Theorem 2.22, one can completely identify the spectra of the representatives of a Weyl representation of CCR:

Corollary 2.12 *Let $(\mathscr{H}, (Q_1, \ldots, Q_N, P_1, \ldots, P_N)) \in \mathsf{CCR}_{\mathrm{W}}(N)$ be a Weyl representation. Then, for all $j = 1, \ldots, N$, Q_j and P_j are absolutely continuous and*

$$\sigma(Q_j) = \sigma_{\mathrm{ac}}(Q_j) = \mathbb{R}, \quad \sigma(P_j) = \sigma_{\mathrm{ac}}(P_j) = \mathbb{R}.$$

Proof This follows from Theorems 2.20, 2.17, and 2.22. ■

2.12 Functional Calculus for a Weak Weyl Relation

For later use, we extend Theorem 2.21 to the case where f is in a more general class of functions on $\mathbb{R}$. We introduce two function spaces:

$$C_{\mathrm{b}}^{1}(\mathbb{R}) := \{f \in C^{1}(\mathbb{R}) | f \text{ and } f' \text{ are bounded}\}, \tag{2.201}$$

$$C_{\mathrm{b},+}(\mathbb{R}) := \{f \in C^{1}(\mathbb{R}) | \text{for some } a \in \mathbb{R}, \sup_{E \geq a} |f(E)| < \infty \text{ and } \sup_{E \geq a} |f'(E)| < \infty\}. \tag{2.202}$$

Theorem 2.23 *Let H be a self-adjoint operator and T be a symmetric operator on $\mathcal{H}$ satisfying (2.187) with (2.188).*

(i) *For all $f \in C_{\mathrm{b}}^{1}(\mathbb{R})$, $f(H)D(T) \subset D(\overline{T})$ and*

$$\overline{T} f(H)\Psi - f(H)T\Psi = i f'(H)C\Psi, \quad \Psi \in D(T). \tag{2.203}$$

(ii) *Let $g \in C^{1}(\mathbb{R})$ be real-valued. Then, for all $\Psi \in D(T) \cap D(g'(H))$, $e^{-itg(H)}\Psi \in D(\overline{T})$, $C\Psi \in D(g'(H))$ and*

$$\overline{T} e^{-itg(H)}\Psi - e^{-itg(H)}T\Psi = t e^{-itg(H)} g'(H)C\Psi. \tag{2.204}$$

(iii) *Suppose that H is bounded from below. Then, for all $f \in C_{\mathrm{b},+}^{1}(\mathbb{R})$, $f(H)D(T) \subset D(\overline{T})$ and (2.203) holds. In particular, for all $z \in \mathbb{C}$ with $\operatorname{Re} z > 0$, $e^{-zH}D(T) \subset D(\overline{T})$ and*

$$\overline{T} e^{-zH} - e^{-zH} T = -i z e^{-zH} C \quad \text{on } D(T). \tag{2.205}$$

Proof

(i) We first consider the case where f is in the set

$$C_{0}^{1}(\mathbb{R}) := \{f \in C^{1}(\mathbb{R}) | \operatorname{supp} f \text{ is compact}\} \subset C_{\mathrm{b}}^{1}(\mathbb{R}).$$

Using the Friedrichs mollifier, one can find a sequence $\{f_n\}_{n=1}^{\infty}$ in $C_0^{\infty}(\mathbb{R})$ such that $\sup_{n \geq 1} \|f_n\|_{\infty} < \infty$, $\sup_{n \geq 1} \|f_n'\|_{\infty} < \infty$ and $\lim_{n \to \infty} f_n(E) = f(E)$, $\lim_{n \to \infty} f_n'(E) = f'(E)$ for all $E \in \mathbb{R}$. Since $C_0^{\infty}(\mathbb{R}) \subset \mathcal{M}_1$, it follows from Theorem 2.21 that

$$\overline{T} f_n(H)\Psi - f_n(H)T\Psi = i f_n'(H)C\Psi, \quad \Psi \in D(T). \tag{2.206}$$

Note that $f_n(H)$, $f(H)$, $f_n'(H)$ and $f'(H)$ are bounded operators. By the functional calculus for H, we have

$$\text{s-}\lim_{n\to\infty} f_n(H) = f(H), \quad \text{s-}\lim_{n\to\infty} f_n'(H) = f'(H).$$

Hence $\overline{T} f_n(H)\Psi \to f(H)T\Psi + if'(H)C\Psi$ as $n \to \infty$. Therefore $f(H)\Psi \in D(\overline{T})$ and (2.203) holds.

We next consider the case where $f \in C_b^1(\mathbb{R})$. Let $\chi \in C_0^\infty(\mathbb{R})$ such that $\chi(0) = 1$ and define $f_n(E) := \chi(E/n)f(E)$, $E \in \mathbb{R}$. Then $f_n \in C_0^1(\mathbb{R})$ and

$$f_n'(E) = \frac{1}{n}\chi'(E/n)f(E) + \chi(E/n)f'(E).$$

By the preceding result, (2.206) holds for the present $\{f_n\}_n$. It is easy to see that

$$\lim_{n\to\infty} f_n(E) = f(E), \quad \lim_{n\to\infty} f_n'(E) = f'(E)$$

with

$$\|f_n\|_\infty \le \|\chi\|_\infty \|f\|_\infty, \quad \|f_n'\|_\infty \le \|\chi'\|_\infty \|f\|_\infty + \|\chi\|_\infty \|f'\|_\infty, \quad \forall n \ge 1.$$

By the functional calculus for H, we have $f_n(H) \to f(H)$ and $f_n'(H) \to f'(H)$ strongly as $n \to \infty$. Then, in the same way as in the preceding paragraph, one can obtain (2.203).

(ii) Let $\chi \in C_0^\infty(\mathbb{R})$ with $\chi(0) = 1$ and, for each $n \in \mathbb{N}$, define

$$f_n(E) := \chi\left(\frac{E}{n}\right) e^{-itg(E)}, \quad E \in \mathbb{R}.$$

Then $f_n \in C_b^1(\mathbb{R})$ with

$$f_n'(E) = \frac{1}{n}\chi'\left(\frac{E}{n}\right) e^{-itg(E)} + \chi\left(\frac{E}{n}\right)(-it)g'(E)e^{-itg(E)}.$$

Hence, by part (i), for all $\Psi \in D(T) \cap D(g'(H))$, $f_n(H)\Psi \in D(\overline{T})$ and

$$\overline{T} f_n(H)\Psi - f_n(H)T\Psi = if_n'(H)C\Psi.$$

It follows from Lemma 2.24 and the functional calculus for H that $Cg'(H) \subset g'(H)C$. Hence, for all $\eta \in D(g'(H))$, $C\eta \in D(g'(H))$ and $g'(H)C\eta = Cg'(H)\eta$. In the same way as above, one can show that

$$\lim_{n\to\infty} f_n(H)\Phi = e^{-itg(H)}\Phi, \quad \forall\Phi \in \mathcal{H},$$

$$\lim_{n\to\infty} f_n'(H)C\Psi = -ite^{-itg(H)}g'(H)C\Psi.$$

Since $\overline{T}$ is closed, it follows that $e^{-itg(H)}\Psi \in D(\overline{T})$ and (2.204) holds.

(iii) Since H is bounded from below, there is a constant $M \geq 0$ such that $H \geq -M$. Let $f \in C^1_{\mathrm{b},+}(\mathbb{R})$. There is a function $g \in C^1_{\mathrm{b}}(\mathbb{R})$ such that $f(E) = g(E)$ for all $E \geq M$. Then, by (i), $g(H)D(T) \subset D(\overline{T})$ and

$$\overline{T}g(H) - g(H)T = ig'(H)C \quad \text{on } D(T).$$

Since the support of the spectral measure of H is included in $[M, \infty)$, it follows that $f(H) = g(H)$ and $f'(H) = g'(H)$. Thus (2.203) holds.

If $\mathrm{Re}\, z > 0$, then the function f on $\mathbb{R}$ defined by

$$f(E) := e^{-zE}, \ E \in \mathbb{R}$$

is in $C^1_{\mathrm{b},+}(\mathbb{R})$ and $f(H) = e^{-zH}$. We have $f'(E) = -ze^{-zE}$. Hence the latter half of part (ii) follows. ■

Chapter 3
Aharonov–Bohm Effect and Inequivalent Representations of CCR

Abstract A two-dimensional quantum system of a charged quantum particle of charge $q \in \mathbb{R} \setminus \{0\}$ with a singular vector potential $\boldsymbol{A}$ is considered. In this system, there appears an irreducible self-adjoint representation $\pi_{\boldsymbol{A}}$ of the CCR with two degrees of freedom. A necessary and sufficient condition is formulated for $\pi_{\boldsymbol{A}}$ to be inequivalent to the Schrödinger representation $\pi_{\mathrm{S}}^{(2)}$ in terms of the magnetic flux of the system. It is shown that the case where $\pi_{\boldsymbol{A}}$ is inequivalent to $\pi_{\mathrm{S}}^{(2)}$ corresponds to the occurrence of the so-called Aharonov–Bohm effect in the present context. Moreover, inequivalence between two representations $\pi_{\boldsymbol{A}}$ and $\pi_{\boldsymbol{A}'}$ for different vector potentials $\boldsymbol{A}$ and $\boldsymbol{A}'$ is discussed. For a vector potential $\boldsymbol{A}$ fixed, $\pi_q := \pi_{\boldsymbol{A}}$ may be regarded as a representation depending on the charge q. A necessary and sufficient condition for π_q to be inequivalent to $\pi_{q'}$ ($q' \in \mathbb{R} \setminus \{0\}$) is given.

3.1 Introduction

We consider a two-dimensional system of a charged quantum particle of a non-zero charge $q \in \mathbb{R} \setminus \{0\}$ moving in the Euclidean plane $\mathbb{R}^2 = \{\boldsymbol{x} = (x_1, x_2)|x_1, x_2 \in \mathbb{R}\}$ under the influence of a classical magnetic field B perpendicular to the plane $\mathbb{R}^2$.

For mathematical generality, we assume that the magnetic field B is a distribution on $\mathbb{R}^2$, which may have isolated singularities at N points $\boldsymbol{a}_n = (a_{n1}, a_{n2}) \in \mathbb{R}^2 (n = 1, \cdots, N,\ N \in \mathbb{N})$ and is continuous on the domain (Fig. 3.1)

$$M := \mathbb{R}^2 \setminus \{\boldsymbol{a}_1, \cdots, \boldsymbol{a}_N\}, \tag{3.1}$$

and that there exists a vector potential

$$\boldsymbol{A} := (A_1, A_2) : M \to \mathbb{R}^2$$

© Springer Nature Singapore Pte Ltd. 2020
A. Arai, *Inequivalent Representations of Canonical Commutation and Anti-Commutation Relations*, Mathematical Physics Studies,
https://doi.org/10.1007/978-981-15-2180-5_3

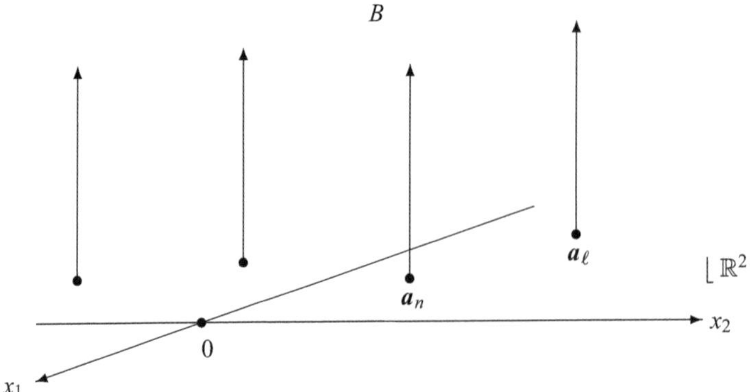

Fig. 3.1 A quantum system with a magnetic field B perpendicular to the plane $\mathbb{R}^2$

such that

$$B = D_1 A_2 - D_2 A_1, \tag{3.2}$$

where D_1 and D_2 are generalized partial (distributional) differential operators in x_1 and x_2 respectively and we assume that

$$A_j \in C(M), \quad j = 1, 2, \tag{3.3}$$

where $C(M)$ denotes the set of continuous functions on M.

An example of B we have in mind is the case where B is concentrated on the points $\boldsymbol{a}_n$ $(n = 1, \ldots, N)$ and hence B is written as a linear combination of two dimensional delta-distributions, a very singular case.

A natural Hilbert space of state vectors for the quantum system under consideration is $L^2(M)$. But the Lebesgue measure of the set $\{\boldsymbol{a}_1, \cdots, \boldsymbol{a}_N\}$ is zero. Hence one has the natural identification $L^2(M) = L^2(\mathbb{R}^2)$. We use this identification.

In what follows, by abuse of notation, we denote by A_j the multiplication operator by the function A_j, acting in $L^2(\mathbb{R}^2)$. Then the multiplication operator A_j is self-adjoint and $\hat{p}_j - qA_j$ is a symmetric operator, where $\hat{p}_j$ is the jth momentum operator acting in $L^2(\mathbb{R}^2)$.[1] One of the main objects in this chapter is the **physical** (or **magnetic**) **momentum operator** $\boldsymbol{P} := (P_1, P_2)$ defined by

$$P_j := \overline{\hat{p}_j - qA_j} \quad (j = 1, 2), \tag{3.4}$$

the closure of $\hat{p}_j - qA_j$, acting in $L^2(\mathbb{R}^2)$.

[1] We use the physical unit system where the reduced Planck constant $\hbar = 1$ and the light speed $c = 1$ unless otherwise stated.

3.2 Essential Self-adjointness of the Physical Momenta

We first prove that P_1 and P_2 can be physical quantities (observables), i.e., they are self-adjoint. For this purpose, we present a series of lemmas. For an open subset D of $\mathbb{R}^d$, we denote by $C_0^\infty(D)$ the set of infinitely differentiable functions on D with compact support in D.

Lemma 3.1 *Let F be a closed subset of $\mathbb{R}^d$ ($d \in \mathbb{N}$) such that its Lebesgue measure $|F|$ is equal to zero. Then $C_0^\infty(\mathbb{R}^d \setminus F)$ is dense in $L^2(\mathbb{R}^d)$.*

Proof Since $\mathbb{R}^d \setminus F$ is an open set, it follows a general theory that $C_0^\infty(\mathbb{R}^d \setminus F)$ is dense in $L^2(\mathbb{R}^d \setminus F)$ (e.g., apply [101, Theorem 6.6]). By the present assumption, $|F| = 0$. Hence $L^2(\mathbb{R}^2)$ is identified with $L^2(\mathbb{R}^d \setminus F)$ in the natural way. Thus the desired result holds. ■

The next lemma easily follows from an application of Lemma 3.1:

Lemma 3.2

(i) *Let $L \in \mathbb{N}$ and $c_1, \cdots, c_L \in \mathbb{R}$ be constants. Then $C_0^\infty\left(\mathbb{R} \setminus \{c_j\}_{j=1}^L\right)$ is dense in $L^2(\mathbb{R})$.*

(ii) *$C_0^\infty(M)$ is dense in $L^2(\mathbb{R}^2)$.*

Let (X, Σ_X, μ_X) and (Y, Σ_Y, μ_Y) be σ-finite measure spaces. For a Σ_X-measurable function $f\colon X \to \mathbb{C} \cup \{\pm\infty\}$ and a Σ_Y-measurable function $g\colon Y \to \mathbb{C} \cup \{\pm\infty\}$, we define a function $f \times g\colon X \times Y \to \mathbb{C} \cup \{\pm\infty\}$ by

$$(f \times g)(x, y) := f(x)g(y), \quad (x, y) \in X \times Y.$$

Let $\mathscr{D}_X$ be a subset of Σ_X-measurable functions on X and $\mathscr{D}_Y$ be a subset of Σ_Y-measurable functions on Y. Then we define a subset of functions on $X \times Y$ by

$$\mathscr{D}_X \overset{\circ}{\otimes} \mathscr{D}_Y := \mathrm{span}\{f \times g | f \in \mathscr{D}_X, \ g \in \mathscr{D}_Y\}.$$

The next lemma is a basic fact:

Lemma 3.3 *Suppose that $L^2(X, d\mu_X)$ and $L^2(Y, d\mu_Y)$ are separable and infinite dimensional. Let $\mathscr{D}_X$ be a dense subspace of $L^2(X, d\mu)$ and $\mathscr{D}_Y$ be a dense subspace of $L^2(Y, d\mu_Y)$. Then $\mathscr{D}_X \overset{\circ}{\otimes} \mathscr{D}_Y$ is dense in $L^2(X \times Y, d\mu_X \otimes d\mu_Y)$.*

Proof By the density of $\mathscr{D}_X$ in $L^2(X, d\mu)$, there exists a CONS $\{\psi_n\}_{n=1}^\infty$ of $L^2(X, d\mu_X)$ such that, for each n, $\psi_n \in \mathscr{D}_X$. Similarly, there exists a CONS $\{\phi_n\}_{n=1}^\infty$ of $L^2(Y, d\mu_Y)$ such that, for each n, $\phi_n \in \mathscr{D}_Y$. Then, by a general theorem (e.g., [32, Theorem 4.2]), $\{\psi_m \times \phi_n | m, \ n \in \mathbb{N}\}$ is a CONS of $L^2(X \times Y, d\mu_X \otimes d\mu_Y)$. Since $\psi_m \times \phi_n \in \mathscr{D}_X \overset{\circ}{\otimes} \mathscr{D}_Y$, it follows that $\mathscr{D}_X \overset{\circ}{\otimes} \mathscr{D}_Y$ is dense in $L^2(X \times Y, d\mu_X \otimes d\mu_Y)$. ■

For convenience, we introduce two subsets of $\mathbb{R}$:

$$X_{\boldsymbol{a}} := \mathbb{R} \setminus \{a_{n1}\}_{n=1}^{N}, \quad Y_{\boldsymbol{a}} := \mathbb{R} \setminus \{a_{n2}\}_{n=1}^{N}. \tag{3.5}$$

By Lemmas 3.2 and 3.3, the subspaces

$$\mathscr{D}_1 := C_0^\infty(\mathbb{R}) \mathring{\otimes} C_0^\infty(Y_{\boldsymbol{a}}), \ \mathscr{D}_2 := C_0^\infty(X_{\boldsymbol{a}}) \mathring{\otimes} C_0^\infty(\mathbb{R}) \tag{3.6}$$

are dense in $L^2(\mathbb{R}^2)$.

Let

$$M_1 := \{\boldsymbol{x} = (x_1, x_2) \in \mathbb{R}^2 | x_2 \in Y_{\boldsymbol{a}}\}, \quad M_2 := \{\boldsymbol{x} = (x_1, x_2) \in \mathbb{R}^2 | x_1 \in X_{\boldsymbol{a}}\}$$

as in Figs. 3.2 and 3.3. Then it is easy to see that

$$\mathscr{D}_1 \subset C_0^\infty(M_1), \quad \mathscr{D}_2 \subset C_0^\infty(M_2).$$

Lemma 3.4 *For each $j = 1, 2$, $\mathscr{D}_j$ is a core for $\hat{p}_j$.*

Proof By a general criterion on essential self-adjointness of a symmetric operator (see Theorem 1.12), we need only to show that $(\hat{p}_j \pm i)\mathscr{D}_j$ are dense in $L^2(\mathbb{R}^2)$. So let $\phi \in ((\hat{p}_1 + i)\mathscr{D}_1)^\perp$. Then, for all $f \in C_0^\infty(\mathbb{R})$ and $g \in C_0^\infty(Y_{\boldsymbol{a}})$

$$\int_{\mathbb{R}^2} \phi(x_1, x_2)^* (-if'(x_1)g(x_2) + if(x_1)g(x_2)) dx_1 dx_2 = 0.$$

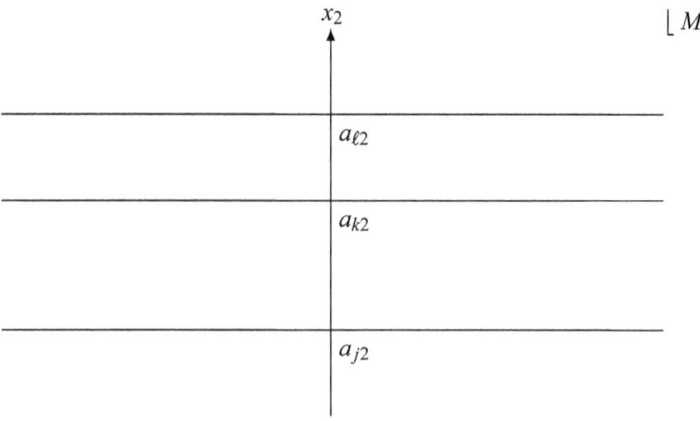

Fig. 3.2 The subset M_1: the straight lines $x_2 = a_{n2}$, $n = 1, \ldots, N$, are removed from $\mathbb{R}^2$

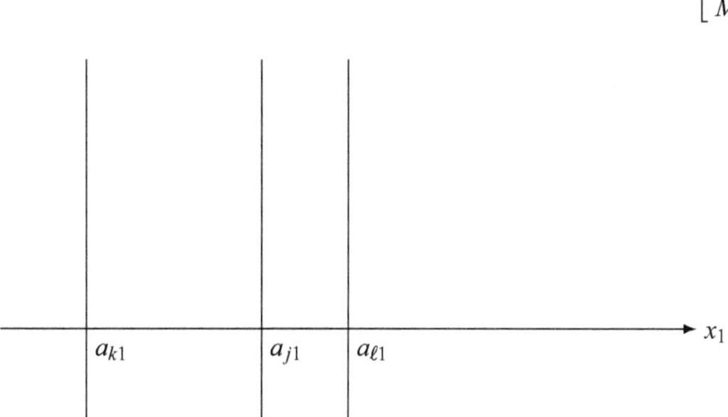

Fig. 3.3 The subset M_2: the straight lines $x_1 = a_{n1}$, $n = 1, \ldots, N$, are removed from $\mathbb{R}^2$

It is easy to see that one can apply Fubini's theorem to the integral on the left hand side to obtain

$$\int_{\mathbb{R}} u(x_1)((-iD_1 + i)f)(x_1)dx_1 = 0$$

with $u(x_1) := \int_{\mathbb{R}} \phi(x_1, x_2)^* g(x_2)dx_2$. By using the Schwarz inequality, one can show that $u \in L^2(\mathbb{R})$. Since $C_0^\infty(\mathbb{R})$ is a core for $-iD_1$, it follows that $u = 0$. Hence $\int_{\mathbb{R}} \phi(x_1, x_2)^* g(x_2)dx_2 = 0$, a.e.$x_1$. By Lemma 3.2, $C_0^\infty(Y_{\boldsymbol{a}})$ is dense in $L^2(\mathbb{R})$. Hence it follows that $\phi = 0$ as an element of $L^2(\mathbb{R}^2)$. Thus $(\hat{p}_1 + i)\mathscr{D}_1$ is dense in $L^2(\mathbb{R}^2)$. Similarly, one can show that $(\hat{p}_1 - i)\mathscr{D}_1$ is dense in $L^2(\mathbb{R}^2)$. The same method of proof applies to $(\hat{p}_2, \mathscr{D}_2)$ too. ■

Remark 3.1 The operator $\hat{p}_1$ (resp. $\hat{p}_2$) as an operator on $L^2(\mathbb{R})$ is not essentially self-adjoint on $C_0^\infty(X_{\boldsymbol{a}})$ (resp. $C_0^\infty(Y_{\boldsymbol{a}})$). Indeed, the function v defined by

$$v(k) := \sum_{n=1}^{N} \frac{1}{k - i} e^{-ika_{n1}}, \quad k \in \hat{\mathbb{R}}$$

is in $L^2(\hat{\mathbb{R}})$ and its inverse Fourier transform $\check{v}$ is in $((\hat{p}_1 + i)C_0^\infty(X_{\boldsymbol{a}}))^\perp$ (use that, for all $f \in C_0^\infty(X_{\boldsymbol{a}})$, $f(a_{n1}) = 0$, $n = 1, \ldots, N$), i.e., $(\hat{p}_1 + i)C_0^\infty(X_{\boldsymbol{a}})$ is not dense in $L^2(\mathbb{R})$. A similar argument applies to $\hat{p}_2$ too. This is a reason why we introduce the subspaces $\mathscr{D}_1$ and $\mathscr{D}_2$ as above.

Let $x_0 \in \mathbb{R}$ and $y_0 \in \mathbb{R}$ be arbitrarily fixed and define

$$\phi_1(\boldsymbol{x}) := \int_{x_0}^{x_1} A_1(x', x_2)dx', \quad \boldsymbol{x} = (x_1, x_2) \in M_1,$$

$$\phi_2(\boldsymbol{x}) := \int_{y_0}^{x_2} A_2(x_1, y')dy', \quad \boldsymbol{x} = (x_1, x_2) \in M_2.$$

Then ϕ_j is continuous and partially differentiable in x_j on M_j with

$$\frac{\partial \phi_j(\boldsymbol{x})}{\partial x_j} = A_j(\boldsymbol{x}), \quad \boldsymbol{x} \in M_j.$$

Since $|\mathbb{R}^2 \setminus M_j| = 0$, ϕ_j can be regarded as a real-valued Borel measurable function on $\mathbb{R}^2$ which is a.e. finite. Therefore the multiplication operator by the function $e^{iq\phi_j}$ is a unitary operator on $L^2(\mathbb{R}^2)$. We denote this unitary operator by U_j:

$$U_j f := e^{iq\phi_j} f, \quad f \in L^2(\mathbb{R}^2). \tag{3.7}$$

Now we can prove the main theorem in this section:

Theorem 3.1 *For each* $j = 1, 2$, P_j *is essentially self-adjoint on* $U_j \mathscr{D}_j$ *and operator equality*

$$P_j = U_j \hat{p}_j U_j^{-1} \quad (j = 1, 2) \tag{3.8}$$

holds.

Proof Let $\eta \in \mathscr{D}_1$ be of the form $\eta = f \times g$ with $f \in C_0^\infty(\mathbb{R})$ and $g \in C_0^\infty(Y_{\boldsymbol{a}})$. Then

$$(U_1\eta)(\boldsymbol{x}) = e^{iq\phi_1(\boldsymbol{x})} f(x_1)g(x_2), \quad \boldsymbol{x} \in M_1.$$

The function on the right hand side is continuous on M_1, partially differentiable in x_1 on M_1 and its support is compact in M_1. Hence it follows that $U_1\eta \in D(\hat{p}_1) \cap D(A_1) \subset D(P_1)$ and

$$(\hat{p}_1 U_1 \eta)(\boldsymbol{x}) = qA_1(\boldsymbol{x})(U_1\eta)(\boldsymbol{x}) + (U_1 \hat{p}_1 \eta)(\boldsymbol{x}).$$

Therefore $P_1 U_1 \eta = U_1 \hat{p}_1 \eta$. Thus $U_1 \mathscr{D}_1 \in D(P_1)$ and, for all $\psi \in \mathscr{D}_1$, $U_1^{-1} P_1 U_1 \psi = \hat{p}_1 \psi$. By this fact and Lemma 3.4, one can conclude that P_1 is essentially self-adjoint on $U_1 \mathscr{D}_1$ and operator equality $U_1^{-1} P_1 U_1 = \hat{p}_1$ holds. This is equivalent to (3.8) with $j = 1$. Similarly, one can prove the statement on P_2. ■

3.3 Strongly Continuous One-Parameter Unitary Groups Generated by the Physical Momenta

Since the self-adjointness of P_j has been established in Theorem 3.1, it is natural to consider the strongly continuous one-parameter unitary group $\{e^{itP_j}\}_{t\in\mathbb{R}}$ generated by P_j. It has in fact an explicit representation:

Theorem 3.2 *For all $t \in \mathbb{R}$ and $\psi \in L^2(\mathbb{R}^2)$,*

$$(e^{itP_j}\psi)(\boldsymbol{x}) = e^{-iq\int_0^t A_j(\boldsymbol{x}+s\boldsymbol{e}_j)ds}\psi(\boldsymbol{x}+t\boldsymbol{e}_j), \quad \boldsymbol{x} \in M_j, \tag{3.9}$$

where $(\boldsymbol{e}_1, \boldsymbol{e}_2)$ is the standard basis of $\mathbb{R}^2$.

Proof By (3.8) and the unitary covariance of functional calculus (Theorem B.3 in Appendix B), we have for all $t \in \mathbb{R}$

$$e^{itP_j} = U_j e^{it\hat{p}_j} U_j^{-1}.$$

By the definition of U_j and formula (1.82), we have for all $\psi \in L^2(\mathbb{R}^2)$ and a.e.$\boldsymbol{x} \in \mathbb{R}^2$

$$(e^{itP_j}\psi)(\boldsymbol{x}) = e^{iq\phi_j(\boldsymbol{x})}e^{-iq\phi_j(\boldsymbol{x}+t\boldsymbol{e}_j)}\psi(\boldsymbol{x}+t\boldsymbol{e}_j).$$

It is easy to see that the right hand side is equal to that of (3.9). ∎

Formula (3.9) has a geometric meaning. In the case where $\boldsymbol{A} = \boldsymbol{0}$, one has $e^{itP_j} = e^{it\hat{p}_j}$ $(j = 1, 2)$. By (1.82), $e^{it\hat{p}_1}$ and $e^{it\hat{p}_2}$ have the function of translating each function $\psi \in L^2(\mathbb{R}^2)$ by $-t$ to the x_1-direction and the x_2-direction respectively. In the case where $q\boldsymbol{A} \neq \boldsymbol{0}$, however, the right hand side of (3.9) means that the translation is distorted by the exponential factor $e^{-iq\int_0^t A_j(\boldsymbol{x}+s\boldsymbol{e}_j)ds}$. Namely $e^{itP_j}\psi$ may be interpreted as a translation of ψ under the existence of a magnetic field. From this viewpoint, e^{itP_j} is called the **magnetic translation** to the jth direction.

3.4 Commutation Relations of Exponential Operators

The contour integral

$$\Phi_C(\boldsymbol{A}) := \int_C \boldsymbol{A}(\boldsymbol{x}) \cdot d\boldsymbol{x} \tag{3.10}$$

of the vector potential $\boldsymbol{A}$ along a closed curve C in M is called the **magnetic flux** penetrating the interior of C.

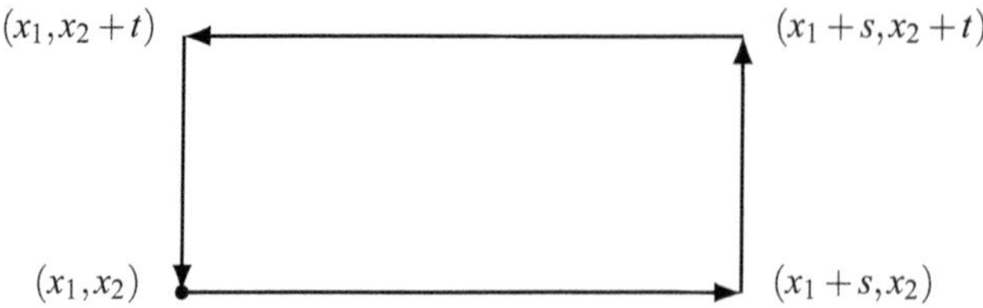

Fig. 3.4 The closed curve $C(x_1, x_2; s, t)$ in the case $s, t > 0$

For each $(x_1, x_2, s, t) \in \mathbb{R}^4$, we denote by $C(x_1, x_2; s, t)$ the rectangular closed curve having the following properties: (1) the point $(x_1, x_2) \in \mathbb{R}^2$ is the initial and the terminal point; (2) $C(x_1, x_2; s, t)$ consists of the following four straight lines (see Fig. 3.4):

$$(x_1, x_2) \to (x_1 + s, x_2) \to (x_1 + s, x_2 + t) \to (x_1, x_2 + t) \to (x_1, x_2).$$

For each $(s, t) \in \mathbb{R}^2$, we introduce subsets $S_1^{(s)}$ and $S_2^{(t)}$ of $\mathbb{R}$ and a subset $M_{s,t}$ of $\mathbb{R}^2$ as follows:

$$S_1^{(s)} = \mathbb{R} \setminus \{a_{n1}, a_{n1} - s\}_{n=1}^N, \quad S_2^{(t)} = \mathbb{R} \setminus \{a_{n2}, a_{n2} - t\}_{n=1}^N,$$
$$M_{s,t} = S_1^{(s)} \times S_2^{(t)}.$$

Note that, if $(x_1, x_2) \in M_{s,t}$, then $C(x_1, x_2; s, t)$ does not pass the points $\boldsymbol{a}_n, n = 1, \cdots, N$. Hence we can define a function $\Phi_{s,t}^{\boldsymbol{A}}$ on $M_{s,t}$ by

$$\Phi_{s,t}^{\boldsymbol{A}}(x_1, x_2) := \Phi_{C(x_1,x_2;s,t)}(\boldsymbol{A}), \quad (x_1, x_2) \in M_{s,t}.$$

Physically, this quantity means the magnetic flux penetrating the interior of $C(x_1, x_2; s, t)$.

Since the two-dimensional Lebesgue measure of the set $\mathbb{R}^2 \setminus M_{s,t}$ is zero, the function $\Phi_{s,t}^{\boldsymbol{A}}$ can be regarded as a real-valued Borel measurable function on $\mathbb{R}^2$ which is a.e. finite. Hence $\Phi_{s,t}^{\boldsymbol{A}}$ defines a self-adjoint multiplication operator on $L^2(\mathbb{R}^2)$. We denote the multiplication operator by the same symbol $\Phi_{s,t}^{\boldsymbol{A}}$.

The next proposition describes a commutation property of e^{isP_1} and e^{itP_2}.

Theorem 3.3 *For all* $s, t \in \mathbb{R}$,

$$e^{isP_1}e^{itP_2} = \exp\left(-iq\Phi_{s,t}^{\boldsymbol{A}}\right) e^{itP_2}e^{isP_1}. \tag{3.11}$$

Proof By Theorem 3.2, for all $\psi \in L^2(\mathbb{R}^2)$,

$$
\begin{aligned}
(e^{isP_1}e^{itP_2}\psi)(\boldsymbol{x}) &= e^{-iq\int_0^s A_1(\boldsymbol{x}+s'\boldsymbol{e}_1)ds'}(e^{itP_2}\psi)(\boldsymbol{x}+s\boldsymbol{e}_1) \\
&= e^{-iq\int_0^s A_1(\boldsymbol{x}+s'\boldsymbol{e}_1)ds'}e^{-iq\int_0^t A_2(\boldsymbol{x}+s\boldsymbol{e}_1+t'\boldsymbol{e}_2)dt'} \\
&\quad \times \psi(\boldsymbol{x}+s\boldsymbol{e}_1+t\boldsymbol{e}_2), \text{ a.e.}\boldsymbol{x}.
\end{aligned}
$$

Similarly, we have

$$
(e^{itP_2}e^{isP_1}\psi)(\boldsymbol{x}) = e^{-iq\int_0^t A_2(\boldsymbol{x}+t'\boldsymbol{e}_2)dt'}e^{-iq\int_0^s A_1(\boldsymbol{x}+t\boldsymbol{e}_2+s'\boldsymbol{e}_1)ds'}\psi(\boldsymbol{x}+t\boldsymbol{e}_2+s\boldsymbol{e}_1), \text{ a.e.}\boldsymbol{x}.
$$

From these two equations, we obtain (3.11). ∎

3.5 Remarks on the Case of Constant Magnetic Fields

As is easily seen, the theory described in the preceding sections includes also the case of ordinary magnetic fields which have no singularity at $\boldsymbol{a}_n$ $(n = 1, \ldots, N)$. A simple example of such magnetic fields is a constant magnetic field. Before going on further, we give here some remarks on the case of a constant magnetic field. So we consider the case where B is a constant function on $\mathbb{R}^2$. As a vector potential $\boldsymbol{A} = (A_1, A_2)$ yielding such a B, one can take the following:

$$
A_1(\boldsymbol{x}) = 0, \quad A_2(\boldsymbol{x}) = Bx_1, \quad \boldsymbol{x} = (x_1, x_2) \in \mathbb{R}^2.
$$

Then we have

$$
P_1 = \hat{p}_1, \quad P_2 = \overline{\hat{p}_2 - qB\hat{q}_1}.
$$

In the present case, one can easily show that, for all $s, t \in \mathbb{R}$,

$$
\Phi_{s,t}^{\boldsymbol{A}}(\boldsymbol{x}) = stB.
$$

Hence (3.11) takes the form

$$
e^{isP_1}e^{itP_2} = e^{-istqB}e^{itP_2}e^{isP_1}. \tag{3.12}
$$

Without loss of generality, we can assume that $qB > 0$. Let

$$
X := \frac{1}{qB}P_1, \quad Y := P_2.
$$

Then (3.12) implies that

$$e^{isX}e^{itY} = e^{-ist}e^{itY}e^{isX}.$$

This means that $(L^2(\mathbb{R}^2), (X, Y))$ is a Weyl representation of the CCR with one degree of freedom: $(L^2(\mathbb{R}^2), (X, Y)) \in \mathsf{CCR}_{\mathrm{W}}(1)$. But this is *not irreducible*, since $e^{is\hat{p}_2}$ commutes with e^{itX} and e^{itY} for all $s, t \in \mathbb{R}$: $e^{is\hat{p}_2} \in \{e^{itX}, e^{itY} | t \in \mathbb{R}\}'$.[2] By von Neumann's uniqueness theorem (Theorem 2.19), $(L^2(\mathbb{R}^2), (X, Y))$ is equivalent to a direct sum of the Schrödinger representation $\pi_{\mathrm{S}}^{(1)} = (L^2(\mathbb{R}), (\hat{q}, \hat{p}))$.[3] Therefore there exist mutually orthogonal closed subspaces $\mathscr{M}_\ell$ of $L^2(\mathbb{R}^2)$ ($\ell = 1, \ldots, L; L \in \mathbb{N}$ or $L = \infty$) such that each $\mathscr{M}_\ell$ reduces X and Y and a unitary transformation $U_\ell \colon \mathscr{M}_\ell \to L^2(\mathbb{R})$ such that

$$U_\ell X_\ell U_\ell^{-1} = \hat{q}, \quad U_\ell Y_\ell U_\ell^{-1} = \hat{p},$$

where X_ℓ (resp. Y_ℓ) is the reduced part of X (resp. Y) to $\mathscr{M}_\ell$.

The result obtained above can be applied to the spectral analysis of the Hamiltonian

$$H_B := \frac{1}{2m}\left(P_1^2 + P_2^2\right) = \frac{1}{2m}\left(\hat{p}_1^2 + \left(\overline{\hat{p}_2 - qB\hat{q}_1}\right)^2\right)$$

of the charged particle, where $m > 0$ denotes the mass of the charged particle. It follows that H_B is reduced by each $\mathscr{M}_\ell$ and

$$U_\ell H_{B,\ell} U_\ell^{-1} = h_B := \frac{1}{2m}\left(\hat{p}^2 + (qB)^2\hat{q}^2\right),$$

where $H_{B,\ell}$ is the reduced part of H_B to $\mathscr{M}_\ell$. Note that the operator h_B is the Hamiltonian of the one-dimensional quantum harmonic oscillator of mass m and angular frequency $\omega = qB/m$ (see Example 2.4). Since h_B is self-adjoint, it follows that $H_{B,\ell}$ is self-adjoint and hence H_B is self-adjoint. By (2.105), one has

$$\sigma(h_B) = \sigma_{\mathrm{d}}(h_B) = \{\varepsilon_n | n \in \mathbb{Z}_+\},$$

where

$$\varepsilon_n := \frac{qB}{m}\left(n + \frac{1}{2}\right).$$

[2]Using the natural isomorphism $L^2(\mathbb{R}^2) \cong L^2(\mathbb{R}) \otimes L^2(\mathbb{R})$ (see (1.58)) with $\hat{p}_1 \cong \hat{p} \otimes I$, $\hat{q}_1 \cong \hat{q} \otimes I$, $\hat{p}_2 \cong I \otimes \hat{p}$, $\hat{q}_2 \cong I \otimes \hat{q}$ (see (2.57)) and (1.63), one can show that $e^{itY} = e^{it\hat{p}_2}e^{-itqB\hat{q}_1}$. Hence $e^{is\hat{p}_2}e^{itY} = e^{itY}e^{is\hat{p}_2}$.

[3]Here $\hat{q} = M_x$, $\hat{p} = -iD_x$ ($x \in \mathbb{R}$).

Hence, by the unitary invariance of spectra and a spectral property of a direct sum operator, we have

$$\sigma(H_B) = \sigma_{\mathrm{p}}(H_B) = \{\varepsilon_n | n \in \mathbb{Z}_+\}.$$

This is a well-known fact in the physics literature (e.g., [103, §111]). But one should note that each eigenvalue ε_n is *degenerate*, since $L \geq 2$. In fact, ε_n is *infinitely degenerate*. A proof of this fact is as follows.

The commutativity of $e^{is\hat{p}_2}$ with e^{itX} and e^{itY} for all $s, t \in \mathbb{R}$ implies that $\hat{p}_2$ strongly commutes with X and Y. Hence $\hat{p}_2$ strongly commutes with H_B (apply Proposition 1.18). Hence, for all Borel sets J, $E_{\hat{p}_2}(J)H_B \subset H_B E_{\hat{p}_2}(J) \cdots (*)$, where $E_{\hat{p}_2}(\cdot)$ is the spectral measure of $\hat{p}_2$. Let $\{\mathbb{I}_r\}_{r=1}^{\infty}$ be a collection of mutually disjoint bounded semi-open intervals such that $\mathbb{R} = \cup_{r=1}^{\infty}\mathbb{I}_r$. Since $\sigma(\hat{p}_2) = \mathbb{R}$ (see (2.40)), it follows that $E_{\hat{p}_2}(\mathbb{I}_r) \neq 0$ for all $r \in \mathbb{N}$. We have $L^2(\mathbb{R}^2) = \oplus_{r=1}^{\infty}\mathscr{E}_r$ with $\mathscr{E}_r := \mathrm{Ran}\,(E_{\hat{p}_2}(\mathbb{I}_r)) \neq \{0\}$. By $(*)$, H_B is reduced by each $\mathscr{E}_r$. Then, in the same way as above, one can show that the reduced part of H_B to $\mathscr{E}_r$ has an eigenvalue ε_n $(n \in \mathbb{N})$. Thus the eigenvalue ε_n is infinitely degenerate.

As this proof shows, the infinite degeneracy of each eigenvalue ε_n of H_B is due to the strong commutativity of $\hat{p}_2$ with H_B, which is equivalent to the translation invariance of H_B in the x_2-direction, i.e., $e^{it\hat{p}_2}H_B e^{-it\hat{p}_2} = H_B$, $t \in \mathbb{R}$.

Remark 3.2 The above description shows that the quantum system of a constant magnetic field in the two-dimensional space $\mathbb{R}^2$ carries a Weyl representation of CCR given by $(L^2(\mathbb{R}^2), (X, Y))$ and the use of it makes very easy the spectral analysis of the Hamiltonian H_B. This is one of the examples in which representations equivalent to Schrödinger representations play important roles in spectral analysis of observables. We have already discussed such an example in Sect. 2.6.3.

3.6 Flatness of Vector Potentials and Existence of Representations of CCR

We now return to the general case where B is not necessarily a constant magnetic field. We denote by $\hat{q}_1$ and $\hat{q}_2$ the position operators on $L^2(\mathbb{R}^2)$ as before. In this section, we show that, for a class of vector potentials $\boldsymbol{A}$, $((\hat{q}_1, \hat{q}_2), (P_1, P_2))$ gives a self-adjoint irreducible representation of the CCR with two degrees of freedom.

To define such a class, we recall a concept from differential geometry:

Definition 3.1 A vector potential $\boldsymbol{A}$ on M is said to be **flat** if $B = 0$ on M.

Note that M is *not simply connected*. Hence, for a flat vector potential $\boldsymbol{A}$, it does not necessarily hold that, for some real-valued function $\phi \in C^1(M)$, $\boldsymbol{A} = \nabla\phi$, where $\nabla := (D_1, D_2)$. This is an important point.

A flat vector potential $\boldsymbol{A}$ of the form $\boldsymbol{A} = \nabla\phi$ for some $\phi \in C^1(M)$ is said to be **trivial**.

We shall see that the non-simply connectedness of M is essential for the theory presented below to be meaningful.

It is easy to see that, for $j, k = 1, 2$,

$$[\hat{q}_j, P_k] = i\delta_{jk} \quad \text{on } C_0^1(M). \tag{3.13}$$

Suppose that $D_1 A_2$ and $D_2 A_1$ are in $L^2_{\mathrm{loc}}(M)$.[4] Then, by a general theorem on generalized partial differentiations, for all $\psi \in C^1(M)$,

$$D_1(A_2\psi) = (D_1A_2)\psi + A_2D_1\psi, \quad D_2(A_1\psi) = (D_2A_1)\psi + A_1D_2\psi.$$

Hence

$$[P_1, P_2] = iqB \quad \text{on } C_0^2(M). \tag{3.14}$$

Commutation relations (3.13) and (3.14) imply the next lemma:

Lemma 3.5 *Assume that $A_j \in C^1(M)$, $j = 1, 2$. Then*

$$\pi_A := (L^2(\mathbb{R}^2), C_0^2(M), (\hat{q}_1, \hat{q}_2, P_1, P_2))$$

is a representation of the CCR with two degrees of freedom if and only $\boldsymbol{A}$ *is flat.*

Example 3.1 The flatness of $\boldsymbol{A}$ means that the magnetic field B as a distribution on $\mathbb{R}^2$ has its support in the set $\{\boldsymbol{a}_1, \cdots, \boldsymbol{a}_N\}$. According to the theory of distributions, such a distribution B is of the form:

$$B(\boldsymbol{x}) = \sum_{n=1}^{N} \sum_{\alpha,\beta=0}^{k_n} \lambda_{\alpha\beta}^{(n)} D_1^\alpha D_2^\beta \delta(\boldsymbol{x} - \boldsymbol{a}_n),$$

where $\delta(\boldsymbol{x} - \boldsymbol{a}_n)$ is the delta distribution with support at $\boldsymbol{a}_n$, $k_n \in \mathbb{Z}_+$ and $\lambda_{\alpha\beta}^{(n)}$'s are real constants.[5] A vector potential yielding such a distribution B is given by the following:

$$A_1(\boldsymbol{x}) = -\frac{1}{2\pi} \sum_{n=1}^{N} \sum_{\alpha,\beta=0}^{k_n} \lambda_{\alpha\beta}^{(n)} D_1^\alpha D_2^\beta \left(\frac{x_2 - a_{n2}}{|\boldsymbol{x} - \boldsymbol{a}_n|^2} \right),$$

$$A_2(\boldsymbol{x}) = \frac{1}{2\pi} \sum_{n=1}^{N} \sum_{\alpha,\beta=0}^{k_n} \lambda_{\alpha\beta}^{(n)} D_1^\alpha D_2^\beta \left(\frac{x_1 - a_{n1}}{|\boldsymbol{x} - \boldsymbol{a}_n|^2} \right).$$

[4] For an open set D in $\mathbb{R}^d$ and $p \in \mathbb{N}$, $f \in L^p_{\mathrm{loc}}(D) \overset{\text{def}}{\Longleftrightarrow}$ for all compact sets K in D, $\int_K |f(x)|^p dx < \infty$.

[5] See, e.g., [78, Chapter II, §4.5] or [166, Example 5 in §4.1].

This is proved by using the well-known formulae[6]:

$$\frac{x_1 - a_{n1}}{|\boldsymbol{x} - \boldsymbol{a}_n|^2} = D_1 \log |\boldsymbol{x} - \boldsymbol{a}_n|, \quad \frac{x_2 - a_{n2}}{|\boldsymbol{x} - \boldsymbol{a}_n|^2} = D_2 \log |\boldsymbol{x} - \boldsymbol{a}_n|,$$
$$(D_1^2 + D_2^2) \log |\boldsymbol{x} - \boldsymbol{a}_n| = 2\pi \delta(\boldsymbol{x} - \boldsymbol{a}_n).$$

A simpler example for $\boldsymbol{A}$ may be given as follows:

$$\widetilde{A}_1(\boldsymbol{x}) = -\frac{1}{2\pi} \sum_{n=1}^{N} \lambda_n \left(\frac{x_2 - a_{n2}}{|\boldsymbol{x} - \boldsymbol{a}_n|^2} \right), \quad \widetilde{A}_2(\boldsymbol{x}) = \frac{1}{2\pi} \sum_{n=1}^{N} \lambda_n \left(\frac{x_1 - a_{n1}}{|\boldsymbol{x} - \boldsymbol{a}_n|^2} \right), \tag{3.15}$$

where λ_n $(n = 1, \cdots, N)$ is a real constant. The magnetic field $\widetilde{B}$ in this case takes the form

$$\widetilde{B}(\boldsymbol{x}) = \sum_{n=1}^{N} \lambda_n \delta(\boldsymbol{x} - \boldsymbol{a}_n).$$

Example 3.2 Using meromorphic functions on $\mathbb{C}$, one can construct a general class of non-trivial flat vector potentials on M as follows. As is well known, the mapping: $\iota : \mathbb{R}^2 \ni \boldsymbol{x} = (x_1, x_2) \mapsto x_1 + i x_2 \in \mathbb{C}$ is a bijection. We set

$$a_n := \iota(\boldsymbol{a}_n) = a_{n1} + i a_{n2} \in \mathbb{C}$$

and

$$M_{\mathbb{C}} := \{\iota(\boldsymbol{x}) | \boldsymbol{x} \in M\}.$$

We denote by $\mathscr{F}_M$ the set of meromorphic functions on $\mathbb{C}$ with poles in $\{a_n\}_{n=1}^N$. Then, for each $f \in \mathscr{F}_M$, one can define a vector potential

$$\boldsymbol{A} := (A_1, A_2)$$

by

$$A_1(\boldsymbol{x}) := \operatorname{Im} f(z), \quad A_2(\boldsymbol{x}) := \operatorname{Re} f(z), \quad z = x_1 + i x_2 \in M_{\mathbb{C}}.$$

It follows from the Cauchy–Riemann equation for f, which is holomorphic in $M_{\mathbb{C}}$, that $\boldsymbol{A}$ is flat on M and divergence free:

$$D_1 A_1 + D_2 A_2 = 0$$

[6] See, e.g., [77, pp. 28–29], [145, Theorem 6] or [166, p. 134].

on M. Hence we see that, with each meromorphic function f in $\mathscr{F}_M$, is associated a flat vector potential. Since $\mathscr{F}_M$ is infinite dimensional as a complex vector space, there may exist plenty of non-trivial flat vector potentials. We remark that this construction of flat vector potentials applies also to the case where isolated singular points of vector potentials are countably infinite.

A simple concrete example is given as follows:

$$f(z) = \frac{1}{2\pi}\sum_{n=1}^{N}\lambda_n \frac{1}{z - a_n},$$

with $\lambda_n \in \mathbb{R}$. In this case, we have $A(x) = (\widetilde{A}_1, \widetilde{A}_2)$ with $\widetilde{A}_j$ given by (3.15) ($j = 1, 2$).

We denote by $\mathscr{A}^1_{\text{flat}}(M)$ the set of flat vector potentials of C^1-class:

$$\mathscr{A}^1_{\text{flat}}(M) := \{A = (A_1, A_2)|A_j \in C^1(M), j = 1, 2 \text{ and } A \text{ is flat on } M\}. \tag{3.16}$$

In view of Lemma 3.5, it is natural to ask if the representation π_A with $A \in \mathscr{A}^1_{\text{flat}}(M)$ is irreducible. The answer is in the affirmative:

Lemma 3.6 *Let $A \in \mathscr{A}^1_{\text{flat}}(M)$. Then π_A is irreducible.*

Proof Let $T \in \{\hat{q}_1, \hat{q}_2, P_1, P_2\}'$. Then $T\hat{q}_j \subset \hat{q}_j T$ and $TP_j \subset P_j T$ ($j = 1, 2$). It follows from the former condition and Lemma 2.7 that $e^{it\hat{q}_j}T = Te^{it\hat{q}_j}$, $j = 1, 2, t \in \mathbb{R}$. Hence, by Lemma 2.11, there exists a function $F \in L^\infty(\mathbb{R}^2)$ such that $T = M_F$ (the multiplication operator by F). Therefore $M_F P_j \subset P_j M_F$ ($j = 1, 2$). Since P_j is self-adjoint and M_F is a bounded operator, it follows from Lemma 2.7 again that, for all $t \in \mathbb{R}$, $e^{itP_j}M_F e^{-itP_j} = M_F$. By this relation and Theorem 3.2, we obtain $F(x) = F(x + te_j)$ (a.e.x). Hence, for all $(t, s) \in \mathbb{R}^2$, $F(x) = F(x + (t, s))$ (a.e.x). Hence, for a constant $c \in \mathbb{C}$, $F(x) = c$ a.e.x. Therefore $T = cI$. Thus $\{\hat{q}_1, \hat{q}_2, P_1, P_2\}$ is irreducible. ■

By Lemmas 3.5 and 3.6, we obtain the next theorem:

Theorem 3.4 *For each $A \in \mathscr{A}^1_{\text{flat}}(M)$, π_A is an irreducible self-adjoint representation of the CCR with two degrees of freedom, i.e., $\pi_A \in \mathsf{CCR}_{\text{sa}}(2)$.*

Thus we have a family $\{\pi_A\}_{A\in\mathscr{A}^1_{\text{flat}}(M)}$ of irreducible representations in $\mathsf{CCR}_{\text{sa}}(2)$ indexed by a class of flat vector potentials. Since $(\hat{q}_1, \hat{q}_2)$ is strongly commuting, π_A is a semi-strong self-adjoint representation, i.e., $\pi_A \in \mathsf{CCR}_{\text{ssa}}(2)$. Then it is interesting to investigate whether or not π_A is equivalent to $\pi_{A'}$ for different vector potentials A and A' in $\mathscr{A}^1_{\text{flat}}(M)$. We next analyze this aspect.

3.7 Inequivalent Representations

Theorem 3.3 implies that P_1 and P_2 strongly commute if and only if

$$e^{-iq\Phi_{s,t}^{A}} = I, \quad \forall (s,t) \in \mathbb{R}^2. \tag{3.17}$$

In this case $(L^2(\mathbb{R}^2), (\hat{q}_1, \hat{q}_2, P_1, P_2))$ becomes a Weyl representation of the CCR with two degrees of freedom. It is obvious that (3.17) holds if and only if, for each $(s,t) \in \mathbb{R}^2$, $q\Phi_{s,t}^{A}(x) \in 2\pi\mathbb{Z}$ for a.e.$x \in \mathbb{R}^2$. Taking this fact into account, we introduce a concept:

Definition 3.2 The magnetic flux of A is said to be **locally quantized** if, for all $(s,t) \in \mathbb{R}^2$, $\Phi_{s,t}^{A}$ is a $2\pi\mathbb{Z}/q$-valued function on $M_{s,t}$.

Theorem 3.5 *Let $A \in \mathscr{A}_{\mathrm{flat}}^{1}(M)$. Then π_A is equivalent to the Schrödinger representation $\pi_{\mathrm{S}}^{(2)}$ if and only if the magnetic flux of A is locally quantized.*

Proof Suppose that $\pi_A \cong \pi_{\mathrm{S}}^{(2)}$. Then $\{e^{it\hat{q}_j}, e^{itP_j} | j = 1,2,\ t \in \mathbb{R}\}$ obeys the Weyl relations with two degrees of freedom. Hence, by Theorem 3.3, (3.17) holds. This means that the magnetic flux of A is locally quantized.

Conversely, suppose that the magnetic flux of A is locally quantized. Then (3.17) holds. Hence, by Theorem 3.3, $\{e^{is\hat{q}_j}, e^{itP_j} | j = 1,2,\ t \in \mathbb{R}\}$ obeys the Weyl relations with two degrees of freedom. Therefore, by Theorem 3.4, $(L^2(\mathbb{R}^2), (\hat{q}_1, \hat{q}_2, P_1, P_2))$ is an irreducible Weyl representation with two degrees of freedom. Thus, by von Neumann's uniqueness theorem (Theorem 2.19), π_A is equivalent to $\pi_{\mathrm{S}}^{(2)}$. ■

We next rewrite the condition of local quantization of the magnetic flux of A in a simpler form. Let

$$\delta_0 := \min\{|a_n - a_m| \mid n \neq m, n, m = 1, \cdots, N\}$$

and, for $0 < \varepsilon < \delta_0$,

$$\gamma_n(A) := \int_{C_\varepsilon(a_n)} A \cdot dx, \tag{3.18}$$

the contour integral of A along the circle

$$C_\varepsilon(a_n) := \{x \in \mathbb{R}^2 : x = a_n + \varepsilon e^{i\theta},\ \theta \in [0, 2\pi]\}$$

with anti-clockwise orientation. We denote by $D(x; s, t)$ ($x = (x_1, x_2),\ (s,t) \in \mathbb{R}^2$) the interior of the closed curve $C(x; s, t)$ and by sgn: $\mathbb{R} \to \mathbb{R}$ the sign function:

$$\mathrm{sgn}(t) := \begin{cases} 1 & \text{if } t \geq 0, \\ -1 & \text{if } t < 0. \end{cases}$$

Lemma 3.7 *Let $\boldsymbol{A} \in \mathscr{A}^1_{\text{flat}}(M)$. Then, each $\gamma_n(\boldsymbol{A})$ is a constant independent of ε and, for all $(s,t) \in \mathbb{R}^2$ and $\boldsymbol{x} \in M_{s,t}$,*

$$\Phi^{\boldsymbol{A}}_{s,t}(\boldsymbol{x}) = \operatorname{sgn}(s)\operatorname{sgn}(t) \sum_{\boldsymbol{a}_n \in D(\boldsymbol{x};s,t)} \gamma_n(\boldsymbol{A}), \tag{3.19}$$

where $\sum_{\boldsymbol{a}_n \in D(\boldsymbol{x};s,t)}$ means the sum over all n such that $\boldsymbol{a}_n \in D(\boldsymbol{x}; s,t)$ (if there is no such n, then $\sum_{\boldsymbol{a}_n \in D(\boldsymbol{x};s,t)}(\cdot) := 0$).

Proof Suppose that $\boldsymbol{a}_{i_1}, \ldots, \boldsymbol{a}_{i_r} \in D(\boldsymbol{x}; s,t)$ for some $\{i_1, \ldots, i_r\} \subset \{1, \ldots, N\}$ with $i_1 < \ldots < i_r$. One can take ε sufficiently small such that $C_\varepsilon(\boldsymbol{a}_{i_j}) \subset D(\boldsymbol{x}; s,t)$ $(j = 1, \ldots, r)$. Let $B_{i_j} := \{\boldsymbol{x} \in \mathbb{R}^2 | \, |\boldsymbol{x} - \boldsymbol{a}_{i_j}| \le \varepsilon\}$, the closed circle with center $\boldsymbol{a}_{i_j}$ and radius ε, and $E := D(\boldsymbol{x}; s,t) \cap (\cup_{j=1}^r B_{i_j})^c$. Then, by the Stokes theorem (or Green's theorem), we have

$$\int_E B dx_1 dx_2 = \int_{\partial E} \boldsymbol{A} \cdot d\boldsymbol{x},$$

where ∂E denotes the boundary of E. But $B = 0$ on E. Hence $\int_{\partial E} \boldsymbol{A} \cdot d\boldsymbol{x} = 0$. On the other hand, we have

$$\int_{\partial E} \boldsymbol{A} \cdot d\boldsymbol{x} = \operatorname{sgn}(s)\operatorname{sgn}(t) \int_{C(\boldsymbol{x};s,t)} \boldsymbol{A} \cdot d\boldsymbol{x} - \sum_{j=1}^r \int_{C_\varepsilon(\boldsymbol{a}_{i_j})} \boldsymbol{A} \cdot d\boldsymbol{x}.$$

Thus we obtain (3.19). ■

The next proposition immediately follows from this lemma.

Proposition 3.1 *Let $\boldsymbol{A} \in \mathscr{A}^1_{\text{flat}}(M)$. Then the magnetic flux of $\boldsymbol{A}$ is locally quantized if and only if, for each $n = 1, \cdots, N$, $\gamma_n(\boldsymbol{A}) \in 2\pi\mathbb{Z}/q$.*

Combining Theorem 3.4 and Proposition 3.1, we arrive at the next theorem which formulates a necessary and sufficient condition for $\pi_{\boldsymbol{A}}$ with $\boldsymbol{A} \in \mathscr{A}^1_{\text{flat}}(M)$ to be equivalent to the Schrödinger representation $\pi_{\text{S}}^{(2)}$:

Theorem 3.6 *Let $\boldsymbol{A} \in \mathscr{A}^1_{\text{flat}}(M)$. Then $\pi_{\boldsymbol{A}}$ is equivalent to the Schrödinger representation $\pi_{\text{S}}^{(2)}$ if and only if, for all $n = 1, \cdots, N$, $\gamma_n(\boldsymbol{A}) \in 2\pi\mathbb{Z}/q$.*

This theorem can be rephrased as follows:

Theorem 3.7 *Let $\boldsymbol{A} \in \mathscr{A}^1_{\text{flat}}(M)$. Then $\pi_{\boldsymbol{A}}$ is inequivalent to $\pi_{\text{S}}^{(2)}$ if and only if $\gamma_n(\boldsymbol{A}) \notin 2\pi\mathbb{Z}/q$ for some $n \in \{1, \ldots, N\}$.*

Remark 3.3 Note that, in the case where the magnetic flux of $\boldsymbol{A} \in \mathscr{A}^1_{\text{flat}}(M)$ is not locally quantized, P_1 and P_2 are *not strongly commuting*. Hence, in this case, $\pi_{\boldsymbol{A}}$ is not a strong self-adjoint representation, but a semi-strong self-adjoint representation.

Example 3.3 Let $\widetilde{\boldsymbol{A}} := (\widetilde{A}_1, \widetilde{A}_2)$ be the vector potential in Example 3.1. Then, by direct computations, we have

$$\gamma_n(\widetilde{\boldsymbol{A}}) = \lambda_n.$$

Suppose that $\lambda_{n_0} \notin 2\pi\mathbb{Z}/q$ with some $n_0 \in \{1, \ldots, N\}$. Then, by Theorem 3.7, $\pi_{\widetilde{\boldsymbol{A}}}$ is inequivalent to $\pi_{\mathrm{S}}^{(2)}$.

As we have seen in Example 3.2, for any meromorphic function f on $M_{\mathbb{C}}$, the vector potential $\boldsymbol{A}_f := (\mathrm{Im}\, f, \mathrm{Re}\, f)$ is in $\mathscr{A}^1_{\mathrm{flat}}(M)$. For any Jordan curve C in M, one has

$$\int_C \boldsymbol{A}_f(\boldsymbol{x}) \cdot d\boldsymbol{x} = \mathrm{Im} \int_C f(z)dz.$$

Using the residue theorem, one obtains

$$\gamma_n(\boldsymbol{A}_f) = \mathrm{Im}\, [2\pi i\, \mathrm{Res}(a_n, f)],$$

where $\mathrm{Res}(a_n, f)$ denotes the residue of f at a_n. This shows that there are infinitely many flat vector potentials of class C^∞ such that $\gamma_n(\boldsymbol{A}) \notin 2\pi\mathbb{Z}/q$.

3.8 Correspondence to Aharonov–Bohm Effect

In this section we show that the situation where $\pi_{\boldsymbol{A}}$ with $\boldsymbol{A} \in \mathscr{A}^1_{\mathrm{flat}}(M)$ is inequivalent to $\pi_{\mathrm{S}}^{(2)}$ physically corresponds to the occurrence of the so-called **Aharonov–Bohm effect** (AB effect for short) [3] in the context under consideration.

For the reader's convenience, we first recall what an AB effect is like. When a split electron beam travels through the space outside a solenoid in which a magnetic field is confined, the interference pattern on a screen behind the solenoid shifts upon alteration of the magnetic field, although the electron beam does not interact with the magnetic field (see Fig. 3.5). This phenomenon is called the AB effect under the present experimental setting. On the other hand, it was only recently that a conclusive experimental confirmation of the AB effect was made (see, e.g., [157]). The AB effect cannot be explained in classical physics. It is a quantum mechanical effect.

In quantum mechanics, what determines the interaction between a charged particle and a magnetic field is not the magnetic field itself, but the vector potentials that produce it. A vector potential can exist beyond the region where the magnetic field is localized (see Example 3.1). Hence an electron can interact with a vector potential even if it does not pass through the region where the magnetic field exists. This is a qualitative picture for the occurrence of the AB effect.

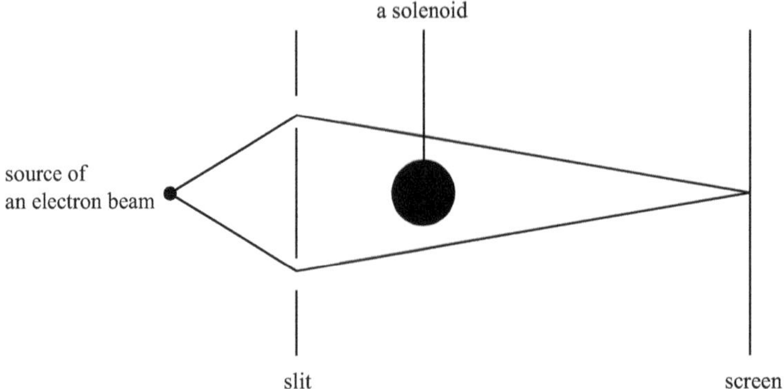

Fig. 3.5 Experimental figure for AB effect

In this section, in the ideal situation where the magnetic field can be regarded as that concentrated on discrete points, we see that there is a beautiful correspondence between the occurrence of the AB effect and the case where π_A is inequivalent to the Schrödinger representation $\pi_S^{(2)}$.

Suppose that the vector potential $\boldsymbol{A}$ is flat. Then the magnetic field B is concentrated on the points $\boldsymbol{a}_1, \cdots, \boldsymbol{a}_N$. Hence the charged particle moving in M does not interact with B directly. As we have already seen, the representation π_A is inequivalent to the Schrödinger representation $\pi_S^{(2)}$ if and only if the magnetic flux of $\boldsymbol{A}$ is not locally quantized. In the case where the magnetic flux is not locally quantized, there exists at least one point $\boldsymbol{a}_n$ such that $\gamma_n(\boldsymbol{A}) \notin 2\pi\mathbb{Z}/q$. Now we take the curve $C(x_1, x_2; s, t)$ such that $\{\boldsymbol{a}_1, \cdots, \boldsymbol{a}_N\} \cap D(x_1, x_2; s, t) = \{\boldsymbol{a}_n\}$. Then we have

$$e^{-isP_1}e^{-itP_2}\psi = \exp\left(-iq\,\mathrm{sgn}(s)\mathrm{sgn}(t)\gamma_n(\boldsymbol{A})\right)e^{-itP_2}e^{-isP_1}\psi, \quad \psi \in L^2(\mathbb{R}^2). \tag{3.20}$$

Hence, for such an s and t, e^{-isP_1} and e^{-itP_2} are not commuting. Following the geometric interpretation described after Theorem 3.2, Eq. (3.20) may be interpreted as follows: Let $s, t > 0$ and

$$C_-\colon (x_1, x_2) \to (x_1 + s, x_2) \to (x_1 + s, x_2 + t),$$

be the lower right part of the curve $C(x_1, x_2; s, t)$ in Fig. 3.4 and

$$C_+\colon (x_1, x_2) \to (x_1, x_2 + t) \to (x_1 + s, x_2 + t)$$

be the upper left part of the curve $C(x_1, x_2; s, t)$ with reversed orientation. Then the left hand side of (3.20) represents the magnetic translation of the state vector ψ along C_+ and the factor $e^{-itP_2}e^{-isP_1}\psi$ on the right hand side of (3.20) represents the magnetic translation of ψ along C_-. Equation (3.20) shows that these magnetic translations do not give the same result and a non-trivial phase factor appears. Physically this means that the phase of the state function of the charged particle starting (x_1, x_2) and arriving at $(x_1 + s, x_2 + t)$ through C_- does not coincide with that through C_+ and a phase shift occurs. This is exactly interpreted as the AB effect in the present context. Thus we see that there is a noteworthy correspondence between the inequivalence of π_A to $\pi_{\mathrm{S}}^{(2)}$ and the occurrence of the AB effect. We emphasize that, in this interpretation, *the AB effect in the present context is purely topological and independent of the details of dynamics of the system (forms of Hamiltonian)*. This suggests that relativistic quantum mechanics also may have aspects related to AB effect.

Remark 3.4 It is well known [157] that, in the case where a magnetic field is confined in a superconductor, the magnetic flux Φ is quantized with $\Phi \in h\mathbb{Z}/2e$, where e is the fundamental charge and h is the Planck constant. Hence, in the unit system where $\hbar = h/2\pi = 1$, Φ is an integer multiple of π/e. Therefore, the quantization of the magnetic flux in a superconductor is a little bit different from the local quantization of the magnetic flux defined in this chapter (by factor two in the case $q = e$). But this difference is very important, because, in the case where the charged particle under consideration is an electron, one has $q = -e$ and, if the magnetic flux penetrating $\boldsymbol{a}_n$ is an odd integer multiple of π/e, then the magnetic flux is not locally quantized. Hence, in this case, in the correspondence described above, the AB effect should occur. In the case where an experiment for the AB effect is made using the magnetic field confined in a superconductor, there are only two patterns of changes of interferences according to the case where the magnetic flux is an even integer multiple of π/e or an odd integer multiple of π/e [157]. This exactly corresponds to the equivalence and the inequivalence of π_A to $\pi_{\mathrm{S}}^{(2)}$ respectively.

3.9 Representations for Different Vector Potentials

In this section, we consider two representations π_A and $\pi_{A'}$ for different vector potentials $\boldsymbol{A}, \boldsymbol{A}' \in \mathscr{A}^1_{\mathrm{flat}}(M)$ and find a necessary and sufficient condition for them to be equivalent.

We first recall an elementary fact. Let $\boldsymbol{A}\colon M \to \mathbb{R}^2$ be continuously differentiable and $\Lambda\colon M \to \mathbb{R}$ be in $C^2(M)$. Then the vector potential

$$\boldsymbol{A}' := \boldsymbol{A} + (\partial_1 \Lambda, \partial_2 \Lambda) \tag{3.21}$$

is called the **gauge transformation** of $\boldsymbol{A}$ with respect to Λ, which is called a **gauge function**. It follows that $\boldsymbol{A}'$ and $\boldsymbol{A}$ yield an identical magnetic field:

$$\partial_1 A_2 - \partial_2 A_1 = \partial_1 A_2' - \partial_2 A_1' \quad (\boldsymbol{A}' = (A_1', A_2')).$$

For $\boldsymbol{A}, \boldsymbol{A}' \in C^1(M)$, we write $\boldsymbol{A} \sim \boldsymbol{A}'$ if there exists a Borel measurable function u on $\mathbb{R}^2$ such that $|u(\boldsymbol{x})| = 1$ a.e. $\boldsymbol{x} \in \mathbb{R}^2$, $D_j u$ $(j = 1, 2)$ is in $L^1_{\text{loc}}(M)$ and

$$A'_j = A_j + iuD_ju^*, \quad j = 1, 2. \tag{3.22}$$

The correspondence $\boldsymbol{A} \mapsto \boldsymbol{A}'$ is a generalization of gauge transformation in the sense that gauge functions are not necessarily in $C^2(M)$.[7]

Taking the complex conjugate of (3.22), we have

$$A'_j = A_j - iu^*D_ju, \quad j = 1, 2. \tag{3.23}$$

Hence (3.22) is equivalent to (3.23).

Lemma 3.8 *The relation* $\sim$ *defined above is an equivalence relation in* $\mathscr{F}^1_{\text{flat}}(M)$.

Proof It is easy to check that the relation $\sim$ satisfies the reflexive and the symmetric relation. To prove that the transitive relation holds for $\sim$, let $\boldsymbol{A} \sim \boldsymbol{A}'$ and $\boldsymbol{A}' \sim \boldsymbol{A}''$. Then there exist Borel measurable functions u, v on $\mathbb{R}^2$ such that $|u| = 1$, $|v| = 1$ and D_ju and D_jv are in $L^1_{\text{loc}}(M)$ satisfying (3.22) and

$$A''_j = A'_j + ivD_jv^*, \quad j = 1, 2.$$

Hence

$$A''_j = A_j + i\left(uD_ju^* + vD_jv^*\right).$$

By a limiting argument using the theory of the Friedrichs mollifier,[8] one can show that $D_j(uv)^*$ is a Borel measurable function with

$$D_j(uv)^* = (D_ju^*)v^* + u^*D_jv^*.$$

Hence $D_j(uv)$ is in $L^1_{\text{loc}}(M)$ and

$$uvD_j(uv)^* = uD_ju^* + vD_jv^*.$$

[7]Note that, if $u = e^{i\Lambda}$ with a real-valued function $\Lambda \in C^2(M)$, then (3.22) gives (3.21).

[8]For each $\varepsilon > 0$, one defines $u_\varepsilon := J_\varepsilon u$, where J_ε is a Friedrichs mollifier. Then $u_\varepsilon \in C^\infty(\mathbb{R}^2)$ and one has $D_j(u_\varepsilon v) = (D_ju_\varepsilon)v + u_\varepsilon D_jv$. Hence, for all $f \in C_0^\infty(M)$,

$$-\int_M u_\varepsilon(\boldsymbol{x})v(\boldsymbol{x})D_jf(\boldsymbol{x})d\boldsymbol{x} = \int_M (D_ju_\varepsilon)(\boldsymbol{x})v(\boldsymbol{x})f(\boldsymbol{x})d\boldsymbol{x} + \int_M u_\varepsilon(\boldsymbol{x})(D_jv)(\boldsymbol{x})f(\boldsymbol{x})d\boldsymbol{x}.$$

Then one takes the limit $\varepsilon \to 0$, where the following facts are used: u_ε and D_ju_ε converge to u and D_ju respectively as $\varepsilon \to 0$ in $L^1(K)$-norm for any compact set K of M.

Obviously $|uv| = 1$. Therefore

$$A''_j = A_j + iuvD_j(uv)^*.$$

Thus $\boldsymbol{A} \sim \boldsymbol{A}''$. ∎

The next theorem characterizes the equivalence between two representations $\pi_{\boldsymbol{A}}$ and $\pi_{\boldsymbol{A}'}$.

Theorem 3.8 *Let $\boldsymbol{A}, \boldsymbol{A}' \in \mathscr{A}^1_{\text{flat}}(M)$. Then $\pi_{\boldsymbol{A}}$ is equivalent to $\pi_{\boldsymbol{A}'}$ if and only if $q\boldsymbol{A} \sim q\boldsymbol{A}'$.*

Proof We denote the physical momentum P_j with vector potential $\boldsymbol{A}$ by $P_j(\boldsymbol{A})$. Suppose that $\pi_{\boldsymbol{A}}$ is equivalent to $\pi_{\boldsymbol{A}'}$: $\pi_{\boldsymbol{A}} \cong \pi_{\boldsymbol{A}'}$. Then there exists a unitary operator U on $L^2(\mathbb{R}^2)$ such that

$$U\hat{q}_jU^{-1} = \hat{q}_j, \quad UP_j(\boldsymbol{A})U^{-1} = P_j(\boldsymbol{A}'), \quad j = 1, 2. \tag{3.24}$$

The first equation and Lemma 2.11 imply that U is a multiplication operator by an essentially bounded Borel measurable function u on $L^2(\mathbb{R}^2)$. By the unitarity of U, $|u| = 1$. It follows from the second equation in (3.24) and the unitary covariance of functional calculus that, for all $f \in L^2(\mathbb{R}^2)$ and $t \in \mathbb{R}$,

$$ue^{itP_j(\boldsymbol{A})}f = e^{itP_j(\boldsymbol{A}')}uf.$$

Using Theorem 3.2 and the arbitrariness of f, we obtain

$$u(\boldsymbol{x} + t\boldsymbol{e}_j) = e^{i\theta_j(t,\boldsymbol{x})}u(\boldsymbol{x}) \quad \text{a.e.}\boldsymbol{x} \tag{3.25}$$

with $\theta_j(t, \boldsymbol{x}) := q\int_0^t (A'_j(\boldsymbol{x} + s\boldsymbol{e}_j) - A_j(\boldsymbol{x} + s\boldsymbol{e}_j))ds$.[9] Hence, by a simple limiting argument, we have for all $f \in C_0^\infty(M)$

$$\int_M u(\boldsymbol{x})\partial_j f(\boldsymbol{x})d\boldsymbol{x} = \lim_{t\to 0}\int_M \frac{e^{i\theta_j(-t,\boldsymbol{x})} - 1}{t}u(\boldsymbol{x})f(\boldsymbol{x})d\boldsymbol{x}.$$

We have

$$\left|\frac{e^{i\theta_j(-t,\boldsymbol{x})} - 1}{t}u(\boldsymbol{x})f(\boldsymbol{x})\right| \le \frac{|\theta_j(-t, \boldsymbol{x})|}{|t|}|f(\boldsymbol{x})| \le |q|C|f(\boldsymbol{x})|, \quad 0 < |t| \le \delta, \quad \boldsymbol{x} \in M$$

---

[9] Note that the set of the points $\boldsymbol{x}$ on which (3.25) holds may depend on t. Hence one cannot conclude from the equation that $u(\boldsymbol{x})$ is partially differentiable in x_j $(j = 1, 2)$ with

$$\partial_j u(\boldsymbol{x}) = iq(A'_j(\boldsymbol{x}) - A_j(\boldsymbol{x}))u(\boldsymbol{x}) \quad \text{a.e.}\boldsymbol{x}.$$

The argument below gives a rigorous proof for the existence of $D_j u$ which is equal to the right hand side of this equation.

with

$$C := \sup_{\boldsymbol{x} \in \operatorname{supp} f,\, |s| \le \delta} (|A_j(\boldsymbol{x} + s\boldsymbol{e}_j)| + |A'_j(\boldsymbol{x} + s\boldsymbol{e}_j)|) < \infty,$$

where $\delta \in (0, 1)$ is taken sufficiently small so that $\boldsymbol{x} + s\boldsymbol{e}_j \in M$ for all $\boldsymbol{x} \in \operatorname{supp} f$ and $|s| \le \delta$, and

$$\lim_{t \to 0} \frac{e^{i\theta_j(-t,\boldsymbol{x})} - 1}{t} u(\boldsymbol{x}) f(\boldsymbol{x}) = -iq(A'_j(\boldsymbol{x}) - A_j(\boldsymbol{x})).$$

Hence, by the Lebesgue dominated convergence theorem, we obtain

$$\int_M u(\boldsymbol{x}) \partial_j f(\boldsymbol{x}) d\boldsymbol{x} = -iq \int_M (A'_j(\boldsymbol{x}) - A_j(\boldsymbol{x})) u(\boldsymbol{x}) f(\boldsymbol{x}) d\boldsymbol{x}.$$

This means that

$$D_j u = iq(A'_j - A_j)u \in L^1_{\mathrm{loc}}(M)$$

and $qA'_j = qA_j - iu^* D_j u$. Hence $q\boldsymbol{A} \sim q\boldsymbol{A}'$ (see (3.23)).

Conversely, suppose that $q\boldsymbol{A} \sim q\boldsymbol{A}'$. Then there exists a Borel measurable function u on $\mathbb{R}^2$ such that $|u| = 1$, $D_j u \in L^1_{\mathrm{loc}}(M)$ and (3.22) holds. Let $U = M_{u^*}$ be the multiplication operator by u^*. Then U is unitary. It is easy to see that $UA'_jU^{-1} = A'_j$ and

$$U\hat{p}_jU^{-1} = \hat{p}_j - iu^*(D_j u) \quad \text{on } U_j\mathscr{D}_j,$$

where U_j is defined by (3.7) (note that each element f of $U_j\mathscr{D}_j$ is continuous on M_j and partially differentiable in x_j with compact support in M_j and hence $D_j(uf) = (D_j u)f + uD_j f$). Hence

$$\begin{aligned} U(\hat{p}_j - qA'_j)U^{-1} &= U\hat{p}_jU^{-1} - qA'_j \\ &\supset (\hat{p}_j - iu^*(D_j u) - qA'_j) \restriction U_j\mathscr{D}_j \\ &= (\hat{p}_j - qA_j) \restriction U_j\mathscr{D}_j. \end{aligned}$$

By Theorem 3.1, $\hat{p}_j - qA_j$ is essentially self-adjoint on $U_j\mathscr{D}_j$. Hence

$$\overline{U(\hat{p}_j - qA'_j)U^{-1}} \supset P_j(\boldsymbol{A}).$$

The left hand side is equal to $UP_j(\boldsymbol{A}')U^{-1}$, which is self-adjoint. Since $P_j(\boldsymbol{A})$ also is self-adjoint, it follows that $UP_j(\boldsymbol{A}')U^{-1} = P_j(\boldsymbol{A})$. It is obvious that $Uq_jU^{-1} = q_j$. Thus $\pi_{\boldsymbol{A}} \cong \pi_{\boldsymbol{A}'}$. ■

The representatives in the representation $\pi_{\boldsymbol{A}}$ with $\boldsymbol{A} = \boldsymbol{0}$ are those of the Schrödinger representation $\pi_{\mathrm{S}}^{(2)}$. Hence Theorem 3.8 yields the following fact:

Corollary 3.1 *Let $\boldsymbol{A} \in \mathscr{A}_{\mathrm{flat}}^{1}(M)$. Then $\pi_{\boldsymbol{A}} \cong \pi_{\mathrm{S}}^{(2)}$ if and only if $q\boldsymbol{A} \sim \boldsymbol{0}$.*

3.10 Charges and Inequivalent Irreducible Representations of CCR

In this section we consider the q-dependence of $\pi_{\boldsymbol{A}}$. Let $\boldsymbol{A} \in \mathscr{A}_{\mathrm{flat}}^{1}(M) \setminus \{0\}$ be fixed ($\boldsymbol{A} \neq \boldsymbol{0}$) and regard $q \in \mathbb{R} \setminus \{0\}$ as a parameter indexing the representation $\pi_{\boldsymbol{A}}$. From this point of view, we write

$$\pi_q := \pi_{\boldsymbol{A}}. \tag{3.26}$$

For convenience, we set

$$\pi_0 := \pi_{\mathrm{S}}^{(2)}.$$

Corollary 3.2 *Let $q, q' \in \mathbb{R} \setminus \{0\}$. Then $\pi_{q'}$ is equivalent to π_q if and only if $(q' - q)\boldsymbol{A} \sim \boldsymbol{0}$.*

Proof By Theorem 3.8, $\pi_{q'}$ is equivalent to π_q if and only if $q'\boldsymbol{A} \sim q\boldsymbol{A}$, which is equivalent to $(q' - q)\boldsymbol{A} \sim \boldsymbol{0}$. ∎

Corollaries 3.1 and 3.2 yield the following result:

Corollary 3.3 *Let $q, q' \in \mathbb{R} \setminus \{0\}$. Then $\pi_{q'}$ is equivalent to π_q if and only if $\pi_{q'-q}$ is equivalent to $\pi_{\mathrm{S}}^{(2)}$.*

By this corollary and Theorem 3.6, one obtains the following result:

Corollary 3.4 *Let $q, q' \in \mathbb{R}$. Then $\pi_{q'}$ is equivalent to π_q if and only if $(q' - q)\gamma_n(\boldsymbol{A}) \in 2\pi\mathbb{Z}$ for all $n = 1, \ldots, N$.*

To rewrite the condition "$(q' - q)\gamma_n(\boldsymbol{A}) \in 2\pi\mathbb{Z}$ for all $n = 1, \ldots, N$" in a simpler form, we recall an equivalence relation in $\mathbb{R}$.

Let $\lambda \in \mathbb{R}$ be a constant and, for two real numbers a, b, define a relation $a \sim b$ by

$$a \sim b \overset{\text{def}}{\Longleftrightarrow} \exists n \in \mathbb{Z} \text{ such that } a - b = n\lambda.$$

It is easy to see that this relation $\sim$ is an equivalence relation in $\mathbb{R}$. We write $a \equiv b \pmod{\lambda}$ to mean that $a \sim b$.

Corollary 3.5 *Let $q, q' \in \mathbb{R}$ and suppose that at least one of $\gamma_1(\boldsymbol{A}), \ldots, \gamma_{N-1}(\boldsymbol{A})$ and $\gamma_N(\boldsymbol{A})$ is not zero. Then $\pi_{q'}$ is inequivalent to π_q if and only if there exists an $n_0 \in \{1, \ldots, N\}$ such that $\gamma_{n_0}(\boldsymbol{A}) \neq 0$ and*

$$q' \not\equiv q \left(\mathrm{mod}\ \frac{2\pi}{\gamma_{n_0}}\right).$$

Proof By Corollary 3.4, $\pi_{q'}$ is equivalent to π_q if and only if $q' \equiv q$ $(\mathrm{mod}\ 2\pi/\gamma_n(\boldsymbol{A}))$ for all $n = 1, \ldots, N$ such that $\gamma_n(\boldsymbol{A}) \neq 0$. Taking the contraposition of this statement, one obtains the desired result. ■

To see a physical meaning of Corollary 3.5 more concretely, we consider the case where $\gamma_{n_j}(\boldsymbol{A}) \neq 0$ for $n_1, \ldots, n_r \in \{1, \ldots, N\}$ $(1 \leq r \leq N)$ and $\gamma_n(\boldsymbol{A}) = 0$ for $n \notin \{n_1, \ldots, n_r\}$. Let

$$\lambda_j := \frac{2\pi}{\gamma_{n_j}(\boldsymbol{A})}, \quad j = 1, \ldots, r$$

and

$$\ell := \min_{j=1,\ldots,r} |\lambda_j|.$$

Then, for all $a \in \mathbb{R}$ and $q, q' \in [a, a+\ell]$ such that $q \neq q'$, $q' \not\equiv q$ $(\mathrm{mod}\ \lambda_j)$, $j = 1, \ldots, r$. Hence, by Corollary 3.5, $\{\pi_q\}_{q\in[a,a+\ell]}$ is a family of mutually inequivalent irreducible self-adjoint representations of the CCR with two degrees of freedom. This shows that electric charges appear as an index set of mutually inequivalent irreducible self-adjoint representations of the CCR with two degrees of freedom. This is an interesting physical correspondence.

3.11 Notes

The contents of this chapter originate from papers [133] and [9] (we refer to [81] as an earlier study of the Aharonov–Bohm effect from a representation-theoretical point of view). Further developments are found in [10–13, 49]. Geometric constructions of the representation $\pi_{\boldsymbol{A}}$ are given by Kurose and Nakazato [102]. An extension to the case where the magnetic field penetrates the interior of a circle is discussed in [90, 91]. Scattering theory in relation to the Aharonov–Bohm effect is discussed in [138]. The contents of Sects. 3.9 and 3.10 are newly written for the present book.

Chapter 4
Time Operators

Abstract A general theory of time operators is presented as part of representation theory of CCR.

4.1 Introduction

A notion of physical time may not be unique depending on regions of physical phenomena and hence on categories of physical theory. For example, the notion of time in Galilean mechanics is different from that in Einstein's theory of relativity. But it is common in both theories that time is physically measured by a laboratory clock which is independent of dynamics of systems studied in experiments. This notion of time is called *external time* [56].[1]

In quantum mechanics, the state vector $\Psi(t)$ at time $t \in \mathbb{R}$ in a quantum system with Hamiltonian H (a self-adjoint operator on a Hilbert space $\mathscr{H}$) is given by $\Psi(t) = e^{-itH}\Psi$, where $\Psi \in \mathscr{H}$ is the initial state vector (the state vector at time zero). But the parameter $t \in \mathbb{R}$ appearing here should be interpreted as an external time (note also that t is not an operator acting in $\mathscr{H}$).

On the other hand, in Einstein's theory of special relativity, each point x in the four-dimensional Minkowski space–time $\mathbb{M}^4$ is represented as $x = (t, x^1, x^2, x^3)$ in a Lorentz coordinate system S_L with t and x^j being the time variable and the jth space variable respectively. Mathematically, a momentum variable p can be regarded as an element of the dual space $(\mathbb{M}^4)^*$ of $\mathbb{M}^4$ and, in its representation $p = (E, p_1, p_2, p_3)$ in S_L, E and p_j denote the energy and the jth momentum variable respectively. The variable p_j is the canonical conjugate variable of x^j. Hence one can infer that E may be the canonical variable of the time variable t.

In quantum mechanics in $\mathbb{R}^d$ ($d \in \mathbb{N}$), as we have seen in Sect. 2.6.1, the position operators $\{\hat{q}_1, \ldots, \hat{q}_d\}$ and the momentum operators $\{\hat{p}_1, \ldots, \hat{p}_d\}$ obey the CCR with degree d. Hence, in view of formal (heuristic) classical–quantum

[1] It is called also *pragmatic* or *laboratory time*.

© Springer Nature Singapore Pte Ltd. 2020

A. Arai, *Inequivalent Representations of Canonical Commutation and Anti-Commutation Relations*, Mathematical Physics Studies,
https://doi.org/10.1007/978-981-15-2180-5_4

correspondence, it may be natural to consider, in quantum mechanics, a symmetric operator T on $\mathscr{H}$ such that

$$[T, H] = i \tag{4.1}$$

on a suitable subspace of $\mathscr{H}$ and to call it a **time operator**.[2]

Some remarks may be necessary on time operators. A germ of a notion of time operators is already found in Heisenberg's paper [87, pp. 177–179] in 1927. But, in 1933, Pauli stated [125, p. 63, footnote 2] that the introduction of a time observable T (a self-adjoint operator) satisfying CCR (4.1) with H being bounded from below and having a discrete eigenvalue is basically forbidden. Since then, studies on time observables seem to have been ruled out for many years without any questions. But *Pauli's statement is false*, as we shall show by an explicit construction of a time operator (see Sect. 4.4). In addition, as we shall demonstrate below, there is not only one class of time operators, but also there are five classes of time operators. It is important to understand this point. Each class of time operators may play different roles in physics from the others.

In standard textbooks in quantum mechanics, the time–energy uncertainty relation

$$\Delta E \Delta t \geq \frac{1}{2}$$

is treated ambiguously, since the definition of the uncertainty ΔE (resp. Δt) of energy (resp. time t) is ambiguous and seems to vary from case to case. In fact, there are some types of time–energy uncertainty relations [56]. As we shall see below, a notion of a time operator gives a mathematically rigorous basis to time–energy uncertainty relation.

Before going into details of a general theory of time operators, here we look at simple examples of time operators.

Example 4.1 Let us consider a quantum system of a non-relativistic quantum particle of mass $m > 0$ under the action of a constant force $F \in \mathbb{R} \setminus \{0\}$ in the one-dimensional space $\mathbb{R}$, which is identified with the x-axis.

An example of such a system is a system of a quantum particle under the influence of the gravitational force $F = mg$, where g is the acceleration of gravity and the x-axis is taken to be the vertical line directed to the center of the earth. Another example is a system of a quantum charged particle of charge $q \in \mathbb{R} \setminus \{0\}$

[2] There is another route to arrive at the notion of a time operator heuristically. In the Hamilton formalism in classical mechanics, the equation of motion for t interpreted as a dynamical variable formally yields $\frac{dt}{dt} = \{t, H_{\rm cl}\}_{\rm PB}$, where $H_{\rm cl}$ is a classical Hamiltonian and $\{\ ,\ \}_{\rm PB}$ is the Poisson bracket. The left hand side of this equation is equal to 1. Hence $\{t, H_{\rm cl}\}_{\rm PB} = 1 \cdots (*)$. One has also $\{x^j, p_k\}_{\rm PB} = \delta_{jk}$ $(j, k = 1, \ldots, d)$. The quantum mechanical correspondent to this relation is $[\hat{q}_j, \hat{p}_k] = i\delta_{jk}$. Hence one infers that the quantum mechanical relation corresponding to $(*)$ should be $[T, H] = i$.

under the influence of a constant electric field $E \in \mathbb{R} \setminus \{0\}$ with direction of the x-axis. In this case, $F = qE$ and this kind of system with an additional scalar potential may cause the so-called **Stark effect**, splittings of energy levels of the original atom by the influence of the electric field.

A natural Hilbert space of state vectors of the system is $L^2(\mathbb{R})$. Then the Hamiltonian of the system is given by

$$H_F := \frac{1}{2m}\hat{p}^2 - F\hat{q}$$

acting in $L^2(\mathbb{R})$, where $\hat{q}$ and $\hat{p}$ are the representatives in the Schrödinger representation $\pi_S^{(1)}$ with one degree of freedom. It is shown that H_F is essentially self-adjoint on $C_0^\infty(\mathbb{R})$.[3] Hence its closure $\overline{H}_F$ is self-adjoint (but, note that $\overline{H}_F$ is bounded neither from below nor from above). The self-adjoint operator

$$T_F = \frac{1}{F}\hat{p} \tag{4.2}$$

satisfies the CCR

$$[T_F, \overline{H}_F] = i$$

on $C_0^\infty(\mathbb{R})$. Hence T_F is a time operator of $\overline{H}_F$.[4]

Note that $(L^2(\mathbb{R}), C_0^\infty(\mathbb{R}), (T_F, \overline{H}_F))$ is a self-adjoint representation of the CCR with one degree of freedom, i.e.,

$$(L^2(\mathbb{R}), C_0^\infty(\mathbb{R}), (T_F, \overline{H}_F)) \in \mathsf{CCR}_{\mathrm{sa}}(1).$$

Hence, by Theorem 2.1(ii), one has the Heisenberg uncertainty relation

$$(\Delta \overline{H}_F)_\Psi (\Delta T_F)_\Psi \geq \frac{1}{2}$$

for all unit vectors $\Psi \in D([T_F, \overline{H}_F])$. This inequality may be interpreted as a form of *time–energy uncertainty relation* in the present model.

[3] Apply the commutator theorem (Theorem 1.13) with $\mathscr{H} = L^2(\mathbb{R})$, $\mathscr{D} = C_0^\infty(\mathbb{R})$, $S = \hat{p}^2 + \hat{q}^2 \geq 1$ and $A = H_F$.

[4] A heuristic argument to find T_F is as follows: in the corresponding model in classical mechanics, the Newton equation of motion takes the form $dp(t)/dt = F$, where $p(t) \in \mathbb{R}$ is the momentum at time t. Hence $p(t) = p(0) + Ft$. Therefore, under the initial condition $p(0) = 0$, one has $t = p(t)/F$.

Example 4.2 The free Hamiltonian of the system in Example 4.1 is given by the non-negative self-adjoint operator

$$H_{\text{NR}} := \frac{1}{2m}\hat{p}^2$$

acting in $L^2(\mathbb{R})$.

A heuristic argument to find a time operator of H_{NR} is as follows: in the corresponding model in classical mechanics, the Newton equation of motion is $dp(t)/dt = 0$. Hence $p(t) = p$ (a constant), implying that the position $q(t)$ of the particle at time t is given by $q(t) = pt/m$ with initial condition $q(0) = 0$ (recall that $dq(t)/dt = p(t)/m$). Hence, for $p \neq 0$, $t = mq(t)p(t)^{-1} = (m/2)(q(t)p(t)^{-1} + p(t)^{-1}q(t))$.

Based on the heuristic argument in the preceding paragraph, one is led to define the operator

$$T_{\text{AB}} := \frac{m}{2}(\hat{q}\hat{p}^{-1} + \hat{p}^{-1}\hat{q}) \tag{4.3}$$

as a candidate of a time operator of H_{NR} (note that $\hat{p}$ is an injective self-adjoint operator and hence $\hat{p}^{-1}$ exists as a self-adjoint operator). Indeed, this is a time operator of H_{NR} as is shown below; it may be called the the **Aharonov–Bohm time operator** [4] for the one-dimensional free quantum system under consideration.

It is easy to see that the subspace

$$\mathscr{D}_{\text{AB}} := \mathscr{F}_1^{-1}C_0^\infty(\hat{\mathbb{R}} \setminus \{0\}) = \{f \in L^2(\mathbb{R}) | \hat{f} \in C_0^\infty(\hat{\mathbb{R}} \setminus \{0\})\} \tag{4.4}$$

is dense in $L^2(\mathbb{R})$ (note that $C_0^\infty(\hat{\mathbb{R}} \setminus \{0\})$ is dense in $L^2(\hat{\mathbb{R}})$ and the inverse Fourier transform $\mathscr{F}_1^{-1}$ is unitary). Moreover, $\mathscr{D}_{\text{AB}} \subset D(\hat{q}) \cap D(\hat{p}^{-1})$ (use that $\mathscr{F}_1\hat{q}\mathscr{F}_1^{-1} = iD_k$ and $\mathscr{F}_1\hat{p}^{-1}\mathscr{F}_1^{-1} = 1/k$) and

$$\hat{q}\mathscr{D}_{\text{AB}} \subset \mathscr{D}_{\text{AB}}, \quad \hat{p}^{-1}\mathscr{D}_{\text{AB}} \subset \mathscr{D}_{\text{AB}}. \tag{4.5}$$

Hence $\mathscr{D}_{\text{AB}} \subset D(T_{\text{AB}})$ and $T_{\text{AB}}\mathscr{D}_{\text{AB}} \subset \mathscr{D}_{\text{AB}}$. Therefore T_{AB} is densely defined and it holds that

$$T_{\text{AB}}^* \supset \frac{m}{2}(\hat{p}^{-1}\hat{q} + \hat{q}\hat{p}^{-1}) = T_{\text{AB}}.$$

Hence T_{AB} is symmetric. Moreover $\hat{p}\mathscr{D}_{\text{AB}} \subset \mathscr{D}_{\text{AB}}$ and hence $H_{\text{NR}}\mathscr{D}_{\text{AB}} \subset \mathscr{D}_{\text{AB}}$. Therefore, for all $f \in \mathscr{D}_{\text{AB}}$,

$$\begin{aligned}[T_{AB}, H_{\text{NR}}]f &= \frac{1}{4}([\hat{q}\hat{p}^{-1}, \hat{p}^2]f + [\hat{p}^{-1}\hat{q}, \hat{p}^2]f) \\ &= \frac{1}{4}([\hat{q}, \hat{p}^2]\hat{p}^{-1}f + \hat{p}^{-1}[\hat{q}, \hat{p}^2]f).\end{aligned}$$

By the CCR for $(\hat{q}, \hat{p})$, we have $[\hat{q}, \hat{p}^2] = 2i\hat{p}$ on $\mathscr{D}_{\mathrm{AB}}$. Hence we obtain

$$[T_{\mathrm{AB}}, H_{\mathrm{NR}}] = i$$

on $\mathscr{D}_{\mathrm{AB}}$. Thus $(L^2(\mathbb{R}), \mathscr{D}_{\mathrm{AB}}, (T_{\mathrm{AB}}, H_{\mathrm{NR}}))$ is in $\mathsf{CCR}(1)$ and T_{AB} is a time operator of H_{NR}.

Example 4.3 We consider a free relativistic particle of mass $m \geq 0$ moving in $\mathbb{R}$. The Hamiltonian of it is given by the free relativistic Schrödinger operator

$$H_{\mathrm{R}} := (\hat{p}^2 + m^2)^{1/2} \tag{4.6}$$

defined by the functional calculus for the self-adjoint operator $\hat{p}$. The operator H_{R} is a non-negative self-adjoint operator with $H_{\mathrm{R}} \geq m$. Let

$$\omega_1(k) := (k^2 + m^2)^{1/2}, \quad k \in \hat{\mathbb{R}}, \tag{4.7}$$

which physically means the energy of the free relativistic particle with momentum k. It follows that

$$\mathscr{F}_1 H_{\mathrm{R}} \mathscr{F}_1^{-1} = \omega_1,$$

the multiplication operator by the function ω_1 on $L^2(\hat{\mathbb{R}})$. In correspondence to classical mechanics as in Example 4.2,[5] we introduce the operator

$$T_{\mathrm{R}} := \frac{1}{2}(H_{\mathrm{R}}\hat{p}^{-1}\hat{q} + \hat{q}\hat{p}^{-1}H_{\mathrm{R}}) \tag{4.8}$$

as a candidate of a time operator of H_{R}. Then, as in the same manner as in the case of the Aharonov–Bohm time operator T_{AB}, one can show that $\mathscr{D}_{\mathrm{AB}} \subset D(T_{\mathrm{R}}) \cap D(H_{\mathrm{R}})$,

[5]The Hamilton equation of motion of the corresponding classical model is the set of the following equations:

$$\frac{dq(t)}{dt} = \frac{\partial H_{\mathrm{cl}}(p(t))}{\partial p(t)} = \frac{p(t)}{H_{\mathrm{cl}}(p(t))},$$

$$\frac{dp(t)}{dt} = -\frac{\partial H_{\mathrm{cl}}(p(t))}{\partial q(t)} = 0,$$

where $H_{\mathrm{cl}}(p) := \sqrt{p^2 + m^2}$ is the classical Hamiltonian corresponding to H_{R}. Hence $p(t) = p$ (a constant) and $q(t) = pt/H_{\mathrm{cl}}(p)$ with $q(0) = 0$. Hence

$$t = H_{\mathrm{cl}}(p)q(t)p^{-1} = \frac{1}{2}\left(H_{\mathrm{cl}}(p)p^{-1}q(t) + q(t)p^{-1}H_{\mathrm{cl}}(p)\right).$$

$T_R \mathscr{D}_{AB} \subset \mathscr{D}_{AB}$, $H_R \mathscr{D}_{AB} \subset \mathscr{D}_{AB}$ and T_R is symmetric. By direct commutations, we have

$$[T_R, H_R] = i$$

on $\mathscr{D}_{AB}$. Hence $(L^2(\mathbb{R}), \mathscr{D}_{AB}, (T_R, H_R))$ is in $\mathsf{CCR}(1)$ and T_R is a time operator of H_R.

Remark 4.1 For each $d \geq 2$, d-dimensional extensions of the above examples are easily constructed, although we do not write them down here.

As is easily noted, all the above examples of the pair (T, H) of a Hamiltonian H and a time operator T (resp. their d-dimensional extensions) are formed out of the Schrödinger representation $\pi_S^{(1)}$ (resp. $\pi_S^{(d)}$). Therefore they are secondary representations of CCR (see (QM3) in Sect. 2.6).

Remark 4.2 There are approaches to time operators using positive operator-valued measures (e.g., [55, 115]) (cf. also [53, 54, 57]). But, in this book, we do not review these approaches. This chapter is intended to describe a mathematical theory of time operators from viewpoints of (inequivalent) representations of CCR. For physical discussions of time operators, we refer the reader to [115, 116] and references therein.

4.2 An Abstract Definition of Time Operator and Time-Energy Uncertainty Relation

From a purely mathematical point of view, a time operator T having property (4.1) is regarded as one of the representatives of a representation of the CCR with degree one. From this point of view, we redefine the notion of a time operator as follows:

Definition 4.1 Let H be a self-adjoint operator on a Hilbert space $\mathscr{H}$. A symmetric operator T on $\mathscr{H}$ is called a **time operator** of H if there exists a (not necessarily dense) subspace $\mathscr{D} \neq \{0\}$ of $\mathscr{H}$ such that $(\mathscr{H}, \mathscr{D}, (T, H))$ is in $\mathsf{CCR}(1)$.

The name "time operator" here is just a sign to indicate that, in applications to quantum mechanics, one has in mind the case where H is the Hamiltonian of a quantum system. But, in the context of pure mathematics, there are no restrictions on H other than self-adjointness.

We also remark on the above definition that a time operator T of H may not be unique. Indeed, if C is a symmetric operator on $\mathscr{H}$ such that $\mathscr{D} \subset D([C, H])$ and $[C, H] = 0$ on $\mathscr{D}$ (i.e., C commutes with H on $\mathscr{D}$), then $[T + C, H] = i$ on $\mathscr{D}$. Hence $T + C$ is also a time operator of H. Therefore it would be natural to introduce the set of time operators of H

$$\mathfrak{T}(H) := \{T \in \mathfrak{L}(\mathscr{H}) | T \text{ is a time operator of } H\}. \tag{4.9}$$

There may be the case $\mathfrak{T}(H) = \emptyset$. But, as the examples in the preceding section show, one can expect that there are some general classes of H such that $\mathfrak{T}(H) \neq \emptyset$. In this section, we show this. But, before doing that, we first note the following fact:

Theorem 4.1 (A Form of Time-Energy Uncertainty Relation) *Let $T \in \mathfrak{T}(H)$ with a CCR-domain $\mathscr{D}$. Then, for all unit vectors $\Psi \in \mathscr{D}$,*

$$(\Delta T)_\Psi (\Delta H)_\Psi \geq \frac{1}{2}. \tag{4.10}$$

Proof This follows from Theorem 2.1(i). ■

Remark 4.3 A theory of a time–energy uncertainty relation without using the notion of a time operator is developed in [127], which includes generalizations of Wigner's time–energy uncertainty relation [163] (cf. also [15, §9]).

In the context of quantum mechanics where H is the Hamiltonian of a quantum system having a time operator, Theorem 4.1 shows that there are no unit state vectors within $\mathscr{D}$ in which the uncertainty (standard deviation) of H is zero.

The set $\mathfrak{T}(H)$ of time operators is closed under convex combinations with a suitable condition:

Proposition 4.1 *Let $N \in \mathbb{N}$ with $N \geq 2$, $T_1, \ldots, T_N \in \mathfrak{T}(H)$ and $\mathscr{D}_j$ be a CCR-domain for (T_j, H) $(j = 1, \ldots, N)$. Suppose that $\cap_{j=1}^N \mathscr{D}_j \neq \{0\}$ and $\cap_{j=}^N D(T_j)$ is dense in $\mathscr{H}$. Then, for all $(a_1, \ldots, a_N) \in \mathbb{R}^N$ satisfying*

$$\sum_{j=1}^N a_j = 1,$$

$\sum_{j=1}^N a_j T_j \in \mathfrak{T}(\mathscr{H})$ with $\cap_{j=1}^N \mathscr{D}_j$ being a CCR-domain for $(\sum_{j=1}^N a_j T_j, H)$.

Proof Since $\cap_{j=1}^N D(T_j)$ is dense, it follows that $T := \sum_{j=1}^N a_j T_j$ is symmetric. Let $\mathscr{D} := \cap_{j=1}^N \mathscr{D}_j$. Then, for all $\Psi \in \mathscr{D}$, $\Psi \in D(TH) \cap D(HT)$ and

$$[T, H]\Psi = \sum_{j=1}^N a_j [T_j, H]\Psi = \sum_{j=1}^N a_j i\Psi = i\Psi.$$

Hence $T \in \mathfrak{T}(H)$ with $\mathscr{D}$ being a CCR-domain for (T, H). ■

For a perturbation of H, here we state only a simple fact:

Proposition 4.2 *Let $T \in \mathfrak{T}(H)$ with a CCR-domain $\mathscr{D}$ and H_1 be a symmetric operator on $\mathscr{H}$ such that $H + H_1$ is essentially self-adjoint. Suppose that $\mathscr{D} \subset D(TH_1) \cap D(H_1 T)$ and T commutes with H_1 on $\mathscr{D}$: $[T, H_1] = 0$ on $\mathscr{D}$. Then T is a time operator of $\overline{H + H_1}$ with $\mathscr{D}$ being a CCR-domain for $(T, \overline{H + H_1})$.*

Proof An easy exercise. ■

4.3 A Structure Generating Pairs of a Hamiltonian and a Time Operator

Let H be a self-adjoint operator on a Hilbert space $\mathscr{H}$ and $T \in \mathfrak{T}(H)$ with a CCR-domain $\mathscr{D}$. We assume that $\mathscr{D}$ is dense in $\mathscr{H}$ and

$$T\mathscr{D} \subset \mathscr{D}, \quad H\mathscr{D} \subset \mathscr{D}. \tag{4.11}$$

Lemma 4.1 *For all* $n \in \mathbb{N}$,

$$[T, H^n] = inH^{n-1} \quad \text{on } \mathscr{D}. \tag{4.12}$$

Proof This can be proved by induction in n using the CCR $[T, H] = i$ on $\mathscr{D}$. ■

We denote by $\mathscr{P}(\mathbb{R})$ the set of all real polynomials of degree more than or equal to 1. Namely, each $P \in \mathscr{P}(\mathbb{R})$ is of the form

$$P(\lambda) = \sum_{j=1}^{n} a_j \lambda^j, \quad \lambda \in \mathbb{R}$$

for some $n \in \mathbb{N}$ with $a_n \in \mathbb{R} \setminus \{0\}$, $a_{n-1}, \ldots, a_1, a_0 \in \mathbb{R}$. For this polynomial P, the operator

$$P(H) := \sum_{j=1}^{n} a_j H^j$$

is self-adjoint with $D(P(H)) = D(H^n)$. We denote by P' the derivative of P. It follows from Lemma 4.1 that

$$[T, P(H)] = iP'(H) \quad \text{on } \mathscr{D}. \tag{4.13}$$

We introduce a subset of $\mathscr{P}(\mathbb{R})$. We say that $P \in \mathscr{P}(\mathbb{R})$ is in the set $\mathscr{P}_H(\mathbb{R})$ if the following conditions are satisfied:

(i) $P'(H)$ is injective;
(ii) $\mathscr{D}$ is an invariant subspace of $P'(H)^{-1}$.

It follows that, for each $P \in \mathscr{P}_H(\mathbb{R})$, the operator

$$T_P := \frac{1}{2}(TP'(H)^{-1} + P'(H)^{-1}T) \tag{4.14}$$

is a symmetric operator on $\mathscr{H}$ with $D(T_P) \supset \mathscr{D}$ and $T_P\mathscr{D} \subset \mathscr{D}$.

Theorem 4.2 *Let $P \in \mathscr{P}_H(\mathbb{R})$. Then T_P is a time operator of $P(H)$, i.e., $T_P \in \mathfrak{T}(P(H))$.*

Proof By the functional calculus for H, $[P'(H)^{-1}, P(H)] = 0$ on $\mathscr{D}$. Hence, for all $\Psi \in \mathscr{D}$,

$$[T_P, P(H)]\Psi = \frac{1}{2}([T, P(H)]P'(H)^{-1}\Psi + P'(H)^{-1}[T, P(H)]\Psi).$$

Using (4.13), one sees that the right hand side is equal to $i\Psi$. Hence $[T_P, P(H)] = i$ on $\mathscr{D}$. Thus T_P is a time operator of $P(H)$. ∎

Theorem 4.2 shows that, for each representation $(\mathscr{H}, \mathscr{D}, (T, H)) \in \mathsf{CCR}(1)$ with $\mathscr{D}$ left invariant by T and H and for each $P \in \mathscr{P}_H(\mathbb{R})$, there exists a representation $(\mathscr{H}, \mathscr{D}, (T_P, P(H))) \in \mathsf{CCR}(1)$ with $\mathscr{D}$ left invariant by T_P and $P(H)$. Therefore we have a sequence $\tau_n = (\mathscr{H}, \mathscr{D}, (T_n, H_n))$ $(n \in \mathbb{Z}_+)$ in $\mathsf{CCR}(1)$ satisfying the following recurrence formula:

$$\begin{array}{ll}
T_0 = T, & H_0 = H, \\
T_1 = T_{P_1}, & H_1 = P_1(H), \quad P_1 \in \mathscr{P}_{H_0}(\mathbb{R}), \\
T_2 = (T_{P_1})_{P_2}, & H_2 = P_2(P_1(H)), \quad P_2 \in \mathscr{P}_{H_1}(\mathbb{R}), \\
\quad\vdots & \quad\vdots \\
T_n = (\cdots((T_{P_1})_{P_2})\cdots)_{P_n}, & H_n = P_n(P_{n-1}(\cdots(P_1(H))\cdots)), \quad P_n \in \mathscr{P}_{H_{n-1}}(\mathbb{R}).
\end{array}$$

This is a structure generating a class of pairs of a Hamiltonian and a time operator. Note that there may be mutually inequivalent representations among τ_n's. For example, if $\{P_1(\lambda)|\lambda \in \sigma(H)\} \neq \sigma(H)$, then $\sigma(P_1(H)) \neq \sigma(H)$ by the spectral mapping theorem. Hence $\tau_0 \not\cong \tau_1$.

Example 4.4 Consider the case where $\mathscr{H} = L^2(\mathbb{R})$, $T = \hat{q}$, $H = \hat{p}$, $\mathscr{D} = \mathscr{D}_{\text{AB}}$ (see Example 4.2). Take $P(\lambda) = \lambda^2/2m$, $\lambda \in \mathbb{R}$. Then $P'(\lambda) = \lambda/m$. Hence $P'(\hat{p}) = \hat{p}/m$. Therefore $P'(\hat{p})$ is injective and $P'(\hat{p})^{-1} = m\hat{p}^{-1}$, which leaves $\mathscr{D}_{\text{AB}}$ invariant. Hence it follows that

$$T_1 = T_{\text{AB}}, \quad H_1 = H_{\text{NR}}.$$

Thus the Aharonov–Bohm time operator is obtained by the above procedure to construct Hamiltonians and time operators. It is obvious that $(\hat{q}, \hat{p})$ is inequivalent to $(T_{\text{AB}}, H_{\text{NR}})$.

4.4 Time Operators of Hamiltonians with Purely Discrete Spectrum

In this section, we introduce a class of time operators of Hamiltonians which are bounded from below and have a purely discrete spectrum.

4.4.1 An Operator Associated with a Self-adjoint Operator with Purely Discrete Spectrum

Let H be a self-adjoint operator on a Hilbert space $\mathscr{H}$ and suppose that H has a purely discrete spectrum

$$\sigma(H) = \sigma_{\mathrm{d}}(H) = \{E_n\}_{n=1}^{\infty} \subset \mathbb{R} \tag{4.15}$$

with $0 < E_1 < \cdots < E_n < E_{n+1} < \cdots$, $E_n \to \infty$ $(n \to \infty)$ satisfying

$$\sum_{n=1}^{\infty} \frac{1}{E_n^2} < \infty. \tag{4.16}$$

Then it is obvious that H is bounded from below with $H \geq E_1$. For simplicity, we assume that each E_n is a simple eigenvalue of H, i.e.,

$$\dim \ker(H - E_n) = 1, \quad n \geq 1.$$

Then there is a CONS $\{e_n\}_{n=1}^{\infty}$ of $\mathscr{H}$ such that

$$He_n = E_n e_n, \quad n \in \mathbb{N}.$$

Condition (4.16) implies that, for each $n \in \mathbb{N}$,

$$\sum_{m \neq n}^{\infty} \frac{1}{|E_n - E_m|^2} < \infty.$$

Hence, by Lemma 1.3, one can define a vector by

$$u_n := \sum_{m \neq n}^{\infty} \frac{1}{E_n - E_m} e_m \tag{4.17}$$

and one has

$$\|u_n\|^2 = \sum_{m \neq n} \frac{1}{|E_n - E_m|^2}. \tag{4.18}$$

Using Lemma 1.3 again, one can define a linear operator $T_{\max}$ on $\mathscr{H}$ as follows:

$$D(T_{\max}) := \left\{ \Psi \in \mathscr{H} \middle| \sum_{n=1}^{\infty} | \langle u_n, \Psi \rangle |^2 < \infty \right\}, \tag{4.19}$$

$$T_{\max}\Psi := i\sum_{n=1}^{\infty}\langle u_n, \Psi\rangle e_n, \quad \Psi \in D(T_{\max}). \tag{4.20}$$

Let

$$\mathscr{D}_0 := \mathrm{span}\{e_n | n \in \mathbb{N}\}.$$

Then $\mathscr{D}_0$ is dense. Note that, for all $n, m \in \mathbb{N}$,

$$\langle u_n, e_m\rangle = \begin{cases} 0 & \text{for } n = m \\ \dfrac{1}{E_n - E_m} & \text{for } n \neq m \end{cases}. \tag{4.21}$$

Hence

$$\sum_{n=1}^{\infty} |\langle u_n, e_m\rangle|^2 = \sum_{n\neq m}^{\infty} \frac{1}{|E_n - E_m|^2} < \infty.$$

Therefore $e_m \in D(T_{\max})$ for all $m \in \mathbb{N}$. Thus

$$\mathscr{D}_0 \subset D(T_{\max}). \tag{4.22}$$

This implies that $D(T_{\max})$ is dense in $\mathscr{H}$. Moreover, we have the following fact:

Proposition 4.3 *The operator $T_{\max}$ is a densely defined closed operator on $\mathscr{H}$.*

Proof We need only to prove the closedness of $T_{\max}$. Let $\Psi_N \in D(T_{\max})$, $N \geq 1$ and $\Psi_N \to \Psi \in \mathscr{H}$ and $T_{\max}\Psi_N \to \Phi \in \mathscr{H}$ as $N \to \infty$. Then $\|T_{\max}\Psi_N\| \to \|\Phi\|$ as $N \to \infty$. Hence there exists a constant $C > 0$ such that $\|T_{\max}\Psi_N\| \leq C$ for all $N \geq 1$. We have

$$\|T_{\max}\Psi_N\|^2 = \sum_{n=1}^{\infty} |\langle u_n, \Psi_N\rangle|^2.$$

Hence, for all $M \in \mathbb{N}$, $\sum_{n=1}^{M} |\langle u_n, \Psi_N\rangle|^2 \leq C^2$. Taking the limit $N \to \infty$ in this inequality, we obtain $\sum_{n=1}^{M} |\langle u_n, \Psi\rangle|^2 \leq C^2$. Then, taking the limit $M \to \infty$, we obtain $\sum_{n=1}^{\infty} |\langle u_n, \Psi\rangle|^2 \leq C^2$. Hence $\Psi \in D(T_{\max})$.

We have for all $n \in \mathbb{N}$

$$\langle e_n, \Phi\rangle = \lim_{N\to\infty}\langle e_n, T_{\max}\Psi_N\rangle = i\lim_{N\to\infty}\langle u_n, \Psi_N\rangle = i\langle u_n, \Psi\rangle.$$

Hence

$$\Phi = \sum_{n=1}^{\infty} \langle e_n, \Phi \rangle \, e_n = i \sum_{n=1}^{\infty} \langle u_n, \Psi \rangle \, e_n = T_{\max} \Psi.$$

Thus $T_{\max}$ is closed. ■

4.4.2 A Time Operator of H

It can be inferred heuristically that $T_{\max}$ may be symmetric. But it seems to be difficult to show without additional conditions that $T_{\max}$ is symmetric. For this reason, we consider the restriction of $T_{\max}$ to $\mathscr{D}_0$:

$$T_0 := T_{\max} \upharpoonright \mathscr{D}_0. \tag{4.23}$$

Proposition 4.4 *The operator T_0 is symmetric.*

Proof It is sufficient to show that, for all $n, m \in \mathbb{N}$, $\langle e_n, T_0 e_m \rangle = \langle T_0 e_n, e_m \rangle$. By the definition of T_0 and (4.21), we have

$$T_0 e_m = i \sum_{n \neq m}^{\infty} \frac{1}{E_n - E_m} e_n. \tag{4.24}$$

Hence

$$\langle e_n, T_0 e_m \rangle = i \sum_{\ell \neq m}^{\infty} \frac{1}{E_\ell - E_m} \langle e_n, e_\ell \rangle = \begin{cases} 0 & \text{for } n = m \\ \dfrac{i}{E_n - E_m} & \text{for } n \neq m \end{cases}.$$

This implies that $\langle e_n, T_0 e_m \rangle = \langle T_0 e_n, e_m \rangle$, $n, m \in \mathbb{N}$. ■

We denote by T_{G} the closure of T_0:

$$T_{\mathrm{G}} := \overline{T}_0. \tag{4.25}$$

Then T_{G} is a closed symmetric operator.

Formula (4.24) shows that $T_{\mathrm{G}} e_m \notin \mathscr{D}_0$. Hence T_{G} does not leave $\mathscr{D}_0$ invariant. By the spectral theorem, we see that

$$D(H) = \left\{ \Psi \in \mathscr{H} \middle| \sum_{n=1}^{\infty} E_n^2 |\, \langle e_n, \Psi \rangle \,|^2 < \infty \right\} \tag{4.26}$$

and

$$H\Psi = \sum_{n=1}^{\infty} E_n \langle e_n, \Psi \rangle e_n, \quad \Psi \in D(H). \tag{4.27}$$

Using (4.26), we see that

$$T_G e_m \notin D(H), \quad m \in \mathbb{N}.$$

Hence $\mathscr{D}_0 \not\subset D(HT_G)$ (it is obvious that $\mathscr{D}_0 \subset D(T_G H)$). Therefore $\mathscr{D}_0 \not\subset D([T_G, H])$. Thus we are led to find another subspace for the commutator $[T_G, H]$.

We introduce

$$\mathscr{D}_c := \text{span}\{e_n - e_m \in \mathscr{H} | n, m \geq 1\}. \tag{4.28}$$

It is obvious that

$$\mathscr{D}_c \subset \mathscr{D}_0.$$

Lemma 4.2 *$\mathscr{D}_c$ is dense in $\mathscr{H}$.*

Proof We need only to show that $\mathscr{D}_c^{\perp} = \{0\}$. So let $\Psi \in \mathscr{D}_c^{\perp}$. Then, for all $n, m \in \mathbb{N}$, $\langle \Psi, e_n - e_m \rangle = 0$. In particular, $\langle \Psi, e_n \rangle = \langle \Psi, e_1 \rangle$, $n \in \mathbb{N} \cdots (*)$. Hence

$$\|\Psi\|^2 = \lim_{N \to \infty} \sum_{k=1}^{N} |\langle e_k, \Psi \rangle|^2 = \lim_{N \to \infty} N |\langle e_1, \Psi \rangle|^2.$$

This implies that $\langle e_1, \Psi \rangle = 0$. Hence, by $(*)$, $\langle e_n, \Psi \rangle = 0$, $n \in \mathbb{N}$. Hence $\Psi = 0$. ■

The next theorem shows that T_G is a time operator of H:

Theorem 4.3

$$\mathscr{D}_c \subset D(T_G H) \cap D(HT_G) \tag{4.29}$$

and

$$[T_G, H]\Psi = i\Psi, \quad \Psi \in \mathscr{D}_c. \tag{4.30}$$

Proof Let $f_{nm} := e_n - e_m$ $(n, m \in \mathbb{N})$. Then we need only to show that, for all $n, m \in \mathbb{N}$ with $n \neq m$, $f_{nm} \in D(T_G H) \cap D(H T_G)$ and $[T_G, H] f_{nm} = i f_{nm}$. By (4.24), we have

$$\begin{aligned} T_G f_{nm} &= i \sum_{k \neq n}^{\infty} \frac{1}{E_k - E_n} e_k - i \sum_{k \neq m}^{\infty} \frac{1}{E_k - E_m} e_k \\ &= i(E_n - E_m) \sum_{k \neq n,m}^{\infty} \frac{1}{(E_k - E_n)(E_k - E_m)} e_k + i \frac{1}{E_m - E_n}(e_n + e_m). \end{aligned}$$

Hence, for each $p \in \mathbb{N}$,

$$\langle e_p, T_G f_{nm} \rangle = \begin{cases} \dfrac{i(E_n - E_m)}{(E_p - E_n)(E_p - E_m)} & \text{for } p \neq m, n \\ \dfrac{i}{E_m - E_n} & \text{for } p = n \text{ or } p = m \end{cases}.$$

Let $p_0 > \max\{m, n\}$ be fixed. Then

$$\sum_{p=p_0}^{\infty} E_p^2 |\langle e_p, T_G f_{nm} \rangle|^2 = (E_n - E_m)^2 \sum_{p=p_0}^{\infty} \frac{E_p^2}{(E_p - E_n)^2 (E_p - E_m)^2} < \infty,$$

where we have used (4.16). Hence $T_G f_{nm} \in D(H)$ and

$$H T_G f_{nm} = i(E_n - E_m) \sum_{k \neq n,m}^{\infty} \frac{E_k}{(E_k - E_n)(E_k - E_m)} e_k + i \frac{1}{E_m - E_n}(E_n e_n + E_m e_m).$$

On the other hand, we have

$$H f_{nm} = E_n e_n - E_m e_m.$$

Hence $H f_{nm} \in D(T_G)$ and

$$\begin{aligned} T_G H f_{nm} &= E_n T_G e_n - E_m T_G e_m \\ &= i(E_n - E_m) \sum_{k \neq n,m}^{\infty} \frac{E_k}{(E_k - E_n)(E_k - E_m)} e_k + i \frac{E_n}{E_m - E_n} e_m - i \frac{E_m}{E_n - E_m} e_n. \end{aligned}$$

Thus $f_{nm} \in D(T_G H) \cap D(H T_G)$ and $[T_G, H] f_{nm} = i f_{nm}$. ∎

We call T_G the **Galapon time operator** [76] associated with H.

4.4.3 A Sufficient Condition for $T_{\max}$ to be Bounded

In this subsection, we first show that, under an additional condition, $T_{\max}$ is a self-adjoint Hilbert–Schmidt operator and $T_{\max} = T_{\mathrm{G}}$.

Theorem 4.4 *Suppose that*

$$\sum_{n=1}^{\infty}\sum_{m\neq n}^{\infty}\frac{1}{|E_n - E_m|^2} < \infty. \tag{4.31}$$

Then $T_{\max}$ is a self-adjoint Hilbert–Schmidt operator and $T_{\mathrm{G}} = T_{\max}$.

Proof By (4.18), we have

$$\sum_{n=1}^{\infty}\|u_n\|^2 = \sum_{n=1}^{\infty}\sum_{m\neq n}^{\infty}\frac{1}{|E_n - E_m|^2} < \infty.$$

Hence, for all $\Psi \in \mathscr{H}$,

$$\sum_{n=1}^{\infty}|\langle u_n, \Psi\rangle|^2 \leq \left(\sum_{n=1}^{\infty}\|u_n\|^2\right)\|\Psi\|^2 < \infty.$$

Hence $\Psi \in D(T_{\max})$. Thus $D(T_{\max}) = \mathscr{H}$. Since $T_{\max}$ is closed by Proposition 4.3, it follows from the closed graph theorem that $T_{\max}$ is bounded.[6]

To prove the self-adjointness of $T_{\max}$, we need only to show the symmetry of $T_{\max}$ which is equivalent to that, for all $\Psi \in \mathscr{H}$, $\langle\Psi, T_{\max}\Psi\rangle \in \mathbb{R}$. Let us prove this. We have

$$\langle\Psi, T_{\max}\Psi\rangle = i\sum_{n=1}^{\infty}\left(\sum_{m\neq n}^{\infty}\frac{1}{E_n - E_m}\langle e_m, \Psi\rangle\langle\Psi, e_n\rangle\right).$$

Hence

$$\langle\Psi, T_{\max}\Psi\rangle^* = -i\sum_{n=1}^{\infty}\left(\sum_{m\neq n}^{\infty}\frac{1}{E_n - E_m}\langle e_n, \Psi\rangle\langle\Psi, e_m\rangle\right). \tag{4.32}$$

By the Cauchy–Schwarz inequality and Bessel's inequality, we have

$$\sum_{n=1}^{\infty}\sum_{m\neq n}^{\infty}\left|\frac{1}{E_n - E_m}\langle e_n, \Psi\rangle\langle\Psi, e_m\rangle\right| \leq \left(\sum_{n=1}^{\infty}\sum_{m\neq n}^{\infty}\frac{1}{|E_n - E_m|^2}\right)\|\Psi\|^2 < \infty.$$

[6] This can be directly proved. Indeed, $\|T_{\max}\Psi\|^2 = \sum_{n=1}^{\infty}|\langle u_n, \Psi\rangle|^2 \leq \left(\sum_{n=1}^{\infty}\|u_n\|^2\right)\|\Psi\|^2$.

Hence we can exchange the summations over n and m on the right hand side of (4.32) to see that the right hand side of (4.32) is equal to $\langle \Psi, T_{\max}\Psi\rangle$. Therefore $\langle \Psi, T_{\max}\Psi\rangle \in \mathbb{R}$. Thus $T_{\max}$ is a bounded self-adjoint operator.

We have by (4.24)

$$T_{\max} e_m = i \sum_{n \neq m}^{\infty} \frac{1}{E_n - E_m} e_n.$$

Hence

$$\|T_{\max}\|_2^2 = \sum_{m=1}^{\infty} \|T_{\max} e_m\|^2 = \sum_{m=1}^{\infty} \sum_{n \neq m}^{\infty} \frac{1}{|E_n - E_m|^2} < \infty. \tag{4.33}$$

Therefore $T_{\max}$ is Hilbert–Schmidt.

Since $T_{\mathrm{G}} \subset T_{\max}$ and T_{G} is densely defined and closed, T_{G} is bounded with $T_G = T_{\max}$. ■

Corollary 4.1 *Assume (4.31). Then T_{G} is a self-adjoint Hilbert–Schmidt operator and there exists a CONS $\{\varphi_n\}_{n=1}^{\infty}$ of $\mathscr{H}$ and real numbers t_n, $n \in \mathbb{N}$ such that*

$$T_{\mathrm{G}}\varphi_n = t_n \varphi_n, \quad n \in \mathbb{N} \tag{4.34}$$

and

$$\sum_{n=1}^{\infty} t_n^2 = \sum_{n=1}^{\infty} \sum_{m \neq n}^{\infty} \frac{1}{(E_n - E_m)^2}. \tag{4.35}$$

Proof By Theorem 4.4, $T_{\mathrm{G}} = T_{\max}$ is a self-adjoint Hilbert–Schmidt operator. Then it follows from Theorem 1.5 that there exist $\{\varphi_n\}_n$ and $\{t_n\}_n$ satisfying (4.34) and $\|T_{\max}\|_2^2 = \sum_{n=1}^{\infty} t_n^2$. By (4.33), we obtain (4.35). ■

Since there are many sequences $\{E_n\}_{n=1}^{\infty}$ satisfying (4.16) and (4.31) (e.g., $E_n = cn^2$ with $c > 0$ a constant; see also Remark 4.6 below), Corollary 4.1 disproves Pauli's statement mentioned in Sect. 4.1 and the belief in non-existence of self-adjoint time operators for a Hamiltonian which is bounded from below and has a purely discrete spectrum. A reason why Pauli was led to a false conclusion is in the lack of careful consideration of the domains of unbounded operators involved and/or a formal (incorrect) identification of an ordinary CCR with a weak Weyl relation or a Weyl relation. The saying that *formal manipulation is a touchy business* ([130, p. 270]) applies to the present case too.

Remark 4.4 Under the assumption of Corollary 4.1, T_{G} may be a new physical quantity related to the usual external time in a suitable sense. Since the spectrum of T_{G} consists of only eigenvalues $\{t_n\}_n$, it may describe a quantization of "time".

Equation (4.35) may be interpreted as a relation between "quantized" times $\{t_n\}_n$ and energy levels $\{E_n\}_n$.

More detailed properties of $T_{\max}$ and T_G have been investigated in [35]. Here we only mention some results.

Theorem 4.5 *Suppose, in addition to the original assumption (4.16) for* $\{E_n\}_{n=1}^\infty$, *that there exist constants* $\alpha > 1, C > 0$ *and* $n_0 \in \mathbb{N}$ *such that*

$$E_n - E_m \geq C(n^\alpha - m^\alpha), \quad n > m > n_0. \tag{4.36}$$

Then T_G *is a bounded self-adjoint operator with* $D(T_G) = \mathscr{H}$ *and* $T_{\max} = T_G$.

We omit the proof of this theorem, since it takes much space (see Theorem 4.5 in [35]).

Remark 4.5 Theorem 4.5 does not cover the case where $E_n = \varepsilon_n := a(n-1) + b$, $n \in \mathbb{N}$ ($a > 0, b > 0$ are constants), i.e., the case where $\{E_n\}_n$ is the spectrum of a one-dimensional quantum harmonic oscillator. But, by using another method, one can prove that T_G with $E_n = \varepsilon_n$ is a bounded self-adjoint operator on $\mathscr{H}$ and $T_G = T_{\max}$ [35, Theorem 4.6]. Putting $\hat{\theta} := aT_G$ and $\hat{N} := a^{-1}(H-b)$ (the number operator), we have

$$[\hat{\theta}, \hat{N}] = i \quad \text{on } \mathscr{D}_c$$

and

$$\sigma(\hat{\theta}) = [-\pi, \pi].$$

This allows one to interpret $\hat{\theta}$ as a *quantum phase operator*. For the details, see [35, Example 4.2].

Remark 4.6 If $\alpha > 3/2$ in Theorem 4.5, then it can be shown that $\{E_n\}_n$ satisfies (4.31) (see [35, Theorem 6.2]). Hence, in this case, T_G is a self-adjoint Hilbert–Schmidt operator by Corollary 4.1.

4.4.4 Unboundedness of T_G

With regard to unboundedness of T_G, the following fact is proved:

Theorem 4.6 ([35, Theorem 5.1]) *Suppose, in addition to the original assumption (4.16) for* $\{E_n\}_{n=1}^\infty$, *that there is a subsequence* $\{E_{p_k}\}_{k=1}^\infty$ *of* $\{E_n\}_n$ *satisfying* $\lim_{k\to\infty}(E_{p_k+1} - E_{p_k}) = 0$. *Then* T_G *is unbounded.*

Proof We have

$$\|T_{\mathrm{G}}e_{p_k}\|^2 = \sum_{n=1}^{\infty}\left|\sum_{m\neq n}\frac{\langle e_m, e_{p_k}\rangle}{E_n - E_m}\right|^2 = \sum_{n\neq p_k}\frac{1}{|E_n - E_{p_k}|^2}$$
$$\geq \frac{1}{|E_{p_k+1} - E_{p_k}|^2} \to \infty \quad (k \to \infty).$$

Since $\|e_{p_k}\| = 1$, this implies that T_{G} is unbounded. ■

Example 4.5 Let $E_n = \lambda n^{\alpha}$ with $1/2 < \alpha < 1$ and a constant $\lambda > 0$. Then $\{E_n\}_{n=1}^{\infty}$ satisfies (4.16) and $\lim_{n\to\infty}(E_{n+1} - E_n) = 0$. Hence, in this case, T_{G} is unbounded.

Remark 4.7 To the author's best knowledge, no examples of T_{G} which is *unbounded* and *self-adjoint* have been constructed.

It is interesting to find a necessary and sufficient condition for a Hamiltonian with discrete eigenvalues to have time operators. For this aspect, we refer the reader to [20]. Recently, Teranishi [155] and Arai and Hiroshima [33] established an existence theorem of a time operator of a *unbounded* self-adjoint operator H with purely discrete spectrum without condition (4.16).

4.4.5 Concluding Remarks

In concluding this section, we give a summary from representation theoretic point of view for the construction of Galapon time operators. One starts with a space of sequences of positive numbers defined by

$$\mathscr{R}_{-2} := \left\{E = \{E_n\}_{n=1}^{\infty}\,\middle|\, 0 < E_n < E_{n+1},\ \forall n \geq 1,\ \sum_{n=1}^{\infty}\frac{1}{E_n^2} < \infty\right\}.$$

Let $\mathscr{H}$ be a separable Hilbert space and fix a CONS $\{e_n\}_{n=1}^{\infty}$ of $\mathscr{H}$. Then there exists a unique unbounded self-adjoint operator H_E on $\mathscr{H}$ such that

$$D(H_E) = \left\{\Psi \in \mathscr{H}\,\middle|\, \sum_{n=1}^{\infty} E_n^2 |\langle e_n, \Psi\rangle|^2 < \infty\right\}, \tag{4.37}$$

$$H_E e_n = E_n e_n, \quad n \in \mathbb{N}. \tag{4.38}$$

It is easy to see that H_E has a purely discrete spectrum with

$$\sigma(H_E) = \sigma_{\mathrm{d}}(H_E) = \{E_n\}_{n=1}^{\infty}$$

and each eigenvalue E_n of H_E is a simple eigenvalue. Hence, by applying Theorem 4.3, H_E has its Galapon time operator; we denote it by T_E. Then

$$\pi_E := (\mathscr{H}, \mathscr{D}_c, (T_E, H_E))$$

is a representation of the CCR with one degree of freedom. Since $\sigma(H_E) \neq \mathbb{R}$, it follows that π_E is inequivalent to any direct sum of the Schrödinger representations $\pi_S^{(1)}$. Moreover, the following fact holds:

Theorem 4.7 *Let $E, E' \in \mathscr{R}_{-2}$ and $E \neq E'$. Then π_E is inequivalent to $\pi_{E'}$.*

Proof If $E \neq E'$, then $\sigma(H_E) \neq \sigma(H_{E'})$. Hence the conclusion follows. ∎

Thus we see that there is a family $\{\pi_E\}_{E \in \mathscr{R}_{-2}}$ of mutually inequivalent representations of the CCR with one degree of freedom. We note that the index set $\mathscr{R}_{-2}$ has a physical meaning, i.e., it is a set of sequences each of which, in the context of quantum mechanics, may be interpreted as a set of energy levels satisfying a condition.

Remark 4.8 It is not known whether π_E is irreducible.

4.5 Strong Time Operators

In this section we introduce a stronger notion of time operators.

4.5.1 Definition and Basic Properties

In what follows, H denotes a self-adjoint operator on a Hilbert space $\mathscr{H}$ unless otherwise stated.

Definition 4.2 A symmetric operator T on $\mathscr{H}$ is called a **strong time operator** of H if $(\mathscr{H}, (T, H))$ is a weak Weyl representation with one degree of freedom, i.e.,

$$e^{itH} T e^{-itH} = T + t, \quad \forall t \in \mathbb{R}. \tag{4.39}$$

In this case, H is said to have a strong time operator T.

We denote by $\mathfrak{T}_s(H)$ the set of strong time operators of H.

By Proposition 2.22, a strong time operator of H is a time operator of H. Hence

$$\mathfrak{T}_s(H) \subset \mathfrak{T}(H). \tag{4.40}$$

But it should be kept in mind that a time operator of H is not necessarily a strong time operator.

To find a physical interpretation for a strong time operator, we consider the case where H is a quantum mechanical Hamiltonian which has a strong time operator T. As is well known, the correspondence: $t \mapsto T(t) := e^{itH}Te^{-itH}$ describes the time development of a quantum mechanical object[7] $\mathscr{O}$ represented by T at time zero, where $T(t)$ represents $\mathscr{O}$ at time t. Equation (4.39) shows that $T(t)$ is given as the shift of T by time t. Moreover, for all unit vectors $\Psi \in D(T)$, (4.39) implies that

$$\langle \Psi(t), T\Psi(t)\rangle = \langle \Psi, T\Psi\rangle + t,$$

where $\Psi(t) := e^{-itH}\Psi$ denotes the state vector at time t. Hence

$$t = \langle \Psi(t), T\Psi(t)\rangle - \langle \Psi, T\Psi\rangle .$$

This shows that an external time t is represented as the difference between the expectation value of T at the time-t state and that at time zero.

We discuss examples of strong time operator later in a separate subsection. For the time being, we continue to study general properties of a strong time operator.

The set $\mathfrak{T}_s(H)$ is invariant under the closure operation:

Proposition 4.5 *For all $T \in \mathfrak{T}_s(\mathscr{H})$, $\overline{T} \in \mathfrak{T}_s(H)$. Moreover, $D(\overline{T}H) \cap D(H\overline{T})$ is dense in $\mathscr{H}$ and*

$$[\overline{T}, H] = i \quad \text{on } D(\overline{T}H) \cap D(H\overline{T}).$$

Proof Taking the closure of both sides of (4.39), we have $\overline{T} + t = \overline{e^{itH}Te^{-itH}} = e^{itH}\overline{T}e^{-itH}$. Hence $\overline{T} \in \mathfrak{T}_s(H)$. The density of $D(\overline{T}H) \cap D(H\overline{T})$ follows from Proposition 2.23. ■

The following fact is interesting:

Theorem 4.8 *If H has a strong time operator, i.e., $\mathfrak{T}_s(H) \neq \emptyset$, then H is absolutely continuous.*

Proof One needs only to apply Theorem 2.22 with $C = I$. ■

This theorem can be used to prove the absolute continuity of Hamiltonians appearing in quantum physics without going into detailed spectral analysis.

Theorem 4.8 is rephrased as follows:

Corollary 4.2 *If H is not absolutely continuous, then H has no strong time operators, i.e., $\mathfrak{T}_s(H) = \emptyset$.*

[7] A quantum mechanical object is not necessarily a quantum mechanical observable in the usual sense and hence it does not need to be represented by a self-adjoint operator.

As in the case of $\mathfrak{T}(H)$, the set $\mathfrak{T}_s(H)$ is closed under convex combinations with a suitable condition:

Proposition 4.6 *Let $T_1, \ldots, T_N \in \mathfrak{T}_s(H)$ such that $\cap_{j=1}^N D(T_j)$ is dense in $\mathscr{H}$. Then, for all $(a_1, \ldots, a_N) \in \mathbb{R}^N$ satisfying*

$$\sum_{j=1}^N a_j = 1,$$

$\sum_{j=1}^N a_j T_j \in \mathfrak{T}_s(H)$.

Proof It is easy to see that $T := \sum_{j=1}^N a_j T_j$ is symmetric. Moreover, for all $t \in \mathbb{R}$, we have by Proposition 1.10

$$e^{itH} T e^{-itH} = \sum_{j=1}^N a_j e^{itH} T_j e^{-itH} = \sum_{j=1}^N a_j (T_j + t) = T + t.$$

Hence $T \in \mathfrak{T}_s(H)$. ■

The next lemma may be used to prove that a symmetric operator T is a strong time operator of H.

Lemma 4.3 *If T is a symmetric operator satisfying*

$$e^{itH} T \subset (T + t) e^{itH}, \quad t \in \mathbb{R}, \tag{4.41}$$

then $T \in \mathfrak{T}_s(H)$.

Proof This is just the case of Lemma 2.22 with $N = 1$. ■

The concept of a strong time operator is unitarily covariant:

Proposition 4.7 *Let $T \in \mathfrak{T}_s(H)$ and $U: \mathscr{H} \to \mathscr{K}$ (a Hilbert space) be a unitary operator. Then UTU^{-1} is a strong time operator of UHU^{-1}, i.e., $UTU^{-1} \in \mathfrak{T}_s(UHU^{-1})$.*

Proof The operator $H' := UHU^{-1}$ is a self-adjoint operator on $\mathscr{K}$ and $T' := UTU^{-1}$ is a symmetric operator on $\mathscr{K}$. By the unitary covariance of functional calculus, we have for all $t \in \mathbb{R}$

$$e^{itH'} T' e^{-itH'} = U e^{itH} U^{-1} T' U e^{-itH} U^{-1} = U e^{itH} T e^{-itH} U = U(T + t)U^{-1} = T' + t.$$

Thus T' is a strong time operator of H'. ■

The next proposition states a simple fact on a perturbation of a strong time operator of H.

Proposition 4.8 *Let $T \in \mathfrak{T}_s(H)$ and T_1 be a symmetric operator on $\mathscr{H}$ strongly commuting with H. Suppose that $D(T) \cap D(T_1)$ is dense in $\mathscr{H}$. Then $T + T_1 \in \mathfrak{T}_s(H)$.*

Proof It is obvious that $T' := T + T_1$ is a symmetric operator. By Proposition 1.10, we have

$$e^{itH}T'e^{-itH} = e^{itH}Te^{-itH} + e^{itH}T_1e^{-itH}.$$

By the assumed properties of T and T', the right hand side is equal to $T + t + T_1 = T' + t$. Hence T' is a strong time operator of H. ■

The next proposition is about non-uniqueness of strong time operators of H.

Proposition 4.9 *Let $T \in \mathfrak{T}_s(H)$ and T' be a symmetric operator on $\mathscr{H}$.*

(i) *If $T' \in \mathfrak{T}_s(H)$, then $T' - T$ strongly commutes with H.*[8]
(ii) *Let $T_0' := T' \upharpoonright D(T') \cap D(T)$. Suppose that $T' - T$ strongly commutes with H and $D(T) \cap D(T')$ is dense. Then $T_0' \in \mathfrak{T}_s(H)$.*

Proof

(i) By the present assumption, we have $e^{itH}T'e^{-itH} = T' + t$, $t \in \mathbb{R}$. Hence, by Proposition 1.10, we obtain

$$e^{itH}(T' - T)e^{-itH} = T' - T. \tag{4.42}$$

This means that $T' - T$ strongly commutes with H.
(ii) It is obvious that T_0' and $T' - T$ are symmetric. The strong commutativity of $T' - T$ with H implies that

$$e^{itH}(T' - T)e^{-itH} = T' - T, \quad t \in \mathbb{R}.$$

Hence $e^{-itH}D(T_0') = D(T_0')$ and, for all $\Psi \in D(T_0')$, $e^{itH}T'e^{-itH}\Psi = (T' + t)\Psi$. Thus $T_0' \in \mathfrak{T}_s(H)$. ■

4.5.2 Self-adjoint Strong Time Operator and Weyl Representation of CCR

We have so far considered strong time operators of H which are not necessarily self-adjoint. Here we examine the case of self-adjoint strong time operators.

[8]See Definition 1.1 for the definition of strong commutativity of a linear operator with a self-adjoint operator.

Theorem 4.9 *Let T be a self-adjoint strong time operator of H. Then $(\mathscr{H}, (T, H))$ is a Weyl representation of the CCR with one degree of freedom, i.e., $(\mathscr{H}, (T, H)) \in$ $\mathsf{CCR}_{\mathrm{W}}(1)$. In particular, T and H are absolutely continuous with*

$$\sigma(T) = \sigma(H) = \mathbb{R}. \tag{4.43}$$

Proof By (4.39) and the unitary covariance of functional calculus, we obtain

$$e^{itH}e^{isT}e^{-itH} = e^{isT}e^{its}, \quad s, t \in \mathbb{R}.$$

Hence

$$e^{isT}e^{itH} = e^{-ist}e^{itH}e^{isT}.$$

Therefore, (T, H) obeys the Weyl relation with one degree of freedom. Thus $(\mathscr{H},$ $(T, H)) \in \mathsf{CCR}_{\mathrm{W}}(1)$. Spectral property (4.43) follows from Corollary 2.12. ■

4.5.3 Non-Self-adjointness of Strong Time Operators

The next theorem is noteworthy:

Theorem 4.10 *Let T be a strong time operator of H, i.e., $T \in \mathfrak{T}_{\mathrm{s}}(H)$. Suppose that $\sigma(H) \neq \mathbb{R}$. Then T is not essentially self-adjoint. In particular, if H is is semi-bounded (i.e., bounded from below or bounded from above). Then T is not essentially self-adjoint.*

Proof Suppose that T were essentially self-adjoint. Then $\overline{T}$ is self-adjoint and $\overline{T} \in \mathfrak{T}_{\mathrm{s}}(H)$. Hence, by Theorem 4.9, $\sigma(H) = \mathbb{R}$. But this contradicts the present assumption. ■

Theorem 4.10 yields the following corollary:

Corollary 4.3 *Suppose that $\sigma(H) \neq \mathbb{R}$. Then H has no self-adjoint strong time operator.*

Corollary 4.4 *Suppose that $\mathscr{H}$ is separable and H has a self-adjoint strong time operator T. Then (T, H) is unitarily equivalent to a direct sum of the Schrödinger representation $\pi_{\mathrm{S}}^{(1)}$.*

Proof This follows from Theorem 4.9 and von Neumann's uniqueness theorem (Theorem 2.19). ■

4.5.4 A Perturbation Theorem

It may be interesting and important to consider conditions for a strong time operator of a self-adjoint operator H_0 to be also a strong time operator of a perturbation of H_0. Here we present only a result in the case of a simple perturbation.

Let H_0 and H_1 be self-adjoint operators on $\mathscr{H}$ satisfying the following conditions:

(H.1) H_0 has a strong time operator T_0.
(H.2) $H := H_0 + H_1$ is essentially self-adjoint.
(H.3) The operator T_0 strongly commutes with H_1, i.e., for all $t \in \mathbb{R}$, $e^{itH_1}T_0 \subset T_0 e^{itH_1}$.

Theorem 4.11 *Under conditions (H.1)–(H.3), T_0 is a strong time operator of $\overline{H}$. In particular, $\overline{H}$ is absolutely continuous.*

Proof By (H.2), we can apply the Trotter product formula [130, Theorem VIII.31] to obtain that, for all $t \in \mathbb{R}$,

$$e^{-it\overline{H}} = \text{s-}\lim_{n\to\infty} \left(e^{-itH_0/n}e^{-itH_1/n}\right)^n.$$

Hence, for all $\Psi, \Phi \in D(T_0)$,

$$\left\langle \Phi, e^{-it\overline{H}}T_0\Psi \right\rangle = \lim_{n\to\infty} \left\langle \Phi, \left(e^{-itH_0/n}e^{-itH_1/n}\right)^n T_0\Psi \right\rangle.$$

By (H.1) and (H.3), we have

$$e^{-itH_0/n}e^{-itH_1/n}T_0\Psi = (T_0 - t/n)e^{-itH_0/n}e^{-itH_1}\Psi.$$

Hence it follows that

$$\left(e^{-itH_0/n}e^{-itH_1/n}\right)^n T_0\Psi = (T_0 - t)\left(e^{-itH_0/n}e^{-itH_1/n}\right)^n \Psi.$$

Therefore

$$\begin{aligned}\left\langle \Phi, e^{-it\overline{H}}T_0\Psi \right\rangle &= \lim_{n\to\infty} \left\langle (T_0 - t)\Phi, \left(e^{-itH_0/n}e^{-itH_1/n}\right)^n \Psi \right\rangle \\ &= \left\langle (T_0 - t)\Phi, e^{-it\overline{H}}\Psi \right\rangle.\end{aligned}$$

This implies that $e^{-it\overline{H}}\Psi \in D(T_0 - t) = D(T_0)$ and

$$(T_0 - t)e^{-it\overline{H}}\Psi = e^{-it\overline{H}}T_0\Psi,$$

i.e.,

$$T_0 e^{-it\overline{H}}\Psi = e^{-it\overline{H}}(T_0 + t)\Psi.$$

Thus T_0 is a strong time operator of $\overline{H}$. The latter half of the theorem follows from Theorem 4.8. ■

Remark 4.9 By Proposition 1.21, (H.3) implies that H_1 and T_0 commute: $[H_1, T_0] = 0$ on $D(H_1 T_0) \cap D(T_0 H_1)$. But the converse does not hold in general. Hence condition (H.3) is a stronger commutativity condition for T_0 and H_1.

4.5.5 Spectral Properties of Strong Time Operators

In this subsection, we investigate spectral properties of a strong time operator. Throughout this subsection, H is a self-adjoint operator on $\mathscr{H}$ and T is a strong time operator of H:

$$T \in \mathfrak{T}_s(H).$$

Lemma 4.4 *Suppose that $\sigma(H) \neq \mathbb{R}$. Then $\sigma(T) = \overline{\Pi}_+$ or $\overline{\Pi}_-$ or $\mathbb{C}$.*

Proof The present assumption and Theorem 4.10 imply that T is not self-adjoint. Hence Theorem 1.17(iv) gives the desired result. ■

Lemma 4.5 *Suppose that H is bounded from below. Then, for all $\beta > 0$, $e^{-\beta H} D(T) \subset D(\overline{T})$ and, for all $\Psi \in D(T)$*

$$\overline{T} e^{-\beta H}\Psi = e^{-\beta H}(T - i\beta)\Psi. \tag{4.44}$$

Proof This is a simple application of Theorem 2.23(iii) with $z = \beta$. ■

Lemma 4.6 *Suppose that H is bounded below. Then, for all $\beta > 0$, $e^{-\beta H} D(T^*) \subset D(T^*)$ and, for all $\Psi \in D(T^*)$*

$$T^* e^{-\beta H}\Psi = e^{-\beta H}(T^* - i\beta)\Psi.$$

Proof Lemma 4.5 implies that $e^{-\beta H}(T - i\beta) \subset \overline{T} e^{-\beta H}$. Hence, by Proposition 1.1,

$$(\overline{T} e^{-\beta H})^* \subset (e^{-\beta H}(T - i\beta))^*.$$

Then, using Proposition 1.4 and the boundedness of $e^{-\beta H}$, one can show that

$$e^{-\beta H}(\overline{T})^* \subset (T^* + i\beta)e^{-\beta H}.$$

This result and the fact $(\overline{T})^* = T^*$ imply the desired result. ■

We are now ready to prove a theorem [17] on spectral properties of T:

Theorem 4.12 *For each $T \in \mathfrak{T}_s(H)$, the following hold:*

(i) *If H is bounded from below, then $\sigma(T)$ is either $\mathbb{C}$ or $\overline{\Pi}_+$.*
(ii) *If H is bounded from above, then $\sigma(T)$ is either $\mathbb{C}$ or $\overline{\Pi}_-$.*
(iii) *If H is bounded, then $\sigma(T) = \mathbb{C}$.*

Proof

(i) By Lemma 4.4, we need only to show that the case $\sigma(T) = \overline{\Pi}_-$ is excluded. For this purpose, suppose that $\sigma(T) = \overline{\Pi}_-$. Then $\Pi_+ = \rho(T)$.

By the Hilbert space decomposition theorem (Theorem 1.2), we have for all $z \in \mathbb{C} \setminus \mathbb{R}$ the orthogonal decomposition

$$\mathscr{H} = \ker(T^* - z^*) \oplus \mathrm{Ran}\,(\overline{T} - z), \tag{4.45}$$

where we have used the fact that, for any closed symmetric operator S, $\mathrm{Ran}\,(S - z)$ is closed.

Since $i \in \rho(T)$, $\mathrm{Ran}\,(\overline{T} - i) = \mathscr{H}$. Hence (4.45) implies that $\ker(T^* + i) = \{0\}$. Since T is not self-adjoint by Theorem 4.10, $\ker(T^* - i) \neq \{0\}$. Hence there exists a non-zero vector $\Psi \in D(T^*)$ such that $T^*\Psi = i\Psi$. Then, by Lemma 4.6, $i(1 - \beta) \in \sigma_{\mathrm{p}}(T^*)$. Since $\beta > 0$ is arbitrary, we can take it to be $1 < \beta$. Then $\gamma := i(1 - \beta) \in \Pi_-$. Taking $z = \gamma^*$ in (4.45), we have the orthogonal decomposition

$$\mathscr{H} = \ker(T^* - \gamma) \oplus \mathrm{Ran}\,(\overline{T} - \gamma^*).$$

Hence $\mathrm{Ran}\,(\overline{T} - \gamma^*)$ is not dense in $\mathscr{H}$. Therefore $\gamma^* \in \sigma(T)$, i.e., $i(\beta - 1) \in \sigma(T)$. But $i(\beta - 1) \in \Pi_+$. This is a contradiction. Thus $\sigma(T) \neq \overline{\Pi}_-$.

(ii) If H is bounded from above, then $\widehat{H} := -H$ is bounded below. It is easy to see that $\widehat{T} := -T$ is a strong time operator of $\widehat{H}$. Hence, by part (i), $\sigma(\widehat{T}) = \mathbb{C}$ or $\overline{\Pi}_+$. On the other hand, $\sigma(T) = \{-\lambda | \lambda \in \sigma(\widehat{T})\}$, which implies that $\sigma(T) = \mathbb{C}$ or $\overline{\Pi}_-$.

(iii) This follows from (i) and (ii). ∎

Example 4.6 Let $\hat{q}_L$ and $\hat{p}_{L,0}$ be as in Sect. 2.6.4. Then it is easy to see that $-\hat{p}_{L,0}$ is a strong time operator of $\hat{q}_L$. Since $\hat{q}_L$ is bounded, it follows from Theorem 4.12 that $\sigma(-\hat{p}_{L,0}) = \mathbb{C}$. Hence $\sigma(\hat{p}_{L,0}) = \mathbb{C}$. This gives another proof of (2.116).

Example 4.7 Let $\pi_{\mathrm{W},a}$ be the weak Weyl representation with degree one defined by (2.179). Then T_a is a strong time operator of $\hat{q}_a$. Since q_a is bounded from below, it follows from Theorem 4.12 that $\sigma(T_a) = \overline{\Pi}_+$ or $\sigma(T_a) = \mathbb{C}$. We have by (2.122)

$$\sigma(T_a) = \overline{\Pi}_+. \tag{4.46}$$

4.5.6 Quasi-Weyl Relation

In this subsection, we remark on an important fact related to the spectral properties of a strong time operator described in Theorem 4.12(i) and (ii).

Let T be a strong time operator of a self-adjoint operator H on a Hilbert space $\mathscr{H}$ and suppose that T is closed with

$$\sigma(T) = \overline{\Pi}_+ \tag{4.47}$$

(see Theorem 4.12(i)). Then, for all $\lambda > 0$, $-i\lambda \in \rho(T)$. The symmetry of T implies that

$$\|(T + i\lambda)\Psi\| \geq \lambda\|\Psi\|, \quad \Psi \in D(T).$$

Hence

$$\|(iT - \lambda)^{-1}\| \leq \frac{1}{\lambda}.$$

Therefore, for all $n \in \mathbb{N}$,

$$\|(iT - \lambda)^{-n}\| \leq \frac{1}{\lambda^n}.$$

Hence, by the Hille–Yosida theorem (Theorem 1.8), iT is the generator of a (C_0)-semi-group $\{V_T(\beta)\}_{\beta\geq 0}$ on $\mathscr{H}$, symbolically written $V_T(\beta) = e^{i\beta T}$, $\beta \geq 0$.

We first note the following fact:

Theorem 4.13 *For all $\beta \geq 0$, $V_T(\beta)$ is an isometry, i.e., $\|V_T(\beta)\Psi\| = \|\Psi\|$, $\Psi \in \mathscr{H}$.*

Proof By the Hille–Yosida theorem, we have

$$\frac{d}{d\beta}V_T(\beta)\Psi = iTV_T(\beta)\Psi = iV_T(\beta)T\Psi, \quad \Psi \in D(iT) = D(T),\ \beta \geq 0.$$

Hence, for all $\Psi \in D(T)$,

$$\frac{d}{d\beta}\|V_T(\beta)\Psi\|^2 = \langle iTV_T(\beta)\Psi, V_T(\beta)\Psi\rangle + \langle V_T(\beta)\Psi, iTV_T(\beta)\Psi\rangle.$$

Since T is symmetric, the right hand side vanishes. Hence $\|V_T(\beta)\Psi\|^2 = \|V_\beta(0)\Psi\|^2 = \|\Psi\|^2$. ■

Let

$$U_H(s) := e^{isH}, \quad s \in \mathbb{R}.$$

Theorem 4.14

$$V_T(\beta)U_H(t) = e^{-i\beta t}U_H(t)V_T(\beta), \quad \beta \geq 0, \, t \in \mathbb{R}. \tag{4.48}$$

Proof We set $V(\beta) := V_T(\beta)$ and $U(t) := U_H(t)$. Let $t \in \mathbb{R}$ be fixed and

$$F(\beta) := e^{i\beta t}U(t)^{-1}V(\beta)U(t).$$

Then it is easy to see that $\{F(\beta)\}_{\beta\geq 0}$ is a (C_0)-semi-group. Let $\Psi \in D(T)$. Then $F(\beta)\Psi$ is strongly differentiable in $\beta \in [0, \infty)$ and

$$\frac{d}{d\beta}F(\beta)\Psi = ie^{i\beta t}U(t)^{-1}(T+t)V(\beta)U(t)\Psi.$$

Note that $U(t)^{-1}(T+t) = TU(t)^{-1}$. Hence $\frac{d}{d\beta}F(\beta)\Psi = iTF(\beta)\Psi$. Hence the generator of $\{F(\beta)\}_{\beta\geq 0}$ is equal to iT. Hence, by the uniqueness of the (C_0)-semi-group stated in the Hille–Yosida theorem, $F(\beta) = V(\beta)$. Thus (4.48) follows. ■

Remark 4.10 In the case where $\sigma(T) = \overline{\Pi}_-$, the operator $-T$ generates a (C_0)-semi-group $\{e^{i\beta(-T)}\}_{\beta\geq 0}$. Putting $V_T(\beta) := e^{i\beta(-T)}$ and $U_H(t) := e^{-itH}$ $t \in \mathbb{R}$, one can show that (4.48) holds.

Relation (4.48) suggests a new algebraic relation. A (C_0)-semi-group $\{V(\beta)\}_{\beta\geq 0}$ and a strongly continuous one-parameter unitary group $\{U(t)\}_{t\in\mathbb{R}}$ are said to obey the **quasi-Weyl relation with one degree of freedom** if

$$V(\beta)U(t) = e^{-i\beta t}U(t)V(\beta), \quad \beta \geq 0, \, t \in \mathbb{R}. \tag{4.49}$$

Let $V(\beta)$ and $U(t)$ be as above. Then, by Stone's theorem, there exists a unique self-adjoint operator H such that

$$U(t) = e^{itH}, \quad t \in \mathbb{R}.$$

We denote by iT the generator of $\{V(\beta)\}_{\beta\geq 0}$.

Theorem 4.15 *Let $V(\beta)$ and $U(t)$ be as above. Then:*

(i) *For all $t \in \mathbb{R}$, $e^{itH}D(T) \subset D(T)$ and*

$$e^{itH}Te^{-itH} = T + t, \quad t \in \mathbb{R}. \tag{4.50}$$

(ii) *If $V(\beta)$ is isometry, then T is symmetric and hence a strong time operator of H. In this case, $\sigma(T) = \overline{\Pi}_+$.*

Proof

(i) Let $\Psi \in D(T)$ and $t \in \mathbb{R}$ be fixed. Then

$$V(\beta)U(t)\Psi = e^{-i\beta t}U(t)V(\beta)\Psi. \tag{4.51}$$

The vector-valued function : $\beta \mapsto V(\beta)\Psi$ is strongly differentiable and

$$\frac{d}{d\beta}V(\beta)\Psi = iTV(\beta)\Psi.$$

Hence the right hand side of (4.51) is strongly differentiable in β and

$$\frac{d}{d\beta}e^{-i\beta t}U(t)V(\beta)\Psi = -ite^{-i\beta t}U(t)V(\beta)\Psi + e^{-i\beta t}U(t)iTV(\beta).$$

Therefore the left hand side of (4.51) is strongly differentiable in β, implying that $U(t)\Psi \in D(T)$ and

$$\frac{d}{d\beta}V(\beta)U(t)\Psi = iTV(\beta)U(t)\Psi.$$

Hence

$$iTV(\beta)U(t)\Psi = -ite^{-i\beta t}U(t)V(\beta)\Psi + e^{-i\beta t}U(t)iTV(\beta).$$

Putting $\beta = 0$, we obtain

$$TU(t)\Psi = -tU(t)\Psi + U(t)T\Psi = U(t)(T - t)\Psi.$$

Hence $U(t)(T - t) \subset TU(t)$. This implies that $U(t)T \subset (T + t)U(t)$. Thus (4.50) follows.

(ii) Let $V(\beta)$ be isometry. Then

$$\frac{d}{d\beta}\|V(\beta)\Psi\|^2 = 0.$$

The left hand side at $\beta = 0$ is equal to $-i\langle T\Psi, \Psi\rangle + i\langle \Psi, T\Psi\rangle$. Hence $\langle T\Psi, \Psi\rangle = \langle \Psi, T\Psi\rangle$. Thus T is symmetric. By this result and (i), T is a strong time operator.

By the Hille–Yosida theorem, there exists a real constant ω such that $\{-i\lambda|\lambda > \omega\} \subset \rho(T)$. Since T is a strong time operator of H, Theorem 4.12 holds. This implies that $\omega \geq 0$ (if $\omega < 0$, then $\rho(T) = \mathbb{C}$, which is a contradiction) and hence $\sigma(T) = \overline{\Pi}_+$. ∎

Corollary 4.5 *Suppose that $(V(\beta), U(t))$ obeys the quasi-Weyl relation (4.49) and iT and H be the generators of $V(\beta)$ and $U(t)$ respectively. Then T is a strong time operator of H with $\sigma(T) = \overline{\Pi}_+$ if and only if $V(\beta)$ is isometry for all $\beta > 0$.*

Remark 4.11 Reducibility of representations $\{V(\beta), U(t) | \beta \geq 0,\ t \in \mathbb{R}\}$ of the quasi-Weyl relation with one degree of freedom is studied in [50].

4.5.7 A Remark on Uniqueness of Weak Weyl Representations

Based on Theorem 4.12, one can classify the set $\mathsf{CCR}_{\mathrm{WW}}(1)$ of weak Weyl representations with degree one into the following mutually disjoint subclasses:

$$\mathscr{W}_{+,1} := \{(\mathscr{H}, (T, H)) \in \mathsf{CCR}_{\mathrm{WW}}(1) |\ \inf \sigma(H) > -\infty \text{ and } \sigma(T) = \overline{\Pi}_+\},$$

$$\mathscr{W}_{+,2} := \{(\mathscr{H}, (T, H)) \in \mathsf{CCR}_{\mathrm{WW}}(1) |\ \sup \sigma(H) = \infty, \inf \sigma(H) > -\infty \text{ and } \sigma(T) = \mathbb{C}\},$$

$$\mathscr{W}_{-,1} := \{(\mathscr{H}, (T, H)) \in \mathsf{CCR}_{\mathrm{WW}}(1) |\ \sup \sigma(H) < \infty \text{ and } \sigma(T) = \overline{\Pi}_-\},$$

$$\mathscr{W}_{-,2} := \{(\mathscr{H}, (T, H)) \in \mathsf{CCR}_{\mathrm{WW}}(1) |\ \sup \sigma(H) < \infty, \inf \sigma(H) = -\infty \text{ and } \sigma(T) = \mathbb{C}\},$$

$$\mathscr{W}_{\mathrm{b}} := \{(\mathscr{H}, (T, H)) \in \mathsf{CCR}_{\mathrm{WW}}(1) |\ H \text{ is bounded}\},$$

$$\mathscr{W}_{\mathrm{ub}} := \{(\mathscr{H}, (T, H)) \in \mathsf{CCR}_{\mathrm{WW}}(1) |\ \sup \sigma(H) = \infty \text{ and } \inf \sigma(H) = -\infty\}.$$

It is easy to see that

$$(\mathscr{H}, (T, H)) \in \mathscr{W}_{+,1} \Longleftrightarrow (\mathscr{H}, (-T, -H)) \in \mathscr{W}_{-,1},$$

$$(\mathscr{H}, (T, H)) \in \mathscr{W}_{+,2} \Longleftrightarrow (\mathscr{H}, (-T, -H)) \in \mathscr{W}_{-,2}.$$

Therefore essentially different subclasses are $\mathscr{W}_{+,1}, \mathscr{W}_{+,2}, \mathscr{W}_{\mathrm{b}}$ and $\mathscr{W}_{\mathrm{ub}}$. It may be natural to ask if, for each of them, there exists a uniqueness theorem like the von Neumann uniqueness theorem. Here we state only a result on uniqueness concerning $\mathscr{W}_{+,1}$

In view of the von Neumann uniqueness theorem, the Schrödinger representation $\pi_{\mathrm{S}}^{(1)}$ is regarded as a reference representation in the theory of self-adjoint representations of the CCR with degree one. Noting that the weak Weyl representation $\pi_{\mathrm{W},a}$ defined by (2.179) is in $\mathscr{W}_{+,1}$ (see (4.46)), one may use it as a reference representation in the theory of weak Weyl representations belonging to $\mathscr{W}_{+,1}$. Indeed, the following theorem is known:

Theorem 4.16 *Let $(\mathscr{H}, (T, H)) \in \mathscr{W}_{+,1}$ with $\mathscr{H}$ separable. Suppose that $\{\overline{T}, H\}' = \mathbb{C}I$.*[9] *Let $a := \inf \sigma(H)$. Then there exists a unitary operator U from $\mathscr{H}$ to $L^2([a, \infty))$ such that*

$$UHU^{-1} = \hat{q}_a, \quad U\overline{T}U^{-1} = \overline{T_a}.$$

For the proof of this theorem, see [19] (cf. also [50, 65, 140, 141] for related aspects).

We now discuss some examples.

4.5.8 Strong Time Operator of $\overline{H}_F$

Theorem 4.17 *Let H_F and T_F be as in Example 4.1. Then T_F is a self-adjoint strong time operator of $\overline{H}_F$.*[10] *Moreover, $\overline{H}_F$ is absolutely continuous with*

$$\sigma(\overline{H}_F) = \sigma_{\text{ac}}(\overline{H}_F) = \mathbb{R}. \tag{4.52}$$

Proof It is obvious that T_F is self-adjoint. Let

$$h_0 := -F\hat{q}.$$

Then $(L^2(\mathbb{R}), (T_F, h_0)) \in \mathsf{CCR}_{\mathrm{W}}(1)$. In particular, T_F is a self-adjoint strong time operator of h_0. We can write

$$H_F = h_0 + h_1, \quad h_1 := H_{\text{NR}}.$$

It follows from functional calculus that T_F strongly commutes with h_1, We already know that H_F is essentially self-adjoint. Hence, by Theorem 4.11, T_F is a strong time operator of $\overline{H}_F$ and $\overline{H}_F$ is absolutely continuous. The spectral property (4.52) follows from Theorem 4.9. ■

By Corollary 4.4, $(T_F, \overline{H}_F)$ is unitarily equivalent to a direct sum of the Schrödinger representation $(\hat{q}, \hat{p})$ with one degree of freedom. In fact, one can construct explicitly a unitary operator U on $L^2(\mathbb{R})$ such that

$$UT_FU^{-1} = \hat{q}, \quad U\overline{H}_FU^{-1} = \hat{p} \tag{4.53}$$

---

[9]Hence $\{\overline{T}, H\}$ is irreducible.

[10]Note that $\overline{H}_F$ is not semi-bounded and hence the existence of a self-adjoint strong time operator of it does not contradict Corollary 4.3.

as follows. Without loss of generality, we can assume that $F > 0$.[11] We introduce a function

$$W_F(k) := \exp\left(i\frac{k^3}{6mF}\right), \quad k \in \hat{\mathbb{R}}.$$

Then the one-dimensional Fourier transform of the multiplication operator by the function W_F

$$U_1 := \mathscr{F}_1^{-1} W_F \mathscr{F}_1$$

is a unitary operator on $L^2(\mathbb{R})$, where $\mathscr{F}_1 \colon L^2(\mathbb{R}) \to L^2(\hat{\mathbb{R}})$ is the one-dimensional Fourier transform (see (1.74)). It is easy to see that

$$U_1 \hat{p} U_1^{-1} = \hat{p}$$

and

$$U_1 \hat{q} U_1^{-1} = \frac{\hat{p}^2}{2mF} + \hat{q} \quad \text{on } \mathscr{S}(\mathbb{R}).$$

Hence

$$U_1 H_F U_1^{-1} = -F\hat{q} \quad \text{on } \mathscr{S}(\mathbb{R})$$

and

$$U_1 T_F U_1^{-1} = \frac{1}{F}\hat{p}.$$

Since $-F\hat{q}$ is essentially self-adjoint on $\mathscr{S}(\mathbb{R})$ and $\overline{H}_F$ is self-adjoint, it follows that

$$U_1 \overline{H}_F U_1^{-1} = -F\hat{q}.$$

---

[11]Let u_s be the **space inversion** on $L^2(\mathbb{R})$: $(u_s f)(x) := f(-x)$, a.e.$x \in \mathbb{R}$. Then u_s is unitary with $u_s^2 = I$ and

$$u_s \hat{q} u_s^{-1} = -\hat{q}, \quad u_s \hat{p} u_s^{-1} = -\hat{p}.$$

Hence

$$u_s H_F u_s^{-1} = H_{-F}, \quad u_s T_F u_s^{-1} = T_{-F}.$$

Let

$$(U_2 f)(k) := \sqrt{F}(\mathscr{F}_1 f)(Fk), \quad f \in L^2(\mathbb{R}), \text{ a.e.} k \in \hat{\mathbb{R}}.$$

Then U_2 is unitary and

$$U_2 T_F U_2^{-1} = k, \quad U_2(-F\hat{q})U_2^{-1} = -iD_k.$$

Hence the operator $U := U_2 U_1$ is unitary and

$$U T_F U^{-1} = k, \quad U \overline{H}_F U^{-1} = -iD_k.$$

Changing the name of the variable from k to x, we obtain (4.53). Thus we arrive at the following fact:

Theorem 4.18 *The pair* $(T_F, \overline{H}_F)$ *is an irreducible Weyl representation of the CCR with one degree of freedom.*

4.5.9 Strong Time Operator of an N-body System

There is a generalization of the preceding example. We consider a system of N non-relativistic quantum particles being in $\mathbb{R}^d$ ($d \in \mathbb{N}$), where the mass of the jth particle is $m_j > 0$ ($j = 1, \ldots, N$) and the jth particle has electric charge $e_j \in \mathbb{R} \setminus \{0\}$. Suppose that a constant electric field $\boldsymbol{E} = (E_1, \ldots, E_d) \in \mathbb{R}^d$ is applied to the system. Then the Hamiltonian of the system is given by

$$H_0(\boldsymbol{E}) := \sum_{j=1}^{N} \left(-\frac{\Delta_{\boldsymbol{x}_j}}{2m_j} - e_j \boldsymbol{E} \cdot \hat{\boldsymbol{q}}_j \right),$$

acting in $L^2((\mathbb{R}^d)^N)$, where $\Delta_{\boldsymbol{x}_j}$ is the generalized Laplacian in the variable $\boldsymbol{x}_j = (x_{j1}, \ldots, x_{jd}) \in \mathbb{R}^d$ and

$$\hat{\boldsymbol{q}}_j = (\hat{q}_{j1}, \ldots, \hat{q}_{jd})$$

with $\hat{q}_{j\ell}$ being the multiplication operator by $x_{j\ell}$ ($\ell = 1, \ldots, d$). Note that

$$H_0(\boldsymbol{E}) = \sum_{j=1}^{N} \sum_{\ell=1}^{d} H_{j,\ell},$$

where

$$H_{j,\ell} := -\frac{\Delta_{x_{j\ell}}}{2m_j} - e_j E_\ell \hat{q}_{j\ell}.$$

Under the natural isomorphism

$$L^2((\mathbb{R}^d)^N) \cong \otimes_{j=1,\dots,N,\,\ell=1,\dots,d} \mathscr{H}_{j,\ell}, \quad \mathscr{H}_{j,\ell} := L^2(\mathbb{R}),$$

the symmetric operator $\overline{H_0(\boldsymbol{E})}$ is unitarily equivalent to the closure of

$$\sum_{j=1}^{N} \sum_{\ell=1}^{d} I \otimes \cdots \otimes I \otimes \overline{H}_{j,\ell} \otimes I \otimes \cdots \otimes I.$$

We already know that $\overline{H}_{j,\ell}$ is self-adjoint (see Example 4.1) and

$$\sigma(\overline{H}_{j,\ell}) = \sigma_{\mathrm{ac}}(\overline{H}_{j,\ell}) = \mathbb{R}.$$

Hence, by a general theory of tensor products of self-adjoint operators, $\overline{H_0(\boldsymbol{E})}$ is self-adjoint (hence $H_0(\boldsymbol{E})$ is essentially self-adjoint[12]) and

$$\sigma(\overline{H_0(\boldsymbol{E})}) = \mathbb{R}. \tag{4.54}$$

Lemma 4.7 *Let $A := (a_{j\ell})$ be a non-zero $N \times d$ real matrix and*

$$T_{A,0} := \sum_{j=1}^{N} \sum_{\ell=1}^{d} a_{j\ell} \hat{p}_{j\ell}. \tag{4.55}$$

Then $T_{A,0}$ is essentially self-adjoint and

$$\sigma(T_A) = \mathbb{R}, \tag{4.56}$$

where

$$T_A := \overline{T}_{A,0}. \tag{4.57}$$

Proof The essential self-adjointness of $T_{A,0}$ follows from applying Theorem 2.18 to the case $\boldsymbol{a} = 0$ and $\boldsymbol{b} = (a_{11}, \dots, a_{N1}, a_{12}, \dots, a_{N2}, \dots, a_{1d}, \dots, a_{Nd})$ (or strong

[12]This can be proved also by applying the commutator theorem (Theorem 1.13) to the case where $S = \sum_{j=1}^{N}(-\Delta_{x_j} + \hat{\boldsymbol{q}}_j^2) + 1$, $A = H_0(\boldsymbol{E})$ and $\mathscr{D} = C_0^\infty((\mathbb{R}^d)^N)$.

commutativity of $(\hat{p}_{j\ell})$). But here we give another proof. Let $\mathscr{F}_{dN}\colon L^2((\mathbb{R}^d)^N) \to L^2((\hat{\mathbb{R}}^d)^N)$ be the dN-dimensional Fourier transform. Then it is easy to see that

$$\mathscr{F}_{dN} T_{A,0} \mathscr{F}_{dN}^{-1} = M_F \quad \text{on } C_0^\infty((\mathbb{R}^d)^N)$$

with $F(\boldsymbol{k}) := \sum_{j=1}^N \sum_{\ell=1}^d a_{j\ell} k_{j\ell}$, $\boldsymbol{k}_1, \ldots, \boldsymbol{k}_N \in \hat{\mathbb{R}}^d$. Since F is a real-valued continuous function, M_F is self-adjoint and essentially self-adjoint on $C_0^\infty((\mathbb{R}^d)^N)$. Hence $T_{A,0}$ is essentially self-adjoint on $\mathscr{F}_{dN}^{-1} C_0^\infty((\mathbb{R}^d)^N)$. Therefore $T_{A,0}$ is essentially self-adjoint and operator equality

$$\mathscr{F}_{dN} T_A \mathscr{F}_{dN}^{-1} = M_F$$

holds. By the unitary invariance of spectrum, $\sigma(T_A) = \sigma(M_F)$. It is easy to see that $F((\mathbb{R}^d)^N) = \mathbb{R}$. Hence $\sigma(M_F) = \mathbb{R}$. Thus (4.56) follows (recall that, for a closable linear operator C, $\sigma(C) = \sigma(\overline{C})$). ■

Theorem 4.19 *Let $\boldsymbol{e} := (e_1, \ldots, e_N) \in \mathbb{R}^N$. Suppose that*

$$\boldsymbol{e} \cdot A\boldsymbol{E} = 1. \tag{4.58}$$

Then T_A is a self-adjoint strong time operator of $\overline{H_0(\boldsymbol{E})}$. In particular, $\overline{H_0(\boldsymbol{E})}$ is absolutely continuous.

Proof Let $T_{j\ell} := \hat{p}_{j\ell}/(e_j E_\ell)$. Then, by Theorem 4.17, we have for all $t \in \mathbb{R}$

$$e^{it\overline{H}_{j\ell}} T_{j\ell} e^{it\overline{H}_{j\ell}} = T_{j\ell} + t.$$

Note that, if $(j', \ell') \neq (j, \ell)$, then $e^{it\overline{H}_{j'\ell'}} T_{j\ell} e^{it\overline{H}_{j'\ell'}} = T_{j\ell}$. Hence

$$e^{it\overline{H_0(\boldsymbol{E})}} T_{j\ell} e^{-it\overline{H_0(\boldsymbol{E})}} = T_{j\ell} + t,$$

which is equivalent to

$$e^{it\overline{H_0(\boldsymbol{E})}} \hat{p}_{j\ell} e^{-it\overline{H_0(\boldsymbol{E})}} = \hat{p}_{j\ell} + e_j E_\ell t.$$

Hence

$$e^{it\overline{H_0(\boldsymbol{E})}} T_{A,0} e^{-it\overline{H_0(\boldsymbol{E})}} = T_{A,0} + \left(\sum_{j=1}^N \sum_{\ell=1}^d a_{j\ell} e_j E_\ell \right) t.$$

By (4.58), $\sum_{j=1}^N \sum_{\ell=1}^d a_{j\ell} e_j E_\ell = 1$. Therefore

$$e^{it\overline{H_0(\boldsymbol{E})}} T_{A,0} e^{-it\overline{H_0(\boldsymbol{E})}} = T_{A,0} + t.$$

Taking the closure of both sides, we obtain $e^{it\overline{H_0(\boldsymbol{E})}}T_A e^{-it\overline{H_0(\boldsymbol{E})}} = T_A + t$. Thus T_A is a strong time operator of $\overline{H_0(\boldsymbol{E})}$. ■

A perturbation of $\overline{H_0(\boldsymbol{E})}$ is given by

$$H_{\boldsymbol{E}} := \overline{H_0(\boldsymbol{E})} + \sum_{1\le i<j\le N} V_{ij}(\boldsymbol{x}_i - \boldsymbol{x}_j), \tag{4.59}$$

where $V_{ij}\colon \mathbb{R}^d \to \mathbb{R} \cup \{\pm\infty\}$ is a potential which is a.e. finite.

For a vector $\boldsymbol{a} \in \mathbb{R}^d$, we define

$$T_{\boldsymbol{a}} := \overline{\sum_{j=1}^{N} \boldsymbol{a} \cdot \hat{\boldsymbol{p}}_j}, \tag{4.60}$$

which is self-adjoint by Lemma 4.7. The total charge of the system is given by

$$Q := \sum_{j=1}^{N} e_j.$$

Theorem 4.20 *Suppose that*

$$Q\boldsymbol{a} \cdot \boldsymbol{E} = 1 \tag{4.61}$$

and that $H_{\boldsymbol{E}}$ is essentially self-adjoint. Then $T_{\boldsymbol{a}}$ is a self-adjoint strong time operator of $\overline{H}_{\boldsymbol{E}}$ and hence $\overline{H}_{\boldsymbol{E}}$ is absolutely continuous and $\sigma(\overline{H}_{\boldsymbol{E}}) = \mathbb{R}$.

Proof The operator $T_{\boldsymbol{a}}$ is the operator T_A with $a_{j\ell} = a_\ell, j = 1, \ldots, N, \ell = 1, \ldots, d$. Then (4.61) implies (4.58). Hence $T_{\boldsymbol{a}}$ is a strong time operator of $\overline{H_0(\boldsymbol{E})}$.

For a Borel measurable function F on $(\mathbb{R}^d)^N$, we have for all $s \in \mathbb{R}$

$$e^{isT_{\boldsymbol{a}}} F e^{-isT_{\boldsymbol{a}}} = G$$

as an operator equation, where

$$G(\boldsymbol{x}_1, \ldots, \boldsymbol{x}_N) := F(\boldsymbol{x}_1 + s\boldsymbol{a}, \ldots, \boldsymbol{x}_N + s\boldsymbol{a}), \quad \text{a.e.}(\boldsymbol{x}_1, \ldots, \boldsymbol{x}_N),$$

Hence, putting $U_{ij}(\boldsymbol{x}_1, \ldots, \boldsymbol{x}_N) := V_{ij}(\boldsymbol{x}_i - \boldsymbol{x}_j)$, we have

$$e^{isT_{\boldsymbol{a}}} U_{ij} e^{-isT_{\boldsymbol{a}}} = U_{ij}.$$

Therefore

$$e^{isT_{\boldsymbol{a}}} V e^{-isT_{\boldsymbol{a}}} = V,$$

where $V := \sum_{1\leq i<j\leq N} U_{ij}$. Since V is a self-adjoint multiplication operator, it follows that $T_{\boldsymbol{a}}$ strongly commutes with V. Thus, by Theorem 4.11, $T_{\boldsymbol{a}}$ is a strong time operator of $\overline{H}_{\boldsymbol{E}}$. ■

Remark 4.12 Theorem 4.20 is a version of a theorem given in [162].

4.5.10 Aharonov–Bohm Time Operators

In this subsection, we show that the Aharonov–Bohm time operator T_{AB} is a strong time operator of H_{NR} and its closure is non-self-adjoint and maximally symmetric.

We first prove a useful formula:

Lemma 4.8 *Let $m > 0$, Δ be the d-dimensional generalized Laplacian acting in $L^2(\mathbb{R}^d)$ and*

$$H_{\mathrm{NR}}^{(d)} := -\frac{1}{2m}\Delta. \tag{4.62}$$

Then, for all $t \in \mathbb{R}$ and $j = 1, \ldots, d$,

$$e^{itH_{\mathrm{NR}}^{(d)}}\hat{q}_j e^{-itH_{\mathrm{NR}}^{(d)}} \supset \hat{q}_j + \frac{t}{m}\hat{p}_j, \tag{4.63}$$

$$e^{itH_{\mathrm{NR}}^{(d)}}\hat{p}_j e^{-itH_{\mathrm{NR}}^{(d)}} = \hat{p}_j. \tag{4.64}$$

Proof Let $A := e^{itH_{\mathrm{NR}}^{(d)}}\hat{q}_j e^{-itH_{\mathrm{NR}}^{(d)}}$ and $\mathscr{F}_d$ be the d-dimensional Fourier transform. Then

$$\mathscr{F}_d A \mathscr{F}_d^{-1} = e^{itE(\boldsymbol{k})} i D_{k_j} e^{-itE(\boldsymbol{k})},$$

where $E(\boldsymbol{k}) := \boldsymbol{k}^2/(2m)$, $\boldsymbol{k} \in \hat{\mathbb{R}}^d$ and D_{k_j} is the generalized partial differential operator in k_j. For all $f \in D(\hat{q}_j) \cap D(\hat{p}_j)$, $\hat{f} \in D(D_{k_j}) \cap D(k_j)$ and

$$D_{k_j} e^{-itE(\boldsymbol{k})}\hat{f} = e^{-itE(\boldsymbol{k})}\left(-\frac{itk_j}{m}\hat{f} + D_{k_j}\hat{f}\right).$$

Hence

$$\mathscr{F}_d A \mathscr{F}_d^{-1}\hat{f} = \frac{tk_j}{m}\hat{f} + iD_{k_j}\hat{f}.$$

Therefore

$$A = \hat{q}_j + \frac{t}{m}\hat{p}_j \quad \text{on } D(\hat{q}_j) \cap D(\hat{p}_j).$$

Thus (4.63) follows.

Formula (4.64) follows from the strong commutativity of $-\Delta = \sum_{\ell=1}^{d} \hat{p}_\ell^2$ with $\hat{p}_j$ or the Fourier analysis. ∎

Lemma 4.9 *For all $t \in \mathbb{R}$ and $j = 1, \ldots, d$, the operator $\hat{q}_j + (t/m)\hat{p}_j$ is essentially self-adjoint and*

$$e^{itH_{\mathrm{NR}}^{(d)}} \hat{q}_j e^{-itH_{\mathrm{NR}}^{(d)}} = \overline{\hat{q}_j + \frac{t}{m}\hat{p}_j}. \tag{4.65}$$

Proof The essential self-adjointness of $\hat{q}_j + (t/m)\hat{p}_j$ follows from an application of Theorem 2.18. Taking the closure of both sides of (4.63) and using the fact that a self-adjoint operator has no non-trivial symmetric extension, one obtains (4.65). ∎

We introduce a subspace of $L^2(\mathbb{R}^d)$:

$$\mathscr{D}_{\mathrm{AB}}^{(d)} := \underbrace{\mathscr{D}_{\mathrm{AB}} \mathring{\otimes} \cdots \mathring{\otimes} \mathscr{D}_{\mathrm{AB}}}_{d \text{ factors}}, \tag{4.66}$$

where $\mathscr{D}_{\mathrm{AB}}$ is defined by (4.4). It follows that $\mathscr{D}_{\mathrm{AB}}^{(d)}$ is dense in $L^2(\mathbb{R}^d)$. It is easy to see that $H_{\mathrm{NR}}^{(d)}$ is injective and that $\mathscr{D}_{\mathrm{AB}}^{(d)} \subset D((H_{\mathrm{NR}}^{(d)})^{-1}) \cap D(\hat{q}_j) \cap D(\hat{p}_j)$, $j = 1, \ldots, d$ and $\hat{q}_j$, $\hat{p}_j$ and $(H_{\mathrm{NR}}^{(d)})^{-1}$ leave $\mathscr{D}_{\mathrm{AB}}^{(d)}$ invariant. Hence one can define

$$T_{\mathrm{AB},0}^{(d)} := \frac{1}{4}\sum_{j=1}^{d} \left((H_{\mathrm{NR}}^{(d)})^{-1} \hat{p}_j \hat{q}_j + \hat{q}_j \hat{p}_j (H_{\mathrm{NR},d})^{-1} \right) \restriction \mathscr{D}_{\mathrm{AB}}^{(d)}. \tag{4.67}$$

It is easy to see that $T_{\mathrm{AB},0}^{(d)}$ is a symmetric operator. Note that

$$T_{\mathrm{AB},0}^{(1)} \subset T_{\mathrm{AB}}.$$

Hence $T_{\mathrm{AB},0}^{(d)}$ is a d-dimensional version of T_{AB}. We set

$$T_{\mathrm{AB}}^{(d)} := \overline{T_{\mathrm{AB},0}^{(d)}}. \tag{4.68}$$

Theorem 4.21 *The operator $T_{\mathrm{AB}}^{(d)}$ is a non-self-adjoint strong time operator of $H_{\mathrm{NR}}^{(d)}$.*

Proof Throughout the proof, we set $H := H_{\mathrm{NR}}^{(d)}$ and $T := T_{\mathrm{AB},0}^{(d)}$. By the functional calculus for H, we have

$$e^{itH} H^{-1} e^{-itH} = H^{-1}.$$

Let $f \in \mathscr{D}_{\mathrm{AB}}^{(d)}$. Then, for all $t \in \mathbb{R}$, $e^{-itH}f \in \mathscr{D}_{\mathrm{AB}}^{(d)}$ (by Fourier analysis). By Lemma 4.8, we have

$$\begin{aligned} e^{itH}Te^{-itH}f &= \frac{1}{4}\sum_{j=1}^{d}\left(H^{-1}\hat{p}_j e^{itH}\hat{q}_j e^{-itH}f + e^{itH}\hat{q}_j e^{-itH}\hat{p}_j H^{-1}f\right) \\ &= \frac{1}{4}\sum_{j=1}^{d}\left(H^{-1}\hat{p}_j(\hat{q}_j + \frac{t}{m}\hat{p}_j)f + (\hat{q}_j + \frac{t}{m}\hat{p}_j)\hat{p}_j H^{-1}f\right) \\ &= (T+t)f. \end{aligned}$$

Hence T is a strong time operator of H. Therefore its closure $T_{\mathrm{AB}}^{(d)}$ is a strong time operator of H. Since H is non-negative, it follows from Theorem 4.10 that $T_{\mathrm{AB}}^{(d)}$ is not self-adjoint. ■

We call $T_{\mathrm{AB}}^{(d)}$ the d-**dimensional Aharonov–Bohm time operator**.

We now consider the case $d = 1$.

Theorem 4.22 *The operator $T_{\mathrm{AB}}^{(1)}$ is non-self-adjoint and maximally symmetric and*

$$\sigma(T_{\mathrm{AB}}^{(1)}) = \overline{\Pi}_+. \tag{4.69}$$

Proof Throughout the proof, we set $T := T_{\mathrm{AB}}^{(1)}$. We first prove the following:

$$\ker(T^* - i) = \{0\}. \tag{4.70}$$

Let $g \in \ker(T^* - i)$. Then, for all $f \in \mathscr{D}_{\mathrm{AB}}$, $\langle (T^* - i)g, f\rangle = 0$. Hence $\langle g, (T+i)f\rangle = 0$. By the Plancherel theorem, we have

$$\frac{m}{2}\int_{\hat{\mathbb{R}}\setminus\{0\}} \hat{g}(k)^* \left(k^{-1}D_k\hat{f}(k) + D_k(k^{-1}\hat{f}(k))\right)dk = -\int_{\hat{\mathbb{R}}} \hat{g}(k)^*\hat{f}(k)dk.$$

Hence

$$m\int_{\hat{\mathbb{R}}\setminus\{0\}} \frac{1}{k}\hat{g}(k)^* D_k\hat{f}(k) - \frac{m}{2}\int_{\hat{\mathbb{R}}\setminus\{0\}} \frac{1}{k^2}\hat{g}(k)^*\hat{f}(k)dk = -\int_{\hat{\mathbb{R}}} \hat{g}(k)^*\hat{f}(k)dk.$$

Since $\hat{f} \in C_0^\infty(\hat{\mathbb{R}} \setminus \{0\})$ is arbitrary, it follows that the L^1_{loc} function $h(k) := k^{-1}g(k)^*$ on $\hat{\mathbb{R}} \setminus \{0\}$ obeys the equation

$$D_k h(k) = -\frac{1}{2k}h(k) + \frac{k}{m}h(k)$$

in the distribution sense on $\hat{\mathbb{R}} \setminus \{0\}$. It is easy to see that the distributional solution to this equation is given by $h(k) = Ce^{k^2/2m}/\sqrt{|k|}$, where $C \in \mathbb{C}$ is a constant. Hence $\hat{g}(k)^* = Ce^{k^2/2m}k/\sqrt{|k|}$. Since $\hat{g} \in L^2(\hat{\mathbb{R}})$, it follows that $C = 0$. Hence $\hat{g} = 0$. Therefore $g = 0$. Thus (4.70) follows. This means that $n_+(T) = 0$. This result and the non-self-adjointness of T imply that $n_-(T) \neq 0$ (see Theorem 1.11(ii)). Hence T is maximally symmetric. By applying Theorem 1.19(i), we obtain (4.69). ■

Remark 4.13 In the same way as in the proof of (4.70), we can show that $n_-(T) = 1$ with

$$\ker(T^* + i) = \{\alpha g_- | \alpha \in \mathbb{C}\}, \quad \hat{g}_-(k) := \frac{k}{\sqrt{|k|}} e^{-k^2/2m}, \ k \in \hat{\mathbb{R}} \setminus \{0\}.$$

Remark 4.14 Unfortunately the present author has been unable to make it clear if $T_{\rm AB}^{(d)}$ with $d \geq 2$ is maximally symmetric.

Remark 4.15 The d-dimensional Aharonov–Bohm time operator $T_{\rm AB}^{(d)}$ is related to time delay in a two-body scattering process; it appears in a formula for the symmetrized time delay introduced by Gérard and Tiedra de Aldecoa (see [79, Theorem 4.6]). This is noteworthy. In the paper [79], however, no mention is made that their operator A_0 ($= T_{\rm AB}^{(d)}$ in our notation) is a time operator of the free Hamiltonian. It is shown in [134] that time operators analogous to $T_{\rm AB}^{(d)}$ in an abstract form are related to localization operators.

4.5.11 *Strong Time Operators of Free Relativistic Schrödinger Operators*

Theorem 4.23 *Let $H_{\rm R}$ and $T_{\rm R}$ be as in Example 4.3. Then*

$$T_{\rm R,0} := T_{\rm R} \upharpoonright \mathscr{D}_{\rm AB}$$

is a strong time operator of $H_{\rm R}$. In particular, $H_{\rm R}$ is absolutely continuous with

$$\sigma(H_{\rm R}) = \sigma_{\rm ac}(H_{\rm R}) = [m, \infty).$$

Proof By Fourier analysis, we have for all $t \in \mathbb{R}$

$$e^{itH_{\rm R}}\hat{q}e^{-itH_{\rm R}} = \hat{q} + t\hat{p}H_{\rm R}^{-1} \quad \text{on } \mathscr{D}_{\rm AB}.$$

By functional calculus,

$$e^{itH_{\rm R}}\hat{p}e^{-itH_{\rm R}} = \hat{p}, \quad e^{itH_{\rm R}}H_{\rm R}e^{-itH_{\rm R}} = H_{\rm R}.$$

Using these relations, one can show in the same way as in the proof of Theorem 4.21 that

$$e^{itH_{\rm R}}T_{{\rm R},0}e^{-itH_{\rm R}} = T_{{\rm R},0} + t, \quad \forall t \in \mathbb{R}.$$

Hence $T_{{\rm R},0}$ is a strong time operator of $H_{\rm R}$. ■

Corollary 4.6 *The operator $\overline{T}_{\rm R}$ is a non-self-adjoint maximally symmetric strong time operator of $H_{\rm R}$ and*

$$\sigma(\overline{T}_{\rm R}) = \overline{\Pi}_+. \tag{4.71}$$

Proof Let

$$\hat{T}_{\rm R} := \overline{T}_{{\rm R},0}.$$

Then, by Theorem 4.23 and Proposition 4.5, $\hat{T}_{\rm R}$ is a strong time operator of $H_{\rm R}$:

$$e^{itH_{\rm R}}\hat{T}_{\rm R}e^{-itH_{\rm R}} = \hat{T}_{\rm R} + t, \quad t \in \mathbb{R}.$$

Since $H_{\rm R}$ is non-negative, it follows from Theorem 4.10 that $\hat{T}_{\rm R}$ is not self-adjoint.

Let $g \in \ker(\hat{T}_{\rm R}^* - i)$. Then, for all $f \in C_0^\infty(\hat{\mathbb{R}} \setminus \{0\})$,

$$\int_{\hat{\mathbb{R}}} \hat{g}(k)^* \Big\{\frac{1}{2}[\omega_1(k)k^{-1}iD_k f(k) + iD_k(k^{-1}\omega_1(k)f(k))] + if(k)\Big\}dk = 0,$$

where ω_1 is defined by (4.7). The distributional solution to this equation is given by

$$\hat{g}(k) = C\frac{k}{\sqrt{|k|\omega_1(k)}}e^{\omega_1(k)}, \quad k \in \hat{\mathbb{R}} \setminus \{0\},$$

where $C \in \mathbb{C}$ is a constant. Since $\hat{g}$ is in $L^2(\hat{\mathbb{R}})$, it follows that $C = 0$ and hence $g = 0$. Therefore $n_+(\hat{T}_{\rm R}) = 0$. By this result and the non-self-adjointness of $\hat{T}_{\rm R}$, $n_-(\hat{T}_{\rm R}) \neq 0$. Hence $\hat{T}_{\rm R}$ is maximally symmetric. It follows from an application of Theorem 1.19(i) that $\sigma(\hat{T}_{\rm R}) = \overline{\Pi}_+$. One has $\hat{T}_{\rm R} \subset \overline{T}_{\rm R}$. Since $\overline{T}_{\rm R}$ is a closed symmetric operator, it follows that $\hat{T}_{\rm R} = \overline{T}_{\rm R}$. Thus $\overline{T}_{\rm R}$ is maximally symmetric and (4.71) holds. ■

Remark 4.16 In the same way as in the above proof, one can show that $n_-(\hat{T}_{\rm R}) = 1$ with

$$\ker(\hat{T}_{\rm R}^* + i) = \{\alpha g | \alpha \in \mathbb{C}\}, \quad \hat{g}(k) := \frac{k}{\sqrt{|k|\omega_1(k)}}e^{-\omega_1(k)}.$$

To make explicit the dependence of H_{R} and $\overline{T}_{\mathrm{R}}$ on the mass m, we write

$$H_{\mathrm{R},m} := H_{\mathrm{R}}, \quad T_{\mathrm{R},m} := \overline{T}_{\mathrm{R}}.$$

We have

$$\sigma(H_{\mathrm{R},m}) = [m, \infty).$$

Hence, if $m_1 \neq m_2$ $(m_1, m_2 \geq 0)$, then $\sigma(H_{\mathrm{R},m_1}) \neq \sigma(H_{\mathrm{R},m_2})$. Thus, if $m_1 \neq m_2$, then $(T_{\mathrm{R},m_1}, H_{\mathrm{R},m_1})$ is inequivalent to $(T_{\mathrm{R},m_2}, H_{\mathrm{R},m_2})$. Thus we arrive at the following fact:

Theorem 4.24 *The set* $\{(T_{\mathrm{R},m}, H_{\mathrm{R},m})\}_{m\geq 0}$ *is a family of mutually inequivalent weak Weyl representations with one degree of freedom.*

This theorem means that the masses of free relativistic quantum particles appear as the index set of a family of mutually inequivalent weak Weyl representations with one degree of freedom consisting of pairs of a free relativistic Schrödinger Hamiltonian and a strong time operator of it. This gives a representation theoretic view for the masses of relativistic quantum particles.

We next consider a d-dimensional version of H_R $(d \in \mathbb{N})$. Let $(\hat{q}_1, \ldots, \hat{q}_d, \hat{p}_1, \ldots, \hat{p}_d)$ be the Schrödinger representation of the CCR with d degrees of freedom and Δ be the d-dimensional generalized Laplacian acting in $L^2(\mathbb{R}^d)$:

$$-\Delta = \sum_{j=1}^{d} \hat{p}_j^2.$$

Then the d-dimensional **free relativistic Schrödinger operator**, acting in $L^2(\mathbb{R}^d)$, is defined by

$$H_{\mathrm{R},m}^{(d)} := (-\Delta + m^2)^{1/2}, \tag{4.72}$$

which is a non-negative self-adjoint operator.

For each $j = 1, \ldots, d$, we define a subspace:

$$\mathscr{D}_j := \{f \in L^2(\mathbb{R}^d) | \hat{f} \in C_0^\infty(\hat{\mathbb{R}}^d), \operatorname{supp} \hat{f} \subset \hat{\mathbb{R}}^d \setminus \{\boldsymbol{k} \in \hat{\mathbb{R}}^d | k_j = 0\}, \tag{4.73}$$

where $\hat{f}$ is the d-dimensional Fourier transform of f. This subspace is dense in $L^2(\mathbb{R}^d)$,

It is easy to see that the operator

$$\frac{1}{2}\left(H_{\mathrm{R},m}^{(d)} \hat{p}_j^{-1} \hat{q}_j + \hat{q}_j \hat{p}_j^{-1} H_{\mathrm{R},m}^{(d)}\right)$$

restricted to $\mathscr{D}_j$ is symmetric and hence closable. Hence one can define

$$T_{\mathrm{R},m,j} := \frac{1}{2}\overline{\left(H_{\mathrm{R},m}^{(d)} \hat{p}_j^{-1} \hat{q}_j + \hat{q}_j \hat{p}_j^{-1} H_{\mathrm{R},m}^{(d)}\right) \upharpoonright \mathscr{D}_j}, \tag{4.74}$$

which is a closed symmetric operator.

In the same way as in the case of $(T_{\mathrm{R},m}, H_{\mathrm{R},m})$, one can prove the following theorem:

Theorem 4.25 *Let $m \geq 0$ and $j = 1, \ldots, d$.*

(i) *The operator $T_{\mathrm{R},m,j}$ is a strong time operator of $H_{\mathrm{R},m}^{(d)}$. In particular, $H_{\mathrm{R},m}^{(d)}$ is absolutely continuous with*

$$\sigma(H_{\mathrm{R},m}^{(d)}) = \sigma_{\mathrm{ac}}(H_{\mathrm{R},m}^{(d)}) = [m, \infty).$$

(ii) *The operator $T_{\mathrm{R},m,j}$ is a non-self-adjoint maximally symmetric operator and*

$$\sigma(T_{\mathrm{R},m,j}) = \overline{\Pi}_{+}. \tag{4.75}$$

(iii) *The set $\{(T_{\mathrm{R},m,j}, H_{\mathrm{R},m}^{(d)})\}_{m\geq 0}$ is a family of mutually inequivalent weak Weyl representations with one degree of freedom.*

4.5.12 Strong Time Operators of a Free Dirac Operator

We next show that the three-dimensional free Dirac operator with mass $m \geq 0$ has a strong time operator. To recall the Dirac operator, let α_j and β be 4×4 Hermitian matrices satisfying the following anti-commutation relations:

$$\{\alpha_j, \alpha_\ell\} = 2\delta_{j\ell} I_4, \quad j, \ell = 1, 2, 3, \tag{4.76}$$

$$\{\alpha_j, \beta\} = 0, \quad \beta^2 = I_4, \quad j = 1, 2, 3, \tag{4.77}$$

where $\{A, B\} := AB + BA$ (anti-commutator) and I_n with $n \in \mathbb{N}$ denotes the $n \times n$ identity matrix. An example of $(\alpha_1, \alpha_2, \alpha_3, \beta)$ is given as follows:

Example 4.8 Let σ_1, σ_2 and σ_3 be the following 2×2 Hermitian matrices:

$$\sigma_1 := \begin{pmatrix} 0 & 1 \\ 1 & 0 \end{pmatrix}, \quad \sigma_2 := \begin{pmatrix} 0 & -i \\ i & 0 \end{pmatrix}, \quad \sigma_3 := \begin{pmatrix} 1 & 0 \\ 0 & -1 \end{pmatrix}. \tag{4.78}$$

These matrices are called the **Pauli matrices**. Using the Pauli matrices, one can define 4×4 Hermitian matrices as follows:

$$\alpha_j = \begin{pmatrix} 0 & \sigma_j \\ \sigma_j & 0 \end{pmatrix}, \quad \beta = \begin{pmatrix} I_2 & 0 \\ 0 & -I_2 \end{pmatrix}. \tag{4.79}$$

It is easy to see that these matrices satisfy (4.77). This representation is called the **standard representation** for α_j and β [156].

In what follows we sometimes omit the identity matrix symbol I_n.

We set

$$\boldsymbol{\alpha} := (\alpha_1, \alpha_2, \alpha_3). \tag{4.80}$$

For a triple of algebraic objects $\boldsymbol{X} = (X_1, X_2, X_3)$, we define

$$\boldsymbol{\alpha} \cdot \boldsymbol{X} := \sum_{j=1}^{3} \alpha_j X_j.$$

With this notation, the **free Dirac operator** H_{D} to be considered is defined by

$$H_{\mathrm{D}} := \boldsymbol{\alpha} \cdot \hat{\boldsymbol{p}} + m\beta, \quad \hat{\boldsymbol{p}} := (\hat{p}_1, \hat{p}_2, \hat{p}_3) \tag{4.81}$$

acting in

$$\begin{aligned} \mathscr{H}_{\mathrm{D}} &:= L^2(\mathbb{R}^3; \mathbb{C}^4) \\ &= \oplus^4 L^2(\mathbb{R}^3) = \{f = (f_1, f_2, f_3, f_4) | f_a \in L^2(\mathbb{R}^3),\ a = 1, 2, 3, 4\}, \end{aligned} \tag{4.82}$$

where, for a 4×4 matrix $M = (M_{ab})_{a,b=1,2,3,4}$ and a linear operator A on $L^2(\mathbb{R}^3)$,

$$(Mf)_a := \sum_{b=1}^{4} M_{ab} f_b, \quad (Ag)_a := Ag_a, \quad f \in \mathscr{H}_{\mathrm{D}},\ g \in \oplus^4 D(A),\ a = 1, 2, 3, 4.$$

The definition of Ag with $g \in \oplus^4 D(A)$ means that we denote the direct sum operator $\oplus^4 A$ on $\mathscr{H}_{\mathrm{D}}$ by A simply. It is easy to see that

$$\mathscr{F}_3 H_{\mathrm{D}} \mathscr{F}_3^{-1} = h_{\mathrm{D}},$$

where h_{D} denotes the multiplication operator by the matrix-valued function

$$h_{\mathrm{D}}(\boldsymbol{k}) := \boldsymbol{\alpha} \cdot \boldsymbol{k} + m\beta, \quad \boldsymbol{k} \in \hat{\mathbb{R}}^3.$$

For each $\boldsymbol{k} \in \hat{\mathbb{R}}^3$, we have

$$\begin{aligned}(\boldsymbol{\alpha} \cdot \boldsymbol{k})^2 &= \sum_{j=1}^{3} \sum_{\ell=1}^{3} k_j k_\ell \alpha_j \alpha_\ell \\ &= \sum_{j=1}^{3} \sum_{\ell=1}^{3} k_j k_\ell (2\delta_{j\ell} - \alpha_\ell \alpha_j) \quad \text{(by (4.76))} \\ &= 2\boldsymbol{k}^2 - (\boldsymbol{\alpha} \cdot \boldsymbol{k})^2.\end{aligned}$$

Hence one obtains

$$(\boldsymbol{\alpha} \cdot \boldsymbol{k})^2 = \boldsymbol{k}^2, \quad \boldsymbol{k} \in \hat{\mathbb{R}}^3. \tag{4.83}$$

This formula and (4.77) imply

$$h_{\mathrm{D}}(\boldsymbol{k})^2 = E_m(\boldsymbol{k})^2, \quad \boldsymbol{k} \in \hat{\mathbb{R}}^3, \tag{4.84}$$

where

$$E_m(\boldsymbol{k}) := \sqrt{\boldsymbol{k}^2 + m^2}, \quad \boldsymbol{k} \in \hat{\mathbb{R}}^3. \tag{4.85}$$

Let

$$u_{\mathrm{D}}(\boldsymbol{k}) := \begin{cases} \dfrac{1}{\sqrt{2E_m(\boldsymbol{k})(m + E_m(\boldsymbol{k}))}}(m + E_m(\boldsymbol{k}) + \beta\boldsymbol{\alpha} \cdot \boldsymbol{k});\ m > 0,\ \boldsymbol{k} \in \hat{\mathbb{R}}^3 \\ \dfrac{1}{\sqrt{2}}\left(1 + \beta\boldsymbol{\alpha} \cdot \dfrac{\boldsymbol{k}}{|\boldsymbol{k}|}\right);\ m = 0,\ \boldsymbol{k} \in \hat{\mathbb{R}}^3 \setminus \{0\} \end{cases}. \tag{4.86}$$

By direct computations using the anti-commutation relations (4.77), one can show that, for each $\boldsymbol{k} \in \hat{\mathbb{R}}^3$ (in the case $m > 0$) or $\boldsymbol{k} \in \hat{\mathbb{R}}^3 \setminus \{0\}$ (in the case $m = 0$), $u_{\mathrm{D}}(\boldsymbol{k})$ is a unitary matrix with

$$u_{\mathrm{D}}(\boldsymbol{k})^{-1} = u_{\mathrm{D}}(\boldsymbol{k})^* = \frac{1}{\sqrt{2E_m(\boldsymbol{k})(m + E_m(\boldsymbol{k}))}}(m + E_m(\boldsymbol{k}) + \boldsymbol{\alpha} \cdot \boldsymbol{k}\beta). \tag{4.87}$$

By direct computations again, we obtain

$$u_{\mathrm{D}}(\boldsymbol{k}) h_{\mathrm{D}}(\boldsymbol{k}) u_{\mathrm{D}}(\boldsymbol{k})^{-1} = \beta E_m(\boldsymbol{k}). \tag{4.88}$$

Hence, letting

$$U_{\rm D} := u_{\rm D}\mathscr{F}_3,$$

we obtain

$$U_{\rm D}H_{\rm D}U_{\rm D}^{-1} = \beta E_m. \tag{4.89}$$

Since βE_m is self-adjoint, it follows that $H_{\rm D}$ is self-adjoint with

$$D(H_{\rm D}) = \mathscr{F}_3^{-1}D(E_m) = \cap_{j=1}^3 D(\hat{p}_j).$$

We now work with the standard representation for $(\boldsymbol{\alpha}, \beta)$ so that β is diagonal (see (4.79)). Then (4.89) implies that

$$U_{\rm D}H_{\rm D}U_{\rm D}^{-1} = \begin{pmatrix} E_m & 0 & 0 & 0 \\ 0 & E_m & 0 & 0 \\ 0 & 0 & -E_m & 0 \\ 0 & 0 & 0 & -E_m \end{pmatrix} = E_m \oplus E_m \oplus (-E_m) \oplus (-E_m). \tag{4.90}$$

Note that the multiplication operator E_m on $L^2(\hat{\mathbb{R}})$ is absolutely continuous with

$$\sigma(E_m) = \sigma_{\rm ac}(E_m) = [m, \infty).$$

Hence, by spectral properties of direct sum operators, we conclude that $H_{\rm D}$ is absolutely continuous with

$$\sigma(H_{\rm D}) = \sigma_{\rm ac}(H_{\rm D}) = (-\infty, -m] \cup [m, \infty). \tag{4.91}$$

Let

$$V_{\rm D} := \mathscr{F}_3^{-1}U_{\rm D} = \mathscr{F}_3^{-1}u_{\rm D}\mathscr{F}_3.$$

Then V_D is unitary on $\mathscr{H}_{\rm D}$. We have

$$\mathscr{F}_3^{-1}E_m\mathscr{F}_3 = H_{{\rm R},m}^{(3)},$$

where $H_{{\rm R},m}^{(3)}$ is defined by (4.72) with $d = 3$. Hence, by (4.90), we obtain

$$V_{\rm D}H_{\rm D}V_{\rm D}^{-1} = H_{{\rm R},m}^{(3)} \oplus H_{{\rm R},m}^{(3)} \oplus (-H_{{\rm R},m}^{(3)}) \oplus (-H_{{\rm R},m}^{(3)}).$$

By Theorem 4.25(i), $T_{\mathrm{R},m,j}$ ($j = 1, 2, 3$) is a strong time operator of $H^{(3)}_{\mathrm{R},m}$. Hence, for each $j = 1, 2, 3$, the operator

$$T_{\mathrm{D},j} := V_{\mathrm{D}}^{-1} \left(T_{\mathrm{R},m,j} \oplus T_{\mathrm{R},m,j} \oplus (-T_{\mathrm{R},m,j}) \oplus (-T_{\mathrm{R},m,j})\right) V_D \tag{4.92}$$

is a strong time operator of H_{D}. Moreover, we obtain the following theorem:

Theorem 4.26 *Let* $j = 1, 2, 3$.

(i) *The operator* $T_{\mathrm{D},j}$ *is a non-self-adjoint symmetric operator with*

$$\sigma(T_{\mathrm{D},j}) = \mathbb{C}. \tag{4.93}$$

(ii) *Write*

$$H_{\mathrm{D},m} := H_{\mathrm{D}}, \quad T_{\mathrm{D},m,j} := T_{\mathrm{D},j}.$$

Then $\{(T_{\mathrm{D},m,j}, H_{\mathrm{D},m})\}_{m\geq 0}$ *is a family of mutually inequivalent weak Weyl representation of the CCR with one degree of freedom.*

Proof

(i) The first half of this part follows from (4.92) and the fact that a direct sum of non-self-adjoint symmetric operators is a non-self-adjoint symmetric operator. The definition (4.92) implies that

$$\sigma(T_{\mathrm{D},m,j}) = \sigma(T_{\mathrm{R},m,j}) \cup \sigma(-T_{\mathrm{R},m,j}).$$

By (4.75), we have

$$\sigma(T_{\mathrm{R},m,j}) \cup \sigma(-T_{\mathrm{R},m,j}) = \overline{\Pi}_+ \cup \overline{\Pi}_- = \mathbb{C}.$$

Thus (4.93) holds.

(ii) By (4.91), $\sigma(H_{\mathrm{D},m}) \neq \sigma(H_{\mathrm{D},m'})$ for $m \neq m'$ ($m, m' \geq 0$). Hence, if $m \neq m'$, then $(T_{\mathrm{D},m,j}, H_{\mathrm{D},m})$ is inequivalent to $(T_{\mathrm{D},m',j}, H_{\mathrm{D},m'})$. ■

The spectral property (4.93) may be interesting. Theorem 4.26(ii) shows that the set of Dirac particle masses has a meaning as the index set for a family of mutually inequivalent weak Weyl representations with one degree of freedom consisting of pairs of a free Dirac Hamiltonian and a strong time operator of it.

4.5.13 A Structure Generating Pairs of a Hamiltonian and a Strong Time Operator

In Theorem 4.2, we have seen a structure which generates pairs of a Hamiltonian and a time operator from one pair of a Hamiltonian and a time operator. In this subsection, we show that a similar structure exists for strong time operators.

We denote by $\mathscr{B}^1_{\mathbb{R}}(\mathbb{R})$ the set of real-valued Borel measurable functions f on $\mathbb{R}$ satisfying the following: there exists a closed Borel set N_f of Lebesgue measure 0 (i.e. $|N_f| = 0$) such that f is continuously differentiable on $\mathbb{R} \setminus N_f$. It follows that, for all $f \in \mathscr{B}^1_{\mathbb{R}}(\mathbb{R})$, $f(\lambda)$ is finite and differentiable for a.e. $\lambda \in \mathbb{R}$ with respect to the one-dimensional Lebesgue measure.

Let H be a self-adjoint operator on a Hilbert space $\mathscr{H}$ and T be a strong time operator of H, i.e., $T \in \mathfrak{T}_s(H)$. Since H is absolutely continuous (Theorem 4.8), it follows that, for all $f \in \mathscr{B}^1_{\mathbb{R}}(\mathbb{R})$

$$E_H(\{\lambda \in \mathbb{R}|\ |f(\lambda)| = \infty\}) = 0, \quad E_H(\{\lambda \in \mathbb{R}||f'(\lambda)| = \infty\}) = 0.$$

Hence, by functional calculus for H, $f(H)$ and $f'(H)$ are self-adjoint.

We introduce a subset of $\mathscr{B}^1_{\mathbb{R}}(\mathbb{R})$:

$$C^1_{T,H}(\mathbb{R}) := \{g \in \mathscr{B}^1_{\mathbb{R}}(\mathbb{R}) | g'(H) \text{ is injective and } D(Tg'(H)^{-1}) \cap D(g'(H)^{-1}T) \text{ is dense}\}. \tag{4.94}$$

Then, for each $g \in C^1_{T,H}(\mathbb{R})$, the operator

$$T_g := \frac{1}{2}\left(Tg'(H)^{-1} + g'(H)^{-1}T\right) \tag{4.95}$$

is a symmetric operator.

Theorem 4.27 ([34, 92]) *Let $T \in \mathfrak{T}_s(H)$ and suppose that T is closed. Then, for all $g \in C^1_{T,H}(\mathbb{R})$, T_g is a strong time operator of $g(H)$, i.e., $T_g \in \mathfrak{T}_s(g(H))$.*

Proof We have

$$\begin{aligned} e^{itg(H)}T_g e^{-itg(H)} &= \frac{1}{2}\left(e^{itg(H)}Tg'(H)^{-1}e^{-itg(H)} + e^{itg(H)}g'(H)^{-1}Te^{-itg(H)}\right) \\ &= \frac{1}{2}\left(e^{itg(H)}Te^{-itg(H)}g'(H)^{-1} + g'(H)^{-1}e^{itg(H)}Te^{-itg(H)}\right). \end{aligned}$$

Theorem 2.23(ii) implies that, for all $t \in \mathbb{R}$, $Te^{-itg(H)} \supset e^{-itg(H)}T + te^{-itg(H)}g'(H)$, where we have used the assumption that T is closed. Hence

$$e^{itg(H)}Te^{-itg(H)} \supset T + tg'(H).$$

Hence

$$e^{itg(H)}Te^{-itg(H)}g'(H)^{-1} \supset Tg'(H)^{-1} + t$$

and

$$g'(H)^{-1}e^{itg(H)}Te^{-itg(H)} \supset g'(H)^{-1}T + t.$$

Therefore

$$e^{itg(H)}T_g e^{-itg(H)} \supset T_g + t.$$

Replacing t with $-t$, we have $e^{-itg(H)}T_g e^{itg(H)} \supset T_g - t$. This implies that

$$T_g + t \supset e^{itg(H)}T_g e^{-itg(H)}.$$

Hence it follows that $e^{itg(H)}T_g e^{-itg(H)} = T + t$. Thus T_g is a strong time operator of $g(H)$. ■

Time operators of the Schrödinger operator H_{NR} and the relativistic Schrödinger operator H_{R} can be understood in a natural way in view of Theorem 4.27:

Example 4.9 Consider the case where $\mathscr{H} = L^2(\mathbb{R})$ and $H = \hat{p}$ and $T = \hat{q}$.

(1) Let $g(\lambda) = \frac{\lambda^2}{2m}$, $\lambda \in \mathbb{R}$ with $m > 0$. Then $g'(\lambda) = \lambda/m$. Hence $g'(H) = \hat{p}/m$ is injective. Hence $D(Tg'(H)^{-1}) \cap D(g'(H)^{-1}T) = D(\hat{q}\hat{p}^{-1}) \cap D(\hat{p}^{-1}\hat{q})$ is dense in $L^2(\mathbb{R})$. Hence $g \in C^1_{\hat{q},\hat{p}}(\mathbb{R})$. In this case we have $g(H) = \hat{p}^2/2m = H_{\mathrm{NR}}$ and

$$T_g = \frac{m}{2}(\hat{q}\hat{p}^{-1} + \hat{p}^{-1}\hat{q}).$$

This is just the Aharonov–Bohm time operator T_{AB} (see (4.3)). Hence, by Theorem 4.27, T_{AB} is a strong time operator of H_{NR}.

(2) Let $g(\lambda) = \sqrt{\lambda^2 + m^2}$, $\lambda \in \mathbb{R}$. Then $g'(\lambda) = \lambda g(\lambda)^{-1}$. Hence $g'(H) = \hat{p}H_{\mathrm{R}}^{-1}$ and it is injective. It follows that $g \in C^1_{\hat{q},\hat{p}}(\mathbb{R})$. In this case $g(H) = H_{\mathrm{R}}$. Hence

$$T_g = \frac{1}{2}(H_{\mathrm{R}}\hat{p}^{-1}\hat{q} + \hat{q}\hat{p}^{-1}H_{\mathrm{R}}) = T_{\mathrm{R}}.$$

Hence T_{R} is a strong time operator of H_{R}.

To make the dependence of T_g on (T, H), we write

$$T_g = \tau(g, T, H).$$

Using Theorem 4.27, one can define a sequence $\{(\mathcal{H}, (T_n, H_n))\}_{n=0}^{\infty}$ of weak Weyl representations with degree one as follows:

$$T_0 := T, \quad H_0 := H,$$

$$T_1 := \overline{\tau(g_1, T_0, H_0)}, \quad H_1 := g_1(H_0) \quad (g_1 \in C^1_{T_0, H_0}(\mathbb{R})),$$

$$T_2 := \overline{\tau(g_2, T_1, H_1)}, \quad H_2 := g_2(H_1) \quad (g_2 \in C^1_{T_1, H_1}(\mathbb{R})),$$

$$\vdots \qquad \vdots$$

$$T_n := \overline{\tau(g_n, T_{n-1}, H_{n-1})}, \quad H_n := g_n(H_{n-1}) \quad (g_n \in C^1_{T_{n-1}, H_{n-1}}(\mathbb{R})),$$

$$\vdots \qquad \vdots$$

This may be an interesting structure.

The following is an example on an abstract level which may be interesting:

Example 4.10 Let T be a closed strong time operator of H. Consider the case where $g(\lambda) = \log|\lambda|$, $\lambda \in \mathbb{R} \setminus \{0\}$ and $g(0) := 0$. Then $g \in \mathscr{B}^1_{\mathbb{R}}(\mathbb{R})$ with $g'(\lambda) = 1/\lambda$, $\lambda \in \mathbb{R} \setminus \{0\}$. Hence $g'(H) = H^{-1}$ (note that $\sigma_{\mathrm{p}}(H) = \emptyset$ by the purely absolute continuity of H and hence H is injective in particular) and $g(H) = \log|H|$. Hence

$$T_g = \frac{1}{2}(TH + HT).$$

By Proposition 4.5, $D(T_g) = D(TH) \cap D(HT)$ is dense. Hence $g \in C^1_{T,H}(\mathbb{R})$. Therefore T_g is a strong time operator of $\log|H|$. For further analysis of this model, see [34].

4.5.14 *Decay-in-Time of Transition Probabilities*

Let H be a self-adjoint operator on $\mathcal{H}$. In the context of quantum mechanics in which H denotes a Hamiltonian, the quantity

$$P_{\Psi,\Phi}(t) := |\left\langle \Phi, e^{-itH}\Psi \right\rangle|^2 \tag{4.96}$$

with $\|\Psi\| = 1 = \|\Phi\|$ is called the **transition probability** from a state Ψ at time 0 to a state Φ at time t (the probability that, when the initial state vector is Ψ, a state vector Φ is found at time t). If H is absolutely continuous, then, by a general theorem (see Theorem D.2),

$$\lim_{t \to \pm\infty} \left\langle \Phi, e^{-itH}\Psi \right\rangle = 0, \quad \Psi, \Phi \in \mathcal{H}. \tag{4.97}$$

Hence the transition probability $P_{\Psi,\Phi}(t)$ decays as the time t tends to infinity. It may be interesting to know the decay order of it. In this subsection, we show that, *if H has a strong time operator T, then T controls the decay order* (recall that, in this case, by Theorem 4.8, $\lim_{t\to\pm\infty} P_{\Psi,\Phi}(t) = 0$). This shows a physical role of a strong time operator.

For each n and $\Psi, \Phi \in D(T^n)$, we define a finite sequence $\{d_\ell(\Phi, \Psi)\}_{\ell=1}^n$ by the following recurrence relations:

$$d_1(\Phi, \Psi) := \|T\Phi\| \, \|\Psi\| + \|\Phi\| \, \|T\Psi\|,$$

$$d_\ell(\Phi, \Psi) := \|T^\ell\Phi\| \, \|\Psi\| + \|\Phi\| \, \|T^\ell\Psi\| + \sum_{k=1}^{\ell-1} {}_\ell C_k d_k(\Phi, T^{\ell-k}\Psi),$$

$$\ell = 2, \ldots, n \ (n \geq 2), \tag{4.98}$$

where ${}_\ell C_k$ is the binomial coefficient with respect to (ℓ, k) with $\ell \geq k$:

$${}_\ell C_k := \frac{\ell!}{(\ell - k)!k!}. \tag{4.99}$$

Theorem 4.28 *Assume that H has a strong time operator T. Let $n \in \mathbb{N}$. Then, for all $\Psi, \Phi \in D(T^n)$ and $t \in \mathbb{R} \setminus \{0\}$,*

$$|\langle \Phi, e^{-itH}\Psi \rangle| \leq \frac{d_n(\Phi, \Psi)}{|t|^n}. \tag{4.100}$$

Proof We prove (4.100) by induction in n. By (4.39), we have for all $t \in \mathbb{R}$

$$te^{-itH}\Psi = (Te^{-itH} - e^{-itH}T)\Psi.$$

Hence

$$t\langle \Phi, e^{-itH}\Psi \rangle = \langle T\Phi, e^{-itH}\Psi \rangle - \langle e^{itH}\Phi, T\Psi \rangle.$$

Therefore, by the Schwarz inequality and the unitarity of e^{-itH}, we obtain

$$|t| \, |\langle \Phi, e^{-itH}\Psi \rangle| \leq \|T\Phi\| \, \|\Psi\| + \|\Phi\| \, \|T\Psi\|.$$

Hence (4.100) with $n = 1$ follows.

We next suppose that (4.100) holds with n replaced by $k = 1, \ldots, \ell$. Then we need only to prove (4.100) with $n = \ell + 1$. So let $\Psi, \Phi \in D(T^{\ell+1})$. It follows from (4.39) and the relation $e^{-itH}e^{itH} = I$ for all $t \in \mathbb{R}$ that

$$e^{itH}T^{\ell+1}e^{-itH} = (T + t)^{\ell+1} = t^{\ell+1} + \sum_{k=0}^{\ell} {}_{\ell+1}C_k T^{\ell+1-k}t^k.$$

Hence

$$t^{\ell+1}\left\langle \Phi, e^{-itH}\Psi\right\rangle = \left\langle T^{\ell+1}\Phi, e^{-itH}\Psi\right\rangle - \sum_{k=0}^{\ell} {}_{\ell+1}C_k \left\langle \Phi, e^{-itH}T^{\ell+1-k}\Psi\right\rangle t^k.$$

Hence

$$|t|^{\ell+1}|\left\langle \Phi, e^{-itH}\Psi\right\rangle| \le \|T^{\ell+1}\Phi\|\,\|\Psi\| + \|\Phi\|\,\|T^{\ell+1}\Psi\|$$
$$+ \sum_{k=1}^{\ell} {}_{\ell+1}C_k |\left\langle \Phi, e^{-itH}T^{\ell+1-k}\Psi\right\rangle||t|^k.$$

Note that $\Phi, T^{\ell+1-k}\Psi \in D(T^k)$ $(k = 1, \ldots, \ell)$. Hence, by the induction hypothesis,

$$|\left\langle \Phi, e^{-itH}T^{\ell+1-k}\Psi\right\rangle||t|^k \le d_k(\Phi, T^{\ell+1-k}\Psi).$$

Therefore

$$t^{\ell+1}|\left\langle \Phi, e^{-itH}\Psi\right\rangle| \le \|T^{\ell+1}\Phi\|\,\|\Psi\| + \|\Phi\|\,\|T^{\ell+1}\Psi\| + \sum_{k=1}^{\ell} {}_{\ell+1}C_k d_k(\Phi, T^{\ell+1-k}\Psi)$$
$$= d_{\ell+1}(\Phi, \Psi).$$

Thus (4.100) with $n = \ell + 1$ holds. ■

Remark 4.17 For all $a \in \mathbb{R}$, $T_a := T - a$ also is a strong time operator of H. Hence Theorem 4.28 holds with T replaced by T_a.

It may be noteworthy that the constant $d_n(\Phi, \Psi)$ in (4.100) depends only on T, Φ and Ψ, not depending on H.

We say that a vector $\Psi \in \mathscr{H}$ has **regularity of order** n **with respect to** T if $\Psi \in D(T^n)$.

Theorem 4.28 tells us that there is a correspondence between the order of regularity of Φ, Ψ with respect to T and the order of decay-in-time of $|\left\langle \Phi, e^{-itH}\Psi\right\rangle|$. This is a very interesting structure, showing a role of strong time operators.

In the case where Ψ and Φ are unit vectors, we have

$$d_1(\Phi, \Psi) = \|T\Phi\| + \|T\Psi\|, \tag{4.101}$$

$$d_\ell(\Phi, \Psi) = \|T^\ell\Phi\| + \|T^\ell\Psi\| + \sum_{k=1}^{\ell-1} {}_\ell C_k d_k(\Phi, T^{\ell-k}\Psi),$$
$$\ell = 2, \ldots, n \ (n \ge 2). \tag{4.102}$$

Theorem 4.28 immediately yields an estimate for $P_{\Psi,\Phi}(t)$:

Corollary 4.7 *Assume that H has a strong time operator T. Let $n \in \mathbb{N}$. Then, for all $\Psi, \Phi \in D(T^n)$ with $\|\Psi\| = \|\Phi\| = 1$ and $t \in \mathbb{R} \setminus \{0\}$,*

$$P_{\Psi,\Phi}(t) \leq \frac{d_n(\Phi, \Psi)^2}{t^{2n}}. \tag{4.103}$$

In particular, for all $\Psi \in D(T)$ with $\|\Psi\| = 1$,

$$|\left\langle \Psi, e^{-itH}\Psi \right\rangle|^2 \leq \frac{4(\Delta T)_\Psi^2}{t^2}. \tag{4.104}$$

Proof Inequality (4.103) follows from (4.100). By Remark 4.17, (4.103) holds with T replaced by $T - \langle \Psi, T\Psi \rangle$. Then (4.103) with $n = 1$ and $\Phi = \Psi$ yields (4.104). ■

The quantity $|\left\langle \Psi, e^{-itH}\Psi \right\rangle|^2$ is called the **survival probability** of the state vector Ψ at time t. It is interesting that the uncertainty $(\Delta T)_\Psi$ of T appears on the right hand side of (4.104) (cf. [111] also).

4.5.15 Existence of Strong Time Operators

As we have already seen in Theorem 4.8, every self-adjoint operator H on a Hilbert space which has a strong time operator is absolutely continuous. A natural question then is: does an absolutely continuous self-adjoint operator H have a strong time operator? In this subsection, we give a partial answer to this question.

A vector $\Psi \in C^\infty(A)$ (see (1.34)) is called a **cyclic vector** of A if $\text{span}\{\Psi, A^n\Psi \mid n \in \mathbb{N}\}$ is dense in $\mathcal{H}$, i.e., Ψ is a cyclic vector for $\{A^n | n \in \mathbb{N}\}$ (see Definition 2.4). In this case we say that A has a cyclic vector Ψ.

Let H be a self-adjoint operator on a Hilbert space $\mathcal{H}$ and $E_H(\cdot)$ be the spectral measure of H. Then, for each unit vector $\Psi \in \mathcal{H}$, a probability measure μ_Ψ on $\mathbb{R}$ is defined by

$$\mu_\Psi(B) := \|E_H(B)\Psi\|^2, \quad B \in \boldsymbol{B},$$

where $\boldsymbol{B}$ is the family of Borel sets of $\mathbb{R}$. We define a function X on $\mathbb{R}$ by

$$X(\lambda) := \lambda, \quad \lambda \in \mathbb{R}.$$

We note the following fact:

Lemma 4.10 *Assume that $\mathcal{H}$ is separable. Suppose that H has a cyclic vector Ψ_0. Then there exists a unitary operator U from $\mathcal{H}$ to $L^2(\mathbb{R}, d\mu_{\Psi_0})$ such that $U\Psi_0 =$*

1 *and* $UHU^{-1} = M_X$, *the multiplication operator by the function* X *acting in* $L^2(\mathbb{R}, d\mu_{\Psi_0})$. *Moreover, the subspace* $\mathrm{span}\{e^{itX}|t \in \mathbb{R}\}$ *is dense in* $L^2(\mathbb{R}, d\mu_{\Psi_0})$.

Proof The first half of the lemma follows from an easy extension of Lemma 1 in [130, §VII.2] to the case of unbounded self-adjoint operators [16, Theorem 1.8]. To prove the second half of the lemma, we note that, by the cyclicity of Ψ_0 for $\{H^n|n \in \mathbb{N}\}$, $\mathrm{span}\{\Psi_0,\ H^n\Psi_0|n \in \mathbb{N}\}$ is dense in $\mathscr{H}$. By the functional calculus for H, we have

$$\lim_{t\to 0}(-i)^n\left(\frac{e^{itH}-1}{t}\right)^n\Psi_0 = H^n\Psi_0.$$

Hence it follows that $\mathrm{span}\{e^{itH}\Psi_0|t \in \mathbb{R}\}$ is dense in $\mathscr{H}$. By the first half of the lemma, we have $Ue^{itH}\Psi_0 = e^{itX}$. Hence $\mathrm{span}\{e^{itX}|t \in \mathbb{R}\}$ is dense in $L^2(\mathbb{R}, d\mu_{\Psi_0})$. ■

Let $\Psi \in \mathscr{H}$. If μ_Ψ is absolutely continuous with respect to the Lebesgue measure on $\mathbb{R}$, then we denote by ρ_Ψ the Radon-Nykodým derivative of μ_Ψ: $\rho_\Psi \geq 0$ and $\mu_\Psi(B) = \int_B \rho_\Psi(\lambda)d\lambda,\ B \in \boldsymbol{B}$.

We introduce a class of self-adjoint operators on $\mathscr{H}$.

Definition 4.3 We say that a self-adjoint operator H on $\mathscr{H}$ is in the class $\mathscr{S}_0(\mathscr{H})$ if it satisfies the following:

(i) H is absolutely continuous.
(ii) H has a cyclic vector Ψ_0 such that ρ_{Ψ_0} is differentiable on $\mathbb{R}$ and

$$\lim_{\lambda\to\pm\infty}\rho_{\Psi_0}(\lambda) = 0, \quad \int_{\rho_{\Psi_0}(\lambda)>0}\frac{\rho'_{\Psi_0}(\lambda)^2}{\rho_{\Psi_0}(\lambda)}d\lambda < \infty.$$

Let $\mathscr{H}$ be separable and $H \in \mathscr{S}_0(\mathscr{H})$ with a cyclic vector Ψ_0 and

$$W_0(\lambda) := \begin{cases} \dfrac{\rho'_{\Psi_0}(\lambda)}{\rho_{\Psi_0}(\lambda)} & \text{for } \rho_{\Psi_0}(\lambda) > 0 \\ 0 & \text{for } \rho_{\Psi_0}(\lambda) = 0 \end{cases}.$$

Then we define an operator Y on $L^2(\mathbb{R}, d\mu_{\Psi_0})$ as follows:

$$D(Y) := \mathrm{span}\{e^{itX}|t \in \mathbb{R}\}, \quad Y := i\frac{d}{d\lambda} + \frac{i}{2}W_0.$$

Lemma 4.11 *The operator* Y *is a symmetric operator.*

Proof By Lemma 4.10, $D(Y)$ is dense in $L^2(\mathbb{R}, d\mu_{\psi_0})$. Using (ii) in Definition 4.3 and integration by parts, we see that, for all $f, g \in D(Y)$,

$$\langle f, Yg\rangle_{L^2(\mathbb{R},d\mu_{\psi_0})} = \langle Yf, g\rangle_{L^2(\mathbb{R},d\mu_{\psi_0})} .$$

Hence Y is a symmetric operator. ■

Lemma 4.12 *The operator Y is a strong time operator of M_X.*

Proof It is obvious that, for all $t \in \mathbb{R}$, $e^{itM_X}D(Y) \subset D(Y)$ and hence $e^{itM_X}D(Y) = D(Y)$. Let $f(\lambda) = e^{is\lambda}$, $s \in \mathbb{R}, \lambda \in \mathbb{R}$. Then, using the fact that $if'(\lambda) = -sf(\lambda)$, we see that

$$(e^{itM_X}Ye^{-itM_X}f)(\lambda) = e^{it\lambda}\left(i\frac{d}{d\lambda} + \frac{i}{2}W_0\right)e^{-i(t-s)\lambda} = tf(\lambda) + (Yf)(\lambda).$$

Hence $e^{itM_X}Ye^{-itM_X} = Y + t$, implying that Y is a strong time operator of M_X. ■

Theorem 4.29 *Assume that $\mathscr{H}$ is separable and that $H \in \mathscr{S}_0(\mathscr{H})$. Then H has a strong time operator T_H given by*

$$T_H := U^{-1}YU. \tag{4.105}$$

Proof We have $U^{-1}M_XU = H$. Then it follows from Proposition 4.7 that T_H defined by (4.105) is a strong time operator of H. ■

It may be instructive to see how the above structure which gives a strong time operator is realized in a simple case:

Example 4.11 Consider the case where $\mathscr{H} = L^2(\mathbb{R})$ and $H = \hat{q}$. Let

$$f_0(x) := \left(\frac{\omega}{\pi}\right)^{1/4} e^{-\omega x^2/2}$$

with $\omega > 0$. Then $\|f_0\| = 1$. It is well known that span$\{f_0, \hat{q}^n f_0 | n \in \mathbb{N}\}$ is dense in $L^2(\mathbb{R})$. Hence f_0 is a cyclic vector of $\hat{q}$. The spectral measure of $\hat{q}$ is given by $E_{\hat{q}}(B) = \chi_B$, $B \in \boldsymbol{B}$. Hence

$$\|E_{\hat{q}}(B)f\|^2 = \int_B |f(x)|^2 dx.$$

Hence $\hat{q}$ is absolutely continuous and

$$\rho_f(\lambda) = |f(\lambda)|^2, \quad \lambda \in \mathbb{R}.$$

In particular,

$$\rho_{f_0}(\lambda) = \left(\frac{\omega}{\pi}\right)^{1/2} e^{-\omega\lambda^2} > 0, \quad \rho'_{f_0}(\lambda) = -2\omega\lambda\rho_{f_0}(\lambda).$$

Hence

$$\lim_{\lambda\to\pm\infty} \rho_{f_0}(\lambda) = 0, \quad \int_{\mathbb{R}} \frac{\rho'_{f_0}(\lambda)^2}{\rho_{f_0}(\lambda)} d\lambda = 4\omega^2 \int_{\mathbb{R}} x^2 f_0(x)^2 dx < \infty.$$

Therefore $\hat{q} \in \mathscr{S}_0(L^2(\mathbb{R}))$. In the present example, the unitary operator U: $L^2(\mathbb{R}_x) \to L^2(\mathbb{R}_\lambda, d\mu_{f_0})$ takes the form

$$Uf_0 = 1, \quad U\hat{q}U^{-1} = \lambda.$$

Let $f(x) = x^n f_0(x)$, $x \in \mathbb{R}$, $n \in \mathbb{N}$. Then $(Uf)(\lambda) = \lambda^n$, $\lambda \in \mathbb{R}$. Hence

$$(UD_x f)(\lambda) = (D_\lambda - \omega\lambda)Uf.$$

In the present case,

$$W_0(\lambda) = -2\omega\lambda.$$

Hence

$$Y = i\frac{d}{d\lambda} - i\omega\lambda$$

Therefore it follows that

$$UiD_xU^{-1} = Y$$

on $\mathrm{span}\{\lambda^n | n \in \mathbb{Z}_+\}$. Thus $T_{\hat{q}} = iD_x$ with $D(T_{\hat{q}}) = \mathrm{span}\{x^n f_0 | n \in \mathbb{Z}_+\}$ is a strong time operator of $\hat{q}$. It is well known that $T_{\hat{q}}$ is essentially self-adjoint on $\mathrm{span}\{x^n f_0 | n \in \mathbb{Z}_+\}$ and $\overline{T_{\hat{q}}} = -\hat{p}$. Hence $-\hat{p}$ is a self-adjoint strong time operator of $\hat{q}$.

Each element of $\mathscr{S}_0(\mathscr{H})$ is multiplicity free (see [130, p. 232, Definition]). In the case where H is not necessarily multiplicity free, we have the following theorem:

Theorem 4.30 *Let H be a self-adjoint operator on a separable Hilbert space $\mathscr{H}$. Suppose that there exist mutually orthogonal closed subspaces $\mathscr{H}_1, \ldots, \mathscr{H}_N$ of $\mathscr{H}$ ($N \in \mathbb{N}$ or $N = \infty$) such that $\mathscr{H} = \oplus_{n=1}^N \mathscr{H}_n$ and H is reduced by each $\mathscr{H}_n$ and the reduced part H_n of H to $\mathscr{H}_n$ is in the class $\mathscr{S}_0(\mathscr{H}_n)$. Then H has a strong time operator T given by $T = \oplus_{n=1}^N T_{H_n}$, where T_{H_n} is defined by (4.105) with H replaced by H_n.*

Proof We have $H = \oplus_{n=1}^N H_n$. By Theorem 4.29, each H_n has a strong time operator T_{H_n}. Then it is easy to see that $T := \oplus_{n=1}^N T_{H_n}$ is a strong time operator of H. ■

Remark 4.18 There is a general structure theorem associated with a self-adjoint operator (see, e.g., [130, p. 226, Lemma 2] for the case of a bounded self-adjoint operator or [16, p. 33, Theorem A.2] for the case of a not necessarily bounded self-adjoint operator).

For any self-adjoint operator H on a separable Hilbert space $\mathscr{H}$, there exists a direct sum decomposition $\mathscr{H} = \oplus_{n=1}^N \mathscr{H}_n$ with $N \in \mathbb{N}$ or $N = \infty$ such that (i) H is reduced by each $\mathscr{H}_n$; (ii) the reduced part of H to $\mathscr{H}_n$ has a cyclic vector Ψ_n.

But one cannot apply this structure theorem to show existence of a strong time operator of H, since it is unclear if Ψ_n can be taken such that H_n is in the class $\mathscr{S}_0(\mathscr{H}_n)$.

4.5.16 Construction of Strong Time Operators of a Self-adjoint Operator from Those of Another Self-adjoint Operator

We consider two self-adjoint operators H and H' acting in Hilbert spaces $\mathscr{H}$ and $\mathscr{H}'$ respectively. If $\mathscr{H} = \mathscr{H}'$, then $H' = H + (H' - H)$ on $D(H) \cap D(H')$ and hence H' can be regarded as a perturbation of H. One can use quantum scattering theory to construct a strong time operator of H'_{ac}, the absolutely continuous part of H' (see Appendix D), from a strong time operator of H_{ac}. This is a basic idea for the subject of this subsection.

We denote by $P_{\mathrm{ac}}(H)$ the orthogonal projection onto the absolutely continuous subspace $\mathscr{H}_{\mathrm{ac}}(H)$ of H.

Lemma 4.13 *Assume the following:*

(A.1) *The wave operator*

$$W := \text{s-}\lim_{t\to\infty} e^{itH'} J e^{-itH} P_{\mathrm{ac}}(H)$$

exists, where s- lim *means strong limit and $J : \mathscr{H} \to \mathscr{H}'$ is a bounded linear operator.*

(A.2) $\lim_{t\to\infty} \|Je^{-itH} P_{\mathrm{ac}}(H)\Psi\| = \|P_{\mathrm{ac}}(H)\Psi\|, \quad \Psi \in \mathscr{H}$.

(A.3) $\mathrm{Ran}(W) = \mathscr{H}_{\mathrm{ac}}(H')$.

Let $U := W \restriction \mathscr{H}_{\mathrm{ac}}(H)$. Then U is a unitary operator from $\mathscr{H}_{\mathrm{ac}}(H)$ to $\mathscr{H}_{\mathrm{ac}}(H')$ such that

$$H'_{\mathrm{ac}} = U H_{\mathrm{ac}} U^{-1}.$$

Proof This lemma is a well-known fact in quantum scattering theory (see, e.g., [5, 100, 132]). ■

Theorem 4.31 *Assume (A.1)–(A.3) in Lemma 4.13. Suppose that H_{ac} has a strong time operator T. Then $T' := UTU^{-1}$ is a strong time operators of H'_{ac}.*

Proof This follows from Lemma 4.13 and an application of Proposition 4.7. ■

Remark 4.19 As is easily seen, one can replace condition "$t \to \infty$" in (A.1) and (A.2) with "$t \to -\infty$".

Theorem 4.31 can be used to construct strong time operators of H'_{ac} from those of H_{ac}.

Remark 4.20 To construct a (not necessarily strong) time operator of a self-adjoint operator H itself (not only a time operator of H_{ac}), one may use the following formulas (see (D.2) and (D.3)):

$$\mathscr{H} = \mathscr{H}_{ac}(H) \oplus \mathscr{H}_{sc}(H) \oplus \mathscr{H}_p(H), \quad H = H_{ac} \oplus H_{sc} \oplus H_p.$$

If H_{ac}, H_{sc} and H_p have time operators T_{ac}, T_{sc} and T_p respectively, then $T := T_{ac} \oplus T_{sc} \oplus T_p$ is a time operator of H.

Example 4.12 We set

$$H_0 := H_{NR}^{(d)},$$

where $H_{NR}^{(d)}$ is defined by (4.62). For each $r \in (0, \infty)$, we denote by χ_r the characteristic function of the set $\{\boldsymbol{x} \in \mathbb{R}^d | \ |\boldsymbol{x}| \ge r\}$. Let $V: \mathbb{R}^d \to \mathbb{R} \cup \{\pm\infty\}$ be a Borel measurable function (potential) satisfying the following:

(V.1) V is H_0-bounded (see (1.3)) and $\|V\|_{H_0} < 1$, where $\|V\|_{H_0}$ is the relative bound of V with respect to H_0 (see (1.4)).

By (V.1), one can define a function h on $(0, \infty)$ by

$$h(r) := \|V(H_0 + 1)^{-1}\chi_r\|, \quad r \in (0, \infty),$$

where χ_r is regarded as the multiplication operator by the function χ_r.

(V.2) $\int_0^\infty h(r)dr < \infty$.

This type of potential V is called an **Enss potential**.[13]

Condition (V.1) implies that there exist constants $a \in [0, 1)$ and $b \ge 0$ such that

$$\|Vf\| \le a\|H_0 f\| + b\|f\|, \quad f \in D(H_0).$$

[13]More generally, an Enss potential is defined to be a (not necessarily multiplication) symmetric operator V on $L^2(\mathbb{R}^d)$ satisfying (V.1) and (V.2).

Hence, by the Kato–Rellich theorem (Theorem 1.15), the symmetric operator

$$H := H_0 + V \tag{4.106}$$

is self-adjoint with $D(H) = D(H_0)$ and bounded from below. The following theorem is known:

Theorem 4.32 (Enss's Theorem) *The wave operators* $W_\pm := \text{s-}\lim_{t\to\pm\infty} e^{itH} e^{-itH_0}$ *exist and*

$$\text{Ran}\,(W_\pm) = \mathscr{H}_{\text{ac}}(H).$$

Moreover, $\sigma_{\text{sc}}(H) = \emptyset$*, the only possible (finite) accumulation point for* $\sigma_{\text{p}}(H)$ *is 0 and any non-zero eigenvalue of* H *has finite multiplicity.*

Proof See textbooks on quantum scattering theory, e.g., [100, Theorem 4.12] or [132, Theorem XI.112]. ∎

By Theorem 4.32, $W_\pm$ are unitary operators from $\mathscr{H}_{\text{ac}}(H_0) = L^2(\mathbb{R}^d)$ to $\mathscr{H}_{\text{ac}}(H)$. We already know that the d-dimensional Aharonov–Bohm time operator $T_{\text{AB}}^{(d)}$ is a strong time operator of H_0 (see Theorem 4.21). Hence, by Theorem 4.31,

$$T_\pm := W_\pm T_{\text{AB}}^{(d)} W_\pm^{-1} \tag{4.107}$$

acting in $\mathscr{H}_{\text{ac}}(H)$ are strong time operators of H_{ac}. Therefore we can apply Theorem 4.28 to obtain estimates for orders of decay-in-time of transition probabilities with respect to H: for all $t \in \mathbb{R} \setminus \{0\}$,

$$|\left\langle g, e^{-itH} f\right\rangle|^2 \le \frac{d_n^\pm(g, f)^2}{t^{2n}}, \quad f, g \in D(T_\pm^n), \quad n \in \mathbb{N}, \tag{4.108}$$

where $d_n^+(g, f)$ (resp. $d_n^-(g, f)$) is the number $d_n(\Phi, \Psi)$ defined by (4.98) with (T, Φ, Ψ) replaced by (T_+, g, f) (resp. (T_-, g, f)).

4.6 Other Classes of Time Operators

So far we have considered two kinds of time operators. In fact, there other classes of time operators. In this section we briefly describe them.

Let H be a self-adjoint operator on a Hilbert space $\mathscr{H}$.

Definition 4.4 A self-adjoint operator T on $\mathscr{H}$ is called an **ultra-strong time operator** of H if (T, H) is a Weyl representation of the CCR with one degree of freedom, i.e., it satisfies

$$e^{itT} e^{isH} = e^{-its} e^{isH} e^{itT}, \quad s, t \in \mathbb{R}.$$

We denote by $\mathfrak{T}_{\mathrm{us}}(H)$ the set of ultra-strong time operators of H.

Proposition 4.10 *An ultra-strong time operator of H is a strong time operator of H.*

Proof This follows from the fact that every Weyl representation is a weak Weyl representation (see Theorem 2.20). ■

By this proposition, we have

$$\mathfrak{T}_{\mathrm{us}}(H) \subset \mathfrak{T}_{\mathrm{s}}(H). \tag{4.109}$$

Proposition 4.11 *A self-adjoint strong time operator of H is an ultra-strong time operator of H.*

Proof Let T be a self-adjoint strong time operator of H. Then, by the proof of Theorem 4.10, (T, H) is a Weyl representation with one degree of freedom. Hence T is an ultra-strong time operator of H. ■

Example 4.13 The time operator T_F in Example 4.1 is an ultra-strong time operator of $\overline{H}_F$, because T_F is a self-adjoint strong time operator of $\overline{H}_F$.

Definition 4.5 A symmetric operator T on $\mathscr{H}$ is called a **weak time operator** of H if there is a non-zero subspace $\mathscr{D}_{\mathrm{w}} \subset D(T) \cap D(H)$ such that

$$\langle T\Psi, H\Phi\rangle - \langle H\Psi, T\Phi\rangle = i\,\langle \Psi, \Phi\rangle\,, \quad \Psi, \Phi \in \mathscr{D}_{\mathrm{w}}. \tag{4.110}$$

We call Eq. (4.110) the **weak CCR** for (T, H) on $\mathscr{D}_{\mathrm{w}}$. The subspace $\mathscr{D}_{\mathrm{w}}$ is called a **weak CCR-domain** for (T, H).

We denote by $\mathfrak{T}_{\mathrm{w}}(H)$ the set of weak time operators of H.

It is easy to see that a time operator of H is a weak time operator of H:

$$\mathfrak{T}(H) \subset \mathfrak{T}_{\mathrm{w}}(H). \tag{4.111}$$

Proposition 4.12 *Let $T \in \mathfrak{T}_{\mathrm{w}}(H)$ with a weak CCR-domain $\mathscr{D}_{\mathrm{w}}$. Then, for all unit vectors $\Psi \in \mathscr{D}_{\mathrm{w}}$, time–energy uncertainty relation (4.10) holds.*

Proof Apply Lemma 2.3. ■

Finally, we define a notion of a generalized time operator which is weaker than that of a weak time operator.

Definition 4.6 ([33]) Let $\mathscr{D}_1$ and $\mathscr{D}_2$ be non-zero subspaces of $\mathscr{H}$. A sesquilinear form $\mathfrak{t}$: $\mathscr{D}_1 \times \mathscr{D}_2 \to \mathbb{C}$ with form domain $Q(\mathfrak{t}) = \mathscr{D}_1 \times \mathscr{D}_2$ (see Sect. 1.11) is called an **ultra-weak time operator** of H if there exists a non-zero subspace $\mathscr{D}$ and $\mathscr{E}$ of $\mathscr{D}_1 \cap \mathscr{D}_2$ such that the following hold:

(i) $\mathscr{E} \subset D(H) \cap \mathscr{D}$.
(ii) (symmetry on $\mathscr{D}$) $\mathfrak{t}(\Phi, \Psi)^* = \mathfrak{t}(\Psi, \Phi)$, $\Phi, \Psi \in \mathscr{D}$.

(iii) (ultra-weak CCR) $H\mathscr{E} \subset \mathscr{D}_1$ and, for all $\Phi,\ \Psi \in \mathscr{E}$,

$$\mathfrak{t}(H\Psi, \Phi)^* - \mathfrak{t}(H\Phi, \Psi) = i\,\langle \Phi, \Psi \rangle .$$

We call $\mathscr{E}$ an **ultra-weak CCR-domain** for $(\mathfrak{t}, H)$ and $\mathscr{D}$ a **symmetry domain** of $\mathfrak{t}$.

We remark that, as is seen in the above definition. An ultra-weak time operator is a sesquilinear form, not an operator. But, by abuse of terminology, we use the name "ultra-weak time operator".

For each $T \in \mathfrak{T}_{\mathrm{w}}(H)$ with $\mathscr{D}_{\mathrm{w}}$ being a weak CCR-domain for (T, H), one can define a sesquilinear form $\mathfrak{t}_T\colon \mathscr{H} \times D(T) \to \mathbb{C}$ by

$$\mathfrak{t}_T(\Phi, \Psi) := \langle \Phi, T\Psi \rangle , \quad \Phi \in \mathscr{H},\ \Psi \in D(T).$$

Proposition 4.13 *The sesquilinear form* $\mathfrak{t}_T$ *is an ultra-weak time operator of* H *with* $\mathscr{D}_{\mathrm{w}}$ *being an ultra-weak CCR-domain for* $(\mathfrak{t}_T, H)$ *and* $D(T)$ *a symmetry domain.*

Proof It is obvious that $\mathscr{D}_{\mathrm{w}} \subset D(H) \cap D(T)$. We have for all $\Phi,\ \Psi \in D(T)$

$$\mathfrak{t}_T(\Phi, \Psi)^* = \langle T\Psi, \Phi \rangle = \langle \Psi, T\Phi \rangle = \mathfrak{t}_T(\Psi, \Phi).$$

Hence the symmetry of $\mathfrak{t}_T$ on $D(T)$ holds. Moreover, for all $\Psi,\ \Phi \in \mathscr{D}_{\mathrm{w}}$

$$\mathfrak{t}_T(H\Psi, \Phi)^* - \mathfrak{t}_T(H\Phi, \Psi) = \langle T\Phi, H\Psi \rangle - \langle H\Phi, T\Psi \rangle = i\,\langle \Phi, \Psi \rangle .$$

Hence $\mathfrak{t}_T$ is an ultra-weak time operator of H with $\mathscr{D}_{\mathrm{w}}$ being an ultra-weak CCR-domain and $D(T)$ a symmetry domain. ■

By Proposition 4.13, there is a mapping f from $\mathfrak{T}_{\mathrm{w}}(H)$ to $\mathfrak{T}_{\mathrm{uw}}(H)$ such that

$$f(T) = \mathfrak{t}_T, \quad T \in \mathfrak{T}_{\mathrm{w}}(H).$$

It is easy to see that f is injective. In this sense, we write

$$\mathfrak{T}_{\mathrm{w}}(H) \subset \mathfrak{T}_{\mathrm{uw}}.$$

Summing up the above considerations, we have found five classes of time operators with the following inclusion relations:

$$\mathfrak{T}_{\mathrm{us}}(H) \subset \mathfrak{T}_{\mathrm{s}}(H) \subset \mathfrak{T}(H) \subset \mathfrak{T}_{\mathrm{w}}(H) \subset \mathfrak{T}_{\mathrm{uw}}(H).$$

For further discussions of ultra-weak time operators, we refer the reader to [33].

In Sect. 4.5.14, we have seen that a strong time operator is related to orders of decay-in-time of transition probabilities. But we remark that it is still unclear if the other time operators play any *proper* roles in relation to physics.

4.7 Generalized Time Operators

The notions of time operators introduced in the previous sections can be generalized. Let H be a self-adjoint operator on a Hilbert space $\mathscr{H}$.

(i) A symmetric operator T on $\mathscr{H}$ is called a **generalized time operator** of H if there exist a non-zero subspace $\mathscr{D} \subset D(TH) \cap D(HT)$ and a non-zero bounded self-adjoint operator C on $\mathscr{H}$ such that

$$[T, H] = iC \quad \text{on } \mathscr{D}. \tag{4.112}$$

(ii) A symmetric operator T on $\mathscr{H}$ is called a **generalized strong time operator** of H if there exists a mapping $K \colon \mathbb{R} \to \mathfrak{B}(\mathscr{H}); \mathbb{R} \ni t \mapsto K(t) \in \mathfrak{B}(\mathscr{H})$ such that, for all $t \in \mathbb{R}$, $K(t)$ is a bounded self-adjoint operator on $\mathscr{H}$ and

$$e^{itH} T e^{-itH} = T + K(t). \tag{4.113}$$

Relation (4.113) is called a **generalized weak Weyl relation** [15].

Commutation relation (4.112) implies a generalized Heisenberg uncertainty relation:

Theorem 4.33 *Let T be a generalized time operator of H satisfying (4.112). Then, for all unit vectors $\Psi \in \mathscr{D} \cap (\ker C)^{\perp}$,*

$$(\Delta H)_{\Psi} (\Delta T)_{\Psi} \geq \frac{1}{2} \delta_C, \tag{4.114}$$

where

$$\delta_C := \inf_{\Psi \in (\ker C)^{\perp},\ \|\Psi\|=1} |\langle \Psi, C\Psi \rangle|.$$

Proof Let $\Psi \in \mathscr{D} \cap (\ker C)^{\perp}$ be a unit vector. Then, applying the Heisenberg–Robertson inequality (2.25) with $A = T$ and $B = H$, we have

$$(\Delta H)_{\Psi} (\Delta T)_{\Psi} \geq \frac{1}{2} |\langle \Psi, C\Psi \rangle| \geq \frac{1}{2} \delta_C.$$

Hence (4.114) holds. ■

Remark 4.21 Since $C \neq 0$ by definition, it follows that $(\ker C)^{\perp} \neq \{0\}$.

Concerning a generalized strong time operator of H, we first note the following fact:

Lemma 4.14 *Let T be a generalized strong time operator of H satisfying (4.113). Suppose that $K(t)$ is strongly differentiable at $t = 0$. Then $K'(0)$ is a bounded self-adjoint operator on $\mathscr{H}$.*

Proof By (4.113) and the density of $D(T)$, we have operator equality

$$K(0) = 0. \tag{4.115}$$

Hence

$$K'(0) = \text{s-}\lim_{t \to 0} \frac{K(t)}{t}.$$

We recall that the strong limit of everywhere defined bounded operators is an everywhere defined bounded operator.[14] Hence $K'(0)$ is in $\mathfrak{B}(\mathscr{H})$. The self-adjointness of $K'(0)$ follows from that of $K(t)/t$. ■

The following proposition shows that, under additional conditions, a generalized strong time operator of H is a generalized time operator of H:

Proposition 4.14 *Let T be a generalized strong time operator of H satisfying (4.113). Suppose that $K(t)$ is strongly differentiable at $t = 0$. Then, for all $\Phi, \Psi \in D(T) \cap D(H)$,*

$$\langle T\Phi, H\Psi\rangle - \langle H\Phi, T\Psi\rangle = \left\langle \Phi, iK'(0)\Psi \right\rangle. \tag{4.116}$$

In particular, if $D(T) \cap D(H)$ is dense in $\mathscr{H}$, then

$$[T, H] = iK'(0) \quad \text{on } D(TH) \cap D(HT). \tag{4.117}$$

Proof Let $\Phi, \Psi \in D(T) \cap D(H)$. Then, by (4.113), we have

$$\left\langle e^{-itH}\Phi, T\Psi \right\rangle = \left\langle T\Phi, e^{itH}\Psi \right\rangle + \left\langle K(t)\Phi, e^{itH}\Psi \right\rangle.$$

It follows that both sides are differentiable in t at $t = 0$ and the differentiations of them at $t = 0$ give

$$\langle (-iH)\Phi, T\Psi \rangle = \langle T\Phi, iH\Psi \rangle + \left\langle K'(0)\Phi, \Psi \right\rangle + \langle K(0)\Phi, iH\Psi \rangle.$$

By (4.115), the last term on the right hand side vanishes. Hence (4.116) follows. ■

For further discussions of generalized (strong) time operators, we refer the reader to [15]. Recently, a generalized strong time operator has appeared in a theory of time crystals [118]. This is very interesting.

[14] This can be proved by applying the Banach–Steinhaus theorem (principle of uniform boundedness) (see, e.g., [130, Theorem III.9] or [32, Theorem 3.49]).

Chapter 5
Representations of Canonical Anti-commutation Relations with Finite Degrees of Freedom

Abstract Basic facts on representations of canonical anti-commutation relations with finite degrees of freedom are described.

5.1 Introduction

Let $\mathfrak{A}$ be an associative algebra over $\mathbb{C}$ and $\mathrm{Map}(\mathfrak{A} \times \mathfrak{A}, \mathfrak{A})$ be the set of mappings from $\mathfrak{A} \times \mathfrak{A}$ to $\mathfrak{A}$.

The mapping $p: \mathfrak{A} \times \mathfrak{A} \to \mathfrak{A} \times \mathfrak{A}$ defined by

$$p(X, Y) := (Y, X), \quad (X, Y) \in \mathfrak{A} \times \mathfrak{A}$$

satisfies $p^2 = I$ (identity). The mapping p induces the mapping R_p: $\mathrm{Map}(\mathfrak{A} \times \mathfrak{A}, \mathfrak{A}) \to \mathrm{Map}(\mathfrak{A} \times \mathfrak{A}, \mathfrak{A})$ given by

$$(R_p f)(X, Y) = f(p(X, Y)) = f(Y, X), \quad f \in \mathrm{Map}(\mathfrak{A}\times\mathfrak{A}, \mathfrak{A}), \quad (X, Y) \in \mathfrak{A}\times\mathfrak{A}.$$

It follows that

$$R_p^2 = I.$$

A mapping $f \in \mathrm{Map}(\mathfrak{A} \times \mathfrak{A}, \mathfrak{A})$ is said to be R_p-symmetric (resp. anti-symmetric) if $R_p f = f$ (resp. $-f$).

Each $f \in \mathrm{Map}(\mathfrak{A} \times \mathfrak{A}, \mathfrak{A})$ is uniquely written as a sum of an R_p**-symmetric** mapping and an R_p**-anti-symmetric** mapping. Indeed, introducing the mappings

$$f_\pm := f \pm R_p f,$$

© Springer Nature Singapore Pte Ltd. 2020

A. Arai, *Inequivalent Representations of Canonical Commutation and Anti-Commutation Relations*, Mathematical Physics Studies,
https://doi.org/10.1007/978-981-15-2180-5_5

one sees that f_+ (resp. f_-) is R_p-symmetric (resp. anti-symmetric) and

$$f = \frac{1}{2}(f_+ + f_-).$$

It is easy to see that this representation is unique.

This general structure may be applied to a subclass of Map($\mathfrak{A} \times \mathfrak{A}$, $\mathfrak{A}$). We define a mapping $B: \mathfrak{A} \times \mathfrak{A} \to \mathfrak{A}$ by

$$B(X, Y) := XY, \quad (X, Y) \in \mathfrak{A} \times \mathfrak{A}.$$

Then B is bilinear. We have

$$B_+(X, Y) = XY + YX = \{X, Y\}, \quad B_-(X, Y) = XY - YX = [X, Y].$$

Hence the anti-commutator $\{\ ,\ \}$ and the commutator $[\ ,\ ]$ can be interpreted as the R_p-symmetric part and the R_p-anti-symmetric part of the bilinear mapping B respectively. Therefore, from symmetry points of view, the commutator and the anti-commutator in fact are natural "components" of the one mapping B.

Based on the above consideration, one may be led to replace the commutator $[\ ,\ ]$ in the Dirac CCR (2.124) by the anti-commutator $\{\ ,\ \}$. Then one obtains anti-commutation relations

$$\{X_j, X_k^*\} = \delta_{jk} I, \quad \{X_j, X_k\} = 0, \quad \{X_j^*, X_k^*\} = 0, \quad j, k = 1, \ldots, N \quad (N \in \mathbb{N}). \tag{5.1}$$

These are called the **canonical anti-commutation relations** (CAR) with N degrees of freedom for $\{X_j\}_{j=1}^N$.

Remark 5.1 For a real constant q, one can define

$$[X, Y]_q := XY - qYX, \quad X, Y \in \mathfrak{A}.$$

This is called the q-commutator. It is obvious that $[X, Y]_1 = [X, Y]$ and $[X, Y]_{-1} = \{X, Y\}$. Hence $[X, Y]_q$ interpolates commutator and anti-commutator. Using q-commutator, one can consider ***q*-deformed CCR**

$$[X_j, Y_k]_{q_{jk}} = \delta_{jk} I, \tag{5.2}$$

$$[X_j, X_k]_{q_{jk}} = 0, \quad [Y_j, Y_k]_{q_{jk}} = 0, \ j, k = 1, \ldots, N \quad (N \in \mathbb{N}), \tag{5.3}$$

where $q_{jk} \in \mathbb{R}$. As is seen, the case $q_{jk} = 1$ (resp. $q_{jk} = -1$) for all $j, k = 1, \ldots, N$ in (5.2) and (5.3) with $Y_j = X_j^*$ gives the Dirac CCR (resp. CAR) with N degrees of freedom. Hence the q-deformed CCR interpolate Dirac CCR and CAR. But, in this book, we do not consider representations of q-deformed CCR. For this subject, see, e.g., [48, 66, 88].

To consider representations of CAR, we first note the following fact:

Lemma 5.1 *Let A be a densely defined closed operator on a Hilbert space $\mathscr{H}$. Suppose that there exists a dense subspace $\mathscr{D} \subset D(A^*A) \cap D(AA^*)$ such that*

$$AA^* + A^*A = I \quad \text{on } \mathscr{D}. \tag{5.4}$$

Then A is bounded with $D(A) = \mathscr{H}$ and $\|A\| \leq 1$. Moreover, operator equality

$$AA^* + A^*A = I \tag{5.5}$$

holds.

Proof For all $\Psi \in \mathscr{D}$, we have by (5.4) $\langle \Psi, (AA^* + A^*A)\Psi \rangle = \|\Psi\|^2$. Hence

$$\|A^*\Psi\|^2 + \|A\Psi\|^2 = \|\Psi\|^2.$$

This implies that $\|A\Psi\| \leq \|\Psi\|$. Hence $A_{\mathscr{D}} := A \restriction \mathscr{D}$ is bounded with $\|A_{\mathscr{D}}\| \leq 1$. Hence $\overline{A}_{\mathscr{D}} \in \mathfrak{B}(\mathscr{H})$ with $\|\overline{A}_{\mathscr{D}}\| \leq 1$. Since $A_{\mathscr{D}} \subset A$ and A is closed, it follows that $\overline{A}_{\mathscr{D}} \subset A$. Hence $A = \overline{A}_{\mathscr{D}}$. Thus $A \in \mathfrak{B}(\mathscr{H})$. Now we know that A^*A and AA^* are in $\mathfrak{B}(\mathscr{H})$. Hence (5.4) and the uniqueness in the extension theorem (Theorem 1.1) imply (5.5). ■

Lemma 5.1 shows that, in contrast to the case of the Dirac CCR, $AA^* - A^*A \subset I$, a densely defined closed operator A satisfying CAR (5.4) is necessarily everywhere defined and bounded. This is an interesting fact.

Based on Lemma 5.1, we define a representation of CAR as follows.

Definition 5.1 Let $\mathscr{H}$ be a Hilbert space and $B_j \in \mathfrak{B}(\mathscr{H})$, $j = 1, \ldots, N$. Then $\rho_B := (\mathscr{H}, \{B_j, B_j^* | j = 1, \ldots, N\})$ is said to be a **representation of the CAR** with N degrees of freedom if

$$\{B_j, B_k^*\} = \delta_{jk} I, \quad \{B_j, B_k\} = 0, \quad j, k = 1, \ldots, N. \tag{5.6}$$

If $\{B_j,\ B_j^* | j = 1, \ldots, N\}$ is irreducible, then ρ_B is said to be **irreducible**.

Remark 5.2 Since each B_j is bounded, the anti-commutativity of B_j and B_k implies that of B_j^* and B_k^* (take the adjoint of the second equation in (5.6)):

$$\{B_j^*, B_k^*\} = 0, \quad j, k = 1, \ldots, N.$$

Example 5.1 Consider the case $\mathscr{H} = \mathbb{C}^2$ and define $b: \mathbb{C}^2 \to \mathbb{C}^2$ by

$$b := \begin{pmatrix} 0 & 1 \\ 0 & 0 \end{pmatrix}.$$

Then it is easy to see that

$$b^* = \begin{pmatrix} 0 & 0 \\ 1 & 0 \end{pmatrix}, \quad b^2 = 0, \quad bb^* = \begin{pmatrix} 1 & 0 \\ 0 & 0 \end{pmatrix}, \quad b^*b = \begin{pmatrix} 0 & 0 \\ 0 & 1 \end{pmatrix}.$$

Hence

$$(b^*)^2 = 0, \quad \{b, b^*\} = I.$$

Hence

$$\rho_b := (\mathbb{C}^2, \{b, b^*\})$$

is a representation of the CAR with one degree of freedom. Moreover, it is easy to check that $\{b, b^*\}' = \mathbb{C}I$. Hence ρ_b is irreducible.

This example has a connection with spin, which is an internal degree of freedom of a quantum particle. As is well known, the **spin degree of freedom** with spin 1/2, is represented by a triple $\boldsymbol{s} = (s_1, s_2, s_3)$ of 2×2 Hermitian matrices given by

$$s_j = \frac{1}{2}\sigma_j, \quad j = 1, 2, 3$$

with the Pauli matrices σ_j, $j = 1, 2, 3$, defined by (4.78). It is easy to see that the following relations hold:

$$s_1 = \frac{1}{2}(b + b^*), \quad s_2 = \frac{i}{2}(b^* - b), \quad s_3 = \frac{1}{2}(bb^* - b^*b). \tag{5.7}$$

Hence the spin matrices s_1, s_2 and s_3 are constructed from the representation ρ_b of CAR.

We recall that the Lie algebra $\mathfrak{su}(2)$ of the two-dimensional special unitary group

$$\mathrm{SU}(2) := \{U \,|\, U \text{ is a } 2 \times 2 \text{ unitary matrix with determinant } \det U = 1\}$$

is given by

$$\mathfrak{su}(2) = \{A \,|\, A \text{ is a } 2 \times 2 \text{ anti-Hermitian matrix with trace zero}\}$$

and $\mathfrak{su}(2)$ is a three dimensional real Lie algebra with a basis $\{e_1, e_2, e_3\}$ of the form

$$e_1 := \frac{1}{2}\begin{pmatrix} 0 & i \\ i & 0 \end{pmatrix}, \quad e_2 := \frac{1}{2}\begin{pmatrix} 0 & -1 \\ 1 & 0 \end{pmatrix}, \quad e_3 := \frac{1}{2}\begin{pmatrix} i & 0 \\ 0 & -i \end{pmatrix}.$$

The basis $\{e_1, e_2, e_3\}$ obeys the following commutation relations:

$$[e_2, e_3] = e_1, \quad [e_3, e_1] = e_2, \quad [e_1, e_2] = e_3.$$

It is obvious that

$$e_1 = is_1, \quad e_2 = -is_2, \quad e_3 = is_3. \tag{5.8}$$

Thus $\mathfrak{su}(2)$ is constructed from the representation ρ_b of the CAR with one degree of freedom.

Conversely, we have by (5.7)

$$b = s_1 + is_2.$$

Hence $\mathfrak{su}(2)$ gives an irreducible representation of the CAR with one degree of freedom.

On the other hand, it is well known (e.g., [164, Chapter III, §4]) that possible values $\{n/2 | n \in \mathbb{Z}_+\} = \{0, 1/2, 1, 3/2, \ldots\}$ of spin angular momentum or orbital angular momentum of a quantum particle appear as the labels of all finite-dimensional irreducible representations of $\mathfrak{su}(2)$, which are mutually inequivalent. In this way, the representation ρ_b is related to physics of angular momentum of a quantum particle.

Example 5.2 Let $N \geq 2$ be a natural number and, for $k \in \mathbb{Z}_+$, $\wedge^k \mathbb{C}^N$ be the k-fold anti-symmetric tensor product (wedge product) of $\mathbb{C}^N$ (see (1.53)) with convention $\wedge^0 \mathbb{C}^N := \mathbb{C}$.[1] Then

$$\wedge(\mathbb{C}^N) := \oplus_{k=0}^N \wedge^k \mathbb{C}^N = \{\psi = (\psi^{(k)})_{k=0}^N | \psi^{(k)} \in \wedge^k \mathbb{C}^N, \, k = 0, \ldots, N\}$$

is called the **fermion Fock space** over $\mathbb{C}^N$. The name comes from that this Hilbert space serves as a Hilbert space of state vectors of fermions whose one-particle Hilbert space is $\mathbb{C}^N$. The subspace $\wedge^k \mathbb{C}^N$ describes the k-particle space of fermions.[2]

For each $z = (z_1, \ldots, z_N) \in \mathbb{C}^N$, there exists a unique linear operator $b(z)$ on $\wedge(\mathbb{C}^N)$ such that its adjoint $b(z)^*$ is of the form

$$(b(z)^* \psi)^{(0)} = 0,$$

$$(b(z)^* \psi)^{(k)} := \sqrt{k} A_k (z \otimes \psi^{(k-1)}), \quad k = 1, \ldots, N, \quad \psi = (\psi^{(k)})_{k=0}^N \in \wedge(\mathbb{C}^N).$$

---

[1]Note that, if $k > N$, then $\wedge^k \mathbb{C}^N = \{0\}$.

[2]Recall that a vector state of k identical fermions is anti-symmetric for any permutations of k fermions. In the present context, this corresponds to the anti-symmetry of $\wedge^k \mathbb{C}^N$.

It is easy to see that, for all $\psi \in \wedge(\mathbb{C}^N)$ such that $\psi^{(\ell)} = 0,\ \ell \neq k$ and

$$\psi^{(k)} = A_k(\boldsymbol{u}_1 \otimes \cdots \otimes \boldsymbol{u}_k) \quad (\boldsymbol{u}_1, \ldots, \boldsymbol{u}_k \in \mathbb{C}^N),$$

one has

$$(b(z)\psi)^{(\ell)} = 0, \quad \ell \neq k-1,$$

$$(b(z)\psi)^{(k-1)} = \frac{1}{\sqrt{k}} \sum_{j=1}^{k} (-1)^{j-1} \langle z, \boldsymbol{u}_j \rangle_{\mathbb{C}^N} A_{k-1}(\boldsymbol{u}_1 \otimes \cdots \otimes \widehat{\boldsymbol{u}_j} \otimes \cdots \otimes \boldsymbol{u}_k),$$

where $\widehat{\boldsymbol{u}_j}$ indicates the omission of $\boldsymbol{u}_j$. Using these representations, one can show that

$$\{b(z), b(\boldsymbol{w})^*\} = \langle z, \boldsymbol{w} \rangle_{\mathbb{C}^N}, \quad \{b(z), b(\boldsymbol{w})\} = 0. \tag{5.9}$$

It follows that, for each $k = 0, 1, \ldots, N$, $b(z)$ maps $\wedge^k \mathbb{C}^N$ to $\wedge^{k-1} \mathbb{C}^N$ with $\wedge^{-1} \mathbb{C}^N := \{0\}$ and $b(z)^*$ maps $\wedge^k \mathbb{C}^N$ to $\wedge^{k+1} \mathbb{C}^N$. For this structure, $b(z)$ (resp. $b(z)^*$) is called the **fermion annihilation** (resp. **creation**) **operator** with test vector z.

We denote by $(\boldsymbol{e}_1, \ldots, \boldsymbol{e}_N)$ the standard basis of $\mathbb{C}^N$: $(\boldsymbol{e}_j)_{j'} = \delta_{jj'},\ j, j' = 1, \ldots, N$. We define

$$b_j := b(\boldsymbol{e}_j), \quad j = 1, \ldots, N.$$

Then (5.9) implies that

$$\{b_j, b_k^*\} = \delta_{jk}, \quad \{b_j, b_k\} = 0, \quad j, k = 1, \ldots, N. \tag{5.10}$$

Thus

$$\rho_{\mathrm{F}}^{(N)} := (\wedge(\mathbb{C}^N), \{b_j, b_j^* | j = 1, \ldots, N\})$$

is a representation of the CAR with N degrees of freedom. Since

$$\dim \wedge(\mathbb{C}^N) = 2^N,$$

this is a finite-dimensional representation. The representation $\rho_{\mathrm{F}}^{(N)}$ is called the **Fock representation of the CAR with N degrees of freedom.**

The Fock representation $\rho_{\mathrm{F}}^{(N)}$ has a simple structure. The vector $\psi_0 \in \wedge(\mathbb{C}^N)$ with

$$\psi_0^{(k)} = \delta_{0k},\ k = 0, \ldots, N$$

is called the **Fock vacuum** in $\wedge(\mathbb{C}^N)$. It follows that

$$b(z)\psi_0 = 0, \quad z \in \mathbb{C}^N$$

and, for all $\boldsymbol{u}_j \in \mathbb{C}^N$ $(j = 1, \ldots, N)$,

$$(b(\boldsymbol{u}_1)^* \cdots b(\boldsymbol{u}_k)^* \psi_0)^{(\ell)} = \delta_{k\ell}\sqrt{k!}A_k(\boldsymbol{u}_1 \otimes \cdots \boldsymbol{u}_k), \quad k, \ell = 1, \ldots, N.$$

Hence

$$\begin{aligned}\wedge^k(\mathbb{C}^N) &= \mathrm{span}\{b(\boldsymbol{u}_1)^* \cdots b(\boldsymbol{u}_k)^* \psi_0 | \boldsymbol{u}_1, \ldots, \boldsymbol{u}_k \in \mathbb{C}^N\} \\ &= \mathrm{span}\{b_{i_1}^* \cdots b_{i_k}^* \psi_0 | i_1 < \cdots < i_k,\ i_1, \ldots, i_k \in \{1, \ldots, N\}\},\end{aligned}$$

where we have used that

$$\{b_j^*, b_k^*\} = 0, \quad (b_j^*)^2 = 0, \quad j, k = 1, \ldots, N.$$

Thus

$$\wedge(\mathbb{C}^N) = \mathrm{span}\{\psi_0,\ b_{i_1}^* \cdots b_{i_k}^* \psi_0 | k = 1, \ldots, N,\ i_1 < \cdots < i_k,\ i_1, \ldots, i_k \in \{1, \ldots, N\}\}.$$

Hence ψ_0 is a cyclic vector for

$$\{b_{i_1}^* \cdots b_{i_k}^* | k = 1, \ldots, N,\ i_1 < \cdots < i_k,\ i_1, \ldots, i_k \in \{1, \ldots, N\}\}.$$

In the same way as in the proof of Proposition 2.16, one can show that $\rho_{\mathrm{F}}^{(N)}$ is irreducible.

In what follows, we first consider representations of the CAR with one degree of freedom and then those of the CAR with arbitrary finite degrees of freedom.

5.2 Representations of the CAR with One Degree of Freedom

Let $(\mathscr{H}, \{B, B^*\})$ be a representation of the CAR with one degree of freedom, i.e., $B \in \mathfrak{B}(\mathscr{H})$,

$$BB^* + B^*B = I, \quad B^2 = 0, \quad (B^*)^2 = 0. \tag{5.11}$$

We define

$$P_B := BB^*, \quad P_{B^*} := B^*B. \tag{5.12}$$

Lemma 5.2 *The operators P_B and P_{B^*} are orthogonal projections and*

$$P_B + P_{B^*} = I. \tag{5.13}$$

Proof It is obvious that P_B and P_{B^*} are bounded self-adjoint operators. By the CAR (5.11), we have

$$P_B^2 = BB^*BB^* = B(I - BB^*)B^* = BB^* = P_B.$$

Hence P_B is an orthogonal projection. Similarly, one can show that $P_{B^*}^2 = P_{B^*}$. Hence P_{B^*} is an orthogonal projection. ■

Let

$$\mathscr{H}_B := P_B\mathscr{H}, \quad \mathscr{H}_{B^*} := P_{B^*}\mathscr{H}.$$

Then, by (5.13), $\mathscr{H}_B$ is orthogonal to $\mathscr{H}_{B^*}$ and

$$\mathscr{H} = \mathscr{H}_B \oplus \mathscr{H}_{B^*}. \tag{5.14}$$

Lemma 5.3

(i) *The operator B is a partial isometry with initial space $\mathscr{H}_{B^*}$ and final space $\mathscr{H}_B$.*
(ii) *The operator B^* is a partial isometry with initial space $\mathscr{H}_B$ and final space $\mathscr{H}_{B^*}$.*

Proof

(i) This is just a rephrase of (5.12).
(ii) This follows from (5.12) and $B = (B^*)^*$. ■

Lemma 5.4 *Assume that $\mathscr{H}$ is separable. Let $d := \dim \mathscr{H}_B$ ($d < \infty$ or $d = \infty$). and $\{e_n^{(0)}\}_{n=1}^d$ be a CONS of $\mathscr{H}_B$. Then there exists a CONS $\{e_n^{(1)}\}_{n=1}^d$ of $\mathscr{H}_{B^*}$ such that, for all $n = 1, \ldots, d$,*

$$B^*e_n^{(0)} = e_n^{(1)}, \quad B^*e_n^{(1)} = 0, \tag{5.15}$$

$$Be_n^{(0)} = 0, \quad Be_n^{(1)} = e_n^{(0)}. \tag{5.16}$$

In particular, if $d < \infty$, then $\mathscr{H}$ is even dimensional with $\dim \mathscr{H} = 2d$.

Proof Let

$$e_n^{(1)} := B^*e_n^{(0)}.$$

Then Lemma 5.3(ii) implies that $\{e_n^{(1)}\}_{n=1}^d$ is a CONS of $\mathscr{H}_{B^*}$. Since $(B^*)^2 = 0$, $B^* e_n^{(1)} = 0$. Since $BP_B = 0$, $Be_n^{(0)} = 0$. Moreover, $Be_n^{(1)} = BB^* e_n^{(0)} = e_n^{(0)}$. The last statement follows from the first half of the lemma and (5.14). ∎

Let $e_n^{(j)}$ $(j = 0, 1)$ be as in Lemma 5.4 and

$$\mathscr{E}_n := \text{span}\{e_n^{(0)}, e_n^{(1)}\}. \tag{5.17}$$

Then, by (5.14), we have

$$\mathscr{H} = \oplus_{n=1}^d \mathscr{E}_n. \tag{5.18}$$

Lemma 5.4 implies that B and B^* are reduced by each $\mathscr{E}_n$. We denote by B_n the reduced part of B to $\mathscr{E}_n$. Then B_n^* is the reduced part of B^* to $\mathscr{E}_n$. Hence we have the direct sum decomposition

$$B = \oplus_{n=1}^d B_n. \tag{5.19}$$

We are now ready to prove a basic theorem on representations of CAR:

Theorem 5.1 *Let $\rho := (\mathscr{H}, \{B, B^*\})$ be a representation of the CAR with one degree of freedom. Assume that $\mathscr{H}$ is separable. Then ρ is equivalent to a direct sum representation of ρ_b. In particular, if ρ is irreducible, then $\rho \cong \rho_b$.*

Proof Let

$$e_1 = \begin{pmatrix} 1 \\ 0 \end{pmatrix}, \quad e_2 = \begin{pmatrix} 0 \\ 1 \end{pmatrix} \in \mathbb{C}^2.$$

Then $\{e_1, e_2\}$ be an orthonormal basis of $\mathbb{C}^2$. Let $\{e_n^{(j)}\}_n$ $(j = 0, 1)$ be as in Lemma 5.4. Then, for each n, there exists a unitary operator U_n from $\mathscr{E}_n$ to $\mathbb{C}^2$ such that $U_n e_n^{(0)} = e_1$ and $U_n e_n^{(1)} = e_2$. It is easy to see that

$$U_n B_n U_n^{-1} = b, \quad U_n B_n^* U_n^{-1} = b^*.$$

The operator $U := \oplus_{n=1}^d U_n$ is a unitary operator from $\mathscr{H}$ to $\oplus^d \mathbb{C}^2$. By (5.19), we have

$$UBU^{-1} = \oplus_{n=1}^d U_n B_n U_n^{-1} = \oplus_{n=1}^d b.$$

This implies that $UB^*U^{-1} = \oplus_{n=1}^d b^*$. Thus ρ is equivalent to a direct sum representation of ρ_b. ∎

Theorem 5.1 shows that, in contrast to representations of the CCR with one degree of freedom, representations of the CAR with one degree of freedom have a very simple structure.

5.3 Representations of the CAR with *N* Degrees of Freedom

We next go on to analysis of representations of the CAR with N degrees of freedom ($N \geq 2$). The following theorem holds:

Theorem 5.2 *Let*

$$\rho_B := (\mathscr{H}, \{B_j, B_j^* | j = 1, \ldots, N\})$$

be a representation of the CAR with N degrees of freedom. Assume that $\mathscr{H}$ is separable. Then ρ_B is equivalent to a direct sum of the Fock representation $\rho_F^{(N)}$ of the CAR with N degrees of freedom:

$$\rho_B \cong \oplus_{n=1}^{M} \rho_F^{(N)} \tag{5.20}$$

with $M \in \mathbb{N}$ or $M = \infty$. In particular, if ρ_B is irreducible, then $\rho_B \cong \rho_F^{(N)}$.

Proof It is obvious that $(\mathscr{H}, \{B_1, B_1^*\})$ is a representation of the CAR with one degree of freedom. Hence, by (5.14),

$$\mathscr{H} = \mathscr{H}_{B_1} \oplus \mathscr{H}_{B_1^*}.$$

Moreover, the mappings $B_1 \colon \mathscr{H}_{B_1^*} \to \mathscr{H}_{B_1}$ and $B_1^* \colon \mathscr{H}_{B_1} \to \mathscr{H}_{B_1^*}$ are unitary operators. By the CAR for $\{B_j, B_j^* | j = 1, \ldots, N\}$, we have

$$B_2 P_\# = P_\# B_2, \quad B_2^* P_\# = P_\# B_2^*, \quad \# = B_1,\ B_1^*.$$

Hence $\mathscr{H}_{B_1}$ and $\mathscr{H}_{B_1^*}$ reduce B_2 and B_2^*. For notational simplicity, we denote the reduced part of B_2 to $\mathscr{H}_\#$ by the same symbol. Hence $(\mathscr{H}_\#, \{B_2, B_2^*\})$ is a representation of the CAR with one degree of freedom. Hence

$$\mathscr{H}_{B_1} = (\mathscr{H}_{B_1})_{B_2} \oplus (\mathscr{H}_{B_1})_{B_2^*}), \quad \mathscr{H}_{B_1^*} = (\mathscr{H}_{B_1^*})_{B_2} \oplus (\mathscr{H}_{B_1^*})_{B_2^*}.$$

Therefore we obtain

$$\mathscr{H} = \left[(\mathscr{H}_{B_1})_{B_2} \oplus (\mathscr{H}_{B_1})_{B_2^*} \right] \oplus \left[(\mathscr{H}_{B_1^*})_{B_2} \oplus (\mathscr{H}_{B_1^*})_{B_2^*} \right].$$

Let $\{u_n\}_n$ be a CONS of $(\mathscr{H}_{B_1})_{B_2}$. Then, by Lemma 5.4, $\{B_2^* u_n\}_n$ be a CONS of $(\mathscr{H}_{B_1})_{B_2^*}$. Hence $\{u_n, B_2^* u_n\}_n$ is a CONS of $\left[(\mathscr{H}_{B_1})_{B_2} \oplus (\mathscr{H}_{B_1})_{B_2^*}\right]$. Hence, Lemma 5.4 again, $\{B_1^* u_n, B_1^* B_2^* u_n\}_n$ is a CONS of $\left[(\mathscr{H}_{B_1^*})_{B_2} \oplus (\mathscr{H}_{B_1^*})_{B_2^*}\right]$. Therefore

$$\{u_n, B_2^* u_n, B_1^* u_n, B_1^* B_2^* u_n\}_n$$

is a CONS of $\mathscr{H}$.

By the CAR for $\{B_j, B_j^* | j = 1, \ldots, N\}$ again, B_3 and B_3^* are reduced by each $(\mathscr{H}_\#)_\natural$ ($\# = B_1, B_1^*$, $\natural = B_2, B_2^*$). Hence each $(\mathscr{H}_\#)_\natural$ has the orthogonal decomposition

$$(\mathscr{H}_\#)_\natural = ((\mathscr{H}_\#)_\natural)_{B_3} \oplus ((\mathscr{H}_\#)_\natural)_{B_3^*}.$$

Repeating this process, we finally arrive at the orthogonal decomposition

$$\mathscr{H} = (\mathscr{H}_1 \oplus \mathscr{H}_2) \oplus (\mathscr{H}_3 \oplus \mathscr{H}_4) \oplus \cdots (\mathscr{H}_{2^N-1} \oplus \mathscr{H}_{2^N})$$

such that the following hold:

- B_N^* is a unitary operator from $\mathscr{H}_{2k-1}$ to $\mathscr{H}_{2k}$ $(k = 1, \ldots, 2^{N-1})$;
- B_{N-1}^* is a unitary operator from $\mathscr{H}_{4k-3} \oplus \mathscr{H}_{4k-2}$ to $\mathscr{H}_{4k-1} \oplus \mathscr{H}_{4k}$ $(k = 1, \ldots, 2^{N-2})$;
- B_{N-2}^* is a unitary operator from $\mathscr{H}_{8k-7} \oplus \mathscr{H}_{8k-6} \oplus \mathscr{H}_{8k-5} \oplus \mathscr{H}_{8k-4}$ to $\mathscr{H}_{8k-3} \oplus \mathscr{H}_{8k-2} \oplus \mathscr{H}_{8k-1} \oplus \mathscr{H}_{8k}$ $(k = 1, \ldots, 2^{N-3})$;

  $\vdots \qquad \vdots$
- B_1^* is a unitary operator from $\oplus_{j=1}^{2^{N-1}} \mathscr{H}_j$ to $\oplus_{j=2^{N-1}+1}^{2^N} \mathscr{H}_j$.

Let $\{v_n\}_n := \{v_n\}_{n=1}^M$ be a CONS of $\mathscr{H}_1$ with $M \in \mathbb{N}$ or $M = \infty$. Then $\{B_N^* v_n\}_n$ is a CONS of $\mathscr{H}_2$. Hence $\{v_n, B_N^* v_n\}_n$ is a CONS of $\mathscr{H}_1 \oplus \mathscr{H}_2$. Then, by the algorithm of the construction, $\{B_{N-1}^* v_n, B_{N-1}^* B_N^* v_n\}_n$ is a CONS of $\mathscr{H}_3 \oplus \mathscr{H}_4$. Hence $\{v_n, B_N^* v_n, B_{N-1}^* v_n, B_{N-1}^* B_N^* v_n\}_n$ is a CONS of $\mathscr{H}_1 \oplus \mathscr{H}_2 \oplus \mathscr{H}_3 \oplus \mathscr{H}_4$. This implies that

$$\{B_{N-2}^* v_n, B_{N-2}^* B_N^* v_n, B_{N-2}^* B_{N-1}^* v_n, B_{N-2}^* B_{N-1}^* B_N^* v_n\}_n$$

is a CONS of $\mathscr{H}_5 \oplus \mathscr{H}_6 \oplus \mathscr{H}_7 \oplus \mathscr{H}_8$. Hence

$$\{v_n, B_N^* v_n, B_{N-1}^* v_n, B_{N-1}^* B_N^* v_n, B_{N-2}^* v_n, B_{N-2}^* B_N^* v_n, B_{N-2}^* B_{N-1}^* v_n, B_{N-2}^* B_{N-1}^* B_N^* v_n\}_n$$

is a CONS of $\oplus_{j=1}^8 \mathscr{H}_j$. Repeating this procedure, one can see that

$$\{v_n, B_{i_1}^* \cdots B_{i_k}^* v_n | k = 1, \ldots, N,\ i_1 < i_2 < \cdots < i_k,\ i_1, \ldots, i_k \in \{1, \ldots, N\}\}_n$$

is a CONS of $\mathscr{H} = \oplus_{j=1}^{2^N}\mathscr{H}_j$. Therefore, putting

$$\mathscr{D}_n := \mathrm{span}\{v_n, B_{i_1}^* \cdots B_{i_k}^* v_n | k = 1, \ldots, N, \ i_1 < i_2 < \cdots < i_k, \ i_1, \ldots, i_k \in \{1, \ldots, N\}\}.$$

we have

$$\mathscr{H} = \oplus_{n=1}^{M} \mathscr{D}_n.$$

It is easy to see that, for each $j = 1, \ldots, N$ and n, B_j and B_j^* leave $\mathscr{D}_n$ invariant. Hence

$$\rho_n := (\mathscr{D}_n, \{B_j, B_j^* | j = 1, \ldots, N\})$$

is a representation of the CAR with N degrees of freedom and

$$\rho_B = \oplus_{n=1}^{M} \rho_n.$$

By Example 5.2 and the isomorphism theorem on Hilbert spaces, there exists a unitary operator $W_n : \mathscr{D}_n \to \wedge(\mathbb{C}^N)$ such that $W_n v_n = \psi_0$ and

$$W_n B_{i_1}^* \cdots B_{i_k}^* v_n = b_{i_1}^* \cdots b_{i_k}^* \psi_0, \quad k = 1, \ldots, N, \ i_1 < \cdots < i_k, \ i_1, \ldots, i_k \in \{1, \ldots, N\}.$$

Then it is easy to see that

$$W_n B_j W_n^{-1} = b_j, \quad j = 1, \ldots, N.$$

Hence each ρ_n is equivalent to $\rho_{\mathrm{F}}^{(N)}$. The operator

$$W := \oplus_{n=1}^{M} W_n$$

is a unitary operator from $\mathscr{H}$ to $\oplus_{n=1}^{M} \wedge (\mathbb{C}^N)$, which gives

$$W B_j W^{-1} = \oplus_{n=1}^{M} b_j, \quad j = 1, \ldots, N.$$

Thus (5.20) holds. ■

Chapter 6
Elements of the Theory of Boson Fock Spaces

Abstract Basic aspects of the theory of boson Fock space are described in connection with studying representations of CCR with infinite degrees of freedom.

6.1 The Boson Fock Space Over a Hilbert Space

Let $\mathscr{H}$ be a Hilbert space. Then, for each $n \in \mathbb{N}$, one has the n-fold symmetric tensor product Hilbert space $\otimes_{\mathrm{s}}^n \mathscr{H}$ of $\mathscr{H}$ (see Sect. 1.5). We set $\otimes_{\mathrm{s}}^0 \mathscr{H} := \mathbb{C}$. The infinite direct sum Hilbert space of $\otimes_{\mathrm{s}}^n \mathscr{H}$, $n = 0, 1, 2, \ldots$

$$\mathscr{F}_{\mathrm{b}}(\mathscr{H}) := \bigoplus_{n=0}^{\infty} \otimes_{\mathrm{s}}^n \mathscr{H} \tag{6.1}$$

$$= \left\{ \Psi = \left\{ \Psi^{(n)} \right\}_{n=0}^{\infty} \mid \Psi^{(n)} \in \otimes_{\mathrm{s}}^n \mathscr{H},\ n \in \mathbb{Z}_+,\ \sum_{n=0}^{\infty} \|\Psi^{(n)}\|^2 < \infty \right\} \tag{6.2}$$

is called the **boson** (or **symmetric**) **Fock space** over $\mathscr{H}$. Following the definition of an infinite direct sum Hilbert space, the inner product of $\mathscr{F}_{\mathrm{b}}(\mathscr{H})$ takes the form

$$\langle \Psi, \Phi \rangle = \sum_{n=0}^{\infty} \left\langle \Psi^{(n)}, \Phi^{(n)} \right\rangle_{\otimes_{\mathrm{s}}^n \mathscr{H}}, \quad \Psi, \Phi \in \mathscr{F}_{\mathrm{b}}(\mathscr{H}).$$

The Hilbert space $\mathscr{F}_{\mathrm{b}}(\mathscr{H})$ is used to describe state vectors of a **Bose field**, i.e., a quantum field of identical bosons. In the context of Bose field theories, the Hilbert space $\otimes_{\mathrm{s}}^n \mathscr{H}$ as a closed subspace of $\mathscr{F}_{\mathrm{b}}(\mathscr{H})$ represents the state vector space of a system of n identical bosons. For this reason, $\otimes_{\mathrm{s}}^n \mathscr{H}$ is called the n**-boson space** or the n**-particle space** of $\mathscr{F}_{\mathrm{b}}(\mathscr{H})$. In particular, $\otimes_{\mathrm{s}}^1 \mathscr{H} = \mathscr{H}$ is called the **one-particle Hilbert space** of $\mathscr{F}_{\mathrm{b}}(\mathscr{H})$, which describes a Hilbert space of state vectors of one boson.

© Springer Nature Singapore Pte Ltd. 2020

A. Arai, *Inequivalent Representations of Canonical Commutation and Anti-Commutation Relations*, Mathematical Physics Studies,
https://doi.org/10.1007/978-981-15-2180-5_6

The subspace

$$\mathscr{F}_{\mathrm{b},0}(\mathscr{H}) := \left\{\Psi \in \mathscr{F}_{\mathrm{b}}(\mathscr{H}) | \exists n_0 \in \mathbb{N} \text{ such that } \Psi^{(n)} = 0, \ \forall n \geq n_0\right\} \tag{6.3}$$

is called the **finite particle subspace** of $\mathscr{F}_{\mathrm{b}}(\mathscr{H})$. It is easy to see that $\mathscr{F}_{\mathrm{b},0}(\mathscr{H})$ is dense in $\mathscr{F}_{\mathrm{b}}(\mathscr{H})$.

For a subspace $\mathscr{D}$ of $\mathscr{H}$, we introduce

$$\mathscr{F}_{\mathrm{b,fin}}(\mathscr{D}) := \left\{\Psi \in \mathscr{F}_{\mathrm{b},0}(\mathscr{H}) | \Psi^{(n)} \in \hat{\otimes}_{\mathrm{s}}^n \mathscr{D}, \ n \in \mathbb{Z}_+\right\}, \tag{6.4}$$

where $\hat{\otimes}_{\mathrm{s}}^n \mathscr{D}$ denotes the n-fold algebraic symmetric tensor product of $\mathscr{D}$. If $\mathscr{D}$ is dense in $\mathscr{H}$, then $\mathscr{F}_{\mathrm{b,fin}}(\mathscr{D})$ is dense in $\mathscr{F}_{\mathrm{b}}(\mathscr{H})$.

Example 6.1 One of the simplest examples of boson Fock space is $\mathscr{F}_{\mathrm{b}}(\mathbb{C})$, the boson Fock space over $\mathbb{C}$. Since $\otimes_{\mathrm{s}}^n \mathbb{C} \cong \mathbb{C}$, we have

$$\mathscr{F}_{\mathrm{b}}(\mathbb{C}) \cong \oplus_{n=0}^{\infty} \mathbb{C} = \ell^2(\mathbb{Z}_+).$$

Example 6.2 For a measure space (X, Σ, μ), one can consider the Hilbert space $L^2(X, d\mu)$ (see Sect. 1.7), which serves as a Hilbert space of state vectors of one boson being in the space X. Then

$$\mathscr{F}_{\mathrm{b}}(L^2(X, d\mu)) = \oplus_{n=0}^{\infty} \otimes_{\mathrm{s}}^n L^2(X, d\mu).$$

By Theorem 1.24(i), we have the natural isomorphism

$$\mathscr{F}_{\mathrm{b}}(L^2(X, d\mu)) \cong \oplus_{n=0}^{\infty} L^2_{\mathrm{sym}}(X^n, d\mu^n),$$

where $\mu^n := \mu \otimes \cdots \otimes \mu$, the n-product measure of μ, and $L^2_{\mathrm{sym}}(X^n, d\mu^n)|_{n=0} := \mathbb{C}$.

Typical examples are given by the case $(X, \Sigma, \mu) = (\mathbb{R}^d, \boldsymbol{B}^d, \mu_{\mathrm{L}}^{(d)})$ or $(\hat{\mathbb{R}}^d, \boldsymbol{B}^d, \mu_{\mathrm{L}}^{(d)})$, where $\mu_{\mathrm{L}}^{(d)}$ denotes the d-dimensional Lebesgue measure.

There are operators acting in the boson Fock space $\mathscr{F}_{\mathrm{b}}(\mathscr{H})$, which play central roles in describing Bose field theories. We next define them.

6.2 Boson Second Quantization Operators

Let T be a densely defined closable operator on $\mathscr{H}$. Then, for each natural number $n \geq 2$ and $j = 1, \ldots, n$, the operator

$$\hat{T}_j := I \otimes \cdots \otimes I \otimes \underbrace{T}_{j\text{-th}} \otimes I \otimes \cdots \otimes I$$

is a densely defined closed operator on the n-fold tensor product Hilbert space $\otimes^n \mathscr{H}$ (see Sect. 1.8). Let

$$T_0^{(n)} := \left(\sum_{j=1}^{n} \hat{T}_j\right) \restriction \hat{\otimes}^n D(T).$$

Then $T_0^{(n)}$ is densely defined. It is easy to see that $D((T_0^{(n)})^*) \supset \hat{\otimes}^n D(T^*)$ and

$$(T_0^{(n)})^* \supset \left(\sum_{j=1}^{n} I \otimes \cdots \otimes I \otimes \underbrace{T^*}_{j\text{-th}} \otimes I \otimes \cdots \otimes I\right) \restriction \hat{\otimes}^n D(T^*).$$

Since $\hat{\otimes}^n D(T^*)$ is dense in $\otimes^n \mathscr{H}$, $(T_0^{(n)})^*$ is densely defined. Hence, by a general theorem, $T_0^{(n)}$ is closable. Hence the closure of it

$$T^{(n)} := \overline{T_0^{(n)}} \quad (n \geq 2) \tag{6.5}$$

is a densely defined closed operator on $\otimes^n \mathscr{H}$. We set

$$T^{(1)} := \overline{T}. \tag{6.6}$$

Lemma 6.1 *For each $n \in \mathbb{N}$, $T^{(n)}$ is reduced by $\otimes_s^n \mathscr{H}$.*

Proof Let $\Psi = \otimes_{j=1}^n \Psi_j \in \hat{\otimes}^n D(T)$, $\Psi_j \in D(T)$ and S_n be the symmetrization operator defined by (1.49). Then

$$S_n \Psi = \frac{1}{n!} \sum_{\sigma \in \mathfrak{S}_n} \otimes_{j=1}^n \Psi_{\sigma(j)}.$$

Hence $S_n \Psi \in D(T^{(n)})$ and

$$T^{(n)} S_n \Psi = \frac{1}{n!} \sum_{\sigma \in \mathfrak{S}_n} T^{(n)} (\otimes_{j=1}^n \Psi_{\sigma(j)}).$$

We have

$$\begin{aligned} T^{(n)} (\otimes_{j=1}^n \Psi_{\sigma(j)}) &= \sum_{k=1}^{n} \Psi_{\sigma(1)} \otimes \cdots \otimes \underbrace{T \Psi_{\sigma(k)}}_{k\text{-th}} \otimes \cdots \otimes \Psi_{\sigma(n)}. \\ &= U_\sigma T^{(n)} \Psi, \end{aligned}$$

where U_σ is defined by (1.48). Hence, by (1.49), we obtain

$$T^{(n)} S_n \Psi = S_n T^{(n)} \Psi.$$

By linearity, this equation extends to all $\Psi \in \hat{\otimes}^n D(T)$. Since $\hat{\otimes}^n D(T)$ is a core for $T^{(n)}$ by the definition of $T^{(n)}$, it follows from a limiting argument that $S_n T^{(n)} \subset T^{(n)} S_n$. Recall that S_n is an orthogonal projection. Hence $T^{(n)}$ is reduced by $\mathrm{Ran}\, S_n = \otimes_{\mathrm{s}}^n \mathscr{H}$. ■

We denote by $T_{\mathrm{b}}^{(n)}$ the reduced part of $T^{(n)}$ to $\otimes_{\mathrm{s}}^n \mathscr{H}$:

$$D(T_{\mathrm{b}}^{(n)}) = D(T^{(n)}) \cap (\otimes_{\mathrm{s}}^n \mathscr{H}), \quad T_{\mathrm{b}}^{(n)} \Psi = T^{(n)} \Psi, \quad \Psi \in D(T_{\mathrm{b}}^{(n)}). \tag{6.7}$$

The following proposition follows from Proposition 1.16:

Proposition 6.1 *Let T be a densely defined closable operator on $\mathscr{H}$.*

(i) *The operator $T_{\mathrm{b}}^{(n)}$ is a densely defined closed operator.*
(ii) *If T is self-adjoint, then so is $T_{\mathrm{b}}^{(n)}$.*

As a convention, we set

$$T_{\mathrm{b}}^{(0)} := 0$$

acting in $\mathbb{C}$. The infinite direct sum operator (see Appendix C)

$$\mathrm{d}\Gamma_{\mathrm{b}}(T) := \oplus_{n=0}^{\infty} T_{\mathrm{b}}^{(n)}, \tag{6.8}$$

which is a densely defined closed operator on $\mathscr{F}_{\mathrm{b}}(\mathscr{H})$, is called the **boson second quantization operator** of T or the boson second quantization of T.

Let $\Omega_{\mathscr{H}} \in \mathscr{F}_{\mathrm{b}}(\mathscr{H})$ be the vector defined by

$$\Omega_{\mathscr{H}}^{(0)} = 1, \quad \Omega_{\mathscr{H}}^{(n)} = 0, \quad n \in \mathbb{N}. \tag{6.9}$$

It is called the **Fock vacuum** in $\mathscr{F}_{\mathrm{b}}(\mathscr{H})$. It is visually written

$$\Omega_{\mathscr{H}} = \{1, 0.0, \ldots\}.$$

It follows that

$$\mathrm{d}\Gamma_{\mathrm{b}}(T) \Omega_{\mathscr{H}} = 0, \tag{6.10}$$

i.e., the Fock vacuum is an eigenvector of $\mathrm{d}\Gamma_{\mathrm{b}}(T)$ with eigenvalue 0.

The boson second quantization operator

$$N_{\mathrm{b}} := \mathrm{d}\Gamma_{\mathrm{b}}(I) \tag{6.11}$$

of $T = I$ is called the **boson number operator** on $\mathscr{F}_{\mathrm{b}}(\mathscr{H})$. The name comes from the easily seen fact

$$N_{\mathrm{b}} \restriction \otimes_{\mathrm{s}}^{n} \mathscr{H} = n, \quad n \in \mathbb{Z}+, \tag{6.12}$$

i.e., N_{b} counts the number of bosons.

In many cases in applications to quantum field theory, the operator T is self-adjoint. In this case, we have the following theorem.

Theorem 6.1 *Let T be a self-adjoint operator on $\mathscr{H}$. Then:*

(i) $\mathrm{d}\Gamma_{\mathrm{b}}(T)$ *is a self-adjoint operator. Moreover, for all $t \in \mathbb{R}$,*

$$e^{it\mathrm{d}\Gamma_{\mathrm{b}}(T)} = \oplus_{n=0}^{\infty} e^{itT_{\mathrm{b}}^{(n)}}. \tag{6.13}$$

(ii) *If T is non-negative, then* $\mathrm{d}\Gamma_{\mathrm{b}}(T)$ *is non-negative and, for all $\beta > 0$,*

$$e^{-\beta\mathrm{d}\Gamma_{\mathrm{b}}(T)} = \oplus_{n=0}^{\infty} e^{-\beta T_{\mathrm{b}}^{(n)}}. \tag{6.14}$$

(iii) *Let $\mathscr{D}$ be a core for T. Then $\mathscr{F}_{\mathrm{b,fin}}(D)$ is a core for* $\mathrm{d}\Gamma_{\mathrm{b}}(T)$.
(iv) *If $\sigma(T) = [m, \infty)$, then*

$$\sigma(\mathrm{d}\Gamma_{\mathrm{b}}(T)) = \{0\} \cup [m, \infty). \tag{6.15}$$

Proof

(i) By Theorem 1.26-(iv), $T^{(n)}$ is self-adjoint. Hence $T_{\mathrm{b}}^{(n)}$ is self-adjoint. Therefore, by a general theorem on the direct sum of self-adjoint operators (see Theorem C.2(i) in Appendix C), $\mathrm{d}\Gamma_{\mathrm{b}}(T)$ is self-adjoint. Formula (6.13) follows from an application of formula (C.3) in Appendix C.
(ii) Let $T \geq 0$. Then, by Theorem 1.26(v), $T^{(n)}$ is non-negative and hence so is $T_{\mathrm{b}}^{(n)}$. Thus, by Theorem C.2(ii), $\mathrm{d}\Gamma_{\mathrm{b}}(T)$ is non-negative. Formula 6.14 follows from an application of formula (C.4) in Appendix C.
(iii) It follows from a limiting argument that, for all $n \in \mathbb{N}$, $\hat{\otimes}^{n} \mathscr{D}$ is a core for $T^{(n)}$. Hence $\hat{\otimes}_{\mathrm{s}}^{n} \mathscr{D}$ is a core for $T_{\mathrm{b}}^{(n)}$. Thus the desired result follows.
(iv) By Theorem C.2(iii) in Appendix C,

$$\sigma(\mathrm{d}\Gamma_{\mathrm{b}}(T)) = \{0\} \cup \sigma(T) \cup \left(\cup_{n=2}^{\infty} \sigma((T^{(n)})_{\mathrm{b}})\right).$$

It is easy to see that, for all $\Psi \in \hat{\otimes}_{\mathrm{s}}^{n} D(T)$, $\langle \Psi, (T^{(n)})_{\mathrm{b}} \Psi \rangle \geq nm \|\Psi\|^2$. By a limiting argument, this extends to all $\Psi \in D((T^{(n)})_{\mathrm{b}})$. Hence $\sigma((T^{(n)})_{\mathrm{b}}) \subset [nm, \infty)$. Thus (6.15) follows. ∎

Remark 6.1 More detailed spectral properties of $\mathrm{d}\Gamma_{\mathrm{b}}(T)$ are known (see [28, Theorem 5.3]).

The following theorem describes a convergent property of a sequence of boson second quantization operators:

Theorem 6.2 *Let T_n $(n \in \mathbb{N})$ and T be self-adjoint operators on $\mathscr{H}$. Suppose that there exists a core $\mathscr{D}$ for T such that $\mathscr{D} \subset \cap_{n=1}^{\infty} D(T_n)$ and, for all $\Psi \in \mathscr{D}$, $\lim_{n\to\infty} T_n\Psi = T\Psi$. Then, for all $t \in \mathbb{R}$,*

$$\text{s-}\lim_{n\to\infty} e^{it\mathrm{d}\Gamma_{\mathrm{b}}(T_n)} = e^{it\mathrm{d}\Gamma_{\mathrm{b}}(T)}. \tag{6.16}$$

Proof It is easy to see that $\mathscr{F}_{\mathrm{b,fin}}(\mathscr{D}) \subset D(\mathrm{d}\Gamma_{\mathrm{b}}(T)) \cap \left(\cap_{n=1}^{\infty} D(\mathrm{d}\Gamma_{\mathrm{b}}(T_n))\right)$ and, for all $\Psi \in \mathscr{F}_{\mathrm{b,fin}}(\mathscr{D})$,

$$\lim_{n\to\infty} \mathrm{d}\Gamma_{\mathrm{b}}(T_n)\Psi = \mathrm{d}\Gamma_{\mathrm{b}}(T)\Psi.$$

By Theorem 6.1, $\mathscr{F}_{\mathrm{b,fin}}(\mathscr{D})$ is a core for $\mathrm{d}\Gamma_{\mathrm{b}}(T)$. Hence, by Theorem 1.29, $\{\mathrm{d}\Gamma_{\mathrm{b}}(T_n)\}_n$ converges to $\mathrm{d}\Gamma_{\mathrm{b}}(T)$ in the strong resolvent sense as $n \to \infty$. Then, by Theorem 1.30, we obtain (6.16). ■

6.3 Boson *Γ*-Operators

Let T be a densely defined closable operator from a Hilbert space $\mathscr{H}$ to a Hilbert space $\mathscr{K}$. Then the n-fold tensor product $\otimes^n T$ of T with $n \geq 2$ is a unique densely defined closed operator from $\otimes^n \mathscr{H}$ to $\otimes^n \mathscr{K}$ such that

$$(\otimes^n T)(\otimes_{j=1}^n \Psi_j) = \otimes_{j=1}^n T\Psi_j, \quad \Psi_j \in D(T),\ j = 1, \ldots, n.$$

In the same way as in the proof of Lemma 6.1, one can prove that, for each $n \geq 2$,

$$S_n(\otimes^n T) \subset (\otimes^n T)S_n. \tag{6.17}$$

Hence we can define a densely defined closed operator $(\otimes^n T)_{\mathrm{b}}$ from $\otimes_{\mathrm{s}}^n \mathscr{H}$ to $\otimes_{\mathrm{s}}^n \mathscr{K}$ as follows: $(\otimes^0 T)_{\mathrm{b}} := 1$, $(\otimes^1 T)_{\mathrm{b}} := \overline{T}$ and, for $n \geq 2$,

$$D((\otimes^n T)_{\mathrm{b}}) := D(\otimes^n T) \cap (\otimes_{\mathrm{s}}^n \mathscr{H}),$$
$$(\otimes^n T)_{\mathrm{b}}\Psi := (\otimes^n T)\Psi, \quad \Psi \in D((\otimes^n T)_{\mathrm{b}}).$$

In the case $\mathscr{H} = \mathscr{K}$, $(\otimes^n T)_{\mathrm{b}}$ is nothing but the reduced part of $\otimes^n T$ to $\otimes_{\mathrm{s}}^n \mathscr{H}$.

One can define an infinite direct sum operator from $\mathscr{F}_{\mathrm{b}}(\mathscr{H})$ to $\mathscr{F}_{\mathrm{b}}(\mathscr{K})$ by

$$\Gamma_{\mathrm{b}}(T) = \oplus_{n=0}^{\infty}(\otimes^n T)_{\mathrm{b}}. \tag{6.18}$$

We call the operator $\Gamma_{\mathrm{b}}(T)$ the **boson Γ-operator** for T.

Theorem 6.3 *Let T be a densely defined closable operator from $\mathscr{H}$ to $\mathscr{K}$.*

(i) *If T is unbounded or bounded with $\|T\| > 1$, then $\Gamma_{\mathrm{b}}(T)$ is unbounded.*
(ii) *If T is a contraction operator, then $\Gamma_{\mathrm{b}}(T)$ is a contraction operator and*

$$\Gamma_{\mathrm{b}}(T)^* = \Gamma_{\mathrm{b}}(T^*). \tag{6.19}$$

(iii) (product law) *Let $S \in \mathfrak{B}(\mathscr{H}, \mathscr{K})$ and $T \in \mathfrak{B}(\mathscr{K}, \mathscr{X})$ ($\mathscr{X}$ is a Hilbert space) be contraction operators. Then,*

$$\Gamma_{\mathrm{b}}(T)\Gamma_{\mathrm{b}}(S) = \Gamma_{\mathrm{b}}(TS). \tag{6.20}$$

(iv) *If $U \colon \mathscr{H} \to \mathscr{K}$ is unitary, then so is $\Gamma_{\mathrm{b}}(U) : \mathscr{F}_{\mathrm{b}}(\mathscr{H}) \to \mathscr{F}_{\mathrm{b}}(\mathscr{K})$ and*

$$\Gamma_{\mathrm{b}}(U)^{-1} = \Gamma_{\mathrm{b}}(U^{-1}). \tag{6.21}$$

Moreover,

$$\Gamma_{\mathrm{b}}(U)\mathscr{F}_{\mathrm{b,fin}}(\mathscr{H}) = \mathscr{F}_{\mathrm{b,fin}}(\mathscr{K}). \tag{6.22}$$

(v) *Let $\mathscr{H} = \mathscr{K}$ and $\mathscr{M}$ be a closed subspace of $\mathscr{H}$. Let $P_{\mathscr{M}}$ be the orthogonal projection onto $\mathscr{M}$. Then $\Gamma_{\mathrm{b}}(P_{\mathscr{M}})$ is the orthogonal projection onto $\mathscr{F}_{\mathrm{b}}(\mathscr{M}) \subset \mathscr{F}_{\mathrm{b}}(\mathscr{H})$.*

Proof

(i) Let T be unbounded. Then $(\otimes^1 T)_{\mathrm{b}} = \overline{T}$ is unbounded. Hence it follows that $\Gamma_{\mathrm{b}}(T)$ is unbounded. Next, let T be bounded with $\|T\| > 1$. Then there exist a constant $c > 0$ and a unit vector $f \in \mathscr{H}$ such that $\|Tf\| \geq 1 + c$. Hence

$$\|(\otimes^n T) \otimes^n f\| = \|Tf\|^n \geq (1 + c)^n.$$

Hence, for $\Psi_n := \{0, 0, \ldots, \otimes^n f, 0, \ldots\}$, we have $\|\Psi_n\| = 1$ and

$$\|\Gamma_{\mathrm{b}}(T)\Psi_n\| = \|(\otimes^n T)_{\mathrm{b}} \otimes^n f\| \geq (1 + c)^n \to \infty \ (n \to \infty).$$

Hence $\Gamma_{\mathrm{b}}(T)$ is unbounded.
(ii) Let T be a contraction operator: $\|T\| \leq 1$. Then

$$\| \otimes_{\mathrm{b}}^n T\| \leq \| \otimes^n T\| = \|T\|^n \leq 1.$$

Hence, for all $\Psi \in \mathscr{F}_{\mathrm{b}}(\mathscr{H})$,

$$\|\Gamma_{\mathrm{b}}(T)\Psi\|^2 = \sum_{n=0}^{\infty} \|(\otimes^n T)_{\mathrm{b}}\Psi^{(n)}\|^2 \leq \sum_{n=0}^{\infty} \|\Psi^{(n)}\|^2 = \|\Psi\|^2.$$

Hence $\|\Gamma_{\mathrm{b}}(T)\| \leq 1$. Thus $\Gamma_{\mathrm{b}}(T)$ is a contraction. We have

$$\Gamma_{\mathrm{b}}(T)^* = \sum_{n=0}^{\infty}((\otimes^n T)_{\mathrm{b}})^*.$$

Since T is bounded, it follows that $((\otimes^n T)_{\mathrm{b}})^* = (\otimes^n T^*)_{\mathrm{b}}$. Thus (6.19) follows.

(iii) This follows from the easily proven fact

$$(\otimes^n T)_{\mathrm{b}}(\otimes^n S)_{\mathrm{b}} = (\otimes^n TS)_{\mathrm{b}}, \quad n \in \mathbb{N}.$$

(iv) Since $UU^* = I$, it follows from (iii) that $\Gamma_{\mathrm{b}}(U)\Gamma_{\mathrm{b}}(U^*) = \Gamma_{\mathrm{b}}(I) = I$. Similarly, $U^*U = I$ implies that $\Gamma_{\mathrm{b}}(U^*)\Gamma_{\mathrm{b}}(U) = I$. By (6.19), we have $\Gamma_{\mathrm{b}}(U^*) = \Gamma_{\mathrm{b}}(U)^*$. Thus $\Gamma_{\mathrm{b}}(U)$ is unitary. Since $U^{-1} = U^*$, we have $\Gamma_{\mathrm{b}}(U^{-1}) = \Gamma_{\mathrm{b}}(U^*) = \Gamma_{\mathrm{b}}(U)^{-1}$. Hence (6.21) holds.
It is easy to see that $(\otimes^n U)_{\mathrm{b}}\hat{\otimes}_{\mathrm{s}}^n\mathscr{H} = \hat{\otimes}_{\mathrm{s}}^n\mathscr{K}$. Hence (6.22) holds.

(v) By (ii), $\Gamma_{\mathrm{b}}(P_{\mathscr{M}})$ is a self-adjoint contraction operator. By (iii),

$$\Gamma_{\mathrm{b}}(P_{\mathscr{M}})^2 = \Gamma_{\mathrm{b}}(P_{\mathscr{M}}^2) = \Gamma_{\mathrm{b}}(P_{\mathscr{M}}).$$

Thus $\Gamma_{\mathrm{b}}(P_{\mathscr{M}})$ is an orthogonal projection. We have $(\otimes^n P_{\mathscr{M}})_{\mathrm{b}}\hat{\otimes}_{\mathrm{b}}^n\mathscr{H} = \hat{\otimes}_{\mathrm{s}}^n\mathscr{M}$. Hence

$$\Gamma_{\mathrm{b}}(P_{\mathscr{M}})\mathscr{F}_{\mathrm{b,fin}}(\mathscr{H}) = \mathscr{F}_{\mathrm{b,fin}}(\mathscr{M}).$$

Since $\mathscr{F}_{\mathrm{b,fin}}(\mathscr{H})$ is dense in $\mathscr{F}_{\mathrm{b}}(\mathscr{H})$ and $\mathscr{F}_{\mathrm{b,fin}}(\mathscr{M})$ is dense in $\mathscr{F}_{\mathrm{b}}(\mathscr{M})$, it follows that $\Gamma_{\mathrm{b}}(P_{\mathscr{M}})\mathscr{F}_{\mathrm{b}}(\mathscr{H}) = \mathscr{F}_{\mathrm{b}}(\mathscr{M})$. Thus $\Gamma_{\mathrm{b}}(P_{\mathscr{M}})$ is the orthogonal projection onto $\mathscr{F}_{\mathrm{b}}(\mathscr{M})$. ■

Basic relations between the boson second quantization operator $\mathrm{d}\Gamma_{\mathrm{b}}(T)$ and the boson Γ-operator $\Gamma_{\mathrm{b}}(T)$ are given in the following theorem:

Theorem 6.4 *Let T be a self-adjoint operator on $\mathscr{H}$. Then:*

(i) *For all $t \in \mathbb{R}$,*

$$\Gamma_{\mathrm{b}}(e^{itT}) = e^{it\mathrm{d}\Gamma_{\mathrm{b}}(T)}. \tag{6.23}$$

(ii) *Let*

$$\mathbb{C}_+ := \{z \in \mathbb{C} | \mathrm{Re}\, z > 0\} \tag{6.24}$$

and $T \geq 0$. *Then* $\mathrm{d}\Gamma_\mathrm{b}(T) \geq 0$ *and, for all* $z \in \overline{\mathbb{C}}_+$,

$$\Gamma_\mathrm{b}(e^{-zT}) = e^{-z\mathrm{d}\Gamma_\mathrm{b}(T)}. \tag{6.25}$$

Proof

(i) By Theorem 1.26(iv), for all $t \in \mathbb{R}$, $\otimes_{j=1}^n e^{itT} = e^{itT^{(n)}}$. Hence

$$(\otimes_{j=1}^n e^{itT})_\mathrm{b} = S_n e^{itT^{(n)}} S_n = e^{itT_\mathrm{b}^{(n)}},$$

where we have used Proposition 1.16(v). Hence $\Gamma_\mathrm{b}(e^{itT}) = \oplus_{n=0}^\infty e^{itT_\mathrm{b}^{(n)}}$. By this fact and formula (6.13), we obtain (6.23).

(ii) In the same manner as in (i), one can show that, for all $\beta \geq 0$, $\Gamma_\mathrm{b}(e^{-\beta T}) = \oplus_{n=0}^\infty e^{-\beta T_\mathrm{b}^{(n)}}$. Hence, by (6.14), we obtain (6.25) with $z = \beta$. We next consider the general case $z = \beta + it$, $\beta \geq 0$, $t \in \mathbb{R}$. In this case we have $e^{-zT} = e^{-\beta T} e^{itT}$. Hence, by Theorem 6.3(iii),

$$\Gamma_\mathrm{b}(e^{-zT}) = \Gamma_\mathrm{b}(e^{-\beta T})\Gamma_\mathrm{b}(e^{itT}) = e^{-\beta\mathrm{d}\Gamma_\mathrm{b}(T)} e^{it\mathrm{d}\Gamma_\mathrm{b}(T)} = e^{-z\mathrm{d}\Gamma_\mathrm{b}(T)}.$$

Thus (6.25) follows. ■

Theorem 6.5 *Let* $U: \mathscr{H} \to \mathscr{K}$ *(a Hilbert space) be a unitary operator and* T *be a densely defined closable operator on* $\mathscr{H}$. *Then*

$$\Gamma_\mathrm{b}(U)\mathrm{d}\Gamma_\mathrm{b}(T)\Gamma_\mathrm{b}(U)^{-1} = \mathrm{d}\Gamma_\mathrm{b}(UTU^{-1}). \tag{6.26}$$

Proof One first proves (6.26) on $\mathscr{F}_{\mathrm{b,fin}}(D(UTU^{-1}))$. Then, by a limiting argument using the fact that $\mathscr{F}_{\mathrm{b,fin}}(D(UTU^{-1}))$ is a core for $\mathrm{d}\Gamma_\mathrm{b}(UTU^{-1})$, one can show that $\mathrm{d}\Gamma_\mathrm{b}(UTU^{-1}) \subset \Gamma_\mathrm{b}(U)\mathrm{d}\Gamma_\mathrm{b}(T)\Gamma_\mathrm{b}(U)^{-1}$.

To prove the converse order relation, let $\mathscr{D} := \Gamma_\mathrm{b}(U)\mathscr{F}_{\mathrm{b,fin}}(D(T))$. Then $\mathscr{D}$ is a core for $\Gamma_\mathrm{b}(U)\mathrm{d}\Gamma_\mathrm{b}(T)\Gamma(U)^{-1}$ and it is shown that (6.26) holds on $\mathscr{D}$. Hence it follows that $\mathrm{d}\Gamma_\mathrm{b}(UTU^{-1}) \supset \Gamma_\mathrm{b}(U)\mathrm{d}\Gamma_\mathrm{b}(T)\Gamma(U)^{-1}$. Thus (6.26) follows. ■

6.4 Creation and Annihilation Operators

For each $f \in \mathscr{H}$, we define an operator $A^\dagger(f)$ on $\mathscr{F}_\mathrm{b}(\mathscr{H})$ as follows:

$$D(A^\dagger(f)) := \left\{ \Psi \in \mathscr{F}_\mathrm{b}(\mathscr{H}) | \sum_{n=1}^{\infty} \|\sqrt{n} S_n(f \otimes \Psi^{(n-1)})\|^2 < \infty \right\}, \tag{6.27}$$

$$(A^\dagger(f)\Psi)^{(0)} = 0, \tag{6.28}$$

$$(A^\dagger(f)\Psi)^{(n)} := \sqrt{n} S_n(f \otimes \Psi^{(n-1)}), \quad n \geq 1, \Psi \in D(A^\dagger(f)). \tag{6.29}$$

It is obvious that

$$\mathscr{F}_{\mathrm{b},0}(\mathscr{H}) \subset D(A^\dagger(f)). \tag{6.30}$$

Hence $A^\dagger(f)$ is densely defined. It follows from (6.29) that $A^\dagger(f)$ maps $\otimes_\mathrm{s}^n \mathscr{H}$ to $\otimes_\mathrm{s}^{n+1} \mathscr{H}$:

$$A^\dagger(f) \otimes_\mathrm{s}^n \mathscr{H} \subset \otimes_\mathrm{s}^{n+1} \mathscr{H}, \quad n \in \mathbb{N}. \tag{6.31}$$

In particular, $A^\dagger(f)$ leaves $\mathscr{F}_{\mathrm{b},0}(\mathscr{H})$ invariant:

$$A^\dagger(f) \mathscr{F}_{\mathrm{b},0}(\mathscr{H}) \subset \mathscr{F}_{\mathrm{b},0}(\mathscr{H}). \tag{6.32}$$

It is easy to see that, for all $\alpha, \beta \in \mathbb{C}$ and $f, g \in \mathscr{H}$,

$$A^\dagger(\alpha f + \beta g) \supset \alpha A^\dagger(f) + \beta A^\dagger(g). \tag{6.33}$$

Lemma 6.2 *For all $f \in \mathscr{H}$, $A^\dagger(f)$ is a densely defined closed operator on $\mathscr{F}_\mathrm{b}(\mathscr{H})$.*

Proof We need only to prove the closedness of $A^\dagger(f)$. Let $\Psi_k \in D(A^\dagger(f))$ $(k \in \mathbb{N})$ be such that $\lim_{k\to\infty} \Psi_k = \Psi \in \mathscr{F}_\mathrm{b}(\mathscr{H})$ and $\lim_{k\to\infty} A^\dagger(f)\Psi_k = \Phi \in \mathscr{F}_\mathrm{b}(\mathscr{H})$. Then, for each $n \in \mathbb{Z}_+$, $\Psi_k^{(n)} \to \Psi^{(n)}$ and $(A^\dagger(f)\Psi_k)^{(n)} \to \Phi^{(n)}$ as $k \to \infty$. The latter implies that $\sqrt{n} S_n(f \otimes \Psi_k^{(n-1)}) \to \Phi^{(n)}$ as $k \to \infty$. We have $f \otimes \Psi_k^{(n-1)} \to f \otimes \Psi^{(n-1)}$ as $k \to \infty$. Since S_n is bounded, it follows that $\sqrt{n} S_n(f \otimes \Psi^{(n-1)}) = \Phi^{(n)}$. We have $\sum_{n=0}^\infty \|\Phi^{(n)}\|^2 < \infty$. Hence $\sum_{n=1}^\infty \|\sqrt{n} S_n(f \otimes \Psi^{(n-1)})\|^2 < \infty$. Therefore $\Psi \in D(A^\dagger(f)$ and $(A^\dagger(f)\Psi)^{(n)} = \Phi^{(n)}$, i.e., $A^\dagger(f)\Psi = \Phi$. Thus $A^\dagger(f)$ is closed. ■

Since $A^\dagger(f)$ is densely defined, one can define

$$A(f) := A^\dagger(f)^*, \tag{6.34}$$

the adjoint of $A^{\dagger}(f)$. By Lemma 6.2, we have

$$A(f)^* = A^{\dagger}(f). \tag{6.35}$$

Property (6.31) implies that

$$A(f) \otimes_{\mathrm{s}}^n \mathscr{H} \subset \otimes_{\mathrm{s}}^{n-1} \mathscr{H}, \quad n \in \mathbb{N} \tag{6.36}$$

with $\otimes_{\mathrm{s}}^{-1} \mathscr{H} := \{0\}$. In particular, $A(f)$ leaves $\mathscr{F}_{\mathrm{b},0}(\mathscr{H})$ invariant:

$$A(f)\mathscr{F}_{\mathrm{b},0}(\mathscr{H}) \subset \mathscr{F}_{\mathrm{b},0}(\mathscr{H}). \tag{6.37}$$

Relation (6.31) means that $A(f)^*$ maps the n-particle space into $(n+1)$-particle space, i.e., it has the function that creates one particle. On the other hand, (6.36) shows that $A(f)$ has the function that annihilates one particle. Based on these properties, $A(f)^*$ and $A(f)$ are called respectively the **boson creation operator** and the **boson annihilation operator** with test vector f.

It follows from (6.33) that, for all $\alpha, \beta \in \mathbb{C}$ and $f, g \in \mathscr{H}$,

$$A(\alpha f + \beta g) = \alpha^* A(f) + \beta^* A(g) \quad \text{on } \mathscr{F}_{\mathrm{b},0}(\mathscr{H}). \tag{6.38}$$

Note that the mapping $\mathscr{H} \ni f \mapsto A(f) \restriction \mathscr{F}_{\mathrm{b},0}(\mathscr{H})$ is *anti-linear*, while the mapping $\mathscr{H} \ni f \mapsto A(f)^* \restriction \mathscr{F}_{\mathrm{b},0}(\mathscr{H})$ is *linear*.

Lemma 6.3 *The Fock vacuum $\Omega_{\mathscr{H}}$ is in $\cap_{f \in \mathscr{H}} D(A(f))$ and*

$$A(f)\Omega_{\mathscr{H}} = 0, \quad f \in \mathscr{H}. \tag{6.39}$$

Proof It follows from (6.28) that, for all $\Psi \in D(A(f)^*)$, $\langle A(f)^*\Psi, \Omega_{\mathscr{H}}\rangle = 0 = \langle \Psi, 0\rangle$. Hence $\Omega_{\mathscr{H}} \in D(A(f)^{**}) = D(A(f))$ and (6.39) holds. ∎

In fact, (6.39) characterizes the Fock vacuum:

Proposition 6.2 *Let $\mathscr{D}$ be a dense subspace of $\mathscr{H}$. Suppose that $\Psi \in \cap_{f \in \mathscr{D}} D(A(f))$ obeys*

$$A(f)\Psi = 0, \quad \forall f \in \mathscr{D}.$$

Then $\Psi = \alpha\Omega_{\mathscr{H}}$ for some $\alpha \in \mathbb{C}$.

Proof For all $\Phi \in \mathscr{F}_{\mathrm{b},0}(\mathscr{H})$,

$$0 = \langle \Phi, A(f)\Psi\rangle = \left\langle A(f)^*\Phi, \Psi\right\rangle = \sum_{n=1}^{\infty} \sqrt{n} \left\langle S_n(f \otimes \Phi^{(n-1)}), \Psi^{(n)}\right\rangle.$$

Let $n \in \mathbb{N}$ be fixed and take $\Phi \in \mathscr{F}_{\mathrm{b},0}(\mathscr{H})$ such that $\Phi^{(m)} = 0$ for $m \neq n-1$. Then

$$\left\langle S_n(f \otimes \Phi^{(n-1)}), \Psi^{(n)} \right\rangle = 0, \quad \forall \Phi^{(n-1)} \in \otimes_{\mathrm{s}}^{n-1} \mathscr{H}.$$

Since $\mathscr{D}$ is dense in $\mathscr{H}$, the subspace

$$\mathrm{span}\{S_n(f \otimes \Phi^{(n-1)}) | f \in \mathscr{D}, \ \Phi^{(n-1)} \in \otimes_{\mathrm{s}}^{n-1} \mathscr{H}\}$$

is dense in $\otimes_{\mathrm{s}}^{n} \mathscr{H}$. Hence $\Psi^{(n)} = 0, \ n \geq 1$. Therefore

$$\Psi = \{\Psi^{(0)}, \ 0, \ 0, \ldots\} = \alpha \Omega_{\mathscr{H}}$$

with $\alpha := \Psi^{(0)}$. ■

The following proposition may have some interest (see Remark 6.2 below):

Proposition 6.3 *If $A(f) = A(g)^*$ on $\mathscr{F}_{\mathrm{b},0}(\mathscr{H})$ ($f, g \in \mathscr{H}$), then $f = g = 0$.*

Proof By (6.39), we have $A(g)^* \Omega_{\mathscr{H}} = 0$. On the other hand, it follows from the definition of $A(\cdot)^*$ that $A(g)^* \Omega_{\mathscr{H}} = \{0, g, 0, 0, \ldots\}$. Hence $g = 0$. Then $A(f) = 0$ on $\mathscr{F}_{\mathrm{b},0}(\mathscr{H})$. Hence, for all $\Psi \in \mathscr{F}_{\mathrm{b},0}(\mathscr{H})$, $\langle \Omega_{\mathscr{H}}, A(f)\Psi \rangle = 0$. This implies that $A(f)^* \Omega_{\mathscr{H}} = 0$. Hence $f = 0$. ■

Remark 6.2 One can define operator-valued functionals $\tilde{A}$ and $\tilde{A}^*$ from $\mathscr{H} \setminus \{0\}$ to $\mathfrak{L}(\mathscr{F}_{\mathrm{b}}(\mathscr{H}))$ by

$$\tilde{A}(f) := A(f), \quad \tilde{A}^*(f) := A(f)^*, \quad f \in \mathscr{H} \setminus \{0\}.$$

Proposition 6.3 tells us that $\mathrm{Ran}\,(\tilde{A}) \cap \mathrm{Ran}\,(\tilde{A}^*) = \emptyset$.

For a class of vectors $\Psi \in D(A(f))$, $A(f)\Psi$ is explicitly written:

Lemma 6.4 *Let $n \geq 1$ be fixed. Then, for all $\Psi \in \mathscr{F}_{\mathrm{b}}(\mathscr{H})$ of the form $\Psi^{(m)} = \delta_{mn} S_n(\otimes^n f_j)$ with $f_j \in \mathscr{H}$ ($m \geq 0$),*

$$(A(f)\Psi)^{(m)} = 0, \quad m \neq n-1, \tag{6.40}$$

$$(A(f)\Psi)^{(n-1)} = \frac{1}{\sqrt{n}} \sum_{j=1}^{n} \langle f, f_j \rangle S_{n-1}(f_1 \otimes \cdots \otimes \hat{f_j} \otimes \cdots \otimes f_n), \tag{6.41}$$

where $\hat{f_j}$ indicates the omission of f_j.

Proof Since $\Psi^{(m)} = 0$ for all $m \neq n$, it follows that $(A(f)\Psi)^{(m)} = 0$ for all $m \neq n-1$. Hence we need only to prove (6.41). Let $\Phi \in \mathscr{F}_{\mathrm{b},0}(\mathscr{H})$ be such that $\Phi^{(m)} = 0$, $m \neq n-1$. Then

$$\begin{aligned}\left\langle \Phi^{(n-1)}, (A(f)\Psi)^{(n-1)} \right\rangle &= \langle \Phi, A(f)\Psi \rangle \\ &= \left\langle A(f)^*\Phi, \Psi \right\rangle = \sqrt{n}\left\langle S_n(f \otimes \Phi^{(n-1)}), \Psi^{(n)} \right\rangle \\ &= \sqrt{n}\left\langle f \otimes \Phi^{(n-1)}, S_n(\otimes_{j=1}^n f_j) \right\rangle \\ &= \frac{1}{\sqrt{n}} \frac{1}{(n-1)!} \sum_{\sigma \in \mathfrak{S}_n} \left\langle f, f_{\sigma(1)} \right\rangle \\ &\qquad \cdot \left\langle \Phi^{(n-1)}, f_{\sigma(2)} \otimes \cdots \otimes f_{\sigma(n)} \right\rangle\end{aligned}$$

One can rewrite $\sum_{\sigma \in \mathfrak{S}_n}$ as $\sum_{j=1}^n \sum_{\sigma \in \mathfrak{S}_n, \sigma(1)=j}$. Then we obtain

$$\left\langle \Phi^{(n-1)}, (A(f)\Psi)^{(n-1)} \right\rangle = \left\langle \Phi^{(n-1)}, \chi \right\rangle$$

with

$$\chi := \frac{1}{\sqrt{n}} \sum_{j=1}^n \left\langle f, f_j \right\rangle S_{n-1}(f_1 \otimes \cdots \hat{f_j} \otimes \cdots \otimes f_n).$$

Since $\Phi^{(n-1)}$ is an arbitrary element in $\otimes_{\mathrm{s}}^{n-1}\mathscr{H}$, we have $(A(f)\Psi)^{(n-1)} = \chi$. Thus we obtain (6.41). ■

Proposition 6.4 *For all $f \in \mathscr{H} \setminus \{0\}$, $A(f)$ and $A(f)^*$ are unbounded.*

Proof For each $N \in \mathbb{N}$, we define a vector $\Psi_N \in \mathscr{F}_{\mathrm{b}}(\mathscr{H})$ by $\Psi_N^{(N)} := \otimes^N f/\|f\|^N$ and $\Psi_N^{(n)} = 0$ for $n \neq N$. Then $\Psi_N \in D(A(f))$ and $\|\Psi_N\| = 1$. Moreover

$$(A(f)\Psi_N)^{(N-1)} = \frac{\sqrt{N}}{\|f\|^{N-2}} \otimes^{N-1} f$$

and $(A(f)\Psi_N)^{(n)} = 0$, $n \neq N-1$. Hence $\|A(f)\Psi_N\| = \sqrt{N}\|f\| \to \infty$ as $N \to \infty$. Thus $A(f)$ is unbounded. Hence $A(f)^*$ also is unbounded.[1] ■

Proposition 6.5 *Let $f \in \mathscr{H}$. Then $\mathscr{F}_{\mathrm{b},0}(\mathscr{H})$ is a core for $A(f)$ and $A(f)^*$.*

[1]In general, if a densely defined closable operator A on a Hilbert space is unbounded, then A^* is unbounded.

Proof Let $\Psi \in D(A(f))$. For $N \in \mathbb{N}$, we define $\Psi_N \in \mathscr{F}_{\mathrm{b},0}(\mathscr{H})$ by

$$\Psi_N^{(n)} = \begin{cases} \Psi^{(n)} & \text{for } 0 \leq n \leq N \\ 0 & \text{for } n \geq N+1 \end{cases}.$$

It is easy to see that $\lim_{N\to\infty} \Psi_N = \Psi$. It follows from the definition of $A(f)$ that $(A(f)\Psi_N)^{(n)} = 0$ for $n \geq N$ and $(A(f)\Psi_N)^{(n)} = (A(f)\Psi)^{(n)}$ for $0 \leq n \leq N-1$. Hence

$$\|A(f)\Psi_N - A(f)\Psi\|^2 = \sum_{n=N}^{\infty} \|(A(f)\Psi)^{(n)}\|^2 \to 0 \ (N \to \infty).$$

Hence $\lim_{N\to\infty} A(f)\Psi_N = A(f)\Psi$. Thus $\mathscr{F}_{\mathrm{b},0}(\mathscr{H})$ is a core for $A(f)$. Similarly, one can show that $\mathscr{F}_{\mathrm{b},0}(\mathscr{H})$ is a core for $A(f)^*$. ■

Using (6.29) repeatedly, one can see that, for all $n \in \mathbb{N}$, and $f_j \in \mathscr{H}$ $(j = 1, \cdots, n)$,

$$(A(f_1)^* \cdots A(f_n)^* \Omega_{\mathscr{H}})^{(m)} = \delta_{nm}\sqrt{n!}S_n(f_1 \otimes \cdots \otimes f_n), \quad m \geq 1. \tag{6.42}$$

This means that $\mathrm{span}\{A(f_1)^* \cdots A(f_n)^*\Omega_{\mathscr{H}} \mid f_j \in \mathscr{H}, j = 1, \ldots, n\}$ is dense in the n boson space $\otimes_{\mathrm{s}}^n \mathscr{H}$. Formula (6.42) also implies the following inequality:

$$\|A(f_1)^* \cdots A(f_n)^* \Omega_{\mathscr{H}}\| \leq \sqrt{n!}\|f_1\| \cdots \|f_n\|. \tag{6.43}$$

It follows from (6.42) that, for each subspace $\mathscr{D}$ of $\mathscr{H}$,

$$\begin{aligned} \mathscr{F}_{\mathrm{b,fin}}(\mathscr{D}) = &\mathrm{span}\big\{\Omega_{\mathscr{H}}, A(f_1)^* \cdots A(f_n)^*\Omega_{\mathscr{H}} \mid n \in \mathbb{N}, \\ &f_j \in \mathscr{D}, \ j = 1, \cdots, n\big\}. \end{aligned} \tag{6.44}$$

Hence, in particular, if $\mathscr{D}$ is dense in $\mathscr{H}$, then

$$\mathrm{span}\{\Omega_{\mathscr{H}}, A(f_1)^* \cdots A(f_n)^*\Omega_{\mathscr{H}} \mid n \in \mathbb{N}, \ f_j \in \mathscr{D}, \ j = 1, \cdots, n\}$$

is dense in $\mathscr{F}_{\mathrm{b}}(\mathscr{H})$. Thus we have the following result:

Theorem 6.6 *If $\mathscr{D}$ is dense in $\mathscr{H}$, $\Omega_{\mathscr{H}}$ is a cyclic vector of the operator set $\{A(f_1)^* \cdots A(f_n)^* | n \in \mathbb{N}, \ f_j \in \mathscr{D}, \ j = 1, \ldots, n\}$.*

This shows a basic role of the creation operators and the Fock vacuum. Moreover, (6.42) implies the following important inner product formula: for all $f_j, g_k \in \mathscr{H}, \ j = 1, \cdots, n, k = 1, \ldots, m$ $(n, m \geq 1)$,

$$\begin{aligned} &\big\langle A(f_1)^* \cdots A(f_n)^*\Omega_{\mathscr{H}}, A(g_1)^* \cdots A(g_m)^*\Omega_{\mathscr{H}}\big\rangle \\ &= \delta_{nm} \sum_{\sigma \in \mathfrak{S}_n} \big\langle f_1, g_{\sigma(1)}\big\rangle \cdots \big\langle f_n, g_{\sigma(n)}\big\rangle. \end{aligned} \tag{6.45}$$

The following corollary follows from (6.45), Theorems 1.21 and 6.6:

Corollary 6.1 *Let $\{e_j\}_{j=1}^{\infty}$ be a CONS of $\mathscr{H}$. Then*

$$\left\{\Omega_{\mathscr{H}}, \frac{C_{j_1\cdots j_n}}{\sqrt{n!}} A(e_{j_1})^* \cdots A(e_{j_n})^* \Omega_{\mathscr{H}} | n \in \mathbb{N}, j_1 \leq \cdots \leq j_n, j_k \in \mathbb{N}, k = 1, \ldots, n\right\}$$

is a CONS of $\mathscr{F}_{\mathrm{b}}(\mathscr{H})$, where $C_{j_1\cdots j_n}$ is defined by (1.55).

Formula (6.42) implies the following proposition:

Proposition 6.6 *Let T be a densely defined closable operator and $n \in \mathbb{N}$. Then, for all $f_1, \ldots, f_n \in D(T)$, $A(f_1)^* \cdots A(f_n)^* \Omega_{\mathscr{H}} \in D(\mathrm{d}\Gamma_{\mathrm{b}}(T)) \cap \Gamma_{\mathrm{b}}(T)$ and*

$$\mathrm{d}\Gamma_{\mathrm{b}}(T) A(f_1)^* \cdots A(f_n)^* \Omega_{\mathscr{H}} = \sum_{j=1}^{n} A(f_1)^* \cdots A(Tf_j)^* \cdots A(f_n)^* \Omega_{\mathscr{H}}, \tag{6.46}$$

$$\Gamma_{\mathrm{b}}(T) A(f_1)^* \cdots A(f_n)^* \Omega_{\mathscr{H}} = A(Tf_1)^* A(Tf_2)^* \cdots A(Tf_n)^* \Omega_{\mathscr{H}}. \tag{6.47}$$

Proof One needs only to recall the definitions of $\mathrm{d}\Gamma_{\mathrm{b}}(T)$ and $\Gamma_{\mathrm{b}}(T)$ and to apply (6.42). ■

In what follows, we denote by $A(\cdot)^{\#}$ either $A(\cdot)$ or $A(\cdot)^*$. A basic relation between $A(f)^{\#}$ and $\Gamma_{\mathrm{b}}(\cdot)$ is given as follows:

Theorem 6.7 *Let U be a unitary operator on $\mathscr{H}$. Then, for all $f \in \mathscr{H}$,*

$$\Gamma_{\mathrm{b}}(U) A(f) \Gamma_{\mathrm{b}}(U)^{-1} = A(Uf), \tag{6.48}$$

$$\Gamma_{\mathrm{b}}(U) A(f)^* \Gamma_{\mathrm{b}}(U)^{-1} = A(Uf)^*. \tag{6.49}$$

Proof Let $\Psi \in D(A(f)^*)$. Then

$$\begin{aligned}
(\Gamma_{\mathrm{b}}(U) A(f)^* \Psi)^{(n)} &= (\otimes^n U) \sqrt{n} S_n (f \otimes \Psi^{(n-1)}) \\
&= \sqrt{n} S_n (Uf \otimes (\otimes^{n-1} U) \Psi^{(n-1)}) \\
&= \sqrt{n} S_n (Uf \otimes (\Gamma_{\mathrm{b}}(U) \Psi)^{(n-1)}) \\
&= (A(Uf)^* \Gamma_{\mathrm{b}}(U) \Psi)^{(n)}.
\end{aligned}$$

Hence $\Psi \in D(A(Uf)^* \Gamma_{\mathrm{b}}(U))$ and $\Gamma_{\mathrm{b}}(U) A(f)^* \Psi = A(Uf)^* \Gamma_{\mathrm{b}}(U) \Psi$. This implies that $\Gamma_{\mathrm{b}}(U) A(f)^* \subset A(Uf)^* \Gamma_{\mathrm{b}}(U)$. Hence $\Gamma_{\mathrm{b}}(U) A(f)^* \Gamma(U)^{-1} \subset A(Uf)^*$. Since $f \in \mathscr{H}$ is arbitrary, one can replace f with $U^{-1} f$ to obtain

$\Gamma_{\mathrm{b}}(U)A(U^{-1}f)^*\Gamma(U)^{-1} \subset A(f)^*$. Note that U is an arbitrary unitary operator. Hence one can replace U^{-1} with U to obtain $\Gamma_{\mathrm{b}}(U)^{-1}A(Uf)^*\Gamma(U) \subset A(f)^*$, where we have used the fact that $\Gamma_{\mathrm{b}}(U^{-1}) = \Gamma_{\mathrm{b}}(U)^{-1}$. Therefore $A(Uf)^* \subset \Gamma_{\mathrm{b}}(U)A(f)^*\Gamma_{\mathrm{b}}(U)^{-1}$. Thus operator equality (6.49) follows. The adjoint of (6.49) yields (6.48). ■

Corollary 6.2 *Let T be a self-adjoint operator on $\mathscr{H}$. Then, for all $f \in \mathscr{H}$ and $t \in \mathbb{R}$,*

$$e^{it\mathrm{d}\Gamma_{\mathrm{b}}(T)}A(f)^{\#}e^{-it\mathrm{d}\Gamma_{\mathrm{b}}(T)} = A(e^{itT}f)^{\#}. \tag{6.50}$$

Proof One needs only to apply Theorem 6.7 with $U = e^{itT}$ and to use Theorem 6.4(i). ■

Taking $T = I$ and $t = \pi$ in (6.50) and using the fact that $e^{i\pi} = -1$, we obtain

$$e^{i\pi N_{\mathrm{b}}}A(f)^{\#}e^{-i\pi N_{\mathrm{b}}} = -A(f)^{\#}. \tag{6.51}$$

This shows that the correspondence $A(f)^{\#} \mapsto -A(f)^{\#}$ is unitarily implementable by $e^{i\pi N_{\mathrm{b}}}$.

6.5 Commutation Relations

In this section we derive commutation relations of $A(f)^{\#}$'s. For each $r \in \mathbb{Z}_+$, we introduce

$$\mathscr{F}_{\mathrm{b},r}(\mathscr{H}) := \{\Psi \in \mathscr{F}_{\mathrm{b}}(\mathscr{H}) \mid \Psi^{(n)} = 0 \text{ for all } n \geq r+1\}.$$

This is a closed subspace of $\mathscr{F}_{\mathrm{b}}(\mathscr{H})$, which is naturally isomorphic to $\oplus_{n=0}^{r} \otimes_{\mathrm{s}}^{n} \mathscr{H}$.

In the context of a Bose field theory, $\mathscr{F}_{\mathrm{b},r}(\mathscr{H})$ is a Hilbert space of state vectors for the subsystem consisting of at most r bosons.

Lemma 6.5 *Let $f \in \mathscr{H}$ and $r \geq 0$. We set $\mathscr{F}_{\mathrm{b},-1}(\mathscr{H}) = \{0\}$.*

(i) *$A(f)^*$ is bounded operator from $\mathscr{F}_{\mathrm{b},r}(\mathscr{H})$ to $\mathscr{F}_{\mathrm{b},r+1}(\mathscr{H})$ with*

$$\|A(f)^* \restriction \mathscr{F}_{\mathrm{b},r}(\mathscr{H})\| \leq \sqrt{r+1}\|f\|.$$

(ii) *$A(f)$ is bounded operator from $\mathscr{F}_{\mathrm{b},r}(\mathscr{H})$ to $\mathscr{F}_{\mathrm{b},r-1}(\mathscr{H})$ with*

$$\|A(f) \restriction \mathscr{F}_{\mathrm{b},r}(\mathscr{H})\| \leq \sqrt{r}\|f\|.$$

(iii) *Let* $f_n, f \in \mathscr{H}$ *and* $\lim_{n\to\infty} f_n = f$. *Then, for all* $\Psi \in \mathscr{F}_{\mathrm{b},0}(\mathscr{H})$,

$$\lim_{n\to\infty} A(f_n)\Psi = A(f)\Psi, \quad \lim_{n\to\infty} A(f_n)^*\Psi = A(f)^*\Psi.$$

Proof Let $\Psi \in \mathscr{F}_{\mathrm{b},r}(\mathscr{H})$. Then it is easy to see that $A(f)\Psi \in \mathscr{F}_{\mathrm{b},r-1}(\mathscr{H})$ and $A(f)^*\mathscr{F}_{\mathrm{b},r+1}(\mathscr{H})$.

(i) We have

$$\|A(f)^*\Psi\|^2 = \sum_{n=1}^{r+1} \|\sqrt{n}S_n(f \otimes \Psi^{(n-1)})\|^2$$
$$\leq (r+1)\|f\|^2 \sum_{n=1}^{r+1} \|\Psi^{(n-1)}\|^2 = (r+1)\|f\|^2\|\Psi\|^2.$$

Hence the desired result follows.

(ii) By a general fact in Hilbert space theory,[2] we have for all $\Psi \in \mathscr{F}_{\mathrm{b},r}(\mathscr{H})$

$$\|A(f)\Psi\| = \sup_{\Phi\in\mathscr{F}_{\mathrm{b},r-1}(\mathscr{H}),\ \|\Phi\|=1} |\langle \Phi, A(f)\Psi\rangle|.$$

We have

$$|\langle \Phi, A(f)\Psi\rangle| = |\langle A(f)^*\Phi, \Psi\rangle| \leq \|A(f)^*\Phi\|\,\|\Psi\|.$$

By (i), $\|A(f)^*\Phi\| \leq \sqrt{r}\|f\|\|\Phi\|$. Hence $\|A(f)\Psi\| \leq \sqrt{r}\|f\|\,\|\Psi\|$. Thus the desired result follows.

(iii) Let $\Psi \in \mathscr{F}_{\mathrm{b},0}(\mathscr{H})$. Then there exists an $r \geq 0$ such that $\Psi \in \mathscr{F}_{\mathrm{b},r}(\mathscr{H})$. By (i), $\|A(f_n)^*\Psi - A(f)^*\Psi\| \leq \sqrt{r+1}\|f_n - f\|\,\|\Psi\|$. Hence $\lim_{n\to\infty} A(f_n)^*\Psi = A(f)^*\Psi$. Similarly, using (ii), we can prove $\lim_{n\to\infty} A(f_n)\Psi = A(f)\Psi$. ■

As an application of Lemma 6.5, one can prove the following fact:

Proposition 6.7 *Let* $\mathscr{D}$ *be a dense subspace of* $\mathscr{H}$ *and* $f \in \mathscr{H}$. *Then* $\mathscr{F}_{\mathrm{b,fin}}(\mathscr{D})$ *is a core for* $A(f)$ *and* $A(f)^*$.

Proof Let $\Psi \in D(A(f))$. By Proposition 6.5, for any $\varepsilon > 0$, there exists a vector $\Psi_\varepsilon \in \mathscr{F}_{\mathrm{b},0}(\mathscr{H})$ such that $\|\Psi_\varepsilon - \Psi\| < \varepsilon$ and $\|A(f)\Psi_\varepsilon - A(f)\Psi\| < \varepsilon$. There exists an $r \in \mathbb{N}$ such that $\Psi_\varepsilon \in \mathscr{F}_{\mathrm{b},r}(\mathscr{H})$. Since $\mathscr{D}$ is dense in $\mathscr{H}$, there exists a vector

[2] Let $\mathscr{K}$ be a Hilbert space and $\mathscr{D}$ be a dense subspace of $\mathscr{K}$. Then, for all $\Psi \in \mathscr{H}$, $\|\Psi\| = \sup_{\Phi\in\mathscr{D},\|\Phi\|=1} |\langle \Phi, \Psi\rangle|$.

$\Phi_\varepsilon \in \mathscr{F}_{\mathrm{b,fin}}(\mathscr{D})$ such that $\Phi_\varepsilon \in \mathscr{F}_{\mathrm{b},r}(\mathscr{H})$ and $\|\Phi_\varepsilon - \Psi_\varepsilon\| < \varepsilon$. By Lemma 6.5, we have

$$\|A(f)\Phi_\varepsilon - A(f)\Psi_\varepsilon\| \le \sqrt{r+1}\|f\|\,\|\Phi_\varepsilon - \Psi_\varepsilon\| \le \sqrt{r+1}\|f\|\varepsilon.$$

Hence, by the triangle inequality, we obtain

$$\|\Phi_\varepsilon - \Psi\| < 2\varepsilon, \quad \|A(f)\Phi_\varepsilon - A(f)\Psi\| < (\sqrt{r+1}\|f\| + 1)\varepsilon.$$

This implies that $\mathscr{F}_{\mathrm{b,fin}}(\mathscr{D})$ is a core for $A(f)$. Similarly, one can prove that $\mathscr{F}_{\mathrm{b,fin}}(\mathscr{D})$ is a core for $A(f)^*$. ■

The operators $A(f)$ and $A(f)^*$ ($f \in \mathscr{H}$) satisfy characteristic commutation relations:

Theorem 6.8 *For all $f, g \in \mathscr{H}$,*

$$[A(f), A(g)^*] = \langle f, g\rangle, \tag{6.52}$$

$$[A(f), A(g)] = 0, \quad [A(f)^*, A(g)^*] = 0. \tag{6.53}$$

on $\mathscr{F}_{\mathrm{b},0}(\mathscr{H})$.

Proof For $n \ge 1$ and $f_j \in \mathscr{H}$ $(j = 1, \ldots, n)$, we define a vector $\Psi_n \in \mathscr{F}_{\mathrm{b},0}(\mathscr{H})$ by

$$\Psi_n^{(m)} := \begin{cases} S_n(f_1 \otimes \cdots \otimes f_n)\ ; & m = n \\ 0 & ; m \ne n \end{cases}.$$

Then

$$(A(f)\Psi_n)^{(m)} = \begin{cases} \dfrac{1}{\sqrt{n}}\displaystyle\sum_{j=1}^n \langle f, f_j\rangle S_{n-1}(f_1 \otimes \cdots \otimes \hat{f_j} \otimes \cdots \otimes f_n)\ ; & m = n-1 \\ 0 & ; m \ne n-1 \end{cases}.$$

By (6.29),

$$(A(g)^*\Psi_n)^{(m)} = \begin{cases} \sqrt{n+1}S_{n+1}(g \otimes f_1 \otimes \cdots \otimes f_n)\ ; & m = n+1 \\ 0 & ; m \ne n+1 \end{cases}.$$

Hence

$$(A(f)A(g)^*\Psi_n)^{(n)} = \langle f, g\rangle\, S_n(f_1 \otimes \cdots \otimes f_n) + \sum_{j=1}^{n} \langle f, f_j\rangle\, S_n(g \otimes f_1 \otimes \cdots \otimes \hat{f}_j \otimes \cdots \otimes f_n),$$

$$(A(g)^*A(f)\Psi_n)^{(n)} = \sum_{j=1}^{n} \langle f, f_j\rangle\, S_n(g \otimes f_1 \otimes \cdots \otimes \hat{f}_j \otimes \cdots \otimes f_n).$$

Therefore

$$([A(f), A(g)^*]\Psi_n)^{(n)} = \langle f, g\rangle\, \Psi_n^{(n)}.$$

If $m \neq n$, then $(A(f)A(g)^*\Psi_n)^{(m)} = 0,\ (A(g)^*A(f)\Psi_n)^{(m)} = 0$. Thus

$$[A(f), A(g)^*]\Psi_n = \langle f, g\rangle\, \Psi_n.$$

It is easy to see that

$$[A(f), A(g)^*]\Omega_{\mathscr{H}} = \langle f, g\rangle\, \Omega_{\mathscr{H}}.$$

Since any vector in $\mathscr{F}_{\mathrm{b,fin}}(\mathscr{H})$ is written as a finite linear combination of $\Omega_{\mathscr{H}}$ and vectors of the form Ψ_n ($n \geq 1$, $f_1, \ldots, f_n \in \mathscr{H}$), it follows that, for all $\Psi \in \mathscr{F}_{\mathrm{b,fin}}(\mathscr{H})$,

$$[A(f), A(g)^*]\Psi = \langle f, g\rangle\, \Psi. \tag{6.54}$$

We next consider the case where $\Psi \in \mathscr{F}_{\mathrm{b},0}(\mathscr{H})$. In this case, there exists an $r \in \mathbb{N}$ such that $\Psi \in \mathscr{F}_{\mathrm{b},r}(\mathscr{H})$. Then there exists a sequence $\{\Psi_N\}_{N=1}^{\infty}$ in $\mathscr{F}_{\mathrm{b,fin}}(\mathscr{H})$ such that, for all $N \geq 1$, $\Psi_N \in \mathscr{F}_{\mathrm{b},r}(\mathscr{H})$ and $\Psi_N \to \Psi$ ($N \to \infty$). By (6.54),

$$A(f)A(g)^*\Psi_N - A(g)^*A(f)\Psi_N = \langle f, g\rangle\, \Psi_N.$$

By Lemma 6.5,

$$A(f)A(g)^*\Psi_N \to A(f)A(g)^*\Psi, \quad A(g)^*A(f)\Psi_N \to A(g)^*A(f)\Psi \quad (N \to \infty).$$

Hence, for all $\Psi \in \mathscr{F}_{\mathrm{b},0}(\mathscr{H})$, (6.54) holds. Namely, (6.52) holds on $\mathscr{F}_{\mathrm{b},0}(\mathscr{H})$. Similarly, one can prove (6.53). ■

Example 6.3 Consider the case where $\mathscr{H}$ is finite-dimensional with $N = \dim \mathscr{H}$ and let $\{e_n\}_{n=1}^{N}$ be a CONS of $\mathscr{H}$. Let

$$A_n := A(e_n), \quad n = 1, \ldots, N.$$

Then it follows that $(\mathscr{F}_{\rm b}(\mathscr{H}), \mathscr{F}_{\rm b,0}(\mathscr{H}), \{A_n, A_n^*|n = 1, \ldots, N\})$ is a representation of the Dirac CCR with N degrees of freedom. Hence, in the case where $\mathscr{H}$ is infinite dimensional, $(\mathscr{F}_{\rm b}(\mathscr{H}), \mathscr{F}_{\rm b,0}(\mathscr{H}), \{A(f), A(f)^*|f \in \mathscr{H}\})$ gives an infinite-dimensional version of representation of the Dirac CCR with finite degrees of freedom.

In view of Example 6.3, we call commutation relations (6.52) and (6.53) the **Dirac canonical commutation relations** (CCR) over $\mathscr{H}$. But, in what follows, we sometimes call them the CCR over $\mathscr{H}$ simply if there is no danger of confusion. We shall discuss in Chap. 8 representation-theoretic aspects of CCR of this type in more details.

With regard to the domain of $A(f)^{\#}$, we have in fact the following:

Corollary 6.3 *For all* $f \in \mathscr{H}$, $D(A(f)) = D(A(f)^*)$ *and, for all* $\Psi \in D(A(f)) (= D(A(f)^*))$,

$$\|A(f)^*\Psi\|^2 = \|A(f)\Psi\|^2 + \|f\|^2\|\Psi\|^2. \tag{6.55}$$

Proof Using (6.52) with $g = f$, we have

$$\|A(f)^*\Psi\|^2 = \|A(f)\Psi\|^2 + \|f\|^2\|\Psi\|^2, \quad \Psi \in \mathscr{F}_{\rm b,0}(\mathscr{H}). \tag{6.56}$$

Let $\Psi \in D(A(f))$. Then there exists a sequence $\{\Psi_n\}_n \subset \mathscr{F}_{\rm b,0}(\mathscr{H})$ such that $\Psi_n \to \Psi$ and $A(f)\Psi_n \to A(f)\Psi$ $(n \to \infty)$. Replacing Ψ by $\Psi_n - \Psi_m$ in (6.56), we see that $\{A(f)^*\Psi_n\}_n$ is a Cauchy sequence. Since $A(f)^*$ is closed, we conclude that $\Psi \in D(A(f)^*)$ and $A(f)^*\Psi = \lim_{n\to\infty} A(f)^*\Psi_n$. Hence $D(A(f)) \subset D(A(f)^*)$ and (6.55) holds. Similarly, one can show that $D(A(f)^*) \subset D(A(f))$. ∎

Theorem 6.8 implies a useful formula:

Corollary 6.4 *Let* $n \in \mathbb{N}$ *and* $f, f_1, \ldots, f_n \in \mathscr{H}$. *Then*

$$\begin{aligned} &A(f)A(f_1)^* \cdots A(f_n)^*\Omega_{\mathscr{H}} \\ &= \sum_{j=1}^n \langle f, f_j\rangle A(f_1)^* \cdots \widehat{A(f_j)^*} \cdots A(f_n)^*\Omega_{\mathscr{H}}, \end{aligned} \tag{6.57}$$

where $\widehat{A(f_j)^*}$ *indicates the omission of* $A(f_j)^*$.

Proof By (6.52), we have

$$\begin{aligned} &A(f)A(f_1)^* \cdots A(f_n)^*\Omega_{\mathscr{H}} \\ &= ([A(f), A(f_1)^*] + A(f_1)^*A(f))A(f_2)^* \cdots A(f_n)^*\Omega_{\mathscr{H}} \\ &= \langle f, f_1\rangle A(f_2)^* \cdots A(f_n)^*\Omega_{\mathscr{H}} + A(f_1)^*A(f)A(f_2)^* \cdots A(f_n)^*\Omega_{\mathscr{H}}. \end{aligned}$$

Iterating this equation and using (6.39), we obtain (6.57). ∎

Corollary 6.5 *Let $\mathscr{D}$ be a subspace of $\mathscr{H}$.*

(i) *For all $f \in \mathscr{D}$, $A(f)^* \mathscr{F}_{\mathrm{b,fin}}(\mathscr{D}) \subset \mathscr{F}_{\mathrm{b,fin}}(\mathscr{D})$.*
(ii) *For all $f \in \mathscr{H}$, $A(f) \mathscr{F}_{\mathrm{b,fin}}(\mathscr{D}) \subset \mathscr{F}_{\mathrm{b,fin}}(\mathscr{D})$.*

Proof

(i) Since $f \in \mathscr{D}$, it follows from (6.44) that $A(f)^* \mathscr{F}_{\mathrm{b,fin}}(\mathscr{D}) \subset \mathscr{F}_{\mathrm{b,fin}}(\mathscr{D})$.
(ii) This follows from (6.57). ■

6.6 Irreducibility

In this section, we prove the following theorem:

Theorem 6.9 *Let $\mathscr{D}$ be a dense subspace of $\mathscr{H}$. Then*

$$\{A(f), A(f)^* \mid f \in \mathscr{D}\}' = \mathbb{C}I. \tag{6.58}$$

In particular, $\{A(f), A(f)^* \mid f \in \mathscr{D}\}$ is irreducible.

Proof Let $T \in \{A(f), A(f)^* \mid f \in \mathscr{D}\}'$. Then $T \in \mathfrak{B}(\mathscr{F}_{\mathrm{b}}(\mathscr{H}))$ and

$$TA(f)^\# \subset A(f)^\# T, \quad f \in \mathscr{D}.$$

In particular, $T\Omega_{\mathscr{H}} \in D(A(f))$ and $A(f)T\Omega_{\mathscr{H}} = TA(f)\Omega_{\mathscr{H}} = 0$ by (6.39). Hence, by Proposition 6.2, there exists a constant $\alpha \in \mathbb{C}$ such that $T\Omega_{\mathscr{H}} = \alpha\Omega_{\mathscr{H}}$. Let

$$\Psi := A(f_1)^* \cdots A(f_n)^* \Omega_{\mathscr{H}}, \quad f_j \in \mathscr{D}\,(j = 1, \ldots, n).$$

Then, using the property that $TA(f)^* \subset A(f)^* T$, we have $T\Psi = \alpha\Psi$. Hence $T = \alpha I$ on $\mathscr{F}_{\mathrm{b,fin}}(\mathscr{D})$. Since $\mathscr{F}_{\mathrm{b,fin}}(\mathscr{D})$ is dense in $\mathscr{F}_{\mathrm{b}}(\mathscr{H})$ and T is bounded, it follows that $T = \alpha I$. Hence (6.58) holds. The irreducibility of $\{A(f), A(f)^* \mid f \in \mathscr{D}\}$ follows from Proposition 2.9(i). ■

6.7 Relative Boundedness of Creation and Annihilation Operators

Let T be a non-negative self-adjoint operator on $\mathscr{H}$ and suppose that T is injective. Then, by Theorem 6.1(ii), $\mathrm{d}\Gamma_{\mathrm{b}}(T)$ is a non-negative self-adjoint operator. Hence one can define the square root $\mathrm{d}\Gamma_{\mathrm{b}}(T)^{1/2}$ of $\mathrm{d}\Gamma_{\mathrm{b}}(T)$ via functional calculus. The purpose of this section is to prove that $A(f)^\#$ with $f \in D(T^{-1/2})$ is relatively bounded with respect to $\mathrm{d}\Gamma_{\mathrm{b}}(T)^{1/2}$. But, to prove this fact, we need some preliminaries.

For each $f \in \mathscr{H}$ and $n \geq 0,\ j = 1, \ldots, n+1$, there exists a unique linear operator $K_j^{(n)}(f)$ from $\otimes^n \mathscr{H}$ to $\otimes^{n+1} \mathscr{H}$ satisfying

$$D(K_j^{(n)}(f)) = \widehat{\otimes}^n \mathscr{H}, \tag{6.59}$$

$$K_1^{(0)}(f)z = zf, \quad z \in \otimes^0 \mathscr{H} = \mathbb{C}, \tag{6.60}$$

$$K_j^{(n)}(f)\left(\otimes_{k=1}^n \Psi_k\right) = \left(\otimes_{k=1}^{j-1} \Psi_k\right) \otimes f \otimes \left(\otimes_{k=j}^n \Psi_k\right), \quad \Psi_k \in \mathscr{H},\ n \geq 1,\ k = 1, \ldots, n, \tag{6.61}$$

where $K_1^{(n)}(f) = f \otimes \left(\otimes_{k=1}^n \Psi_k\right)$ and $K_{n+1}^{(n)} = \left(\otimes_{k=1}^n \Psi_k\right) \otimes f$. Indeed, one first defines the operation of $K_j^{(n)}(f)$ to vectors of the form $\otimes_{k=1}^n \Psi_k$ ($\Psi_k \in \mathscr{H},\ k = 1, \ldots, n$) by (6.61) and then extends it to all vectors $\Psi \in \widehat{\otimes}^n \mathscr{H}$ by linearity. This is well-defined, i.e., the extension $K_j^{(n)}(f)\Psi$ is independent of the way of representing Ψ as a linear combination of tensors of the form $\otimes_{k=1}^n \Psi_k,\ \Psi_k \in \mathscr{H},\ k = 1, \ldots, n$.

Lemma 6.6 *Let $f \in \mathscr{H}$ and $n \geq 0,\ j = 1, \ldots, n+1$. Then, for all $\Psi \in \widehat{\otimes}^n \mathscr{H}$,*

$$\|K_j^{(n)}(f)\Psi\| = \|f\|\|\Psi\|. \tag{6.62}$$

In particular, $K_j^{(n)}(f)$ is bounded with $\|K_j^{(n)}(f)\| = \|f\|$.

Proof Each vector Ψ in $\widehat{\otimes}^n \mathscr{H}$ is represented as follows:

$$\Psi = \sum_{m=1}^M \otimes_{k=1}^n \Psi_k^{(m)}$$

with $M \in \mathbb{N},\ \Psi_k^{(m)} \in \mathscr{H},\ k = 1, \ldots, n,\ m = 1, \ldots, M$. Let

$$D_k := \mathrm{span}\{\Psi_k^{(m)} \mid m = 1, \ldots, M\}.$$

Then $d_k := \dim D_k \leq M$. Hence we can take a CONS $\{e_i^{(k)} \mid i = 1, \ldots, d_k\}$ of D_k. Hence we have $\Psi_k^{(m)} = \sum_{i=1}^{d_k} c_i^{(km)} e_i^{(k)}$ with $c_i^{(km)} := \left\langle e_i^{(k)}, \Psi_k^{(m)} \right\rangle$. Therefore

$$\Psi = \sum_{i_1=1}^{d_1} \cdots \sum_{i_n=1}^{d_n} c_{i_1 \cdots i_n} e_{i_1}^{(1)} \otimes \cdots \otimes e_{i_n}^{(n)},$$

where $c_{i_1 \cdots i_n} := \sum_{m=1}^M \left(\prod_{k=1}^n c_{i_k}^{(km)}\right)$. Hence

$$K_j^{(n)}(f)\Psi = \sum_{i_1=1}^{d_1} \cdots \sum_{i_n=1}^{d_n} c_{i_1 \cdots i_n} e_{i_1}^{(1)} \otimes \ldots \otimes e_{i_{j-1}}^{(j-1)} \otimes f \otimes e_{i_j}^{(j)} \otimes \cdots \otimes e_{i_n}^{(n)}.$$

By direct computation, we have

$$\|K_j^{(n)}(f)\Psi\|^2 = \|f\|^2 \sum_{i_1=1}^{d_1} \cdots \sum_{i_n=1}^{d_n} |c_{i_1\cdots i_n}|^2 = \|f\|^2\|\Psi\|^2.$$

Thus (6.62) holds and hence $K_j^{(n)}(f)$ is bounded with $\|K_j^{(n)}(f)\| = \|f\|$. ■

Lemma 6.7 *Let $f \in \mathscr{H}$ and $n \in \mathbb{Z}_+$. Then*

$$S_{n+1}(f \otimes \Psi) = \frac{1}{n+1}\sum_{j=1}^{n+1} K_j^{(n)}(f)\Psi, \quad \Psi \in S_n(\widehat{\otimes}^n \mathscr{H}). \tag{6.63}$$

Proof For each vector $\eta = S_n(\Psi_1 \otimes \cdots \otimes \Psi_n)$ $(\Psi_j \in \mathscr{H},\ j = 1, \ldots, n)$,

$$\begin{aligned}
S_{n+1}(f \otimes \eta) &= S_{n+1}(f \otimes \Psi_1 \otimes \cdots \otimes \Psi_n) \\
&= \frac{1}{(n+1)!}\sum_{\sigma \in \mathfrak{S}_n} (f \otimes \Psi_{\sigma(1)} \otimes \cdots \otimes \Psi_{\sigma(n)} \\
&\quad + \Psi_{\sigma(1)} \otimes f \otimes \Psi_{\sigma(2)} \otimes \cdots \otimes \Psi_{\sigma(n)} \\
&\quad \cdots + \Psi_{\sigma(1)} \otimes \cdots \otimes \Psi_{\sigma(n)} \otimes f) \\
&= \frac{1}{n+1}\sum_{j=1}^{n+1} K_j^{(n)}(f)\eta.
\end{aligned}$$

Hence, by linearity, (6.63) follows. ■

We are now ready to prove the $\mathrm{d}\Gamma_{\mathrm{b}}(T)^{1/2}$-boundedness of $A(f)^{\#}$ with $f \in D(T^{-1/2})$, which plays a basic role in analysis on boson Fock spaces.

Theorem 6.10 *Let T be a non-negative and injective self-adjoint operator on $\mathscr{H}$. Then, for all $f \in D(T^{-1/2})$,*

$$D(\mathrm{d}\Gamma_{\mathrm{b}}(T)^{1/2}) \subset D(A(f)) \cap D(A(f)^*)$$

and, for all $\Psi \in D(\mathrm{d}\Gamma_{\mathrm{b}}(T)^{1/2})$,

$$\|A(f)\Psi\| \le \|T^{-1/2}f\|\,\|\mathrm{d}\Gamma_{\mathrm{b}}(T)^{1/2}\Psi\|, \tag{6.64}$$

$$\|A(f)^*\Psi\|^2 \le \|T^{-1/2}f\|^2\,\|\mathrm{d}\Gamma_{\mathrm{b}}(T)^{1/2}\Psi\|^2 + \|f\|^2\|\Psi\|^2. \tag{6.65}$$

Proof For each $n \in \mathbb{N},\ j = 1, \ldots, n$, we define a linear operator $T_{n,j}^{(1/2)}$ on $\otimes^n \mathscr{H}$ as follows:

$$T_{n,j}^{(1/2)} := I \otimes \cdots \otimes I \otimes \underbrace{T^{1/2}}_{j\text{th}} \otimes I \otimes \cdots \otimes I.$$

Then, for all $\Psi^{(n)} \in S_n(\widehat{\otimes}^n \mathscr{H})$,

$$K_j^{(n)}(f)\Psi^{(n)} = T_{n+1,j}^{(1/2)} K_j^{(n)}(T^{-1/2} f)\Psi^{(n)}. \tag{6.66}$$

Let $\Phi \in \mathscr{F}_{\mathrm{b,fin}}(D(T))$. Then, by Lemma 6.7 and (6.66), we have

$$\begin{aligned}
\left\langle (A(f)\Phi)^{(n)}, \Psi^{(n)} \right\rangle &= \left\langle \Phi^{(n+1)}, \sqrt{n+1} S_n(f \otimes \Psi^{(n)}) \right\rangle \\
&= \frac{1}{\sqrt{n+1}} \sum_{j=1}^{n+1} \left\langle \Phi^{(n+1)}, K_n^{(j)}(f)\Psi^{(n)} \right\rangle \\
&= \frac{1}{\sqrt{n+1}} \sum_{j=1}^{n+1} \left\langle T_{n+1,j}^{(1/2)} \Phi^{(n+1)}, K_j^{(n)}(T^{-1/2} f)\Psi^{(n)} \right\rangle.
\end{aligned}$$

Hence, by the Schwarz inequality and (6.62),

$$\begin{aligned}
&\left| \left\langle (A(f)\Phi)^{(n)}, \Psi^{(n)} \right\rangle \right| \\
&\leq \frac{1}{\sqrt{n+1}} \sum_{j=1}^{n+1} \| T_{n+1,j}^{(1/2)} \Phi^{(n+1)} \| \, \| K_j^{(n)}(T^{-1/2} f)\Psi^{(n)} \| \\
&\leq \frac{1}{\sqrt{n+1}} \left(\sum_{j=1}^{n+1} \| T_{n+1,j}^{(1/2)} \Phi^{(n+1)} \|^2 \right)^{1/2} \left(\sum_{j=1}^{n+1} \| K_j^{(n)}(T^{-1/2} f)\Psi^{(n)} \|^2 \right)^{1/2} \\
&= \| T^{-1/2} f \| \left(\sum_{j=1}^{n+1} \| T_{n+1,j}^{(1/2)} \Phi^{(n+1)} \|^2 \right)^{1/2} \| \Psi^{(n)} \|.
\end{aligned}$$

On the other hand,

$$\sum_{j=1}^{n+1} \| T_{n+1,j}^{(1/2)} \Phi^{(n+1)} \|^2 = \left\langle \Phi^{(n+1)}, T_{\mathrm{b}}^{(n+1)} \Phi^{(n+1)} \right\rangle.$$

Hence

$$\left|\left\langle (A(f)\Phi)^{(n)}, \Psi^{(n)}\right\rangle\right| \le \|T^{-1/2}f\| \left\langle \Phi^{(n+1)}, T_{\mathrm{b}}^{(n+1)}\Phi^{(n+1)}\right\rangle^{1/2} \|\Psi^{(n)}\|.$$

Since $S_n(\widehat{\otimes}^n \mathscr{H})$ is dense in $\otimes_{\mathrm{s}}^n \mathscr{H}$, it follows[3] that

$$\|(A(f)\Phi)^{(n)}\| \le \|T^{-1/2}f\| \left\langle \Phi^{(n+1)}, T_{\mathrm{b}}^{(n+1)}\Phi^{(n+1)}\right\rangle^{1/2}.$$

Hence

$$\|A(f)\Phi\|^2 \le \|T^{-1/2}f\|^2 \langle \Phi, \mathrm{d}\Gamma_{\mathrm{b}}(T)\Phi\rangle = \|T^{-1/2}f\|^2 \|\mathrm{d}\Gamma_{\mathrm{b}}(T)^{1/2}\Phi\|^2,$$

i.e.,

$$\|A(f)\Phi\| \le \|T^{-1/2}f\| \, \|\mathrm{d}\Gamma_{\mathrm{b}}(T)^{1/2}\Phi\|. \tag{6.67}$$

Since $\mathscr{F}_{\mathrm{b,fin}}(D(T))$ is a core for $\mathrm{d}\Gamma_{\mathrm{b}}(T)$ by Theorem 6.1(iii), it is a core for $\mathrm{d}\Gamma_{\mathrm{b}}(T)^{1/2}$ (apply Proposition 1.13). Hence, by a limiting argument, (6.67) extends to all $\Phi \in D(\mathrm{d}\Gamma_{\mathrm{b}}(T)^{1/2})$ with $D(\mathrm{d}\Gamma_{\mathrm{b}}(T)^{1/2}) \subset D(A(f))$. Thus (6.64) holds. Inequality (6.65) follows from (6.64) and Corollary 6.3. ■

Remark 6.3 It should be noted that, in Theorem 6.10, separability of $\mathscr{H}$ is not assumed. In the case where $\mathscr{H}$ is separable, there is another (easier) proof of Theorem 6.10 using the multiplication operator form of spectral theorem (see [28, Remark 5.8]).

Taking $T = I$ in Theorem 6.10, we obtain an estimate showing the relative boundedness of $A(\cdot)^{\#}$ with respect to $N_{\mathrm{b}}^{1/2}$:

Corollary 6.6 *For all $f \in \mathscr{H}$, $D(N_{\mathrm{b}}^{1/2}) \subset D(A(f)) \cap D(A(f)^*)$ and*

$$\|A(f)\Psi\| \le \|f\| \, \|N_{\mathrm{b}}^{1/2}\Psi\|, \tag{6.68}$$

$$\|A(f)^*\Psi\| \le \|f\| \, \|(N_{\mathrm{b}}+1)^{1/2}\Psi\|, \quad \Psi \in D(N_{\mathrm{b}}^{1/2}). \tag{6.69}$$

Proof Estimate (6.68) is a direct consequence of (6.64). As for (6.69), we need only to note that $\|N_{\mathrm{b}}^{1/2}\Psi\|^2 + \|\Psi\|^2 = \|(N_{\mathrm{b}}+1)\Psi\|^2, \quad \Psi \in D(N_{\mathrm{b}}^{1/2})$. ■

Corollary 6.7 *Let T be a non-negative and injective self-adjoint operator on $\mathscr{H}$.*

[3] For any vector Φ in a Hilbert space $\mathscr{X}$, $\|\Phi\| = \sup_{\Psi \in \mathscr{D}, \Psi \ne 0} |\langle \Phi, \Psi\rangle|/\|\Psi\|$, where $\mathscr{D}$ is a dense subspace of $\mathscr{X}$.

(i) *Let* $f, f_n \in D(T^{-1/2})$ *and* $\lim_{n\to\infty} T^{-1/2} f_n = T^{-1/2} f$. *Then, for all* $\Psi \in D(\mathrm{d}\Gamma_{\mathrm{b}}(T)^{1/2})$,

$$\lim_{n\to\infty} A(f_n)\Psi = A(f)\Psi. \tag{6.70}$$

(ii) *Let* $f, f_n \in D(T^{-1/2})$ *and* $\lim_{n\to\infty} f_n = f$, $\lim_{n\to\infty} T^{-1/2} f_n = T^{-1/2} f$. *Then, for all* $\Psi \in D(\mathrm{d}\Gamma_{\mathrm{b}}(T)^{1/2})$,

$$\lim_{n\to\infty} A(f_n)^*\Psi = A(f)^*\Psi. \tag{6.71}$$

(iii) *Let* $f \in D(T^p) \cap D(T^{-1/2})$ *with* $p \in \mathbb{N}$ *and* $\Psi \in D(\mathrm{d}\Gamma_{\mathrm{b}}(T)^{1/2})$. *Then the mapping:* $\mathbb{R} \ni t \mapsto A(e^{itT} f)^{\#}\Psi$ *is* p *times strongly differentiable and the strong derivative of the* p*-th order is given by*

$$\frac{d^p}{dt^p} A(e^{itT} f)^{\#}\Psi = A(i^p e^{itT} T^p f)^{\#}\Psi. \tag{6.72}$$

Proof

(i) By (6.64), we have

$$\|A(f_n)\Psi - A(f)\Psi\| \le \|T^{-1/2} f_n - T^{-1/2} f\| \, \|\mathrm{d}\Gamma_{\mathrm{b}}(T)^{1/2}\Psi\| \to 0 \ (n \to \infty).$$

Thus (6.70) holds. (ii) By (6.65), we have

$$\begin{aligned}\|A(f_n)^*\Psi - A(f)^*\Psi\| &\le \|T^{-1/2} f_n - T^{-1/2} f\| \, \|\mathrm{d}\Gamma_{\mathrm{b}}(T)^{1/2}\Psi\| \\ &\quad + \|f_n - f\| \, \|\Psi\| \to 0 \ (n \to \infty).\end{aligned}$$

Thus (6.71) holds.

(iii) Let $\Psi \in D(\mathrm{d}\Gamma_{\mathrm{b}}(T)^{1/2})$. We first consider the case $p = 1$. Hence $f \in D(T) \cap D(T^{-1/2})$. For $\varepsilon \in \mathbb{R} \setminus \{0\}$, we define

$$B_\varepsilon := \frac{1}{\varepsilon}(A(e^{i(t+\varepsilon)T} f)^{\#}\Psi - A(e^{itT} f)^{\#}\Psi).$$

Then $B_\varepsilon = A(g_\varepsilon)^{\#}\Psi$ with $g_\varepsilon := \varepsilon^{-1}(e^{i\varepsilon T} - 1)e^{itT} f$. Since $e^{itT} f \in D(T)$, it follows that $\lim_{\varepsilon\to 0} g_\varepsilon = iTe^{itT} f = ie^{itT} Tf$. By the functional calculus for T, $g_\varepsilon \in D(T^{-1/2})$ with $T^{-1/2} g_\varepsilon = \varepsilon^{-1}(e^{i\varepsilon T} - 1)e^{itT} T^{-1/2} f$. It is obvious that $T^{-1/2} f \in D(T)$. Hence $\lim_{\varepsilon\to 0} T^{-1/2} g_\varepsilon = iTe^{itT} T^{-1/2} f = T^{-1/2}(iTe^{itT} f)$. Hence, by (i) and (ii), $\lim_{\varepsilon\to 0} B_\varepsilon = A(ie^{itT} Tf)^{\#}\Psi$. Hence $A(e^{itT} f)^{\#}\Psi$ is strongly differentiable in t and (6.72) with $p = 1$ holds. Suppose that the statement of (iii) holds for some p. Then we can apply the preceding argument to f replaced by $i^p T^p f$ to conclude that the statement of (iii) holds for p replaced by $p + 1$. ■

Corollary 6.8

(i) *Let* $f,\ f_n \in \mathscr{H},\ \lim_{n\to\infty} f_n = f$. *Then, for all* $\Psi \in D(N_{\rm b}^{1/2})$,

$$\lim_{n\to\infty} A(f_n)^{\#}\Psi = A(f)^{\#}\Psi. \tag{6.73}$$

(ii) *Let* T *be a self-adjoint operator on* $\mathscr{H}$, $f \in D(T^p)$ *with* $p \in \mathbb{N}$ *and* $\Psi \in D(N_{\rm b}^{1/2})$. *Then the mapping:* $\mathbb{R} \ni t \mapsto A(e^{itT} f)^{\#}\Psi$ *is* p *times strongly differentiable and the strong derivative of the* p*-th order is given by (6.72).*

Proof

(i) This follows from (i) and (ii) with $T = I$ in Corollary 6.7.
(ii) This follows from (i) and the proof of Corollary 6.7(iii). ■

Theorem 6.10 also implies infinitesimal smallness of $A(f)^{\#}$ ($f \in D(T^{-1/2})$) with respect to $\mathrm{d}\Gamma_{\rm b}(T)$:

Corollary 6.9 *Let* T *be a non-negative and injective self-adjoint operator on* $\mathscr{H}$ *and* $f \in D(T^{-1/2})$. *Then, for all* $\varepsilon > 0$ *and* $\Psi \in D(\mathrm{d}\Gamma_{\rm b}(T))$,

$$\|A(f)\Psi\| \le \varepsilon\|\mathrm{d}\Gamma_{\rm b}(T)\Psi\| + \frac{1}{4\varepsilon}\|T^{-1/2}f\|^2\|\Psi\|, \tag{6.74}$$

$$\|A(f)^*\Psi\| \le \varepsilon\|\mathrm{d}\Gamma_{\rm b}(T)\Psi\| + \left(\frac{1}{4\varepsilon}\|T^{-1/2}f\|^2 + \|f\|\right)\|\Psi\|. \tag{6.75}$$

Proof We estimate the right hand side of (6.64) and (6.65). Let $\Psi \in D(\mathrm{d}\Gamma_{\rm b}(T))$. Then, by the Schwarz inequality, we have

$$\|\mathrm{d}\Gamma_{\rm b}(T)^{1/2}\Psi\| = \sqrt{\langle \Psi, \mathrm{d}\Gamma_{\rm b}(T)\Psi\rangle} \le \|\Psi\|^{1/2}\|\mathrm{d}\Gamma_{\rm b}(T)\Psi\|^{1/2}.$$

Using the elementary inequality

$$ab \le \varepsilon a^2 + \frac{1}{4\varepsilon}b^2 \quad (a > 0,\ b > 0, \varepsilon > 0), \tag{6.76}$$

we obtain

$$\|T^{-1/2}f\|\,\|\mathrm{d}\Gamma_{\rm b}(T)^{1/2}\Psi\| \le \varepsilon\|\mathrm{d}\Gamma_{\rm b}(T)\Psi\| + \frac{1}{4\varepsilon}\|T^{-1/2}f\|^2\|\Psi\|.$$

This estimate and (6.64) yield (6.74).

Inequality (6.65) implies that

$$\|A(f)^*\Psi\| \le \|T^{-1/2}f\|\,\|\mathrm{d}\Gamma_{\rm b}(T)^{1/2}\Psi\| + \|f\|\|\Psi\|.$$

Hence we obtain (6.75). ■

6.8 Application to Self-adjointness

Corollary 6.9 has applications to proving self-adjointness of symmetric operators defined as a perturbation of $\mathrm{d}\Gamma_{\mathrm{b}}(T)$. Here we present a basic example. For a self-adjoint operator T on $\mathscr{H}$ and a vector $g \in \mathscr{H}$, we define

$$H_{\mathrm{v}} := \mathrm{d}\Gamma_{\mathrm{b}}(T) + \frac{1}{\sqrt{2}}(A(g)^* + A(g)). \tag{6.77}$$

It is easy to see that $(A(g) + A(g)^*)/\sqrt{2}$ is a symmetric operator with

$$D(A(g)) \cap D(A(g)^*) \supset \mathscr{F}_{\mathrm{b},0}(\mathscr{H}) \supset \mathscr{F}_{\mathrm{b,fin}}(D(T)).$$

Hence $\mathscr{F}_{\mathrm{b,fin}}(D(T)) \subset D(H_{\mathrm{v}})$. Since $\mathscr{F}_{\mathrm{b,fin}}(D(T))$ is dense in $\mathscr{F}_{\mathrm{b}}(\mathscr{H})$, it follows that H_{v} is a symmetric operator.

Theorem 6.11 *Assume that T is a non-negative injective self-adjoint operator and*

$$g \in D(T^{-1/2}). \tag{6.78}$$

Then H_{v} is self-adjoint with $D(H_{\mathrm{v}}) = D(\mathrm{d}\Gamma_{\mathrm{b}}(T))$ and bounded from below. Moreover, any core for $\mathrm{d}\Gamma_{\mathrm{b}}(T)$ is a core for H_{v}.

Proof Let $H_{\mathrm{I}} := (A(g)^* + A(g))/\sqrt{2}$. Then, by Corollary 6.9, $D(\mathrm{d}\Gamma_{\mathrm{b}}(T)^{1/2}) \subset D(H_{\mathrm{I}})$. Hence $D(\mathrm{d}\Gamma_{\mathrm{b}}(T)) \subset D(H_{\mathrm{I}})$ and $D(H_{\mathrm{v}}) = D(\mathrm{d}\Gamma_{\mathrm{b}}(T))$. Let $\Psi \in D(\mathrm{d}\Gamma_{\mathrm{b}}(T))$. Then, by (6.74) and (6.75), we have

$$\|H_{\mathrm{I}}\Psi\| \le \frac{1}{\sqrt{2}}(\|A(g)^*\Psi\| + \|A(g)\Psi\|) \le \varepsilon\|\mathrm{d}\Gamma_{\mathrm{b}}(T)\Psi\| + c_\varepsilon\|\Psi\|,$$

where $\varepsilon > 0$ is arbitrary and $c_\varepsilon > 0$ is a constant depending on ε, $\|T^{-1/2}g\|$ and $\|g\|$. Take $\varepsilon < 1$. Then, by the Kato–Rellich theorem (see Theorem 1.15), we obtain the desired result. ■

In the context of quantum field theory, the operator H_{v} denotes the Hamiltonian of an abstract quantum field model, called an **abstract van Hove model** or an **abstract van Hove–Miyatake model** (see Sect. 10.9 in the present book).

6.9 Representations of Boson Second Quantization Operators in Terms of $A(\cdot)^{\#}$

In this section we derive some formulae for boson second quantization operators in terms of creation and annihilation operators.

Lemma 6.8 *Let $T \in \mathfrak{L}(\mathcal{H})$ be densely defined and $S \in \mathfrak{L}(\mathcal{H})$. Then, for all $f \in D(T^*) \cap D(S)$ and $\Psi \in \mathcal{F}_{\mathrm{b},0}(\mathcal{H})$,*

$$A(Sf)^* A(T^* f)\Psi = \mathrm{d}\Gamma_{\mathrm{b}}(P_{T^* f, Sf})\Psi, \tag{6.79}$$

where $P_{f,g}$ $(f, g \in \mathcal{H})$ is defined by (1.22).

Proof If $\Psi = \alpha\Omega_{\mathcal{H}}$ $(\alpha \in \mathbb{C})$, then (6.79) obviously holds with both sides being the zero vector. Let Ψ be of the form

$$\Psi = A(f_1)^* \cdots A(f_p)^* \Omega_{\mathcal{H}} \quad (f_1, \ldots, f_p \in \mathcal{H}) \tag{6.80}$$

with $p \geq 1$. Then we have by CCR

$$\begin{aligned} A(Sf)^* A(T^* f)\Psi &= \sum_{j=1}^{p} A(f_1)^* \cdots \underbrace{A(\langle T^* f, f_j\rangle Sf)^*}_{j\text{th}} \cdots A(f_p)^* \Omega_{\mathcal{H}} \\ &= \sum_{j=1}^{p} A(f_1)^* \cdots \underbrace{A(P_{T^* f, Sf} f_j)^*}_{j\text{th}} \cdots A(f_p)^* \Omega_{\mathcal{H}} \\ &= \mathrm{d}\Gamma_{\mathrm{b}}(P_{T^* f, Sf})\Psi_p, \end{aligned}$$

where we have used (6.46). Hence (6.79) holds. Then, by linearity, (6.79) holds for all $\Psi \in \mathcal{F}_{\mathrm{b,fin}}(\mathcal{H})$. Recall that $\mathcal{F}_{\mathrm{b,fin}}(\mathcal{H})$ is dense in $\mathcal{F}_{\mathrm{b},0}(\mathcal{H})$ and $A(Sf)^* A(T^* f)$ is a bounded operator on $\otimes_{\mathrm{s}}^{p} \mathcal{H}$ for each $p \geq 1$. Moreover $P_{T^* f, Sf}$ is a bounded operator. Hence it follows from a limiting argument that (6.79) can be extended to all $\Psi \in \mathcal{F}_{\mathrm{b},0}(\mathcal{H})$. ∎

Theorem 6.12 *Let $S, T \in \mathfrak{B}(\mathcal{H})$ and $\{e_n\}_{n=1}^{\infty}$ be a CONS of $\mathcal{H}$. Then, for all $\Psi \in \mathcal{F}_{\mathrm{b},0}(\mathcal{H})$,*

$$\sum_{n=1}^{\infty} A(Se_n)^* A(T^* e_n)\Psi = \mathrm{d}\Gamma_{\mathrm{b}}(ST)\Psi. \tag{6.81}$$

Proof There exists an integer p_0 such that, for all $p > p_0$, $\Psi^{(p)} = 0$. Hence we need only to prove (6.81) for $\Psi^{(p)}$ with $p \leq p_0$. For each $N \in \mathbb{N}$, we define an operator P_N by

$$P_N := \sum_{n=1}^{N} P_{T^* e_n, Se_n}.$$

We have by Lemma 6.8

$$\sum_{n=1}^{N} A(Se_n)^* A(T^* e_n)\Psi^{(p)} = L_N \Psi^{(p)},$$

where

$$L_N = P_N \otimes I \otimes \cdots \otimes I + I \otimes P_N \otimes \cdots \otimes I + \cdots + I \otimes \cdots \otimes I \otimes P_N.$$

For all $g \in \mathscr{H}$, we have by the Cauchy–Schwarz inequality and the Bessel inequality

$$\begin{aligned} |\langle g, P_N f\rangle| &\leq \sum_{n=1}^{N} |\langle e_n, Tf\rangle \langle g, Se_n\rangle| \\ &\leq (\sum_{n=1}^{N} |\langle e_n, Tf\rangle|^2)^{1/2} (\sum_{n=1}^{N} |\langle S^* g, e_n\rangle|^2)^{1/2} \\ &\leq \|Tf\| \, \|S^* g\| \leq \|T\| \, \|S^*\| \, \|f\| \, \|g\|. \end{aligned}$$

Hence the operator norm $\|P_N\|$ of P_N is uniformly bounded with

$$\|P_N\| \leq \|T\| \, \|S^*\|.$$

It is easy to see that $\text{s-}\lim_{N\to\infty} P_N = ST$. Hence, by an application of Proposition 1.25, we obtain

$$\lim_{N\to\infty} L_N \Psi^{(p)} = (ST)_{\mathrm{b}}^{(p)} \Psi^{(p)} = (\mathrm{d}\Gamma_{\mathrm{b}}(ST)\Psi)^{(p)}.$$

Thus (6.81) follows. ∎

Corollary 6.10 *Let S be a bounded self-adjoint operator on $\mathscr{H}$ and $\{e_n\}_{n=1}^{\infty}$ be a CONS of $\mathscr{H}$. Then, for all $\Psi \in D(\mathrm{d}\Gamma_{\mathrm{b}}(S^2)^{1/2})$, $\Psi \in \cap_{n=1}^{\infty} D(A(Se_n))$ and*

$$\sum_{n=1}^{\infty} \|A(Se_n)\Psi\|^2 = \|\mathrm{d}\Gamma_{\mathrm{b}}(S^2)^{1/2}\Psi\|^2. \tag{6.82}$$

In particular, for all $\Psi \in D(N_{\mathrm{b}}^{1/2})$,

$$\sum_{n=1}^{\infty} \|A(e_n)\Psi\|^2 = \|N_{\mathrm{b}}^{1/2}\Psi\|^2. \tag{6.83}$$

Proof We apply (6.81) with $T^* = S$ to obtain

$$\sum_{n=1}^{\infty} \|A(Se_n)\Psi\|^2 = \|\mathrm{d}\Gamma_{\mathrm{b}}(S^2)^{1/2}\Psi\|^2$$

for all $\Psi \in \mathscr{F}_{\mathrm{b},0}(\mathscr{H})$. Let $\Psi \in D(\mathrm{d}\Gamma_{\mathrm{b}}(S^2)^{1/2})$ next. For each $p \in \mathbb{Z}_+$, we define $\Psi_p \in \mathscr{F}_{\mathrm{b},0}(\mathscr{H})$ by $\Psi_p^{(k)} := \delta_{pk}\Psi^{(p)}$ $(k \in \mathbb{Z}_+)$. Then we have $\Psi = \sum_{p=0}^{\infty} \Psi_p$ and

$$\|\mathrm{d}\Gamma_{\mathrm{b}}(S^2)^{1/2}\Psi\|^2 = \sum_{p=1}^{\infty} \|\mathrm{d}\Gamma_{\mathrm{b}}(S^2)^{1/2}\Psi_p\|^2,$$

where we have used that $\mathrm{d}\Gamma_{\mathrm{b}}(S^2)^{1/2}\Psi_0 = 0$. By the preceding result, we have

$$\|\mathrm{d}\Gamma_{\mathrm{b}}(S^2)^{1/2}\Psi_p\|^2 = \sum_{n=1}^{\infty} \|A(Se_n)\Psi_p\|^2.$$

Hence

$$\|\mathrm{d}\Gamma_{\mathrm{b}}(S^2)^{1/2}\Psi\|^2 = \sum_{p=1}^{\infty}\sum_{n=1}^{\infty} \|A(Se_n)\Psi_p\|^2.$$

This implies that $\sum_{p=1}^{\infty} \|A(Se_n)\Psi_p\|^2 < \infty$ and

$$\|\mathrm{d}\Gamma_{\mathrm{b}}(S^2)^{1/2}\Psi\|^2 = \sum_{n=1}^{\infty}\sum_{p=0}^{\infty} \|A(Se_n)\Psi_p\|^2.$$

Note that $(A(Se_n)\Psi_p)^{(k)} = \delta_{k,p-1}(A(Se_n)\Psi)^{(k)}$. Hence, for all $M \in \mathbb{N}$,

$$\sum_{p=1}^{M} \|A(Se_n)\Psi_p\|^2 = \sum_{q=0}^{M-1} \|(A(Se_n)\Psi)^{(q)}\|^2 = \|A(Se_n)\Psi_M\|^2,$$

where $\Psi_M^{(q)} := \Psi^{(q)}$ for $0 \le q \le M$ and $\Psi_M^{(q)} = 0$ for $q \ge M + 1$. Hence

$$\|\mathrm{d}\Gamma_{\mathrm{b}}(S^2)^{1/2}\Psi_M\|^2 \ge \|A(Se_n)\Psi_M|^2,$$

$$\|\mathrm{d}\Gamma_{\mathrm{b}}(S^2)^{1/2}(\Psi_M - \Psi_N)\|^2 \ge \|A(Se_n)(\Psi_M - \Psi_N)\|^2, \quad M > N.$$

Since $\|\mathrm{d}\Gamma_{\mathrm{b}}(S^2)^{1/2}(\Psi_M - \Psi_N)\| \to 0$ as $M, N \to \infty$, it follows that, for each $n \in \mathbb{N}$, $\{A(Se_n)\Psi_M\}_M$ is a Cauchy sequence. Since $A(Se_n)$ is closed, one can conclude that $\Psi \in D(A(Se_n))$ and $A(Se_n)\Psi_M \to A(Se_n)\Psi$ as $M \to \infty$. In particular, $\Psi \in \cap_{n=1}^{\infty} D(A(Se_n))$. Then we have

$$\sum_{p=0}^{\infty} \|A(Se_n)\Psi_p\|^2 = \|A(Se_n)\Psi\|^2.$$

Thus (6.82) follows. Formula (6.83) is just a special case of (6.82) with $S = I$. ■

We next consider the case where S or T is unbounded.

Theorem 6.13 *Let T be a densely defined closable operator on $\mathscr{H}$ and $S \in \mathfrak{B}(\mathscr{H})$ such that ST is closable. Let $\{e_n\}_{n=1}^{\infty} \subset D(T^*)$ be a CONS of $\mathscr{H}$. Then*

$$d\Gamma_b(ST)\Psi = \sum_{n=1}^{\infty} A(Se_n)^* A(T^* e_n)\Psi, \quad \Psi \in \mathscr{F}_{b,\text{fin}}(D(T)). \tag{6.84}$$

Proof Similar to the proof of Theorem 6.12. But, in the present case, vectors Ψ on which $d\Gamma_b(ST)$ and $A(Se_n)^* A(T^* e_n)$ act are restricted to those in $\mathscr{F}_{b,\text{fin}}(D(T))$. ■

A shortcoming in formula (6.84) may be in that the approximating operators $\sum_{n=1}^{N} A(Se_n)^* A(T^* e_n)$ $(N = 1, 2, \ldots)$ are not necessarily symmetric even if $\overline{ST}$ is self-adjoint. For a class of non-negative self-adjoint operators T on $\mathscr{H}$, this point can be improved:

Theorem 6.14 *Let T be a non-negative self-adjoint operator on $\mathscr{H}$ such that there exists a CONS $\{e_n\}_{n=1}^{\infty} \subset D(T^{1/2})$ of $\mathscr{H}$ satisfying*

$$\lim_{N\to\infty} \sum_{n=1}^{N} \langle e_n, f \rangle T^{1/2} e_n = T^{1/2} f, \quad f \in D(T^{1/2}). \tag{6.85}$$

Then, for all $\Psi \in \mathscr{F}_{b,\text{fin}}(D(T))$,

$$d\Gamma_b(T)\Psi = \sum_{n=1}^{\infty} A(T^{1/2} e_n)^* A(T^{1/2} e_n)\Psi. \tag{6.86}$$

Proof Let Ψ be given by (6.80) with $f_j \in D(T)$ $(j = 1, \ldots, p)$. Then, in the same way as in the proof of Theorem 6.13, one can show that, for each $N \in \mathbb{N}$,

$$\sum_{n=1}^{N} A(T^{1/2} e_n)^* A(T^{1/2} e_n)\Psi_p = \sum_{j=1}^{p} A(f_1)^* \cdots A(F_{j,N})^* \cdots A(f_p)^* \Omega_{\mathscr{H}},$$

where $F_{j,N} := \sum_{n=1}^{N} \left\langle e_n, T^{1/2} f_j \right\rangle T^{1/2} e_n$. By (6.85), $\lim_{N\to\infty} F_{j,N} = T f_j$. Thus (6.86) holds. ■

Example 6.4 Let T be a non-negative unbounded self-adjoint operator on $\mathscr{H}$. Suppose that $\sigma(T) = \sigma_{\mathrm{d}}(T)$ (the discrete spectrum of T) $= \{\lambda_n\}_{n=1}^{\infty}$ with $0 \le \lambda_1 \le \lambda_2 \le \cdots \le \lambda_n \le \lambda_{n+1} \le \cdots$, counting multiplicities, and $\lambda_n \to \infty$ as $n \to \infty$. Then there exists a CONS $\{\varphi_n\}_{n=1}^{\infty}$ of $\mathscr{H}$ such that $T\varphi_n = \lambda_n\varphi_n$, $n \in \mathbb{N}$ (φ_n's are eigenvectors of T). We have $T^{1/2}\varphi_n = \lambda_n^{1/2}\varphi_n$. Hence, for all $f \in D(T^{1/2})$,

$$\sum_{n=1}^{N} \langle \varphi_n, f\rangle T^{1/2}\varphi_n = \sum_{n=1}^{N} \left\langle \varphi_n, T^{1/2} f\right\rangle \varphi_n \to T^{1/2} f.$$

Hence (6.85) with $e_n = \varphi_n$ holds. Thus, in this case too, (6.86) holds with $e_n = \varphi_n$.

A weak form of (6.84) is given as follows:

Theorem 6.15 *Let S and T be densely defined closable operators on $\mathscr{H}$ such that $D(S) \cap D(T^*)$ is dense in $\mathscr{H}$ and ST is a densely defined closable operator. Let $\{e_n\}_{n=1}^{\infty} \subset D(S) \cap D(T^*)$ be a CONS of $\mathscr{H}$. Then, for all $\Psi \in \mathscr{F}_{\mathrm{b,fin}}(D(ST))$ and $\Phi \in \mathscr{F}_{\mathrm{fin}}(D(S^*))$,*

$$\langle \Phi, \mathrm{d}\Gamma_{\mathrm{b}}(ST)\Psi\rangle = \lim_{N\to\infty} \left\langle \Phi, \sum_{n=1}^{N} A(Se_n)^* A(T^*e_n)\Psi \right\rangle. \tag{6.87}$$

In particular,

$$\|\mathrm{d}\Gamma_{\mathrm{b}}(T^*T)^{1/2}\Psi\|^2 = \sum_{n=1}^{\infty} \|A(T^*e_n)\Psi\|^2, \quad \Psi \in \mathscr{F}_{\mathrm{b,fin}}(D(T^*T)). \tag{6.88}$$

Proof Let $\Psi = A(f_1)^* \cdots A(f_p)^*\Omega_{\mathscr{H}}$ and $\Phi = A(g_1)^* \cdots A(g_p)^*\Omega_{\mathscr{H}}$ with $f_j \in D(ST)$ and $g_k \in D(S^*)$ $(j, k = 1, \ldots, p)$. Then, as in the proof of Theorem 6.13, we have

$$\Theta_N := \sum_{n=1}^{N} A(Se_n)^* A(T^*e_n)\Psi = \sum_{j=1}^{p} A(f_1)^* \cdots A(F_{j,N})^* \cdots A(f_p)^*\Omega_{\mathscr{H}},$$

where $F_{j,N} := \sum_{n=1}^{N} \left\langle e_n, Tf_j\right\rangle Se_n$. Hence

$$\langle \Phi, \Theta_N\rangle = \sum_{j=1}^{p} \sum_{\sigma\in\mathfrak{S}_p} \left\langle g_{\sigma(1)}, f_1\right\rangle \cdots \left\langle g_{\sigma(j)}, F_{j,N}\right\rangle \cdots \left\langle g_{\sigma(p)}, f_p\right\rangle.$$

We have

$$\lim_{N\to\infty} \left\langle g, F_{j,N}\right\rangle = \left\langle S^*g, Tf_j\right\rangle = \left\langle g, STf_j\right\rangle, \quad g \in D(S^*).$$

Hence

$$\lim_{N\to\infty}\langle\Phi,\Theta_N\rangle=\sum_{j=1}^{p}\sum_{\sigma\in\mathfrak{S}_p}\left\langle g_{\sigma(1)},f_1\right\rangle\cdots\left\langle g_{\sigma(j)},STf_j\right\rangle\cdots\left\langle g_{\sigma(p)},f_p\right\rangle$$
$$=\langle\Phi,\mathrm{d}\Gamma_{\mathrm{b}}(ST)\Psi\rangle\,.$$

By linearity, this extends to all $\Psi\in\mathscr{F}_{\mathrm{b,fin}}(D(ST))$ and $\Phi\in\mathscr{F}_{\mathrm{b,fin}}(D(S^*))$. Hence (6.87) holds. Note that $\overline{T^*T}=\overline{T}^*\overline{T}$ is a non-negative self-adjoint operator on $\mathscr{H}$. Hence $\mathrm{d}\Gamma_{\mathrm{b}}(T^*T)$ is a non-negative self-adjoint operator on $\mathscr{F}_{\mathrm{b}}(\mathscr{H})$. Therefore $\mathrm{d}\Gamma_{\mathrm{b}}(T^*T)^{1/2}$ is defined via functional calculus. Taking $\Phi=\Psi$ and $S=T^*$ in (6.87), we obtain (6.88). ■

Corollary 6.11 *Let T be a non-negative self-adjoint operator on $\mathscr{H}$ and $\{e_n\}_{n=1}^{\infty}\subset D(T^{1/2})$ be a CONS of $\mathscr{H}$. Then, for all $\Psi\in D(\mathrm{d}\Gamma_{\mathrm{b}}(T)^{1/2})$,*

$$\sum_{n=1}^{\infty}\|A(T^{1/2}e_n)\Psi\|^2\le\|\mathrm{d}\Gamma_{\mathrm{b}}(T)^{1/2}\Psi\|^2. \tag{6.89}$$

Proof Since $\mathscr{F}_{\mathrm{b,fin}}(D(T))$ is a core for $\mathrm{d}\Gamma_{\mathrm{b}}(T)^{1/2}$ by Theorem 6.1(iii), for each $\Psi\in D(\mathrm{d}\Gamma_{\mathrm{b}}(T)^{1/2})$, there exists a sequence $\{\Psi_k\}_k$ in $\mathscr{F}_{\mathrm{b,fin}}(D(T))$ such that $\lim_{k\to\infty}\Psi_k=\Psi$ and $\lim_{k\to\infty}\mathrm{d}\Gamma_{\mathrm{b}}(T)^{1/2}\Psi_k=\mathrm{d}\Gamma_{\mathrm{b}}(T)^{1/2}\Psi$. By (6.88) with T replaced by $T^{1/2}$, we have

$$\sum_{n=1}^{\infty}\|A(T^{1/2}e_n)\Psi_k\|^2=\|\mathrm{d}\Gamma_{\mathrm{b}}(T)^{1/2}\Psi_k\|^2$$

and

$$\|A(T^{1/2}e_n)\Psi_k-A(T^{1/2}e_n)\Psi\|\le\|\mathrm{d}\Gamma_{\mathrm{b}}(T)^{1/2}\Psi_k-\mathrm{d}\Gamma_{\mathrm{b}}(T)^{1/2}\Psi\|\to 0\ \ (k\to\infty).$$

In particular, $\lim_{k\to\infty}\|A(T^{1/2}e_n)\Psi_k\|=\|A(T^{1/2}e_n)\Psi\|$. Hence, by applying Fatou's theorem, we obtain (6.89). ■

Remark 6.4 In fact, the equality in (6.89) holds:

$$\sum_{n=1}^{\infty}\|A(T^{1/2}e_n)\Psi\|^2=\|\mathrm{d}\Gamma_{\mathrm{b}}(T)^{1/2}\Psi\|^2,\quad \Psi\in D(\mathrm{d}\Gamma_{\mathrm{b}}(T)^{1/2}). \tag{6.90}$$

Hence, by the polarization identity, we have

$$\left\langle \mathrm{d}\Gamma_{\mathrm{b}}(T)^{1/2}\Phi, \mathrm{d}\Gamma_{\mathrm{b}}(T)^{1/2}\Psi \right\rangle = \sum_{n=1}^{\infty} \left\langle A(T^{1/2}e_n)\Phi, A(T^{1/2}e_n)\Psi \right\rangle,$$

$$\Phi, \Psi \in D(\mathrm{d}\Gamma_{\mathrm{b}}(T)^{1/2}). \tag{6.91}$$

But, to prove (6.90), we need to pass to another representation of $\{A(f), A(f)^* | f \in \mathscr{H}\}$ where T is represented by a multiplication operator on an L^2-space. This representation is obtained by employing the multiplication operator form of the spectral theorem. For the details, see [28, Theorem 5.21].

6.10 Commutation Relations Between $A(f)^{\#}$ and $\mathrm{d}\Gamma_{\mathrm{b}}(T)$

We next consider commutation properties of boson second quantization operators with boson creation and annihilation operators.

Proposition 6.8 *Let T be a densely defined closable operator on $\mathscr{H}$.*

(i) *For all $f \in D(T)$,*

$$\mathscr{F}_{\mathrm{b,fin}}(D(T)) \subset D([\mathrm{d}\Gamma_{\mathrm{b}}(T), A(f)^*])$$

and

$$[\mathrm{d}\Gamma_{\mathrm{b}}(T), A(f)^*] = A(Tf)^* \text{ on } \mathscr{F}_{\mathrm{b,fin}}(D(T)).$$

(ii) *For all $f \in D(T^*)$,*

$$\mathscr{F}_{\mathrm{b,fin}}(D(T)) \subset D([\mathrm{d}\Gamma_{\mathrm{b}}(T), A(f)])$$

and

$$[\mathrm{d}\Gamma_{\mathrm{b}}(T), A(f)] = -A(T^* f) \text{ on } \mathscr{F}_{\mathrm{b,fin}}(D(T)).$$

Proof

(i) Let $\Psi = \{\Psi^{(n)}\}_{n=0}^{\infty} \in \mathscr{F}_{\mathrm{b,fin}}(D(T))$. Then, by (6.44), each $\Psi^{(n)}$ with $n \geq 1$ is a linear combination of vectors of the form

$$\eta = A(f_1)^* \cdots A(f_n)^* \Omega_{\mathscr{H}}$$

with $f_j \in D(T)$, $j = 1, \ldots, n$. By Proposition 6.6, $\eta \in D(\mathrm{d}\Gamma_{\mathrm{b}}(T))$ and

$$\mathrm{d}\Gamma_{\mathrm{b}}(T)\eta = \sum_{j=1}^{n} A(f_1)^* \cdots A(Tf_j)^* \cdots A(f_n)^* \Omega_{\mathscr{H}}.$$

Hence $\mathrm{d}\Gamma_{\mathrm{b}}(T)\eta \in D(A(f)^*)$ and

$$A(f)^*\mathrm{d}\Gamma_{\mathrm{b}}(T)\eta = \sum_{j=1}^{n} A(f)^* A(f_1)^* \cdots A(Tf_j)^* \cdots A(f_n)^* \Omega_{\mathscr{H}}.$$

Since f is in $D(T)$, $A(f)^*\eta \in D(\mathrm{d}\Gamma_{\mathrm{b}}(T))$ and

$$\mathrm{d}\Gamma_{\mathrm{b}}(T)A(f)^*\eta = A(Tf)^*\eta + \sum_{j=1}^{n} A(f)^* A(f_1)^* \cdots A(Tf_j)^* \cdots A(f_n)^* \Omega_{\mathscr{H}}.$$

Hence

$$[\mathrm{d}\Gamma_{\mathrm{b}}(T), A(f)^*]\eta = A(Tf)^*\eta.$$

It is easy to see that $[\mathrm{d}\Gamma_{\mathrm{b}}(T), A(f)^*]\Omega_{\mathscr{H}} = A(Tf)^*\Omega_{\mathscr{H}}$. Thus the desired result follows.

(ii) Let $f \in D(T^*)$. Let $\Psi = A(f_1)^* \cdots A(f_n)^* \Omega_{\mathscr{H}}$ with $f_j \in D(T)$ ($j = 1, \ldots, n$). Then, by (6.57), we see as in (i) that $\Psi \in D([\mathrm{d}\Gamma_{\mathrm{b}}(T), A(f)])$ and

$$[\mathrm{d}\Gamma_{\mathrm{b}}(T), A(f)]\Psi = -A(T^* f)\Psi$$

Thus the desired result follows. ■

In the case where T is self-adjoint, non-negative and injective, Proposition 6.8 can be more refined:

Theorem 6.16 *Let T be a non-negative and injective self-adjoint operator on $\mathscr{H}$.*

(i) *For all $\Psi, \Phi \in D(\mathrm{d}\Gamma_{\mathrm{b}}(T))$ and $f \in D(T) \cap D(T^{-1/2})$,*

$$\langle \mathrm{d}\Gamma_{\mathrm{b}}(T)\Psi, A(f)\Phi\rangle - \left\langle A(f)^*\Psi, \mathrm{d}\Gamma_{\mathrm{b}}(T)\Phi\right\rangle = -\langle \Psi, A(Tf)\Phi\rangle, \tag{6.92}$$

$$\left\langle \mathrm{d}\Gamma_{\mathrm{b}}(T)\Psi, A(f)^*\Phi\right\rangle - \langle A(f)\Psi, \mathrm{d}\Gamma_{\mathrm{b}}(T)\Phi\rangle = \left\langle \Psi, A(Tf)^*\Phi\right\rangle. \tag{6.93}$$

(ii) *For each* $f \in D(T) \cap D(T^{-1/2})$, $A(f)^{\#}$ *maps* $D(\mathrm{d}\Gamma_{\mathrm{b}}(T)^{3/2})$ *to* $D(\mathrm{d}\Gamma_{\mathrm{b}}(T))$ *and satisfies commutation relations*

$$[\mathrm{d}\Gamma_{\mathrm{b}}(T), A(f)^*] = A(Tf)^*,$$
$$[\mathrm{d}\Gamma_{\mathrm{b}}(T), A(f)] = -A(Tf)$$

on $D(\mathrm{d}\Gamma_{\mathrm{b}}(T)^{3/2})$.

Proof

(i) By Proposition 6.8, for all $\Psi \in \mathscr{F}_{\mathrm{b,fin}}(D(T))$ and $\Phi \in D(\mathrm{d}\Gamma_{\mathrm{b}}(T))$, (6.92) holds. Since $\mathscr{F}_{\mathrm{b,fin}}(D(T))$ is a core for $\mathrm{d}\Gamma_{\mathrm{b}}(T)$, for any $\Psi \in D(\mathrm{d}\Gamma_{\mathrm{b}}(T))$, there exists a sequence $\{\Psi_n\}_n$ in $\mathscr{F}_{\mathrm{b,fin}}(D(T))$ such that $\Psi_n \to \Psi$ and $\mathrm{d}\Gamma_{\mathrm{b}}(T)\Psi_n \to \mathrm{d}\Gamma_{\mathrm{b}}(T)\Psi$ $(n \to \infty)$. We have

$$\langle \mathrm{d}\Gamma_{\mathrm{b}}(T)\Psi_n, A(f)\Phi\rangle - \left\langle A(f)^*\Psi_n, \mathrm{d}\Gamma_{\mathrm{b}}(T)\Phi\right\rangle = -\langle \Psi_n, A(Tf)\Phi\rangle .$$

By (6.75), $A(f)^*\Psi_n \to A(f)^*\Psi$ $(n \to \infty)$. Hence, taking the limit $n \to \infty$ in the above equation, we obtain (6.92). Formula (6.93) follows from taking the complex conjugate of (6.92).

(ii) Let $\Psi \in D(\mathrm{d}\Gamma_{\mathrm{b}}(T)^{3/2})$ and $\Phi \in D(\mathrm{d}\Gamma_{\mathrm{b}}(T))$. Then

$$\mathrm{d}\Gamma_{\mathrm{b}}(T)\Psi \in D(\mathrm{d}\Gamma_{\mathrm{b}}(T)^{1/2}) \subset D(A(f)^*) \cap D(A(f)).$$

Hence, by (i),

$$\left\langle \left(A(f)^*\mathrm{d}\Gamma_{\mathrm{b}}(T) + A(Tf)^*\right)\Psi, \Phi\right\rangle = \left\langle A(f)^*\Psi, \mathrm{d}\Gamma_{\mathrm{b}}(T)\Phi\right\rangle .$$

This means that $A(f)^*\Psi \in D(\mathrm{d}\Gamma_{\mathrm{b}}(T))$ and

$$\mathrm{d}\Gamma_{\mathrm{b}}(T)A(f)^*\Psi = \left(A(f)^*\mathrm{d}\Gamma_{\mathrm{b}}(T) + A(Tf)^*\right)\Psi.$$

Hence the statement on $A(f)^*$ is proved. Similarly, one can prove the statement on $A(f)$. ■

6.11 Segal Field Operators

Operators on $\mathscr{F}_{\mathrm{b}}(\mathscr{H})$ can be constructed from $A(f)$ and $A(f)^*$ $(f \in \mathscr{H})$. A simple example of such operators is given by $A(f) + A(f)^*$. As already seen in Sect. 6.8, it is a symmetric operator. Hence one can define a closed symmetric operator

$$\Phi_{\mathrm{S}}(f) := \frac{1}{\sqrt{2}}\overline{(A(f) + A(f)^*)}, \quad f \in \mathscr{H}, \tag{6.94}$$

where the constant $1/\sqrt{2}$ is just for convenience. The operator $\Phi_S(f)$ is called the **Segal field operator** with test vector f, which plays a fundamental role in applications to Bose field theory. The purpose of this section is to investigate basic properties of Segal field operators.

6.11.1 Basic Properties

Proposition 6.9 *Let T be a non-negative and injective self-adjoint operator on $\mathscr{H}$.*

(i) *For all $f \in D(T^{-1/2})$, $D(\mathrm{d}\Gamma_b(T)^{1/2}) \subset D(A(f)) \cap D(A(f)^*)$ and, for all $\Psi \in D(\mathrm{d}\Gamma_b(T)^{1/2})$,*

$$\Phi_S(f)\Psi = \frac{1}{\sqrt{2}}(A(f) + A(f)^*)\Psi \tag{6.95}$$

and

$$\|\Phi_S(f)\Psi\| \leq \sqrt{2}\|T^{-1/2}f\| \, \|\mathrm{d}\Gamma_b(T)^{1/2}\Psi\| + \frac{1}{\sqrt{2}}\|f\| \, \|\Psi\|. \tag{6.96}$$

(ii) *For all $f \in \mathscr{H}$, $D(N_b^{1/2}) \subset D(\Phi_S(f))$ and, for all $\Psi \in D(N_b^{1/2})$, (6.95) holds and*

$$\|\Phi_S(f)\Psi\| \leq \sqrt{2}\|f\| \left(\|N_b^{1/2}\Psi\| + \frac{1}{2}\|\Psi\| \right). \tag{6.97}$$

Proof

(i) Let $f \in D(T^{-1/2})$. Then Theorem 6.10 implies that $D(\mathrm{d}\Gamma_b(T)^{1/2}) \subset D(\Phi_S(f))$ and (6.95) holds. Hence

$$\|\Phi_S(f)\Psi\| \leq \frac{1}{\sqrt{2}}(\|A(f)\Psi\| + \|A(f)^*\Psi\|).$$

Then, using (6.64) and (6.65), we obtain (6.96).

(ii) This follows from (i) in the case $T = I$. ■

Let

$$\mathscr{D}_b := \cap_{f \in \mathscr{H}} D(A(f)) \cap D(A(f)^*). \tag{6.98}$$

Then

$$D(N_{\mathrm{b}}^{1/2}) \subset \mathscr{D}_{\mathrm{b}}.$$

The real linearity of $A(f)^{\#}$ in f implies that of the correspondence $f \mapsto \Phi_{\mathrm{S}}(f)\Psi$ ($\Psi \in \mathscr{D}_{\mathrm{b}}$):

$$\Phi_{\mathrm{S}}(af + bg)\Psi = (a\Phi_{\mathrm{S}}(f) + b\Phi_{\mathrm{S}}(g))\,\Psi,\ a, b \in \mathbb{R},\ f, g \in \mathscr{H},\ \Psi \in \mathscr{D}_{\mathrm{b}}. \tag{6.99}$$

Remark 6.5 For any non-real α, if $f \neq 0$ ($f \in \mathscr{H}$), then $\Phi_{\mathrm{S}}(\alpha f) \neq \alpha\Phi_{\mathrm{S}}(f)$. Namely, the correspondence $f \mapsto \Phi_{\mathrm{S}}(f)\Psi$ ($\Psi \in \mathscr{D}_{\mathrm{b}}$) is *not complex linear*. This is because the correspondence $f \mapsto A(f)\Psi$ ($\Psi \in \mathscr{D}_{\mathrm{b}}$) is anti-linear: $A(\alpha f)\Psi = \alpha^* A(f)\Psi$, $\alpha \in \mathbb{C}$ (see (6.38)).

Proposition 6.10 *For all $f \in \mathscr{H} \setminus \{0\}$, $\Phi_{\mathrm{S}}(f)$ is neither bounded from below nor bounded from above.*

Proof For each $N \in \mathbb{N}$, we define $\Psi_N(f) \in \mathscr{F}_{\mathrm{b,fin}}(\mathscr{H})$ by $\Psi_N^{(n)}(f) = (\sqrt{2}\|f\|^n)^{-1} \otimes^n f$ for $n = N, N+1$ and $\Psi_N^{(n)}(f) = 0$ for $n \neq N, N+1$. Then $\|\Psi_N(f)\| = 1$ and

$$\begin{aligned}\langle\Psi_N, A(f)^*\Psi_N\rangle &= \left\langle \frac{1}{\sqrt{2}\|f\|^{N+1}} \otimes^{N+1} f, \frac{\sqrt{N+1}}{\sqrt{2}\|f\|^N} \otimes^{N+1} f \right\rangle \\ &= \frac{\|f\|}{2}\sqrt{N+1}.\end{aligned}$$

Hence $\langle\Psi_N, A(f)\Psi_N\rangle = \langle\Psi_N, A(f)^*\Psi_N\rangle^* = \sqrt{N+1}\|f\|/2$. Therefore

$$\langle\Psi_N(f), \Phi_{\mathrm{S}}(f)\Psi_N(f)\rangle = \frac{\sqrt{N+1}\|f\|}{\sqrt{2}} \to \infty\ (N \to \infty).$$

Hence $\Phi_{\mathrm{S}}(f)$ is not bounded from above. Since $\Phi_{\mathrm{S}}(-f) = -\Phi_{\mathrm{S}}(f)$, it follows that $\|\Psi_N(-f)\| = 1$ and

$$\langle\Psi_N(-f), \Phi_{\mathrm{S}}(f)\Psi_N(-f)\rangle = -\frac{\sqrt{N+1}}{\sqrt{2}}\|f\| \to -\infty\ (N \to \infty).$$

Hence $\Phi_{\mathrm{S}}(f)$ is not bounded from below. ■

We have for all $f \in \mathscr{H}$ and $\Psi \in D(A(f)) \cap D(A(f)^*)$

$$\Phi_{\mathrm{S}}(if)\Psi = \frac{i}{\sqrt{2}}(A(f)^* - A(f))\Psi. \tag{6.100}$$

By this formula and (6.94), we obtain the following relations: for all $\Psi \in D(A(f)) \cap D(A(f)^*)$,

$$A(f)^*\Psi = \frac{1}{\sqrt{2}}(\Phi_S(f) - i\Phi_S(if))\Psi, \tag{6.101}$$

$$A(f)\Psi = \frac{1}{\sqrt{2}}(\Phi_S(f) + i\Phi_S(if))\Psi. \tag{6.102}$$

These are formulae recovering $A(\cdot)^\#$ from Segal field operators.

Proposition 6.11

(i) *For each* $f \in \mathscr{H}$, $\Phi_S(f)$ *leaves* $\mathscr{F}_{b,0}(\mathscr{H})$ *invariant and obeys commutation relations*

$$[\Phi_S(f), \Phi_S(g)] = i\,\mathrm{Im}\,\langle f, g\rangle, \quad f, g \in \mathscr{H} \tag{6.103}$$

on $\mathscr{F}_{b,0}(\mathscr{H})$.

(ii) *Let* $\mathscr{D}$ *be a subspace of* $\mathscr{H}$. *Then, for each* $f \in \mathscr{D}$, $\Phi_S(f)$ *leaves* $\mathscr{F}_{b,\mathrm{fin}}(\mathscr{D})$ *invariant.*

(iii) *Let* $\mathscr{D}$ *be a subspace of* $\mathscr{H}$ *and*

$$\mathscr{P}_{\Phi_S}(\mathscr{D}) := \mathrm{span}\Big\{\Omega_{\mathscr{H}}, \Phi_S(f_1)\cdots\Phi_S(f_n)\Omega_{\mathscr{H}} \,\Big|\, n \geq 1, \\ f_j \in \mathscr{D}, j = 1, \ldots, n\Big\}.$$

Then

$$\mathscr{F}_{b,\mathrm{fin}}(\mathscr{D}) = \mathscr{P}_{\Phi_S}(\mathscr{D}). \tag{6.104}$$

In particular, if $\mathscr{D}$ *is dense, then* $\mathscr{P}_{\Phi_S}(\mathscr{D})$ *is dense in* $\mathscr{F}_b(\mathscr{H})$.

(iv) (continuity) *Let* T *be a non-negative and injective self-adjoint operator on* $\mathscr{H}$. *Suppose that* $f_n, f \in D(T^{-1/2})$ *and* $f_n \to f, T^{-1/2}f_n \to T^{-1/2}f$ $(n \to \infty)$. *Then, for all* $\Psi \in D(\mathrm{d}\Gamma_b(T)^{1/2})$, $\Phi_S(f_n)\Psi \to \Phi_S(f)\Psi$ $(n \to \infty)$.

Proof

(i) This follows from Theorem 6.8.

(ii) By Corollary 6.5, for all $f \in \mathscr{D}$, $A(f)^\#\mathscr{F}_{b,\mathrm{fin}}(\mathscr{D}) \subset \mathscr{F}_{b,\mathrm{fin}}(\mathscr{D})$. Hence

$$\Phi_S(f)\mathscr{F}_{b,\mathrm{fin}}(\mathscr{D}) \subset \mathscr{F}_{b,\mathrm{fin}}(\mathscr{D}).$$

(iii) Let $f_j \in \mathscr{D}$ $(j = 1, \ldots, n)$. Then, by direct computations, we have

$$\Phi_{\rm S}(f_1)\cdots\Phi_{\rm S}(f_n)\Omega_{\mathscr{H}} = \sum_{2^n \text{ terms}} \frac{1}{2^{n/2}} A(f_1)^{\#}\cdots A(f_n)^{\#}\Omega_{\mathscr{H}},$$

Using (6.52), (6.53) and (6.39), we see that the vector on the right hand side is in $\mathscr{F}_{\rm b,fin}(\mathscr{D})$. Hence $\mathscr{P}_{\Phi_{\rm S}}(\mathscr{D}) \subset \mathscr{F}_{\rm b,fin}(\mathscr{D})$.

Conversely, let $\Psi = A(f_1)^* \cdots A(f_n)^*\Omega_{\mathscr{H}} \in \mathscr{F}_{\rm b,fin}(\mathscr{D})$ $(f_j \in \mathscr{D}, j = 1, \ldots, n)$. Then, using (6.101), we see that the vector on the right hand side is in $\mathscr{P}_{\Phi_{\rm S}}(\mathscr{D})$. Hence $\mathscr{F}_{\rm b,fin}(\mathscr{D}) \subset \mathscr{P}_{\Phi_{\rm S}}(\mathscr{D})$. Thus $\mathscr{F}_{\rm b,fin}(\mathscr{D}) = \mathscr{P}_{\Phi_{\rm S}}(\mathscr{D})$.

(iv) This follows from (6.96). ■

6.11.2 Self-Adjointness of the Segal Field Operator

Lemma 6.9 *Let $f \in \mathscr{H}$. Then every vector in $\mathscr{F}_{\rm b,0}(\mathscr{H})$ is an entire analytic vector of $\Phi_{\rm S}(f)$.*

Proof Let $n \geq 1$. Then we have

$$\Phi_{\rm S}(f)^n = \frac{1}{2^{n/2}} \sum_{2^n \text{ terms}} \underbrace{A(f)^{\#}\cdots A(f)^{\#}}_{n \text{ factors}}$$

on $\mathscr{F}_{\rm b,0}(\mathscr{H})$. Let $\Psi_k = \{0, \cdots, 0, \Psi^{(k)}, 0, \cdots\} \in \mathscr{F}_{\rm b,0}(\mathscr{H})$ $(k = 0, 1, 2, \cdots)$. Then

$$\underbrace{A(f)^{\#}\cdots A(f)^{\#}}_{n \text{ factors}}\Psi_k \in \mathscr{F}_{{\rm b},n+k}(\mathscr{H}).$$

Hence, by Corollary 6.6,

$$\begin{aligned}
\|\underbrace{A(f)^{\#}\cdots A(f)^{\#}}_{n \text{ factors}}\Psi_k\| &\leq \|f\|\,\|(N_{\rm b}+1)^{1/2}\underbrace{A(f)^{\#}\cdots A(f)^{\#}}_{n-1 \text{ factors}}\Psi_k\| \\
&\leq (n+k)^{1/2}\|f\|\,\|\underbrace{A(f)^{\#}\cdots A(f)^{\#}}_{n-1 \text{ factors}}\Psi_k\| \\
&\leq (n+k)^{1/2}(n-1+k)^{1/2}\cdots(1+k)^{1/2} \\
&\qquad \times\|f\|^n\|\Psi_k\|. \qquad (6.105)
\end{aligned}$$

Hence, for all $t > 0$,

$$\sum_{n=0}^{\infty} \frac{\|\Phi_S(f)^n \Psi_k\|}{n!} t^n$$
$$\leq \sum_{n=0}^{\infty} \frac{2^{n/2} t^n \|f\|^n}{n!} (n+k)^{1/2}(n-1+k)^{1/2} \cdots (1+k)^{1/2} \|\Psi_k\|.$$

By d'Alembert's ratio test, the infinite series on the right hand side converges. Hence Ψ_k is an entire analytic vector of $\Phi_S(f)$. Since any vector Ψ in $\mathscr{F}_{b,0}(\mathscr{H})$ is a linear combination of vectors of the form Ψ_k ($k \geq 0$), it follows that Ψ is an entire analytic vector of $\Phi_S(f)$. ∎

Theorem 6.17 *Let $f \in \mathscr{H}$. Then $\Phi_S(f)$ is self-adjoint. Moreover, for any dense subspace $\mathscr{D}$ of $\mathscr{H}$, $\mathscr{F}_{b,\mathrm{fin}}(\mathscr{D})$ is a core for $\Phi_S(f)$.*

Proof It is obvious that $\Phi_S(f)\mathscr{F}_{b,0}(\mathscr{H}) \subset \mathscr{F}_{b,0}(\mathscr{H})$. Hence, by Lemma 6.9 and Theorem 1.14, $\Phi_S(f)$ is essentially self-adjoint on $\mathscr{F}_{b,0}(\mathscr{H})$ and hence $\Phi_S(f)$ is self-adjoint.

By the preceding result, for each $\Psi \in D(\Phi_S(f))$ and any $\varepsilon > 0$, there exists a vector $\Psi_\varepsilon \in \mathscr{F}_{b,0}(\mathscr{H})$ such that $\|\Psi_\varepsilon - \Psi\| < \varepsilon$ and $\|\Phi_S(f)\Psi_\varepsilon - \Phi_S(f)\Psi\| < \varepsilon$. One has $\mathscr{F}_{b,\mathrm{fin}}(\mathscr{D}) \subset \mathscr{F}_{b,0}(\mathscr{H})$ and $\mathscr{F}_{b,\mathrm{fin}}(\mathscr{D})$ is dense in $\mathscr{F}_b(\mathscr{H})$. There exists an $r \in \mathbb{N}$ such that $\Psi_\varepsilon \in \mathscr{F}_{b,r}(\mathscr{H})$. By Lemma 6.5, $\Phi_S(f)$ can be regarded as a bounded operator from $\mathscr{F}_{b,r}(\mathscr{H})$ to $\mathscr{F}_{b,r+1}(\mathscr{H})$. Hence there exists a vector $\Theta_\varepsilon \in \mathscr{F}_{b,\mathrm{fin}}(\mathscr{D})$ such that $\|\Theta_\varepsilon - \Psi_\varepsilon\| < \varepsilon$ and $\|\Phi_S(f)\Theta_\varepsilon - \Phi_S(f)\Psi_\varepsilon\| < \varepsilon$. Therefore, using the triangle inequality, we obtain

$$\|\Theta_\varepsilon - \Psi\| < 2\varepsilon, \quad \|\Phi_S(f)\Theta_\varepsilon - \Phi_S(f)\Psi\| < 2\varepsilon.$$

This means that $\mathscr{F}_{b,\mathrm{fin}}(\mathscr{D})$ is a core for $\Phi_S(f)$. ∎

By Theorem 6.17, for all $f \in \mathscr{H}$, $e^{i\Phi_S(f)}$ is a unitary operator on $\mathscr{F}_b(\mathscr{H})$. This unitary operator is called the **Weyl operator** with test vector f.

Theorem 6.18

$$\left\langle \Omega_{\mathscr{H}}, e^{i\Phi_S(f)}\Omega_{\mathscr{H}} \right\rangle = e^{-\|f\|^2/4}, \tag{6.106}$$

$$\left\langle A(f_1)^* \cdots A(f_n)^*\Omega_{\mathscr{H}}, e^{i\Phi_S(f)}\Omega_{\mathscr{H}} \right\rangle = \frac{i^n}{2^{n/2}} \langle f_1, f\rangle \cdots \langle f_n, f\rangle \, e^{-\|f\|^2/4}. \tag{6.107}$$

Proof By Lemma 6.9, we have

$$\left\langle \Omega_{\mathscr{H}}, e^{i\Phi_S(f)}\Omega_{\mathscr{H}} \right\rangle = \sum_{n=0}^{\infty} \frac{i^n}{n!} C_n,$$

where

$$C_n := \left\langle \Omega_{\mathscr{H}}, \Phi_{\mathrm{S}}(f)^n \Omega_{\mathscr{H}} \right\rangle.$$

Using (6.39) and the commutation relation

$$[A(f), \Phi_{\mathrm{S}}(g)] = \frac{1}{\sqrt{2}} \langle f, g \rangle \quad \text{on } \mathscr{F}_{\mathrm{b},0}(\mathscr{H}) \ (f, g \in \mathscr{H}), \tag{6.108}$$

we see that

$$C_n = \frac{(n-1)}{2} \|f\|^2 C_{n-2}.$$

Note that $C_1 = 0$. Hence $C_{2n-1} = 0$, $n \in \mathbb{N}$. On the other hand, we have by iteration

$$C_{2n} = \frac{\|f\|^{2n}(2n)!}{2^{2n} n!}.$$

Hence (6.106) follows.

By Lemma 6.9 again, we have

$$\left\langle A(f_1)^* \cdots A(f_n)^* \Omega_{\mathscr{H}}, e^{i\Phi_{\mathrm{S}}(f)} \Omega_{\mathscr{H}} \right\rangle = \sum_{m=0}^{\infty} \frac{i^m}{m!} a_m,$$

where

$$a_m := \left\langle A(f_1)^* \cdots A(f_n)^* \Omega_{\mathscr{H}}, \Phi_{\mathrm{S}}(f)^m \Omega_{\mathscr{H}} \right\rangle.$$

It is easy to see that, if $m < n$, then $a_m = 0$. So let $m \geq n$. Then, using (6.108) and (6.39), one can show that

$$a_m = \frac{m \langle f_1, f \rangle}{\sqrt{2}} \left\langle A(f_2)^* \cdots A(f_n)^* \Omega_{\mathscr{H}}, \Phi_{\mathrm{S}}(f)^{m-1} \Omega_{\mathscr{H}} \right\rangle.$$

Iterating the right hand side similarly, we obtain

$$a_m = \frac{m!}{2^{n/2}(m-n)!} \langle f_1, f \rangle \cdots \langle f_n, f \rangle \left\langle \Omega_{\mathscr{H}}, \Phi_{\mathrm{S}}(f)^{m-n} \Omega_{\mathscr{H}} \right\rangle.$$

Hence

$$\sum_{m=0}^{\infty} \frac{i^m}{m!} a_m = \frac{i^n}{2^{n/2}} \left(\prod_{j=1}^{n} \langle f_j, f \rangle \right) \sum_{k=0}^{\infty} \frac{i^k}{k!} \left\langle \Omega_{\mathscr{H}}, \Phi_{\mathrm{S}}(f)^k \Omega_{\mathscr{H}} \right\rangle$$
$$= \frac{i^n}{2^{n/2}} \left(\prod_{j=1}^{n} \langle f_j, f \rangle \right) \left\langle \Omega_{\mathscr{H}}, e^{i\Phi_{\mathrm{S}}(f)} \Omega_{\mathscr{H}} \right\rangle.$$

Thus (6.107) is obtained. ∎

6.11.3 Vacuum Expectation Values

In general, for a linear operator X on $\mathscr{F}_{\mathrm{b}}(\mathscr{H})$ such that $\Omega_{\mathscr{H}} \in D(X)$, the number $\langle \Omega_{\mathscr{H}}, X\Omega_{\mathscr{H}} \rangle$ is called the **vacuum expectation value** of X.

A slight modification of the Segal field operator is given by

$$\Psi(f, g) := A(f) + A(g)^*, \quad f, g \in \mathscr{H}. \tag{6.109}$$

For each $n \in \mathbb{N}$ and $f_j, g_j \in \mathscr{H}$ $(j = 1, \ldots, n)$, we consider the vacuum expectation value

$$W_n((f_1, g_1), \cdots, (f_n, g_n)) := \langle \Omega_{\mathscr{H}}, \Psi(f_1, g_1) \cdots \Psi(f_n, g_n) \Omega_{\mathscr{H}} \rangle, \quad n \in \mathbb{N}. \tag{6.110}$$

of $\Psi(f_1, g_1) \cdots \Psi(f_n, g_n)$.

Using (6.39) and the CCR for $A(\cdot)^{\#}$, one can easily show that

$$W_2((f_1, g_1), (f_2, g_2)) = \langle f_1, g_2 \rangle. \tag{6.111}$$

Theorem 6.19 *Let $n \in \mathbb{N}$ and $f_j, g_j \in \mathscr{H}$, $j = 1, \ldots, n$. Then*

$$W_{2n-1}((f_1, g_1), \cdots, (f_{2n-1}, g_{2n-1})) = 0, \tag{6.112}$$

$$W_{2n}((f_1, g_1), \cdots, (f_{2n}, g_{2n})) = \sum_{\text{pairings}} \langle f_{i_1}, g_{j_1} \rangle \langle f_{i_2}, g_{j_2} \rangle \cdots \langle f_{i_n}, g_{j_n} \rangle, \tag{6.113}$$

where $\sum_{\text{pairings}}$ means the sum over all $(2n)!/2^n n!$ ways of writing $1, \ldots, 2n$ as n distinct (unordered) pairs $(i_1, j_1), \ldots, (i_n, j_n)$.

Proof By (6.39) and the CCR for $A(\cdot)^{\#}$, we have

$$W_n((f_1, g_1), \cdots, (f_n, g_n))$$
$$= \sum_{j=2}^{n} \langle f_1, g_j \rangle \left\langle \Omega_{\mathscr{H}}, \Psi(f_2, g_2) \cdots \widehat{\Psi(f_j, g_j)} \cdots \Psi(f_n, g_n) \Omega_{\mathscr{H}} \right\rangle. \tag{6.114}$$

If n is odd, then a successive iteration of this equation finally gives

$$W_n((f_1, g_1), \cdots, (f_n, g_n)) = \sum_{i=1}^{n} c_i \langle \Omega_{\mathscr{H}}, \Psi(f_i, g_i) \Omega_{\mathscr{H}} \rangle,$$

where c_i is a constant. But, by (6.39), $\langle \Omega_{\mathscr{H}}, \Psi(f_i, g_i) \Omega_{\mathscr{H}} \rangle = 0$. Hence (6.112) follows. Formula (6.113) can be proved by (6.114) and induction in n. ■

Since

$$\Phi_S(f) = \frac{1}{\sqrt{2}} \Psi(f, f) \quad \text{on } D(A(f)) \cap D(A(f)^*),$$

Theorem 6.19 implies the following corollary:

Corollary 6.12 *Let $n \in \mathbb{N}$ and $f_j \in \mathscr{H}$, $j = 1, \ldots, 2n$. Then*

$$\langle \Omega_{\mathscr{H}}, \Phi_S(f_1) \cdots \Phi_S(f_{2n-1}) \Omega_{\mathscr{H}} \rangle = 0,$$
$$\langle \Omega_{\mathscr{H}}, \Phi_S(f_1) \cdots \Phi_S(f_{2n}) \Omega_{\mathscr{H}} \rangle = \frac{1}{2^n} \sum_{\text{pairings}} \langle f_{i_1}, f_{j_1} \rangle \cdots \langle f_{i_n}, f_{j_n} \rangle.$$

6.11.4 *Irreducibility of Segal Field Operators*

We next consider irreducibility of Segal field operators.

Theorem 6.20 *For any dense subspace $\mathscr{D}$ of $\mathscr{H}$, $\{\Phi_S(f) | f \in \mathscr{D}\}' = \mathbb{C}I$. In particular, $\{\Phi_S(f) | f \in \mathscr{D}\}$ is irreducible.*

Proof Let $T \in \{\Phi_S(f) | f \in \mathscr{D}\}'$. Then, for all $f \in \mathscr{D}$, $T\Phi_S(f) \subset \Phi_S(f)T$. Let $\mathscr{D}_f := D(A(f)^*) = D(A(f))$ (see Corollary 6.3). Hence, for all $\Psi, \Phi \in \mathscr{D}_f$,

$$\left\langle \Psi, T(A(f)^* + A(f))\Phi \right\rangle = \left\langle (A(f) + A(f)^*)\Psi, T\Phi \right\rangle.$$

Replacing f by if, one has

$$\left\langle \Psi, T(A(f)^* - A(f))\Phi \right\rangle = \left\langle (A(f) - A(f)^*)\Psi, T\Phi \right\rangle.$$

From these two equations, one obtains

$$\left\langle \Psi, TA(f)^*\Phi \right\rangle = \langle A(f)\Psi, T\Phi \rangle, \quad \langle \Psi, TA(f)\Phi \rangle = \left\langle A(f)^*\Psi, T\Phi \right\rangle.$$

The first equation implies that $T\Phi \in D(A(f)^*)$ and $TA(f)^*\Phi = A(f)^*T\Phi$, i.e., $TA(f)^* \subset A(f)^*T$. Similarly, the second equation implies that $TA(f) \subset A(f)T$. Hence $T \in \{A(f), A(f)^*|f \in \mathscr{D}\}' = \mathbb{C}I$ (by Theorem 6.9), i.e., $T = cI$ for some $c \in \mathbb{C}$. Thus $\{\Phi_S(f)|f \in \mathscr{D}\}' = \mathbb{C}I$ and hence $\{\Phi_S(f)|f \in \mathscr{D}\}$ is irreducible. ■

6.11.5 Some Formulae and Spectrum of $\Phi_S(f)$

Let $\mathfrak{A}$ be an algebra and $A \in \mathfrak{A}$. Then, for each $n \in \mathbb{Z}_+$, one can define a mapping $\mathrm{ad}^n(A)\colon \mathfrak{A} \to \mathfrak{A}$ by the following recurrence relations: for all $B \in \mathfrak{A}$,

$$\mathrm{ad}^0(A)B := B, \quad \mathrm{ad}^n(A)B := [A, \mathrm{ad}^{n-1}(A)B], \quad n \geq 1. \tag{6.115}$$

The mapping $\mathrm{ad}^n(A)$ is called the **adjoint operator** associated with A.

Lemma 6.10 *For all $A,\ B \in \mathfrak{A}$ and $N \in \mathbb{N}$,*

$$\sum_{n=0}^{N} {}_N C_n (-1)^n A^{N-n} B A^n = \mathrm{ad}^N(A)B, \tag{6.116}$$

where ${}_N C_n$ is the binomial coefficient with respect to (N, n) (see (4.99)).

Proof This can be proved by induction in N, where the following formula is used:

$${}_{n+1}C_k = {}_n C_k + {}_n C_{k-1}, \quad k = 1, \ldots, n,\ n \in \mathbb{N}. \tag{6.117}$$

■

Theorem 6.21 *Let $f, g \in \mathscr{H}$. Then the following operator equalities hold:*

$$e^{i\Phi_S(f)} A(g) e^{-i\Phi_S(f)} = A(g) - \frac{i}{\sqrt{2}} \langle g, f \rangle. \tag{6.118}$$

$$e^{i\Phi_S(f)} A(g)^* e^{-i\Phi_S(f)} = A(g)^* + \frac{i}{\sqrt{2}} \langle f, g \rangle. \tag{6.119}$$

Proof Let $\Psi, \Phi \in \mathscr{F}_{\mathrm{b},0}(\mathscr{H})$. Then, in the same way as in the proof of Lemma 6.9, one can show that $e^{-i\Phi_S(f)}\Psi \in D(A(f)^\#)$ and

$$A(g)^\# e^{-i\Phi_S(f)}\Psi = \sum_{n=0}^{\infty} \frac{(-i)^n}{n!} A(g)^\# \Phi_S(f)^n \Psi.$$

Hence

$$\left\langle e^{-i\Phi_S(f)}\Phi, A(g)e^{-i\Phi_S(f)}\Psi\right\rangle = \sum_{n=0}^{\infty}\sum_{m=0}^{\infty}\frac{i^m}{m!}\frac{(-i)^n}{n!}\left\langle\Phi, \Phi_S(f)^m A(g)\Phi_S(f)^n\Psi\right\rangle.$$

Since the series on the right hand side is absolutely convergent, one can rearrange the summation $\sum_{m=0}^{\infty}\sum_{n=0}^{\infty}$ in the form $\sum_{N=0}^{\infty}\sum_{n+m=N}$ so that

$$\left\langle e^{-i\Phi_S(f)}\Phi, A(g)e^{-i\Phi_S(f)}\Psi\right\rangle = \sum_{N=0}^{\infty}\frac{i^N}{N!}\langle\Phi, A_N\Psi\rangle$$

with

$$A_N := \sum_{n=0}^{N} {}_NC_n(-1)^n\Phi_S(f)^{N-n}A(g)\Phi_S(f)^n.$$

By (6.116), we have

$$A_N = \mathrm{ad}^N(\Phi_S(f))A(g) \quad \text{on } \mathscr{F}_{\mathrm{b},0}(\mathscr{H}).$$

We have

$$A_0\Psi = A(g)\Psi,$$

$$A_1\Psi = [\Phi(f), A(g)]\Psi = -\frac{1}{\sqrt{2}}\langle g, f\rangle\Psi,$$

$$A_N\Psi = 0, \quad N \geq 2.$$

Hence

$$\left\langle e^{-i\Phi_S(f)}\Phi, A(g)e^{-i\Phi_S(f)}\Psi\right\rangle = \left\langle\Phi, \left(A(g) - \frac{i}{\sqrt{2}}\langle g, f\rangle\right)\Psi\right\rangle.$$

Therefore

$$e^{i\Phi_S(f)}A(g)e^{-i\Phi_S(f)}\Psi = \left(A(g) - \frac{i}{\sqrt{2}}\langle g, f\rangle\right)\Psi.$$

Since $\mathscr{F}_{\mathrm{b},0}(\mathscr{H})$ is a core for $A(f)$ by Proposition 6.5, it follows from a limiting argument that

$$e^{i\Phi_S(f)}A(g)e^{-i\Phi_S(f)} \supset A(g) - \frac{i}{\sqrt{2}}\langle g, f\rangle.$$

Since $f \in \mathscr{H}$ is arbitrary, replacing f with $-f$ gives

$$e^{-i\Phi_S(f)}A(g)e^{i\Phi_S(f)} \supset A(g) + \frac{i}{\sqrt{2}}\langle g, f\rangle$$

which is equivalent to

$$A(g) - \frac{i}{\sqrt{2}}\langle g, f\rangle \supset e^{i\Phi_S(f)}A(g)e^{-i\Phi_S(f)}.$$

Thus operator equality (6.118) holds. Operator equality (6.119) follows from taking the adjoint of (6.118). ■

Corollary 6.13 *For all $f, g \in \mathscr{H}$,*

$$e^{i\Phi_S(f)}\Phi_S(g)e^{-i\Phi_S(f)} = \Phi_S(g) - \operatorname{Im}\langle f, g\rangle . \tag{6.120}$$

Proof By (6.118) and (6.119), we have

$$e^{i\Phi_S(f)}\Phi_S(g)e^{-i\Phi_S(f)} = \Phi_S(g) - \operatorname{Im}\langle f, g\rangle$$

on $D(A(g)) \cap D(A(g)^*)$. Since $D(A(g)) \cap D(A(g)^*)$ is a core for $\Phi_S(g)$ and $\Phi_S(g)$ is self-adjoint, operator equality (6.120) follows. ■

Corollary 6.13 implies the following result on the spectrum of $\Phi_S(f)$.

Corollary 6.14 *For all $f \in \mathscr{H} \setminus \{0\}$,*

$$\sigma(\Phi_S(f)) = \mathbb{R}. \tag{6.121}$$

Proof Let $t \in \mathbb{R}$ and $U(t) := e^{i\Phi_S(itf)}$. Then, by (6.120), we have

$$U(t)\Phi_S(f)U(t)^{-1} = \Phi_S(f) = \Phi_S(f) + t\|f\|^2.$$

By the unitary invariance of spectrum, we have

$$\sigma(\Phi_S(f)) = \sigma(\Phi_S(f) + t\|f\|^2) = \{\lambda + t\|f\|^2 | \lambda \in \sigma(\Phi_S(f))\}.$$

Since $t \in \mathbb{R}$ is arbitrary, we obtain (6.121). ■

6.11.6 Properties of Weyl Operators

Theorem 6.22 *Let $f, g \in \mathscr{H}$.*

(i)

$$e^{i\Phi_S(f)}e^{i\Phi_S(g)} = e^{-i\mathrm{Im}\langle f,g\rangle}e^{i\Phi_S(g)}e^{i\Phi_S(f)}. \tag{6.122}$$

(ii)

$$e^{i\Phi_S(f+g)} = e^{i\mathrm{Im}\langle f,g\rangle/2}e^{i\Phi_S(f)}e^{i\Phi_S(g)}, \tag{6.123}$$

(iii) (strong continuity) *Let $f_n \in \mathscr{H}$ and $f_n \to f$ $(n \to \infty)$. Then*

$$\text{s-}\lim_{n\to\infty} e^{i\Phi_S(f_n)} = e^{i\Phi_S(f)}. \tag{6.124}$$

(iv) (irreducibility) *For any dense subspace $\mathscr{D} \subset \mathscr{H}$, $\{e^{i\Phi_S(f)} | f \in \mathscr{D}\}' = \mathbb{C}I$. In particular, $\{e^{i\Phi_S(f)} | f \in \mathscr{D}\}$ is irreducible.*

Proof

(i) By Corollary 6.13 and the unitary covariance of functional calculus, we have

$$e^{i\Phi_S(f)}e^{i\Phi_S(g)}e^{-i\Phi_S(f)} = e^{i(\Phi_S(g)-\mathrm{Im}\,\langle f,g\rangle} = e^{-i\mathrm{Im}\,\langle f,g\rangle}e^{i\Phi_S(g)}.$$

Multiplying $e^{i\Phi_S(f)}$ from the right side, we obtain (6.122).

(ii) Let

$$U(t) := e^{it^2\mathrm{Im}\,\langle f,g\rangle/2}e^{it\Phi_S(f)}e^{it\Phi_S(g)}, \quad t \in \mathbb{R}.$$

Then, in the same way as in the proof of Theorem 2.18, one can show that $\{U(t)\}_{t\in\mathbb{R}}$ is a strongly continuous one-parameter unitary group. Hence, by the Stone theorem, there exists a unique self-adjoint operator S on $\mathscr{F}_b(\mathscr{H})$ such that $U(t) = e^{itS}$, $t \in \mathbb{R}$. It is easy to see that, for all $\Psi \in D(\Phi_S(f)) \cap D(\Phi_S(g))$,

$$\lim_{t\to 0}\frac{1}{t}(U(t)-1)\Psi = i(\Phi_S(f)+\Phi_S(g))\Psi.$$

Hence $\Psi \in D(S)$ and $(\Phi_S(f)+\Phi_S(g))\Psi = S\Psi$. Therefore $\Phi_S(f)+\Phi_S(g)) \subset S$. Note that

$$\Phi_S(f)+\Phi_S(g) \supset \Phi_S(f+g) \upharpoonright \mathscr{F}_{b,0}(\mathscr{H}).$$

Since $\Phi_S(f+g)$ is essentially self-adjoint on $\mathscr{F}_{b,0}(\mathscr{H})$, it follows that $\Phi(f+g) = S$. Thus (6.123) holds.

(iii) Let $\Psi \in \mathscr{F}_{\mathrm{b},0}(\mathscr{H})$. Then, by direct computation, we have

$$\|e^{i\Phi_{\mathrm{S}}(f_n)}\Psi - e^{i\Phi_{\mathrm{S}}(f)}\Psi\|^2 = 2\|\Psi\|^2 - 2\mathrm{Re}\left\langle \Psi, e^{-i\Phi_{\mathrm{S}}(f_n)}e^{i\Phi_{\mathrm{S}}(f)}\Psi \right\rangle.$$

By (6.123),

$$e^{-i\Phi_{\mathrm{S}}(f_n)}e^{i\Phi_{\mathrm{S}}(f)}\Psi = e^{i\mathrm{Im}\,\langle f_n, f\rangle/2}e^{i\Phi_{\mathrm{S}}(f-f_n)}\Psi.$$

Hence

$$\|e^{i\Phi_{\mathrm{S}}(f_n)}\Psi - e^{i\Phi_{\mathrm{S}}(f)}\Psi\|^2 = 2\|\Psi\|^2 - 2\mathrm{Re}\left\{e^{i\mathrm{Im}\,\langle f_n, f\rangle/2}\left\langle \Psi, e^{i\Phi_{\mathrm{S}}(f-f_n)}\Psi \right\rangle\right\}.$$

We have $\lim_{n\to\infty} \mathrm{Im}\,\langle f_n, f\rangle = \mathrm{Im}\,\|f\|^2 = 0$. By Proposition 1.14 and Proposition 6.11(iv) with $T = I$,

$$\|e^{i\Phi_{\mathrm{S}}(f-f_n)}\Psi - \Psi\| \le \|\Phi_{\mathrm{S}}(f - f_n)\Psi\| \to 0\ (n \to \infty).$$

Hence the right hand side of the above equation converges to 0 as $n \to \infty$. Thus

$$\lim_{n\to\infty} e^{i\Phi_{\mathrm{S}}(f_n)}\Psi = e^{i\Phi_{\mathrm{S}}(f)}\Psi.$$

Since $\mathscr{F}_{\mathrm{b},0}(\mathscr{H})$ is dense in $\mathscr{F}_{\mathrm{b}}(\mathscr{H})$ and $\sup_{n\ge 1}\|e^{i\Phi_{\mathrm{S}}(f_n)}\| = 1 < \infty$, a limiting argument gives (6.124).

(iv) Let $T \in \{e^{i\Phi_{\mathrm{S}}(f)} | f \in \mathscr{D}\}'$. Then, for all $f \in \mathscr{D}$, $t \in \mathbb{R}$ and $\Psi \in D(\Phi_{\mathrm{S}}(f))$, $Te^{it\Phi_{\mathrm{S}}(f)}\Psi = e^{it\Phi_{\mathrm{S}}(f)}T\Psi$. Since $e^{it\Phi_{\mathrm{S}}(f)}\Psi$ is strongly differentiable in t and T is bounded, it follows that $e^{it\Phi_{\mathrm{S}}(f)}T\Psi$ also is strongly differentiable in t. Hence, considering the strong differentiation of the above equation at $t = 0$, one obtains that $T\Psi \in D(\Phi_{\mathrm{S}}(f))$ and $T\Phi_{\mathrm{S}}(f)\Psi = \Phi_{\mathrm{S}}(f)T\Psi$. This means that $T\Phi_{\mathrm{S}}(f) \subset \Phi_{\mathrm{S}}(f)T$. Therefore $T \in \{\Phi_{\mathrm{S}}(f) | f \in \mathscr{D}\}'$. Hence, by Theorem 6.20, $T = cI$ for some $c \in \mathbb{C}$. Thus $\{e^{i\Phi_{\mathrm{S}}(f)} | f \in \mathscr{D}\}$ is irreducible.

■

Formula (6.122) for the Weyl operators $e^{i\Phi_{\mathrm{S}}(f)}$ and $e^{i\Phi_{\mathrm{S}}(g)}$ is called the **Weyl relation** for $\Phi_{\mathrm{S}}(f)$ and $\Phi_{\mathrm{S}}(g)$.

As a corollary to Theorem 6.22, we obtain an important result:

Corollary 6.15 (Strong Commutativity) *Let $f, g \in \mathscr{H}$. Then $\Phi_{\mathrm{S}}(f)$ and $\Phi_{\mathrm{S}}(g)$ strongly commute if and only if* $\mathrm{Im}\,\langle f, g\rangle = 0$.

Proof Suppose that $\Phi_{\mathrm{S}}(f)$ and $\Phi_{\mathrm{S}}(g)$ strongly commute. Then, by Proposition 1.17, for all $s, t \in \mathbb{R}$, $e^{it\Phi_{\mathrm{S}}(f)}e^{is\Phi_{\mathrm{S}}(g)} = e^{is\Phi_{\mathrm{S}}(g)}e^{it\Phi_{\mathrm{S}}(f)}$. Hence, by (6.122), $e^{-its\mathrm{Im}\,\langle f,g\rangle} = 1$ for all $s, t \in \mathbb{R}$. Hence $\mathrm{Im}\,\langle f, g\rangle = 0$.

Conversely, suppose that $\mathrm{Im}\,\langle f, g\rangle = 0$. Then $\mathrm{Im}\,\langle tf, sg\rangle = 0$, $s, t \in \mathbb{R}$. Using the real linearity of $\Phi_S(f)$ in f and (6.122), we obtain $e^{it\Phi_S(f)}e^{is\Phi_S(g)} = e^{is\Phi_S(g)}e^{it\Phi_S(f)}$. Hence, by Proposition 1.17 again, $\Phi_S(f)$ and $\Phi_S(g)$ strongly commute. ■

6.12 Canonical Free Bose Field and Canonical Conjugate Momentum

Let J be a conjugation on $\mathscr{H}$ and $\mathscr{H}_J$ be the real part of $\mathscr{H}$ with respect to J (see (1.68)). For each $f \in \mathscr{H}_J$, we define

$$\phi_F(f) := \Phi_S(f), \quad \pi_F(f) := \Phi_S(if), \tag{6.125}$$

where $\Phi_S(f)$ is the Segal field operator with test vector f. By Proposition 6.11(i), we have for all $f, g \in \mathscr{H}_J$

$$[\phi_F(f), \pi_F(g)] = i\,\langle f, g\rangle\,, \tag{6.126}$$

$$[\phi_F(f), \phi_F(g)] = 0, \quad [\pi_F(f), \pi_F(g)] = 0 \tag{6.127}$$

on $\mathscr{F}_{b,0}(\mathscr{H})$. It follows from (6.99) that, for all $f, g \in \mathscr{H}_J$ and $a, b \in \mathbb{R}$,

$$\phi_F(af + bg) = a\phi_F(f) + b\phi_F(g), \tag{6.128}$$

$$\pi_F(af + bg) = a\pi_F(f) + b\pi_F(g) \quad \text{on } \mathscr{D}_b. \tag{6.129}$$

The mapping $f \mapsto \phi_F(f)$ is called the **canonical free Bose field** over $\mathscr{H}_J$ and the mapping $f \mapsto \pi_F(f)$ is called the **canonical conjugate momentum** of $\phi_F(f)$.

Theorem 6.22 is rewritten in terms of $\phi_F(f)$ and $\pi_F(f)$ as follows:

Theorem 6.23 *Let $f, g \in \mathscr{H}_J$ and $a, b \in \mathbb{R}$.*

(i) *(Weyl relations)*

$$e^{i\phi_F(f)}e^{i\pi_F(g)} = e^{-i\langle f,g\rangle}e^{i\pi_F(g)}e^{i\phi_F(f)}, \tag{6.130}$$

$$e^{i\phi_F(f)}e^{i\phi_F(g)} = e^{i\phi_F(g)}e^{i\phi_F(f)}, \tag{6.131}$$

$$e^{i\pi_F(f)}e^{i\pi_F(g)} = e^{i\pi_F(g)}e^{i\pi_F(f)}. \tag{6.132}$$

(ii)

$$e^{i\phi_F(af+bg)} = e^{ia\phi_F(f)}e^{ib\phi_F(g)}, \quad e^{i\pi_F(af+bg)} = e^{ia\pi_F(f)}e^{ib\pi_F(g)}. \tag{6.133}$$

(iii) (strong continuity) *Let* $f_n \in \mathscr{H}_J$ *and* $f_n \to f \in \mathscr{H}_J$ $(n \to \infty)$. *Then*

$$\text{s-}\lim_{n\to\infty} e^{i\phi_{\mathrm{F}}(f_n)} = e^{i\phi_{\mathrm{F}}(f)}, \quad \text{s-}\lim_{n\to\infty} e^{i\pi_{\mathrm{F}}(f_n)} = e^{i\pi_{\mathrm{F}}(f)}. \tag{6.134}$$

(iv) (irreducibility) *For any dense subspace* $\mathscr{W} \subset \mathscr{H}_J$, $\{e^{i\phi_{\mathrm{F}}(f)}, e^{i\pi_{\mathrm{F}}(f)} | f \in \mathscr{W}\}' = \mathbb{C}I$. *In particular,* $\{e^{i\phi_{\mathrm{F}}(f)}, e^{i\pi_{\mathrm{F}}(f)} | f \in \mathscr{W}\}$ *is irreducible.*

Proof We prove only (iv). Let $T \in \{e^{i\phi_{\mathrm{F}}(f)}, e^{i\pi_{\mathrm{F}}(f)} | f \in \mathscr{W}\}'$. Then, for all $f, g \in \mathscr{W}$, $Te^{i\phi_{\mathrm{F}}(f)} = e^{i\phi_{\mathrm{F}}(f)}T$ and $Te^{i\pi_{\mathrm{F}}(g)} = e^{i\pi_{\mathrm{F}}(g)}T$. Hence $Te^{i\phi_{\mathrm{F}}(f)}e^{i\pi_{\mathrm{F}}(g)} = e^{i\phi_{\mathrm{F}}}e^{i\pi_{\mathrm{F}}(g)}T$. Using (6.123), we obtain $Te^{i\Phi_{\mathrm{S}}(f+ig)} = e^{i\Phi_{\mathrm{S}}(f+ig)}T$. Note that $\{f + ig | f, g \in \mathscr{W}\}$ is dense in $\mathscr{H}$. Hence, by Theorem 6.22(iv), $T = \alpha I$ for some $\alpha \in \mathbb{C}$. Thus the desired result follows. ■

6.13 Symplectic Spaces and Generalization of Segal Field Operators

As is easily seen, the real number

$$s(f, g) := \mathrm{Im}\,\langle f, g\rangle\,, \quad f, g \in \mathscr{H}$$

appearing in the commutation relation (6.103) of Segal field operators has the following properties:

$$s(af_1 + bf_2, g) = as(f_1, g) + bs(f_2, g), \quad f_1, f_2, g \in \mathscr{H},\ a, b \in \mathbb{R}, \tag{6.135}$$

$$s(f, g) = -s(g, f), \quad f, g \in \mathscr{H}, \tag{6.136}$$

$$\{f \in \mathscr{H} | s(f, g) = 0,\ \forall g \in \mathscr{H}\} = \{0\}. \tag{6.137}$$

Equations (6.135) and (6.136) imply that the correspondence $\mathscr{H} \times \mathscr{H} \ni (f, g) \mapsto s(f, g)$ is a real-bilinear form. Note that the complex Hilbert space $\mathscr{H}$ can be viewed as a real Hilbert space by replacing the coefficient field $\mathbb{C}$ with $\mathbb{R}$.

These properties can be abstracted to obtain a general concept of space. Let $\mathscr{X}$ be a vector space over $\mathbb{K}$ ($\mathbb{K} = \mathbb{R}$ or $\mathbb{C}$). Let ω be a mapping from $\mathscr{X} \times \mathscr{X}$ to $\mathbb{K}$ satisfying the following:

(i) ($\mathbb{K}$-bilinearity) For all $f, g, h \in \mathscr{X}$ and $a, b \in \mathbb{K}$,

$$\omega(af + bg, h) = a\,\omega(f, h) + b\,\omega(g, h),$$

$$\omega(h, af + bg) = a\,\omega(h, f) + b\,\omega(h, g).$$

(ii) (anti-symmetry) For all $f, g \in \mathscr{X}$, $\omega(f, g) = -\omega(g, f)$.

(iii) (non-degeneracy) $\{f \in \mathscr{X} | \omega(f, g) = 0, \forall g \in \mathscr{X}\} = \{0\}$.

The mapping ω is called a $\mathbb{K}$-**symplectic form** on $\mathscr{X}$ and $(\mathscr{X}, \omega)$ is called a $\mathbb{K}$-symplectic space. If $\mathbb{K} = \mathbb{R}$ (resp. $\mathbb{C}$), then $(\mathscr{X}, \omega)$ is called a **real** (resp. **complex**) **symplectic space**.

Example 6.5 The pair $(\mathscr{H}, s)$ described above is a real symplectic space.

Example 6.6 Let T be an injective bounded self-adjoint operator on $\mathscr{X}$ and

$$\omega(f, g) = \mathrm{Im}\, \langle f, Tg \rangle_{\mathscr{X}}, \; f, g \in \mathscr{X}.$$

Then ω is an $\mathbb{R}$-symplectic form on $\mathscr{X}$.

Example 6.7 Let $\mathscr{X}$ be a complex Hilbert space and C be a conjugation on $\mathscr{X}$ (see Definition 1.2). Suppose that there exists a bounded linear operator W on $\mathscr{X}$ such that $\|Wf\| = \|f\|$, $f \in \mathscr{X}$ and

$$W^2 = -I, \quad CW = WC.$$

Let

$$\omega_{C,W}(f, g) := \langle Cf, Wg \rangle_{\mathscr{X}}, \quad f, g \in \mathscr{X}.$$

Then $\omega_{C,W}$ is a $\mathbb{C}$-symplectic form on $\mathscr{X}$.[4] Hence $(\mathscr{X}, \omega_{C,W})$ is a complex symplectic space.

Let $(\mathscr{X}, \omega)$ be a $\mathbb{K}$-symplectic space. Then a linear operator S: $\mathscr{X} \to \mathscr{X}$ is called an **iso-symplectic transformation** if

$$\omega(Sf, Sg) = \omega(f, g), \quad f, g \in \mathscr{X}.$$

It follows from the non-degeneracy of ω that any iso-symplectic transformation is injective.

If S is an iso-symplectic transformation and surjective, then S is called a **symplectic transformation**.

---

[4]It is straightforward to prove the $\mathbb{C}$-bilinearity of $\omega_{C,W}$. With regard to anti-symmetry, one can proceed as follows, where properties (1.67), $C^2 = I$, $CW = WC$,$\langle Wf, Wg \rangle = \langle f, g \rangle$, $f, g \in \mathscr{X}$ (this follows from the polarization identity and norm-preserving property $\|Wf\| = \|f\|$, $f \in \mathscr{X}$) and $W^2 = -I$ are used:

$$\omega_{C,W}(f, g) = \langle CWg, f \rangle = \langle WCg, f \rangle = \langle W(WCg), Wf \rangle = -\langle Cg, Wf \rangle = -\omega_{C,W}(g, f).$$

To prove non-degeneracy of $\omega_{C,W}$, let $g \in \mathscr{H}$ be such that, for all $f \in \mathscr{H}$, $\omega_{C,W}(f, g) = 0$. Then $0 = \langle CWg, f \rangle$. Hence $CWg = 0$. Since $C^2 = I$, it follows that $Wg = 0$, implying $g = 0$. Thus $\omega_{C,W}$ is non-degenerate.

If S is a symplectic transformation, then so is S^{-1}. Moreover, if S and T are symplectic transformations, then so is ST. It follows that the set

$$\mathrm{Sp}(\mathscr{X}) := \{S : \mathscr{X} \to \mathscr{X} | S \text{ is symplectic}\}$$

of all symplectic transformations on $\mathscr{X}$ forms a group. This group is called the **symplectic group** on $(\mathscr{X}, \omega)$.

With the above generalization of $s(f, g)$, it may be natural to generalize the concept of Segal field operators. Let $(\mathscr{X}, \omega)$ be a real symplectic space and $\{\Phi_\omega(f) | f \in \mathscr{X}\}$ be a family of self-adjoint operators on a complex Hilbert space $\mathscr{F}$. Then $\Phi_\omega(f)$ is called an **ω-Segal field operator** with test vector $f \in \mathscr{X}$ if there exists a dense subspace $\mathscr{D}$ of $\mathscr{F}$ left invariant by $\Phi_\omega(f)$ $(f \in \mathscr{X})$ such that, for all $f, g \in \mathscr{D}$ and $a, b \in \mathbb{R}$,

$$\Phi_\omega(af + bg) = a\Phi_\omega(f) + b\Phi_\omega(g) \quad \text{on } \mathscr{D}$$

and

$$[\Phi_\omega(f), \Phi_\omega(g)] = i\omega(f, g), \quad f, g \in \mathscr{X}$$

on $\mathscr{D}$.

6.14 Quadratic Operators

For later use (see Sect. 8.6), we introduce operators quadratic in the annihilation operator $A(\cdot)$ and the creation operator $A(\cdot)^*$. Let K be a Hilbert–Schmidt operator on a Hilbert space $\mathscr{H}$ and J be a conjugation on $\mathscr{H}$. Then, by Theorem 1.4, there exist ONS's $\{\phi_n\}_{n=1}^{N_K}$, $\{\psi_n\}_{n=1}^{N_K}$ of $\mathscr{H}$ $(N_K < \infty$ or $N_K = \infty)$ and $\lambda_n > 0$ $(n = 1, \ldots, N_K)$ such that

$$\sum_{n=1}^{\infty} \lambda_n^2 < \infty \quad \text{(in the case } N_K = \infty)$$

and

$$K = \sum_{n=1}^{N_K} \lambda_n \langle \psi_n, \cdot \rangle \phi_n. \tag{6.138}$$

If $N_K < \infty$, then we define operators $\langle A^\# | K | A^\# \rangle$ by

$$\langle A^* | K | A^* \rangle := \sum_{n=1}^{N_K} \lambda_n A(J\psi_n)^* A(\phi_n)^*, \tag{6.139}$$

$$\langle A | K | A \rangle := \sum_{n=1}^{N_K} \lambda_n A(\psi_n) A(J\phi_n). \tag{6.140}$$

In the case $N_K = \infty$, for each $N \in \mathbb{N}$, we first define an operator K_N on $\mathscr{H}$ by

$$K_N := \sum_{n=1}^{N} \lambda_n \langle \psi_n, \cdot \rangle \phi_n. \tag{6.141}$$

and set

$$\langle A^*|K_N|A^* \rangle := \sum_{n=1}^{N} \lambda_n A(J\psi_n)^* A(\phi_n)^*, \tag{6.142}$$

$$\langle A|K_N|A \rangle := \sum_{n=1}^{N} \lambda_n A(\psi_n) A(J\phi_n). \tag{6.143}$$

Lemma 6.11

(i) *For all* $\Psi \in \mathscr{F}_{\mathrm{b},0}(\mathscr{H})$, *the limit* $\lim_{N\to\infty} \langle A^*|K_N|A^* \rangle \Psi$ *exists.*
(ii) *For all* $\Psi \in \mathscr{F}_{\mathrm{b,fin}}(\mathscr{H})$, *the limit* $\lim_{N\to\infty} \langle A|K_N|A \rangle \Psi$ *exists.*

Proof For notational simplicity, we set $T_N := \langle A^*|K_N|A^* \rangle$ and $S_N := \langle A|K_N|A \rangle$.

(i) Let $\Psi \in \mathscr{F}_{\mathrm{b},0}(\mathscr{H})$. Then $(T_N\Psi)^{(0)} = 0,\ (T_N\Psi)^{(1)} = 0$ and

$$(T_N\Psi)^{(p+2)} = \sqrt{(p+2)(p+1)} S_{p+2}(\eta_N \otimes \Psi^{(p)}), \quad p \geq 0,$$

where $\eta_N := \sum_{n=1}^{N} \lambda_n J\psi_n \otimes \phi_n \in \mathscr{H} \otimes \mathscr{H}$. For all $M > N$, we have

$$\|\eta_M - \eta_N\|^2 = \sum_{n=N+1}^{M} \lambda_n^2 \longrightarrow 0\ (M, N \to \infty).$$

Hence $\{\eta_N\}_N$ is a Cauchy sequence in $\mathscr{H} \otimes \mathscr{H}$. Therefore the limit

$$\eta(K) := \lim_{N\to\infty} \eta_N = \sum_{n=1}^{\infty} \lambda_n J\psi_n \otimes \phi_n \in \mathscr{H} \otimes \mathscr{H} \tag{6.144}$$

exists. By the continuity of tensor product operation and the boundedness of S_{p+2}, we see that

$$\lim_{N\to\infty} (T_N\Psi)^{(p+2)} = \sqrt{(p+2)(p+1)} S_{p+2}(\eta(K) \otimes \Psi^{(p)}). \tag{6.145}$$

Thus $\lim_{N\to\infty} T_N\Psi$ exists.

(ii) It is sufficient to prove that, for all vectors Ψ of the form

$$\Psi_p = A(f_1)^* \cdots A(f_p)^* \Omega_{\mathscr{H}}, \quad p \geq 0,\ f_j \in \mathscr{H},\ j = 1, \ldots, p, \tag{6.146}$$

$\lim_{N\to\infty} S_N \Psi$ exists. It is obvious that

$$S_N \Psi_p = 0, \quad p = 0, 1.$$

Hence $\lim_{N\to\infty} S_N \Psi_p = 0,\ p = 0, 1$.

Let $p \geq 2$. Then, using the CCR for $A(\cdot)^\#$, one can show that

$$S_N \Psi_p = \sum_{\substack{j,k=1 \\ j\neq k}}^{p} \Phi_N^{(j,k)} A(f_1)^* \cdots \widehat{A(f_j)^*} \cdots \widehat{A(f_k)^*} \cdots A(f_p)^* \Omega_{\mathscr{H}},$$

where $\Phi_N^{(j,k)} := \sum_{n=1}^N \lambda_n \langle J\phi_n, f_j \rangle \langle \psi_n, f_k \rangle$. Using (6.138), one can show that

$$\lim_{N\to\infty} \Phi_N^{(j,k)} = \langle K^* J f_j, f_k \rangle.$$

Hence

$$\lim_{N\to\infty} S_N \Psi_p = \sum_{\substack{j,k=1 \\ j\neq k}}^{p} \langle K^* J f_j, f_k \rangle A(f_1)^* \cdots \widehat{A(f_j)^*} \cdots \widehat{A(f_k)^*} \cdots A(f_p)^* \Omega_{\mathscr{H}}. \tag{6.147}$$

Thus the desired result follows. ■

Remark 6.6 By Corollary 1.4, $\eta(K) = U_J K$.

By Lemma 6.11, one can introduce densely defined operators $\langle A|K|A\rangle_0^\#$ as follows:

$$D(\langle A^*|K|A^*\rangle_0) := \mathscr{F}_{\mathrm{b},0}(\mathscr{H}), \tag{6.148}$$

$$\langle A^*|K|A^*\rangle_0 \Psi := \lim_{N\to\infty} \langle A^*|K_N|A^*\rangle \Psi, \quad \Psi \in D(\langle A^*|K|A^*\rangle_0) \tag{6.149}$$

and

$$D(\langle A|K|A\rangle_0) := \mathscr{F}_{\mathrm{b,fin}}(\mathscr{H}), \tag{6.150}$$

$$\langle A|K|A\rangle_0 \Psi := \lim_{N\to\infty} \langle A|K_N|A\rangle \Psi, \quad \Psi \in D(\langle A|K|A\rangle_0). \tag{6.151}$$

Recall that K^* also is Hilbert–Schmidt with

$$K^* = \sum_{n=1}^{N_K} \lambda_n \langle \phi_n, \cdot \rangle \psi_n.$$

Hence $\langle A^\# | K^* | A^\# \rangle_0$ is defined by replacing K with K^* in the above definitions. Then it is easy to see that

$$(\langle A^* | K | A^* \rangle_0)^* \supset \langle A | K^* | A \rangle_0, \tag{6.152}$$

$$(\langle A | K | A \rangle_0)^* \supset \langle A^* | K^* | A^* \rangle_0. \tag{6.153}$$

Hence $\langle A^\# | K | A^\# \rangle_0^*$ is densely defined. Therefore $\langle A^\# | K | A^\# \rangle_0$ is closable. Based on this fact, we define closed operators by

$$\left\langle A^\# | K | A^\# \right\rangle := \overline{\langle A^\# | K | A^\# \rangle_0}, \tag{6.154}$$

the closure of $\langle A^\# | K | A^\# \rangle_0$. We call operators of the forms $\langle A^* | K | A^* \rangle$ and $\langle A | K | A \rangle$ **quadratic operators** in A^* and A respectively.

It follows from (6.145) and (6.147) that

$$(\langle A^* | K | A^* \rangle \Psi)^{(0)} = 0, \quad (\langle A^* | K | A^* \rangle \Psi)^{(1)} = 0, \tag{6.155}$$

$$(\langle A^* | K | A^* \rangle \Psi)^{(p+2)} = \sqrt{(p+2)(p+1)} S_{p+2}(\eta(K) \otimes \Psi^{(p)}),$$
$$p \geq 0, \ \Psi \in D(\langle A^* | K | A^* \rangle) \tag{6.156}$$

and

$$\langle A | K | A \rangle = 0 \quad \text{on span}\{\Omega_{\mathscr{H}}, \ A(f)^* \Omega_{\mathscr{H}} | f \in \mathscr{H}\}, \tag{6.157}$$

$$\langle A | K | A \rangle \Psi_p = \sum_{\substack{j,k=1 \\ j \neq k}}^{p} \langle K^* J f_j, f_k \rangle A(f_1)^* \cdots \widehat{A(f_j)^*} \cdots \widehat{A(f_k)^*} \cdots A(f_p)^* \Omega_{\mathscr{H}} \tag{6.158}$$

for vectors Ψ_p of the form (6.146).

The following lemma summarizes basic properties of quadratic operators.

Lemma 6.12

(i) $\mathscr{F}_{\mathrm{b},0}(\mathscr{H}) \subset D(\langle A^* | K | A^* \rangle)$. *Moreover, for all* $\Psi \in D(\langle A^* | K | A^* \rangle)$ *and* $p \in \mathbb{Z}_+$,

$$\|(\langle A^* | K | A^* \rangle \Psi)^{(p+2)}\| \leq \sqrt{(p+2)(p+1)} \|\eta(K)\| \, \|\Psi^{(p)}\|. \tag{6.159}$$

In particular, $\langle A^*|K|A^*\rangle$ is a bounded operator from the p-particle space $\otimes_{\mathrm{s}}^p \mathscr{H}$ to the $(p+2)$-particle space $\otimes_{\mathrm{s}}^{p+2} \mathscr{H}$.

(ii) *$D(N_{\mathrm{b}}) \subset D(\langle A^*|K|A^*\rangle)$ and*

$$\left\| \left\langle A^*|K|A^*\right\rangle \Psi \right\| \leq \|\eta(K)\| \, \|(N_{\mathrm{b}}+2)\Psi\|, \quad \Psi \in D(N_{\mathrm{b}}). \tag{6.160}$$

(iii) *$\mathscr{F}_{\mathrm{b},0}(\mathscr{H}) \subset D(\langle A|K|A\rangle)$. Moreover, for each $p \in \mathbb{Z}_+$ and $\Psi \in D(\langle A|K|A\rangle)$,*

$$\|(\langle A|K|A\rangle \Psi)^{(p-2)}\| \leq \sqrt{(p-1)p}\|\eta(K^*)\| \, \|\Psi^{(p)}\|. \tag{6.161}$$

In particular, $\langle A|K|A\rangle$ is a bounded operator from the p-particle space $\otimes_{\mathrm{s}}^p \mathscr{H}$ to the $(p-2)$-particle space $\otimes_{\mathrm{s}}^{p-2} \mathscr{H}$, where $\otimes_{\mathrm{s}}^{-2} \mathscr{H} := \{0\}$ and $\otimes_{\mathrm{s}}^{-1} \mathscr{H} := \{0\}$.

(iv) *$D(N_{\mathrm{b}}) \subset D(\langle A|K|A\rangle)$ and*

$$\| \langle A|K|A\rangle \Psi \| \leq \|\eta(K)\| \, \|N_{\mathrm{b}}\Psi\|, \quad \Psi \in D(N_{\mathrm{b}}). \tag{6.162}$$

Proof

(i) It is obvious from the definition of $\langle A^*|K|A^*\rangle$ that $\mathscr{F}_{\mathrm{b},0}(\mathscr{H})$ is a subset of $D(\langle A^*|K|A^*\rangle)$. Inequality (6.159) follows from (6.156). This implies the desired boundedness property of $\langle A^*|K|A^*\rangle$.

(ii) By (6.159), we have

$$\begin{aligned}\|(\left\langle A^*|K|A^*\right\rangle \Psi)^{(p+2)}\|^2 &\leq (p+2)(p+1)\|\eta(K)\|^2\|\Psi^{(p)}\|^2 \\ &= \|\eta(K)\|^2\|((N_{\mathrm{b}}+1)^{1/2}(N_{\mathrm{b}}+2)^{1/2}\Psi)^{(p)}\|^2, \quad \Psi \in D(N_{\mathrm{b}}).\end{aligned}$$

Taking the summation over p from 0 to ∞, we have

$$\sum_{p=0}^{\infty} \|(\left\langle A^*|K|A^*\right\rangle \Psi)^{(p+2)}\|^2 \leq \|\eta(K)\|^2\|(N_{\mathrm{b}}+1)^{1/2}(N_{\mathrm{b}}+2)^{1/2}\Psi\|^2.$$

Hence $\Psi \in D(\langle A^*|K|A^*\rangle)$ and (6.160) holds, where we have used the easily proved fact that $\|(N_{\mathrm{b}}+1)^{1/2}\Phi\| \leq \|(N_{\mathrm{b}}+2)^{1/2}\Phi\|$, $\Phi \in D(N_{\mathrm{b}}^{1/2})$.

(iii) By (6.152), we have $\langle A^*|K^*|A^*\rangle^* \supset \langle A|K|A\rangle_0$. Since $\langle A^*|K^*|A^*\rangle$ is a bounded operator from $\otimes_{\mathrm{s}}^p \mathscr{H}$ to $\otimes_{\mathrm{s}}^{p+2} \mathscr{H}$ by (i) with K replaced by K^*, it follows that $\langle A|K|A\rangle_0$ is a bounded operator from $\hat{\otimes}_{\mathrm{s}}^{p+2} \mathscr{H}$ to $\hat{\otimes}_{\mathrm{s}}^p \mathscr{H}$. Hence

$$\mathscr{F}_{\mathrm{b},0}(\mathscr{H}) \subset D(\langle A|K|A\rangle)$$

and

$$\left\langle A^*|K^*|A^*\right\rangle^* \supset \langle A|K|A\rangle .$$

This means that, for each $p \in \mathbb{Z}_+$, $\langle A|K|A\rangle$ is a bounded operator from $\otimes_s^{p+2}\mathcal{H}$ to $\otimes_s^p\mathcal{H}$ with

$$\|(\langle A|K|A\rangle\Psi)^{(p)}\| \le (p+1)(p+2)\|\eta(K^*)\|\,\|\Psi^{(p+2)}\|, \quad \Psi \in D(\langle A|K|A\rangle).$$

Hence (6.161) holds.

(iv) Since we have proved (iii), this part can proved in quite the same way as in the proof of (ii). ■

Lemma 6.13 *Let K_1 be a Hilbert–Schmidt operator on $\mathcal{H}$ and $K_2 \in \mathfrak{B}(\mathcal{H})$. Let $\{e_n\}_{n=1}^\infty$ be a CONS of $\mathcal{H}$. Then*

$$\lim_{N\to\infty}\sum_{n=1}^N A(K_1e_n)^*A(JK_2e_n)^*\Psi = \left\langle A^*|K_1K_2^*|A^*\right\rangle\Psi, \quad \Psi \in \mathcal{F}_{\mathrm{b},0}(\mathcal{H}), \tag{6.163}$$

$$\lim_{N\to\infty}\sum_{n=1}^N A(K_1e_n)A(JK_2e_n)\Psi = \left\langle A|K_2K_1^*|A\right\rangle\Psi, \quad \Psi \in \mathcal{F}_{\mathrm{b,fin}}(\mathcal{H}). \tag{6.164}$$

Proof Let $\Sigma_N := \sum_{n=1}^N A(K_1e_n)^*A(JK_2e_n)^*$. Then

$$\Sigma_N\Psi = \sum_{n=1}^N A(JK_2e_n)^*A(K_1e_n)^*\Psi.$$

Hence

$$(\Sigma_N\Psi)^{(p+2)} = \sqrt{(p+2)(p+1)}S_{p+2}(u_N \otimes \Psi^{(p)})$$

with $u_N := \sum_{n=1}^\infty JK_2e_n \otimes K_1e_n$. By Corollary 1.4, we have $U_JT_N = u_N$ with $T_N := \sum_{n=1}^N \langle K_2e_n, \cdot\rangle\, K_1e_n$. Let $P_N := \sum_{n=1}^N \langle e_n, \cdot\rangle\, e_n$. Then $T_N = K_1P_NK_2^*$. It is easy to see that

$$\|K_1P_N - K_1\|_2^2 = \sum_{n=N+1}^\infty \|K_1e_n\|^2 \to 0 \ (N \to \infty).$$

Hence, by (1.25), we obtain

$$\|T_N - K_1K_2^*\|_2 \le \|K_1P_N - K_1\|_2\|K_2^*\| \to 0 \ (N \to \infty).$$

Hence $\lim_{N\to\infty} u_N = U_JK_1K_2^*$ in $\mathcal{H} \otimes \mathcal{H}$, implying that

$$\lim_{N\to\infty}(\Sigma_N\Psi)^{(p+2)} = \sqrt{(p+2)(p+1)}S_{p+2}(U_JK_1K_2^* \otimes \Psi^{(p)}).$$

By this result and (6.156), we see that (6.163) holds.

To prove (6.164), let Ψ_p be a vector of the form (6.146). Then

$$\sum_{n=1}^{N} A(K_1 e_n) A(J K_2 e_n) \Psi = \sum_{\substack{j,k=1 \\ j \neq k}}^{p} \Psi_N^{(j,k)} A(f_1)^* \cdots \widehat{A(f_j)^*} \cdots \widehat{A(f_k)^*} \cdots A(f_p)^* \Omega_{\mathscr{H}},$$

where $\Psi_N^{(j,k)} := \sum_{n=1}^{N} \langle J K_2 e_n, f_j \rangle \langle K_1 e_n, f_k \rangle$. It is easy to see that

$$\lim_{N \to \infty} \Psi_N^{(j,k)} = \langle K_1 K_2^* J f_j, f_k \rangle = \langle (K_2 K_1^*)^* J f_j, f_k \rangle .$$

Thus (6.164) holds. ■

6.15 The Boson Fock Space Over a Direct Sum Hilbert Space

In this section we consider the boson Fock space $\mathscr{F}_{\mathrm{b}}(\mathscr{H} \oplus \mathscr{K})$ over the direct sum Hilbert space

$$\mathscr{H} \oplus \mathscr{K} := \{(f, g) | f \in \mathscr{H}, g \in \mathscr{K}\}$$

of two Hilbert spaces $\mathscr{H}$ and $\mathscr{K}$. We prove that $\mathscr{F}_{\mathrm{b}}(\mathscr{H} \oplus \mathscr{K})$ is isomorphic to $\mathscr{F}_{\mathrm{b}}(\mathscr{H}) \otimes \mathscr{F}_{\mathrm{b}}(\mathscr{K})$ in a natural way.

To avoid confusions, we denote the annihilation operator on $\mathscr{F}_{\mathrm{b}}(\mathscr{H})$ by $A_{\mathscr{H}}(\cdot)$.

Theorem 6.24 *Assume that $\mathscr{H}$ and $\mathscr{K}$ are separable. Then there exists a unique unitary operator $U\colon \mathscr{F}_{\mathrm{b}}(\mathscr{H} \oplus \mathscr{K}) \to \mathscr{F}_{\mathrm{b}}(\mathscr{H}) \otimes \mathscr{F}_{\mathrm{b}}(\mathscr{K})$ which has the following properties:*

(i)

$$U \Omega_{\mathscr{H} \oplus \mathscr{K}} = \Omega_{\mathscr{H}} \otimes \Omega_{\mathscr{K}} \tag{6.165}$$

(ii)

$$U \mathscr{F}_{\mathrm{b,fin}}(\mathscr{H} \oplus \mathscr{K}) = \mathscr{F}_{\mathrm{b,fin}}(\mathscr{H}) \hat{\otimes} \mathscr{F}_{\mathrm{b,fin}}(\mathscr{K}). \tag{6.166}$$

(iii) *For all $f \in \mathscr{H}$ and $g \in \mathscr{K}$,*

$$U A_{\mathscr{H} \oplus \mathscr{K}}(f, g)^{\#} U^{-1} = \overline{A_{\mathscr{H}}(f)^{\#} \otimes I + I \otimes A_{\mathscr{K}}(g)^{\#}}. \tag{6.167}$$

Proof Throughout the proof, we set $A(f, g) := A_{\mathscr{H} \oplus \mathscr{K}}(f, g)$, $(f, g) \in \mathscr{H} \oplus \mathscr{K}$. Let $\{e_n\}_{n=1}^{\infty}$ and $\{\xi_n\}_{n=1}^{\infty}$ be CONSs of $\mathscr{H}$ and $\mathscr{K}$ respectively. Then

$\{(e_n, 0), (0, \xi_n)|n \in \mathbb{N}\}$ is a CONS of $\mathscr{H} \oplus \mathscr{K}$. By Corollary 6.1, the set

$$\big\{N_{i_1\cdots i_n, j_1\cdots j_m} A(e_{i_1}, 0)^* \cdots A(e_{i_n}, 0)^* A(0, \xi_{j_1})^* \cdots A(0, \xi_{j_m})^* \Omega_{\mathscr{H}\oplus\mathscr{K}} \mid i_1 \le \cdots \le i_n,\ j_1 \le \cdots \le j_m,\ i_k, j_\ell \in \mathbb{N}, n, m \in \{0\} \cup \mathbb{N}\big\}$$

is a CONS of $\mathscr{F}_{\mathrm{b}}(\mathscr{H} \oplus \mathscr{K})$, where

$$N_{i_1\cdots i_n, j_1\cdots j_m} := \|A(e_{i_1}, 0)^* \cdots A(e_{i_n}, 0)^* A(0, \xi_{j_1})^* \cdots A(0, \xi_{j_m})^* \Omega_{\mathscr{H}\oplus\mathscr{K}}\|^{-1}.$$

Similarly,

$$\big\{N_{i_1\cdots i_n; j_1\cdots j_m} A_{\mathscr{H}}(e_{i_1})^* \cdots A_{\mathscr{H}}(e_{i_n})^* \Omega_{\mathscr{H}} \otimes A_{\mathscr{K}}(\xi_{j_1})^* \cdots A_{\mathscr{K}}(\xi_{j_m})^* \Omega_{\mathscr{K}} \mid i_1 \le \cdots \le i_n,\ j_1 \le \cdots \le j_m,\ i_k, j_\ell \in \mathbb{N}, n, m \in \{0\} \cup \mathbb{N}\big\}$$

is a CONS of $\mathscr{F}_{\mathrm{b}}(\mathscr{H}) \otimes \mathscr{F}_{\mathrm{b}}(\mathscr{K})$. Hence, by the isomorphism theorem (Theorem 1.3), there exists a unique unitary operator U: $\mathscr{F}_{\mathrm{b}}(\mathscr{H} \oplus \mathscr{K}) \to \mathscr{F}_{\mathrm{b}}(\mathscr{H}) \otimes \mathscr{F}_{\mathrm{b}}(\mathscr{K})$ such that, for all $m, n \in \{0\} \cup \mathbb{N}$, $i_k, j_\ell \in \mathbb{N}$,

$$\begin{aligned} &U A(e_{i_1}, 0)^* \cdots A(e_{i_n}, 0)^* A(0, \xi_{j_1})^* \cdots A(0, \xi_{j_m})^* \Omega_{\mathscr{H}\oplus\mathscr{K}} \\ &= A_{\mathscr{H}}(e_{i_1})^* \cdots A_{\mathscr{H}}(e_{i_n})^* \Omega_{\mathscr{H}} \otimes A_{\mathscr{K}}(\xi_{j_1})^* \cdots A_{\mathscr{K}}(\xi_{j_m})^* \Omega_{\mathscr{K}}. \end{aligned}$$

In particular, (6.165) holds.

For all $f_j \in \mathscr{H}$, $g_j \in \mathscr{K}$ $(j \in \mathbb{N})$, we have expansions

$$f_j = \sum_{n=1}^{\infty} a_{nj} e_n, \quad g_k = \sum_{n=1}^{\infty} b_{nk} \xi_n$$

with $a_{nj} := \langle e_n, f_j \rangle$ and $b_{nk} := \langle \xi_n, g_k \rangle$. Hence, for all $q, r \ge 0$,

$$\begin{aligned} \Phi &:= A(f_1, 0)^* \cdots A(f_r, 0)^* A(0, g_1)^* \cdots A(0, g_q)^* \Omega_{\mathscr{H}\oplus\mathscr{K}} \\ &= \sum_{\substack{n_1,\cdots,n_r \ge 1 \\ m_1,\cdots,m_q \ge 1}}^{\infty} a_{n_1 1} \cdots a_{n_r r} b_{m_1 1} \cdots b_{m_q q} A(e_{n_1}, 0)^* \cdots A(e_{n_r}, 0)^* \\ &\qquad \times A(0, \xi_{m_1})^* \cdots A(0, \xi_{m_q})^* \Omega_{\mathscr{H}\oplus\mathscr{K}}. \end{aligned}$$

By the continuity of U,

$$U\Phi = \sum_{\substack{n_1,\cdots,n_r\geq 1\\ m_1,\cdots,m_q\geq 1}}^{\infty} a_{n_1 1}\cdots a_{n_r r} b_{m_1 1}\cdots b_{m_q q} U A(e_{n_1},0)^*\cdots A(e_{n_r},0)^*$$
$$\times A(0,\xi_{m_1})^*\cdots A(0,\xi_{m_q})^*\Omega_{\mathscr{H}\oplus\mathscr{K}}$$
$$= \sum_{\substack{n_1,\cdots,n_r\geq 1\\ m_1,\cdots,m_q\geq 1}}^{\infty} a_{n_1 1}\cdots a_{n_r r} b_{m_1 1}\cdots b_{m_q q} A_{\mathscr{H}}(e_{n_1})^*\cdots A_{\mathscr{H}}(e_{n_r})^*\Omega_{\mathscr{H}}$$
$$\otimes A_{\mathscr{K}}(\xi_{m_1})^*\cdots A_{\mathscr{K}}(\xi_{m_q})^*\Omega_{\mathscr{K}}$$
$$= A_{\mathscr{H}}(f_1)^*\cdots A_{\mathscr{H}}(f_r)^*\Omega_{\mathscr{H}}\otimes A_{\mathscr{K}}(g_1)^*\cdots A_{\mathscr{K}}(g_q)^*\Omega_{\mathscr{K}}.$$

On the other hand, for all $f\in\mathscr{H}$, $g\in\mathscr{K}$,

$$A(f,g)^* = A(f,0)^* + A(0,g)^*$$

on $\mathscr{F}_{\mathrm{b,fin}}(\mathscr{H}\oplus\mathscr{K})$. Hence $\mathscr{F}_{\mathrm{b,fin}}(\mathscr{H}\oplus\mathscr{K})$ is algebraically spanned by vectors of the form Φ. Thus (6.166) holds.

To prove (iii), let

$$\Psi = A_{\mathscr{H}}(f_1)^*\cdots A_{\mathscr{H}}(f_r)^*\Omega_{\mathscr{H}}\otimes A_{\mathscr{K}}(g_1)^*\cdots A_{\mathscr{K}}(g_q)^*\Omega_{\mathscr{K}}. \tag{6.168}$$

Then

$$UA(f,g)^*U^{-1}\Psi$$
$$= U[A(f,0)+A(0,g)]^*A(f_1,0)^*\cdots A(f_r,0)^*A(0,g_1)^*\cdots A(0,g_q)^*\Omega_{\mathscr{H}\oplus\mathscr{K}}$$
$$= UA(f,0)^*A(f_1,0)^*\cdots A(f_r,0)^*A(0,g_1)^*\cdots A(0,g_q)^*\Omega_{\mathscr{H}\oplus\mathscr{K}}$$
$$+UA(0,g)^*A(f_1,0)^*\cdots A(f_r,0)^*A(0,g_1)^*\cdots A(0,g_q)^*\Omega_{\mathscr{H}\oplus\mathscr{K}}$$
$$= A_{\mathscr{H}}(f)^*A_{\mathscr{H}}(f_1)^*\cdots A_{\mathscr{H}}(f_r)^*\Omega_{\mathscr{H}}\otimes A_{\mathscr{K}}(g_1)^*\cdots A_{\mathscr{K}}(g_q)^*\Omega_{\mathscr{K}}$$
$$+A_{\mathscr{H}}(f_1)^*\cdots A_{\mathscr{H}}(f_r)^*\Omega_{\mathscr{H}}\otimes A_{\mathscr{K}}(g)^*A_{\mathscr{K}}(g_1)^*\cdots A_{\mathscr{K}}(g_q)^*\Omega_{\mathscr{K}}$$
$$= [A_{\mathscr{H}}(f)^*\otimes I + I\otimes A_{\mathscr{K}}(g)^*]\Psi.$$

Hence (6.167) with $\# = *$ holds on $\mathscr{F}_{\mathrm{b,fin}}(\mathscr{H})\hat{\otimes}\mathscr{F}_{\mathrm{b,fin}}(\mathscr{K})$. Since $\mathscr{F}_{\mathrm{b,fin}}(\mathscr{X})$ ($\mathscr{X}$ is a Hilbert space) is a core for $A_{\mathscr{X}}(f)^*$, it follows that $\mathscr{F}_{\mathrm{b,fin}}(\mathscr{H})\hat{\otimes}\mathscr{F}_{\mathrm{b,fin}}(\mathscr{K})$ is a core for $A(f)^*\otimes I + I\otimes A(g)^*$. Hence operator equality (6.167) with $\# = *$ holds. In general, for a densely defined closable operator A, $A^* = \left(\bar{A}\right)^*$. Hence

$$UA(f,g)U^{-1} = \left(A_{\mathscr{H}}(f)^*\otimes I + I\otimes A_{\mathscr{K}}(g)^*\right)^* \supset A_{\mathscr{H}}(f)\otimes I + I\otimes A_{\mathscr{K}}(g).$$

Hence, for all $\Psi \in \mathscr{F}_{\mathrm{b,fin}}(\mathscr{H})\hat{\otimes}\mathscr{F}_{\mathrm{b,fin}}(\mathscr{K})$,

$$UA(f, g)U^{-1}\Psi = (A_{\mathscr{H}}(f) \otimes I + I \otimes A_{\mathscr{K}}(g))\Psi.$$

Then, in the same way as in the case of $UA(f, g)^*U^{-1}$, one can conclude that (6.167) holds for $A(f, g)^\# = A(f, g)$. ■

We call the isomorphism U in Theorem 6.24 the **natural isomorphism** between $\mathscr{F}_{\mathrm{b}}(\mathscr{H} \oplus \mathscr{K})$ and $\mathscr{F}_{\mathrm{b}}(\mathscr{H}) \otimes \mathscr{F}_{\mathrm{b}}(\mathscr{K})$ and write

$$\mathscr{F}_{\mathrm{b}}(\mathscr{H} \oplus \mathscr{K}) \cong \mathscr{F}_{\mathrm{b}}(\mathscr{H}) \otimes \mathscr{F}_{\mathrm{b}}(\mathscr{K}).$$

Example 6.8 We give an important example of the natural isomorphism established above, which is useful in applications to quantum field theory. We consider the boson Fock space $\mathscr{F}_{\mathrm{b}}(L^2(\mathbb{R}^d))$ over $L^2(\mathbb{R}^d)$ ($d \in \mathbb{N}$). Let σ and κ be constants satisfying $0 \le \sigma < \kappa$ and $A := \{\boldsymbol{p} \in \mathbb{R}^d \,|\, \sigma \le |\boldsymbol{p}| \le \kappa\}$. Then each element $f \in L^2(\mathbb{R}^d)$ has a unique orthogonal decomposition $f = f_1 + f_2$ with $f_1 \in L^2(A)$ and $f_2 \in L^2(A^c)$ (explicitly $f_1 = \chi_A f$, $f_2 = \chi_{A^c} f$). Hence

$$L^2(\mathbb{R}^d) = L^2(A) \oplus L^2(A^c).$$

Therefore, by Theorem 6.24,

$$\mathscr{F}_{\mathrm{b}}(L^2(\mathbb{R}^d)) \cong \mathscr{F}_{\mathrm{b}}(L^2(A)) \otimes \mathscr{F}_{\mathrm{b}}(L^2(A^c)). \tag{6.169}$$

Remark 6.7 As is easily seen, isomorphism (6.169) holds for any Borel set $A \in \boldsymbol{B}^d$.

Theorem 6.25 *Let T and S be densely defined closable operators on $\mathscr{H}$ and $\mathscr{K}$ respectively. Let U be the unitary operator in Theorem 6.24. Then*

$$U\mathrm{d}\Gamma_{\mathrm{b}}(T \oplus S)U^{-1} = \overline{\mathrm{d}\Gamma_{\mathrm{b}}(T) \otimes I + I \otimes \mathrm{d}\Gamma_{\mathrm{b}}(S)}. \tag{6.170}$$

Proof Let Ψ be a vector of the form (6.168) with $f_j \in D(T)$, $g_k \in D(S)$. Then

$$\begin{aligned} &U\mathrm{d}\Gamma_{\mathrm{b}}(T \oplus S)U^{-1}\Psi \\ &= \sum_{j=1}^{r} UA(f_1, 0)^* \cdots A(Tf_j, 0)^* \cdots A(f_r, 0)^* A(0, g_1)^* \cdots A(0, g_q)^* \Omega_{\mathscr{H}\oplus\mathscr{K}} \\ &\quad + \sum_{k=1}^{q} UA(f_1, 0)^* \cdots A(f_r, 0)^* A(0, g_1)^* \cdots A(0, Sg_k) \cdots A(0, g_q)^* \Omega_{\mathscr{H}\oplus\mathscr{K}} \end{aligned}$$

$$= \sum_{j=1}^{r} A(f_1, 0)^* \cdots A(Tf_j, 0)^* \cdots A(f_r, 0)^* \Omega_{\mathcal{H}} \otimes A(0, g_1)^* \cdots A(0, g_q)^* \Omega_{\mathcal{K}}$$

$$+ \sum_{k=1}^{q} A(f_1, 0)^* \cdots A(f_r, 0)^* \Omega_{\mathcal{H}} \otimes A(0, g_1)^* \cdots A(0, Sg_k) \cdots A(0, g_q)^* \Omega_{\mathcal{K}}$$

$$= [\mathrm{d}\Gamma_{\mathrm{b}}(T) \otimes I + I \otimes \mathrm{d}\Gamma_{\mathrm{b}}(S)]\Psi.$$

It follows that, for all $\Psi \in \mathcal{F}_{\mathrm{b,fin}}(D(T)) \hat{\otimes} \mathcal{F}_{\mathrm{b,fin}}(D(S))$,

$$U \mathrm{d}\Gamma_{\mathrm{b}}(T \oplus S) U^{-1} \Psi = [\mathrm{d}\Gamma_{\mathrm{b}}(T) \otimes I + I \otimes \mathrm{d}\Gamma_{\mathrm{b}}(S)]\Psi.$$

Since $\mathcal{F}_{\mathrm{b,fin}}(D(T))$ (resp. $\mathcal{F}_{\mathrm{b,fin}}(D(S))$) is a core for $\mathrm{d}\Gamma_{\mathrm{b}}(T)$ (resp. $\mathrm{d}\Gamma_{\mathrm{b}}(S)$), it follows that $\mathcal{F}_{\mathrm{b,fin}}(D(T)) \hat{\otimes} \mathcal{F}_{\mathrm{b,fin}}(D(S))$ is a core for $\mathrm{d}\Gamma_{\mathrm{b}}(T) \otimes I + I \otimes \mathrm{d}\Gamma_{\mathrm{b}}(S)$. We also have

$$U \mathcal{F}_{\mathrm{b,fin}}(D(T) \oplus D(S)) = \mathcal{F}_{\mathrm{b,fin}}(D(T)) \hat{\otimes} \mathcal{F}_{\mathrm{b,fin}}(D(S)).$$

Hence (6.170) holds. ■

Example 6.9 Let $A \subset \mathbb{R}^d$ be as in Example 6.8. Let $F: \mathbb{R}^d \to \mathbb{R}$ be a Borel measurable function a.e. finite. Then the multiplication operator by the function F on $L^2(\mathbb{R}^d)$ is self-adjoint. We denote the multiplication operator by the same symbol. Let p_A be the orthogonal projection onto $L^2(A)$. Then we have $p_A = \chi_A$, the multiplication operator by the characteristic function of A. It is easy to see that, for all $f \in D(F)$, $p_A f \in D(F)$ and $F p_A f = p_A F f$. Hence F is reduced by $L^2(A)$ and hence by $L^2(A^c)$. We have $F = F_A \oplus F_{A^c}$ with $F_A = \chi_A F$ and $F_{A^c} = \chi_{A^c} F$. Therefore, by Theorem 6.25,

$$\mathrm{d}\Gamma_{\mathrm{b}}(F) \cong \overline{\mathrm{d}\Gamma_{\mathrm{b}}(F_A) \otimes I + I \otimes \mathrm{d}\Gamma_{\mathrm{b}}(F_{A^c})}$$

under the natural isomorphism (6.169).

Remark 6.8 The contents of this section can be easily extended to the case where $\mathcal{H} \oplus \mathcal{K}$ is replaced by the direct sum $\oplus_{n=1}^{N} \mathcal{H}_n$ of N Hilbert spaces $\mathcal{H}_1, \ldots, \mathcal{H}_N$ with $N \geq 3$.

Chapter 7
Elements of the Theory of Fermion Fock Spaces

Abstract Fundamental aspects of the theory of fermion Fock space are described.

7.1 Definitions and Basic Properties

In Example 5.2 in Chap. 5, we mentioned the fermion Fock space $\wedge(\mathbb{C}^N)$ over the N-dimensional complex Euclidean vector space $\mathbb{C}^N$. In this chapter, we consider an abstract (infinite dimensional) version of this space and describe its basic structures.

Let $\mathscr{K}$ be a complex Hilbert space and, for each $p \in \mathbb{N}$, $\wedge^p(\mathscr{K})$ be the p-fold anti-symmetric tensor product of $\mathscr{K}$ (see Sect. 1.5). We set $\wedge^0(\mathscr{K}) := \mathbb{C}$. The infinite direct sum Hilbert space of $\wedge^p(\mathscr{K})$, $p = 0, 1, 2, \ldots$

$$\begin{aligned} \mathscr{F}_{\mathrm{f}}(\mathscr{K}) &:= \bigoplus_{p=0}^{\infty} \wedge^p(\mathscr{K}) \\ &= \left\{ \Psi = \{\Psi^{(p)}\}_{p=0}^{\infty} | \Psi^{(p)} \in \wedge^p(\mathscr{K}),\ p \geq 0,\ \sum_{p=0}^{\infty} \|\Psi^{(p)}\|^2 < \infty \right\} \end{aligned} \tag{7.1}$$

is called the **fermion Fock space over** $\mathscr{K}$ or the **anti-symmetric Fock space over** $\mathscr{K}$. This Hilbert space is used to describe state vectors of a **Fermi field**, i.e., a quantum field of identical fermions.

Remark 7.1 Let $\mathscr{K}$ be finite-dimensional with $n = \dim \mathscr{K}$. Then, for all $p > n$, $\wedge^p(\mathscr{K}) = \{0\}$. Hence $\mathscr{F}_{\mathrm{f}}(\mathscr{K}) = \oplus_{p=0}^{n} \wedge^p (\mathscr{K})$. Therefore $\mathscr{F}_{\mathrm{f}}(\mathscr{K})$ is finite-dimensional with $\dim \mathscr{F}_{\mathrm{f}}(\mathscr{K}) = \sum_{p=0}^{n} \dim \wedge^p(\mathscr{K}) = 2^n$. Hence analysis of $\mathscr{F}_{\mathrm{f}}(\mathscr{K})$ is reduced to finite dimensional analysis. This is a big difference from the case of boson Fock spaces.

© Springer Nature Singapore Pte Ltd. 2020

A. Arai, *Inequivalent Representations of Canonical Commutation and Anti-Commutation Relations*, Mathematical Physics Studies,
https://doi.org/10.1007/978-981-15-2180-5_7

Based on the fact mentioned in Remark 7.1, *throughout this chapter, we assume that $\mathscr{K}$ is infinite dimensional*, unless otherwise stated.

In the context of concrete quantum field models, $\mathscr{K}$ represents the state vector space of a fermion, called the **one-particle space** of a fermion or the **one-fermion space**, and the Hilbert space $\wedge^p(\mathscr{K})$ with $p \geq 2$ as a closed subspace of $\mathscr{F}_{\mathrm{f}}(\mathscr{K})$ represents the state vector space consisting of p identical fermions. For this reason, $\wedge^p(\mathscr{K})$ is called the p**-particle space of fermions**.

The subspace of the Fermion Fock space $\mathscr{F}_{\mathrm{f}}(\mathscr{K})$ corresponding to the finite particle subspace of a boson Fock space is given by

$$\mathscr{F}_{\mathrm{f},0}(\mathscr{K}) := \left\{ \Psi \in \mathscr{F}_{\mathrm{f}}(\mathscr{K}) | \exists p_0 \in \mathbb{N} \text{ such that } \Psi^{(p)} = 0, \ \forall p \geq p_0 \right\}. \tag{7.2}$$

It is called the **finite particle subspace** of $\mathscr{F}_{\mathrm{f}}(\mathscr{K})$. It follows that $\mathscr{F}_{\mathrm{f},0}(\mathscr{K})$ is dense in $\mathscr{F}_{\mathrm{f}}(\mathscr{K})$.

The vector

$$\Omega_{\mathscr{K}} := \{1, 0, 0, \ldots\} \in \mathscr{F}_{\mathrm{f}}(\mathscr{K}) \tag{7.3}$$

is called the **fermion Fock vacuum**.

As in the case of boson Fock space, for a subspace $\mathscr{D}$ of $\mathscr{K}$, we introduce a subspace $\wedge_{\mathrm{fin}}(\mathscr{D})$ of $\mathscr{F}_{\mathrm{f}}(\mathscr{K})$ by

$$\wedge_{\mathrm{fin}}(\mathscr{D}) := \widehat{\oplus}_{p=0}^{\infty} A_p(\hat{\otimes}^p D), \tag{7.4}$$

where $\hat{\otimes}^p \mathscr{D}$ denotes the p-fold algebraic tensor product of $\mathscr{D}$. It follows that, if $\mathscr{D}$ is dense in $\mathscr{K}$, then $\wedge_{\mathrm{fin}}(\mathscr{D})$ is dense in $\mathscr{F}_{\mathrm{f}}(\mathscr{K})$.

Example 7.1 Let (X, Σ, μ) be the measure space as in Sect. 1.7 and consider the case $\mathscr{K} = L^2(X, d\mu)$. Then, by Theorem 1.24(ii), we have

$$\mathscr{F}_{\mathrm{f}}(L^2(X, d\mu)) = \oplus_{p=0}^{\infty} \wedge^p (L^2(X, d\mu)) \cong \oplus_{p=0}^{\infty} L^2_{\mathrm{as}}(X^p, d\mu^p),$$

We call this isomorphism the **natural isomorphism** between $\mathscr{F}_{\mathrm{f}}(L^2(X, d\mu))$ and $\oplus_{p=0}^{\infty} L^2_{\mathrm{as}}(X^p, d\mu^p)$.

7.2 Fermion Second Quantization Operators

In the same manner as in the case of second quantization operators on a boson Fock space, one can define second quantization operators on the fermion Fock space $\mathscr{F}_{\mathrm{f}}(\mathscr{K})$.

Let T be a densely defined closable operator on $\mathscr{K}$. Then, in the same way as in the proof of Lemma 6.1, one can show that the operator $T^{(p)}$ on $\otimes^p \mathscr{K}$ (see (6.5)) is reduced by $\wedge^p(\mathscr{K})$. We denote the reduced part by $T_{\mathrm{f}}^{(p)}$:

$$T_{\mathrm{f}}^{(p)} := (T^{(p)})_{\wedge^p(\mathscr{K})}, \quad p \geq 2. \tag{7.5}$$

As a convention, we set $T_{\mathrm{f}}^{(0)} = 0$ and $T_{\mathrm{f}}^{(1)} := \overline{T}$. It follows from Proposition 1.16(iii) that $T_{\mathrm{f}}^{(p)}$ is a densely defined closed operator. The infinite direct sum operator of $T_{\mathrm{f}}^{(p)}$, $p = 0, 1, 2, \ldots$

$$\mathrm{d}\Gamma_{\mathrm{f}}(T) := \oplus_{p=0}^{\infty} T_{\mathrm{f}}^{(p)} \tag{7.6}$$

on $\mathscr{F}_{\mathrm{f}}(\mathscr{K})$ is called the **fermion second quantization operator of** T. In the same way as in the proof of Theorem 6.1, one can prove the following theorem:

Theorem 7.1 *Let T be a self-adjoint operator on $\mathscr{K}$.*

(i) $\mathrm{d}\Gamma_{\mathrm{f}}(T)$ *is a self-adjoint operator and, for all $t \in \mathbb{R}$,*

$$e^{it\mathrm{d}\Gamma_{\mathrm{f}}(T)} = \oplus_{p=0}^{\infty} e^{itT_{\mathrm{f}}^{(p)}}.$$

(ii) *If T is non-negative, then $\mathrm{d}\Gamma_{\mathrm{f}}(T)$ is non-negative and, for all $z \in \overline{\mathbb{C}}_+$,*

$$e^{-z\mathrm{d}\Gamma_{\mathrm{f}}(T)} = \oplus_{p=0}^{\infty} e^{-zT_{\mathrm{f}}^{(p)}}.$$

(iii) *Let $\mathscr{D}$ be a core for T. Then $\wedge_{\mathrm{fin}}(D)$ is a core for $\mathrm{d}\Gamma_{\mathrm{f}}(T)$.*
(iv) *If $\sigma(T) = [m, \infty)$, then*

$$\sigma(\mathrm{d}\Gamma_{\mathrm{f}}(T)) = \{0\} \cup [m, \infty). \tag{7.7}$$

It follows from (7.3) that

$$\mathrm{d}\Gamma_{\mathrm{f}}(T)\Omega_{\mathscr{K}} = 0. \tag{7.8}$$

Hence

$$0 \in \sigma_{\mathrm{p}}(\mathrm{d}\Gamma_{\mathrm{f}}(T)). \tag{7.9}$$

The fermion second quantization operator

$$N_{\mathrm{f}} := \mathrm{d}\Gamma_{\mathrm{f}}(I) \tag{7.10}$$

of the identity I on $\mathscr{K}$ is called the **fermion number operator**. It is obvious that N_{f} is reduced by each $\wedge^p(\mathscr{K})$ and

$$N_{\mathrm{f}} \upharpoonright \wedge^p(\mathscr{K}) = p$$

7.3 Fermion Γ-Operators

Let T be a densely defined closable operator from a Hilbert space $\mathscr{K}$ to a Hilbert space $\mathscr{H}$ and $p \in \mathbb{N}$. Then the p-fold tensor product $\otimes^p T$ of T is a unique densely defined closed operator from $\otimes^p \mathscr{K}$ to $\otimes^p \mathscr{H}$ such that

$$(\otimes^p T)(\otimes_{j=1}^p \Psi_j) = \otimes_{j=1}^p T\Psi_j, \quad \Psi_j \in D(T),\ j = 1, \ldots, p.$$

In the same way as in the proof of Lemma 6.1, one can prove that, for each $p \geq 2$,

$$A_{\mathscr{H},p}(\otimes^p T) \subset (\otimes^p T)A_{\mathscr{K},p}, \tag{7.11}$$

where $A_{\mathscr{X},p}$ ($\mathscr{X} = \mathscr{K}, \mathscr{H}$) is the anti-symmetrization operator on $\otimes^p \mathscr{X}$. Hence we can define a densely defined closed operator $\wedge^p T$ from $\wedge^p(\mathscr{K})$ to $\wedge^p(\mathscr{H})$ as follows: $\wedge^0 T := 1$, $\wedge^1 T := \overline{T}$ and, for $p \geq 2$,

$$D(\wedge^p T) := D(\otimes^p T) \cap \wedge^p(\mathscr{K}), \tag{7.12}$$

$$(\wedge^p T)\Psi := (\otimes^p T)\Psi, \quad \Psi \in D(\wedge^p T). \tag{7.13}$$

In the case $\mathscr{H} = \mathscr{K}$, $\wedge^p T$ is nothing but the reduced part of $\otimes^p T$ to $\wedge^p(\mathscr{K})$.

One can define an infinite direct sum operator $\Gamma_{\mathrm{f}}(T)$ from $\mathscr{F}_{\mathrm{f}}(\mathscr{K})$ to $\mathscr{F}_{\mathrm{f}}(\mathscr{H})$ by

$$\Gamma_{\mathrm{f}}(T) := \oplus_{p=0}^{\infty} \wedge^p T. \tag{7.14}$$

We call the operator $\Gamma_{\mathrm{f}}(T)$ the **fermion Γ-operator** for T.

It obvious that

$$\Gamma_{\mathrm{f}}(T)\Omega_{\mathscr{K}} = \Omega_{\mathscr{H}}. \tag{7.15}$$

The following theorems can be proved in quite the same manner as in the proof of Theorem 6.3.

Theorem 7.2 *Let T be a densely defined closable operator from $\mathscr{K}$ to $\mathscr{H}$.*

(i) *If T is unbounded, then $\Gamma_{\mathrm{f}}(T)$ is unbounded.*
(ii) *If T is a contraction operator, then $\Gamma_{\mathrm{f}}(T)$ is a contraction operator and*

$$\Gamma_{\mathrm{f}}(T)^* = \Gamma_{\mathrm{f}}(T^*). \tag{7.16}$$

(iii) *If $\mathscr{D}$ is a core for T, then $\wedge_{\mathrm{fin}}(\mathscr{D})$ is a core for $\Gamma_{\mathrm{f}}(T)$.*

Theorem 7.3 *Let $\mathscr{K}$, $\mathscr{H}$ and $\mathscr{X}$ be complex Hilbert spaces.*

(i) *If T is self-adjoint on $\mathscr{K}$, then so is $\Gamma_{\mathrm{f}}(T)$.*
(ii) *If $U : \mathscr{K} \to \mathscr{H}$ is unitary, then so is $\Gamma_{\mathrm{f}}(U) : \mathscr{F}_{\mathrm{f}}(\mathscr{K}) \to \mathscr{F}_{\mathrm{f}}(\mathscr{H})$ and*

$$\Gamma_{\mathrm{f}}(U)^{-1} = \Gamma_{\mathrm{f}}(U^{-1}). \tag{7.17}$$

Moreover,

$$\Gamma_{\mathrm{f}}(U) \wedge_{\mathrm{fin}} (\mathscr{K}) = \wedge_{\mathrm{fin}}(\mathscr{H}). \tag{7.18}$$

(iii) *Let $T \in \mathfrak{B}(\mathscr{K}, \mathscr{H})$ and $S \in \mathfrak{B}(\mathscr{H}, \mathscr{X})$ be contraction operators. Then,*

$$\Gamma_{\mathrm{f}}(S)\Gamma_{\mathrm{f}}(T) = \Gamma_{\mathrm{f}}(ST).$$

(iv) *Let T_n $(n \in \mathbb{N})$ and T be contraction operators from $\mathscr{K}$ to $\mathscr{H}$ satisfying $\text{s-}\lim_{n\to\infty} T_n = T$. Then*

$$\text{s-}\lim_{n\to\infty} \Gamma_{\mathrm{f}}(T_n) = \Gamma_{\mathrm{f}}(T).$$

The next theorem gives relations between fermion second quantization operators and fermion Γ-operators:

Theorem 7.4 *Let T be a self-adjoint operator on $\mathscr{K}$.*

(i) *For all $t \in \mathbb{R}$,*

$$\Gamma_{\mathrm{f}}(e^{itT}) = e^{it\mathrm{d}\Gamma_{\mathrm{f}}(T)}.$$

(ii) *If $T \geq 0$, then $\mathrm{d}\Gamma_{\mathrm{f}}(T) \geq 0$ and, for all $z \in \overline{\mathbb{C}}_+$,*

$$\Gamma_{\mathrm{f}}(e^{-zT}) = e^{-z\mathrm{d}\Gamma_{\mathrm{f}}(T)}.$$

Proof Similar to the proof of Theorem 6.4. ∎

7.4 Fermion Annihilation and Creation Operators

For each $u \in \mathscr{K}$, one can define an operator $B^\dagger(u)$ on $\mathscr{F}_\mathrm{f}(\mathscr{K})$ as follows:

$$D(B^\dagger(u)) = \left\{ \Psi = \{\Psi^{(p)}\}_{p=0}^\infty \in \mathscr{F}_\mathrm{f}(\mathscr{K}) \,\Big|\, \sum_{p=1}^\infty \|\sqrt{p} A_p(u \otimes \Psi^{(p-1)})\|^2 < \infty \right\}, \tag{7.19}$$

$$(B^\dagger(u)\Psi)^{(0)} = 0, \tag{7.20}$$

$$(B^\dagger(u)\Psi)^{(p)} = \sqrt{p} A_p(u \otimes \Psi^{(p-1)}), \quad p \geq 1,\ \Psi \in D(B^\dagger(u)). \tag{7.21}$$

It is easy to see that $B^\dagger(u)$ is a densely defined closed operator with

$$\mathscr{F}_{\mathrm{f},0}(\mathscr{K}) \subset D(B^\dagger(u)).$$

We set

$$B(u) := (B^\dagger(u))^*. \tag{7.22}$$

Lemma 7.1 *For all $u \in \mathscr{K}$, $\Omega_\mathscr{K} \in D(B(u))$ and*

$$B(u)\Omega_\mathscr{K} = 0. \tag{7.23}$$

Proof Property (7.20) implies that, for all $\Psi \in D(B^\dagger(u))$, $\langle \Omega_\mathscr{K}, B^\dagger(u)\Psi \rangle = 0 = \langle 0, \Psi \rangle$. Hence $\Omega_\mathscr{K} \in D(B(u))$ and $B(u)\Omega_\mathscr{K} = 0$. ■

Lemma 7.2 *The operator $B(u)$ is a densely defined closed operator with*

$$D(B(u)) \supset \wedge_\mathrm{fin}(\mathscr{K})$$

and, for all $p \geq 1$ and $u_j \in \mathscr{K}$, $j = 1, \ldots, p$,

$$(B(u)u_1 \wedge \cdots \wedge u_p)^{(q)} = 0, \quad q \neq p-1, \tag{7.24}$$

$$(B(u)u_1 \wedge \cdots \wedge u_p)^{(p-1)} = \sum_{j=1}^p (-1)^{j-1} \langle u, u_j \rangle u_1 \wedge \cdots \wedge \hat{u}_j \wedge \cdots \wedge u_p, \tag{7.25}$$

where $\hat{u}_j$ indicates the omission of u_j.

Proof Let $\Psi = u_1 \wedge \cdots \wedge u_p$. Since $B^\dagger(u)$ maps $\wedge^p(\mathscr{K})$ to $\wedge^{p+1}(\mathscr{K})$, it follows that, for all $\Phi \in D(B^\dagger(u))$,

$$\left\langle \Psi, B^\dagger(u)\Phi \right\rangle = \left\langle \Psi, \sqrt{p} A_p(u \otimes \Phi^{(p-1)}) \right\rangle = \left\langle \sqrt{p}\Psi, u \otimes \Phi^{(p-1)} \right\rangle = \langle \Psi_{p-1}, \Phi \rangle,$$

where $(\Psi_{p-1})^{(q)} = 0$ for $q \neq p-1$ and

$$(\Psi_{p-1})^{(p-1)} = \frac{1}{\sqrt{(p-1)!}} \sum_{\sigma \in \mathfrak{S}_p} \operatorname{sgn}(\sigma) \langle u, u_{\sigma(1)} \rangle u_{\sigma(2)} \otimes \cdots \otimes u_{\sigma p}. \tag{7.26}$$

Hence $\Psi \in D(B(u))$ and $B(u)\Psi = \Psi_{p-1}$. In particular, (7.24) holds. The summation $\sum_{\sigma \in \mathfrak{S}_p}$ on the right hand side of (7.26) is rewritten as

$$\sum_{\sigma \in \mathfrak{S}_p} = \sum_{j=1}^{p} \sum_{\sigma \in \mathfrak{S}_p, \sigma(1)=j}.$$

Using this structure, one can easily see that the right hand side of (7.26) is equal to that of (7.25). ∎

By Lemma 7.2, the adjoint $B(u)^*$ of $B(u)$ exists and

$$B(u)^* = B^\dagger(u). \tag{7.27}$$

In what follows, we mainly use the symbol $B(u)^*$ instead of $B^\dagger(u)$.

As remarked in the proof of Lemma 7.2, $B(u)^*$ maps $\wedge^p(\mathscr{K})$ to $\wedge^{p+1}(\mathscr{K})$ for all $p \geq 0$. This implies that $B(u)$ maps $\wedge^p(\mathscr{K})$ to $\wedge^{p-1}(\mathscr{K})$ with $\wedge^{-1}(\mathscr{K}) := \{0\}$. Based on these properties, $B(u)^*$ and $B(u)$ are called respectively the **fermion creation operator** and the **fermion annihilation operator** with test vector u.

Proposition 7.1 *Let $n \in \mathbb{N}$ and $u_j \in \mathscr{K}$, $j = 1, \ldots, n$. Then,*

$$\Omega_{\mathscr{K}} \in D(B(u_1)^* \cdots B(u_n)^*)$$

and, for all $p \geq 1$,

$$(B(u_1)^* \cdots B(u_n)^* \Omega_{\mathscr{K}})^{(p)} = \delta_{np} u_1 \wedge \cdots \wedge u_p. \tag{7.28}$$

Proof We need only to apply (7.21) repeatedly. ∎

It follows from Proposition 7.1 that, for every subspace $\mathscr{D} \subset \mathscr{K}$,

$$\wedge_{\text{fin}}(\mathscr{D}) = \text{span}\{\Omega_{\mathscr{K}}, B(u_1)^* \cdots B(u_p)^* \Omega_{\mathscr{K}} | p \geq 1, u_j \in \mathscr{D}, j = 1, \cdots, p\}. \tag{7.29}$$

Formula (7.28) and Theorem 1.22 imply the following theorem:

Theorem 7.5 *Let* $\{e_j\}_{j=1}^{\infty}$ *be a CONS of* $\mathscr{K}$. *Then*

$$\{\Omega_{\mathscr{K}},\ B(e_{i_1})^* \cdots B(e_{i_p})^* \Omega_{\mathscr{K}} | p \geq 1,\ i_1, \ldots, i_p \in \mathbb{N},\ i_1 < i_2 < \ldots < i_p\}$$

is a CONS of $\mathscr{F}_{\mathrm{f}}(\mathscr{K})$.

For each $u \in \mathscr{K}$, we denote by $B(u)^{\#}$ either $B(u)$ or $B(u)^*$. Formulae (7.25) and (7.28) give the following result:

Corollary 7.1 *For all* $p \in \mathbb{N}$ *and* $u_1, \ldots, u_p \in \mathscr{K}$, $B(u_1)^* \cdots B(u_p)^* \in D(B(u)^{\#})$ *and*

$$B(u)B(u_1)^* \cdots B(u_p)^* \Omega_{\mathscr{K}} = \sum_{j=1}^{p} (-1)^{j-1} B(u_1)^* \cdots \widehat{B(u_j)^*} \cdots B(u_p)^* \Omega_{\mathscr{K}}. \tag{7.30}$$

We now prove a basic theorem:

Theorem 7.6 *For every* $u \in \mathscr{K}$, $B(u)$ *and* $B(u)^*$ *are everywhere defined bounded operators on* $\mathscr{F}_{\mathrm{f}}(\mathscr{K})$ *with*

$$\|B(u)\| = \|B(u)^*\| = \|u\|. \tag{7.31}$$

Moreover, the following anti-commutation relations hold:

$$\{B(u), B(v)^*\} = \langle u, v \rangle, \tag{7.32}$$

$$\{B(u), B(v)\} = 0, \quad \{B(u)^*, B(v)^*\} = 0, \quad u, v \in \mathscr{K}. \tag{7.33}$$

Proof Let $u, v \in \mathscr{K}$ and $\Psi_p := B(u_1)^* \cdots B(u_p)^* \Omega_{\mathscr{K}}$, $p \in \mathbb{N}$. Then, by Corollary 7.1, we have

$$\begin{aligned} B(u)B(v)^* \Psi_p &= \langle u, v \rangle \Psi_p \\ &\quad - \sum_{j=1}^{p} (-1)^{j-1} \langle u, u_j \rangle B(v)^* B(u_1)^* \cdots \widehat{B(u_j)^*} \cdots B(u_p)^* \Omega_{\mathscr{K}}, \end{aligned}$$

$$B(v)^* B(u) \Psi_p = \sum_{j=1}^{p} (-1)^{j-1} \langle u, u_j \rangle B(v)^* B(u_1)^* \cdots \widehat{B(u_j)^*} \cdots B(u_p)^* \Omega_{\mathscr{K}}.$$

Hence

$$(B(u)B(v)^* + B(v)^* B(u)) \Psi_p = \langle u, v \rangle \Psi_p.$$

By (7.29), we obtain

$$B(u)B(v)^* + B(v)^*B(u) = \langle u, v \rangle \quad \text{on } \wedge_{\text{fin}}(\mathscr{K}). \tag{7.34}$$

In particular,

$$B(u)B(u)^* + B(u)^*B(u) = \|u\|^2 \quad \text{on } \wedge_{\text{fin}}(\mathscr{K}).$$

Hence, by applying Lemma 5.1 to the case $A = \|u\|^{-1}B(u)$ with $u \neq 0$, we conclude that $\|u\|^{-1}B(u)$ is everywhere defined bounded operator with $\|B(u)\| \leq \|u\|$ and operator equality

$$B(u)B(u)^* + B(u)^*B(u) = \|u\|^2$$

holds (for $u = 0$, this is trivial). Hence $B(u)^*$ also is everywhere defined bounded operator and $\|B(u)^*\| = \|B(u)\| \leq \|u\|$. Since $\wedge_{\text{fin}}(\mathscr{K})$ is dense in $\mathscr{F}_{\text{f}}(\mathscr{K})$, (7.34) extends to (7.32).

Taking $u = v$ in (7.32) and using (7.23), we have

$$\|B(u)^*\Omega_{\mathscr{K}}\| = \|u\|. \tag{7.35}$$

But $\|B(u)^*\Omega_{\mathscr{H}}\| \leq \|B(u)^*\| \, \|\Omega_{\mathscr{K}}\| = \|B(u)^*\|$. Hence $\|B(u)^*\| \geq \|u\|$. Thus it follows that $\|B(u)^*\| = \|u\|$. Since $\|B(u)\| = \|B(u)^*\|$, (7.31) holds.

We have

$$(B(u)^*B(v)^*\Psi_p)^{(p+2)} = u \wedge v \wedge u_1 \wedge \cdots \wedge u_p \tag{7.36}$$

$$= -v \wedge u \wedge u_1 \wedge \cdots \wedge u_p \tag{7.37}$$

$$= -(B(v)^*B(u)^*\Psi_p)^{(p+2)}. \tag{7.38}$$

Hence it follows that $\{B(u)^*, B(v)^*\} = 0$ on $\wedge_{\text{fin}}(\mathscr{K})$. Since $\wedge_{\text{fin}}(\mathscr{K})$ is dense in $\mathscr{F}_{\text{f}}(\mathscr{K})$ and $B(\cdot)^*$ is bounded as already proved, we obtain operator equality $\{B(u)^*, B(v)^*\} = 0$. Taking the adjoint of this equation, we have operator equality $\{B(u), B(v)\} = 0$. ∎

Anti-commutation relations (7.32) and (7.33) are called the **canonical anti-commutation relations** (CAR) over $\mathscr{K}$ or indexed by $\mathscr{K}$.

Putting $u = v$ in (7.33), we have

$$B(u)^2 = 0, \quad B(u)^{*2} = 0, \quad u \in \mathscr{K}. \tag{7.39}$$

Hence $B(u)$ and $B(u)^*$ are nilpotent operators.

As we have already seen, for all $u \in \mathscr{K}$, $B(u)$ vanishes the fermion Fock vacuum $\Omega_{\mathscr{K}}$ (see (7.23)). In fact, the converse is true:

Theorem 7.7 *Let $\mathscr{D}$ be a dense subspace of $\mathscr{K}$ and $\Psi \in \mathscr{F}_{\mathrm{f}}(\mathscr{K})$ be a vector such that, for all $u \in \mathscr{D}$, $B(u)\Psi = 0$. Then there exists a constant $\alpha \in \mathbb{C}$ such that $\Psi = \alpha\Omega_{\mathscr{K}}$.*

Proof For all $\Phi \in \mathscr{F}_{\mathrm{f}}(\mathscr{K})$, we have

$$0 = \langle \Phi, B(u)\Psi \rangle = \left\langle B(u)^*\Phi, \Psi \right\rangle.$$

Let $\Phi \in \wedge^p(\mathscr{K})$ $(p \geq 0)$. Then $\left\langle \sqrt{p+1}A_{p+1}(u \otimes \Phi), \Psi^{(p+1)} \right\rangle = 0$. The subspace algebraically spanned by $\{A_{p+1}(u \otimes \Phi) | u \in \mathscr{D}, \Phi \in \wedge^p(\mathscr{K})\}$ is dense in $\wedge^{p+1}(\mathscr{K})$. Hence $\Psi^{(p+1)} = 0$. Since $p \geq 0$ is arbitrary, it follows that $\Psi = \{\Psi^{(0)}, 0, 0, \ldots\} = \alpha\Omega_{\mathscr{K}}$ with $\alpha := \Psi^{(0)}$. ■

The following theorem is important:

Theorem 7.8 *Let $\mathscr{D}$ be a dense subspace of $\mathscr{K}$. Then $\{B(u), B(u)^* | u \in \mathscr{D}\}' = \mathbb{C}I$. In particular, $\{B(u), B(u)^* | u \in \mathscr{D}\}$ is irreducible.*

Proof Let $T \in \{B(u), B(u)^* | u \in \mathscr{D}\}'$ so that, for all $u \in \mathscr{D}$, $B(u)^\# T = TB(u)^\#$. Then, by (7.23), $B(u)T\Omega_{\mathscr{K}} = 0$. Hence, by Theorem 7.7, there exists a constant $\alpha \in \mathbb{C}$ such that $T\Omega_{\mathscr{K}} = \alpha\Omega_{\mathscr{K}}$. Let $\Psi := B(u_1)^* \cdots B(u_p)^*\Omega_{\mathscr{K}}$ $(u_j \in \mathscr{D}, j = 1, \ldots, p, p \in \mathbb{N})$. Then it follows from the commutativity of T and $B(u_j)^*$ that $T\Psi = \alpha\Psi$. Since the subspace algebraically spanned by vectors of the form Ψ is dense in $\mathscr{F}_{\mathrm{f}}(\mathscr{K})$ and T is bounded, it follows that $T = \alpha I$. Thus $\{B(u), B(u)^* | u \in \mathscr{D}\}' = \mathbb{C}I$. By this fact and Proposition 2.9(i), $\{B(u), B(u)^* | u \in \mathscr{D}\}$ is irreducible. ■

7.5 Commutation Relations Between $B(\cdot)^\#$ and $d\Gamma_{\mathrm{f}}(\cdot)$

Theorem 7.9 *Let T be a densely defined closable operator on $\mathscr{K}$.*

(i) *For every $u \in D(T)$, $B(u)^*$ leaves $D(d\Gamma_{\mathrm{f}}(T))$ invariant and*

$$[d\Gamma_{\mathrm{f}}(T), B(u)^*] = B(Tu)^* \quad \text{on } D(\mathrm{d}\Gamma_{\mathrm{f}}(T)). \tag{7.40}$$

(ii) *For every $u \in D(T^*)$, $B(u)$ leaves* $\mathrm{d}\Gamma_{\mathrm{f}}(T)$ *invariant and*

$$[d\Gamma_{\mathrm{f}}(T), B(u)] = -B(T^*u) \quad \text{on } D(\mathrm{d}\Gamma_{\mathrm{f}}(T)). \tag{7.41}$$

Proof

(i) Let $\Psi \in \wedge_{\mathrm{fin}}(D(T))$. Then $\Psi \in D(\mathrm{d}\Gamma_{\mathrm{f}}(T))$ and $B(u)^*\Psi \in D(\mathrm{d}\Gamma_{\mathrm{f}}(T))$. Moreover,

$$(\mathrm{d}\Gamma_{\mathrm{f}}(T)B(u)^*\Psi)^{(p)} = \sqrt{p}A_p(Tu \otimes \Psi^{(p-1)}) + \sqrt{p}A_p(u \otimes (\mathrm{d}\Gamma_{\mathrm{f}}(T)\Psi)^{p-1}),$$
$$(B(u)^*\mathrm{d}\Gamma_{\mathrm{f}}(T)\Psi)^{(p)} = \sqrt{p}A_p(u \otimes (\mathrm{d}\Gamma_{\mathrm{f}}(T)\Psi)^{(p-1)}).$$

Hence

$$([\mathrm{d}\Gamma_{\mathrm{f}}(T), B(u)^*]\Psi)^{(p)} = \sqrt{p}A_p(Tu \otimes \Psi^{(p-1)}) = (B(Tu)^*\Psi)^{(p)}.$$

Therefore

$$[\mathrm{d}\Gamma_{\mathrm{f}}(T), B(u)^*]\Psi = B(Tu)^*\Psi.$$

Let $\Psi \in D(\mathrm{d}\Gamma_{\mathrm{f}}(T))$. Since $\wedge_{\mathrm{fin}}(D(T))$ is a core for $d\Gamma_{\mathrm{f}}(T)$, there exists a sequence $\{\Psi_n\}_n$ in $\wedge_{\mathrm{fin}}(D(T))$ such that $\Psi_n \to \Psi, d\Gamma_{\mathrm{f}}(T)\Psi_n \to d\Gamma_{\mathrm{f}}(T)\Psi$ $(n \to \infty)$. By the result obtained above, we have

$$d\Gamma_{\mathrm{f}}(T)B(u)^*\Psi_n = B(u)^*d\Gamma_{\mathrm{f}}(T)\Psi_n + B(Tu)^*\Psi_n.$$

By the boundedness of $B(\cdot)^*$, the right hand side of this equation converges to $B(u)^*d\Gamma_{\mathrm{f}}(T)B(u)^*\Psi + B(Tu)^*\Psi$ as $n \to \infty$. Hence, by the closedness of $d\Gamma_{\mathrm{f}}(T)$, $B(u)^*\Psi \in D(d\Gamma_{\mathrm{f}}(T))$ and $d\Gamma_{\mathrm{f}}(T)B(u)^*\Psi = B(u)^*d\Gamma_{\mathrm{f}}(T)\Psi + B(Tu)^*\Psi$. Thus, $B(u)^*$ leaves $D(d\Gamma_{\mathrm{f}}(T))$ invariant and (7.40) holds on $D(\mathrm{d}\Gamma_{\mathrm{f}}(T))$.

(ii) Let $u \in D(T^*)$. Then, by considering the adjoint of (7.40) with T replaced by T^*, we have

$$[\mathrm{d}\Gamma_{\mathrm{f}}(T), B(u)] = -B(T^*u) \quad \text{on } \wedge_{\mathrm{fin}}(D(T)).$$

Then, by a limiting argument similar to that in (i), one can see that $B(u)$ leaves $D(\mathrm{d}\Gamma_{\mathrm{f}}(T))$ invariant and (7.41) holds. ■

7.6 Tranformations of $B(\cdot)^{\#}$ by $\Gamma_{\mathrm{f}}(\cdot)$

Theorem 7.10 *Let W be a unitary operator on $\mathscr{K}$. Then*

$$\Gamma_{\mathrm{f}}(W)B(u)^{\#}\Gamma_{\mathrm{f}}(W)^{-1} = B(Wu)^{\#}, \quad u \in \mathscr{K}. \tag{7.42}$$

Proof Let $\Psi = B(u_1)^* \cdots B(u_n)^* \Omega_{\mathscr{K}}$ ($n \geq 0$, $u_j \in \mathscr{K}$, $j = 1, \ldots, n$). Then it is easy to see that $\Gamma_{\mathrm{f}}(W) B(u)^* \Psi = B(Wu)^* \Gamma_{\mathrm{f}}(W) \Psi$. Since the subspace algebraically spanned by vectors of the form Ψ is dense in $\mathscr{F}_{\mathrm{f}}(\mathscr{K})$ and $\Gamma_{\mathrm{f}}(W)$ and $B(u)^*$ are bounded, it follows that $\Gamma_{\mathrm{f}}(W) B(u)^* = B(Wu)^* \Gamma_{\mathrm{f}}(W)$. Hence (7.42) with $B(u)^\# = B(u)^*$ holds. Similarly (or by taking the adjoint of (7.42) with $B(u)^\# = B(u)^*$), we obtain (7.42) with $B(u)^\# = B(u)$. ■

By this theorem and Theorem 7.4(i), we obtain the following theorem:

Theorem 7.11 *Let T be a self-adjoint operator on $\mathscr{K}$. Then, for all $t \in \mathbb{R}$ and $u \in \mathscr{K}$,*

$$e^{itd\Gamma_{\mathrm{f}}(T)} B(u)^\# e^{-itd\Gamma_{\mathrm{f}}(T)} = B(e^{itT} u)^\#,$$

Corollary 7.2 *For all $u \in \mathscr{K}$,*

$$e^{i\pi N_{\mathrm{f}}} B(u)^\# e^{-i\pi N_{\mathrm{f}}} = -B(u)^\#,$$

We introduce a bounded self-adjoint operator $(-1)^{N_{\mathrm{f}}}$ on $\mathscr{F}_{\mathrm{f}}(\mathscr{K})$ by

$$((-1)^{N_{\mathrm{f}}} \Psi)^{(p)} := (-1)^p \Psi^{(p)}, \quad \Psi \in \mathscr{F}_{\mathrm{f}}(\mathscr{K}),\ p \geq 0. \tag{7.43}$$

By functional calculus, we have

$$(-1)^{N_{\mathrm{f}}} = e^{i\pi N_{\mathrm{f}}} = e^{-i\pi N_{\mathrm{f}}}. \tag{7.44}$$

Corollary 7.2 is rephrased as the anti-commutativity of $(-1)^{N_{\mathrm{f}}}$ and $B(u)^\#$:

$$\{(-1)^{N_{\mathrm{f}}}, B(u)^\#\} = 0. \tag{7.45}$$

7.7 Representations of Fermion Second Quantization Operators in Terms of $B(\cdot)^\#$

As in the case of boson second quantization operators, fermion second quantization operators also have representations in terms of $B(\cdot)^\#$. We first derive a basic formula:

Lemma 7.3 *Let $T \in \mathfrak{L}(\mathscr{K})$ be densely defined and $S \in \mathfrak{L}(\mathscr{K})$. Then, for all $u \in D(S) \cap D(T^*)$,*

$$B(Su)^* B(T^* u) = \mathrm{d}\Gamma_{\mathrm{f}}(P_{T^*u, Su}), \tag{7.46}$$

where $P_{v,w}$ ($v, w \in \mathscr{K}$) is defined by (1.23).

Proof As in Lemma 6.8, one first proves (7.46) on $\wedge_{\mathrm{fin}}(\mathscr{K})$ (see (7.29)). Since $B(Su)^* B(T^*u)$ is bounded (this is different from the bosonic case), one obtains (7.46) as an operator equality. ■

Theorem 7.12 *Let $S, T \in \mathfrak{B}(\mathscr{K})$ and $\{e_n\}_{n=1}^{\infty}$ be a CONS of $\mathscr{K}$. Then, for all $\Psi \in \mathscr{F}_{\mathrm{f},0}(\mathscr{K})$,*

$$\sum_{n=1}^{\infty} B(Se_n)^* B(T^*e_n)\Psi = \mathrm{d}\Gamma_{\mathrm{f}}(ST)\Psi. \tag{7.47}$$

Proof Since we have Lemma 7.3, one can prove (7.47) in the same way as in the proof of Theorem 6.12. ■

Corollary 7.3 *Let S be a bounded self-adjoint operator on $\mathscr{K}$ and $\{e_n\}_{n=1}^{\infty}$ be a CONS of $\mathscr{K}$. Then, for all $\Psi \in D(\mathrm{d}\Gamma_{\mathrm{f}}(S^2)^{1/2})$,*

$$\sum_{n=1}^{\infty} \|B(Se_n)\Psi\|^2 = \|\mathrm{d}\Gamma_{\mathrm{f}}(S^2)^{1/2}\Psi\|^2. \tag{7.48}$$

In particular, for all $\Psi \in D(N_{\mathrm{f}}^{1/2})$,

$$\sum_{n=1}^{\infty} \|B(e_n)\Psi\|^2 = \|N_{\mathrm{f}}^{1/2}\Psi\|^2. \tag{7.49}$$

Proof We apply (7.47) with $T^* = S$ to obtain

$$\sum_{n=1}^{\infty} \|B(Se_n)\Psi\|^2 = \|\mathrm{d}\Gamma_{\mathrm{f}}(S^2)^{1/2}\Psi\|^2$$

for all $\Psi \in \mathscr{F}_{\mathrm{f},0}(\mathscr{H})$. Then, in the same way as in the proof of Corollary 6.10, one obtains (7.48). Formula (7.49) is just a special case of (7.48) with $S = I$. ■

7.8 Uniform Differentiability of Basic Operator-Valued Functions

For a self-adjoint operator T on $\mathscr{K}$ and a vector $u \in \mathscr{K}$, one has bounded operator-valued functions $B(e^{itT}u)$ and $B(e^{itT}u)^*$ of t. In this section we consider differentiability of them.

To state a theorem on this aspect, we recall a definition of differentiability of an operator-valued function on an interval $\mathbb{I}$ of $\mathbb{R}$. Let $\mathscr{H}$ be a Hilbert space and $a \in \mathbb{I}$. Then an operator-valued function $T(\cdot) : \mathbb{I} \to \mathfrak{B}(\mathscr{H}); \mathbb{I} \ni t \mapsto T(t) \in \mathfrak{B}(\mathscr{H})$ is said to be **uniformly differentiable at** $t = a$ if the uniform limit

$$T'(a) := \text{u-}\lim_{t \to a} \frac{1}{t-a}(T(t) - T(a)) \in \mathfrak{B}(\mathscr{H})$$

exists (i.e., $\lim_{t\to a} \|T'(a) - (t-a)^{-1}(T(t) - T(a))\| = 0$). If $T(\cdot)$ is uniformly differentiable at any point t in $\mathbb{I}$, then $T(\cdot)$ is said to be uniformly differentiable on $\mathbb{I}$. In this case, we define

$$\frac{dT(t)}{dt} := \text{u-}\lim_{\varepsilon \to 0} \frac{1}{\varepsilon}(T(t+\varepsilon) - T(t)) \in \mathfrak{B}(\mathscr{H})$$

and call it the **uniform derivative** of $T(\cdot)$. The uniform derivative $dT(t)/dt$ of $T(\cdot)$ is written also $T'(t)$.

If $dT(t)/dt$ is uniformly differentiable on $\mathbb{I}$, then $T(\cdot)$ is said to be twice uniformly differentiable on $\mathbb{I}$. In this case, the second uniform derivative of $T(\cdot)$ is defined by

$$\frac{d^2T(t)}{dt^2} := \frac{dT'(t)}{dt}.$$

In a similar manner, for any natural number $n \geq 3$, one can define n times uniform differentiability of $T(\cdot)$ and the nth uniform derivative $T^{(n)}(t) = d^nT(t)/dt^n$ inductively:

$$\frac{d^nT(t)}{dt^n} := \frac{dT^{(n-1)}(t)}{dt}.$$

Theorem 7.13 *Let $u \in D(T^n)$ and $n \in \mathbb{N}$. Then $B(e^{itT}u)^\#$ is n times uniformly differentiable in $t \in \mathbb{R}$ with uniform derivative*

$$\frac{d^nB(e^{itT}u)^\#}{dt^n} = B(i^ne^{itT}T^nu)^\#.$$

Proof We first consider the case $n = 1$. Let $t \in \mathbb{R}$, $\varepsilon \in \mathbb{R} \setminus \{0\}$, $u \in D(T)$ and

$$A_\varepsilon := \frac{B(e^{i(t+\varepsilon)T}u)^\# - B(e^{itT}u)^\#}{\varepsilon} - B(ie^{itT}Tu)^\#.$$

Then we have $A_\varepsilon = B(e^{itT}u_\varepsilon)^\#$ with

$$u_\varepsilon := \frac{e^{i\varepsilon T}u - u}{\varepsilon} - iTu.$$

Hence $\|A_\varepsilon\| = \|e^{itT}u_\varepsilon\| = \|u_\varepsilon\|$. It is well known that $\lim_{\varepsilon\to} \|u_\varepsilon\| = 0$. Thus the present theorem with $n = 1$ holds. Similarly, by induction in n, one can prove the theorem with any n. ∎

7.9 The Fermion Fock Space Over a Direct Sum Hilbert Space

Let $\mathscr{K}_1$ and $\mathscr{K}_2$ be complex Hilbert spaces. Then one can consider the fermion Fock space $\mathscr{F}_{\mathrm{f}}(\mathscr{K}_1 \oplus \mathscr{K}_2)$ over the direct sum Hilbert space $\mathscr{K}_1 \oplus \mathscr{K}_2$. As in the boson Fock space case, there exists a natural isomorphism between $\mathscr{F}_{\mathrm{f}}(\mathscr{K}_1 \oplus \mathscr{K}_2)$ and $\mathscr{F}_{\mathrm{f}}(\mathscr{K}_1) \otimes \mathscr{F}_{\mathrm{f}}(\mathscr{K}_2)$ if $\mathscr{K}_1$ and $\mathscr{K}_2$ are separable. To be definite, for $i = 1, 2$, we denote by $B_{\mathscr{K}_i}(u)$ the annihilation operator on $\mathscr{F}_{\mathrm{f}}(\mathscr{K}_i)$ with test vector $u \in \mathscr{K}_i$. On the other hand, we simply denote by $B(u, v)$ the annihilation operator on $\mathscr{F}_{\mathrm{f}}(\mathscr{K}_1 \oplus \mathscr{K}_2)$ with test vector $(u, v) \in \mathscr{K}_1 \oplus \mathscr{K}_2$.

Theorem 7.14 *Suppose that $\mathscr{K}_1$ and $\mathscr{K}_2$ are separable. Then there exists a unique unitary operator $U : \mathscr{F}_{\mathrm{f}}(\mathscr{K}_1 \oplus \mathscr{K}_2) \to \mathscr{F}_{\mathrm{f}}(\mathscr{K}_1) \otimes \mathscr{F}_{\mathrm{f}}(\mathscr{K}_2)$ such that*

$$U\Omega_{\mathscr{K}_1\oplus\mathscr{K}_2} = \Omega_{\mathscr{K}_1} \otimes \Omega_{\mathscr{K}_2} \tag{7.50}$$

and, for all $n, m \in \mathbb{Z}_+$ and $u_i \in \mathscr{K}_1, v_j \in \mathscr{K}_2$ ($i = 1, \ldots, n$, $j = 1, \ldots, m$),

$$UB(u_1, 0)^* \cdots B(u_n, 0)^* B(0, v_1)^* \cdots B(0, v_m)^* \Omega_{\mathscr{K}_1\oplus\mathscr{K}_2} \tag{7.51}$$

$$= B_{\mathscr{K}_1}(u_1)^* \cdots B_{\mathscr{K}_1}(u_n)^* \Omega_{\mathscr{K}_1} \otimes B_{\mathscr{K}_2}(v_1)^* \cdots B_{\mathscr{K}_2}(v_m)^* \Omega_{\mathscr{K}_2} \tag{7.52}$$

Moreover, the following hold:

(i) *For all $u \in \mathscr{K}_1, v \in \mathscr{K}_2$,*

$$UB(u, v)^{\#}U^{-1} = B_{\mathscr{K}_1}(u)^{\#} \otimes I + (-1)^{N_{\mathrm{f}}} \otimes B_{\mathscr{K}_2}(v)^{\#}. \tag{7.53}$$

(ii) *Let T_1 and T_2 be densely defined closable operators on $\mathscr{K}_1$ and $\mathscr{K}_2$ respectively. Then*

$$U\mathrm{d}\Gamma_{\mathrm{f}}(T_1 \oplus T_2)U^{-1} = \overline{\mathrm{d}\Gamma_{\mathrm{f}}(T_1) \otimes I + I \otimes \mathrm{d}\Gamma_{\mathrm{f}}(T_2)}. \tag{7.54}$$

(iii) *Let A_1 and A_2 be contraction operators on $\mathscr{K}_1$ and $\mathscr{K}_2$ respectively. Then*

$$U\Gamma(A_1 \oplus A_2)U^{-1} = \Gamma(A_1) \otimes \Gamma(A_2). \tag{7.55}$$

Proof Let $\{e_n\}_{n=1}^\infty$ and $\{f_n\}_{n=1}^\infty$ be CONS's of $\mathcal{K}_1$ and $\mathcal{K}_2$ respectively. Then, by Theorem 1.20,

$$\{B(e_{i_1})^* \cdots B(e_{i_p})^* \Omega_{\mathcal{K}_1} \otimes B(f_{j_1})^* \cdots B(f_{j_q})^* \Omega_{\mathcal{K}_2} | \, p, q \geq 0, i_1 < \ldots < i_p, \\ j_1 < \ldots < j_q, i_k, j_\ell \in \mathbb{N}, k = 1, \ldots, p, \ell = 1, \ldots, q\}$$

is a CONS of $\mathcal{F}_{\mathrm{f}}(\mathcal{K}_1) \otimes \mathcal{F}_{\mathrm{f}}(\mathcal{K}_2)$. The set $\{(e_n, 0), (0, f_n) | \, n \in \mathbb{N}\}$ is a CONS of $\mathcal{K}_1 \oplus \mathcal{K}_2$. Hence, by Theorem 7.5,

$$\{B(e_{i_1}, 0)^* \cdots B(e_{i_p}, 0)^* B(0, f_{j_1})^* \cdots B(0, f_{j_q})^* \Omega_{\mathcal{K}_1 \oplus \mathcal{K}_2} | \, p, q \geq 0, \\ i_1 < \ldots < i_p, \; j_1 < \ldots < j_q, i_k, j_\ell \in \mathbb{N}, k = 1, \ldots, p, \ell = 1, \ldots, q\}$$

is a CONS of $\mathcal{F}_{\mathrm{f}}(\mathcal{K}_1 \oplus \mathcal{K}_2)$. Hence, by the isomorphism theorem, there exists a unique unitary operator U from $\mathcal{F}_{\mathrm{f}}(\mathcal{K}_1 \oplus \mathcal{K}_2)$ to $\mathcal{F}_{\mathrm{f}}(\mathcal{K}_1) \otimes \mathcal{F}_{\mathrm{f}}(\mathcal{K}_2)$ such that, for all $p, q \geq 0$ and $i_1 < \ldots < i_p, j_1 < \ldots < j_q, i_k, j_\ell \in \mathbb{N}$,

$$\begin{aligned} & U B(e_{i_1}, 0)^* \cdots B(e_{i_p}, 0)^* B(0, f_{j_1})^* \cdots B(0, f_{j_q})^* \Omega_{\mathcal{K}_1 \oplus \mathcal{K}_2} \\ &= B(e_{i_1})^* \cdots B(e_{i_p})^* \Omega_{\mathcal{K}_1} \otimes B(f_{j_1})^* \cdots B(f_{j_q})^* \Omega_{\mathcal{K}_2} \end{aligned}$$

Hence, in particular, (7.50) holds. It is easy to see that, for each $u \in \mathcal{K}_1$,

$$B(u)^* = \sum_{n=1}^\infty \langle e_n, u \rangle \, B(e_n)^* \tag{7.56}$$

in the sense of operator norm. Using this fact, one can prove (7.52).

(i) Let $\Psi = B(u_1, 0)^* \cdots B(u_n, 0)^* B(0, v_1)^* \cdots B(0, v_m)^* \Omega_{\mathcal{K}_1 \oplus \mathcal{K}_2}$. Then

$$B(u, v)^* \Psi = B(u, 0)^* \Psi + B(0, v)^* \Psi = B(u, 0)^* \Psi + (-1)^n \Psi',$$

where $\Psi' := B(u_1, 0)^* \cdots B(u_n, 0)^* B(0, v)^* B(0, v_1)^* \cdots B(0, v_m)^* \Omega_{\mathcal{K}_1 \oplus \mathcal{K}_2}$. Hence, by the preceding result, we have

$$\begin{aligned} U B(u, v)^* \Psi &= (B(u)^* \otimes I) U \Psi + (-1)^n (I \otimes B(v)^*) U \Psi \\ &= (B(u)^* \otimes I) U \Psi + ((-1)^{N_{\mathrm{f}}} \otimes B(v)^*) U \Psi. \end{aligned}$$

Since the subspace algebraically spanned by vectors of the form Ψ is dense in $\mathcal{F}_{\mathrm{f}}(\mathcal{K} \oplus \mathcal{K}_2)$ and $B(u, v)^*$, $B(u)^*$ and $(-1)^{N_{\mathrm{f}}} \otimes B(v)^*$ are bounded, it follows that

$$U B(u, v)^* = (B(u) \otimes I) U + ((-1)^{N_{\mathrm{f}}} \otimes B(v)^*) U.$$

Thus (7.53) with $\# = *$ holds. Taking the adjoint of the operator equality just obtained, we have (7.53) with $\# = \emptyset$.

Propositions (ii) and (iii) can be proved in the same way as in the proof of Theorem 6.25. ■

Remark 7.2 Note that the operator $(-1)^{N_{\mathrm{f}}}$ appears on the right hand side of (7.53). This is a big difference from the case of boson creation and annihilation operators.

Remark 7.3 Theorem 7.14 can be easily extended to the case where $\mathscr{K}_1 \oplus \mathscr{K}_2$ is replaced by the direct sum $\oplus_{j=1}^N \mathscr{K}_j$ of N Hilbert spaces $\mathscr{K}_1, \ldots, \mathscr{K}_N$ with $N \geq 3$.

Chapter 8
Representations of CCR with Infinite Degrees of Freedom

Abstract A general theory of representations of CCR with infinite degrees of freedom is presented. A basic class of such representations, called Fock representations, is described. Moreover, deformations of Fock representations are discussed.

8.1 Representation of CCR

In Chap. 6, we have seen that boson creation and annihilation operators on the boson Fock space $\mathscr{F}_{\mathrm{b}}(\mathscr{H})$ obey the CCR over $\mathscr{H}$ (Theorem 6.8) and the Fock vacuum $\Omega_{\mathscr{H}}$ is a cyclic vector for $\{A(f_1)^*A(f_2)^* \cdots A(f_n)^* | n \in \mathbb{N},\ f_j \in \mathscr{D},\ j = 1, \ldots, n\}$ with $\mathscr{D}$ being a dense subspace of $\mathscr{H}$ (Theorem 6.6). Moreover, it has been suggested that various operators on the boson Fock space $\mathscr{F}_{\mathrm{b}}(\mathscr{H})$ can be constructed from $A(\cdot)$ and $A(\cdot)^*$. One such operator is the Segal field operator $\Phi_{\mathrm{S}}(f)$ with test vector $f \in \mathscr{H}$, which is a simple linear combination of creation operator $A(f)^*$ and annihilation operator $A(f)$ and has interesting properties as shown in Sect. 6.11. Noting these aspects and taking a representation theoretic point of view, one may regard the boson Fock space $\mathscr{F}_{\mathrm{b}}(\mathscr{H})$ as a Hilbert space in which algebraic objects obeying CCR are realized as $A(f)^{\#}$'s ($f \in \mathscr{H}$). This point of view may lead one to introduce a representation theoretical concept of CCR indexed by a vector space.

Definition 8.1 Let $\mathscr{F}$ be a Hilbert space and $\mathscr{D}$ be a dense subspace in $\mathscr{F}$. Let $\mathscr{V}$ be a complex inner product space and $C\colon \mathscr{V} \to \mathfrak{L}(\mathscr{F}); \mathscr{V} \ni f \mapsto C(f) \in \mathfrak{L}(\mathscr{F})$ such that, for all $f \in \mathscr{V}$, $C(f)$ is densely defined and closed. The triplet $(\mathscr{F}, \mathscr{D}, \{C(f) | f \in \mathscr{V}\})$ (or $(\mathscr{F}, \mathscr{D}, \{C(f), C(f)^* | f \in \mathscr{V}\})$) is called a **representation of the CCR over** $\mathscr{V}$ if the following properties are satisfied:

(C.1) For all $f \in \mathscr{V}$, $\mathscr{D} \subset D(C(f)) \cap D(C(f)^*)$ and $C(f)\mathscr{D} \subset \mathscr{D}$, $C(f)^*\mathscr{D} \subset \mathscr{D}$.
(C.2) (anti-linearity in test vectors) For all $f, g \in \mathscr{V}$ and $\alpha, \beta \in \mathbb{C}$, $C(\alpha f + \beta g) = \alpha^* C(f) + \beta^* C(g)$ on $\mathscr{D}$.

© Springer Nature Singapore Pte Ltd. 2020

A. Arai, *Inequivalent Representations of Canonical Commutation and Anti-Commutation Relations*, Mathematical Physics Studies,
https://doi.org/10.1007/978-981-15-2180-5_8

(C.3) (CCR) For all $f, g \in \mathscr{V}$,

$$[C(f), C(g)^*] = \langle f, g \rangle_{\mathscr{V}}, \tag{8.1}$$

$$[C(f), C(g)] = 0, \quad [C(f)^*, C(g)^*] = 0 \tag{8.2}$$

on $\mathscr{D}$.

The Hilbert space $\mathscr{F}$ is called a **representation space** of the CCR under consideration. The subspace $\mathscr{D}$ is called a **CCR-domain** for $\{C(f), C(f)^* | f \in \mathscr{V}\}$.

If $\{C(f), C(f)^* \mid f \in \mathscr{V}\}$ is irreducible (resp. reducible), then the representation $(\mathscr{F}, \mathscr{D}, \{C(f) | f \in \mathscr{V}\})$ is said to be **irreducible** (resp. **reducible**).

Remark 8.1 The second commutation relation in (8.2) follows from taking the adjoint of the first one.

Remark 8.2 Property (C.2) implies the linearity of $C(f)^*$ in f: for all $f, g \in \mathscr{V}$ and $\alpha, \beta \in \mathbb{C}$, $C(\alpha f + \beta g)^* = \alpha C(f)^* + \beta C(g)^*$ on $\mathscr{D}$.

Remark 8.3 By Lemma 2.1(iii), (8.1) with $g = f$ implies that, for $f \neq 0$, at least one of $C(f)$ and $C(f)^*$ is unbounded. Hence, in fact, both $C(f)$ and $C(f)^*$ are unbounded.

Example 8.1 Let $(\mathscr{F}, \mathscr{D}, \{C(f) | f \in \mathscr{V}\})$ be a representation of the CCR over $\mathscr{V}$ and $U: \mathscr{F} \to \mathscr{F}'$ ($\mathscr{F}'$ is a Hilbert space) be a unitary operator. Then $(\mathscr{F}', U\mathscr{D}, \{UC(f)U^{-1} | f \in \mathscr{V}\})$ is a representation of the CCR over $\mathscr{V}$.

Example 8.2 Let $\mathscr{H}$ be a Hilbert space and $\mathscr{V}$ be a dense subspace of $\mathscr{H}$. Then the triplet

$$\rho_{\mathrm{F}}(\mathscr{V}) := \{\mathscr{F}_{\mathrm{b}}(\mathscr{H}), \mathscr{F}_{\mathrm{b},0}(\mathscr{H}), \{A(f) | f \in \mathscr{V}\}\}$$

is an irreducible representation of the CCR over $\mathscr{V}$ (the irreducibility is due to Theorem 6.9). This representation is called the **Fock representation of the CCR over $\mathscr{V}$**.

Example 8.3 Let $\mathscr{V}$ be a dense subspace of a Hilbert space $\mathscr{H}$, $L: \mathscr{V} \to \mathbb{C}$ be a linear functional and

$$C_L(f) := A(f) + L(f)^*, \quad f \in \mathscr{V}. \tag{8.3}$$

Then it is easy to see that $\{C_L(f), C_L(f)^* | f \in \mathscr{V}\}$ satisfies the CCR over $\mathscr{V}$ on $\mathscr{F}_{\mathrm{b},0}(\mathscr{H})$. Let $T \in \{C_L(f), C_L(f)^* | f \in \mathscr{V}\}'$. Then it follows that $T \in \{A(f), A(f)^* \mid f \in \mathscr{V}\}'$. Hence $T = \alpha I$ for some $\alpha \in \mathbb{C}$. Therefore $\{C_L(f), C_L(f)^* | f \in \mathscr{V}\}' = \mathbb{C}I$. Thus

$$\beta_L(\mathscr{V}) := (\mathscr{F}_{\mathrm{b}}(\mathscr{H}), \mathscr{F}_{\mathrm{b},0}(\mathscr{H}), \{C_L(f) | f \in \mathscr{V}\}) \tag{8.4}$$

is an irreducible representation of the CCR over $\mathscr{V}$. The correspondence $(A(\cdot), A(\cdot)^*) \mapsto (C_L(\cdot), C_L(\cdot)^*)$ of operator-valued functionals on $\mathscr{V}$ is called the **Bogoliubov translation** by the translation functional $L(\cdot)$.

Example 8.4 Let $W: \mathscr{H} \to \mathscr{H}$ be an isometry (i.e., W is linear and $\|Wf\| = \|f\|$, $f \in \mathscr{H}$) and

$$A_W(f) := A(Wf).$$

Let $\mathscr{V}$ be a dense subspace of $\mathscr{H}$. Then $(\mathscr{F}_{\mathrm{b}}(\mathscr{H}), \mathscr{F}_{\mathrm{b},0}(\mathscr{H}), \{A_W(f)|f \in \mathscr{V}\})$ is a representation of the CCR over $\mathscr{V}$. It may depend on properties of W if this representation is irreducible.

(1) The case where W is unitary

In this case, $W\mathscr{V}$ is dense in $\mathscr{H}$. Hence, by Theorem 6.9, $\{A_W(f), A_W(f)^*| f \in \mathscr{V}\}$ is irreducible.

(2) The case where W is not unitary

In this case, $(W\mathscr{V})^\perp \neq \{0\}$. Hence, for all $g \in (W\mathscr{V})^\perp \setminus \{0\}$ and $f \in \mathscr{V}$, $\langle Wf, g\rangle = 0$. Therefore, by (6.118),

$$e^{i\Phi_{\mathrm{S}}(g)} A_W(f) e^{-i\Phi_{\mathrm{S}}(g)} = A_W(f) - \frac{i}{\sqrt{2}}\langle Wf, g\rangle = A_W(f),$$

i.e., $e^{i\Phi_{\mathrm{S}}(g)} A_W(f) = A_W(f) e^{i\Phi_{\mathrm{S}}(g)}$. This implies also that

$$e^{i\Phi_{\mathrm{S}}(g)} A_W(f)^* = A_W(f)^* e^{i\Phi_{\mathrm{S}}(g)}.$$

For every $g \neq 0$, $e^{i\Phi_{\mathrm{S}}(g)} \neq I$.[1] Thus $\{A_W(f), A_W(f)^*| f \in \mathscr{V}\}' \neq \mathbb{C}I$. The set $\{A_W(f), A_W(f)^*|f \in \mathscr{V}\}$ is $*$-invariant. Hence $\{A_W(f), A_W(f)^*|f \in \mathscr{V}\}$ is not irreducible.

As in the case of representations of CCR with finite degrees of freedom, we define a concept of equivalence among representations of CCR in the sense defined above.

Definition 8.2 Let $(\mathscr{F}, \mathscr{D}, \{C(f)|f \in \mathscr{V}\})$ and $(\mathscr{F}', \mathscr{D}', \{C'(f)|f \in \mathscr{V}\})$ be two representations of the CCR over $\mathscr{V}$. They are said to be **equivalent** if there exists a unitary operator $U: \mathscr{F} \to \mathscr{F}'$ such that, for all $f \in \mathscr{V}$, $C'(f) = UC(f)U^{-1}$ (operator equality). In that case, we often say simply that $\{C(f)|f \in \mathscr{V}\}$ is equivalent to $\{C'(f)|f \in \mathscr{V}\}$.

[1] In general, for a self-adjoint operator A on a Hilbert space, $\sigma(e^{iA}) = \{e^{i\lambda}|\lambda \in \sigma(A)\}$ by the spectral mapping theorem. Recall that, if $g \neq 0$, then $\sigma(\Phi_{\mathrm{S}}(g)) = \mathbb{R}$ by (6.121).

Remark 8.4 The operator equality $C'(f) = UC(f)U^{-1}$ implies the operator equality

$$C'(f)^* = UC(f)^*U^{-1}.$$

Remark 8.5 Suppose that $\rho := (\mathscr{F}, \mathscr{D}, \{C(f)|f \in \mathscr{V}\})$ is equivalent to $\rho' := (\mathscr{F}', \mathscr{D}', \{C'(f)|f \in \mathscr{V}\})$. Then, by Proposition 2.8, ρ is reducible (resp. irreducible) if and only if ρ' is reducible (resp. irreducible). Hence, if ρ is reducible and ρ' is irreducible, then ρ is inequivalent to ρ'. But this type of inequivalence is a trivial inequivalence. Non-trivial inequivalences may occur among irreducible representations.

Example 8.5 Let us consider the representation $(\mathscr{F}_{\mathrm{b}}(\mathscr{H}), \mathscr{F}_{\mathrm{b},0}(\mathscr{H}), \{A_W(f)|f \in \mathscr{V}\})$ in Example 8.4. Suppose that W is unitary. Then, $\Gamma_{\mathrm{b}}(W)$ is a unitary operator on $\mathscr{F}_{\mathrm{b}}(\mathscr{H})$ and, by Theorem 6.7,

$$\Gamma_{\mathrm{b}}(W)A(f)\Gamma_{\mathrm{b}}(W)^{-1} = A_W(f), \quad f \in \mathscr{V}.$$

Hence $(\mathscr{F}_{\mathrm{b}}(\mathscr{H}), \mathscr{F}_{\mathrm{b},0}(\mathscr{H}), \{A_W(f)|f \in \mathscr{V}\})$ is equivalent to the Fock representation $\rho_{\mathrm{F}}(\mathscr{V})$.

Example 8.6 We consider the case where W is not unitary in Example 8.4. The reducibility of the representation in this case can be understood structurally as follows. Let $\mathscr{M} := \overline{W\mathscr{V}}$, which is not equal to $\mathscr{H}$ in the present case. Hence we have a non-trivial orthogonal decomposition

$$\mathscr{H} = \mathscr{M} \oplus \mathscr{M}^{\perp}.$$

Then, by Theorem 6.24, there exists a unitary operator $U\colon \mathscr{F}_{\mathrm{b}}(\mathscr{H}) \to \mathscr{F}_{\mathrm{b}}(\mathscr{M}) \otimes \mathscr{F}_{\mathrm{b}}(\mathscr{M}^{\perp})$ such that

$$UA_W(f)U^{-1} = A_{\mathscr{M}}(Wf) \otimes I, \quad f \in \mathscr{V},$$

where $A_{\mathscr{M}}(\cdot)$ denotes the annihilation operator on $\mathscr{F}_{\mathrm{b}}(\mathscr{M})$. Let $\{\Psi_n\}_{n=1}^{\infty}$ be a CONS of $\mathscr{F}_{\mathrm{b}}(\mathscr{M}^{\perp})$ and $\mathscr{L}_n := \{\alpha\Psi_n|\alpha \in \mathbb{C}\}$. Then we have the orthogonal decomposition

$$U\mathscr{F}_{\mathrm{b}}(\mathscr{H}) = \oplus_{n=1}^{\infty}\mathscr{F}_{\mathrm{b}}(\mathscr{M}) \otimes \mathscr{L}_n.$$

Since $\mathscr{L}_n$ is one dimensional, there exists a unitary operator $V_n\colon \mathscr{F}_{\mathrm{b}}(\mathscr{M}) \otimes \mathscr{L}_n \to \mathscr{F}_{\mathrm{b}}(\mathscr{M})$ such that $V_n(\Psi \otimes (\alpha\Psi_n) = \alpha\Psi$, $\alpha \in \mathbb{C}$, $\Psi \in \mathscr{F}_{\mathrm{b}}(\mathscr{M})$. Thus, putting $V := \oplus_{n=1}^{\infty} V_n$ and $Y := VU$, we obtain

$$Y\mathscr{F}_{\mathrm{b}}(\mathscr{H}) = \oplus_{n=1}^{\infty}\mathscr{F}_{\mathrm{b}}^{(n)}(\mathscr{M})$$

with $\mathscr{F}_{\mathrm{b}}^{(n)}(\mathscr{M}) := \mathscr{F}_{\mathrm{b}}(\mathscr{M})$, $n \in \mathbb{N}$. We denote by $A_{\mathscr{M}}^{(n)}(\cdot)$ the annihilation operator on $\mathscr{F}_{\mathrm{b}}^{(n)}(\mathscr{M})$. Since $W\mathscr{V}$ is dense in $\mathscr{M}$, $\rho_n := (\mathscr{F}_{\mathrm{b}}^{(n)}(\mathscr{M}), \mathscr{F}_{\mathrm{b},0}(\mathscr{M}), \{A_{\mathscr{M}}^{(n)}(Wf)$ $f \in \mathscr{V}\})$ is an irreducible representation of the CCR over $\mathscr{V}$. It is not difficult to show that

$$YA_W(f)Y^{-1} = \oplus_{n=1}^{\infty} A_{\mathscr{M}}^{(n)}(Wf).$$

Thus $(\mathscr{F}_{\mathrm{b}}(\mathscr{H}), \mathscr{F}_{\mathrm{b},0}(\mathscr{H}), \{A_W(f)|f \in \mathscr{V}\})$ is equivalent to $\oplus_{n=1}^{\infty}\rho_n$, the direct sum of irreducible representations ρ_n. As in Example 8.5, we have

$$A_{\mathscr{M}}^{(n)}(Wf) = \Gamma_{\mathrm{b}}(W)A(f)\Gamma_{\mathrm{b}}(W)^{-1}, \quad f \in \mathscr{V},$$

where $\Gamma_{\mathrm{b}}(W)$ is considered as a unitary operator from $\mathscr{F}_{\mathrm{b}}(\mathscr{H})$ onto $\mathscr{F}_{\mathrm{b}}(\mathscr{M})$. Hence ρ_n is equivalent to the Fock representation $\rho_{\mathrm{F}}(\mathscr{V})$. Thus $(\mathscr{F}_{\mathrm{b}}(\mathscr{H}), \mathscr{F}_{\mathrm{b},0}(\mathscr{H})$, $\{A_W(f)\,|f \in \mathscr{V}\})$ is equivalent to an infinite direct sum of the Fock representation $\rho_{\mathrm{F}}(\mathscr{V})$. This is a simple example of completely reducible representations.[2]

8.2 Cyclic Representations

We first discuss representations of CCR which can be treated more easily.

Definition 8.3 Let

$$\rho := (\mathscr{F}, \mathscr{D}, \{C(f)|f \in \mathscr{V}\}) \tag{8.5}$$

be a representation of the CCR over $\mathscr{V}$.

(i) If there exists a unit vector $\Omega \in \mathscr{D}$ ($\|\Omega\| = 1$) such that $C(f)\Omega = 0$, $\forall f \in \mathscr{V}$, then Ω is called a **vacuum** for ρ or $\{C(f)|f \in \mathscr{V}\}$ and ρ is said to have a vacuum Ω. In this case, one can define a subspace:

$$\mathscr{F}_{\rho,\mathrm{fin}} := \mathrm{span}\{\Omega, C(f_1)^* \cdots C(f_n)^*\Omega|n \in \mathbb{N},\ f_j \in \mathscr{V},\ j = 1, \ldots, n\}. \tag{8.6}$$

(ii) Representation ρ is called a **cyclic representation** if it has a vacuum Ω which is cyclic for $\{C(f_1)^{\#} \cdots C(f_n)^{\#})|n \in \mathbb{N},\ f_j \in \mathscr{V},\ j = 1, \ldots, n\}$. We write this representation as $(\mathscr{F}, \mathscr{D}, \{C(f)|f \in \mathscr{V}\}, \Omega)$.

[2] A reducible representation of the CCR over an inner product space $\mathscr{V}$ is said to be **completely reducible** if it is equivalent to a direct sum representation of irreducible representations of the CCR over $\mathscr{V}$.

Example 8.7 The Fock representation $\rho_{\mathrm{F}}(\mathscr{V})$ (Example 8.2) is a cyclic representation with a vacuum being equal to the Fock vacuum $\Omega_{\mathscr{H}}$.

Lemma 8.1 *Let ρ be a representation of the CCR over $\mathscr{V}$ of the form (8.5) having a vacuum Ω. Then, for all $f, g \in \mathscr{V}$, CCR (8.1) and (8.2) hold on $\mathscr{F}_{\rho,\mathrm{fin}}$.*

Proof Let $\Phi \in \mathscr{D}$ and $\Psi \in \mathscr{F}_{\rho,\mathrm{fin}}$. Then

$$\begin{aligned}\left\langle \Phi, [C(f), C(g)^*]\Psi \right\rangle &= \left\langle [C(g), C(f)^*]\Phi, \Psi \right\rangle = \langle \langle g, f\rangle \Phi, \Psi\rangle \\ &= \langle \Phi, \langle f, g\rangle \Psi\rangle .\end{aligned}$$

Since $\mathscr{D}$ is dense in $\mathscr{F}$, it follows that $[C(f), C(g)^*]\Psi = \langle f, g\rangle \Psi$. Similarly, one can prove (8.2) on $\mathscr{F}_{\rho,\mathrm{fin}}$. ■

Let ρ be a representation of the CCR over $\mathscr{V}$ given by (8.5) having a vacuum Ω. Then one can define a linear operator $C_{\mathrm{fin}}(f)$ $(f \in \mathscr{V})$ on the Hilbert space

$$\mathscr{F}_\rho := \overline{\mathscr{F}_{\rho,\mathrm{fin}}} \quad \text{(the closure of } \mathscr{F}_{\rho,\mathrm{fin}}) \tag{8.7}$$

as follows:

$$D(C_{\mathrm{fin}}(f)) := \mathscr{F}_{\rho,\mathrm{fin}}, \quad C_{\mathrm{fin}}(f)\Psi := C(f)\Psi, \quad \Psi \in \mathscr{F}_{\rho,\mathrm{fin}}.$$

Then it easy to see that $C_{\mathrm{fin}}(f)$ is densely defined and $\mathscr{F}_{\rho,\mathrm{fin}} \subset D(C_{\mathrm{fin}}(f)^*)$ with $C_{\mathrm{fin}}(f)^*\Psi = C(f)^*\Psi$, $\Psi \in \mathscr{F}_{\rho,\mathrm{fin}}$. Hence $C_{\mathrm{fin}}(f)$ is closable. We denote its closure by $C_\Omega(f)$.

Lemma 8.1 shows that

$$\pi_\Omega(\mathscr{V}) := \{\mathscr{F}_\rho, \mathscr{F}_{\rho,\mathrm{fin}}, \{C_\Omega(f) | f \in \mathscr{V}\}, \Omega\}$$

is a cyclic representation of the CCR over $\mathscr{V}$.

Theorem 8.1 *Let ρ be a representation of the CCR over $\mathscr{V}$ given by (8.5) having a vacuum Ω. Then there exists a unitary operator $U\colon \mathscr{F}_\rho \to \mathscr{F}_{\mathrm{b}}(\mathscr{H})$ such that $U\Omega = \Omega_{\mathscr{H}}$ and $UC_\Omega(f)U^{-1} = A(f)$, $f \in \mathscr{V}$. In particular, $\pi_\Omega(\mathscr{V})$ is equivalent to the Fock representation $\rho_{\mathrm{F}}(\mathscr{V})$.*

Proof We define a mapping $U_0\colon \mathscr{F}_{\rho,\mathrm{fin}} \to \mathscr{F}_{\mathrm{b},0}(\mathscr{H})$ by

$$\begin{gathered} U_0\Omega := \Omega_{\mathscr{H}}, \\ U_0C(f_1)^* \cdots C(f_n)^*\Omega := A(f_1)^* \cdots A(f_n)^*\Omega_{\mathscr{H}} \end{gathered}$$

$(f_j \in \mathscr{V}, j = 1, \cdots, n,\ n \in \mathbb{N})$. For any element Ψ in $\mathscr{F}_{\rho,\mathrm{fin}}$, $U_0\Psi$ is defined by linear extension (this is well-defined). It is easy to see that U_0 is isometry and $\mathrm{Ran}\,(U_0) = \mathscr{F}_{\mathrm{b,fin}}(\mathscr{V})$. Since $\mathscr{F}_{\rho,\mathrm{fin}}$ is dense in $\mathscr{F}_\rho$, by the extension theorem on densely defined bounded linear operators, the closure $U := \overline{U_0}$ is a unitary operator from $\mathscr{F}_\rho$ to $\mathscr{F}_b(\mathscr{H})$.

Let $f \in \mathscr{V}$. Then, by CCR and the fact that $C(f)\Omega = 0$, we have

$$C_\Omega(f)C(f_1)^* \cdots C(f_n)^*\Omega = \sum_{j=1}^{n} \langle f, f_j \rangle C(f_1)^* \cdots \overbrace{C(f_j)^*}^{j\text{th}} \cdots C(f_n)^*\Omega.$$

Hence

$$\begin{aligned} UC_\Omega(f)C(f_1)^* \cdots C(f_n)^*\Omega &= \sum_{j=1}^{n} \langle f, f_j \rangle A(f_1)^* \cdots \overbrace{A(f_j)^*}^{j\text{th}} \cdots A(f_n)^*\Omega_{\mathscr{H}} \\ &= A(f)A(f_1)^* \cdots A(f_n)^*\Omega_{\mathscr{H}}. \end{aligned}$$

Hence, for all $\Psi \in \mathscr{F}_{b,\text{fin}}(\mathscr{H})$, $UC_\Omega(f)U^{-1}\Psi = A(f)\Psi$. Recall that $\mathscr{F}_{b,\text{fin}}(\mathscr{V})$ is a core for $A(f)$ (see Proposition 6.7). Hence it follows from a limiting argument that $A(f) \subset UC_\Omega(f)U^{-1}$. Exchanging the role of $A(f)$ and $C_\Omega(f)$, we see that $C_\Omega(f) \subset U^{-1}A(f)U$, i.e., $UC_\Omega(f)U^{-1} \subset A(f)$. Hence we obtain operator equality $UC_\Omega(f)U^{-1} = A(f)$. Thus the first half of the theorem follows. With regard to the second half of the theorem, we need only to note that $UC_\Omega(f)U^{-1} = A(f)$ implies $UC_\Omega(f)^*U^{-1} = A(f)^*$. ■

Corollary 8.1 *Let $(\mathscr{F}, \mathscr{D}, \{C(f)|f \in \mathscr{V}\}, \Omega)$ be a cyclic representation of the CCR over $\mathscr{V}$. Suppose that, for all $f \in \mathscr{V}$, $\mathscr{F}_{\rho,\text{fin}}$ is a core for $C(f)$. Then $(\mathscr{F}, \mathscr{D}, \{C(f)|f \in \mathscr{V}\}, \Omega)$ is equivalent to the Fock representation $\rho_F(\mathscr{V})$.*

Proof By the cyclicity of Ω, we have $\mathscr{F}_\rho = \mathscr{F}$. Hence the operator U in the proof of Theorem 8.1 is a unitary operator from $\mathscr{F}$ to $\mathscr{F}_b(\mathscr{H})$ By the assumption on a core of $C(f)$, $C_\Omega(f) = C(f)$, $f \in \mathscr{V}$. Thus, by Theorem 8.1, we obtain $UC(f)^\# U^{-1} = A(f)^\#$ for all $f \in \mathscr{V}$. ■

8.3 Second Quantization Operator Associated with a Representation of CCR

Formula (6.91) may be regarded as a form in which the Fock representation $\rho_F(\mathscr{V})$ of the CCR over $\mathscr{V}$ (see Example 8.2) yields the boson second quantization operator $d\Gamma_b(T)$ for a class of non-negative self-adjoint operators T. We generalize this structure to introduce a concept of second quantization operator associated with a representation of CCR.

8.3.1 A General Fact on a Sesquilinear Form

Let $\mathscr{A} := \{A_n\}_{n=1}^{\infty}$ be a sequence of densely defined closed operators on a Hilbert space $\mathscr{X}$ and

$$D_{\mathscr{A}} := \left\{ \Psi \in \cap_{n=1}^{\infty} D(A_n) | \sum_{n=1}^{\infty} \|A_n \Psi\|^2 < \infty \right\}.$$

Then one can define a sesquilinear form $\mathfrak{s}$: $D_{\mathscr{A}} \times D_{\mathscr{A}}$ (see Sect. 1.11) by

$$\mathfrak{s}(\Phi, \Psi) := \sum_{n=1}^{\infty} \langle A_n \Phi, A_n \Psi \rangle , \quad \Phi, \Psi \in D_{\mathscr{A}}. \tag{8.8}$$

Lemma 8.2 *The sesquilinear form $\mathfrak{s}$ is closed and non-negative.*

Proof Using the Schwarz inequality, one can easily see that the right hand side of (8.8) is absolutely convergent and that $\mathfrak{s}$ is a non-negative sesquilinear form on $D_{\mathscr{A}}$.

We next show that $\mathfrak{s}$ is closed. Let $\Psi_k \in D_{\mathscr{A}}$ $(k \in \mathbb{N})$ and $\Psi_k \to \Psi \in \mathscr{X}$ $(k \to \infty)$, $\mathfrak{s}(\Psi_k - \Psi_\ell, \Psi_k - \Psi_\ell) \to 0$ $(k, \ell \to \infty)$. Hence, for any $\varepsilon >$, there exists a number $k_0 \in \mathbb{N}$ such that, for all $k, \ell \geq k_0$,

$$\sum_{n=1}^{\infty} \|A_n \Psi_k - A_n \Psi_\ell\|^2 < \varepsilon. \tag{8.9}$$

Hence, for all $n \in \mathbb{N}$, $\|A_n \Psi_k - A_n \Psi_\ell\|^2 < \varepsilon$, $k, \ell \geq k_0$. This means that $\{A_n \Psi_k\}_{k=1}^{\infty}$ is a Cauchy sequence in $\mathscr{X}$. Hence $\Phi_n := \lim_{k \to \infty} A_n \Psi_k \in \mathscr{X}$ exists. Since A_n is closed, it follows that $\Psi \in D(A_n)$ and $A_n \Psi = \Phi_n$. In particular, $\Psi \in \cap_{n=1}^{\infty} D(A_n)$. Taking the limit $\ell \to \infty$ in (8.9) and using Fatou's lemma, we obtain

$$\sum_{n=1}^{\infty} \|A_n \Psi_k - A_n \Psi\|^2 \leq \varepsilon, \quad k \geq k_0. \tag{8.10}$$

Since

$$\|A_n \Psi\|^2 \leq 2(\|A_n \Psi - A_n \Psi_k\|^2 + \|A_n \Psi_k\|^2),$$

we have

$$\sum_{n=1}^{\infty} \|A_n \Psi\|^2 \leq 2(\varepsilon + \sum_{n=1}^{\infty} \|A_n \Psi_k\|^2) < \infty, \quad k \geq k_0.$$

Hence $\Psi \in D_{\mathscr{A}}$. Then (8.10) means that $\lim_{k\to\infty} \mathfrak{s}(\Psi_k - \Psi, \Psi_k - \Psi) = 0$. Thus $\mathfrak{s}$ is closed. ■

Under an additional condition, we obtain an existence theorem of a self-adjoint operator associated with $\mathscr{A}$:

Theorem 8.2 *Suppose that $D_{\mathscr{A}}$ is dense in $\mathscr{X}$. Then there exists a unique non-negative self-adjoint operator H on $\mathscr{X}$ such that $D_{\mathscr{A}} = D(H^{1/2})$ and, for all $\Phi, \Psi \in D_{\mathscr{A}}$,*

$$\left\langle H^{1/2}\Phi, H^{1/2}\Psi \right\rangle = \sum_{n=1}^{\infty} \langle A_n\Phi, A_n\Psi\rangle . \tag{8.11}$$

Proof Since we have Lemma 8.2, we can apply Theorem 1.31 to conclude that there exists a unique non-negative self-adjoint operator H on $\mathscr{X}$ such that $D_{\mathscr{A}} = D(H^{1/2})$ and $\left\langle H^{1/2}\Phi, H^{1/2}\Psi\right\rangle = \mathfrak{s}(\Phi, \Psi)$ for all $\Phi, \Psi \in D_{\mathscr{A}}$. Thus (8.11) follows. ■

By von Neumann's theorem, $A_n^* A_n$ is a non-negative self-adjoint operator. We introduce a subspace

$$\mathscr{D}_{\mathscr{A},2} := \{\Psi \in \cap_{n=1}^{\infty} D(A_n^* A_n) | \sum_{n=1}^{\infty} A_n^* A_n \Psi \text{ strongly converges}\}.$$

It follows that $\mathscr{D}_{\mathscr{A},2} \subset \mathscr{D}_{\mathscr{A}}$.

Corollary 8.2 *Suppose that $D_{\mathscr{A}}$ is dense in $\mathscr{X}$. Let H be the self-adjoint operator in Theorem 8.2. Then $\mathscr{D}_{\mathscr{A},2} \subset D(H)$ and*

$$H\Psi = \sum_{n=1}^{\infty} A_n^* A_n \Psi, \quad \Psi \in \mathscr{D}_{\mathscr{A},2}. \tag{8.12}$$

Proof Let $\Psi \in \mathscr{D}_{\mathscr{A},2}$ and $\Phi \in \mathscr{D}_{\mathscr{A}} = D(H^{1/2})$. Then, by (8.11), we have

$$\left\langle H^{1/2}\Phi, H^{1/2}\Psi\right\rangle = \sum_{n=1}^{\infty} \left\langle \Phi, A_n^* A_n \Psi\right\rangle = \left\langle \Phi, \sum_{n=1}^{\infty} A_n^* A_n \Psi \right\rangle,$$

where we have used the strong convergence of $\sum_{n=1}^{\infty} A_n^* A_n \Psi$. This means that $H^{1/2}\Psi \in D(H^{1/2})$ (i.e., $\Psi \in D(H)$) and (8.12) holds. ■

8.3.2 A Second Quantization Operator Associated with a Representation of CCR

We fix a representation

$$\rho := (\mathscr{F}, \mathscr{D}, \{C(f)|f \in \mathscr{V}\}) \tag{8.13}$$

of the CCR over $\mathscr{V}$. Throughout this subsection, we assume that *$\mathscr{V}$ is a dense subspace of a Hilbert space $\mathscr{K}$.*

Theorem 8.3 *Let T be a non-negative self-adjoint operator on $\mathscr{K}$. Suppose that $D(T^{1/2})$ includes a CONS $\{e_n\}_{n=1}^\infty$ of $\mathscr{K}$ with $T^{1/2}e_n \in \mathscr{V}$, $n \in \mathbb{N}$ and*

$$\mathscr{Q}_C := \left\{ \Psi \in \cap_{n=1}^\infty D(C(T^{1/2}e_n)) \left| \sum_{n=1}^\infty \|C(T^{1/2}e_n)\Psi\|^2 < \infty \right. \right\} \tag{8.14}$$

is dense in $\mathscr{F}$. Then there exists a unique non-negative self-adjoint operator $H_C(T)$ on $\mathscr{F}$ such that

$$D(H_C(T)^{1/2}) = \mathscr{Q}_C, \tag{8.15}$$

$$\left\langle H_C(T)^{1/2}\Phi, H_C(T)^{1/2}\Psi \right\rangle = \sum_{n=1}^\infty \left\langle C(T^{1/2}e_n)\Phi, C(T^{1/2}e_n)\Psi \right\rangle, \quad \Phi, \Psi \in \mathscr{Q}_C. \tag{8.16}$$

Proof We need only apply Theorem 8.2 to the case where $\mathscr{X} = \mathscr{F}$ and $A_n = C(T^{1/2}e_n)$. ■

We call the non-negative self-adjoint operator $H_C(T)$ a **second quantization operator of T associated with** $\{C(f)|f \in \mathscr{V}\}$. The operator $H_C(T)$ may depend on the choice of the CONS $\{e_n\}_{n=1}^\infty$.

In order to show that $\mathscr{Q}_C$ is dense in concrete representations of CCR on the boson Fock space $\mathscr{F}_\mathrm{b}(\mathscr{H})$ over a Hilbert space $\mathscr{H}$, the following lemma may be useful:

Lemma 8.3 *Let L be a densely defined linear operator on a Hilbert space $\mathscr{H}$ and $\{e_n\}_{n=1}^\infty \subset D(L)$ be a CONS of $\mathscr{H}$. Let $\Psi \in \mathscr{F}_{\mathrm{b,fin}}(D(L^*)) \subset \mathscr{F}_\mathrm{b}(\mathscr{H})$. Then:*

(i)

$$\sum_{n=1}^\infty \|A(Le_n)\Psi\|^2 < \infty. \tag{8.17}$$

(ii) *Suppose that* $\sum_{n=1}^{\infty} \|Le_n\|^2 < \infty$. *Then*

$$\sum_{n=1}^{\infty} \|A(Le_n)^* \Psi\|^2 < \infty. \tag{8.18}$$

Proof We need only to show (8.17) and (8.18) for Ψ of the form

$$\Psi = A(f_1)^* \cdots A(f_p)^* \Omega_{\mathscr{H}}, \quad p \geq 0,\ f_j \in D(L^*),\ j = 1, \ldots, p.$$

(i) If $\Psi = \Omega_{\mathscr{H}}$ (the case $p = 0$), then $A(Le_n)\Psi = 0$, $n \in \mathbb{N}$. Hence (8.17) trivially holds. Let $p \geq 1$ next. Then we have

$$A(Le_n)\Psi = \sum_{j=1}^{p} \langle Le_n, f_j \rangle \Psi^{(j)},$$

where $\Psi^{(j)} := A(f_1)^* \cdots \widehat{A(f_j)^*} \cdots A(f_p)^* \Omega_{\mathscr{H}}$. Hence

$$\sum_{n=1}^{\infty} \|A(Le_n)\Psi\|^2 = \sum_{j,k=1}^{p} \left\langle \Psi^{(j)}, \Psi^{(k)} \right\rangle \sum_{n=1}^{\infty} \langle L^* f_j, e_n \rangle \langle e_n, L^* f_k \rangle$$

$$= \sum_{j,k=1}^{p} \left\langle \Psi^{(j)}, \Psi^{(k)} \right\rangle \langle L^* f_j, L^* f_k \rangle.$$

Therefore (8.17) holds.

(ii) By (6.55), we have

$$\|A(Le_n)^* \Psi\|^2 = \|A(Le_n)\Psi\|^2 + \|Le_n\|^2.$$

Hence, by (i) and the present assumption, (8.18) holds. ■

Example 8.8 Let us consider the case $\rho = \rho_{\mathrm{F}}(\mathscr{V})$, the Fock representation of the CCR over $\mathscr{V} \subset \mathscr{H}$. We suppose that $\mathscr{V} \cap D(T)$ is dense in $\mathscr{H}$. Then there exists a CONS $\{e_n\}_{n=1}^{\infty}$ of $\mathscr{H}$ with $e_n \in \mathscr{V} \cap D(T)$ ($n \in \mathbb{N}$) so that $T^{1/2} e_n \in D(T^{1/2})$. Let

$$\mathscr{Q}_A := \left\{ \Psi \in \cap_{n=1}^{\infty} D(A(T^{1/2} e_n)) \,\middle|\, \sum_{n=1}^{\infty} \|A(T^{1/2} e_n)\Psi\|^2 < \infty \right\}. \tag{8.19}$$

Then it follows from Lemma 8.3 with $L = T^{1/2}$ that

$$\mathscr{F}_{\mathrm{b,fin}}(D(T^{1/2})) \subset \mathscr{Q}_A.$$

Hence $\mathscr{Q}_A$ is dense in $\mathscr{F}_{\mathrm{b}}(\mathscr{H})$. Therefore, by Theorem 8.3, there exists a unique non-negative self-adjoint operator $H_A(T)$ on $\mathscr{F}_{\mathrm{b}}(\mathscr{H})$ such that $D(H_A(T)^{1/2}) = \mathscr{Q}_A$ and, for all $\Phi, \Psi \in \mathscr{Q}_A$,

$$\left\langle H_A(T)^{1/2}\Phi, H_A(T)^{1/2}\Psi \right\rangle = \sum_{n=1}^{\infty} \left\langle A(T^{1/2}e_n)\Phi, A(T^{1/2}e_n)\Psi \right\rangle. \tag{8.20}$$

There are three cases to be considered separately.

(1) The case where T is bounded

In this case, (6.85) holds. Hence, by (6.86), for all $\Psi \in \mathscr{F}_{\mathrm{b,fin}}(D(T))$ and $\Phi \in D(H_A(T)^{1/2})$, the right hand side of (8.20) is equal to $\langle \Phi, \mathrm{d}\Gamma_{\mathrm{b}}(T)\Psi \rangle$. Hence $\Psi \in D(H_A(T))$ and $H_A(T)\Psi = \mathrm{d}\Gamma_{\mathrm{b}}(T)\Psi$. Since $\mathscr{F}_{\mathrm{b,fin}}(D(T))$ is a core for $\mathrm{d}\Gamma_{\mathrm{b}}(T)$, it follows that $\mathrm{d}\Gamma_{\mathrm{b}}(T) \subset H_A(T)$. Hence

$$H_A(T) = \mathrm{d}\Gamma_{\mathrm{b}}(T). \tag{8.21}$$

Thus $H_A(T)$ coincides with $\mathrm{d}\Gamma_{\mathrm{b}}(T)$. This implies also that $H_A(T)^{1/2} = \mathrm{d}\Gamma_{\mathrm{b}}(T)^{1/2}$. Hence, in the present case,

$$\mathscr{Q}_A = D(\mathrm{d}\Gamma_{\mathrm{b}}(T)^{1/2}). \tag{8.22}$$

(2) The case where T is unbounded and satisfies (6.85)

In this case too, the arguments in (1) work without alterations. Hence (8.21) and (8.22) hold.

(3) The case where T is unbounded and $\{e_n\}_{n=1}^{\infty}$ does not satisfy (6.85)

In this case, by Corollary 6.11, $D(\mathrm{d}\Gamma_{\mathrm{b}}(T)^{1/2}) \subset \mathscr{Q}_A = D(H_A(T)^{1/2})$ and

$$\left\langle H_A(T)^{1/2}\Phi, H_A(T)^{1/2}\Psi \right\rangle = \left\langle \mathrm{d}\Gamma_{\mathrm{b}}(T)^{1/2}\Phi, \mathrm{d}\Gamma_{\mathrm{b}}(T)^{1/2}\Psi \right\rangle, \quad \Phi, \Psi \in D(\mathrm{d}\Gamma_{\mathrm{b}}(T)^{1/2}). \tag{8.23}$$

This means that the sesquilinear form associated with $H_A(T)$ is an extension of that associated with $\mathrm{d}\Gamma_{\mathrm{b}}(T)$.[3] But it does not mean operator equality (8.21). It follows from (8.23) that, for all $\Psi \in D(H_A(T)) \cap D(\mathrm{d}\Gamma_{\mathrm{b}}(T)^{1/2})$, $\Psi \in D(\mathrm{d}\Gamma_{\mathrm{b}}(T))$ and $H_A(T)\Psi = \mathrm{d}\Gamma_{\mathrm{b}}(T)\Psi$.

[3] A sesquilinear form $\mathfrak{s}'$ is called an extension of a sesquilinear form $\mathfrak{s}$ if $\mathscr{Q}(\mathfrak{s}) \subset \mathscr{Q}(\mathfrak{s}')$ and $\mathfrak{s}(\Psi, \Phi) = \mathfrak{s}'(\Psi, \Phi)$, $\Psi, \Phi \in \mathscr{Q}(\mathfrak{s})$.

Any CONS $\{e_n\}_{n=1}^{\infty}$ of $\mathscr{H}$ satisfies (6.85) with $T = I$. Hence, by the fact in (1) given above, we obtain the following result:

Corollary 8.3 *Let $\{e_n\}_{n=1}^{\infty}$ be a CONS of $\mathscr{H}$. Then*

$$D(N_{\mathrm{b}}^{1/2}) = \left\{ \Psi \in \cap_{n=1}^{\infty} D(A(e_n)) \left| \sum_{n=1}^{\infty} \|A(e_n)\Psi\|^2 < \infty \right. \right\}, \tag{8.24}$$

$$\left\langle N_{\mathrm{b}}^{1/2}\Phi, N_{\mathrm{b}}^{1/2}\Psi \right\rangle = \sum_{n=1}^{\infty} \langle A(e_n)\Phi, A(e_n)\Psi \rangle, \quad \Phi, \Psi \in D(N_{\mathrm{b}}^{1/2}), \tag{8.25}$$

independently of the choice of $\{e_n\}_{n=1}^{\infty}$.

In what follows, we assume that T obeys the assumption of Theorem 8.3.

Proposition 8.1

(i) *If ρ has a vacuum Ω, then $H_C(T)$ has a ground state with ground state energy being zero: $H_C(T)\Omega = 0$.*
(ii) *Conversely, if $H_C(T)$ has a zero-energy ground state Ω, then $C(f)\Omega = 0$ for all $f \in \mathrm{span}\{T^{1/2}e_n | n \in \mathbb{N}\}$.*

Proof

(i) The present assumption implies that $C(T^{1/2}e_n)\Omega = 0$, $n \in \mathbb{N}$. Hence, by (8.16), Hence $\|H_C(T)^{1/2}\Omega\|^2 = 0$. Therefore $H_C(T)^{1/2}\Omega = 0$. This implies that $H_C(T)\Omega = 0$.
(ii) By the present assumption, $\Omega \in \ker H_C(T)$. Hence, by Lemma 1.1, $\Omega \in \ker H_C(T)^{1/2}$. Hence $\left\langle H_C(T)^{1/2}\Omega, H_C(T)^{1/2}\Omega \right\rangle = 0$. Then, by (8.16),

$$\sum_{n=1}^{\infty} \|C(T^{1/2}e_n)\Omega\|^2 = 0.$$

Hence $\|C(T^{1/2}e_n)\Omega\|^2 = 0$, $n \in \mathbb{N}$, which implies that $C(T^{1/2}e_n)\Omega = 0$, $n \in \mathbb{N}$. Hence the desired result follows. ■

Proposition 8.2 *Let $N \in \mathbb{N}$ and*

$$f_{T,N} := \sum_{n=1}^{N} \left\langle T^{1/2}e_n, f \right\rangle T^{1/2}e_n, \quad f \in \mathscr{V}.$$

Suppose that the CCR-domain $\mathscr{D}$ in the representation ρ is a subset of $D(H_C(T)^{1/2})$ and, for all $f \in \mathscr{V} \cap D(T)$ such that $Tf \in \mathscr{V}$,

$$\lim_{N\to\infty} \left\langle \Phi, C(f_{T,N})^*\Psi \right\rangle = \left\langle \Phi, C(Tf)^{*}{}'\Psi \right\rangle, \quad \Phi, \Psi \in \mathscr{D}.$$

Then, for all $\Phi \in \mathscr{D}$, $\Psi \in \mathscr{D} \cap D(H_C(T))$ and $f \in \mathscr{V} \cap D(T)$ such that $Tf \in \mathscr{V}$,

$$\left\langle H_C(T)^{1/2}\Phi, H_C(T)^{1/2}C(f)^*\Psi\right\rangle - \langle C(f)\Phi, H_C(T)\Psi\rangle = \left\langle \Phi, C(Tf)^*\Psi\right\rangle, \tag{8.26}$$

$$\left\langle H_C(T)^{1/2}\Phi, H_C(T)^{1/2}C(f)\Psi\right\rangle - \left\langle C(f)^*\Phi, H_C(T)\Psi\right\rangle = -\langle \Phi, C(Tf)\Psi\rangle . \tag{8.27}$$

Proof Throughout the proof, we set $H := H_C(T)$. Let $\Phi \in \mathscr{D}$ and $\Psi \in \mathscr{D} \cap D(H)$. Then, by the present assumption, $\Psi \in D(C(f)^*)$ and $C(f)^*\Psi \in \mathscr{D}$. Hence, by (8.16), we have

$$\left\langle H^{1/2}\Phi, H^{1/2}C(f)^*\Psi\right\rangle = \sum_{n=1}^{\infty}\left\langle C(T^{1/2}e_n)\Phi, C(T^{1/2}e_n)C(f)^*\Psi\right\rangle.$$

By CCR on $\mathscr{D}$, we have

$$C(T^{1/2}e_n)C(f)^*\Psi = C(f)^*C(T^{1/2}e_n)\Psi + \left\langle T^{1/2}e_n, f\right\rangle\Psi$$

and $C(f)C(T^{1/2}e_n)\Phi = C(T^{1/2}e_n)C(f)\Phi$. Hence we obtain

$$\left\langle H^{1/2}\Phi, H^{1/2}C(f)^*\Psi\right\rangle - \langle C(f)\Phi, H\Psi\rangle = \sum_{n=1}^{\infty}\left\langle T^{1/2}e_n, f\right\rangle\left\langle \Phi, C(T^{1/2}e_n)^*\Psi\right\rangle.$$

The right hand side is equal to $\lim_{N\to\infty}\left\langle \Phi, C(f_{T,N})^*\Psi\right\rangle = \langle \Phi, C(Tf)^*\Psi\rangle$. Thus (8.26) holds. Similarly, one can prove (8.27). ∎

Note that (8.26) and (8.27) have a structure similar to the structure of (6.92) and (6.93) apart from domains of commutation relations.

Remark 8.6 A slightly different concept of second quantization operator associated with a representation of CCR is introduced in [28, §5.22]. But, in this book, we use the term "second quantization operator associated with $\{C(f)|f \in \mathscr{V}\}$" in the sense defined above.

Example 8.9 Let us find a second quantization operator associated with $\{C_L(f)|f \in \mathscr{V}\}$ in Example 8.3. Let T be a non-negative and injective self-adjoint operator on $\mathscr{H}$ and consider the case where $\mathscr{V} = D(T^{-1/2})$ and

$$L(f) = \frac{1}{\sqrt{2}}\left\langle T^{-1/2}g, T^{-1/2}f\right\rangle, \quad f \in D(T^{-1/2})$$

for some fixed $g \in D(T^{-1/2})$. In this case, we write

$$\beta_L(\mathscr{V}) = \beta_L$$

simply. Let $\{e_n\}_{n=1}^{\infty} \subset D(T^{1/2})$ be a CONS of $\mathscr{H}$. Then $T^{1/2}e_n \in D(T^{-1/2})$, $n \in \mathbb{N}$. Hence, by Theorem 8.3, there exists a unique non-negative self-adjoint operator H_{C_L} on $\mathscr{F}_{\mathrm{b}}(\mathscr{H})$ such that

$$\left\langle H_{C_L}^{1/2}\Phi, H_{C_L}^{1/2}\Psi\right\rangle = \sum_{n=1}^{\infty}\left\langle C_L(T^{1/2}e_n)\Phi, C_L(T^{1/2}e_n)\Psi\right\rangle, \quad \Phi, \Psi \in \mathscr{Q}_{C_L}.$$

For all $\Psi, \Phi \in D(C_L(T^{1/2}e_n)) = D(A(T^{1/2}e_n))$, we have

$$\begin{aligned}\left\langle C_L(T^{1/2}e_n)\Phi, C_L(T^{1/2}e_n)\Psi\right\rangle =& \left\langle A(T^{1/2}e_n)\Phi, A(T^{1/2}e_n)\Psi\right\rangle \\ &+ L(T^{1/2}e_n)\left\langle \Phi, A(T^{1/2}e_n)\Psi\right\rangle \\ &+ L(T^{1/2}e_n)^*\left\langle A(T^{1/2}e_n)\Phi, \Psi\right\rangle \\ &+ |L(T^{1/2}e_n)|^2 \langle \Phi, \Psi\rangle . \end{aligned} \tag{8.28}$$

In particular,

$$\begin{aligned}\|C_L(T^{1/2}e_n)\Psi\|^2 =& \|A(T^{1/2}e_n)\Psi\|^2 + L(T^{1/2}e_n)\left\langle \Psi, A(T^{1/2}e_n)\Psi\right\rangle \\ &+ L(T^{1/2}e_n)^*\left\langle A(T^{1/2}e_n)\Psi, \Psi\right\rangle \\ &+ |L(T^{1/2}e_n)|^2\|\Psi\|^2. \end{aligned} \tag{8.29}$$

Let

$$a_n := \|A(T^{1/2}e_n)\Psi\|, \quad b_n := |L(T^{1/2}e_n)|\,\|\Psi\|.$$

Then, using the Schwarz inequality, we obtain

$$\|C_L(T^{1/2}e_n)\Psi\|^2 \le a_n^2 + 2a_nb_n + b_n^2 \le 2(a_n^2 + b_n^2).$$

By the Parseval equality, we have

$$\sum_{n=1}^{\infty} b_n^2 = \frac{1}{2}\|T^{-1/2}g\|^2\|\Psi\|^2 < \infty. \tag{8.30}$$

Therefore it follows that

$$\mathscr{Q}_A \subset \mathscr{Q}_{C_L}.$$

Conversely, let $\Psi \in \mathscr{Q}_{C_L}$. Then $\Psi \in \cap_{n=1}^{\infty} D(A(T^{1/2}e_n))$ and

$$\sum_{n=1}^{\infty} \|C_L(T^{1/2}e_n)\Psi\|^2 < \infty.$$

By (8.29), we have

$$\|C_L(T^{1/2}e_n)\Psi\|^2 \geq (a_n - b_n)^2.$$

Hence $\sum_{n=1}^{\infty}(a_n - b_n)^2 < \infty$. We have

$$a_n^2 \leq 2(a_n - b_n)^2 + 2b_n^2.$$

Therefore $\sum_{n=1}^{\infty} a_n^2 < \infty$, implying that $\Psi \in \mathscr{Q}_A$. Thus we obtain

$$\mathscr{Q}_{C_L} = \mathscr{Q}_A. \tag{8.31}$$

Let $\Phi, \Psi \in \mathscr{F}_{\mathrm{b,fin}}(D(T^{1/2}))$. Then, in the same way as in the proof of Theorem 6.14, one can show that

$$\lim_{N\to\infty} \sum_{n=1}^{N} L(T^{1/2}e_n)\left\langle \Phi, A(T^{1/2}e_n)\Psi \right\rangle = \frac{1}{\sqrt{2}} \langle \Phi, A(g)\Psi \rangle .$$

Using this fact and (8.30), we obtain

$$\left\langle H_{C_L}^{1/2}\Phi, H_{C_L}^{1/2}\Psi \right\rangle = \left\langle \Phi, (H_{\mathrm{v}} + \frac{1}{2}\|T^{-1/2}g\|^2)\Psi \right\rangle, \tag{8.32}$$

where H_{v} is defined by (6.77). By Theorem 6.11, the operator

$$\hat{H}_{\mathrm{v}} := H_{\mathrm{v}} + \frac{1}{2}\|T^{-1/2}g\|^2$$

is self-adjoint with $D(\hat{H}_{\mathrm{v}}) = D(\mathrm{d}\Gamma_{\mathrm{b}}(T))$ and bounded from below. The subspace $\mathscr{F}_{\mathrm{b,fin}}(D(T^{1/2}))$ is a core for $\mathrm{d}\Gamma_{\mathrm{b}}(T)$. Hence, by Theorem 6.11, it is a core for $\hat{H}_{\mathrm{v}}$ too. Hence, by a limiting argument, one can show that (8.32) extends to all $\Psi, \Phi \in D(\mathrm{d}\Gamma_{\mathrm{b}}(T)) = D(\hat{H}_{\mathrm{v}})$ with $D(\mathrm{d}\Gamma_{\mathrm{b}}(T)) \subset D(H_{C_L}^{1/2})$. In particular,

$$\hat{H}_{\mathrm{v}} \geq 0 \tag{8.33}$$

and

$$\left\langle H_{C_L}^{1/2}\Phi, H_{C_L}^{1/2}\Psi\right\rangle = \left\langle \hat{H}_{\mathrm{v}}^{1/2}\Phi, \hat{H}_{\mathrm{v}}^{1/2}\Psi\right\rangle, \quad \Psi, \Phi \in D(\mathrm{d}\Gamma_{\mathrm{b}}(T)). \tag{8.34}$$

Since $D(\mathrm{d}\Gamma_{\mathrm{b}}(T))$ is a core for $\mathrm{d}\Gamma_{\mathrm{b}}(T)^{1/2}$, it follows that $D(\mathrm{d}\Gamma_{\mathrm{b}}(T)^{1/2}) \subset D(H_{C_L}^{1/2}) = \mathscr{Q}_{C_L} = \mathscr{Q}_A$ (see (8.31)) and (8.34) extends to all $\Psi, \Phi \in D(\mathrm{d}\Gamma_{\mathrm{b}}(T)^{1/2})$.

Let us consider the case where T is bounded. Then, by (8.22), we have

$$D(H_{C_L}^{1/2}) = D(\mathrm{d}\Gamma_{\mathrm{b}}(T)^{1/2}).$$

Hence it follows that

$$H_{C_L} = \hat{H}_{\mathrm{v}}. \tag{8.35}$$

Thus, in the present case, H_{v} is a second quantization operator of T associated with $\{C_L(f)|f \in D(T^{-1/2})\}$ up to an additive constant.

8.4 Diagonalization of $H_C(T)$

We continue to study the self-adjoint operator $H_C(T)$ defined by (8.16).

Theorem 8.4 *Assume that (6.85) holds with $T^{1/2}e_n \in \mathscr{V}$, $n \in \mathbb{N}$ and suppose that the representation ρ is equivalent to the Fock representation $\rho_{\mathrm{F}}(\mathscr{V})$, i.e., there exists a unitary operator $U: \mathscr{F} \to \mathscr{F}_{\mathrm{b}}(\mathscr{H})$ such that*

$$UC(f)U^{-1} = A(f), \quad f \in \mathscr{V}. \tag{8.36}$$

Then

$$UH_C(T)U^{-1} = \mathrm{d}\Gamma_{\mathrm{b}}(T). \tag{8.37}$$

Proof By (8.36), for all $\Psi \in \mathscr{Q}_C$, we have

$$\sum_{n=1}^{\infty} \|A(T^{1/2}e_n)U\Psi\|^2 = \sum_{n=1}^{\infty} \|C(T^{1/2}e_n)\Psi\|^2 < \infty.$$

Hence $U\Psi \in \mathscr{Q}_A$. Conversely, if $\Phi \in \mathscr{Q}_A$, then

$$\sum_{n=1}^{\infty} \|C(T^{1/2}e_n)U^{-1}\Phi\|^2 = \sum_{n=1}^{\infty} \|A(T^{1/2}e_n)\Phi\|^2 < \infty.$$

Hence $U^{-1}\Phi \in \mathscr{Q}_C$. Thus $U\mathscr{Q}_C = \mathscr{Q}_A$. Since (6.85) holds, $\mathscr{Q}_A = D(\mathrm{d}\Gamma_{\mathrm{b}}(T)^{1/2})$. Therefore, for all $\Psi, \Phi \in D(\mathrm{d}\Gamma_{\mathrm{b}}(T)^{1/2}) = \mathscr{Q}_A\ (= U\mathscr{Q}_C)$,

$$\begin{aligned}\left\langle H_C(T)^{1/2}\Phi, H_C(T)^{1/2}\Psi\right\rangle &= \sum_{n=1}^{\infty}\left\langle A(T^{1/2}e_n)U\Phi, A(T^{1/2}e_n)U\Psi\right\rangle \\ &= \left\langle \mathrm{d}\Gamma_{\mathrm{b}}(T)^{1/2}U\Phi, \mathrm{d}\Gamma_{\mathrm{b}}(T)^{1/2}U\Psi\right\rangle \\ &= \left\langle U^{-1}\mathrm{d}\Gamma_{\mathrm{b}}(T)^{1/2}U\Phi, U^{-1}\mathrm{d}\Gamma_{\mathrm{b}}(T)^{1/2}U\Psi\right\rangle.\end{aligned}$$

Note that $U^{-1}\mathrm{d}\Gamma_{\mathrm{b}}(T)U$ is a non-negative self-adjoint operator and, by functional calculus,

$$(U^{-1}\mathrm{d}\Gamma_{\mathrm{b}}(T)U)^{1/2} = U^{-1}\mathrm{d}\Gamma_{\mathrm{b}}(T)^{1/2}U.$$

Hence

$$\left\langle H_C(T)^{1/2}\Phi, H_C(T)^{1/2}\Psi\right\rangle = \left\langle (U^{-1}\mathrm{d}\Gamma_{\mathrm{b}}(T)U)^{1/2}\Phi, (U^{-1}\mathrm{d}\Gamma_{\mathrm{b}}(h)U)^{1/2}\Psi\right\rangle. \tag{8.38}$$

By this fact and the uniqueness of the self-adjoint operator associated with a densely defined non-negative closed sesquilinear form, we obtain that $H_C(T) = U^{-1}\mathrm{d}\Gamma_{\mathrm{b}}(T)U$. This is equivalent to (8.37). ■

Remark 8.7 As is seen from the proof given above, in the case where T is unbounded and does not satisfy (6.85), one can prove only the following properties unless additional conditions are assumed: $D(\mathrm{d}\Gamma_{\mathrm{b}}(T^{1/2})) \subset U\mathscr{Q}_C$ and (8.38) holds for all $\Psi, \Phi \in D(\mathrm{d}\Gamma_{\mathrm{b}}(T)^{1/2})$.

Remark 8.8 For a sufficient condition for the representation ρ to be equivalent to $\rho_{\mathrm{F}}(\mathscr{V})$, see Corollary 8.1.

In the context of quantum field theory, a Bose field Hamiltonian H, which is a self-adjoint operator on a Hilbert space, is said to be **diagonalized** if it is unitarily equivalent to a boson second quantization operator $\mathrm{d}\Gamma_{\mathrm{b}}(h)$ of a self-adjoint operator h (not necessarily bounded) up to an additive constant. In such a case, the spectrum of the Hamiltonian H is completely identified if the spectrum of the one-particle Hamiltonian h is known.

Theorem 8.4 shows that, for any non-negative self-adjoint operator T satisfying (6.85) with $T^{1/2}e_n \in \mathscr{V}$, $n \in \mathbb{N}$, the operator $H_C(T)$ can be diagonalized if the representation ρ is equivalent to the Fock representation $\rho_{\mathrm{F}}(\mathscr{V})$.

8.5 Analysis of Bogoliubov Translations

In this section, we give a detailed analysis of the Bogoliubov translation (C_L, C_L^*) in Example 8.3.

Lemma 8.4 *Let $\mathscr{D}$ be a dense subspace of $\mathscr{H}$. Suppose that there exists a non-zero vector $\Omega \in \cap_{f\in\mathscr{D}} D(A(f))$ such that, for all $f \in \mathscr{D}$, $C_L(f)\Omega = 0$. Then, there exists a vector $h \in \mathscr{H}$ such that $L(f) = \langle h, f\rangle$, $f \in \mathscr{D}$.*

Proof Without loss of generality, we can assume that $\|\Omega\| = 1$. The present assumption implies that $A(f)\Omega = -L(f)^*\Omega$, $f \in \mathscr{D}$. There exists an $n \in \mathbb{Z}_+$ such that $\Omega^{(n)} \neq 0$. We define $\Psi_n \in \mathscr{F}_{\mathrm{b},0}(\mathscr{H})$ by $\Psi_n^{(n)} := \Omega^{(n)}$ and $\Psi^{(m)} = 0$, $m \neq n$ and put $\omega_n := \langle \Omega, \Psi_n\rangle = \|\Omega^{(n)}\|^2 \neq 0$. Then

$$L(f) = -\frac{1}{\omega_n}\langle A(f)\Omega, \Psi_n\rangle = -\frac{1}{\omega_n}\left\langle \Omega, A(f)^*\Psi_n\right\rangle.$$

By the Schwarz inequality and Lemma 6.5(i), we obtain

$$|L(f)| \leq \frac{1}{\omega_n}\|A(f)^*\Psi_n\| \leq C_n\|f\|, \quad f \in \mathscr{D},$$

with $C_n :== \sqrt{(n+1)/\omega_n}$. Hence L is bounded. Since $\mathscr{D}$ is dense, it follows from the extension theorem that there exists a unique $\widetilde{L} \in \mathscr{H}^*$ such that $\widetilde{L}(f) = L(f)$, $f \in \mathscr{D}$. Therefore, by Riesz's representation theorem, there exists a unique vector $h \in \mathscr{H}$ such that $\widetilde{L}(f) = \langle h, f\rangle$, $f \in \mathscr{H}$. Hence we have $L(f) = \langle h, f\rangle$, $f \in \mathscr{D}$. ■

Theorem 8.5 *There exists a unitary operator U on $\mathscr{F}_{\mathrm{b}}(\mathscr{H})$ such that*

$$UC_L(f)U^{-1} = A(f), \quad f \in \mathscr{V} \tag{8.39}$$

if and only if there exists a vector $h \in \mathscr{H}$ such that $L(f) = \langle h, f\rangle_{\mathscr{H}}$, $f \in \mathscr{V}$.

Proof Suppose that there exists a unitary operator U on $\mathscr{F}_{\mathrm{b}}(\mathscr{H})$ such that (8.39) holds. Then, letting $\Omega := U^{-1}\Omega_{\mathscr{H}}$, we have $C_L(f)\Omega = 0$ and $\|\Omega\| = 1$AD Hence, by Lemma 8.4 with $\mathscr{D} = \mathscr{V}$, there exists a vector $h \in \mathscr{H}$ such that $L(f) = \langle h, f\rangle_{\mathscr{H}}$, $f \in \mathscr{V}$.

Conversely, suppose that there exists a vector $h \in \mathscr{H}$ such that $L(f) = \langle h, f\rangle$, $f \in \mathscr{V}$. Then, by (6.118), we have

$$e^{i\sqrt{2}\Phi_{\mathrm{S}}(ih)}A(f)e^{-i\sqrt{2}\Phi_{\mathrm{S}}(ih)} = C_L(f), \quad f \in \mathscr{V}.$$

Hence, letting $U := e^{-i\sqrt{2}\Phi_{\mathrm{S}}(ih)}$, we obtain $UC_L(f)U^{-1} = A(f)$, $f \in \mathscr{V}$. ■

Corollary 8.4 *The representation* $(\mathscr{F}_b(\mathscr{H}), \mathscr{F}_{b,0}(\mathscr{H}), \{C_L(f)|f \in \mathscr{V}\})$ *of CCR is equivalent to the Fock representation* $\rho_F(\mathscr{V})$ *if and only if there exists a vector* $h \in \mathscr{H}$ *such that* $L(f) = \langle h, f \rangle$, $f \in \mathscr{V}$.

8.6 Bogoliubov Transformations

A more general class of representations of the CCR over a subspace $\mathscr{V}$ of a Hilbert space $\mathscr{H}$ may be found by considering operators of the form $A(W_1 f) + A(W_2 f)^*$ $(f \in \mathscr{V})$ with suitable mappings W_1 and W_2 from $\mathscr{V}$ to $\mathscr{H}$. This is an idea for what follows in this section. To describe a general structure behind the idea, we first present a complex symplectic space structure of $\mathscr{H} \oplus \mathscr{H}$ (see Sect. 6.13).

We fix a conjugation J on $\mathscr{H}$ (see Definition 1.2). Let $\gamma : \mathscr{H} \oplus \mathscr{H} \to \mathscr{H} \oplus \mathscr{H}$ be defined by

$$\gamma(f, g) := (g, -f), \quad (f, g) \in \mathscr{H} \oplus \mathscr{H}. \tag{8.40}$$

Then it is easy to see that γ is a bounded linear operator on $\mathscr{H} \oplus \mathscr{H}$ with

$$\|\gamma(f, g)\| = \|(f, g)\|, \quad (f, g) \in \mathscr{H} \oplus \mathscr{H} \tag{8.41}$$

and

$$\gamma^* = -\gamma, \quad \gamma^2 = -I. \tag{8.42}$$

Moreover

$$\tilde{J} := J \oplus J \tag{8.43}$$

is a conjugation on $\mathscr{H} \oplus \mathscr{H}$ and

$$\tilde{J}\gamma = \gamma\tilde{J}. \tag{8.44}$$

Hence, one can apply Example 6.7 to the case $\mathscr{X} = \mathscr{H} \oplus \mathscr{H}$, $C = \tilde{J}$ and $W = \gamma$ to conclude that the mapping $\omega_J : \mathscr{H} \oplus \mathscr{H} \to \mathbb{C}$ defined by

$$\omega_J(\Psi, \Phi) := \langle (\tilde{J}\Psi, \gamma\Phi \rangle, \quad \Psi, \Phi \in \mathscr{H} \oplus \mathscr{H} \tag{8.45}$$

is a $\mathbb{C}$-symplectic form on $(\mathscr{H} \oplus \mathscr{H})$. Hence $(\mathscr{H} \oplus \mathscr{H}, \omega_J)$ is a complex symplectic space. Explicitly, we have

$$\omega_J((f, g), (f', g')) = \langle Jf, g' \rangle - \langle Jg, f' \rangle, \quad (f, g), (f', g') \in \mathscr{H} \oplus \mathscr{H}. \tag{8.46}$$

Note that $\widetilde{J}$ and γ have the following operator matrix representations:

$$\widetilde{J} = \begin{pmatrix} J & 0 \\ 0 & J \end{pmatrix}, \quad \gamma = \begin{pmatrix} 0 & I \\ -I & 0 \end{pmatrix},$$

where an operator matrix

$$A := \begin{pmatrix} A_{11} & A_{12} \\ A_{21} & A_{22} \end{pmatrix}$$

with $A_{ij} \in \mathfrak{L}(\mathscr{H})$ $(D(A) := (D(A_{11}) \cap D(A_{21})) \oplus (D(A_{12}) \cap D(A_{22})))$ acts on $(f_1, f_2) \in D(A)$ as

$$A \begin{pmatrix} f_1 \\ f_2 \end{pmatrix} = \begin{pmatrix} A_{11} f_1 + A_{12} f_2 \\ A_{21} f_1 + A_{22} f_2 \end{pmatrix}.$$

Lemma 8.5 *Let $\mathbb{S}$ be a bounded linear operator on $\mathscr{H} \oplus \mathscr{H}$. Then $\mathbb{S}$ is an iso-symplectic transformation if and only if*

$$(\widetilde{J}\mathbb{S}^* \widetilde{J})\gamma\ \mathbb{S} = \gamma. \tag{8.47}$$

Proof Let $\Psi, \Phi \in \mathscr{H} \oplus \mathscr{H}$. Then

$$\begin{aligned} \omega_J(\mathbb{S}\Psi, \mathbb{S}\Phi) = \omega_J(\Psi, \Phi) &\iff \langle \widetilde{J}\mathbb{S}\Psi, \gamma\mathbb{S}\Phi\rangle = \langle \widetilde{J}\Psi, \gamma\Phi\rangle \\ &\iff \langle \mathbb{S}^*\gamma\widetilde{J}\mathbb{S}\Psi, \Phi\rangle = \langle \gamma\widetilde{J}\Psi, \Phi\rangle \\ &\iff \langle \widetilde{J}\Phi, \widetilde{J}\mathbb{S}^*\gamma\widetilde{J}\mathbb{S}\Psi\rangle = \langle \widetilde{J}\Phi, \gamma\Psi\rangle, \\ &\iff \langle \widetilde{J}\Phi, \widetilde{J}\mathbb{S}^*\widetilde{J}\gamma\ \mathbb{S}\Psi\rangle = \langle \widetilde{J}\Phi, \gamma\Psi\rangle, \end{aligned}$$

where we have used (8.44). Since $\mathrm{Ran}\,\widetilde{J} = \mathscr{H} \oplus \mathscr{H}$, it follows that the last statement is equivalent to (8.47). ■

For each $(f, g) \in \mathscr{H} \oplus \mathscr{H}$, we define an operator $\Phi(f, g)$ on $\mathscr{F}_{\mathrm{b}}(\mathscr{H})$ by

$$\Phi(f, g) := A(Jf) + A(g)^*. \tag{8.48}$$

Using CCR of $A(\cdot)^\#$ and (8.46), one can see that

$$[\Phi(f, g), \Phi(f', g')] = \omega_J((f, g), (f', g')), \quad (f, g), (f', g') \in \mathscr{H} \oplus \mathscr{H} \tag{8.49}$$

on $\mathscr{F}_{\mathrm{b},0}(\mathscr{H})$. Note that

$$A(f) = \Phi(Jf, 0), \quad f \in \mathscr{H}. \tag{8.50}$$

The next proposition immediately follows from (8.49).

Proposition 8.3 *A linear operator* $\mathbb{S}$ *on* $\mathcal{H} \oplus \mathcal{H}$ *is an iso-symplectic transformation on* $(\mathcal{H} \oplus \mathcal{H}, \omega_J)$ *if and only if*

$$[\Phi(\mathbb{S}(f,g)), \mathbb{S}(f',g'))] = [\Phi(f,g), \Phi(f',g')], \quad (f,g), (f',g') \in \mathcal{H} \oplus \mathcal{H},$$

on $\mathcal{F}_{\mathrm{b},0}(\mathcal{H})$.

In view of Proposition 8.3, it is interesting to ask if $\Phi(\mathbb{S}(\cdot))$ is unitarily implementable, i.e., there exists a unitary operator U on $\mathcal{F}_{\mathrm{b}}(\mathcal{H})$ such that

$$\overline{\Phi(\mathbb{S}(f,g))} = U\overline{\Phi(f,g)}U^{-1}, \quad (f,g) \in \mathcal{H} \oplus \mathcal{H}.$$

This is a basic motivation of the theory presented below.

For each $T \in \mathfrak{B}(\mathcal{H})$, we define an operator $T_J \in \mathfrak{B}(\mathcal{H})$ by

$$T_J := JTJ. \tag{8.51}$$

The operator T_J is called the **conjugation of T with respect to J**. Note that

$$(T_J)^* = (T^*)_J.$$

Let $\mathbb{S}$ be an iso-symplectic transformation on $(\mathcal{H} \oplus \mathcal{H}, \omega_J)$ and

$$\mathbb{S} := \begin{pmatrix} S_1 & S_2 \\ S_3 & S_4 \end{pmatrix}$$

be the operator matrix representation of $\mathbb{S}$ ($S_i \in \mathfrak{B}(\mathcal{H})$, $i = 1, 2, 3, 4$). Then (8.47) is equivalent to the following set of operator equations:

$$(S_{1J})^* S_4 - (S_{3J})^* S_2 = I, \tag{8.52}$$

$$(S_{4J})^* S_1 - (S_{2J})^* S_3 = I, \tag{8.53}$$

$$(S_{1J})^* S_3 - (S_{3J})^* S_1 = 0, \tag{8.54}$$

$$(S_{2J})^* S_4 - (S_{4J})^* S_2 = 0, \tag{8.55}$$

where $S_{iJ} := (S_i)_J$ ($i = 1, 2, 3, 4$). Note that (8.52) is equivalent to (8.53).

To find a subclass of $\mathbb{S}$'s, we consider the subset

$$\mathcal{H}^{(J)} := \{(f, Jf) \mid f \in \mathcal{H}\} \subset \mathcal{H} \oplus \mathcal{H},$$

which may be a natural subset. Indeed, $\tilde{J}$ leaves $\mathcal{H}^{(J)}$ invariant bijectively:

$$\tilde{J}\mathcal{H}^{(J)} = \mathcal{H}^{(J)}.$$

We note the following fact:

Lemma 8.6 *Let $\mathbb{S}$ be an iso-symplectic transformation on $(\mathscr{H} \oplus \mathscr{H}, \omega_J)$. Then $\mathbb{S}$ leaves $\mathscr{H}^{(J)}$ invariant if and only if*

$$S_2 = S_{3J}, \quad S_4 = S_{1J}. \tag{8.56}$$

Proof We have for all $f \in \mathscr{H}$

$$\mathbb{S}(f, Jf) = (S_1 f + S_2 Jf, S_3 f + S_4 Jf).$$

Hence $\mathbb{S}(f, Jf) \in \mathscr{H}^{(J)}$ if and only if

$$JS_1 f + JS_2 Jf = S_3 f + S_4 Jf. \tag{8.57}$$

Suppose that $\mathbb{S}(f, Jf) \in \mathscr{H}^{(J)}$. Then (8.57) holds. Replacing f by if, we obtain

$$-JS_1 f + S_{2J} f = S_3 f - S_4 Jf.$$

Hence it follows that $S_{2J} f = S_3 f$ and $JS_1 f = S_4 Jf$. Therefore (8.56) holds.
Conversely, assume (8.56). Then (8.57) holds. Hence $\mathbb{S}(f, Jf) \in \mathscr{H}^{(J)}$. ■

Let $\mathbb{S}$ be an iso-symplectic transformation $(\mathscr{H} \oplus \mathscr{H}, \omega_J)$ such that

$$\mathbb{S}\mathscr{H}^{(J)} \subset \mathscr{H}^{(J)}. \tag{8.58}$$

Then, by Lemma 8.6, $\mathbb{S}$ is of the form

$$\mathbb{S} = \begin{pmatrix} S_J & T \\ T_J & S \end{pmatrix} \tag{8.59}$$

with relations

$$S^* S - T^* T = I, \tag{8.60}$$

$$S^* T_J = T^* S_J. \tag{8.61}$$

We have

$$\Phi(\mathbb{S}(Jf, 0)) = A(Sf) + A(JTf)^*, \quad f \in \mathscr{H}.$$

This operator is closable. We define

$$B(f) := \overline{A(Sf) + A(JTf)^*}, \quad f \in \mathscr{H}. \tag{8.62}$$

It follows that

$$B(f)^* = (A(Sf) + A(JTf)^*)^* \supset A(JTf) + A(Sf)^*.$$

Lemma 8.7 *Let S and T be in $\mathfrak{B}(\mathscr{H})$ satisfying (8.60) and (8.61). Then the triple*

$$\rho_{S,T}(\mathscr{H}) := (\mathscr{F}_{\mathrm{b}}(\mathscr{H}), \mathscr{F}_{\mathrm{b},0}(\mathscr{H}), \{B(f) \mid f \in \mathscr{H}\}) \tag{8.63}$$

is a representation of the CCR over $\mathscr{H}$.

Proof The anti-linearity of $B(f) \restriction \mathscr{F}_{\mathrm{b},0}(\mathscr{H})$ in f is easily seen. It is easy to see that, for all $f \in \mathscr{H}$, $B(f)^{\#}\mathscr{F}_{\mathrm{b},0}(\mathscr{H}) \subset \mathscr{F}_{\mathrm{b},0}(\mathscr{H})$ and, for all $\Psi \in \mathscr{F}_{\mathrm{b},0}(\mathscr{H})$,

$$\begin{aligned}[B(f), B(g)^*]\Psi &= [A(Sf) + A(JTf)^*, A(Sg)^* + A(JTg)]\Psi \\ &= (\langle Sf, Sg\rangle - \langle JTg, JTf\rangle)\Psi \\ &= \left\langle f, (S^*S - T^*T)g\right\rangle \Psi = \langle f, g\rangle \Psi,\end{aligned}$$

where we have used (8.60). Similarly, using (8.61), we can show that $[B(f), B(g)]\Psi = 0$. ■

The correspondence $\mathscr{B}_{S,T}$: $(A(\cdot), A(\cdot)^*) \mapsto (B(\cdot), B(\cdot)^*)$ is called the **bosonic Bogoliubov transformation** associated with (S, T). Note that this comes from $\Phi(\mathbb{S}(Jf, 0))$ with an iso-symplectic transformation $\mathbb{S}$ such that $\mathbb{S}\mathscr{H}^{(J)} \subset \mathscr{H}^{(J)}$.

The transformation $\mathscr{B}_{S,T}$ is said to be **proper** if there exists a unitary operator U on $\mathscr{F}_{\mathrm{b}}(\mathscr{H})$ such that $B(f) = UA(f)U^{-1}$, $f \in \mathscr{H}$ (hence $B(f)^* = UA(f)^*U^{-1}$), while $\mathscr{B}_{S,T}$ is said to be **improper** if it is not proper.

Remark 8.9 Conditions (8.60) and (8.61) are weakened. Namely, one can consider the case where S and T are not necessarily bounded operators on $\mathscr{H}$ satisfying the weak forms of (8.60) and (8.61):

$$\langle Sf, Sg\rangle - \langle Tf, Tg\rangle = \langle f, g\rangle\,, \tag{8.64}$$

$$\langle Sf, T_J g\rangle = \langle Tf, S_J g\rangle \tag{8.65}$$

for all f, g in a dense subspace $\mathscr{V} \subset D(T) \cap D(S) \cap D(T_J) \cap D(S_J)$. In this case, we define $B(f)$ by (8.62) with $f \in \mathscr{V}$. Then it follows that

$$\rho_{S,T}(\mathscr{V}) := (\mathscr{F}_{\mathrm{b}}(\mathscr{H}), \mathscr{F}_{\mathrm{b},0}(\mathscr{H}), \{B(f) | f \in \mathscr{V}\})$$

is a representation of the CCR over $\mathscr{V}$. Formula (8.64) implies that

$$\|Sf\|^2 - \|Tf\|^2 = \|f\|^2, \quad f \in \mathscr{V}.$$

Hence S is bounded if and only if T is bounded. In this case, (8.64) and (8.65) extend to all $f, g \in \mathscr{H}$ with S and T replaced by $\overline{S}$ and $\overline{T}$ respectively. Then

(8.60) and (8.61) hold with S and T replaced by $\overline{S}$ and $\overline{T}$ respectively. But, in the case where S or T is *unbounded* (then both S and T are unbounded), $\mathscr{B}_{S,T}$ may be considered as a new type of bosonic Bogoliubov transformation. We call it a **singular bosonic Bogoliubov transformation**. For a study on singular bosonic Bogoliubov transformation, see [30].

We first consider a condition under which the representation $\rho_{S,T}(\mathscr{H})$ is not equivalent to any direct sum of $\rho_{\mathrm{F}}(\mathscr{H})$.

Lemma 8.8 *Equation (8.60) implies that S is injective and S^* is surjective.*

Proof Equation (8.60) is equivalent to the equation $S^*S = I + T^*T$. Since T^*T is a non-negative self-adjoint operator, S^*S is bijective. This implies the statement of the present lemma. ∎

The following proposition is a key fact in considering inequivalence of representations of CCR.

Proposition 8.4 *Assume that $\mathscr{H}$ is separable. Let $L \in \mathfrak{L}(\mathscr{H})$ and $\mathscr{D}$ be a dense subspace of $\mathscr{H}$ such that $\mathscr{D} \subset D(L)$. Suppose that there exists a non-zero vector $\Omega \in \mathscr{F}_{\mathrm{b}}(\mathscr{H})$ such that, for all $\Psi \in \mathscr{F}_{\mathrm{b},0}(\mathscr{H})$ and $f \in \mathscr{D}$,*

$$\left\langle A(f)^*\Psi, \Omega\right\rangle = \langle A(JLf)\Psi, \Omega\rangle . \tag{8.66}$$

Then L is bounded and $\overline{L}$ is Hilbert–Schmidt.

Proof For each $n \geq 0$, we define a vector Ω_n in the n-particle space $\otimes_{\mathrm{s}}^n \mathscr{H}$ by $\Omega_n^{(n)} := \Omega^{(n)}$ and $\Omega_n^{(m)} := 0$, $m \neq n$. Let $\Psi \in \otimes_{\mathrm{s}}^n \mathscr{H}$. Then, for all $f \in \mathscr{D}$, $A(f)^*\Psi \in \otimes_{\mathrm{s}}^{n+1}\mathscr{H}$ and $A(JLf)\Psi \in \otimes_{\mathrm{s}}^{n-1}\mathscr{H}$ ($\otimes_{\mathrm{s}}^{-1}\mathscr{H} := \{0\}$). Hence we have from (8.66) $\langle A(f)^*\Psi, \Omega_{n+1}\rangle = \langle A(JLf)\Psi, \Omega_{n-1}\rangle$. Therefore

$$\langle \Psi, A(f)\Omega_{n+1}\rangle = \left\langle \Psi, A(JLf)^*\Omega_{n-1}\right\rangle .$$

Thus

$$A(f)\Omega_{n+1} = A(JLf)^*\Omega_{n-1}, \quad n \geq 0,\ f \in \mathscr{D}, \tag{8.67}$$

where $\Omega_{-1} := 0$. In particular, $A(f)\Omega_1 = 0$. Hence, by Proposition 6.2, $\Omega_1 = \langle \Omega_{\mathscr{H}}, \Omega_1\rangle\, \Omega_{\mathscr{H}} = 0$. Then, by (8.67), $A(f)\Omega_3 = 0$ for all $f \in \mathscr{D}$. Hence, by Proposition 6.2 again, $\Omega_3 = 0$. By repeating this process with (8.67), we can show by induction in n that $\Omega_{2n-1} = 0$ for all $n \geq 1$.

By (8.67), we have $A(f)\Omega_2 = cA(JLf)^*\Omega_{\mathscr{H}}$, where $c := \Omega^{(0)} \in \mathbb{C}$. If $c = 0$, then $A(f)\Omega_2 = 0$ for all $f \in \mathscr{D}$. Hence, by Proposition 6.2 again, $\Omega_2 = 0$. Then, in the same way as above, we obtain $\Omega_{2n} = 0$ for all $n \geq 0$. Thus $\Omega = 0$. But this is a contradiction. Hence $c \neq 0$. Therefore $\|A(JLf)^*\Omega_{\mathscr{H}}\|^2 = \|A(f)\Omega_2\|^2/|c|^2$.

Hence $\|Lf\|^2 = \|A(f)\Omega_2\|^2/|c|^2$. Let $\{e_n|n \in \mathbb{N}\} \subset \mathscr{D}$ be a CONS of $\mathscr{H}$. Then, by (6.83),

$$\sum_{n=1}^{\infty} \|Le_n\|^2 = \frac{1}{|c|^2} \sum_{n=1}^{\infty} \|A(e_n)\Omega_2\|^2 \leq \frac{1}{|c|^2} \|N_{\mathrm{b}}^{1/2}\Omega_2\|^2 < \infty.$$

In particular, $\|Le_n\| \leq C$, $n \in \mathbb{N}$ with $C := |c|^{-1}\|N_{\mathrm{b}}^{1/2}\Omega_2\|$ and C is independent of the choice of $\{e_n|n \in \mathbb{N}\}$. Hence, for each $\Psi \in \mathscr{D} \setminus \{0\}$, $\|L(\Psi/\|\Psi\|)\| \leq C$, i.e., $\|L\Psi\| \leq C\|\Psi\|$. Therefore $L \restriction \mathscr{D}$ is a bounded operator. Hence L is bounded and $\sum_{n=1}^{\infty} \|\overline{L}e_n\|^2 < \infty$. Thus $\overline{L}$ is Hilbert–Schmidt. ■

Theorem 8.6 *Assume that $\mathscr{H}$ is separable. Suppose that* Ran $S = \mathscr{H}$ *and T is not Hilbert–Schmidt. Then $\rho_{S,T}(\mathscr{H})$ is not equivalent to any direct sum of $\rho_{\mathrm{F}}(\mathscr{H})$.*

Proof We prove the contraposition. Hence suppose that there exists a unitary operator U from $\mathscr{F}_{\mathrm{b}}(\mathscr{H})$ to $\oplus_{p=1}^{N}\mathscr{F}_{\mathrm{b}}(\mathscr{H})$ ($N < \infty$ or $N = \infty$) such that $UB(f)U^{-1} = \oplus_{p=1}^{N} A(f)$, $f \in \mathscr{H}$. Let $\Omega := U^{-1}\{\Omega_{\mathscr{H}}, 0, 0, \ldots\}$. Then $\|\Omega\| = 1$ and $B(f)\Omega = 0$, i.e.,

$$\overline{(A(Sf) + A(JTf)^*)}\Omega = 0.$$

Hence, for all $\Psi \in \mathscr{F}_{\mathrm{b},0}(\mathscr{H})$, $\langle A(Sf)^*\Psi, \Omega\rangle = -\langle A(JTf)\Psi, \Omega\rangle$. By Lemma 8.8 and the present assumption on S, S is bijective. Hence, putting $K := TS^{-1}$, we have for all $f \in \mathscr{H}$, $\langle A(f)^*\Psi, \Omega\rangle = -\langle A(JKf)\Psi, \Omega\rangle$. Hence, by Proposition 8.4, K is Hilbert–Schmidt. Thus $T = KS$ is Hilbert–Schmidt.[4] ■

To see a physical meaning of Theorem 8.6, for a CONS $\{e_n\}_{n=1}^{\infty}$ of $\mathscr{H}$, we introduce a quantity

$$n_B := \sum_{n=1}^{\infty} \|B(e_n)\Omega_{\mathscr{H}}\|^2, \tag{8.68}$$

which is a finite non-negative number or $+\infty$. Intuitively, n_B is interpreted as an expectation value of the number operator "$\sum_{n=1}^{\infty} B(e_n)^*B(e_n)$" associated with $B(\cdot)$ with respect to the bare vacuum $\Omega_{\mathscr{H}}$. By (8.62), we have

$$n_B = \sum_{n=1}^{\infty} \|Te_n\|^2 = \operatorname{Tr} T^*T. \tag{8.69}$$

Hence $n_B = +\infty$ if and only if T is not Hilbert–Schmidt. Therefore *the physical phenomenon $n_B = +\infty$ is associated with the inequivalence stated in Theorem 8.6.*

---

[4]Recall that, if T is Hilbert–Schmidt on a Hilbert space $\mathscr{X}$, then, for all $A \in \mathfrak{B}(\mathscr{X})$, AT and TA are Hilbert–Schmidt.

We next consider the Bogoliubov transformation $\mathscr{B}_{S,T}$ under a stronger condition for (S, T).

Proposition 8.5 *Let $\mathbb{S}$ be given by (8.59) satisfying (8.60), (8.61). Then $\mathbb{S}$ is symplectic if and only if the following operator equations hold:*

$$SS^* - T_J T_J^* = I, \tag{8.70}$$

$$T_J S_J^* = ST^*. \tag{8.71}$$

Proof Let $\mathbb{S}$ be symplectic. Then $\mathbb{S}$ is bijective. Hence the inverse $\mathbb{S}^{-1}$ exists. By (8.47), we have

$$-\gamma(\widetilde{J}\mathbb{S}^*\widetilde{J})\gamma\mathbb{S} = I.$$

Hence

$$\mathbb{S}^{-1} = -\gamma(\widetilde{J}\mathbb{S}^*\widetilde{J})\gamma = \begin{pmatrix} S_J^* & -T_J^* \\ -T^* & S^* \end{pmatrix}.$$

Then, by direct computations, one can see that $\mathbb{S}\mathbb{S}^{-1} = I$ is equivalent to (8.70) and (8.71).

Conversely, assume (8.70) and (8.71). Let

$$\mathbb{T} := \begin{pmatrix} S_J^* & -T_J^* \\ -T^* & S^* \end{pmatrix}.$$

Then, by using (8.70) and (8.71), one can show that $\mathbb{S}\mathbb{T} = I$. Hence $\mathbb{S}$ is bijective. Therefore $\mathbb{S}$ is symplectic. ■

Lemma 8.9 *Assume (8.60), (8.61), (8.70) and (8.71). Then $\{B(f), B(f)^* | f \in \mathscr{H}\}' = \mathbb{C}I$. In particular, $\{B(f), B(f)^* | f \in \mathscr{H}\}$ is irreducible.*

Proof Let $X \in \{B(f), B(f)^* | f \in \mathscr{H}\}'$. Then $XB(f)^\# \subset B(f)^\# X$, $f \in \mathscr{H}$. Using (8.70) and (8.71), one can express $A(f)$ in terms of $B(\cdot)^\#$. Namely, for all $f \in \mathscr{H}$,

$$A(f) = B(S^* f) - B(T^* J f)^*, \quad A(f)^* = B(S^* f)^* - B(T^* J f) \tag{8.72}$$

on $\mathscr{F}_{\mathrm{b},0}(\mathscr{H})$. Hence, for all $\Phi, \Psi \in \mathscr{F}_{\mathrm{b},0}(\mathscr{H})$, we have

$$\langle XA(f)\Phi, \Psi\rangle = \langle X\Phi, A(f)^*\Psi\rangle.$$

Since $\mathscr{F}_{\mathrm{b},0}(\mathscr{H})$ is a core for $A(f)^*$, it follows that $X\Phi \in D((A(f)^*)^*) = D(A(f))$ and $A(f)X\Phi = XA(f)\Phi$. Since $\mathscr{F}_{\mathrm{b},0}(\mathscr{H})$ is a core for $A(f)$, it follows from a limiting argument that $XA(f) \subset A(f)X$. Similarly, one can show that

$XA(f)^* \subset A(f)^*X$. Hence $X \in \{A(f), A(f)^*|f \in \mathscr{H}\}' = \mathbb{C}I$. Therefore $X = cI$ for some $c \in \mathbb{C}$. Hence $\{B(f), B(f)^*|f \in \mathscr{H}\}' = \mathbb{C}I$. In particular, $\{B(f), B(f)^*|f \in \mathscr{H}\}$ is irreducible. ■

We remark the following fact:

Lemma 8.10 *Assume (8.60) and (8.70). Then S and S^* are bijective.*

Proof By (8.70), one can apply Lemma 8.8 with S replaced by S^* to conclude that S^* is injective and $S^{**} = S$ is surjective. By this result and Lemma 8.8, we obtain the desired result. ■

With regard to the converse of Theorem 8.6, the following result is known:

Theorem 8.7 *Assume (8.60), (8.61), (8.70) and (8.71). Suppose that T is Hilbert–Schmidt. Then $\rho_{S,T}(\mathscr{H})$ is equivalent to $\rho_{\mathrm{F}}(\mathscr{H})$.*

For proofs of this theorem, see, e.g., [123, 135–137, 146], [132, Theorems XI.106 and XI.108], [7] or [93, Chapter 2].

In summary, we have obtained the following characterization on proper Bogoliubov transformations:

Corollary 8.5 *Assume (8.60), (8.61), (8.70) and (8.71). Then $\rho_{S,T}(\mathscr{H})$ is equivalent to $\rho_{\mathrm{F}}(\mathscr{H})$ if and only if T is Hilbert–Schmidt.*

8.7 Second Quantization Operator Associated with $\{B(f)|f \in \mathscr{H}\}$

Let h be a non-negative self-adjoint operator on $\mathscr{H}$ and $\{e_n\}_{n=1}^{\infty} \subset D(h^{1/2})$ be a CONS of $\mathscr{H}$. Then, according to the arguments in Sect. 8.3.2, we define a subspace

$$\mathscr{D}_B := \left\{ \Psi \in \cap_{n=1}^{\infty} D(B(h^{1/2}e_n)) \middle| \sum_{n=1}^{\infty} \|B(h^{1/2}e_n)\Psi\|^2 < \infty \right\}, \tag{8.73}$$

where $B(f)$ is defined by (8.62). In this section, we consider only the case where h is bounded for technical simplicity. We assume the following:

Assumption (B)

(B.1) Equations (8.60) and (8.61) hold.
(B.2) The self-adjoint operator h is bounded and $Jh \subset hJ$.
(B.3) The operator T is Hilbert–Schmidt.

Under Assumption (B),

$$T_h := Th^{1/2} \tag{8.74}$$

is Hilbert–Schmidt. Hence $(T_h)_J = JT_hJ$ is Hilbert–Schmidt. We set

$$S_h := Sh^{1/2}, \tag{8.75}$$

which also is bounded under Assumption (B).

Lemma 8.11 *Under Assumption (B), the subspace $\mathscr{Q}_B$ is dense in $\mathscr{F}_{\mathrm{b}}(\mathscr{H})$ with*

$$\mathscr{F}_{\mathrm{b,fin}}(\mathscr{H})) \subset \mathscr{Q}_B. \tag{8.76}$$

Proof Let $\Psi \in \mathscr{F}_{\mathrm{b},0}(\mathscr{H})$. Then

$$a_n := \|B(h^{1/2}e_n)\Psi\|^2 \le b_n := 2(\|A(S_h e_n)\Psi\|^2 + \|A(JT_h e_n)^*\Psi\|^2).$$

By the Hilbert–Schmidtness of T_h, we have $\sum_{n=1}^{\infty} \|JT_h e_n\|^2 < \infty$. Let $\Psi \in \mathscr{F}_{\mathrm{b,fin}}(\mathscr{H})$. Then, by the present assumption and Lemma 8.3, $\sum_{n=1}^{\infty} b_n < \infty$. Hence $\sum_{n=1}^{\infty} a_n < \infty$. Therefore $\Psi \in \mathscr{Q}_B$. Thus (8.76) holds. ∎

By Lemma 8.11, we can apply Theorem 8.3 to obtain the following theorem:

Theorem 8.8 *Suppose that Assumption (B) holds. Then there exists a unique non-negative self-adjoint operator $H_B(h)$ on $\mathscr{F}_{\mathrm{b}}(\mathscr{H})$ such that*

$$D(H_B(h)^{1/2}) = \mathscr{Q}_B, \tag{8.77}$$

$$\left\langle H_B(h)^{1/2}\Phi, H_B(h)^{1/2}\Psi\right\rangle = \sum_{n=1}^{\infty}\left\langle B(h^{1/2}e_n)\Phi, B(h^{1/2}e_n)\Psi\right\rangle,$$
$$\Phi, \Psi \in D(H_B(h)^{1/2}). \tag{8.78}$$

The operator $H_B(h)$ is a second quantization operator of h associated with $\{B(f)|f \in \mathscr{H}\}$.

To find a concrete form of $H_B(h)$, we compute the right hand side of (8.78) for $\Phi \in D(H_B(h)^{1/2})$ and $\Psi \in \mathscr{F}_{\mathrm{b,fin}}(\mathscr{H})$. We have

$$\begin{aligned} B(h^{1/2}e_n)^* B(h^{1/2}e_n) = {} & A(S_h e_n)^* A(S_h e_n) + A((T_h)_J J e_n)^* A((T_h)_J J e_n) \\ & + A(S_h e_n)^* A((T_h)_J J e_n)^* + A((T_h)_J J e_n) A(S_h e_n) \\ & + \|T_h e_n\|^2 \end{aligned}$$

on $\mathscr{F}_{\mathrm{b},0}(\mathscr{H})$.

Let $\Psi \in \mathscr{F}_{\mathrm{b,fin}}(\mathscr{H})$. Then, by Theorem 6.13, we have

$$\lim_{N\to\infty} \sum_{n=1}^{N} A(S_h e_n)^* A(S_h e_n)\Psi = \mathrm{d}\Gamma_{\mathrm{b}}(S_h S_h^*)\Psi, \tag{8.79}$$

$$\lim_{N\to\infty}\sum_{n=1}^{N}A((T_h)_J Je_n)^* A((T_h)_J Je_n)\Psi = \mathrm{d}\Gamma_{\mathrm{b}}((T_h)_J(T_h)_J^*)\Psi. \tag{8.80}$$

By Lemma 6.13, we have

$$\begin{aligned}\lim_{N\to\infty}\sum_{n=1}^{N}A(S_he_n)^* A((T_h)_J Je_n)^*\Psi &= \lim_{N\to\infty}\sum_{n=1}^{N}A((T_h)_J Je_n)^* A(J(S_h)_J e_n)^*\Psi \\ &= \left\langle A^*|(T_h)_J(S_h)_J^*|A^*\right\rangle \Psi, \end{aligned} \tag{8.81}$$

$$\lim_{N\to\infty}\sum_{n=1}^{N}A((T_h)_J Je_n)A(S_he_n)\Psi = \left\langle A|(S_h)_J(T_h)_J^*|A\right\rangle \Psi. \tag{8.82}$$

It is obvious that

$$\sum_{n=1}^{\infty}\|T_he_n\|^2 = \|T_h\|_2^2.$$

Hence, introducing the operator

$$\begin{aligned}H_{\mathrm{qd}} :=& \,\mathrm{d}\Gamma_{\mathrm{b}}(S_hS_h^* + (T_h)_J(T_h)_J^*) \\ &+ \left\langle A^*|(T_h)_J(S_h)_J^*|A^*\right\rangle + \left\langle A|(S_h)_J(T_h)_J^*|A\right\rangle, \end{aligned} \tag{8.83}$$

we obtain

$$\begin{aligned}&\left\langle H_B(h)^{1/2}\Phi, H_B(h)^{1/2}\Psi\right\rangle = \left\langle \Phi, (H_{\mathrm{qd}} + \|T_h\|_2^2)\Psi\right\rangle, \\ &\qquad\qquad \Phi \in D(H_B(h)^{1/2}),\ \Psi \in \mathscr{F}_{\mathrm{b,fin}}(\mathscr{H}). \end{aligned} \tag{8.84}$$

It follows from the definition of H_{qd} and (8.84) that $H_{\mathrm{qd}} \restriction \mathscr{F}_{\mathrm{b,fin}}(\mathscr{H})$ is a symmetric operator bounded from below with

$$H_{\mathrm{qd}} \restriction \mathscr{F}_{\mathrm{b,fin}}(\mathscr{H}) \geq -\|T_h\|_2^2.$$

Theorem 8.9 *Suppose that Assumption (B) holds. Then H_{qd} is essentially self-adjoint on $\mathscr{F}_{\mathrm{b,fin}}(\mathscr{H})$ and $\overline{H}_{\mathrm{qd}} \geq -\|T_h\|_2^2$.*

Proof We write $H := H_{\mathrm{qd}} = A + B$ with

$$A := \mathrm{d}\Gamma_{\mathrm{b}}(S_hS_h^* + (T_h)_J(T_h)_J^*), \quad B := \left\langle A^*|(T_h)_J(S_h)_J^*|A^*\right\rangle + \left\langle A|(S_h)_J(T_h)_J^*|A\right\rangle.$$

Let $\mathscr{D} := D(A) \cap \mathscr{F}_{\mathrm{b},0}(\mathscr{H})$ and $\Psi \in \mathscr{D}$. Since $\mathscr{F}_{\mathrm{b,fin}}(\mathscr{H})$ is a core for A (note that $D(S_hS_h^*) + (T_h)_J(T_h)_J^*) = \mathscr{H}$), there exists a sequence $\{\Psi_n\}_{n=1}^{\infty}$ in $\mathscr{F}_{\mathrm{b,fin}}(\mathscr{H})$ such that $\Psi_n \to \Psi$, $A\Psi_n \to A\Psi$ as $n \to \infty$. By Lemma 6.12, $B\Psi_n \to B\Psi$ as $n \to \infty$. Therefore $H\Psi_n \to H\Psi$ as $n \to \infty$. Thus

$$H \upharpoonright \mathscr{D} \subset H' := \overline{H \upharpoonright \mathscr{F}_{\mathrm{b,fin}}(\mathscr{H})}.$$

Hence it is sufficient to show that H is essentially self-adjoint on $\mathscr{D}$. For this purpose, we apply Theorem C.3 in Appendix C to the case where $\mathscr{H}_n = \otimes_{\mathrm{s}}^n \mathscr{H}$, $\mathscr{D}_0 = \mathscr{F}_{\mathrm{b},0}(\mathscr{H})$, $\hat{N} = N_{\mathrm{b}}$ and (A, B) given above. By Lemma 6.12, there exist constants $c_i \geq 0$, $i = 1, 2$ such that, for all $\Psi \in \mathscr{F}_{\mathrm{b},0}(\mathscr{H})$,

$$\begin{aligned}\|B\Psi\| &\leq c_1\|(N_{\mathrm{b}}+2)\Psi\| + c_2\|N_{\mathrm{b}}\Psi\|) \\ &\leq (2c_1 + c_2)\|(N_{\mathrm{b}} + 1)\Psi\| \\ &\leq (2c_1 + c_2)\|(N_{\mathrm{b}} + 1)^2\Psi\|,\end{aligned}$$

where we have used the following facts (which are easily proved): for all $\Psi \in D(N_{\mathrm{b}})$,

$$\|\Psi\| \leq \|(N_{\mathrm{b}} + 1)\Psi\|, \quad \|N_{\mathrm{b}}\Psi\| \leq \|(N_{\mathrm{b}} + 1)\Psi\|$$

and, for any non-negative self-adjoint operator X on a Hilbert space, $\|(X + 1)\| \leq \|(X + 1)^2 f\|$, $f \in D(X^2)$. Hence condition (i) in Sect. C.3 in Appendix C is satisfied in the present case.

Lemma 6.12 implies also that, for $|m - n| > 2$, $\langle\Psi^{(m)}, B\Psi^{(n)}\rangle = 0$. Hence condition (ii) also in Sect. C.3 in Appendix C is satisfied in the present case. Since $H' \geq -\|T_h\|_2^2$, it follows that $H \geq -\|T_h\|_2^2$ on $\mathscr{D}$. Thus we can apply Theorem C.3 to conclude that H is essentially self-adjoint on $\mathscr{D}$. Hence H' is a self-adjoint operator. Thus H is essentially self-adjoint on $\mathscr{F}_{\mathrm{b,fin}}(\mathscr{H})$. ∎

Remark 8.10 Let K be a Hilbert–Schmidt operator on $\mathscr{H}$ and ξ be a non-negative self-adjoint operator on $\mathscr{H}$ such that

$$H_{\xi,K} := \mathrm{d}\Gamma_{\mathrm{b}}(\xi) + \langle A^*|K|A^*\rangle + \langle A|K^*|A\rangle$$

is bounded from below on $D(\mathrm{d}\Gamma_{\mathrm{b}}(\xi)) \cap \mathscr{F}_{\mathrm{b},0}(\mathscr{H})$. Then, in quite the same way as in the proof of Theorem 8.9, one can prove that $H_{\xi,K}$ is essentially self-adjoint on $D(\mathrm{d}\Gamma_{\mathrm{b}}(\xi)) \cap \mathscr{F}_{\mathrm{b},0}(\mathscr{H})$ and the closure $\overline{H}_{\xi,K}$ is a self-adjoint operator bounded from below.

We now come back to relation (8.84). By Theorem 8.9 and a limiting argument, one can show that (8.84) extends to all $\Psi \in D(\overline{H}_{\mathrm{qd}})$. Hence it follows that $D(\overline{H}_{\mathrm{qd}}) \subset D(H_B(h))$ and $H_B(h)\Psi = (\overline{H}_{\mathrm{qd}} + \|T_h\|_2^2)\Psi$, $\Psi \in D(\overline{H}_{\mathrm{qd}})$. This means that $\overline{H}_{\mathrm{qd}} + \|T_h\|_2^2 \subset H_B(h)$. Thus

$$H_B(h) = \overline{H}_{\mathrm{qd}} + \|T_h\|_2^2. \tag{8.85}$$

Theorem 8.10 *Assume (8.60), (8.61), (8.70) and (8.71). Suppose, in addition, that Assumption (B) holds. Then there is a unitary operator* U_B *on* $\mathscr{F}_{\mathrm{b}}(\mathscr{H})$ *such that*

$$U_B B(f) U_B^{-1} = A(f), \quad f \in \mathscr{H}, \tag{8.86}$$

$$U_B(\overline{H}_{\mathrm{qd}} + \|T_h\|_2^2)U_B^{-1} = \mathrm{d}\Gamma_{\mathrm{b}}(h). \tag{8.87}$$

Proof Since T is Hilbert–Schmidt, it follows from Corollary 8.5 that there exists a unitary operator U_B on $\mathscr{F}_{\mathrm{b}}(\mathscr{H})$ such that (8.86) holds. Hence, by Theorem 8.4, $U_B H_B(h) U_B^{-1} = \mathrm{d}\Gamma_{\mathrm{b}}(h)$. By this fact and (8.85), we obtain (8.87). ∎

Remark 8.11 Suppose that S and T are independent of h. Then, by construction, the unitary operator U_B is independent of h. Hence, under additional conditions, (8.87) may be extended to the case where h is unbounded through a limiting argument.

Remark 8.12 There is another method for proof of (8.87). See, e.g. [7, §IV] or [93, §3.5]. For recent studies on diagonalizations of Hamiltonians quadratic in annihilation and creation operators, see [52, 60, 61, 119].

Theorem 8.10 immediately yields the following corollary:

Corollary 8.6 *Under the same assumption as in Theorem 8.10,* $\overline{H}_{\mathrm{qd}}$ *has a unique ground state given by*

$$\Psi_B := U_B^{-1}\Omega_{\mathscr{H}},$$

up to constant multiples, with ground state energy $-\|T_h\|_2^2$*:*

$$\overline{H}_{\mathrm{qd}}\Psi_B = -\|T_h\|_2^2\Psi_B.$$

8.8 Representations of Heisenberg CCR

As in the case of finite degrees of freedom, one can also consider representations of CCR of Heisenberg type.

Definition 8.4 Let $\mathscr{W}$ be a real inner product space. Let $\mathscr{F}$ be a Hilbert space and $\mathscr{D}$ be a dense subspace of $\mathscr{F}$. Suppose that, for each $f \in \mathscr{W}$, symmetric operators $\phi(f)$ and $\pi(f)$ on $\mathscr{F}$ are given, satisfying the following conditions:

(i) $\mathscr{D} \subset \cap_{f\in\mathscr{W}} D(\phi(f)) \cap D(\pi(f))$ and $\phi(f)\mathscr{D} \subset \mathscr{D}, \pi(f)\mathscr{D} \subset \mathscr{D}$.
(ii) For all $f, g \in \mathscr{W}$ and $a, b \in \mathbb{R}$, $\phi(af + bg) = a\phi(f) + b\phi(g)$ and $\pi(af + bg) = a\pi(f) + b\pi(g)$ on $\mathscr{D}$.

(iii) (Heisenberg CCR over $\mathscr{W}$) For all $f, g \in \mathscr{W}$,

$$[\phi(f), \pi(g)] = i \langle f, g \rangle,$$
$$[\phi(f), \phi(g)] = 0, \quad [\pi(f), \pi(g)] = 0$$

on $\mathscr{D}$.

Then $(\mathscr{F}, \mathscr{D}, \{\phi(f), \pi(f) \mid f \in \mathscr{W}\})$ is called a **representation of the Heisenberg CCR over** $\mathscr{W}$.[5] We call $\mathscr{D}$ a **CCR-domain** for $\{\phi(f), \pi(f) | f \in \mathscr{W}\}$.

Two representations $(\mathscr{F}, \mathscr{D}, \{\phi(f), \pi(f) \mid f \in \mathscr{W}\})$ and $(\mathscr{F}', \mathscr{D}', \{\phi'(f), \pi'(f) \mid f \in \mathscr{W}\})$ of the Heisenberg CCR over $\mathscr{W}$ are said to be **equivalent** if there exists a unitary operator $U: \mathscr{F} \to \mathscr{F}'$ such that, for all $f \in \mathscr{W}$, $U\overline{\phi(f)}U^{-1} = \overline{\phi'(f)}$ and $U\overline{\pi(f)}U^{-1} = \overline{\pi'(f)}$.

The representation $(\mathscr{F}, \mathscr{D}, \{\phi(f), \pi(f) \mid f \in \mathscr{W}\})$ is said to be **irreducible** (resp. **reducible**) if $\{\phi(f), \pi(f) \mid f \in \mathscr{W}\}$ is irreducible (resp. reducible).

Remark 8.13 In this definition, self-adjointness of $\phi(f)$ (resp. $\pi(f)$) is not required. The subspace $\mathscr{D}$ is not necessarily unique.

Remark 8.14 Let $f \neq 0$ ($f \in \mathscr{W}$). By Definition 8.4(iii), we have

$$[\phi(f), \pi(f)] = i \|f\|^2 \quad \text{on } \mathscr{D}.$$

Hence $Q := \|f\|^{-1}\phi(f)$ and $P := \|f\|^{-1}\pi(f)$ obey the CCR with one degree of freedom. Therefore, at least one of $\phi(f)$ and $\pi(f)$ is unbounded.

Remark 8.15 Remark 8.5 applies also to representations of Heisenberg CCR.

Example 8.10 With notation in Sect. 6.12, let $\mathscr{W}$ be a subspace of $\mathscr{H}_J$. Then formulas (6.126)–(6.129) show that

$$\pi_F(\mathscr{W}) := (\mathscr{F}_b(\mathscr{H}), \mathscr{F}_{b,0}(\mathscr{H}), \{\phi_F(f), \pi_F(f) \mid f \in \mathscr{W}\}) \tag{8.88}$$

is a representation of the Heisenberg CCR over $\mathscr{W}$. This representation is called the **Fock representation of the Heisenberg CCR over** $\mathscr{W}$.

There is a simple deformation of a given representation of Heisenberg CCR:

Proposition 8.6 *Let $(\mathscr{F}, \mathscr{D}, \{\phi(f), \pi(f) \mid f \in \mathscr{W}\})$ be a representation of the Heisenberg CCR over $\mathscr{W}$ as above and two linear functionals $\mu, \nu : \mathscr{W} \to \mathbb{R}$ be given. Define*

$$\phi_\mu(f) := \phi(f) + \mu(f), \tag{8.89}$$
$$\pi_\nu(f) := \pi(f) + \nu(f), \quad f \in \mathscr{W}, \tag{8.90}$$

[5] We sometimes omit "Heisenberg" if there is no danger of confusion.

Then:

(i) $(\mathscr{F}, \mathscr{D}, \{\phi_\mu(f), \pi_\nu(f) \mid f \in \mathscr{W}\})$ *is a representation of the Heisenberg CCR over* $\mathscr{W}$.
(ii) *If* $(\mathscr{F}, \mathscr{D}, \{\phi(f), \pi(f) \mid f \in \mathscr{W}\})$ *is irreducible, then so is* $(\mathscr{F}, \mathscr{D}, \{\phi_\mu(f), \pi_\nu(f) \mid f \in \mathscr{W}\})$.
(iii) *If* $\phi(f)$ *and* $\pi(f)$ *are essentially self-adjoint, then so are* $\phi_\mu(f)$ *and* $\pi_\nu(f)$ *with*

$$\overline{\phi_\mu(f)} = \overline{\phi(f)} + \mu(f), \tag{8.91}$$

$$\overline{\pi_\nu(f)} = \overline{\pi(f)} + \nu(f), \quad f \in \mathscr{W}. \tag{8.92}$$

Proof An easy exercise. ∎

We call $(\mathscr{F}, \mathscr{D}, \{\phi_\mu(f), \pi_\nu(f) \mid f \in \mathscr{W}\})$ the **translation** of $(\mathscr{F}, \mathscr{D}, \{\phi(f), \pi(f) \mid f \in \mathscr{W}\})$ by (μ, ν).

The following proposition shows that a representation of the CCR over a complex inner product space with an additional condition yields a representation of the Heisenberg CCR over a real inner product space.

Proposition 8.7 *Let* $\mathscr{V}$ *be a complex inner product space and* J *be a conjugation on* $\mathscr{V}$. *Suppose that there exists a representation* $(\mathscr{F}, \mathscr{D}, \{C(f) \mid f \in \mathscr{V}\})$ *of the CCR over* $\mathscr{V}$ *and let*

$$\phi(f) := \frac{1}{\sqrt{2}}(C(f)^* + C(f)),$$

$$\pi(f) := \frac{i}{\sqrt{2}}(C(f)^* - C(f)), \quad f \in \mathscr{V}_J,$$

where $\mathscr{V}_J$ *is the real part of* $\mathscr{V}$ *with respect to* J *(see (1.68)). Then:*

(i) $(\mathscr{F}, \mathscr{D}, \{\phi(f), \pi(f) \mid f \in \mathscr{V}_J\})$ *is a representation of the Heisenberg CCR over* $\mathscr{V}_J$.
(ii) *If* $\mathscr{D}$ *is a core for* $C(f)$ *and* $C(f)^*$ *for all* $f \in \mathscr{V}$ *and* $\{C(f), C(f)^* | f \in \mathscr{V}\}' = \mathbb{C}I$, *then* $\{\phi(f), \pi(f) \mid f \in \mathscr{V}_J\}' = \mathbb{C}I$. *In particular,* $\{\phi(f), \pi(f) \mid f \in \mathscr{V}_J\}$ *is irreducible.*

Proof

(i) An easy exercise.
(ii) Let $T \in \{\phi(f), \pi(f) \mid f \in \mathscr{V}_J\}'$. Then $T\phi(f) \subset \phi(f)T$ and $T\pi(f) \subset \pi(f)T$ for all $f \in \mathscr{V}_J$. It follows that

$$TC(f)^* = C(f)^*T, \quad TC(f) = C(f)T \quad \text{on } \mathscr{D}.$$

By the linearity of $C(f)^*$ in f and the anti-linearity of $C(f)$ in f, these relations extend to all $f \in \mathscr{V}$. Since $\mathscr{D}$ is a core for $C(f)^*$ and $C(f)$

by the present assumption, it follows from a simple limiting argument that $TC(f)^* \subset C(f)^*T$ and $TC(f) \subset C(f)T$, $f \in \mathscr{V}$. This means that $T \in \{C(f), C(f)^* | f \in \mathscr{V}\}' = \mathbb{C}I$. Hence $T = \alpha I$ for some $\alpha \in \mathbb{C}$. Therefore $\{\phi(f), \pi(f) \mid f \in \mathscr{V}_J\}' = \mathbb{C}I$. Hence $\{\phi(f), \pi(f) \mid f \in \mathscr{V}_J\}$ is irreducible. ■

We call $(\mathscr{F}, \mathscr{D}, \{\phi(f), \pi(f) \mid f \in \mathscr{V}_J\})$ the **representation of the Heisenberg CCR induced by** $(\mathscr{F}, \mathscr{D}, \{C(f) \mid f \in \mathscr{V}\})$ with conjugation J.

Example 8.11 The Fock representation $\pi_{\mathrm{F}}(\mathscr{W})$ (Example 8.10) is the representation of the Heisenberg CCR induced by $(\mathscr{F}_{\mathrm{b}}(\mathscr{H}), \mathscr{F}_{\mathrm{b},0}(\mathscr{H}), \{A(f) \mid f \in \mathscr{W}_{\mathbb{C}}\})$ with conjugation J.

To consider the converse of Proposition 8.7, we recall a basic concept in Hilbert space theory. Let $\mathscr{R}$ be a real Hilbert space with inner product $\langle\ ,\ \rangle_{\mathscr{R}}$. Then the product space $\mathscr{R} \times \mathscr{R} := \{(f, g) \mid f, g \in \mathscr{R}\}$ becomes a complex vector space with the following operations of addition and scalar multiplication: for all $(f, g), (f', g') \in \mathscr{R} \times \mathscr{R}$ and $\alpha = a + ib \in \mathbb{C}\,(a, b \in \mathbb{R})$,

$$(f, g) + (f', g') := (f + g, f' + g'),$$

$$\alpha(f, g) := (af - bg, ag + bf).$$

This complex vector space is called the **complexification** of $\mathscr{R}$ and denoted by $\mathscr{R}_{\mathbb{C}}$.

Note that $\psi := (f, g) = (f, 0) + i(g, 0)\ (f, g \in \mathscr{R})$. Hence, identifying $(f, 0)$ with f, we have $\psi = f + ig$. The vector f (resp. g) is called the **real** (resp. **imaginary**) **part** of ψ and written as $f = \operatorname{Re}\psi$ (resp. $g = \operatorname{Im}\psi$).

A natural inner product is introduced in $\mathscr{R}_{\mathbb{C}}$ by

$$\langle f + ig, f' + ig'\rangle_{\mathscr{R}_{\mathbb{C}}} := \langle f, f'\rangle_{\mathscr{R}} + \langle g, g'\rangle_{\mathscr{R}} - i\,\langle g, f'\rangle_{\mathscr{R}} + i\,\langle f, g'\rangle_{\mathscr{R}}$$

$(f, f', g, g' \in \mathscr{R})$. Then the inner product space $(\mathscr{R}_{\mathbb{C}}, \langle\ ,\ \rangle_{\mathscr{R}_{\mathbb{C}}})$ becomes a complex Hilbert space. This Hilbert space is called the **complexification** of the real Hilbert space $\mathscr{R}$ and denoted by the same symbol $\mathscr{R}_{\mathbb{C}}$ for simplicity. Henceforth the subscript $\mathscr{R}_{\mathbb{C}}$ in $\langle\ ,\ \rangle_{\mathscr{R}_{\mathbb{C}}}$ is omitted, provided that there would be no danger of confusion.

We define a mapping $J_{\mathscr{R}}: \mathscr{R}_{\mathbb{C}} \to \mathscr{R}_{\mathbb{C}}$ by

$$J_{\mathscr{R}}(f + ig) := f - ig, \quad f, g \in \mathscr{R}.$$

Then it is easy to see that $J_{\mathscr{R}}$ is a conjugation. This conjugation is called the **complex conjugation** on $\mathscr{R}_{\mathbb{C}}$. It immediately follows that

$$\mathscr{R} = (\mathscr{R}_{\mathbb{C}})_{J_{\mathscr{R}}}.$$

Example 8.12 Let $\mathscr{H}$ be a complex Hilbert space and J be a conjugation on $\mathscr{H}$. Then $(\mathscr{H}_J)_{\mathbb{C}} \cong \mathscr{H}$.

Proposition 8.8 *Let $(\mathscr{F}, \mathscr{D}, \{\phi(f), \pi(f) \mid f \in \mathscr{W}\})$ be a representation of the Heisenberg CCR over $\mathscr{W}$ as in Definition 8.4 and $\mathscr{W}_{\mathbb{C}}$ be the complexification of $\mathscr{W}$.*

(i) *Let*

$$\Phi(f) := \phi(\mathrm{Re}\, f) + \pi(\mathrm{Im}\, f), \quad f \in \mathscr{W}_{\mathbb{C}}. \tag{8.93}$$

Then $\Phi(f)$ is symmetric and, for all $f, g \in \mathscr{W}_{\mathbb{C}}$,

$$[\Phi(f), \Phi(g)] = i\, \mathrm{Im}\, \langle f, g\rangle \tag{8.94}$$

on $\mathscr{D}$.

(ii) *Let*

$$C(f) := \frac{1}{\sqrt{2}}(\Phi(f) + i\Phi(if)), \quad f \in \mathscr{W}_{\mathbb{C}}.$$

Then $C(f)$ is closable and $(\mathscr{F}, \mathscr{D}, \{\overline{C(f)} \mid f \in \mathscr{W}_{\mathbb{C}}\})$ is a representation of the CCR over $\mathscr{W}_{\mathbb{C}}$.

Proof An easy exercise. ∎

Propositions 8.7 and 8.8 show that considering a representation of the Heisenberg CCR over a real inner product space $\mathscr{W}$ is essentially equivalent to considering a representation of the CCR over $\mathscr{W}_{\mathbb{C}}$.

Remark 8.16 Let $(\mathscr{X}, \omega)$ be an $\mathbb{R}$-symplectic space. Suppose that a set $\{\Phi(f)|f \in (\mathscr{X}, \omega)\}$ of symmetric (not necessarily self-adjoint) operators on a Hilbert space $\mathscr{F}$ is given and $\mathscr{D}$ be a dense subspace of $\mathscr{F}$ left invariant by each $\Phi(f),\ f \in (\mathscr{X}, \omega)$. Then the triple $(\mathscr{F}, \mathscr{D}, \{\Phi(f)|f \in (\mathscr{X}, \omega)\})$ is called a **representation of the generalized Heisenberg CCR** over $(\mathscr{X}, \omega)$ if, for all $f, g \in \mathscr{X}, a, b \in \mathbb{R}$ and $\Psi \in \mathscr{D}$,

$$\Phi(af + bg)\Psi = [a\,\Phi(f) + b\,\Phi(g)]\Psi$$

and

$$[\Phi(f), \Phi(g)]\Psi = i\omega(f, g)\Psi. \tag{8.95}$$

The ω-Segal field operators introduced in Sect. 6.13 give a representation of the generalized Heisenberg CCR over $(\mathscr{X}, \omega)$.

8.9 Weyl Representations of CCR Over Real Inner Product Spaces

The concept of Weyl representations of CCR with finite degrees of freedom can be naturally extended to the case of infinite degrees of freedom.

Definition 8.5 Let $\mathscr{W}$ be a real inner product space and $\mathscr{F}$ be a complex Hilbert space. Suppose that, for each $f \in \mathscr{W}$, two self-adjoint operators $\phi(f)$ and $\pi(f)$ on $\mathscr{F}$ are defined. The pair $(\mathscr{F}, \{\phi(f), \pi(f) \mid f \in \mathscr{W}\})$ is called a **Weyl representation** of the CCR over $\mathscr{W}$ on $\mathscr{F}$ if the following hold:

(i) For all $f, g \in \mathscr{W}$ and $a, b \in \mathbb{R}$,

$$e^{i\phi(af+bg)} = e^{ia\phi(f)}e^{ib\phi(g)}, \tag{8.96}$$

$$e^{i\pi(af+bg)} = e^{ia\pi(f)}e^{ib\pi(g)}. \tag{8.97}$$

(ii) (Weyl relations over $\mathscr{W}$) For all $f, g \in \mathscr{W}$ and $t \in \mathbb{R}$

$$e^{i\phi(f)}e^{i\pi(g)} = e^{-i\langle f,g\rangle}e^{i\pi(g)}e^{i\phi(f)}, \tag{8.98}$$

$$e^{i\phi(f)}e^{i\phi(g)} = e^{i\phi(g)}e^{i\phi(f)}, \tag{8.99}$$

$$e^{i\pi(f)}e^{i\pi(g)} = e^{i\pi(g)}e^{i\pi(f)}. \tag{8.100}$$

Remark 8.17 A more general definition of a Weyl representation is possible as given in [37].

A Weyl representation $(\mathscr{F}, \{\phi(f), \pi(f) \mid f \in \mathscr{W}\})$ is said to be **irreducible** (resp. **reducible**) if $\{e^{i\phi(f)}, e^{i\pi(f)} \mid f \in \mathscr{W}\}$ is irreducible (resp. reducible).

Two Weyl representations $(\mathscr{F}, \{\phi(f), \pi(f) \mid f \in \mathscr{W}\})$ and $(\mathscr{F}', \{\phi'(f), \pi'(f) \mid f \in \mathscr{W}\})$ are said to be **equivalent** if there exists a unitary operator $U : \mathscr{F} \to \mathscr{F}'$ such that, for all $f \in \mathscr{W}$, $U\phi(f)U^{-1} = \phi'(f)$ and $U\pi(f)U^{-1} = \pi'(f)$.

If $\dim \mathscr{W} = \infty$, then $(\mathscr{F}, \{\phi(f), \pi(f) \mid f \in \mathscr{W}\})$ is said to be a **Weyl representation with infinite degrees of freedom**.

Remark 8.18 Relations (8.96) and (8.97) imply that, for all $a, t \in \mathbb{R}$,

$$e^{it\phi(af)} = e^{ita\phi(f)}, \quad e^{it\pi(af)} = e^{ita\pi(f)}.$$

Hence it follows that

$$\phi(af) = a\phi(f), \quad \pi(af) = a\pi(f), \quad a \in \mathbb{R}, \ f \in \mathscr{W}. \tag{8.101}$$

Remark 8.19 Relations (8.99) and (8.100) imply that $\{\phi(f) \mid f \in \mathscr{W}\}$ (resp. $\{\pi(f) \mid f \in \mathscr{W}\}$) is a family of strongly commuting self-adjoint operators on $\mathscr{F}$. Hence, for all f, g, $\phi(f) + \phi(g)$ is essentially self-adjoint and, for all $t \in \mathbb{R}$,

$$e^{it\left(\overline{\phi(f)+\phi(g)}\right)} = e^{it\phi(f)}e^{it\phi(g)}.$$

Similarly, $\pi(f) + \pi(g)$ is essentially self-adjoint and

$$e^{it\left(\overline{\pi(f)+\pi(g)}\right)} = e^{it\pi(f)}e^{it\pi(g)}.$$

Then, it follows from (i) that, for all $t \in \mathbb{R}$,

$$e^{it\phi(f+g)} = e^{it\left(\overline{\phi(f)+\phi(g)}\right)}, \quad e^{it\pi(f+g)} = e^{it\left(\overline{\pi(f)+\pi(g)}\right)}.$$

Hence

$$\phi(f+g) = \overline{\phi(f)+\phi(g)}, \quad \pi(f+g) = \overline{\pi(f)+\pi(g)}, \quad f, g \in \mathscr{W}. \tag{8.102}$$

Formulas (8.101) and (8.102) show that $\phi(f)$ and $\pi(f)$ are real linear in f with additivity being of the form (8.102).

Remark 8.20 Remark 8.5 applies also to Weyl representations of CCR.

Remark 8.21 Properties (8.98)–(8.100) imply that, for all $f, g \in \mathscr{W}$, $\phi(f)$, $\phi(g)$, $\pi(f)$ and $\pi(g)$ obey CCR:

$$[\phi(f), \pi(g)] = i\langle f, g\rangle, \quad [\phi(f), \phi(g)] = 0, \quad [\pi(f), \pi(g)] = 0 \tag{8.103}$$

on the Gårding domain associated with $(\phi(f), \phi(g), \pi(f), \pi(g))$ (see (2.140) and Theorem 2.16). Hence, by Proposition 2.2, each commutation relation in (8.103) holds on its maximal domain. In this sense, $\{\phi(f), \pi(f) | f \in \mathscr{W}\}$ is a representation of the Heisenberg CCR over $\mathscr{W}$. In contrast to the case of Weyl representations of CCR with finite degrees of freedom, a uniqueness theorem similar to the von Neumann uniqueness theorem does not hold anymore for Weyl representations of CCR with *infinite degrees of freedom*. Namely, there are mutually (non-trivially) inequivalent Weyl representations of the CCR over an infinite dimensional real inner product space. See Corollary 8.7 below and Chap. 10.

We say that a representation $(\mathscr{F}, \mathscr{D}, \{\phi(f), \pi(f) | f \in \mathscr{W}\})$ of the Heisenberg CCR over $\mathscr{W}$ is a Weyl representation of the CCR over $\mathscr{W}$ if $(\mathscr{F}, \{\phi(f), \pi(f) | f \in \mathscr{W}\})$ is a Weyl representation of the CCR over $\mathscr{W}$.

The Fock representation $\pi_{\mathrm{F}}(\mathscr{W})$ of the Heisenberg CCR over $\mathscr{W}$ in Example 8.10 is in fact an irreducible Weyl representation:

Theorem 8.11 *The Fock representation $\pi_{\mathrm{F}}(\mathscr{W})$ is an irreducible Weyl representation of the CCR over $\mathscr{W}$.*

Proof This follows from Theorem 6.23. ■

Proposition 8.9 *Let $(\mathscr{F}, \{\phi(f), \pi(f) \mid f \in \mathscr{W}\})$ be a Weyl representation of the CCR over a real inner product space $\mathscr{W}$ and $\phi_\mu(f)$ and $\pi_\nu(f)$ be defined by (8.89) and (8.90) respectively. Then $(\mathscr{F}, \{\phi_\mu(f), \pi_\nu(f) \mid f \in \mathscr{W}\})$ is a Weyl representation of the CCR over $\mathscr{W}$. Moreover, if $(\mathscr{F}, \{\phi(f), \pi(f) \mid f \in \mathscr{W}\})$ is irreducible, then so is $(\mathscr{F}, \{\phi_\mu(f), \pi_\nu(f) \mid f \in \mathscr{W}\})$.*

Proof It is obvious that $\phi_\mu(f)$ and $\pi_\nu(f)$ are self-adjoint and

$$e^{i\phi_\mu(f)} = e^{i\phi(f)}e^{i\mu(f)}, \quad e^{i\pi_\nu(f)} = e^{i\pi(f)}e^{i\nu(f)}. \tag{8.104}$$

These relations imply that $\{e^{i\phi_\mu(f)}, e^{i\pi_\nu(f)} | f \in \mathscr{W}\}$ obeys the Weyl relations over $\mathscr{W}$. Hence $(\mathscr{F}, \{\phi_\mu(f), \pi_\nu(f) \mid f \in \mathscr{W}\})$ is a Weyl representation of the CCR over $\mathscr{W}$.

Suppose that $(\mathscr{F}, \{\phi(f), \pi(f) \mid f \in \mathscr{W}\})$ is irreducible. Let $T \in \{e^{i\phi_\mu(f)}, e^{i\pi_\nu(f)} \mid f \in \mathscr{W}\}'$. Then, by (8.104), $T \in \{e^{i\phi(f)}, e^{i\pi(f)} \mid f \in \mathscr{W}\}' = \mathbb{C}I$. Hence $T = \alpha I$ for some $\alpha \in \mathbb{C}$. Thus $(\mathscr{F}, \{\phi_\mu(f), \pi_\nu(f) \mid f \in \mathscr{W}\})$ is irreducible. ■

The following proposition is often useful in applications to quantum field theory:

Proposition 8.10 *Let $(\mathscr{F}, \{\phi(f), \pi(f) \mid f \in \mathscr{W}\})$ be a Weyl representation of the CCR over a real inner product space $\mathscr{W}$. Let $\overline{\mathscr{W}}$ be the completion of $\mathscr{W}$ and T be an injective linear mapping on $\overline{\mathscr{W}}$ such that, for all $f, g \in D(T)$, $\langle f, Tg\rangle_{\overline{\mathscr{W}}} = \langle Tf, g\rangle_{\overline{\mathscr{W}}}$ and define*

$$q_T(f) := \phi(T^{-1}f), \quad p_T(g) := \pi(Tg) \tag{8.105}$$

for $f \in D(T^{-1})$ and $g \in D(T)$ such that $T^{-1}f \in \mathscr{W}$ and $Tg \in \mathscr{W}$. Let $\mathscr{W}_0 \subset D(T) \cap D(T^{-1})$ be a subspace such that $T^{\pm 1}(\mathscr{W}_0) \subset \mathscr{W}$. Then:

(i) *$(\mathscr{F}, \{q_T(f), p_T(f) | f \in \mathscr{W}_0\})$ is a Weyl representation of the CCR over $\mathscr{W}_0$.*
(ii) *Suppose that the mappings $f \mapsto e^{i\phi(f)}$ and $f \mapsto e^{i\pi(f)}$ are weakly continuous: if $f_n, f \in \mathscr{W}$ ($n \in \mathbb{N}$) and $\lim_{n\to\infty} f_n = f$, then $\text{w-}\lim_{n\to\infty} e^{i\phi(f_n)} = e^{i\phi(f)}$ and $\text{w-}\lim_{n\to\infty} e^{i\pi(f_n)} = e^{i\pi(f)}$, where w-lim means weak limit (see Sect. 1.1.2), and $T^{\pm 1}(\mathscr{W}_0)$ are dense in $\mathscr{W}$. Let $\{\phi(f), \pi(f) | f \in \mathscr{W}\}' = \mathbb{C}I$. Then $\{q_T(f), p_T(f) | f \in \mathscr{W}_0\}' = \mathbb{C}I$. In particular, $(\mathscr{F}, \{q_T(f), p_T(f) | f \in \mathscr{W}_0\})$ is irreducible.*

Proof

(i) By the assumption for T, we have $\langle T^{-1}f, Tg\rangle = \langle f, g\rangle$, $f, g \in \mathscr{W}_0$. Hence $e^{iq_T(f)}e^{ip_T(g)} = e^{-i\langle f,g\rangle}e^{ip_T(g)}e^{iq_T(f)}$. It is easy to show that, for $(\phi(f), \pi(f)) = (q_T(f), p_T(f))$ $(f \in \mathscr{W}_0)$, (8.96), (8.97), (8.99) and (8.100) hold. Thus $(\mathscr{F}, \{q_T(f), p_T(f)|f \in \mathscr{W}_0\})$ is a Weyl representation of the CCR over $\mathscr{W}_0$.

(ii) Let $B \in \{e^{iq_T(f)}, e^{ip_T(f)}|f \in \mathscr{W}_0\}'$. Then, for all $f \in T^{-1}(\mathscr{W}_0)$ and $g \in T(\mathscr{W}_0)$, $Be^{i\phi(f)} = e^{i\phi(f)}B$ and $Be^{i\pi(g)} = e^{i\pi(g)}B$. Let $f \in \mathscr{W}$. Then, by the density of $T^{-1}(\mathscr{W}_0)$ in $\mathscr{W}$, there exists a sequence $\{f_n\}_{n=1}^{\infty}$ in $T^{-1}(\mathscr{W})$ such that $f_n \to f$ as $n \to \infty$. We have $Be^{i\phi(f_n)} = e^{i\phi(f_n)}B$. Taking the weak limit of both sides as $n \to \infty$, we obtain $Be^{i\phi(f)} = e^{i\phi(f)}B$. Similarly, one can show that $Be^{i\pi(f)} = e^{i\pi(f)}B$. Therefore $B \in \{e^{i\phi(f)}, e^{i\pi(f)}\}' = \mathbb{C}I$. Hence $B = \alpha I$ for some $\alpha \in \mathbb{C}$. Thus $\{e^{iq_T(f)}, e^{ip_T(f)}|f \in \mathscr{W}_0\}' = \mathbb{C}I$. ■

8.10 Translations of Fock Representation of Heisenberg CCR

Let $\pi_F(\mathscr{W})$ be the Fock representation of the Heisenberg CCR over $\mathscr{W}$ given by (8.88) and consider a translation of it:

$$\phi_\mu(f) := \phi_F(f) + \mu(f), \tag{8.106}$$

$$\pi_\nu(f) := \pi_F(f) + \nu(f), \quad f \in \mathscr{W}, \tag{8.107}$$

where $\mu,\ \nu : \mathscr{W} \to \mathbb{R}$ are real-valued linear functionals. By Proposition 8.9,

$$\pi_{\mu,\nu}(\mathscr{W}) := (\mathscr{F}_b(\mathscr{H}), \{\phi_\mu(f), \pi_\nu(f)|f \in \mathscr{W}\}) \tag{8.108}$$

is an irreducible Weyl representation of the CCR over $\mathscr{W}$. A natural question is: under what conditions is it equivalent or inequivalent to the Fock representation $\pi_F(\mathscr{W})$? The answer is given in Theorem 8.12 below.

We extend μ and ν to linear functionals on $\mathscr{W}_\mathbb{C}$ as follows: for $f = f_1 + if_2 \in \mathscr{W}_\mathbb{C}$ $(f_1, f_2 \in \mathscr{W})$,

$$\mu(f) := \mu(f_1) + i\mu(f_2), \quad \nu(f) := \nu(f_1) + i\nu(f_2). \tag{8.109}$$

Let

$$L(f) := \frac{1}{\sqrt{2}}(\mu(f) + i\nu(f)), \quad f \in \mathscr{W}_\mathbb{C}. \tag{8.110}$$

Theorem 8.12 *The representation $\pi_{\mu,\nu}(\mathscr{W})$ is equivalent to $\pi_F(\mathscr{W})$ if and only if there exists a vector $h \in \mathscr{H}$ such that $L(f) = \langle h, f\rangle$, $f \in \mathscr{W}_\mathbb{C}$.*

Proof Suppose that $\pi_{\mu,\nu}(\mathscr{W})$ is equivalent to $\pi_{\mathrm{F}}(\mathscr{W})$. Then there exists a unitary operator U on $\mathscr{F}_{\mathrm{b}}(\mathscr{H})$ such that, for all $f \in \mathscr{W}$, $U\phi_\mu(f)U^{-1} = \phi_{\mathrm{F}}(f)$ and $U\pi_\nu(f)U^{-1} = \pi_{\mathrm{F}}(f)$. Hence $U\phi_\mu(f) = \phi_{\mathrm{F}}(f)U$ and $U\pi_\nu(f) = \pi_{\mathrm{F}}(f)U$. In particular, we have $U(\phi_\mu(f) + i\pi_\nu(f)) = (\phi_{\mathrm{F}}(f) + i\pi_{\mathrm{F}}(f))U$. Let $\Phi, \Psi \in \mathscr{F}_{\mathrm{b},0}(\mathscr{H})$. Then

$$(\phi_\mu(f) + i\pi_\nu(f))\Psi = \sqrt{2}(A(f) + L(f)^*)\Psi,$$

$$(\phi_{\mathrm{F}}(f) + i\pi_{\mathrm{F}}(f))^*\Phi = \sqrt{2}A(f)^*\Phi.$$

Hence

$$\langle \Phi, U(A(f) + L(f)^*)\Psi \rangle = \langle A(f)^*\Phi, U\Psi \rangle.$$

Since $\mathscr{F}_{\mathrm{b},0}(\mathscr{H})$ is a core for $A(f)^*$, it follows that $U\Psi \in D((A(f)^*)^*) = D(A(f))$ and $A(f)U\Psi = U(A(f) + L(f)^*)\Psi$, i.e., $U^{-1}A(f)U\Psi = (A(f) + L(f)^*)\Psi$. It is easy to see that this extends to all $f \in \mathscr{W}_{\mathbb{C}}$. Recall that $\mathscr{F}_{\mathrm{b},0}(\mathscr{H})$ is a core for $A(g)$, $g \in \mathscr{H}$. Hence we obtain operator equality

$$U^{-1}A(f)U = A(f) + L(f)^* = C_L(f), \quad f \in \mathscr{W}_{\mathbb{C}}. \tag{8.111}$$

This means that the Bogoliubov translation $C_L(\cdot)$ by L is equivalent to the Fock representation of the CCR over $\mathscr{W}_{\mathbb{C}}$. Hence, by Theorem 8.5, there exists a vector $h \in \mathscr{H}$ such that $L(f) = \langle h, f \rangle$, $f \in \mathscr{W}_{\mathbb{C}}$.

Conversely, suppose that there exists a vector $h \in \mathscr{H}$ such that $L(f) = \langle h, f \rangle$, $f \in \mathscr{W}_{\mathbb{C}}$. Then one can define the unitary operator

$$U = e^{-i\sqrt{2}\Phi_{\mathrm{S}}(ih)}.$$

By a general formula, (8.111) holds. Hence, for all $f \in \mathscr{W}$,

$$U^{-1}\phi_{\mathrm{F}}(f)U = \phi_\mu(f), \quad U^{-1}\pi_{\mathrm{F}}(f)U = \pi_\nu(f),$$

on $D(A(f)) \cap D(A(f)^*)$. Since $\phi_\mu(f)$ and $\pi_\nu(f)$ are essentially self-adjoint on $D(A(f)) \cap D(A(f)^*)$, we obtain operator equalities

$$U^{-1}\phi_{\mathrm{F}}(f)U = \phi_\mu(f), \quad U^{-1}\pi_{\mathrm{F}}(f)U = \pi_\nu(f).$$

Thus $\pi_{\mu,\nu}(\mathscr{W})$ is equivalent to $\pi_{\mathrm{F}}(\mathscr{W})$. ∎

Corollary 8.7 *Suppose that L is not continuous on $\mathscr{W}_{\mathbb{C}}$ in the strong topology of $\mathscr{H}$. Then $\pi_{\mu,\nu}(\mathscr{W})$ is inequivalent to $\pi_{\mathrm{F}}(\mathscr{W})$.*

Proof By Riesz's representation theorem, there exists a vector $h \in \mathscr{H}$ such that $L(f) = \langle h, f \rangle$, $f \in \mathscr{W}_{\mathbb{C}}$ if and only if L is continuous on $\mathscr{W}_{\mathbb{C}}$ in the strong topology of $\mathscr{H}$. Hence, by the present assumption, there is no such vector h. Therefore the conclusion follows from Theorem 8.12. ∎

There are classes of non-continuous linear functionals on a dense subspace of $\mathscr{H}$. For example, let T be a densely defined closable *unbounded* linear operator on $\mathscr{H}$. Then $D(T^*) \neq \mathscr{H}$ (if $D(T^*) = \mathscr{H}$, then T^* is bounded by the closed graph theorem and hence $\overline{T} = (T^*)^*$ is bounded, a contradiction). Hence there exists a non-zero vector $g \notin D(T^*)$. Then the linear functional $L_{T,g}$ defined by

$$L_{T,g}(f) := \langle g, Tf \rangle, \quad f \in D(T)$$

is not continuous. (If $L_{T,g}$ is continuous, then, by Riesz's representation theorem, there exists a vector $h \in \mathscr{H}$ such that $L_{T,g}(f) = \langle h, f \rangle$, $f \in D(T)$. Hence $\langle g, Tf \rangle = \langle h, f \rangle$, $f \in D(T)$. This means that $g \in D(T^*)$ and $T^* g = h$. But this is a contradiction.) If $\mathscr{H}$ is infinite dimensional, then there are infinitely many densely defined closed unbounded operators. Thus we see that *there are infinitely many irreducible Weyl representations which are inequivalent to the Fock representation* $\pi_{\mathrm{F}}(\mathscr{W})$.

8.11 A Class of Irreducible Weyl Representations of CCR (I)

In this section we show that one can obtain a simple class of irreducible Weyl representations of the CCR over a real Hilbert space by deforming the Segal field operators.

Let $\mathscr{H}$ be a Hilbert space and J be a conjugation on $\mathscr{H}$. Suppose that there exist bounded operators S and T on $\mathscr{H}$ satisfying the following conditions: for all $f, g \in \mathscr{H}_J$,

$$\operatorname{Re} \langle Sf, Tg \rangle = \langle f, g \rangle, \tag{8.112}$$

$$\operatorname{Im} \langle Sf, Sg \rangle = 0, \quad \operatorname{Im} \langle Tf, Tg \rangle = 0. \tag{8.113}$$

We define the following self-adjoint operators on the boson Fock space $\mathscr{F}_{\mathrm{b}}(\mathscr{H})$ over $\mathscr{H}$:

$$\phi^{(S)}(f) := \Phi_{\mathrm{S}}(Sf), \tag{8.114}$$

$$\pi^{(T)}(f) := \Phi_{\mathrm{S}}(iTf), \quad f \in \mathscr{H}_J, \tag{8.115}$$

where $\Phi_{\mathrm{S}}(g)$ ($g \in \mathscr{H}$) is the Segal field operator with test vector g. Using (6.122), one can show that $\{e^{i\phi^{(S)}(f)}, e^{i\pi^{(T)}(f)} | f \in \mathscr{H}_J\}$ obeys the Weyl relations over $\mathscr{H}_J$. Hence we have:

Lemma 8.12 *The pair $(\mathscr{F}_{\mathrm{b}}(\mathscr{H}), \{\phi^{(S)}(f), \pi^{(T)}(f) | f \in \mathscr{H}_J\})$ is a Weyl representation of the CCR over $\mathscr{H}_J$.*

One needs an additional condition for $\{e^{\phi^{(S)}(f)}, e^{i\pi^{(T)}(f)}|f \in \mathscr{H}_J\}$ to be irreducible:

Theorem 8.13 *Let $\mathscr{W}$ be a dense subspace of $\mathscr{H}_J$. Suppose that $S^* - iT^*$ is injective. Then $(\mathscr{F}_b(\mathscr{H}), \{\phi^{(S)}(f), \pi^{(T)}(f)|f \in \mathscr{W}\})$ is an irreducible Weyl representation of the CCR over $\mathscr{W}$.*

Proof We need only to show that $\mathfrak{A} := \{e^{i\phi^{(S)}(f)}, e^{i\pi^{(T)}(f)}|f \in \mathscr{W}\}$ is irreducible. To prove this, let $B \in \mathfrak{A}'$. Then, for all $f, g \in \mathscr{W}$,

$$Be^{i\Phi_S(Sf)} = e^{i\Phi_S(Sf)}B, \quad Be^{i\Phi_S(iTf)} = e^{i\Phi_S(iTf)}B.$$

Let $K := S + iT$. Then, by (6.123), we have

$$e^{i\Phi_S(Kf)} = ce^{i\Phi_S(Sf)}e^{i\Phi_S(iTf)}, \quad f \in \mathscr{W},$$

where $c := e^{i\|f\|^2/2}$. Hence $Be^{i\Phi_S(Kf)} = e^{i\Phi_S(Kf)}B$ for all $f \in \mathscr{W}$. Since $K^* = S^* - iT^*$ is injective by the present assumption, $\mathrm{Ran}\,((K^*)^*) = \mathrm{Ran}\,(K)$ is dense in $\mathscr{H}$ by Corollary 1.1. Hence it follows that $K\mathscr{H}_J$ is dense in $\mathscr{H}$ and hence $K\mathscr{W}$ is dense in $\mathscr{H}$. Hence, by Theorem 6.22(iv), $\{e^{i\Phi_S(Kf)}|f \in \mathscr{W}\}$ is irreducible. It is obvious that the set $\{e^{i\Phi_S(Kf)}|f \in \mathscr{W}\}$ is $*$-invariant. Therefore, by Proposition 2.9, $\{e^{i\Phi_S(Kf)}|f \in \mathscr{W}\}' = \mathbb{C}I$. Hence $B = \alpha I$ for some constant $\alpha \in \mathbb{C}$. Thus $\{e^{i\phi^{(S)}(f)}, e^{i\pi^{(T)}(f)}|f \in \mathscr{W}\}$ is irreducible. ■

We note the following fact:

Lemma 8.13 *Equation (8.112) implies that $S + T$ is injective.*

Proof Let $f \in \ker(S+T)$, so that $Sf = -Tf$. Hence $\langle Sf, Tf\rangle = -\|Tf\|^2$. Hence, by (8.112), $\|f\|^2 = -\|Tf\|^2 \le 0$. Therefore $\|f\|^2 = 0$. Thus $f = 0$. ■

Concerning equivalence of $(\mathscr{F}_b(\mathscr{H}), \phi^{(S)}(f), \pi^{(T)}(f)|f \in \mathscr{H}_J\})$ to the Fock representation $\pi_F(\mathscr{H}_J)$, one has the following theorem:

Theorem 8.14 *Assume the following:*

(a) $\mathrm{Ran}\,(S + T)$ *is dense.*
(b) $S - T$ *is not Hilbert–Schmidt.*

Let $\mathscr{W}$ be a dense subspace of $\mathscr{H}_J$. Then $(\mathscr{F}_b(\mathscr{H}), \{\phi^{(S)}(f), \pi^{(T)}(f)|f \in \mathscr{W}\})$ is inequivalent to the Fock representation $\pi_F(\mathscr{W})$.

Proof Suppose that the representation $(\mathscr{F}_b(\mathscr{H}), \{\phi^{(S)}(f), \pi^{(T)}(f)|f \in \mathscr{W}\})$ were equivalent to $\pi_F(\mathscr{W})$. Then there exists a unitary operator U on $\mathscr{F}_b(\mathscr{H})$ such that

$$\phi_F(f) = U\Phi_S(Sf)U^{-1}, \quad \pi_F(f) = U\Phi_S(iTf)U^{-1}, \quad f \in \mathscr{W}. \tag{8.116}$$

Let $\Omega := U^{-1}\Omega_{\mathscr{H}}$. Then it follows from (8.116) and (6.39) that

$$\frac{1}{\sqrt{2}}A(f)^*\Omega_{\mathscr{H}} = U\Phi_S(Sf)\Omega, \quad \frac{i}{\sqrt{2}}A(f)^*\Omega_{\mathscr{H}} = U\Phi_S(iTf)\Omega, \ f \in \mathscr{W}.$$

Hence $\Phi_S(Sf)\Omega = -i\Phi_S(iTf)\Omega$. This implies that, for all $\Psi \in \mathscr{F}_{b,0}(\mathscr{H})$,

$$\left\langle A((S+T)f)^*\Psi, \Omega\right\rangle = -\left\langle A((S-T)f)\Psi, \Omega\right\rangle, \quad f \in \mathscr{W}.$$

This can be extended as

$$\left\langle A((S+T)f)^*\Psi, \Omega\right\rangle = -\left\langle A((S-T)Jf)\Psi, \Omega\right\rangle, \quad f \in \mathscr{W}_{\mathbb{C}}.$$

Let $\mathscr{D} := (S+T)\mathscr{W}_{\mathbb{C}}$, which is dense in $\mathscr{H}$ by condition (a). Then, for all $g \in \mathscr{D}$,

$$\left\langle A(g)^*\Psi, \Omega\right\rangle = -\left\langle A(J(S-T)_J(S+T)^{-1}g)\Psi, \Omega\right\rangle.$$

Hence, by Proposition 8.4, $W := \overline{(S-T)_J(S+T)^{-1}}$ is Hilbert–Schmidt. Therefore $W(S+T)$ also is Hilbert–Schmidt. But $W(S+T) = (S-T)_J$. Hence it follows that $S-T$ is Hilbert–Schmidt. But this contradicts (b). ■

Remark 8.22 Let

$$B_0(f) := \frac{1}{\sqrt{2}}(\phi^{(S)}(f) + i\pi^{(T)}(f)) \restriction \mathscr{F}_{b,0}(\mathscr{H}), \quad f \in \mathscr{H}_J,$$

and $B_0(f) := B_0(f_1) - iB_0(f_2)$ for $f = f_1 + if_2$, $f_1, f_2 \in \mathscr{H}_J$. Since $D(B_0(f)^*)$ is dense, $B_0(f)$ is closable; we denote the closure of it by $B(f)$. Then the Heisenberg CCR of $\{\phi^{(S)}(f), \pi^{(T)}(f) | f \in \mathscr{H}_J\}$ imply that, for all $f, g \in \mathscr{H}$,

$$[B(f), B(g)^*] = \langle f, g\rangle, \quad [B(f), B(g)] = 0 \quad \text{on } \mathscr{F}_{b,0}(\mathscr{H}).$$

Hence $(\mathscr{F}_b(\mathscr{H}), \mathscr{F}_{b,0}(\mathscr{H}), \{B(f), B(f)^* | f \in \mathscr{H}\})$ is a representation of the CCR over $\mathscr{H}$. On the other hand, one has

$$B(f) = \frac{1}{2}(A((S+T)f) + A((S-T)Jf)^*) \quad \text{on } \mathscr{F}_{b,0}(\mathscr{H}).$$

Hence $(A(\cdot), A(\cdot)^*) \mapsto (B(\cdot), B(\cdot)^*)$ is a Bogoliubov transformation. Under additional conditions (see Theorem 8.7), one can prove that, if $S-T$ is Hilbert–Schmidt, then $(\mathscr{F}_b(\mathscr{H}), \mathscr{F}_{b,0}(\mathscr{H}), \{B(f), B(f)^* | f \in \mathscr{H}\})$ is equivalent to the Fock representation $\rho_F(\mathscr{H})$, implying that $(\mathscr{F}_b(\mathscr{H}), \{\phi^{(S)}(f), \pi^{(T)}(f) | f \in \mathscr{H}_J\})$ is equivalent to $\pi_F(\mathscr{H}_J)$.

8.12 A Class of Irreducible Weyl Representations of CCR (II)

Let T be a symmetric operator on $\mathscr{H}$ (not necessarily bounded) such that

$$JT \subset TJ. \tag{8.117}$$

Then T is reduced by $\mathscr{H}_J$ in the sense that, for all $f \in D(T)$, the real part $\mathrm{Re}\, f$ of f (see (1.69)) is in $\mathscr{H}_J \cap D(T)$ and $\mathrm{Re}\,(Tf) = T\mathrm{Re}\, f$. In particular, $D(T) \cap \mathscr{H}_J$ is dense in $\mathscr{H}_J$ and $T(D(T) \cap \mathscr{H}_J) \subset \mathscr{H}_J$.

Lemma 8.14 *Suppose that T is injective. Then $JT^{-1} \subset T^{-1}J$.*

Proof Let $f \in D(T^{-1}) = \mathrm{Ran}\, T$. Then $g := T^{-1}f$ is in $D(T)$. Hence Jg is in $D(T)$ and $TJg = JTg = Jf$. Hence $Jf \in \mathrm{Ran}\, T = D(T^{-1})$ and $Jg = T^{-1}Jf$. We have $Jg = JT^{-1}f$. Thus $JT^{-1} \subset T^{-1}J$. ■

For an injective symmetric operator T on $\mathscr{H}$ satisfying (8.117), we define self-adjoint operators

$$\phi_T(f) := \phi_{\mathrm{F}}(T^{-1}f), \quad f \in D(T^{-1}) \cap \mathscr{H}_J, \tag{8.118}$$

$$\pi_T(g) := \pi_{\mathrm{F}}(Tg), \quad g \in D(T) \cap \mathscr{H}_J, \tag{8.119}$$

acting in $\mathscr{F}_{\mathrm{b}}(\mathscr{H})$.

Let $\mathscr{W}$ be a dense subspace of $\mathscr{H}_J$ and $\mathfrak{S}_{\mathscr{W}}(\mathscr{H})$ be the set of injective symmetric operators T on $\mathscr{H}$ satisfying the following conditions:

(i) (8.117) holds.
(ii) $\mathscr{W} \subset D(T) \cap D(T^{-1})$ and $T^{\pm 1}\mathscr{W}$ are dense in $\mathscr{H}_J$.

For each $T \in \mathfrak{S}_{\mathscr{W}}(\mathscr{H})$, we define a triple

$$\Pi_T(\mathscr{W}) := (\mathscr{F}_{\mathrm{b}}(\mathscr{H}), \mathscr{F}_{\mathrm{b},0}(\mathscr{H}), \{\phi_T(f), \pi_T(f) | f \in \mathscr{W}\}). \tag{8.120}$$

Proposition 8.11 *The triple $\Pi_T(\mathscr{W})$ is a representation of the Heisenberg CCR over $\mathscr{W}$.*

Proof For all $f, g \in \mathscr{W}$, the CCR for $\phi_T(f)$ and $\pi_T(g)$ on $\mathscr{F}_{\mathrm{b},0}(\mathscr{H})$ follow from (6.53) and the elementary fact that $\langle T^{-1}f, Tg\rangle = \langle f, g\rangle$. ■

Theorem 8.15 *The representation $\Pi_T(\mathscr{W})$ is an irreducible Weyl representation of the CCR over $\mathscr{W}$.*

Proof This follows from an application of Proposition 8.10. ■

It is interesting to ask under what condition $\Pi_{T_1}(\mathscr{W})$ is equivalent or inequivalent to $\Pi_{T_2}(\mathscr{W})$ for $T_1,\ T_2 \in \mathfrak{S}_{\mathscr{W}}(\mathscr{H})$ with $T_1 \neq T_2$. To answer this question, we need some preliminaries.

For $T_1, T_2 \in \mathfrak{S}_{\mathscr{W}}(\mathscr{H})$ such that $D(T_2^{-1}T_1) \cap D(T_2T_1^{-1}) \cap D(T_1^{-1}T_2) \cap D(T_1T_2^{-1})$ is dense in $\mathscr{H}$, we define

$$S_{\pm} := \frac{1}{2}(T_2^{-1}T_1 \pm T_2T_1^{-1}). \tag{8.121}$$

Note that

$$\mathscr{D}_{21} := D(S_+) = D(S_-) = D(T_2T_1^{-1}) \cap D(T_2^{-1}T_1).$$

Hence $S_{\pm}$ are densely defined and we have

$$S_{\pm}^* \supset \frac{1}{2}(T_1T_2^{-1} \pm T_1^{-1}T_2).$$

Hence $S_{\pm}^*$ are densely defined. Therefore $S_{\pm}$ are closable. We set

$$\mathscr{D}_{12} := D(T_1T_2^{-1}) \cap D(T_1^{-1}T_2) \subset D(S_{\pm}^*).$$

It follows that

$$\mathscr{D}_{12} \cap \mathscr{D}_{21} = D(T_1^{-1}T_2) \cap D(T_1T_2^{-1}) \cap D(T_2^{-1}T_1) \cap D(T_2T_1^{-1}). \tag{8.122}$$

Lemma 8.15 *The following hold:*

(i) *For all* $f, g \in D_{21}$,

$$\langle S_+ f, S_+ g\rangle - \langle S_- f, S_- g\rangle = \langle f, g\rangle, \tag{8.123}$$

$$\langle S_- f, S_+ g\rangle = \langle S_+ f, S_-\rangle \tag{8.124}$$

(ii) *For all* $f, g \in \mathscr{D}_{12}$,

$$\left\langle S_+^* f, S_+^* g\right\rangle - \left\langle S_-^* f, S_-^* g\right\rangle = \langle f, g\rangle, \tag{8.125}$$

$$\left\langle S_-^* f, S_+^* g\right\rangle = \left\langle S_+^* f, S_-^* g\right\rangle. \tag{8.126}$$

Proof

(i) Let $f, g \in \mathscr{D}_{21}$. Then

$$\langle S_+ f, S_+ g\rangle = \frac{1}{4}(\left\langle T_2^{-1}T_1 f, T_2^{-1}T_1\right\rangle + 2\langle f, g\rangle + \left\langle T_2T_1^{-1} f, T_2T_1^{-1} g\right\rangle),$$

$$\langle S_- f, S_- g\rangle = \frac{1}{4}(\left\langle T_2^{-1}T_1 f, T_2^{-1}T_1\right\rangle - 2\langle f, g\rangle + \left\langle T_2T_1^{-1} f, T_2T_1^{-1} g\right\rangle).$$

Hence (8.123) holds.. Similarly, one can prove (8.124).

(ii) This also can be proved by direct computations as in (i). ∎

Relations (8.123)–(8.126) remind us of the theory of bosonic Bogoliubov transformations (see Remark 8.9). Thus we are led to proceed as follows.

We fix $T_1, T_2 \in \mathfrak{S}_{\mathscr{W}}(\mathscr{H})$ and set

$$\mathscr{V} := \mathscr{D}_{21}. \tag{8.127}$$

For each $f \in \mathscr{V}$, we introduce a new operator $B(f)$ by

$$B(f) := \overline{A(\overline{S}_+ f) + A(J\overline{S}_- f)^*}. \tag{8.128}$$

We have

$$B(f)^* \supset A(\overline{S}_+ f)^* + A(J\overline{S}_- f). \tag{8.129}$$

Note that $\{B(f), B(f)^* | f \in \mathscr{V}\}$ satisfy

$$[B(f), B(g)^*] = \langle f, g \rangle ,$$

$$[B(f), B(g)] = 0, \quad [B(f)^*, B(g)^*] = 0, \quad f, g \in \mathscr{V},$$

on $\mathscr{F}_{\mathrm{b},0}(\mathscr{H})$. Hence $(\mathscr{F}_{\mathrm{b}}(\mathscr{H}), \{B(f), B(f)^* | f \in \mathscr{V}\})$ is a representation of the CCR over $\mathscr{V}$.

Remark 8.23 If S_+ or S_- is unbounded (then both S_+ and S_- are unbounded), then the correspondence $(A(\cdot), A(\cdot)^*) \mapsto (B(\cdot), B(\cdot)^*)$ is a singular bosonic Bogoliubov transformation in the sense defined in Remark 8.9. An analysis of this singular Bogoliubov transformation is given in [30].

In the present book, we discuss only the case where $S_\pm$ are bounded. Thus, in what follows, we assume the following:

(T) The operators $T_2^{-1} T_1$ and $T_2 T_1^{-1}$ are bounded.

Under this assumption, $S_\pm$ are bounded. Hence, for all $f \in \mathscr{H}$, $B(f)$ is defined by (8.128). Moreover, (8.123)–(8.126) imply the following operator equations:

$$\overline{S}_+^* \overline{S}_+ - \overline{S}_-^* \overline{S}_- = I, \tag{8.130}$$

$$\overline{S}_-^* \overline{S}_+ = \overline{S}_+^* \overline{S}_-, \tag{8.131}$$

$$\overline{S}_+ \overline{S}_+^* - \overline{S}_- \overline{S}_-^* = I, \tag{8.132}$$

$$\overline{S}_- \overline{S}_+^* = \overline{S}_+ \overline{S}_-^*. \tag{8.133}$$

Lemma 8.16 *Let $\mathscr{H}$ be separable and assume (T). Then there exists a unitary operator U on $\mathscr{F}_{\mathrm{b}}(\mathscr{H})$ such that*

$$B(f) = UA(f)U^{-1}, \quad f \in \mathscr{H} \tag{8.134}$$

if and only if $\overline{S}_-$ is Hilbert–Schmidt.

Proof This follows from an application of Corollary 8.5. ■

Theorem 8.16 *Let $\mathscr{H}$ be separable and assume (T). Suppose that $\overline{S}_-$ is Hilbert–Schmidt. Then there exists a unitary operator U on $\mathscr{F}_{\mathrm{b}}(\mathscr{H})$ such that operator equations*

$$U\phi_{T_1}(f)U^{-1} = \phi_{T_2}(f), \tag{8.135}$$

$$U\pi_{T_1}(f)U^{-1} = \pi_{T_2}(f), \quad f \in \mathscr{W}, \tag{8.136}$$

hold. Namely, $\Pi_{T_1}(\mathscr{W}) \cong \Pi_{T_2}(\mathscr{W})$.

Proof By Lemma 8.16, there exists a unitary operator U on $\mathscr{F}_{\mathrm{b}}(\mathscr{H})$ satisfying (8.134). This implies also

$$B(f)^* = UA(f)^*U^{-1}, \quad f \in \mathscr{H}. \tag{8.137}$$

Hence, for all $f \in \mathscr{W}$, we have

$$B(T_1^{\pm 1}f) = UA(T_1^{\pm 1}f)U^{-1}, \quad B(T_1^{\pm 1}f)^* = UA(T_1^{\pm 1}f)U^{-1}.$$

Hence

$$\frac{1}{\sqrt{2}}(B(T_1^{-1}f)^* + B(T_1^{-1}f)) = U\phi_{T_1}(f)U^{-1},$$

$$\frac{i}{\sqrt{2}}(B(T_1f)^* - B(T_1f)) = U\pi_{T_1}(f)U^{-1}.$$

By (8.128) and (8.129), we have

$$B(T_1^{-1}f)^* + B(T_1^{-1}f) = A(\overline{S}_+T_1^{-1}f + \overline{S}_-T_1^{-1}f)^* + A(\overline{S}_-T_1^{-1}f + \overline{S}_+T_1^{-1}f),$$

$$B(T_1f)^* - B(T_1f) = A(\overline{S}_+T_1f - \overline{S}_-T_1f)^* + A(\overline{S}_-T_1f - \overline{S}_+T_1f).$$

on $\mathscr{F}_{\mathrm{b},0}(\mathscr{H})$. For all $g \in \mathscr{D}_{12}$, one can show that

$$\left\langle g, \overline{S}_+T_1^{-1}f + \overline{S}_-T_1^{-1}f\right\rangle = \left\langle g, T_2^{-1}f\right\rangle,$$

$$\left\langle g, \overline{S}_+T_1f - \overline{S}_-T_1f\right\rangle = \langle g, T_2f\rangle.$$

Since $\mathscr{D}_{12}$ is dense by the original assumption, it follows that

$$\overline{S}_+T_1^{-1}f + \overline{S}_-T_1^{-1}f = T_2^{-1}f, \quad \overline{S}_+T_1f - \overline{S}_-T_1f = T_2f.$$

Hence

$$B(T_1^{-1}f)^* + B(T_1^{-1}f) = A(T_2^{-1}f)^* + A(T_2^{-1}f),$$
$$B(T_1f)^* - B(T_1f) = A(T_2f)^* - A(T_2f)$$

on $\mathscr{F}_{\mathrm{b},0}(\mathscr{H})$. Therefore

$$\phi_{T_2}(f) = U\phi_{T_1}(f)U^{-1}, \ \pi_{T_2}(f) = U\pi_{T_1}(f)U^{-1} \quad \text{on } \mathscr{F}_{\mathrm{b},0}(\mathscr{H}). \tag{8.138}$$

This implies that

$$\phi_{T_2}(f) \restriction \mathscr{F}_{\mathrm{b},0}(\mathscr{H}) \subset U\phi_{T_1}(f)U^{-1}, \ \pi_{T_2}(f) \restriction \mathscr{F}_{\mathrm{b},0}(\mathscr{H}) \subset U\pi_{T_1}(f)U^{-1}. \tag{8.139}$$

Recall that $\mathscr{F}_{\mathrm{b},0}(\mathscr{H})$ is a core for $\phi_{T_2}(f)$ and $\pi_{T_2}(f)$. Moreover, $U\phi_{T_1}(f)U^{-1}$ and $U\pi_{T_1}(f)U^{-1}$ are self-adjoint. Hence (8.139) yields operator equalities (8.135) and (8.136). ■

Theorem 8.17 *Let $\mathscr{H}$ be separable and assume (T). Suppose that there exists a unitary operator U on $\mathscr{F}_{\mathrm{b}}(\mathscr{H})$ satisfying (8.135) and (8.136). Then $\overline{S}_-$ is Hilbert–Schmidt.*

Proof It is easy to see that, for all $\Psi \in \mathscr{F}_{\mathrm{b},0}(\mathscr{H})$ and $f \in \mathscr{W}$,

$$A(f)\Psi = \frac{1}{\sqrt{2}}(\phi_{T_1}(T_1f) + i\pi_{T_1}(T_1^{-1}f))\Psi.$$

Hence

$$A(f)\Psi = \frac{1}{\sqrt{2}}(U^{-1}\phi_{T_2}(T_1f)U\Psi + iU^{-1}\pi_{T_2}(T_1^{-1}f)U\Psi),$$

which implies that

$$UA(f)\Psi = \frac{1}{\sqrt{2}}(\phi_{T_2}(T_1f)U\Psi + i\pi_{T_2}(T_1^{-1}f)U\Psi).$$

Hence, for all $\Phi \in \mathscr{F}_{\mathrm{b},0}(\mathscr{H})$,

$$\left\langle U^{-1}\Phi, A(f)\Psi \right\rangle = \left\langle U^{-1}(A(S_+f)^*\Phi + A(S_-f)\Phi), \Psi \right\rangle.$$

Since $\mathscr{F}_{\mathrm{b},0}(\mathscr{H})$ is a core for $A(f)$, it follows that $U^{-1}\Phi \in D(A(f)^*)$ and

$$A(f)^*U^{-1}\Phi = U^{-1}(A(S_+f)^* + A(S_-f))\Phi.$$

By the density of $\mathscr{W}$ in $\mathscr{H}_J$ and the strong continuity of the mapping $\mathscr{H} \ni g \mapsto A(g)^{\#}\Phi$, this equation can be extended to all $f \in \mathscr{H}_J$ with $S_\pm$ replaced by $\overline{S}_\pm$:

$$A(f)^* U^{-1}\Phi = U^{-1}(A(\overline{S}_+ f)^* + A(\overline{S}_- f))\Phi, \quad f \in \mathscr{H}_J.$$

By (6.39), we have $\left\langle A(f)^* U^{-1}\Phi, \Omega_{\mathscr{H}} \right\rangle = 0$. Hence, putting $\Omega := U\Omega_{\mathscr{H}}$, we obtain

$$\left\langle A(\overline{S}_+ f)^*\Phi, \Omega \right\rangle = -\left\langle A(\overline{S}_- f)\Phi, \Omega \right\rangle, \quad f \in \mathscr{H}_J. \tag{8.140}$$

By (8.130), we have

$$\overline{S}_+^* \overline{S}_+ = 1 + \overline{S}_-^* \overline{S}_-.$$

This implies that $\overline{S}_+^* \overline{S}_+$ is bijective on $\mathscr{H}_J$ (note that $\overline{S}_\pm$ are reduced by $\mathscr{H}_J$). In particular, $\overline{S}_+$ is injective on $\mathscr{H}_J$. On the other hand, (8.132) gives

$$\overline{S}_+ \overline{S}_+^* = 1 + \overline{S}_- \overline{S}_-^*.$$

This implies that $\overline{S}_+ \overline{S}_+^*$ is bijective on $\mathscr{H}_J$. In particular, $\overline{S}_+$ is surjective on $\mathscr{H}_J$. Hence $\overline{S}_+$ is bijective on $\mathscr{H}_J$. Hence, putting $K := \overline{S}_- \overline{S}_+^{-1}$, we have for all $f \in \mathscr{H}_J$, $\langle A(f)^*\Phi, \Omega \rangle = -\langle A(Kf)\Phi, \Omega \rangle$. This equation can be extended to all $f \in \mathscr{H}$:

$$\left\langle A(f)^*\Phi, \Omega \right\rangle = -\langle A(JKf)\Phi, \Omega \rangle, \quad f \in \mathscr{H}.$$

Hence, by Proposition 8.4, K is Hilbert–Schmidt. Thus $\overline{S}_- = K\overline{S}_+$ is Hilbert–Schmidt. ■

In summary, we have proved the following theorem:

Theorem 8.18 *Let $\mathscr{H}$ be separable. Let $T_1, T_2 \in \mathfrak{S}_{\mathscr{W}}(\mathscr{H})$ and assume (T). Then $\Pi_{T_1}(\mathscr{W}) \cong \Pi_{T_2}(\mathscr{W})$ if and only if $\overline{S}_-$ is Hilbert–Schmidt.*

In view of finding inequivalent irreducible Weyl representations of CCR, it may be convenient to rephrase this theorem as follows:

Corollary 8.8 *Let $\mathscr{H}$ is separable. Let $T_1, T_2 \in \mathfrak{S}_{\mathscr{W}}(\mathscr{H})$ and assume (T). Then $\Pi_{T_1}(\mathscr{W})$ is inequivalent to $\Pi_{T_2}(\mathscr{W})$ if and only if $\overline{S}_-$ is not Hilbert–Schmidt.*

Remark 8.24 The conditions for T_1 and T_2 in Theorem 8.18 are related to an equivalence relation in a subset of $\mathfrak{S}_{\mathscr{W}}(\mathscr{H})$. Let

$$\mathfrak{S}_{\mathscr{W}}^{\times} := \{T \in \mathfrak{S}_{\mathscr{W}}(\mathscr{H}) | T \text{ is surjective}\}.$$

Then, for all $T \in \mathfrak{S}_{\mathscr{W}}^{\times}$, $T^{-1} \in \mathfrak{B}(\mathscr{H})$. For $T_1, T_2 \in \mathfrak{S}_{\mathscr{W}}$, we write $T_1 \sim T_2$ if the following conditions are satisfied: (1) $T_2 T_1^{-1}$ and $T_2^{-1} T_1$ are bounded; (2)

$\overline{T_2T_1^{-1} - T_2^{-1}T_1}$ is Hilbert–Schmidt. It is easy to see that the relation $\sim$ is an equivalence relation in $\mathfrak{S}_{\mathscr{W}}^{\times}$.[6] Hence Theorem 8.18 implies the following: Let $T_1, T_2 \in \mathfrak{S}_{\mathscr{W}}^{\times}$. Then, if $T_1 \sim T_2$, then $\Pi_{T_1}(\mathscr{W}) \cong \Pi_{T_2}(\mathscr{W})$. Conversely, if $\Pi_{T_1}(\mathscr{W}) \cong \Pi_{T_2}(\mathscr{W})$ for $T_1, T_2 \in \mathfrak{S}_{\mathscr{W}}^{\times}$ with $T_2T_1^{-1}$ and $T_2^{-1}T_1$ being bounded, then $T_1 \sim T_2$. This may be an interesting structure.

Remark 8.25 In the case where at least one of $T_2^{-1}T_1$ and $T_2T_1^{-1}$ is unbounded, we need a separate consideration (see [24, Theorem 5.6]). Such an example appears in relation to Casimir effect in quantum electrodynamics [29, 30].

A simple consequence of Theorem 8.18 is as follows.

Corollary 8.9 (Shale [146]) *Let S be a bijective bounded self-adjoint operators on $\mathscr{H}$ such that $SJ = JS$. Then the Weyl representation $(\mathscr{F}_{\rm b}(\mathscr{H}), \{\phi_S(f), \pi_S(f) | f \in \mathscr{H}_J\})$ is equivalent to the Fock representation $\rho_{\rm F}(\mathscr{H})$ if and only if $S^2 - I$ is Hilbert–Schmidt.*

Proof We need only to apply Theorem 8.18 to the case where $T_1 = S$ and $T_2 = I$. In this case, we have

$$\overline{S_-} = S_- = \frac{1}{2}\left(S - S^{-1}\right) = \frac{1}{2}S^{-1}(S^2 - I).$$

Hence $\overline{S_-}$ is Hilbert–Schmidt if and only $S^2 - I$ is Hilbert–Schmidt. Thus the desired result follows. ∎

Example 8.13 A simple example of (T_1, T_2) such that $\overline{S_-}$ is not Hilbert–Schmidt is given as follows. Let $T \in \mathfrak{S}_{\mathscr{W}}(\mathscr{H})$ and

$$T_1 := T, \quad T_2 = \alpha T$$

with $\alpha \in \mathbb{R} \setminus \{0, \pm 1\}$. Then, $T_1, T_2 \in \mathfrak{S}_{\mathscr{W}}(\mathscr{H})$ and we have

$$\overline{S_-} = \frac{1}{2}\left(\frac{1}{\alpha} - \alpha\right) \neq 0.$$

It is easy to see that, in the present example, condition (T) is trivially is satisfied. Suppose that $\mathscr{H}$ is infinite dimensional. Then any non-zero scalar operator on $\mathscr{H}$ is not Hilbert–Schmidt. Hence $\overline{S_-}$ is not Hilbert–Schmidt. Thus, by Corollary 8.8, $\Pi_T(\mathscr{W})$ is inequivalent to $\Pi_{\alpha T}(\mathscr{W})$. Note that, for this inequivalence, the infinite dimensionality of $\mathscr{H}$, which makes $\Pi_T(\mathscr{W})$ and $\Pi_{\alpha T}(\mathscr{W})$ to be representations of CCR with *infinite degrees of freedom*, is essential.

[6] Condition that $T^{-1} \in \mathfrak{B}(\mathscr{H})$ ($T \in \mathfrak{S}_{\mathscr{W}}^{\times}$) is needed for $\sim$ to be transitive.

8.13 Functional Schrödinger Representation

In concluding this chapter, we briefly mention a representation of the Heisenberg CCR over a real Hilbert space which can be regarded as an infinite-dimensional version of Schrödinger representations of CCR, although it will be not used in the rest of this book (but it is important to know the existence of the representation). For this purpose, we first recall a basic existence theorem on a random process over a real Hilbert space:

Theorem 8.19 *Let $\mathscr{H}$ be a separable real Hilbert space. Then there exist a probability space (Q, Σ, μ) and a family $\{\phi(f)|f \in \mathscr{H}\}$ of random variables on (Q, Σ) (real-valued Σ-measurable functions on Q which are finite μ-a.e.) such that the following hold:*

(i) *(real linearity in test vectors) For all $f, g \in \mathscr{H}$ and $a, b \in \mathbb{R}$,*

$$\phi(af + bg) = a\phi(f) + b\phi(g), \quad \mu\text{-a.e.}$$

(ii) *(fullness of the random variables) The σ-field Σ is the smallest σ-field such that, for all $f \in \mathscr{H}$, $\phi(f)$ is measurable.*

(iii) *(Gaussian measure)*

$$\int_Q e^{i\phi(f)} d\mu = e^{-\|f\|^2_{\mathscr{H}}/4}, \quad f \in \mathscr{H}.$$

For a proof of this theorem, see, e.g., [147, §1.1], [148, Theorem 2.3.4], [72, §3.2], [105, §5.4] or [21, §2.4].

Remark 8.26 Theorem 8.19 is concerned with existence of infinite-dimensional Gaussian measures. To explain an idea behind Theorem 8.19, let $\dot{\mathbb{R}}$ be the one-point compactification of $\mathbb{R}$ and

$$Q := \prod_{n=1}^{\infty} \dot{\mathbb{R}} = \{q = (q_n)_{n=1}^{\infty} | n \in \mathbb{N},\ q_n \in \dot{\mathbb{R}}\},$$

the infinite direct product of $\dot{\mathbb{R}}$, and Σ be the σ-algebra generated by the coordinate functions $c_n\colon Q \mapsto \dot{\mathbb{R}}$; $c_n(q) := q_n \in \dot{\mathbb{R}}$, $q \in Q$, $n \in \mathbb{N}$. Let μ_{G} be the Gaussian measure on $(\mathbb{R}, \boldsymbol{B}^1)$ given by

$$\mu_{\mathrm{G}}(B) = \frac{1}{\sqrt{\pi}} \int_B e^{-x^2} dx, \quad B \in \boldsymbol{B}^1.$$

Then the measure μ in Theorem 8.19 in the present context is the infinite product measure of μ_{G}: $\mu = \otimes_{n=1}^{\infty} \mu_{\mathrm{G}}$ (symbolically written $d\mu(q) = \prod_{n=1}^{\infty} d\mu_{\mathrm{G}}(q_n)$); the existence of it can be proved by applying Kolmogorov's theorem (e.g., [148,

Theorem 2.1]). This is an infinite-dimensional Gaussian measure. Let $\{e_n\}_{n=1}^{\infty}$ be a CONS of $\mathscr{H}$. Then each $f \in \mathscr{H}$ is expanded as

$$f = \sum_{n=1}^{\infty} \langle e_n f \rangle e_n.$$

One can show that

$$\phi_N(f) := \sum_{n=1}^{N} \langle e_n, f \rangle c_n \quad (N \in \mathbb{N})$$

converges in $L^2(Q, d\mu)$ as $N \to \infty$. Hence

$$\phi(f) := \lim_{N \to \infty} \phi_N(f)$$

in the L^2-sense is an element of $L^2(Q, d\mu)$. Note that

$$\phi(e_n) = c_n. \tag{8.141}$$

It is easy to see that

$$\int_Q e^{i\phi_N(f)} d\mu = e^{-\sum_{n=1}^{N} \langle e_n, f \rangle^2 / 4}, \quad f \in \mathscr{H}.$$

The right hand side converges to $e^{-\|f\|^2/4}$ as $N \to \infty$. Hence it follows that

$$\int_Q e^{i\phi(f)} d\mu = e^{-\|f\|^2/4}, \quad f \in \mathscr{H}.$$

In this way, one obtains a model for $((Q, \Sigma, \mu), \{\phi(f) | f \in \mathscr{H}\})$. We remark also that there are other models for $((Q, \Sigma, \mu), \{\phi(f) | f \in \mathscr{H}\})$.

The family $\{\phi(f) | f \in \mathscr{H}\}$ of random variables in Theorem 8.19 is called the **Gaussian random process** indexed by $\mathscr{H}$ with probability space (Q, Σ, μ). In the case where $\mathscr{H}$ is infinite dimensional, the measure μ is called an **infinite-dimensional Gaussian measure** on (Q, Σ). It follows that, for all $n \in \mathbb{N}$ and $f_j \in \mathscr{H}$, $j = 1, \ldots, n$,

$$\phi(f_1) \cdots \phi(f_n) \in L^2(Q, d\mu)$$

(for a proof of this fact, see references cited above).

The following theorem is significant (for a proof, see references cited above):

Theorem 8.20 *There exists a unitary operator U from the boson Fock space $\mathscr{F}_{\mathrm{b}}(\mathscr{H}_{\mathbb{C}})$ over $\mathscr{H}_{\mathbb{C}}$ to $L^2(Q, d\mu)$ such that the following hold:*

(i) $U\Omega_{\mathscr{H}} = 1 \in L^2(Q, d\mu)$.
(ii) $U\mathscr{F}_{\mathrm{b,fin}}(\mathscr{H}_{\mathbb{C}}) = \mathrm{span}\{1, \phi(f_1)\cdots\phi(f_n)|n \in \mathbb{N}, f_j \in \mathscr{H}, j = 1, \ldots, n\}$.
(iii) $U\phi_{\mathrm{F}}(f)U^{-1} = \phi(f)$, $f \in \mathscr{H}$, *where $\phi_F(f)$ is defined by (6.125) with $\mathscr{H}_J = \mathscr{H}$ and the right hand side is the multiplication operator by the function $\phi(f)$ on Q.*

The Hilbert space $L^2(Q, d\mu)$ is called a ***Q*-space representation** of the boson Fock space $\mathscr{F}_{\mathrm{b}}(\mathscr{H}_{\mathbb{C}})$. Let $\mathscr{H}$ be infinite dimensional and

$$\pi(f) := U\pi_{\mathrm{F}}(f)U^{-1}, \quad f \in \mathscr{H}. \tag{8.142}$$

Then

$$\pi_{\mathrm{S}}^{\infty} := (L^2(Q, d\mu), \{\phi(f), \pi(f)|f \in \mathscr{H}\})$$

is an irreducible Weyl representation of the CCR over $\mathscr{H}$ with infinite degrees of freedom, which is equivalent to the Fock one $(\mathscr{F}_{\mathrm{b}}(\mathscr{H}_{\mathbb{C}}), \{\phi_{\mathrm{F}}(f), \pi_{\mathrm{F}}(f)|f \in \mathscr{H}\})$.

To write down the form of operation of $\pi(f)$, we introduce an operator D_f with $f \in \mathscr{H}$ by

$$D_f := \sqrt{2}UA(f)U^{-1}. \tag{8.143}$$

Let

$$\mathscr{P}(\mathscr{H}) := \mathrm{span}\{1, P_n(\phi(f_1), \ldots, \phi(f_n))|n \in \mathbb{N}, P_n \text{ is a polynomial of } n \text{ real variables with complex coefficients}, f_1, \ldots, f_n \in \mathscr{H}\}.$$

Then, by Theorem 8.20(ii), $\mathscr{P}(\mathscr{H})$ is dense in $L^2(Q, d\mu)$. One can show that, for all $f \in \mathscr{H}$, $\mathscr{P}(\mathscr{H}) \subset D(D_f)$ and

$$D_f P_n(\phi(f_1), \ldots, \phi(f_n)) = \sum_{j=1}^{n} \langle f, f_j \rangle (\partial_j P_n)(\phi(f_1), \ldots, \phi(f_n)) \tag{8.144}$$

and

$$\pi(f) = -iD_f + i\phi(f) \quad \text{on } \mathscr{P}(\mathscr{H})$$

(for a proof of these facts, see [21, §6.11], [28, §5.26]).

Formula (8.144) shows that $D_f \restriction \mathscr{P}(\mathscr{H})$ can be regarded as a directional differential operator with the direction of f. We note also that, in the representation π_S^∞, $\phi(f)$ is a multiplication operator and each $\phi(e_n)$ plays a role of a coordinate function in an infinite-dimensional space (see (8.141)). Thus π_S^∞ can be regarded as an infinite-dimensional version of the Schrödinger representation.[7] We call π_S^∞ the **functional Schrödinger representation of the Heisenberg CCR over** $\mathscr{H}$ on $L^2(Q, d\mu)$. For further details of this representation, see [28, §5.26].

The functional Schrödinger representation π_S^∞ is very useful in constructions and analysis of models of quantum scalar fields. For example, in this representation, an interaction of a self-interacting neutral quantum scalar field with cutoffs can be defined as a multiplication operator on $L^2(Q, d\mu)$ with a suitable space Q, and the heat semi-group ((C_0)-semi-group) defined by the Hamiltonian of a neutral quantum scalar field with such an interaction can be represented in terms of functional integrations (Feynman–Kac–Nelson-type formulae). It is known that functional integral methods make it possible to derive detailed properties of quantum field models. For further details of functional integral methods, see, e.g., [21, 62, 72, 80, 105, 147, 148]. We remark also that the representation π_S^∞ is related to white noise analysis (see, e.g., [122]).

[7]One can easily show that, if $\mathscr{H}$ is finite-dimensional with $\dim \mathscr{H} = N < \infty$, then $(L^2(Q, d\mu), \{\phi(f), \pi(f) \mid f \in \mathscr{H}\})$ is equivalent to the Schrödinger representation $\pi_S^{(N)}$ (see [28, §5.26]).

Chapter 9
Representations of CAR with Infinite Degrees of Freedom

Abstract Some basic representations of CAR with infinite degrees of freedom are described in the framework of fermion Fock spaces.

9.1 Definitions

Definition 9.1 Let $\mathfrak{F}$ and $\mathscr{H}$ be complex Hilbert spaces, and

$$\mathfrak{A}_{\mathrm{CAR}}(\mathscr{H}) := \{\psi(f), \psi(f)^* | f \in \mathscr{H}\}$$

be a subset of $\mathfrak{B}(\mathfrak{F})$, the Banach space of everywhere defined bounded linear operators on $\mathfrak{F}$. The pair $(\mathfrak{F}, \mathfrak{A}_{\mathrm{CAR}}(\mathscr{H}))$ is called a **representation of the canonical anti-commutation relations** (CAR) over $\mathscr{H}$ if the following hold:

(i) (anti-linearity in test vectors) For all $f, g \in \mathscr{H}$ and $\alpha, \beta \in \mathbb{C}$, $\psi(\alpha f + \beta g) = \alpha^* \psi(f) + \beta^* \psi(g)$.
(ii) (CAR) For all $f, g \in \mathscr{H}$,

$$\{\psi(f), \psi(g)\} = 0, \quad \{\psi(f), \psi(g)^*\} = \langle f, g \rangle . \tag{9.1}$$

Remark 9.1 Taking the adjoint of the first equation in (9.1), we have

$$\{\psi(f)^*, \psi(g)^*\} = 0, \quad f, g \in \mathscr{H}.$$

Definition 9.2 A representation $(\mathfrak{F}, \mathfrak{A}_{\mathrm{CAR}}(\mathscr{H}))$ of the CAR over $\mathscr{H}$ is said to be **reducible** (resp. **irreducible**) if $\mathfrak{A}_{\mathrm{CAR}}(\mathscr{H})$ is reducible (resp. irreducible).

Definition 9.3 Let $(\mathfrak{F}', \mathfrak{A}'_{\mathrm{CAR}}(\mathscr{H}))$ with $\mathfrak{A}'_{\mathrm{CAR}}(\mathscr{H}) := \{\psi'(f), \psi'(f)^* | f \in \mathscr{H}\})$ be another representation of the CAR over $\mathscr{H}$. Then the two representations $(\mathfrak{F}', \mathfrak{A}'_{\mathrm{CAR}}(\mathscr{H}))$ and $(\mathfrak{F}, \mathfrak{A}_{\mathrm{CAR}}(\mathscr{H}))$ are said to be **equivalent** if there exists a unitary operator $U: \mathfrak{F} \to \mathfrak{F}'$ such that, for all $f \in \mathscr{H}$, $\psi'(f) = U\psi(f)U^{-1}$.

© Springer Nature Singapore Pte Ltd. 2020

A. Arai, *Inequivalent Representations of Canonical Commutation and Anti-Commutation Relations*, Mathematical Physics Studies,
https://doi.org/10.1007/978-981-15-2180-5_9

Example 9.1 Let $\mathscr{F}_{\mathrm{f}}(\mathscr{K})$ be the fermion Fock space over a Hilbert space $\mathscr{K}$ (see Chap. 7) and $B(u)$ be the fermion annihilation operator on $\mathscr{F}_{\mathrm{f}}(\mathscr{K})$ with test vector $u \in \mathscr{K}$. Then, by Theorems 7.6 and 7.8,

$$(\mathscr{F}_{\mathrm{f}}(\mathscr{K}), \{B(u), B(u)^* | u \in \mathscr{K}\}) \tag{9.2}$$

is an irreducible representation of the CAR over $\mathscr{K}$. This representation is called the **Fock representation of the CAR over** $\mathscr{K}$.

Example 9.2 Let $S \in \mathfrak{B}(\mathscr{K})$ such that $S^*S = I$, i.e., S is an isometry on $\mathscr{K}$, and let

$$B_S(u) := B(Su), \quad u \in \mathscr{K}.$$

Then it is easy to see that $(\mathscr{F}_{\mathrm{f}}(\mathscr{K}), \{B_S(u), B_S(u)^* | u \in \mathscr{K}\})$ is a representation of the CAR over $\mathscr{K}$. There are two cases to be considered:

(1) The case where S is unitary. In this case, we have

$$S^*S = I, \quad SS^* = I.$$

By Theorem 7.10,

$$B_S(u) = \Gamma_{\mathrm{f}}(S)B(u)\Gamma_{\mathrm{f}}(S)^{-1}, \quad u \in \mathscr{K}.$$

Hence $(\mathscr{F}_{\mathrm{f}}(\mathscr{K}), \{B_S(u), B_S(u)^* | u \in \mathscr{K}\})$ is equivalent to the Fock representation $(\mathscr{F}_{\mathrm{f}}(\mathscr{K}), \{B(u), B(u)^* | u \in \mathscr{K}\})$.

(2) The case where S is not unitary. In this case $(\mathrm{Ran}\, S)^\perp \neq \{0\}$. Let $v \in (\mathrm{Ran}\, S)^\perp$ and $v \neq 0$. Then, by direct computations using the CAR for $B(\cdot)^\#$, we have

$$[B_S(u), B(v)^* B(v)] = \langle Su, v \rangle\, B(v) = 0.$$

Hence $B(v)^* B(v) \in \{B_S(u), B_S(u)^* | u \in \mathscr{K}\}'$ and $B(v)^* B(v)$ is not a scalar operator. Therefore $\{B_S(u), B_S(u)^* | u \in \mathscr{K}\}$ is reducible. Thus $(\mathscr{F}_{\mathrm{f}}(\mathscr{K}), \{B_S(u), B_S(u)^* | u \in \mathscr{K}\})$ is inequivalent to $(\mathscr{F}_{\mathrm{f}}(\mathscr{K}), \{B(u), B(u)^* | u \in \mathscr{K}\})$ (but this is a trivial inequivalence).

To analyze the structure of the reducibility of $\{B_S(u), B_S(u)^* | u \in \mathscr{K}\}$, we note the orthogonal decomposition

$$\mathscr{K} = \mathrm{Ran}\, S \oplus \ker S^*.$$

Hence, by Theorem 7.14,

$$\mathscr{F}_{\mathrm{f}}(\mathscr{K}) \cong \mathscr{F}_{\mathrm{f}}(\mathrm{Ran}\, S) \otimes \mathscr{F}_{\mathrm{f}}(\ker S^*).$$

Under this isomorphism, $B_S(u)^\#$ is unitarily equivalent to $B(Su)^\# \otimes I$. Let $\{\Psi_n\}_{n=1}^N$ ($N \in \mathbb{N}$ or $N = \infty$) be a CONS of $\mathscr{F}_\mathrm{f}(\ker S^*)$ and $\mathscr{H}_n = \{\alpha\Psi_n | \alpha \in \mathbb{C}\}$. Then $\mathscr{F}_\mathrm{f}(\ker S^*) = \oplus_{n=1}^N \mathscr{H}_n$. Hence

$$\mathscr{F}_\mathrm{f}(\mathscr{K}) \cong \oplus_{n=1}^N \mathscr{F}_\mathrm{f}(\mathrm{Ran}\, S) \otimes \mathscr{H}_n \cong \oplus_{n=1}^N \mathscr{M}_n,$$

where $\mathscr{M}_n := \mathscr{F}_\mathrm{f}(\mathrm{Ran}\, S)$. Under this isomorphism, $B_S(u)^\#$ is reduced by $\mathscr{M}_n$ and $(\mathscr{M}_n, \{B_S(u), B_S(u)^* | u \in \mathscr{K}\})$ gives an irreducible representation of the CAR over $\mathscr{K}$. Since $\Gamma_\mathrm{f}(S)$ is a unitary operator from $\mathscr{F}_\mathrm{f}(\mathscr{K})$ to $\mathscr{F}_\mathrm{f}(\mathrm{Ran}\, S)$, we have

$$\Gamma_\mathrm{f}(S) B(u)^\# \Gamma_\mathrm{f}(S)^{-1} = B_S(u)^\#$$

on $\mathscr{F}_\mathrm{f}(\mathrm{Ran}\, S)$. Hence, for each n, $(\mathscr{M}_n, \{B_S(u), B_S(u)^* | u \in \mathscr{K}\})$ is equivalent to the Fock representation $(\mathscr{F}_\mathrm{f}(\mathscr{K}), \{B(u), B(u)^* | u \in \mathscr{K}\})$. Thus $(\mathscr{F}_\mathrm{f}(\mathscr{K}), \{B_S(u), B_S(u)^* | u \in \mathscr{K}\})$ is equivalent to a direct sum representation of the Fock representation of the CAR over $\mathscr{K}$.

Remark 9.2 A general theory of representations of CAR is given in [41].

9.2 Fermionic Bogoliubov Transformations

For the fermion annihilation and creation operators too, one can consider analogues of Bogoliubov transformations of the boson annihilation and creation operators. Let C be a conjugation on $\mathscr{K}$ and $S, T \in \mathfrak{B}(\mathscr{K})$ satisfying

$$S^*S + T^*T = I, \quad S^*T_C + T^*S_C = 0, \tag{9.3}$$

where, for $A \in \mathfrak{B}(\mathscr{K})$, $A_C \in \mathfrak{B}(\mathscr{K})$ is the conjugation of A with respect to C (see (8.51)). Then one defines

$$B_{S,T}(u) := B(Su) + B(CTu)^*, \quad u \in \mathscr{K}. \tag{9.4}$$

It is easy to see that $(\mathscr{F}_\mathrm{f}(\mathscr{K}), \{B_{S,T}(u), B_{S,T}(u)^* | u \in \mathscr{K}\})$ is a representation of the CAR over $\mathscr{K}$. The correspondence $(B(\cdot), B(\cdot)^*) \mapsto (B_{S,T}(\cdot), B_{S,T}(\cdot)^*)$ is called the **fermionic Bogoliubov transformation** associated with (S, T).

For the fermionic Bogoliubov transformation too, one has a result corresponding to Theorem 8.6 on a bosonic Bogoliubov transformation. To state and prove it, we need two lemmas.

Lemma 9.1 *Assume that* $S \in \mathfrak{B}(\mathscr{K})$ *and* $T \in \mathfrak{B}(\mathscr{K})$ *satisfy (9.3). Suppose that* Ran S *is dense in* $\mathscr{K}$. *Then* S *is injective.*

Proof Let $u \in \ker S$. Then, by (9.3),

$$T^*Tu = u, \quad S^*CTu = 0. \tag{9.5}$$

Since $\mathscr{K} = \ker S^* \oplus \overline{\operatorname{Ran} S}$ by the Hilbert space decomposition theorem (Theorem 1.2), it follows that $\ker S^* = \{0\}$. Hence, by the second equation in (9.5), $CTu = 0$. Hence $Tu = 0$. Then the first equation in (9.5) yields $u = 0$. Hence $\ker S = \{0\}$, implying that S is injective. ■

Lemma 9.2 *Assume that $\mathscr{K}$ is separable. Let $\mathscr{D}$ be a dense subspace of $\mathscr{K}$ and $L \in \mathfrak{L}(\mathscr{K})$ with $D(L) \supset \mathscr{D}$. Suppose that there exists a non-zero vector $\Omega \in \mathscr{F}_{\mathrm{f}}(\mathscr{K})$ such that*

$$B(CLu)^*\Omega = -B(u)\Omega. \quad u \in \mathscr{D}, \tag{9.6}$$

Then L is bounded and $\overline{L}$ is Hilbert–Schmidt.

Proof For each $n \geq 0$, we define a vector Ω_n in the n-particle space by $\Omega_n^{(n)} := \Omega^{(n)}$ and $\Omega_n^{(m)} := 0$, $m \neq n$. Then we have $\Omega = \sum_{n=0}^{\infty} \Omega_n$. Since $B(\cdot)^{\#}$ is bounded, it follows that

$$\sum_{n=0}^{\infty} B(CLu)^*\Omega_n = -\sum_{n=1}^{\infty} B(u)\Omega_n, \quad u \in \mathscr{D},$$

where we have used that $B(u)\Omega_0 = 0$. Hence

$$B(u)\Omega_1 = 0, \tag{9.7}$$

$$B(CLu)^*\Omega_{n-1} = -B(u)\Omega_{n+1}, \quad n \geq 1. \tag{9.8}$$

It follows from (9.7) and Theorem 7.7 that $\Omega_1 = 0$. Then, by (9.8) with $n = 2$, $B(u)\Omega_3 = 0$, $u \in \mathscr{D}$. Hence, by Theorem 7.7 again, $\Omega_3 = 0$. By induction in n, one can show that $\Omega_{2n-1} = 0$, $n \geq 1$. Taking $n = 1$ in (9.8), one has $B(CLu)^*\Omega_0 = -B(u)\Omega_2$. If $\Omega_0 = 0$, then $B(u)\Omega_2 = 0$, $u \in \mathscr{D}$. Hence, by Theorem 7.7, $\Omega_2 = 0$. Then, by induction, one has $\Omega_{2n} = 0$ for all $n \in \mathbb{Z}_+$. Thus $\Omega = 0$. But this is a contradiction. Therefore $\Omega_0 \neq 0$. Hence $\Omega_0 = \alpha\Omega_{\mathscr{K}}$ with some $\alpha \in \mathbb{C} \setminus \{0\}$. Then $B(CLu)^*\Omega_{\mathscr{K}} = -\alpha^{-1}B(u)\Omega_2$. Taking the norm of both sides, we obtain $\|Lu\| = |\alpha|^{-1}\|B(u)\Omega_2\|$. Let $\{e_n\}_{n=1}^{\infty}$ be a CONS of $\mathscr{K}$ with $e_n \in \mathscr{D}$, $n \in \mathbb{N}$. Then

$$\sum_{n=1}^{\infty} \|Le_n\|^2 = |\alpha|^{-2} \sum_{n=1}^{\infty} \|B(e_n)\Omega_2\|^2 = c_0 := |\alpha|^{-2}\|N_{\mathrm{f}}^{1/2}\Omega_2\|^2 < \infty,$$

where we have used (7.49). This implies that $\sum_{n=1}^{\infty} \|Le_n\|^2$ is independent of the choice of $\{e_n\}_{n=1}^{\infty}$ and $\|Le_n\| \le \sqrt{c_0}$, $n \in \mathbb{N}$. Hence, for all $\Psi \in \mathscr{D} \setminus \{0\}$,

$$\|L(\|\Psi\|^{-1}\Psi)\| \le \sqrt{c_0}.$$

Therefore $\|L\Psi\| \le \sqrt{c_0}\|\Psi\|$. Thus L is bounded and $\overline{L}$ is Hilbert–Schmidt. ■

Theorem 9.1 *Assume that $\mathscr{K}$ is separable. Suppose that* Ran S *is dense and T is not Hilbert–Schmidt. Then $(\mathscr{F}_{\mathrm{f}}(\mathscr{K}), \{B_{S,T}(u), B_{S,T}(u)^* | \, u \in \mathscr{K}\})$ is inequivalent to any direct sum representation of the Fock representation $(\mathscr{F}_{\mathrm{f}}(\mathscr{K}), \{B(u), B(u)^* | u \in \mathscr{K}\})$.*

Proof The method of proof is similar to that of Theorem 8.6. We prove the contraposition of the above statement. Suppose that $(\mathscr{F}_{\mathrm{f}}(\mathscr{K}), \{B_{S,T}(u), B_{S,T}(u)^* | \, u \in \mathscr{K}\})$ is equivalent to a direct sum representation

$$(\oplus_{n=1}^{N} \mathscr{F}_{\mathrm{f}}(\mathscr{K}), \{\oplus_{n=1}^{N} B(u), \oplus_{n=1}^{N} B(u)^* | u \in \mathscr{K}\})$$

of the Fock representation $(\mathscr{F}_{\mathrm{f}}(\mathscr{K}), \{B(u), B(u)^* | u \in \mathscr{K}\})$, where $N < \infty$ or $N = \infty$. Then there exists a unitary operator U from $\mathscr{F}_{\mathrm{f}}(\mathscr{K})$ to $\oplus_{n=1}^{N} \mathscr{F}_{\mathrm{f}}(\mathscr{K})$ such that $U B_{S,T}(u) U^{-1} = \oplus_{n=1}^{N} B(u)$, $u \in \mathscr{K}$. Let

$$\Omega := U^{-1}(\Omega_{\mathscr{K}}, 0, 0, \dots).$$

Then $\|\Omega\| = 1$ (hence $\Omega \ne 0$) and $B_{S,T}(u)\Omega = 0$, i.e.,

$$B(CTu)^*\Omega = -B(Su)\Omega. \tag{9.9}$$

Since Ran S is dense, it follows from Lemma 9.1 that S is injective. Hence we have

$$B(CLu)^*\Omega = -B(u)\Omega, \quad u \in \text{Ran}\, S,$$

where $L := TS^{-1}$. Then, by Lemma 9.2, $\overline{L}$ is Hilbert–Schmidt and hence $T = \overline{L}S$ also is Hilbert–Schmidt. ■

With regard to irreducibility of $\{B_{S,T}(u), B_{S,T}(u)^* | u \in \mathscr{K}\}$, the following theorem holds.

Theorem 9.2 *Assume (9.3) and*

$$SS^* + T_C(T_C)^* = I, \quad ST^* + T_C(S_C)^* = 0. \tag{9.10}$$

Then $(\mathscr{F}_{\mathrm{f}}(\mathscr{K}), \{B_{S,T}(u), B_{S,T}(u)^* | u \in \mathscr{K}\})$ is an irreducible representation of the CAR over $\mathscr{K}$.

Proof Using (9.10), one can express $B(u)$ in terms of $B_{S,T}(\cdot)^{\#}$ as follows:

$$B(u) = B_{S,T}(S^* u) + B_{S,T}(T^* C u)^*, \quad u \in \mathscr{K}. \tag{9.11}$$

Hence, for all $A \in \{B_{S,T}(u), B_{S,T}(u)^* | u \in \mathscr{K}\}'$, $A \in \{B(u), B(u)^* | u \in \mathscr{K}\}' = \mathbb{C}I$ (Example 9.1). Therefore $\{B_{S,T}(u), B_{S,T}(u)^* | u \in \mathscr{K}\}' = \mathbb{C}I$. Thus the desired result follows. ■

The following theorem is a standard result in the theory of fermionic Bogoliubov transformations.

Theorem 9.3 *Let $\mathscr{K}$ be separable. Suppose that (9.3) and (9.10) hold. Then $(\mathscr{F}_{\mathrm{f}}(\mathscr{K}), \{B_{S,T}(u), B_{S,T}(u)^* | u \in \mathscr{K}\})$ is equivalent to the Fock representation $(\mathscr{F}_{\mathrm{f}}(\mathscr{K}), \{B(u), B(u)^* | u \in \mathscr{K}\})$ if and only if T is Hilbert–Schmidt.*

For a proof of this fact, see, e.g, [137], [123, §2.4] or [156, §10.3]. Cf. also [38] for related aspects.

9.3 A Class of Representations of CAR

We now fix an orthogonal decomposition

$$\mathscr{K} = \mathscr{K}_+ \oplus \mathscr{K}_- = \{f = (f_+, f_-) | f_+ \in \mathscr{K}_+, f_- \in \mathscr{K}_-\}$$

of $\mathscr{K}$ with $\mathscr{K}_+$ and $\mathscr{K}_-$ being mutually orthogonal non-trivial closed subspaces ($\mathscr{K}_+ \neq \{0\}, \mathscr{K}$). We sometimes identify $f_+ \in \mathscr{K}_+$ (resp. $f_- \in \mathscr{K}_-$) with $(f_+, 0)$ (resp. $(0, f_-)$) so that f is written $f = f_+ + f_-$.

Let C be a conjugation on $\mathscr{K}$. We denote by

$$\mathfrak{B}_\pm := \mathfrak{B}(\mathscr{K}, \mathscr{K}_\pm)$$

the Banach space of everywhere defined bounded linear operators from $\mathscr{K}$ to $\mathscr{K}_\pm$ and introduce a subset of the direct product space $\mathfrak{B}_+ \times \mathfrak{B}_-$:

$$\mathfrak{T}(\mathscr{K}) := \{T = (T_+, T_-) \mid T_\pm \in \mathfrak{B}_\pm, \ T_+^* T_+ + (T_-^* T_-)_C = I\},$$

where

$$(T_-^* T_-)_C := C T_-^* T_- C.$$

Let $P_\pm$ be the orthogonal projections onto $\mathscr{K}_\pm$. We assume that

$$P_\pm C = C P_\pm. \tag{9.12}$$

Then $C\mathscr{K}_\pm = \mathscr{K}_\pm$.

Each $T \in \mathfrak{T}(\mathscr{K})$ defines an element of $\mathfrak{B}(\mathscr{K})$ by

$$Tf := (T_+ f, T_- f), \quad f \in \mathscr{K}. \tag{9.13}$$

For each $T \in \mathfrak{T}(\mathscr{K})$, we define an anti-linear mapping $\psi_T : \mathscr{K} \to \mathfrak{B}(\mathscr{F}_{\mathrm{f}}(\mathscr{K}))$ by

$$\psi_T(f) := B(T_+ f, 0) + B(0, T_- Cf)^*, \quad f \in \mathscr{K}. \tag{9.14}$$

It is obvious that

$$\psi_T(f)^* = B(T_+ f, 0)^* + B(0, T_- Cf).$$

Let

$$\mathfrak{A}_T(\mathscr{K}) := \{\psi_T(f), \psi_T(f)^* | f \in \mathscr{K}\}. \tag{9.15}$$

Lemma 9.3 *For all $f, g \in \mathscr{K}$, the following anti-commutation relations hold:*

$$\{\psi_T(f), \psi_T(g)\} = \{\psi_T(f)^*, \psi_T(g)^*\} = 0, \tag{9.16}$$

$$\{\psi_T(f), \psi_T(g)^*\} = \langle f, g\rangle . \tag{9.17}$$

Proof By direct computations using (9.1), we have

$$\begin{aligned}\{\psi_T(f), \psi_T(g)^*\} &= \langle T_+ f, T_+ g\rangle + \langle T_- Cg, T_- Cf\rangle = \langle f, T_+^* T_+ g\rangle + \langle Cg, T_-^* T_- f\rangle \\ &= \langle f, T_+^* T_+ g + (T_-^* T_-)_C g\rangle \\ &= \langle f, g\rangle .\end{aligned}$$

Hence we obtain (9.17). Similarly, one can prove (9.16). ■

Lemma 9.3 shows that $(\mathscr{F}_{\mathrm{f}}(\mathscr{K}), \mathfrak{A}_T(\mathscr{K}))$ is a representation of the CAR over $\mathscr{K}$.

Remark 9.3 The standard choice of $T = (T_+, T_-)$ in the literature is given by $T_+ = P$ and $T_- = C(I - P)C$, where P is the orthogonal projection onto $\mathscr{K}_+$(see, e.g., [123, pp. 22–23] and [156, §10.1.3]). In this case, the representation is called a quasi-free representation. Hence the representation $(\mathscr{F}_{\mathrm{f}}(\mathscr{K}), \mathfrak{A}_T(\mathscr{K}))$ gives a generalization of the quasi-free representation.

To find more detailed properties of the representation $(\mathscr{F}_{\mathrm{f}}(\mathscr{K}), \mathfrak{A}_T(\mathscr{K}))$, we introduce the following additional conditions for $T = (T_+, T_-) \in \mathfrak{T}(\mathscr{K})$:

(T.1) $T_+ T_+^* = I$,
(T.2) $T_- T_-^* = I$,
(T.3) $T_- C T_+^* = 0$.

Multiplying (T.3) by C from the right and taking the adjoint of the resulting equation, one sees that (T.3) is equivalent to

(T.4) $T_+ C T_-^* = 0.$

We define a subset of $\mathfrak{T}(\mathscr{K})$:

$$\mathfrak{T}_*(\mathscr{K}) := \{T \in \mathfrak{T}(\mathscr{K}) | \text{(T.1)–(T.3) hold}\}. \tag{9.18}$$

Lemma 9.4 *Let $T \in \mathfrak{T}_*(\mathscr{K})$. Then*

$$B(f) = \psi_T(T_+^* f_+) + \psi_T(C T_-^* f_-)^*, \quad f = (f_+, f_-),\ f_\pm \in \mathscr{K}_\pm. \tag{9.19}$$

Proof By (T.1) and (T.3), we have

$$\psi_T(T_+^* f_+) = B(f_+, 0), \quad f_+ \in \mathscr{K}_+.$$

Property (T.3) implies that $T_+ C T_-^* = 0$. By this fact and (T.2), we obtain

$$\psi_T(C T_-^* f_-) = B(0, f_-)^*, \quad f_- \in \mathscr{K}_-.$$

We have $B(f) = B(f_+, 0) + B(0, f_-)$. Thus we obtain (9.19). ■

Lemma 9.5 *For all $T \in \mathfrak{T}_*(\mathscr{K})$, $(\mathscr{F}_{\mathrm{f}}(\mathscr{K}), \mathfrak{A}_T(\mathscr{K}))$ is irreducible.*

Proof Let $A \in \{\psi_T(f), \psi_T(f)^* | \mathscr{K}\}'$. Then $A\psi_T(f) = \psi_T(f)A$ and $A\psi_T(f)^* = \psi_T(f)^* A$. Hence, by (9.19), $AB(f) = B(f)A$ and $AB(f)^* = B(f)^* A$ for all $f \in \mathscr{K}$. Therefore $A \in \{B(f), B(f)^* | f \in \mathscr{K}\}'$. As we have seen in Example 9.1, $\{B(f), B(f)^* | f \in \mathscr{K}\}' = \mathbb{C}I$. Hence $A = \alpha I$ for some $\alpha \in \mathbb{C}$. Thus $\mathfrak{A}_T(\mathscr{K})$ is irreducible. ■

Lemmas 9.3 and 9.5 immediately yield the following theorem:

Theorem 9.4 *For each $T \in \mathfrak{T}_*(\mathscr{K})$, $(\mathscr{F}_{\mathrm{f}}(\mathscr{K}), \mathfrak{A}_T(\mathscr{K}))$ is an irreducible representation of the CAR over $\mathscr{K}$.*

Thus we have a family $\{(\mathscr{F}_{\mathrm{f}}(\mathscr{K}), \mathfrak{A}_T(\mathscr{K})\}_{T \in \mathfrak{T}_*(\mathscr{K})}$ of irreducible representations of the CAR over $\mathscr{K}$.

We next consider equivalence or inequivalence of two representations $(\mathscr{F}_{\mathrm{f}}(\mathscr{K}), \mathfrak{A}_T(\mathscr{K}))$ and $(\mathscr{F}_{\mathrm{f}}(\mathscr{K}), \mathfrak{A}_S(\mathscr{K}))$ with $S \neq T$ $(S, T \in \mathfrak{T}_*(\mathscr{K}))$.

For each pair $(S, T) \in \mathfrak{T}_*(\mathscr{K}) \times \mathfrak{T}_*(\mathscr{K})$, we define linear operators V and W on $\mathscr{K}$ as follows:

$$Vf := (S_+ T_+^* f_+, S_- T_-^* f_-), \tag{9.20}$$

$$Wf := (C S_+ C T_-^* f_-, C S_- C T_+^* f_+), \quad f = (f_+, f_-) \in \mathscr{K}. \tag{9.21}$$

Let

$$V_C := CVC, \quad W_C := CWC.$$

Lemma 9.6 *The following equations hold:*

$$V^*V + W^*W = I, \tag{9.22}$$

$$V^*W_C + W^*(V_C)^* = 0, \tag{9.23}$$

$$VV^* + W_C W_C^* = I, \tag{9.24}$$

$$VW^* + W_C V_C^* = 0. \tag{9.25}$$

Proof The operators V and W have the following operator matrix representations:

$$V = \begin{pmatrix} S_+T_+^* & 0 \\ 0 & S_-T_-^* \end{pmatrix}, \quad W = \begin{pmatrix} 0 & CS_+CT_-^* \\ CS_-CT_+^* & 0 \end{pmatrix}.$$

Using these representations and properties of S and T, one can easily prove (9.22)–(9.25) by direct computations. ■

We define

$$C(f) := B(Vf) + B(CWf)^*, \quad f \in \mathscr{K}. \tag{9.26}$$

Then it is easy to see that $\{C(f), C(f)^* | f \in \mathscr{K}\}$ is a representation of the CAR over $\mathscr{K}$.

Lemma 9.7 *There exists a unitary operator* $\mathbb{U}$ *on* $\mathscr{F}_{\mathrm{f}}(\mathscr{K})$ *such that*

$$B(f) = \mathbb{U}C(f)\mathbb{U}^*, \quad f \in \mathscr{K}, \tag{9.27}$$

if and only if W *is Hilbert–Schmidt.*

Proof Since we have (9.22)–(9.25), the lemma follows from an application of Theorem 9.3 to the case $S = V$ and $T = W$. ■

Now we are ready to state and prove the main theorem in this section:

Theorem 9.5 *Let* T *and* S *be in* $\mathfrak{T}_*(\mathscr{K})$ *with* $T \neq S$. *Then the two representations* $(\mathscr{F}_{\mathrm{f}}(\mathscr{K}), \mathfrak{A}_T(\mathscr{K}))$ *and* $(\mathscr{F}_{\mathrm{f}}(\mathscr{K}), \mathfrak{A}_S(\mathscr{K}))$ *are equivalent if and only if* $CS_+CT_-^*$ *and* $CS_-CT_+^*$ *are Hilbert–Schmidt.*

Proof Suppose that $(\mathscr{F}_{\mathrm{f}}(\mathscr{K}), \mathfrak{A}_T(\mathscr{K}))$ and $(\mathscr{F}_{\mathrm{f}}(\mathscr{K}), \mathfrak{A}_S(\mathscr{K}))$ are equivalent. Then there exists a unitary operator $\mathbb{U}$ on $\mathscr{F}_{\mathrm{f}}(\mathscr{K})$ such that

$$\mathbb{U}\psi_S(f)\mathbb{U}^* = \psi_T(f), \quad f \in \mathscr{K}. \tag{9.28}$$

Then, by (T.1)–(T.3), we have

$$B(f_+, 0) = \mathbb{U}\left(B(S_+T_+^* f_+, 0) + B(0, S_-CT_+^* f_+)^*\right)\mathbb{U}^*, \quad f_+ \in \mathscr{K}_+. \tag{9.29}$$

and

$$B(0, f_-) = \mathbb{U}\left(B(S_+CT_-^* f_-, 0)^* + B(0, S_-T_-^* f_-)\right)\mathbb{U}^*, \quad f_- \in \mathscr{K}_-. \tag{9.30}$$

Hence, adding the first equation to the second one, we obtain (9.27). Therefore, by Lemma 9.7, W is Hilbert–Schmidt, which is equivalent to that $CS_+CT_-^*$ and $CS_-CT_+^*$ are Hilbert–Schmidt.

Conversely, suppose that $CS_+CT_-^*$ and $CS_-CT_+^*$ are Hilbert–Schmidt. Then W is Hilbert–Schmidt. Hence, by Lemma 9.7, there exists a unitary operator $\mathbb{U}$ on $\mathscr{F}_{\mathrm{f}}(\mathscr{K})$ such that (9.27) holds. Then (9.29) and (9.30) hold. Using the property that $T_+^*T_+ + C(T_-^*T_-)C = I$, we obtain (9.28). ■

The contraposition of Theorem 9.5 gives an inequivalence theorem on the two representations $(\mathscr{F}_{\mathrm{f}}(\mathscr{K}), \mathfrak{A}_T(\mathscr{K}))$ and $(\mathscr{F}_{\mathrm{f}}(\mathscr{K}), \mathfrak{A}_S(\mathscr{K}))$:

Theorem 9.6 *Let T and S be in $\mathfrak{T}_*(\mathscr{K})$ with $T \neq S$. Then the two representations $(\mathscr{F}_{\mathrm{f}}(\mathscr{K}), \mathfrak{A}_T(\mathscr{K}))$ and $(\mathscr{F}_{\mathrm{f}}(\mathscr{K}), \mathfrak{A}_S(\mathscr{K}))$ are inequivalent if and only if $CS_+CT_-^*$ or $CS_-CT_+^*$ is not Hilbert–Schmidt.*

Chapter 10
Physical Correspondences in Quantum Field Theory

Abstract It is shown that there are correspondences between characteristic physical situations or phenomena in quantum field theory and inequivalent representations of CCR or CAR.

10.1 Quantum Field Models in Hamiltonian Formalism

In what follows, we use the physical unit system where the reduced Planck constant $\hbar$ and the speed of light c are equal to 1.

We first give the definition of abstract quantum field models in Hamiltonian formalism.[1]

Definition 10.1 (Bose Field Models)

(1) An **abstract relativistic Bose field model** is defined to be a triple $(\mathscr{F}, H, \{\phi(f), \pi(f)|f \in \mathscr{W}\})$ consisting of a Hilbert space $\mathscr{F}$, a self-adjoint operator H on $\mathscr{F}$ and an irreducible representation $(\mathscr{F}, \mathscr{D}, \{\phi(f), \pi(f)|f \in \mathscr{W}\})$ of the Heisenberg CCR over a real inner product space $\mathscr{W}$ such that, for all $f \in \mathscr{W}$, $\phi(f)$ and $\pi(f)$ are closed symmetric operators on $\mathscr{F}$ and $\{\phi(f), \pi(f)|f \in \mathscr{W}\}' = \mathbb{C}I$.[2] The operator-valued functionals $f \mapsto \phi(f)$ and $f \mapsto \pi(f)$ are called the **time-zero fields** of the model.

(2) An **abstract non-relativistic Bose field model** is defined to be a triple $(\mathscr{F}, H, \{C(f)|f \in \mathscr{V}\})$ consisting of a Hilbert space $\mathscr{F}$, a self-adjoint operator H on $\mathscr{F}$ and an irreducible representation $(\mathscr{F}, \mathscr{D}, \{C(f), C(f)^*|f \in \mathscr{V}\})$ of the CCR over a complex inner product space $\mathscr{V}$. The operator-valued functional $f \mapsto C(f)$ is called the **time-zero field** of the model.

[1] There are other ways to define quantum field theories axiomatically. But, in the present book, we do not go into the details (see, e.g., [28, 39, 45, 149, 149, 152]).

[2] If $\phi(f)$ or $\pi(f)$ is not self-adjoint, then $\{\phi(f), \pi(f)|f \in \mathscr{W}\}$ is not $*$-invariant. For this reason, we assume this stronger irreducible condition. One may replace the condition with that $\{\phi(f), \pi(f), \phi(f)^*, \pi(f)^*|f \in \mathscr{W}\}$, which is $*$-invariant, is irreducible.

© Springer Nature Singapore Pte Ltd. 2020

A. Arai, *Inequivalent Representations of Canonical Commutation and Anti-Commutation Relations*, Mathematical Physics Studies,
https://doi.org/10.1007/978-981-15-2180-5_10

In each model, the operator H is called the **Hamiltonian** of the model.

In an abstract relativistic Bose field model $(\mathscr{F}, H, \{\phi(f), \pi(f)|f \in \mathscr{W}\})$, the **time-$t$ fields** $\phi(t,\cdot)$ and $\pi(t,\cdot)$ are defined by

$$\phi(t,f) := e^{itH}\phi(f)e^{-itH}, \quad \pi(t,f) := e^{itH}\pi(f)e^{-itH}, \quad t \in \mathbb{R},\ f \in \mathscr{W}, \tag{10.1}$$

while, in an abstract non-relativistic Bose field model $(\mathscr{F}, H, \{C(f)|f \in \mathscr{V}\})$, the **time-$t$ field** is defined by

$$C(t,f) := e^{itH}C(f)e^{-itH}, \quad f \in \mathscr{V}. \tag{10.2}$$

Remark 10.1 The words "relativistic" and "non-relativistic" here should be understood only as symbolical ones to distinguish two types of Bose field models (only the difference of types of time-zero field(s) is taken into account here). More differences of the two models can be described by taking into account relativistic symmetries and non-relativistic ones as unitary (or projective) representations on $\mathscr{F}$ (for more details, see [28, Chapters 8 and 9]). In what follows, we may sometimes omit the word "relativistic" or "non-relativistic".

The irreducibility of the time-zero fields of a Bose field model implies the irreducibility of the time-t fields of it:

Proposition 10.1 *Let $(\mathscr{F}, H, \{\phi(f), \pi(f)|f \in \mathscr{W}\})$ and $(\mathscr{F}, H, \{C(f)|f \in \mathscr{V}\})$ be as in Definition 10.1. Then:*

(i) *For all $t \in \mathbb{R}$, $\{\phi(t,f), \pi(t,f)|f \in \mathscr{W}\}$ is irreducible with $\{\phi(t,f), \pi(t,f)|f \in \mathscr{W}\}' = \mathbb{C}I$.*
(ii) *For all $t \in \mathbb{R}$, $\{C(t,f), C(t,f)^*|f \in \mathscr{V}\}$ is irreducible.*

Proof Since $\{\phi(t,f), \pi(t,f)|f \in \mathscr{W}\}$ is unitarily equivalent to $\{\phi(f), \pi(f)|f \in \mathscr{W}\}$ by the unitary operator e^{itH} and the latter is irreducible with $\{\phi(f), \pi(f)|f \in \mathscr{W}\}' = \mathbb{C}I$, it follows from an application of Proposition 2.8 that $\{\phi(t,f), \pi(t,f)| f \in \mathscr{W}\}$ is irreducible with $\{\phi(t,f), \pi(t,f)|f \in \mathscr{W}\}' = \mathbb{C}I$. Similarly, one can prove the irreducibility of $\{C(t,f), C(t,f)^*|f \in \mathscr{V}\}$. ■

Definition 10.2

(1) An abstract relativistic Bose field model $(\mathscr{F}, H, \{\phi(f), \pi(f)|f \in \mathscr{W}\})$ is said to be **equivalent** to an abstract relativistic Bose field model $(\mathscr{F}', H', \{\phi(f)', \pi(f)'\ |f \in \mathscr{W}\})$ if there exist a unitary operator $U: \mathscr{F} \to \mathscr{F}'$ and a real constant E such that

$$UHU^{-1} = H' + E, \quad U\phi(f)U^{-1} = \phi(f)', \quad U\pi(f)U^{-1} = \pi(f)', \quad f \in \mathscr{W}. \tag{10.3}$$

(2) An abstract non-relativistic Bose field model $(\mathscr{F}, H, \{C(f)|f \in \mathscr{V}\})$ is said to be **equivalent** to an abstract non-relativistic Bose field model $(\mathscr{F}', H', \{C(f)'$

$|f \in \mathscr{V}\})$ if there exist a unitary operator $U: \mathscr{F} \to \mathscr{F}'$ and a real constant E such that

$$UHU^{-1} = H' + E, \quad UC(f)U^{-1} = C(f)', \quad f \in \mathscr{V}.$$

Remark 10.2 One can define a relation $(\mathscr{F}, H, \{\phi(f), \pi(f)|f \in \mathscr{W}\}) \sim (\mathscr{F}', H', \{\phi(f)', \pi(f)' \mid f \in \mathscr{W}\})$ if there exist a unitary operator $U: \mathscr{F} \to \mathscr{F}'$ and a real constant E satisfying (10.3). Then it is easy to see that this relation $\sim$ is an equivalence relation. Therefore the above definition on equivalence between two Bose field models is well-defined. The same applies to abstract non-relativistic Bose field models.

Remark 10.3 A detailed analysis of Hamiltonian formalism using Weyl representations of CCR is given by Araki [36].

Remark 10.4 A concrete relativistic (resp. non-relativistic) Bose field model is defined by giving algebraic relations between H and $\{\phi(f), \pi(f)|f \in \mathscr{W}\}$ (resp. $\{C(f)|f \in \mathscr{V}\}$), which include relations implying field equations for time-t fields. A Hilbert space operator realization of the algebraic relations yields a representation of the model. Therefore, for a model, there may exist inequivalent representations of it.

Remark 10.5 Only a unitary equivalence between H and $H' + E$ does not imply the equivalence of the models.

One can define also an abstract Fermi field model:

Definition 10.3 (Fermi Field Model) An **abstract Fermi field model** is defined to be a triple $(\mathscr{F}, H, \{\psi(f)|f \in \mathscr{H}\})$ consisting of a Hilbert space $\mathscr{F}$, a self-adjoint operator H on $\mathscr{F}$ and an irreducible representation $(\mathscr{F}, \mathscr{D}, \{\psi(f), \psi(f)^*|f \in \mathscr{H}\})$ of the CAR over a complex Hilbert space $\mathscr{H}$. The operator-valued functional $f \mapsto \psi(f)$ is called the **time-zero field** of the model.

The time t-fields of an abstract Fermi field model $(\mathscr{F}, H, \{\psi(f)|f \in \mathscr{H}\})$ are defined by $\{\psi(t, f), \psi(t, f)^*|f \in \mathscr{H}\}$ with

$$\psi(t, f) := e^{itH}\psi(f)e^{-itH}, \quad f \in \mathscr{H}. \tag{10.4}$$

As in the case of Bose field models, the time-t fields of an abstract Fermi field model also are irreducible:

Proposition 10.2 *Let $(\mathscr{F}, H, \{\psi(f)|f \in \mathscr{H}\})$ be an abstract Fermi field model. Then, for all $t \in \mathbb{R}$, $\{\psi(t, f), \psi(t, f)^*|f \in \mathscr{H}\}$ is irreducible.*

Proof Similar to the proof of Proposition 10.1. ■

The notion of equivalence between two abstract Fermi field models is defined in the same way as in the case of abstract Bose field models.

Example 10.1 Let $\mathscr{H}$ be a Hilbert space and $\mathscr{F}_{\rm b}(\mathscr{H})$ be the boson Fock space over $\mathscr{H}$. Let h be a self-adjoint operator on $\mathscr{H}$ and $d\Gamma_{\rm b}(h)$ be the boson second quantization operator of h. Then $(\mathscr{F}_{\rm b}(\mathscr{H}), d\Gamma_{\rm b}(h), \{A(f)|f \in \mathscr{H}\})$ is an abstract non-relativistic Bose field model with $d\Gamma_{\rm b}(h)$ being the Hamiltonian of the model. The time-t field is given by

$$A(t, f) = e^{itd\Gamma_{\rm b}(h)}A(f)e^{-itd\Gamma_{\rm b}(h)} = A(e^{ith}f),$$

where we have used Corollary 6.2. Using Corollary 6.8(ii), one can show that the operator $A(t, f)$ obeys the field equation

$$i\frac{d}{dt}A(t, f)\Psi = A(t, hf)\Psi, \quad f \in D(h),\ \Psi \in D(N_{\rm b}^{1/2}),$$

where the time derivative is taken in the sense of strong derivative. This is an abstract free **de Broglie equation** [28, Chapter 9].[3]

Example 10.2 Let T be a non-negative and injective self-adjoint operator on a Hilbert space $\mathscr{H}$ such that $T^{1/2} \in \mathfrak{S}_{\mathscr{W}}(\mathscr{H})$ (see Sect. 8.12). Let

$$H_{\rm b} := d\Gamma_{\rm b}(T)$$

and $\phi_{T^{1/2}}(f)$ and $\pi_{T^{1/2}}(f)$ be defined by (8.118) and (8.119) with T replaced by $T^{1/2}$ respectively. Then

$$\mathsf{M}(T, \mathscr{W}) := (\mathscr{F}_{\rm b}(\mathscr{H}), H_{\rm b}, \{\phi_{T^{1/2}}(f), \pi_{T^{1/2}}(f)|f \in \mathscr{W}\})$$

is an abstract relativistic Bose field model. This type of abstract quantum field model is called a **free relativistic Bose field model**.

By Theorem 6.16(ii), the commutation relations hold on $D(H_{\rm b}^{3/2})$:

$$[H_{\rm b}, \phi_{T^{1/2}}(f)] = -i\pi_{T^{1/2}}(f), \tag{10.5}$$

$$[H_{\rm b}, \pi_{T^{1/2}}(f)] = i\phi_{T^{1/2}}(T^2 f), \quad f \in D(T^2) \cap \mathscr{W}. \tag{10.6}$$

In view of Remark 10.4, the model $\mathsf{M}(T, \mathscr{W})$ can be regarded as an operator realization of the following algebraic relations:

$$[\phi(f), \pi(g)] = i\langle f, g\rangle,\ [\phi(f), \phi(g)] = 0,\ [\pi(f), \pi(g)] = 0, \quad f, g \in \mathscr{W}, \tag{10.7}$$

$$[H, \phi(f)] = -i\pi(f), \quad f \in \mathscr{W}, \tag{10.8}$$

$$[H, \pi(f)] = i\phi(T^2 f), \quad f \in D(T^2) \cap \mathscr{W} \tag{10.9}$$

[3]The free bosonic quantum de Broglie field theory described in [28, Chapter 9] is a concrete realization of the present abstract model.

with correspondence $(H, \phi(f), \pi(f)) \mapsto (H_{\mathrm{b}}, \phi_{T^{1/2}}(f), \pi_{T^{1/2}}(f))$.

For a concrete realization of the model $\mathsf{M}(T, \mathscr{W})$, see Sect. 10.7 below.

Let T_1 and T_2 be non-negative injective self-adjoint operators on $\mathscr{H}$ such that (i) $T_1^{1/2}, T_2^{1/2} \in \mathfrak{S}_{\mathscr{W}}(\mathscr{H})$; (ii) $T_2^{-1/2}T_1^{1/2}$ and $T_2^{1/2}T_1^{-1/2}$ are bounded. Suppose that

$$A_{21} := \overline{T_2^{-1/2}T_1^{1/2} - T_2^{1/2}T_1^{-1/2}} \tag{10.10}$$

is not Hilbert–Schmidt. Then, by Theorem 8.18, $\mathsf{M}(T_1, \mathscr{W})$ is not equivalent to $\mathsf{M}(T_2, \mathscr{W})$.

As a special case, we consider the case where $T_1 := T^2$ and $T_2 = \alpha^2 T^2$ with $T \in \mathfrak{S}_{\mathscr{W}}(\mathscr{H})$ $(\alpha > 0, \alpha \neq 1)$. Then

$$A_{21} = \frac{1}{\alpha} - \alpha \neq 0.$$

Suppose that $\mathscr{H}$ is infinite dimensional. Then A_{21} is not Hilbert–Schmidt. Hence, by Theorem 8.18, $\mathsf{M}(T^2, \mathscr{W})$ is not equivalent to $\mathsf{M}(\alpha^2 T^2, \mathscr{W})$.

If $\sigma(T^2) = [0, \infty)$, then $\sigma(\alpha^2 T^2) = [0, \infty)$. Then we have by Theorem 6.1(iv)

$$\sigma(\mathrm{d}\Gamma_{\mathrm{b}}(T^2)) = \sigma(\mathrm{d}\Gamma_{\mathrm{b}}(\alpha^2 T^2)) = [0, \infty).$$

Hence the spectrum of the Hamiltonian $\mathrm{d}\Gamma_{\mathrm{b}}(T_2)$ is equal to that of $\mathrm{d}\Gamma_{\mathrm{b}}(\alpha^2 T^2)$. Therefore we have an example of inequivalent abstract free relativistic Bose field models in which the spectra of Hamiltonians coincide.

Example 10.3 Let $\mathscr{K}$ be a Hilbert space and $\mathscr{F}_{\mathrm{f}}(\mathscr{K})$ be the fermion Fock space over $\mathscr{K}$. We denote by $B(u)$ the fermion annihilation operator with test vector $u \in \mathscr{K}$ on $\mathscr{F}_{\mathrm{f}}(\mathscr{K})$. Let h be a self-adjoint operator on $\mathscr{K}$ and $\mathrm{d}\Gamma_{\mathrm{f}}(h)$ be the fermion second quantization operator of h. Then $(\mathscr{F}_{\mathrm{f}}(\mathscr{K}), \mathrm{d}\Gamma_{\mathrm{f}}(h), \{B(u)|u \in \mathscr{K}\})$ is an abstract Fermi field model. The time-t field is given by

$$B(t, u) = e^{it\mathrm{d}\Gamma_{\mathrm{f}}(h)} B(f) e^{-i\mathrm{d}\Gamma_{\mathrm{f}}(h)} = B(e^{ith}u),$$

where we have used Theorem 7.11. By Theorem 7.13, $B(t, u)$ obeys the field equation

$$i\frac{d}{dt}B(t, u) = B(t, hu), \quad u \in D(h),$$

where the time derivative is taken in the sense of uniform derivative. The present model is an abstract form of the free fermionic quantum de Broglie field theory [28, Chapter 9, §9.8].

Equivalence of quantum field models can be rephrased in terms of time-t fields:

Proposition 10.3 *Let* $\mathsf{M} = (\mathscr{F}, H, \{\phi(f), \pi(f) | f \in \mathscr{W}\})$ *and* $\mathsf{M}' = (\mathscr{F}', H', \{\phi(f)', \pi(f)' | f \in \mathscr{W}\})$ *be abstract relativistic Bose field models. Then* M *is equivalent to* M' *if and only if there exists a unitary operator* $U: \mathscr{F} \to \mathscr{F}'$ *such that, for all* $t \in \mathbb{R}$ *and* $f \in \mathscr{W}$,

$$U\phi(t, f)U^{-1} = \phi'(t, f), \quad U\pi(t, f)U^{-1} = \pi'(t, f), \tag{10.11}$$

where $\phi'(t, f)$ *and* $\pi'(t, f)$ *are the time-zero fields of the model* M'.

Proof Suppose that M is equivalent to M'. Then there exist a unitary operator U: $\mathscr{F} \to \mathscr{F}'$ and a real constant E satisfying (10.3). Then we have for all $t \in \mathbb{R}$ and $f \in \mathscr{W}$

$$\begin{aligned} U\phi(t, f)U^{-1} &= Ue^{itH}U^{-1}U\phi(f)U^{-1}Ue^{-itH}U^{-1} = e^{itUHU^{-1}}\phi'(f)e^{-itUHU^{-1}} \\ &= e^{it(H'+E)}\phi'(f)e^{-it(H'+E)} = e^{itH'}\phi'(f)e^{-itH'} \\ &= \phi'(t, f). \end{aligned}$$

Similarly, one can show that $U\pi(t, f)U^{-1} = \pi'(t, f)$.

Conversely, suppose that there exists a unitary operator $U: \mathscr{F} \to \mathscr{F}'$ such that, for all $t \in \mathbb{R}$ and $f \in \mathscr{W}$, (10.11) holds. Then the case $t = 0$ gives

$$U\phi(f)U^{-1} = \phi'(f), \quad U\pi(f)U^{-1} = \pi'(f).$$

Using the first of these relations, we have

$$Ue^{itH}\phi(f)e^{-itH}U^{-1} = e^{itH'}\phi'(f)e^{-itH'} = e^{itH'}U\phi(f)U^{-1}e^{itH'}.$$

This implies that $V(t)\phi(f)V(t)^{-1} = \phi(f)$, where

$$V(t) := U^{-1}e^{-itH'}Ue^{itH}.$$

Similarly, we have $V(t)\pi(f)V(t)^{-1} = \pi(f)$. Hence $V(t) \in \{\phi(f), \pi(f) | f \in \mathscr{W}\}' = \mathbb{C}I$. Therefore $V(t) = c(t)I$ for some $c(t) \in \mathbb{C}$. It follows that $c(t)e^{itH'} = e^{itUHU^{-1}}$. It is easy to see that $V(t)$ is a strongly continuous semi-group. Hence it follows that $c(t) = e^{itE}$ for some $E \in \mathbb{R}$. Therefore $e^{it(H'+E)} = e^{itUHU^{-1}}$. This implies that $H' + E = UHU^{-1}$. Thus M is equivalent to M'. ■

Remark 10.6 Proposition 10.3 holds also for abstract non-relativistic Bose field models and abstract Fermi field models.

Remark 10.7 In this book, we do not present a general theory of quantum fields. For this topic, see, e.g., [28, 39, 45, 62, 63, 67, 85, 147, 149, 152].

Remark 10.8 In what follows, we shall consider models whose time-zero fields depend on parameters. In such cases, we are interested in investigating equivalence or inequivalence between time-zero fields of different parameters. In relation to this subject, the following is an elementary fact that one should keep in mind: if time-zero fields of different parameters are mutually inequivalent, then the time-t fields of different parameters also are mutually inequivalent. This easily follows from the fact that time-t fields are unitarily equivalent to time-zero fields (see (10.1)–(10.4)). We shall not mention this structure in any of the quantum field models which are discussed below.

10.2 Scale Transformations of Time-Zero Fields

Let $\phi_T(f)$ and $\pi_T(f)$ be defined by (8.118) and (8.119) respectively with $T \in \mathfrak{S}_{\mathscr{W}}(\mathscr{H})$ and, for $\lambda > 0$, define $\phi_{T,\lambda}(f)$ and $\pi_{T,\lambda}(f)$ by

$$\phi_{T,\lambda}(f) := \lambda\phi_T(f), \quad \pi_{T,\lambda}(f) := \lambda^{-1}\pi_T(f), \quad f \in \mathscr{W}. \tag{10.12}$$

The correspondence : $\lambda \mapsto \phi_{T,\lambda}(f)$ is called the **scale transformation** of $\phi_T(f)$ with scaling parameter λ. Note that

$$\phi_{T,\lambda}(f) = \phi_{T/\lambda}(f), \quad \pi_{T,\lambda}(f) = \pi_{T/\lambda}(f). \tag{10.13}$$

Hence, by Theorem 8.15,

$$\Pi_{T,\lambda}(\mathscr{W}) := (\mathscr{F}_{\mathrm{b}}(\mathscr{H}), \{\phi_{T,\lambda}(f), \pi_{T,\lambda}(f) | f \in \mathscr{W}\}) \tag{10.14}$$

is an irreducible Weyl representation of the CCR over $\mathscr{W}$ and

$$\Pi_{T,\lambda}(\mathscr{W}) = \Pi_{T/\lambda}(\mathscr{W}). \tag{10.15}$$

Theorem 10.1 *Assume that $\mathscr{H}$ is separable and infinite dimensional. Let $\lambda_1 \neq \lambda_2$ ($\lambda_1, \lambda_2 > 0$). Then $\Pi_{T,\lambda_1}(\mathscr{W})$ is inequivalent to $\Pi_{T,\lambda_2}(\mathscr{W})$.*

Proof We apply Theorem 8.18 with the following case:

$$T_1 = \frac{T}{\lambda_1}, \quad T_2 = \frac{T}{\lambda_2}.$$

Then it is not so difficult to show that the assumption of Theorem 8.18 is satisfied. We have

$$S := \overline{T_2^{-1}T_1 - T_2T_1^{-1}} = \frac{\lambda_2}{\lambda_1} - \frac{\lambda_1}{\lambda_2}.$$

Since $\lambda_1 \neq \lambda_2$, S is a non-zero scalar operator. By the present assumption, $\mathscr{H}$ is infinite dimensional. Hence S is not Hilbert–Schmidt. Thus, by Theorem 8.18, $\Pi_{T,\lambda_1}(\mathscr{W}) \not\cong \Pi_{T,\lambda_2}(\mathscr{W})$. ■

We emphasize that, in Theorem 10.1, the infinite dimensionality of $\mathscr{H}$ is essential.

Let $\mathscr{H}$ be infinite dimensional. Then Theorem 10.1 shows that $\{\Pi_{T,\lambda}(\mathscr{W})\}_{\lambda>0}$ is a family of mutually inequivalent irreducible Weyl representations of the CCR over $\mathscr{W}$. This gives a representation theoretic meaning to scale parameters λ as indices of a family of mutually inequivalent irreducible Weyl representations of CCR.

10.3 Bogoliubov Translations in Finite and Infinite Volume Theories

In a theory of Bose gas, a class of Bogoliubov translations (see Example 8.3) appears.[4] In this section, we show that, in view of representations of CCR, there is an essential difference between the Bogoliubov translations in a finite volume theory and an infinite volume one.

10.3.1 *Finite Volume Case*

Let M be a bounded open set in the position space $\mathbb{R}^d$ ($d \in \mathbb{N}$) and $\mathscr{F}_{\mathrm{b}}(L^2(M))$ be the boson Fock space over $L^2(M)$. We denote by $a_M(\cdot)$ the annihilation operator on $\mathscr{F}_{\mathrm{b}}(L^2(M))$. We consider a system of Bose gas in M whose particle density is given by a constant $\rho > 0$. For each $f \in L^2(M)$, one has by the Schwarz inequality

$$\int_M |f(\boldsymbol{x})|d\boldsymbol{x} \leq \|f\|_{L^2(M)}|M|^{1/2} < \infty,$$

where $|M|$ denotes the Lebesgue measure of M. Hence the operator

$$b_M(f) := a_M(f) + \sqrt{\rho}\int_M f(\boldsymbol{x})^* d\boldsymbol{x}, \quad f \in L^2(M). \tag{10.16}$$

is well-defined. This is a Bogoliubov translation whose translation functional is given by

$$L_M(f) := \sqrt{\rho}\int_M f(\boldsymbol{x})d\boldsymbol{x}, \quad f \in L^2(M).$$

---

[4] Here we do not present a systematic description of a theory of Bose gas; we restrict our considerations to the form and properties of CCR only. For more details, see [40, 70]. For physical discussions on the subject, see, e.g., [109, 153, 154].

By Example 8.3,

$$\rho_M := (\mathscr{F}_{\mathrm{b}}(L^2(M)), \mathscr{F}_{\mathrm{b},0}(L^2(M)), \{b_M(f) | f \in L^2(M)\})$$

is an irreducible representation of the CCR over $L^2(M)$.

Theorem 10.2 *The representation ρ_M is equivalent to the Fock representation $\rho_{\mathrm{F}}(L^2(M)) = (\mathscr{F}_{\mathrm{b}}(L^2(M)), \mathscr{F}_{\mathrm{b},0}(L^2(M)), \{a_M(f) | f \in L^2(M)\})$.*

Proof Since M is bounded, $\sqrt{\rho} \in L^2(M)$. Hence one can rewrite

$$L_M(f) = \langle \sqrt{\rho}, f \rangle_{L^2(M)}, \quad f \in L^2(M).$$

Hence, by Theorem 8.5, $b_M(f)$ is unitarily equivalent to $a_M(f)$. ■

10.3.2 Infinite Volume Case

We next consider the case where a Bose gas is placed in an unbounded open set $X \subset \mathbb{R}^d$ with $|X| = \infty$. In this case, the Hilbert space of state vectors of the system under consideration is taken to be the boson Fock space $\mathscr{F}_{\mathrm{b}}(L^2(X))$ over $L^2(X)$. We denote by $a_X(\cdot)$ the annihilation operator on $\mathscr{F}_{\mathrm{b}}(L^2(X))$. We assume that the particle density $\rho > 0$ is a constant. For each $f \in L^1(X) \cap L^2(X)$, one can define an operator $b_{X,\rho}(f)$ by

$$b_{X,\rho}(f) := a_X(f) + \sqrt{\rho} \int_X f(\boldsymbol{x})^* d\boldsymbol{x}.$$

This is a Bogoliubov translation. Since $L^1(X) \cap L^2(X)$ is dense in $L^2(X)$, it follows from Example 8.3, $(\mathscr{F}_{\mathrm{b}}(L^2(X)), \mathscr{F}_{\mathrm{b},0}(L^2(X)), \{b_{X,\rho}(f) | f \in L^1(X) \cap L^2(X)\})$ is an irreducible representation of the CCR over $L^1(X) \cap L^2(X)$.

Theorem 10.3 *The representation*

$$(\mathscr{F}_{\mathrm{b}}(L^2(X)), \mathscr{F}_{\mathrm{b},0}(L^2(X)), \{b_{X,\rho}(f) | f \in L^1(X) \cap L^2(X)\})$$

is inequivalent to the Fock representation

$$(\mathscr{F}_{\mathrm{b}}(L^2(X)), \mathscr{F}_{\mathrm{b},0}(L^2(X)), \{a_X(f) | f \in L^1(X) \cap L^2(X)\}).$$

Proof In the present case, $\sqrt{\rho} \notin L^2(X)$. Hence it follows from the fundamental lemma of the calculus of variations (du Bois–Reymond lemma) that there is no vector $h \in L^2(X)$ such that

$$\sqrt{\rho} \int_X f(\boldsymbol{x})^* d\boldsymbol{x} = \langle f, h \rangle_{L^2(X)}, \quad f \in L^1(X) \cap L^2(X).$$

Therefore, by Theorem 8.5, $b_{X,\rho}(f)$ is not unitarily equivalent to $a_X(f)$. ■

Physically the inequivalence in Theorem 10.3 is related to the occurrence of a Bose–Einstein condensation (BEC).[5]

Corollary 10.1 *Let $\rho_1 > 0$, $\rho_2 > 0$ and $\rho_1 \neq \rho_2$. Then $\{b_{X,\rho_1}(f)|f \in L^1(X) \cap L^2(X)\}$ is inequivalent to $\{b_{X,\rho_2}(f)|f \in L^1(X) \cap L^2(X)\}$.*

Proof Suppose that $\{b_{X,\rho_1}(f)|f \in L^1(X) \cap L^2(X)\}$ were equivalent to $\{b_{X,\rho_2}(f)|f \in L^1(X) \cap L^2(X)\}$. Then there exists a unitary operator U on $\mathscr{F}_{\mathrm{b}}(L^2(X))$ such that $b_{X,\rho_2}(f) = Ub_{X,\rho_1}(f)U^{-1}$, $f \in L^1(X) \cap L^2(X)$. Without loss of generality, one can assume that $\rho_1 > \rho_2$. Then the equation implies that $a_X(f) = Ub_{X,\rho}(f)U^{-1}$ with $\rho = (\sqrt{\rho_1} - \sqrt{\rho_2})^2 > 0$. But this contradicts Theorem 10.3. Thus $\{b_{X,\rho_1}(f)|f \in L^1(X) \cap L^2(X)\}$ is not equivalent to $\{b_{X,\rho_2}(f)|f \in L^1(X) \cap L^2(X)\}$. ■

Thus we have a family $\{\{b_{X,\rho}(f)|f \in L^1(X) \cap L^2(X)\}|\rho > 0\}$ of mutually inequivalent irreducible representations of the CCR over $L^1(X)\cap L^2(X)$. This shows that *the particle densities in an infinite volume system have a meaning as the labels of a family of mutually inequivalent irreducible representations of CCR.*

10.4 Translations of Heisenberg CCR and BEC

We consider the case where $X = \mathbb{R}^d$ in the notation in Sect. 10.3.2 and define

$$\hat{\phi}_{\mathrm{F}}(f) := \frac{1}{\sqrt{2}}\overline{(a_{\mathbb{R}^d}(f)^* + a_{\mathbb{R}^d}(f))}, \tag{10.17}$$

$$\hat{\pi}_{\mathrm{F}}(f) := \frac{i}{\sqrt{2}}\overline{(a_{\mathbb{R}^d}(f)^* - a_{\mathbb{R}^d}(f))}, \quad f \in L^2_{\mathbb{R}}(\mathbb{R}^d), \tag{10.18}$$

where $L^2_{\mathbb{R}}(\mathbb{R}^d)$ denotes the set of real elements in $L^2(\mathbb{R}^d)$:

$$L^2_{\mathbb{R}}(\mathbb{R}^d) := \{f \in L^2(\mathbb{R}^d)|f^* = f\}. \tag{10.19}$$

Then, for any dense subspace $\mathscr{W}$ of $L^2_{\mathbb{R}}(\mathbb{R}^d)$,

$$\pi_{\mathrm{F}}^{(d)}(\mathscr{W}) := (\mathscr{F}_{\mathrm{b}}(L^2(\mathbb{R}^d)), \mathscr{F}_{\mathrm{b},0}(L^2(\mathbb{R}^d)), \{\hat{\phi}_{\mathrm{F}}(f), \hat{\pi}_{\mathrm{F}}(f)|f \in \mathscr{W}\}) \tag{10.20}$$

is an example of Fock representations of Heisenberg CCR (see Example 8.10). Hence it is an irreducible Weyl representation of the CCR over $\mathscr{W}$. In a theory of free

[5]For physical aspects of this phenomenon, see, e.g., [71, 99, 159].

Bose gas in $\mathbb{R}^d$ with a constant particle density $\rho > 0$, a translation of Heisenberg CCR of the following type appears [40, 70]:

$$\hat{\phi}_\theta(f) := \hat{\phi}_{\mathrm{F}}(f) + \sqrt{2\rho}(\cos\theta)\int_{\mathbb{R}^d} f(\boldsymbol{x})d\boldsymbol{x}, \tag{10.21}$$

$$\hat{\pi}_\theta(f) := \hat{\pi}_{\mathrm{F}}(f) + \sqrt{2\rho}(\sin\theta)\int_{\mathbb{R}^d} f(\boldsymbol{x})d\boldsymbol{x}, \tag{10.22}$$

where $\theta \in \mathbb{R}$ and $f \in L^1(\mathbb{R}^d) \cap L^2_{\mathbb{R}}(\mathbb{R}^d)$.

We denote by $\mathscr{S}_{\mathbb{R}}(\mathbb{R}^d)$ the space of real elements in $\mathscr{S}(\mathbb{R}^d)$, the space of rapidly decreasing C^∞-functions on $\mathbb{R}^d$:

$$\mathscr{S}_{\mathbb{R}}(\mathbb{R}^d) := \{f \in \mathscr{S}(\mathbb{R}^d) | f^* = f\} \tag{10.23}$$

By Proposition 8.9,

$$\pi_\theta(\mathscr{S}_{\mathbb{R}}(\mathbb{R}^d)) := (\mathscr{F}_{\mathrm{b}}(L^2(\mathbb{R}^d)), \{\hat{\phi}_\theta(f), \hat{\pi}_\theta(f) | f \in \mathscr{S}_{\mathbb{R}}(\mathbb{R}^d)\}) \tag{10.24}$$

is an irreducible Weyl representation of the CCR over $\mathscr{S}_{\mathbb{R}}(\mathbb{R}^d)$. By the periodicity of $\cos\theta$ and $\sin\theta$, we can assume that $\theta \in (0, 2\pi]$ without loss of generality. We first note the following fact:

Theorem 10.4 *For all $\theta \in [0, 2\pi)$, $\pi_\theta(\mathscr{S}_{\mathbb{R}}(\mathbb{R}^d))$ is inequivalent to $\pi_{\mathrm{F}}^{(d)}(\mathscr{S}_{\mathbb{R}}(\mathbb{R}^d))$.*

Proof The representation $\pi_\theta(\mathscr{S}_{\mathbb{R}}(\mathbb{R}^d))$ is the representation $\pi_{\mu,\nu}(\mathscr{W})$ given by (8.108) with the following choice of $(\mathscr{H}, , \mathscr{W}, \mu, \nu)$:

$$\mathscr{H} = L^2(\mathbb{R}^d), \quad \mathscr{W} = \mathscr{S}_{\mathbb{R}}(\mathbb{R}^d),$$

$$\mu(f) = \sqrt{2}(\cos\theta)\int_{\mathbb{R}^d} f(\boldsymbol{x})d\boldsymbol{x}, \quad \nu(f) = \sqrt{2}(\sin\theta)\int_{\mathbb{R}^d} f(\boldsymbol{x})d\boldsymbol{x}, \quad f \in \mathscr{S}_{\mathbb{R}}(\mathbb{R}^d).$$

Hence, in the present case, we have

$$L(f) = \frac{1}{\sqrt{2}}(\mu(f) + i\nu(f)) = e^{i\theta}\int_{\mathbb{R}^d} f(\boldsymbol{x})d\boldsymbol{x}.$$

It is easy to see that there is no vector $h \in L^2(\mathbb{R}^d)$ such that, for all $f \in \mathscr{S}(\mathbb{R}^d)$, $L(f) = \langle h, f\rangle_{L^2(\mathbb{R}^d)}$. Hence, by Theorem 8.12, the representation $\pi_\theta(\mathscr{S}_{\mathbb{R}}(\mathbb{R}^d))$ is inequivalent to $\pi_{\mathrm{F}}^{(d)}(\mathscr{S}_{\mathbb{R}}(\mathbb{R}^d))$. ■

We next consider the family $\{\pi_\theta(\mathscr{S}_{\mathbb{R}}(\mathbb{R}^d))\}_{\theta\in[0,2\pi)}$,

Theorem 10.5 *Let $\theta_1, \theta_2 \in [0, 2\pi)$ and $\theta_1 \neq \theta_2$. Then $\pi_{\theta_1}(\mathscr{S}_{\mathbb{R}}(\mathbb{R}^d))$ is inequivalent to $\pi_{\theta_2}(\mathscr{S}_{\mathbb{R}}(\mathbb{R}^d))$.*

Proof Suppose that $\pi_{\theta_1}(\mathscr{S}_{\mathbb{R}}(\mathbb{R}^d))$ were equivalent to $\pi_{\theta_2}(\mathscr{S}_{\mathbb{R}}(\mathbb{R}^d))$. Then there exists a unitary operator U on $\mathscr{F}_{\mathrm{b}}(L^2(\mathbb{R}^d))$ such that

$$U\hat{\phi}_{\theta_1}(f)U^{-1} = \hat{\phi}_{\theta_2}(f), \quad U\hat{\pi}_{\theta_1}(f)U^{-1} = \hat{\pi}_{\theta_2}(f), \quad f \in \mathscr{S}_{\mathbb{R}}(\mathbb{R}^d).$$

Hence

$$U\hat{\phi}_{\mathrm{F}}(f)U^{-1} = \hat{\phi}_{\mathrm{F}}(f) + \mu_{12}(f), \tag{10.25}$$

$$U\hat{\pi}_{\mathrm{F}}(f)U^{-1} = \hat{\pi}_{\mathrm{F}}(f) + \nu_{12}(f). \tag{10.26}$$

where

$$\mu_{12}(f) := \sqrt{2}(\cos\theta_2 - \cos\theta_1)\int_{\mathbb{R}^d} f(\boldsymbol{x})d\boldsymbol{x},$$

$$\nu_{12}(f) := \sqrt{2}(\sin\theta_2 - \sin\theta_1)\int_{\mathbb{R}^d} f(\boldsymbol{x})d\boldsymbol{x}.$$

We have

$$L_{12}(f) := \frac{1}{\sqrt{2}}(\mu_{12}(f) + i\nu_{12}(f)) = (e^{i\theta_2} - e^{i\theta_1})\int_{\mathbb{R}^d} f(\boldsymbol{x})d\boldsymbol{x}.$$

Since $\theta_1, \theta_2 \in [0, 2\pi)$ and $\theta_1 \neq \theta_2$, it follows that $e^{i\theta_2} - e^{i\theta_1} \neq 0$. Hence there is no vector $h \in L^2(\mathbb{R}^d)$ such that, for all $f \in \mathscr{S}(\mathbb{R}^d)$, $L_{12}(f) = \langle h, f\rangle_{L^2(\mathbb{R}^d)}$. Hence, by Theorem 8.12, (10.25) and (10.26) do not hold. Thus we arrive at a contradiction. ∎

By Theorem 10.5, $\{\pi_\theta(\mathscr{S}_{\mathbb{R}}(\mathbb{R}^d))\}_{\theta\in[0,2\pi)}$ is a family of inequivalent irreducible Weyl representations of the CCR over $\mathscr{S}_{\mathbb{R}}(\mathbb{R}^d)$. This family is related to the occurrence of a BEC with a spontaneous symmetry breaking (see [40, 70], [150, Part II, Chapter 13] or [18, Chapter 10] for details).

10.5 Improper Bogoliubov Transformation and Renormalization

In this section, we reconsider the representation $\rho_{S,T}(\mathscr{H})$ of the CCR over $\mathscr{H}$ defined by (8.63). We take over the notation in Sects. 8.6 and 8.7. As we have seen in Sect. 8.7, a self-adjoint operator $\overline{H}_{\mathrm{qd}}$ associated with (S, T, h) is defined under Assumption (B) (see (8.83)). Hence, if h is injective in addition, then we have a Bose field model

$$\mathsf{M}_{\mathrm{qd}} := (\mathscr{F}_{\mathrm{b}}(\mathscr{H}), \overline{H}_{\mathrm{qd}}, \{\phi_{h^{1/2}}(f), \pi_{h^{1/2}}(f) | f \in D(h^{-1/2}) \cap \mathscr{H}_J\}),$$

where $\phi_{h^{1/2}}(f)$ and $\pi_{h^{1/2}}(f)$ are defined by (8.118) and (8.119) with $T = h^{1/2}$ respectively.

Suppose that T_h is not Hilbert–Schmidt. Then Assumption (B) does not hold. In addition, $(T_h)_J(S_h)_J^*$ may not be Hilbert–Schmidt (e.g., if $(S_h)_J^*$ is bijective, then $(T_h)_J(S_h)_J^*$ is not Hilbert–Schmidt). If $(T_h)_J(S_h)_J^*$ is not Hilbert–Schmidt, then $\langle A^*|(T_h)_J(S_h)_J^*|A^*\rangle$ cannot be defined and hence H_{qd} loses mathematical meaning, implying that the model M_{qd} cannot be defined.

The main purpose of this section is to show that, even in the case where T_h and $(T_h)_J(S_h)_J^*$ are not Hilbert–Schmidt, there exists a model which may be regarded as a correct or "renormalized" form of M_{qd} in such a case.

We assume the following:

Assumption (ST)

(ST.1) Relations (8.60), (8.61), (8.70) and (8.71) hold.

(ST.2) The operator h is a bounded non-negative injective self-adjoint operator on $\mathscr{H}$ and $Jh \subset hJ$.

We first describe a model equivalent to M_{qd} in the case where T is Hilbert–Schmidt.

10.5.1 A Model Equivalent to M_{qd}

In this subsection, we assume that T is Hilbert–Schmidt in addition to Assumption (ST). Then we have Theorem 8.10. Therefore the model M_{qd} is equivalent to the model

$$\mathsf{M}'_{\mathrm{qd}} := (\mathscr{F}_{\mathrm{b}}(\mathscr{H}), H'_{\mathrm{qd}}, \{\phi'_{h^{1/2}}(f), \pi'_{h^{1/2}}(f) | f \in D(h^{-1/2}) \cap \mathscr{H}_J\}),$$

where

$$H'_{\mathrm{qd}} := \mathrm{d}\Gamma_{\mathrm{b}}(h) - \|T_h\|_2^2$$

and

$$\phi'_{h^{1/2}}(f) := U_B \phi_{h^{1/2}}(f) U_B^{-1}, \quad \pi'_{h^{1/2}}(f) := U_B \pi_{h^{1/2}}(f) U_B^{-1}.$$

To compute operators $\phi'_{h^{1/2}}(f)$ and $\pi'_{h^{1/2}}(f)$, we use (8.72). By (8.72) and (8.86), we have

$$U_B A(f) U_B^{-1} = A(S^* f) - A(T^* J f)^*,$$
$$U_B A(f)^* U_B^{-1} = A(S^* f)^* - A(T^* J f)$$

on $U_B\mathscr{F}_{\mathrm{b},0}(\mathscr{H})$. Hence, for all $f \in D(h^{-1/2}) \cap \mathscr{H}_J$,

$$\phi'_{h^{1/2}}(f) = \Phi_{\mathrm{S}}((S^* - T^*)h^{-1/2}f), \tag{10.27}$$

$$\pi'_{h^{1/2}}(f) = \Phi_{\mathrm{S}}(i(S^* + T^*)h^{1/2}f) \tag{10.28}$$

on $U_B\mathscr{F}_{\mathrm{b},0}(\mathscr{H})$, where $\Phi_{\mathrm{S}}(\cdot)$ is the Segal field operator (see (6.94)).

10.5.2 A Renormalized Model

We next consider the case where T_h is not Hilbert–Schmidt. In this case, H_{qd} may not be defined as we have already discussed. But we note that, in the form of model $\mathsf{M}'_{\mathrm{qd}}$, the informally renormalized Hamiltonian "$H'_{\mathrm{qd}} + \|T_h\|_2^2$" (this expression is meaningless, since $\|T_h\|_2^2 = \infty$ by the non-Hilbert–Schmidtness of T_h) may be redefined as $\mathrm{d}\Gamma_{\mathrm{b}}(h)$. Moreover the right hand sides of (10.27) and (10.28) are well-defined even in the case where T_h is not Hilbert–Schmidt (note that, if T_h is not Hilbert–Schmidt, then T is not Hilbert–Schmidt, since $h^{1/2}$ is bounded under the present assumption). Therefore we introduce new operators

$$\phi_{S,T}(f) := \Phi_{\mathrm{S}}((S^* - T^*)h^{-1/2}f), \tag{10.29}$$

$$\pi_{S,T}(f) := \Phi_{\mathrm{S}}(i(S^* + T^*)h^{1/2}f), \quad f \in D(h^{-1/2}) \cap \mathscr{H}_J. \tag{10.30}$$

Theorem 10.6 *Suppose that T is not Hilbert–Schmidt in addition to Assumption (ST). Then $(\mathscr{F}_{\mathrm{b}}(\mathscr{H}), \{\phi_{S,T}(f), \pi_{S,T}(f) | f \in D(h^{-1/2}) \cap \mathscr{H}_J\})$ is an irreducible Weyl representation of the CCR over $D(h^{-1/2}) \cap \mathscr{H}_J$. Moreover, it is inequivalent to the Fock representation $(\mathscr{F}_{\mathrm{b}}(f), \{\phi_{h^{1/2}}(f), \pi_{h^{1/2}}(f) | f \in D(h^{-1/2}) \cap \mathscr{H}_J\})$.*

Proof Throughout the proof, we set

$$\phi(f) := \Phi_{\mathrm{S}}((S^* - T^*)f),$$
$$\pi(f) := \Phi_{\mathrm{S}}(i(S^* + T^*)f), \quad f \in \mathscr{H}_J.$$

Then

$$\phi_{S,T}(f) = \phi(h^{-1/2}f), \quad \pi_{S,T}(f) = \pi(h^{1/2}f), \quad f \in D(h^{-1/2}) \cap \mathscr{H}_J.$$

Let

$$\hat{S} := S^* - T^*, \quad \hat{T} := S^* + T^*.$$

Then, using (8.70) and (8.71), one can show that

$$\operatorname{Im}\left\langle \hat{S}f, \hat{S}g\right\rangle = 0, \quad \operatorname{Im}\left\langle \hat{T}f, \hat{T}g\right\rangle = 0,$$

$$\operatorname{Re}\left\langle \hat{S}f, \hat{T}g\right\rangle = \langle f, g\rangle, \quad f, g \in \mathscr{H}_J.$$

We have

$$\hat{S}^* - i\hat{T}^* = (1-i)S - (1+i)T.$$

Let $g \in \ker(\hat{S}^* - \hat{T}^*)$. Then $(1-i)Sg = (1+i)Tg$. Hence $\|Sg\|^2 = \|Tg\|^2$. By (8.60), $\|Sg\|^2 = \|Tg\|^2 + \|g\|^2$. Hence $\|g\|^2 = 0$. Hence $g = 0$. Therefore $\ker(\hat{S}^* - \hat{T}^*) = \{0\}$, implying that $\hat{S}^* - i\hat{T}^*$ is injective. Hence, by an application of Theorem 8.13, $(\mathscr{F}_{\mathrm{b}}(\mathscr{H}), \{\phi(f), \pi(f)|f \in \mathscr{H}_J\})$ is an irreducible Weyl representation of the CCR over $\mathscr{H}_J$.

We have

$$\hat{S} + \hat{T} = 2S^*.$$

Hence Ran $(\hat{S} + \hat{T}) = \mathscr{H}$. The easily seen equation

$$\hat{S} - \hat{T} = -2T^*$$

implies that $\hat{S} - \hat{T}$ is not Hilbert–Schmidt. Therefore, by an application of Theorem 8.14, $\{\phi(f), \pi(f)|f \in \mathscr{W}\}$ is inequivalent to $\{\phi_{\mathrm{F}}(f), \pi_{\mathrm{F}}(f)|f \in \mathscr{W}\}$ for any dense subspace $\mathscr{W}$ of $\mathscr{H}_J$. This implies that $\{\phi(h^{-1/2}f), \pi(h^{1/2}f)|f \in D(h^{-1/2}) \cap \mathscr{H}_J\}$ is inequivalent to $\{\phi_{\mathrm{F}}(h^{-1/2}f), \pi_{\mathrm{F}}(h^{1/2}f)|f \in D(h^{-1/2}) \cap \mathscr{H}_J\}$. ■

Let

$$H_{\mathrm{ren}}(h) := \mathrm{d}\Gamma_{\mathrm{b}}(h)$$

and define a renormalized model for M_{qd} as

$$\mathsf{M}_{\mathrm{qd,\,ren}} := (\mathscr{F}_{\mathrm{b}}(\mathscr{H}), H_{\mathrm{ren}}(h), \{\phi_{S,T}(f), \pi_{S,T}(f)|f \in \mathscr{H}_J\}).$$

By Theorem 10.6, *the time-zero fields $\phi_{S,T}(f)$ and $\pi_{S,T}(f)$ in the renormalized model are inequivalent to the Fock ones $\phi_{h^{1/2}}(f)$ and $\pi_{h^{1/2}}(f)$*. This is an interesting structure.

10.6 Representations of CCR in a Theory of Weakly Interacting Bosons

In this section, we consider representations of CCR appearing in a theory of weakly interacting bosons. This theory is related to the superfluidity [44]. For physical discussions of the theory, we refer the reader to [44], [153, §3.6], [154, §4-6] or [109, §7.4].

10.6.1 Finite Volume Theory

We first consider a Bose field theory on the cubic box $\mathbb{I}_L^d$ (see (1.114)). We recall that the momentum space corresponding to this configuration space is Γ_L^d given by (1.115). We work with momentum representation. Then the Hilbert space for this system is taken to be the boson Fock space

$$\mathscr{F}_{\mathrm{b}}(\ell^2(\Gamma_L^d)) = \oplus_{n=0}^{\infty} \otimes_{\mathrm{s}}^n \ell^2(\Gamma_L^d) \tag{10.31}$$

over $\ell^2(\Gamma_L^d)$.

For each $\boldsymbol{p} \in \Gamma_L^d$, we define a vector $e_{\boldsymbol{p}} \in \ell^2(\Gamma_L^d)$ by

$$(e_{\boldsymbol{p}})_{\boldsymbol{k}} := \delta_{\boldsymbol{pk}}, \quad \boldsymbol{k} \in \Gamma_L^d. \tag{10.32}$$

Then $\{e_{\boldsymbol{p}}\}_{\boldsymbol{p}\in\Gamma_L^d}$ is a CONS of $\ell^2(\Gamma_L^d)$.

We denote by $a_L(f)$ the annihilation operator with test vector $f \in \ell^2(\Gamma_L^d)$, acting in $\mathscr{F}_{\mathrm{b}}(\ell^2(\Gamma_L^d))$. For each $\boldsymbol{k} \in \Gamma_L^d$, we define an operator $a_{\boldsymbol{k}}$ by

$$a_{\boldsymbol{k}} := a_L(e_{\boldsymbol{k}}). \tag{10.33}$$

It follows from the CCR of $a_L(\cdot)$ and $a_L(\cdot)^*$ that $\{a_{\boldsymbol{k}}, a_{\boldsymbol{k}}^* | \boldsymbol{k} \in \Gamma_L^d\}$ obeys the CCR indexed by Γ_L^d: for all $\boldsymbol{k},\ \boldsymbol{p} \in \Gamma_L^d$,

$$[a_{\boldsymbol{k}}, a_{\boldsymbol{p}}^*] = \delta_{\boldsymbol{kp}}, \quad [a_{\boldsymbol{k}}, a_{\boldsymbol{k}}] = 0 \quad \text{on } \mathscr{F}_{\mathrm{b},0}(\ell^2(\Gamma_L^d)).$$

We consider a system of non-relativistic bosons interacting weakly. For each $\kappa > 0$, an approximate Hamiltonian H_κ for such a system is defined by

$$H_\kappa := \sum_{\substack{\boldsymbol{k}\in\Gamma_L^d \\ 0<|\boldsymbol{k}|\le\kappa}} \left(\frac{1}{2m}\boldsymbol{k}^2 + g\rho\right) a_{\boldsymbol{k}}^* a_{\boldsymbol{k}} + \sum_{\substack{\boldsymbol{k}\in\Gamma_L^d \\ 0<|\boldsymbol{k}|\le\kappa}} \frac{g\rho}{2}(a_{-\boldsymbol{k}}^* a_{\boldsymbol{k}}^* + a_{\boldsymbol{k}} a_{-\boldsymbol{k}}), \tag{10.34}$$

where $m > 0$ is the mass of a bare boson, $\rho > 0$ denotes the density of bare bosons with momentum $\boldsymbol{k} = 0$ and $g > 0$ is a coupling constant. As is seen from the definition of H_κ, the parameter $\kappa > 0$ physically means a momentum cutoff for bosons, called an **ultraviolet cutoff** for boson momenta[6] (note that $\sum_{\boldsymbol{k}\in\Gamma_L^d} a_{-\boldsymbol{k}}^* a_{\boldsymbol{k}}^*$ and $\sum_{\boldsymbol{k}\in\Gamma_L^d} a_{-\boldsymbol{k}} a_{-\boldsymbol{k}}$ have no mathematical meaning). It is easy to see that H_κ is a symmetric operator with $D(H_\kappa) \supset \mathscr{F}_{\mathrm{b},0}(\ell^2(\Gamma_L^d))$.

Since H_κ is an operator quadratic in $a_{\boldsymbol{k}}$ and $a_{\boldsymbol{k}}^*$, it may be diagonalized through a proper Bogoliubov transformation. Indeed this is the case as shown below. Let

$$\epsilon(\boldsymbol{k}) := \frac{1}{2m}\boldsymbol{k}^2 + g\rho, \quad \boldsymbol{k} \in \hat{\mathbb{R}}^d. \tag{10.35}$$

Then $\epsilon(\boldsymbol{k}) \geq g\rho$, implying that $\epsilon(\boldsymbol{k})^2 - (g\rho)^2 \geq 0$. Hence one can define a function on $\hat{\mathbb{R}}^d$ by

$$\eta(\boldsymbol{k}) := \sqrt{\epsilon(\boldsymbol{k})^2 - (g\rho)^2}, \quad \boldsymbol{k} \in \hat{\mathbb{R}}^d. \tag{10.36}$$

Note that

$$\eta(\boldsymbol{0}) = 0.$$

Taking this fact into account, we introduce a function on $\hat{\mathbb{R}}^d \setminus \{\boldsymbol{0}\}$:

$$\gamma(\boldsymbol{k}) := \frac{\epsilon(\boldsymbol{k})}{\eta(\boldsymbol{k})}, \quad \boldsymbol{k} \in \hat{\mathbb{R}}^d \setminus \{\boldsymbol{0}\}. \tag{10.37}$$

It obvious that

$$\eta(\boldsymbol{k}) < \epsilon(\boldsymbol{k}), \quad \boldsymbol{k} \in \hat{\mathbb{R}}^d. \tag{10.38}$$

Hence

$$\gamma(\boldsymbol{k}) > 1, \quad \boldsymbol{k} \in \hat{\mathbb{R}}^d \setminus \{\boldsymbol{0}\}. \tag{10.39}$$

We have

$$\gamma(\boldsymbol{k}) = \frac{1}{\sqrt{1 - \dfrac{(g\rho)^2}{\epsilon(\boldsymbol{k})^2}}}, \quad \boldsymbol{k} \in \hat{\mathbb{R}}^d \setminus \{\boldsymbol{0}\}.$$

[6] A momentum space region in which $|\boldsymbol{k}|$ is very large is called an ultraviolet region, since the wavelength $2\pi/|\boldsymbol{k}|$ is very short.

For all $\boldsymbol{k} \in \Gamma_L^d \setminus \{\boldsymbol{0}\}$,

$$\epsilon(\boldsymbol{k}) \geq \varepsilon_L := \frac{1}{2m}\left(\frac{2\pi}{L}\right)^2 + g\rho.$$

Hence

$$\gamma(\boldsymbol{k}) \leq \gamma_L := \frac{1}{\sqrt{1 - \dfrac{(g\rho)^2}{\varepsilon_L^2}}}, \quad \boldsymbol{k} \in \Gamma_L^d \setminus \{\boldsymbol{0}\}. \tag{10.40}$$

Therefore $\gamma(\cdot)$ is bounded on $\Gamma_L^d \setminus \{\boldsymbol{0}\}$ (but, unbounded on $\hat{\mathbb{R}}^d \setminus \{\boldsymbol{0}\}$).

By (10.39), one can define two sequences $\{u_{\boldsymbol{k}}\}_{\boldsymbol{k}\in\Gamma_L^d}$ and $\{v_{\boldsymbol{k}}\}_{\boldsymbol{k}\in\Gamma_L^d}$ by

$$u_{\boldsymbol{k}} := \sqrt{\frac{1+\gamma(\boldsymbol{k})}{2}}, \quad v_{\boldsymbol{k}} := \sqrt{\frac{\gamma(\boldsymbol{k})-1}{2}}, \quad \boldsymbol{k} \in \Gamma_L^d \setminus \{\boldsymbol{0}\}, \tag{10.41}$$

$$u_{\boldsymbol{0}} := 1, \quad v_{\boldsymbol{0}} := 0. \tag{10.42}$$

It is easy to show that the following equations hold:

$$u_{\boldsymbol{k}}^2 - v_{\boldsymbol{k}}^2 = 1, \quad u_{\boldsymbol{k}} = u_{-\boldsymbol{k}}, \quad v_{\boldsymbol{k}} = v_{-\boldsymbol{k}}, \quad \boldsymbol{k} \in \Gamma_L^d. \tag{10.43}$$

$$u_{\boldsymbol{k}} v_{\boldsymbol{k}} = \frac{g\rho}{2\eta(\boldsymbol{k})}, \quad \boldsymbol{k} \neq \boldsymbol{0}. \tag{10.44}$$

It follows from the boundedness of γ that $\{u_{\boldsymbol{k}}\}_{\boldsymbol{k}\in\Gamma_L^d}$ and $\{v_{\boldsymbol{k}}\}_{\boldsymbol{k}\in\Gamma_L^d}$ are bounded.

We now define

$$b_{\boldsymbol{k}} := u_{\boldsymbol{k}} a_{\boldsymbol{k}} + v_{\boldsymbol{k}} a_{-\boldsymbol{k}}^*, \quad \boldsymbol{k} \in \Gamma_L^d. \tag{10.45}$$

Note that

$$b_{\boldsymbol{0}} = a_{\boldsymbol{0}}.$$

Using the CCR for $a_{\boldsymbol{k}}^{\#}$ and (10.43), one can show that $\{b_{\boldsymbol{k}}, b_{\boldsymbol{k}}^* | \boldsymbol{k} \in \Gamma_L^d\}\}$ obeys the CCR indexed by Γ_L^d: for all $\boldsymbol{k}, \boldsymbol{p} \in \Gamma_L^d$,

$$[b_{\boldsymbol{k}}, b_{\boldsymbol{p}}^*] = \delta_{\boldsymbol{k}\boldsymbol{p}}, \quad [b_{\boldsymbol{k}}, b_{\boldsymbol{k}}] = 0 \quad \text{on } \mathscr{F}_{\mathrm{b},0}(\ell^2(\Gamma_L^d)). \tag{10.46}$$

Equation (10.45) implies that

$$b_{\boldsymbol{k}}^* \supset u_{\boldsymbol{k}} a_{\boldsymbol{k}}^* + v_{\boldsymbol{k}} a_{-\boldsymbol{k}}, \quad \boldsymbol{k} \in \Gamma_L^d. \tag{10.47}$$

Using (10.45) and (10.47), one can represent $a_{\boldsymbol{k}}$ and $a_{\boldsymbol{k}}^*$ in terms of $b_{\boldsymbol{k}}$ and $b_{\boldsymbol{k}}^*$: for all $\boldsymbol{k} \in \Gamma_L^d$,

$$a_{\boldsymbol{k}} = u_{\boldsymbol{k}} b_{\boldsymbol{k}} - v_{\boldsymbol{k}} b_{-\boldsymbol{k}}^*, \tag{10.48}$$

$$a_{\boldsymbol{k}}^* = u_{\boldsymbol{k}} b_{\boldsymbol{k}}^* - v_{\boldsymbol{k}} b_{-\boldsymbol{k}} \tag{10.49}$$

on $\mathscr{F}_{\mathrm{b},0}(\ell^2(\Gamma_L^d))$. Substituting these representations for $a_{\boldsymbol{k}}$ and $a_{\boldsymbol{k}}^*$ on the right hand side of (10.34), we obtain

$$H_\kappa = \sum_{\boldsymbol{k} \in \Gamma_L^d, |\boldsymbol{k}| \le \kappa} \eta(\boldsymbol{k}) b_{\boldsymbol{k}}^* b_{\boldsymbol{k}} + E_\kappa \quad \text{on } \mathscr{F}_{\mathrm{b},0}(\ell^2(\Gamma_L^d)), \tag{10.50}$$

where

$$E_\kappa := - \sum_{\boldsymbol{k} \in \Gamma_L^d, |\boldsymbol{k}| \le \kappa} \eta(\boldsymbol{k}) v_{\boldsymbol{k}}^2 = -\frac{(g\rho)^2}{2} \sum_{\boldsymbol{k} \in \Gamma_L^d, |\boldsymbol{k}| \le \kappa} \frac{1}{\epsilon(\boldsymbol{k}) + \eta(\boldsymbol{k})} < 0. \tag{10.51}$$

Note that the operator $\sum_{\boldsymbol{k} \in \Gamma_L^d, |\boldsymbol{k}| \le \kappa} \eta(\boldsymbol{k}) b_{\boldsymbol{k}}^* b_{\boldsymbol{k}}$ is non-negative on $\mathscr{F}_{\mathrm{b},0}(\ell^2(\Gamma_L^d))$. Hence H_κ is bounded from below on $\mathscr{F}_{\mathrm{b},0}(\ell^2(\Gamma_L^d))$ with $H_\kappa \ge E_\kappa$ on $\mathscr{F}_{\mathrm{b},0}(\ell^2(\Gamma_L^d))$. Then, in the same way as in the proof of Theorem 8.9, one can prove the following fact:

Theorem 10.7 *The operator H_κ is essentially self-adjoint on $\mathscr{F}_{\mathrm{b},0}(\ell^2(\Gamma_L^d)$ and $\overline{H}_\kappa \ge E_\kappa$.*

We note that (10.45) defines a Bogoliubov transformation in the following way. We define a mapping $j_d \colon \ell^2(\Gamma_L^d) \to \ell^2(\Gamma_L^d)$ by

$$(j_d f)(\boldsymbol{k}) := f(-\boldsymbol{k})^*, \quad f \in \ell^2(\Gamma_L^d),\ \boldsymbol{k} \in \Gamma_L^d. \tag{10.52}$$

Then it is easy to see that j_d is a conjugation on $\ell^2(\Gamma_L^d)$.

For each $\kappa > 0$, we define functions (sequences) $u^{(\kappa)}$ and $v^{(\kappa)}$ on Γ_L^d as follows:

$$(u^{(\kappa)})_{\boldsymbol{k}} := \begin{cases} u_{\boldsymbol{k}} & \text{for } |\boldsymbol{k}| \le \kappa \\ 1 & \text{for } |\boldsymbol{k}| > \kappa \end{cases}, \qquad (v^{(\kappa)})_{\boldsymbol{k}} := \begin{cases} v_{\boldsymbol{k}} & \text{for } |\boldsymbol{k}| \le \kappa \\ 0 & \text{for } |\boldsymbol{k}| > \kappa \end{cases}.$$

Then $u^{(\kappa)}$ and $v^{(\kappa)}$ are real-valued bounded sequences satisfying

$$(u^{(\kappa)})_{\boldsymbol{k}}^2 - (v^{(\kappa)})_{\boldsymbol{k}}^2 = 1,\ (u^{(\kappa)})_{\boldsymbol{k}} = (u^{(\kappa)})_{-\boldsymbol{k}},\ (v^{(\kappa)})_{\boldsymbol{k}} = (v^{(\kappa)})_{-\boldsymbol{k}},\ \boldsymbol{k} \in \Gamma_L^d. \tag{10.53}$$

Let

$$b^{(\kappa)}(f) := \overline{a_L(u^{(\kappa)}f) + a_L(j_d(v^{(\kappa)}f))^*} \quad f \in \ell^2(\Gamma_L^d), \tag{10.54}$$

where, for a sequence w on Γ_L^d, $(wf)_{\boldsymbol{k}} := w_{\boldsymbol{k}} f_{\boldsymbol{k}}$, $\boldsymbol{k} \in \Gamma_L^d$. Then

$$b^{(\kappa)}(e_{\boldsymbol{k}}) = \begin{cases} b_{\boldsymbol{k}} & \text{for } |\boldsymbol{k}| \leq \kappa \\ a_{\boldsymbol{k}} & \text{for } |\boldsymbol{k}| > \kappa \end{cases}.$$

It is easy to see the correspondence $(a_L(\cdot), a_L(\cdot)^*) \mapsto (b^{(\kappa)}(\cdot), b^{(\kappa)}(\cdot)^*)$ is a Bogoliubov transformation with conjugation $J = j_d$. Note that (8.70) and (8.71) hold for $S = u^{(\kappa)}$ and $T = v^{(\kappa)}$ with $J = j_d$. Hence, by Lemma 8.9, we obtain the following:

Lemma 10.1 *The triple*

$$\tau_L := (\mathscr{F}_{\mathrm{b}}(\ell^2(\Gamma_L^d)), \mathscr{F}_{\mathrm{b},0}(\ell^2(\Gamma_L^d)), \{b^{(\kappa)}(f) \mid f \in \ell^2(\Gamma_L^d)\}) \tag{10.55}$$

is an irreducible representation of the CCR over $\ell^2(\Gamma_L^d)$.

Let

$$\chi_\kappa(\boldsymbol{k}) := \chi_{[0,\kappa]}(|\boldsymbol{k}|), \quad \boldsymbol{k} \in \hat{\mathbb{R}}^d. \tag{10.56}$$

Then $\chi_\kappa \eta$ can be regarded as a non-negative bounded, multiplication self-adjoint operator on $\ell^2(\Gamma_L^d)$ satisfying $j_d(\chi_\kappa \eta) \subset (\chi_\kappa \eta) j_d$.

Theorem 10.8 *There exists a unitary operator* U *on* $\mathscr{F}_{\mathrm{b}}(\ell^2(\Gamma_L^d))$ *such that*

$$U b^{(\kappa)}(f) U^{-1} = a_L(f), \quad f \in \ell^2(\Gamma_L^d), \tag{10.57}$$

$$U(\overline{H}_\kappa - E_\kappa)U^{-1} = \mathrm{d}\Gamma_{\mathrm{b}}^{(L)}(\chi_\kappa \eta), \tag{10.58}$$

where $\mathrm{d}\Gamma_{\mathrm{b}}^{(L)}(\chi_\kappa \eta)$ *denotes the boson second quantization operator of the multiplication operator* $\chi_\kappa \eta$, *acting in* $\mathscr{F}_{\mathrm{b}}(\ell^2(\Gamma_L^d))$.

Proof Let $h = \chi_\kappa \eta$, $S = u^{(\kappa)}$, $T = v^{(\kappa)}$ and $J = j_d$. Then the pair (h, T) satisfies Assumption (B) in Sect. 8.7 and (S, T, J) obeys (8.60), (8.61), (8.70) and (8.71). Hence, by an application of Theorem 8.10, the result follows. ■

Theorem 10.8 implies the existence of a unique ground state of $\overline{H}_\kappa$:

Corollary 10.2 *The Hamiltonian* $\overline{H}_\kappa$ *has a unique ground state* Ω_κ *given by*

$$\Omega_\kappa := U^{-1} \Omega_{\ell^2(\Gamma_L^d)},$$

up to constant multiples, with ground state energy E_κ:

$$\overline{H}_\kappa \Omega_\kappa = E_\kappa \Omega_\kappa.$$

Formula (10.57) implies that

$$b^{(\kappa)}(f)\Omega_\kappa = 0, \quad f \in \ell^2(\Gamma_L^d).$$

By (10.48) and (10.49), we have

$$a_L(f) = b^{(\kappa)}(u^{(\kappa)} f) - b^{(\kappa)}(v^{(\kappa)} j_d f)^*. \tag{10.59}$$

on $\mathscr{F}_{\mathrm{b},0}(\ell^2(\Gamma_L^d))$. Hence

$$\|a_L(f)\Omega_\kappa\|^2 = \|b^{(\kappa)}(v^{(\kappa)} j_d f)^* \Omega_\kappa\|^2 = \|v^{(\kappa)} f\|^2.$$

Hence, by Corollary 8.3, we have

$$\|N_{\mathrm{b}}^{1/2} \Omega_\kappa\|^2 = \sum_{\boldsymbol{k} \in \Gamma_L^d,\, |\boldsymbol{k}| \le \kappa} v_{\boldsymbol{k}}^2, \tag{10.60}$$

where N_{b} is the boson number operator on $\mathscr{F}_{\mathrm{b}}(\ell^2(\Gamma_L^d))$. This gives a physical meaning of the sequence $\{v_{\boldsymbol{k}}\}_{\boldsymbol{k}}$. By (8.69) with $B = b^{(\kappa)}$, we have also

$$n_{b^{(\kappa)}} = \sum_{\boldsymbol{k} \in \Gamma_L^d} \|b_{\boldsymbol{k}}^{(\kappa)} \Omega_{\ell^2(\Gamma_L^d)}\|^2 = \sum_{\boldsymbol{k} \in \Gamma_L^d,\, |\boldsymbol{k}| \le \kappa} v_{\boldsymbol{k}}^2. \tag{10.61}$$

We have by (10.38)

$$\frac{1}{\epsilon(\boldsymbol{k}) + \eta(\boldsymbol{k})} > \frac{1}{2\varepsilon(\boldsymbol{k})} = \frac{m}{\boldsymbol{k}^2 + 2mg\rho}.$$

It is easy to see that, if $d \ge 2$, then

$$\sum_{\boldsymbol{k} \in \Gamma_L^d} \frac{m}{\boldsymbol{k}^2 + 2mg\rho} = \infty$$

and hence

$$\sum_{\boldsymbol{k} \in \Gamma_L^d} \frac{1}{\epsilon(\boldsymbol{k}) + \eta(\boldsymbol{k})} = \infty.$$

By this fact and (10.51),

$$\lim_{\kappa\to\infty} E_\kappa = -\infty$$

for $d \geq 2$. This is an example of ultraviolet divergences[7] in quantum field models. Hence, to construct a model without ultraviolet cutoff as a suitable limit of the model

$$\mathsf{M}_{L,\kappa} := (\mathscr{F}_{\mathrm{b}}(\ell^2(\Gamma_L^d)), \overline{H}_\kappa, \{a_L(f)|f \in \ell^2(\Gamma_L^d)\})$$

with $d \geq 2$, one needs an energy renormalization.

Remark 10.9 In non-relativistic quantum field theory (quantum theory of de Broglie field), the time-zero field in the Fock representation is given by the annihilation operator (see [28, Chapter 9]).

By Theorem 10.8, the model $\mathsf{M}_{L,\kappa}$ is equivalent to the model

$$\mathsf{M}'_{L,\kappa} := (\mathscr{F}_{\mathrm{b}}(\ell^2(\Gamma_L^d)), \mathrm{d}\Gamma_{\mathrm{b}}^{(L)}(\chi_\kappa\eta), \{Ua_L(f)U^{-1}|f \in \ell^2(\Gamma_L^d)\}).$$

By (10.59)

$$Ua_L(f)U^{-1} = a_L(u^{(\kappa)}f) - a_L(v^{(\kappa)}j_d f)^* \tag{10.62}$$

on $\mathscr{F}_{\mathrm{b},0}(\ell^2(\Gamma_L^d))$. Note that the right hand side is well-defined even if $\kappa = \infty$, since u and v are bounded on Γ_L^d. The operator $\mathrm{d}\Gamma_{\mathrm{b}}^{(L)}(\eta)$ is a non-negative self-adjoint operator. Hence we define

$$c_L(f) := \overline{a_L(uf) - a_L(vj_d f)^*}, \quad f \in \ell^2(\Gamma_L^d), \tag{10.63}$$

$$\hat{H}_L := \mathrm{d}\Gamma_{\mathrm{b}}^{(L)}(\eta). \tag{10.64}$$

It is easy to see that, for all $f \in \ell^2(\Gamma_L^d)$,

$$a_L(f) = c_L(uf) + c_L(vj_d f)^* \tag{10.65}$$

on $\mathscr{F}_{\mathrm{b},0}(\ell^2(\Gamma_L^d))$. Hence it follows that $\{c_L(f)|f \in \ell^2(\Gamma_L^d)\}$ is an irreducible representation of the CCR over $\ell^2(\Gamma_L^d)$. Thus we have a renormalized model

$$\mathsf{M}_{\mathrm{ren},L} := (\mathscr{F}_{\mathrm{b}}(\ell^2(\Gamma_L^d), \hat{H}_L, \{c_L(f)|f \in \ell^2(\Gamma_L^d)\})$$

[7] A quantity Q_κ (a scalar, a vector or an operator) which depends on κ is said to be ultraviolet divergent if Q_κ diverges in a suitable sense as $\kappa \to \infty$ (removal of ultraviolet cutoff).

without ultraviolet cutoff, which may be regarded as a genuine form of a fictitious model corresponding to the symbolical expression "$\lim_{\kappa\to\infty} \mathsf{M}_{L,\kappa}$". In this model, the ground state (physical vacuum) is given by

$$\Omega_L := \Omega_{\ell^2(\Gamma_L^d)}. \tag{10.66}$$

We have

$$\sum_{\boldsymbol{k}\in\Gamma_L^d} v_{\boldsymbol{k}}^2 = \frac{(g\rho)^2}{2} \sum_{\boldsymbol{k}\in\Gamma_L^d\setminus\{\boldsymbol{0}\}} \frac{1}{\eta(\boldsymbol{k})(\epsilon(\boldsymbol{k})+\eta(\boldsymbol{k}))}.$$

Note that, for all $\boldsymbol{k} \in \Gamma_L^d \setminus \{\boldsymbol{0}\}$,

$$\frac{1}{2\epsilon(\boldsymbol{k})^2} < \frac{1}{\eta(\boldsymbol{k})(\epsilon(\boldsymbol{k})+\eta(\boldsymbol{k}))} \le \frac{\gamma(\boldsymbol{k})^2}{1+\gamma(\boldsymbol{k})}\frac{1}{\epsilon(\boldsymbol{k})^2} < \frac{\gamma(\boldsymbol{k})}{\epsilon(\boldsymbol{k})^2} < \frac{\gamma_L}{\epsilon(\boldsymbol{k})^2}.$$

Since $\sum_{\boldsymbol{k}\in\Gamma_L^d\setminus\{\boldsymbol{0}\}} 1/\epsilon(\boldsymbol{k})^2 < \infty$ if and only if $d \le 3$, it follows that the multiplication operator v is Hilbert–Schmidt if and only $d \le 3$. Therefore the representation $\{c_L(f)|f \in \ell^2(\Gamma_L^d)\}$ is equivalent to the Fock representation $\{a_L(f)|f \in \ell^2(\Gamma_L^d)\}$ if and only $d \le 3$.

10.6.2 Infinite Volume Theory

We next consider an infinite volume theory, corresponding heuristically to the case $L \to \infty$ in the model $\mathsf{M}_{\mathrm{ren},L}$. A standard way to construct such a theory is to apply the idea of the Wightman reconstruction theorem in axiomatic quantum field theory (e.g., [149], [45, Chapter 3, §3]) or Gel'fand–Naĭmark–Segal (GNS) construction in representation theory of $*$-algebras (e.g., [85, Chapter III, §III.2.2], [18, Chapter 2, §2.5]). The vacuum expectation values in the model $\mathsf{M}_{\mathrm{ren},L}$ are defined by

$$W_n^{(L)}(f_1,\ldots,f_n) := \left\langle \Omega_L, \tilde{c}_L(f_1)^\# \cdots \tilde{c}_L(f_n)^\# \Omega_L \right\rangle,$$
$$f_1,\ldots,f_n \in \mathscr{E}_d \ (n \in \mathbb{Z}_+),$$

where $\mathscr{E}_d$ is a suitable subspace of $L^2(\hat{\mathbb{R}}^d)$ (the restriction of f_j on Γ_L^d is regarded as an element of $\ell^2(\Gamma_L^d)$) and

$$\tilde{c}_L(f) := \sqrt{\frac{(2\pi)^d}{L^d}} c_L(f), \quad f \in \mathscr{E}_d.$$

The multiplicative factor operator $\sqrt{(2\pi)^d/L^d}$ is used in order that the infinite volume limit $\lim_{L\to\infty} W_n^{(L)}(f_1,\ldots,f_n)$ exists.

Let ϵ, η and γ be as in the preceding subsection. We define functions u and v on $\hat{\mathbb{R}}^d$ as follows:

$$u(\boldsymbol{k}) := \sqrt{\frac{1+\gamma(\boldsymbol{k})}{2}}, \quad v(\boldsymbol{k}) := \sqrt{\frac{\gamma(\boldsymbol{k})-1}{2}}, \quad \boldsymbol{k} \in \hat{\mathbb{R}}^d \setminus \{\boldsymbol{0}\}, \tag{10.67}$$

$$u(\boldsymbol{0}) := 1, \quad v(\boldsymbol{0}) := 0. \tag{10.68}$$

Then

$$u(\boldsymbol{k})^2 - v(\boldsymbol{k})^2 = 1, \quad u(\boldsymbol{k}) = u(-\boldsymbol{k}), \quad v(\boldsymbol{k}) = v(-\boldsymbol{k}), \quad \boldsymbol{k} \in \hat{\mathbb{R}}^d.$$

We have

$$u(\boldsymbol{k}) = \frac{g\rho}{\sqrt{2\eta(\boldsymbol{k})(\epsilon(\boldsymbol{k}) - \eta(\boldsymbol{k}))}}, \tag{10.69}$$

$$v(\boldsymbol{k}) = \frac{g\rho}{\sqrt{2\eta(\boldsymbol{k})(\epsilon(\boldsymbol{k}) + \eta(\boldsymbol{k}))}}, \quad \boldsymbol{k} \neq \boldsymbol{0}. \tag{10.70}$$

Note that $\lim_{|\boldsymbol{k}|\to 0} \eta(\boldsymbol{k}) = 0$. Hence $\lim_{|\boldsymbol{k}|\to 0} u(\boldsymbol{k}) = +\infty$ and $\lim_{|\boldsymbol{k}|\to 0} v(\boldsymbol{k}) = +\infty$. Therefore u and v are unbounded near the origin $\boldsymbol{0} \in \hat{\mathbb{R}}^d$. In the context of quantum field theory, this kind of singularity is called an **infrared singularity**.[8] To avoid this singularity, we take as $\mathscr{E}_d$ the following:

$$\mathscr{E}_d := \{f \in C(\hat{\mathbb{R}}^d) | \operatorname{supp} f \text{ is bounded with } \operatorname{supp} f \subset \hat{\mathbb{R}}^d \setminus \{\boldsymbol{0}\}\}. \tag{10.71}$$

As is easily seen, for all $f \in \mathscr{E}_d$, uf and vf are bounded continuous functions on $\hat{\mathbb{R}}^d$ with bounded support and hence $uf, vf \in L^2(\hat{\mathbb{R}}^d)$.

Lemma 10.2 *For all* $f_1, f_2 \in \mathscr{E}_d$,

$$\lim_{L\to\infty} \left\langle \Omega_L, \tilde{c}_L(f_1)^* \tilde{c}_L(f_2)^* \Omega_L \right\rangle = -\int_{\hat{\mathbb{R}}^d} u(\boldsymbol{k}) v(\boldsymbol{k}) f_1(-\boldsymbol{k}) f_2(\boldsymbol{k}) d\boldsymbol{k}, \tag{10.72}$$

$$\lim_{L\to\infty} \langle \Omega_L, \tilde{c}_L(f_1) \tilde{c}_L(f_2) \Omega_L \rangle = -\int_{\hat{\mathbb{R}}^d} u(\boldsymbol{k}) v(\boldsymbol{k}) f_1(-\boldsymbol{k})^* f_2(\boldsymbol{k})^* d\boldsymbol{k}, \tag{10.73}$$

$$\lim_{L\to\infty} \left\langle \Omega_L, \tilde{c}_L(f_1)^* \tilde{c}_L(f_2) \Omega_L \right\rangle = \int_{\hat{\mathbb{R}}^d} v(\boldsymbol{k})^2 f_1(\boldsymbol{k}) f_2(\boldsymbol{k})^* d\boldsymbol{k}, \tag{10.74}$$

$$\lim_{L\to\infty} \left\langle \Omega_L, \tilde{c}_L(f_1) \tilde{c}_L(f_2)^* \Omega_L \right\rangle = \int_{\hat{\mathbb{R}}^d} u(\boldsymbol{k})^2 f_1(\boldsymbol{k})^* f_2(\boldsymbol{k}) d\boldsymbol{k}. \tag{10.75}$$

[8] A momentum space region in which the modulus $|\boldsymbol{k}|$ of a wave number vector $\boldsymbol{k} \in \hat{\mathbb{R}}^d$ is very small (hence the wave length $2\pi/|\boldsymbol{k}|$ is very long) is called an **infrared region**. The word "infrared" comes from this correspondence.

Proof By (10.63) and (6.39), we have

$$\begin{aligned}\langle \Omega_L, \tilde{c}_L(f_1)^*\tilde{c}_L(f_2)^*\Omega_L\rangle &= -\frac{(2\pi)^d}{L^d}\langle \Omega_L, a_L(vj_d f_1)a_L(uf_2)^*\Omega_L\rangle \\ &= -\frac{(2\pi)^d}{L^d}\sum_{\boldsymbol{k}\in\Gamma_L^d} u(\boldsymbol{k})v(\boldsymbol{k})f_1(-\boldsymbol{k})f_2(\boldsymbol{k}).\end{aligned}$$

Hence, by a theorem on Riemann integrals, the right hand side converges to that of (10.72) as $L \to \infty$. Formula (10.73) follows from taking the complex conjugate of (10.72). In the same manner as in the proof of (10.72), one can prove (10.74) and (10.75). ■

Theorem 10.9 *For all $n \in \mathbb{Z}_+$ and $f_1, \ldots, f_n \in \mathscr{E}_d$, the infinite volume limit*

$$W_n(f_1, \ldots, f_n) := \lim_{L\to\infty} W_n^{(L)}(f_1, \ldots, f_n)$$

exists.

Proof If n is odd, then, by Theorem 6.19, $W_n^{(L)}(\cdot) = 0$. Hence $\lim_{L\to\infty} W_n^{(L)}(\cdot) = 0$. In the case where n is even with $n = 2p$, by Theorem 6.19 again, $W_n^{(L)}$ is written as a combinatorial sum of

$$\left\langle \Omega_L, c_L(f_{i_1})^\# c_L(f_{j_1})^\# \Omega_L\right\rangle \cdots \left\langle \Omega_L, c_L(f_{i_p})^\# c_L(f_{j_p})^\# \Omega_L\right\rangle$$

with $i_1 < \cdots < i_p, i_k < j_k$. By Lemma 10.2, this quantity converges as $L \to \infty$. Thus the result follows. ■

We now construct a representation $\{c(f)|f \in \mathscr{E}_d\}$ of the CCR over $\mathscr{E}_d$ on the boson Fock space

$$\mathscr{F}_\mathrm{b}(L^2(\hat{\mathbb{R}}^d)) = \oplus_{n=0}^\infty \otimes_\mathrm{s}^n L^2(\hat{\mathbb{R}}^d) \tag{10.76}$$

over $L^2(\hat{\mathbb{R}}^d)$ such that, for all $n \in \mathbb{Z}_+$ and $f_1, \ldots, f_n \in \mathscr{E}_d$,

$$W_n(f_1, \ldots, f_n) = \left\langle \Omega, c(f_1)^\# \cdots c(f_n)^\# \Omega\right\rangle \tag{10.77}$$

with

$$\Omega := \Omega_{L^2(\hat{\mathbb{R}}^d)}.$$

This corresponds to a concrete form of construction by the Wightman reconstruction theorem or the GNS construction in the present context. Intuitively, the infinite space $\mathbb{R}^d$ (resp. $\hat{\mathbb{R}}^d$) here is regarded as the infinite volume limit of $\mathbb{I}_L^d$ (resp. $\ell^2(\Gamma_L^d)$).

We denote by $a(f)$ the annihilation operator with test vector $f \in L^2(\hat{\mathbb{R}}^d)$ acting in $\mathscr{F}_{\mathrm{b}}(L^2(\hat{\mathbb{R}}^d))$. Let $J_d \colon L^2(\hat{\mathbb{R}}^d) \to L^2(\hat{\mathbb{R}}^d)$ be the conjugation defined by

$$(J_d f)(\boldsymbol{k}) := f(-\boldsymbol{k})^*, \quad f \in L^2(\hat{\mathbb{R}}L^d),\ \boldsymbol{k} \in \hat{\mathbb{R}}^d. \tag{10.78}$$

Corresponding to the case of the renormalized model $\mathsf{M}_{\mathrm{ren},L}$ in the finite volume theory, we define a densely defined closed operator

$$c(f) := \overline{a(uf) - a(J_d(vf))^*}, \quad f \in \mathscr{E}_d \tag{10.79}$$

on $\mathscr{F}_{\mathrm{b}}(L^2(\hat{\mathbb{R}}^d))$. It is easy to see that

$$\tau_\infty := (\mathscr{F}_{\mathrm{b}}(L^2(\hat{\mathbb{R}}^d)), \mathscr{F}_{\mathrm{b},0}(L^2(\hat{\mathbb{R}}^d)), \{c(f) \mid f \in \mathscr{E}_d)\}) \tag{10.80}$$

is a representation of the CCR over $\mathscr{E}_d$.

Remark 10.10 Note that u and v are unbounded self-adjoint multiplication operators on $L^2(\hat{\mathbb{R}}^d)$. Hence the correspondence $(a(\cdot), a(\cdot)^*) \mapsto (c(\cdot), c(\cdot)^*)$ is a singular Bogoliubov transformation (see Sect. 8.6).

It follows that, for all $f \in \mathscr{E}_d$,

$$a(f) = c(uf) + c(vJ_d f)^* \tag{10.81}$$

on $\mathscr{F}_{\mathrm{b},0}(L^2(\hat{\mathbb{R}}^d)$, which, together with the density of $\mathscr{E}_d$ in $L^2(\hat{\mathbb{R}}^d)$, implies that $\{\{c(f), c(f)^* \mid f \in \mathscr{E}_d\}$ is irreducible. Therefore τ_∞ is an irreducible representation of the CCR over $\mathscr{E}_d$. Moreover, by using Theorem 6.19, one can show that (10.77) holds.

A big difference between the finite volume theory and the infinite one is stated in the following theorem:

Theorem 10.10 *The representation τ_∞ is inequivalent to the Fock representation*

$$\rho_{\mathrm{F}}(\mathscr{E}_d) = (\mathscr{F}_{\mathrm{b}}(L^2(\hat{\mathbb{R}}^d)), \mathscr{F}_{\mathrm{b},0}(L^2(\hat{\mathbb{R}}^d)), \{a(f) \mid f \in \mathscr{E}_d\})$$

over $\mathscr{E}_d$.

Proof Suppose that τ_∞ were equivalent to the Fock representation $\rho_{\mathrm{F}}(\mathscr{E}_d)$. Then there exists a unitary operator U on $\mathscr{F}_{\mathrm{b}}(L^2(\hat{\mathbb{R}}^d))$ such that $Uc(f)U^{-1} = a(f),\ f \in \mathscr{E}_d$. Hence, letting $\Omega := U^{-1}\Omega_{L^2(\hat{\mathbb{R}}^d)}$, we have that $c(f)\Omega = 0$, $\forall f \in \mathscr{E}_d$. Then, for all $\Psi \in \mathscr{F}_{\mathrm{b},0}(L^2(\hat{\mathbb{R}}^d))$ and $f \in \mathscr{E}_d$,

$$\left\langle a(uf)^*\Psi, \Omega \right\rangle = \langle a(vJ_d f)\Psi, \Omega \rangle .$$

This implies that $\overline{\langle a(f)^*\Psi, \Omega\rangle} = \left\langle a(J_d vu^{-1} f)\Psi, \Omega\right\rangle$, $f \in \mathscr{E}_d$. Hence, by Proposition 8.4, $\overline{vu^{-1}}$ is a Hilbert–Schmidt operator. But $T := \overline{vu^{-1}}$ is the multiplication operator by the continuous function $v(\boldsymbol{k})u(\boldsymbol{k})^{-1}$ on $\hat{\mathbb{R}}^d \setminus \{\boldsymbol{0}\}$. Hence the spectrum $\sigma(T)$ of T is not discrete. This is a contradiction. ■

The model corresponding to the infinite volume limit of the model $\mathsf{M}_{\mathrm{ren},L}$ is defined by

$$\mathsf{M}_{\mathrm{ren},\infty} := (\mathscr{F}_{\mathrm{b}}(L^2(\hat{\mathbb{R}}^d)), \hat{H}_\infty, \{c(f)|f \in \mathscr{E}_d\})$$

with

$$\hat{H}_\infty := \mathrm{d}\Gamma_{\mathrm{b}}(\eta).$$

It is easy to see that the time-t field in the present model is given by

$$c(t, f) := e^{it\hat{H}_\infty} c(f) e^{-it\hat{H}_\infty} = c(e^{it\eta} f), \quad f \in \mathscr{E}_d,\ t \in \mathbb{R}.$$

The Hamiltonian $\hat{H}_\infty$ has a unique ground state

$$\Omega_\infty := \Omega_{L^2(\hat{\mathbb{R}}^d)}$$

with $\hat{H}_\infty \Omega_\infty = 0$.

By Theorem 10.10, the model $\mathsf{M}_{\mathrm{ren},\infty}$ is inequivalent to any free Bose field model of the form $(\mathscr{F}_{\mathrm{b}}(L^2(\hat{\mathbb{R}}^d)), \mathrm{d}\Gamma_{\mathrm{b}}(T), \{a(f)|f \in \mathscr{E}_d\})$ with T being a self-adjoint operator on $L^2(\hat{\mathbb{R}}^d)$.

In concluding this section, we consider the dependence of $c(f)$ on the parameters m, g and ρ. We set

$$\lambda := (m, g, \rho) \in \mathbb{R}^3_+ \ \ (\mathbb{R}_+ := (0, \infty)) \tag{10.82}$$

and define

$$p(\lambda) := mg\rho. \tag{10.83}$$

To indicate the dependence of $c(f)$ on λ, we write

$$c_\lambda(f) := c(f), \quad \tau_{\infty,\lambda} := \tau_\infty. \tag{10.84}$$

Theorem 10.11 *If* $p(\lambda_1) \neq p(\lambda_2)$ $(\lambda_i := (m_i, g_i, \rho_i) \in \mathbb{R}^3_+,\ i = 1, 2)$, *then* τ_{∞,λ_1} *is inequivalent to* τ_{∞,λ_2}.

Proof Suppose that τ_{∞,λ_1} were equivalent to τ_{∞,λ_2}. Then there exists a unitary operator U on $\mathscr{F}_{\rm b}(L^2(\hat{\mathbb{R}}^d))$ such that $Uc_{\lambda_1}(f)U^{-1} = c_{\lambda_2}(f),\ f \in \mathscr{E}_d$. By (10.81), we have for all $\Psi \in \mathscr{F}_{\rm b,0}(L^2(\hat{\mathbb{R}}^d))$

$$\begin{aligned} a(f)\Psi &= c_{\lambda_1}(u_1 f)\Psi - c_{\lambda_1}(v_1 J_d f)^*\Psi \\ &= U^{-1}c_{\lambda_2}(u_2 f)U\Psi - U^{-1}c_{\lambda_2}(v_2 J_d f)^*U\Psi, \end{aligned}$$

where $u_i := u$ and $v_i := v$ with $(m, g, \rho) = \lambda_i$ $(i = 1, 2)$. Hence

$$Ua(f)\Psi = c_{\lambda_2}(u_2 f)U\Psi - c_{\lambda_2}(v_2 J_d f)^*U\Psi.$$

Hence, for all $\Phi \in \mathscr{F}_{\rm b,0}(L^2(\hat{\mathbb{R}}^d))$,

$$\langle \Phi, Ua(f)\Psi\rangle = \big\langle (c_{\lambda_2}(u_2 f)^* - c_{\lambda_2}(v_2 J_d f))\Phi, U\Psi\big\rangle.$$

Using (10.81) again, we see that the right hand side is equal to $\langle a(f)^*\Phi, U\Psi\rangle$. Hence $\langle \Phi, Ua(f)\Psi\rangle = \langle a(f)^*\Phi, U\Psi\rangle$. Since $\mathscr{F}_{\rm b,0}(L^2(\hat{\mathbb{R}}^d))$ is a core for $a(f)$, this equation extends to all $\Psi \in D(a(f))$. The subspace $\mathscr{F}_{\rm b,0}(L^2(\hat{\mathbb{R}}^d))$ is a core for $a(f)^*$ too. Hence $U\Psi \in D((a(f)^*)^*) = D(a(f))$ and $a(f)U\Psi = Ua(f)\Psi$. Therefore $Ua(f) \subset a(f)U$, which implies that $Ua(f)^* \subset a(f)^*U$. Hence $U \in \{a(f), a(f)^* | f \in \mathscr{E}_d\}'$. By this fact and Theorem 6.9, we conclude that $U = e^{i\theta}$ for some $\theta \in \mathbb{R}$. Thus

$$c_{\lambda_1}(f) = c_{\lambda_2}(f),\ f \in \mathscr{E}_d.$$

This implies

$$a((u_1 - u_2)f) = a((v_1 - v_2)J_d f)^* \quad \text{on } \mathscr{F}_{\rm b,0}(L^2(\hat{\mathbb{R}}^d)).$$

Then, by Proposition 6.3, $(u_1 - u_2)f = 0$ and $(v_1 - v_2)J_d f = 0$ for all $f \in \mathscr{E}_d$. Hence $u_1(\boldsymbol{k}) = u_2(\boldsymbol{k})$ and $v_1(\boldsymbol{k}) = v_2(\boldsymbol{k}),\ \boldsymbol{k} \in \hat{\mathbb{R}}^d \setminus \{\boldsymbol{0}\}$. These equations imply that $m_1 g_1 \rho_1 = m_2 g_2 \rho_2$. ■

Theorem 10.11 tells us the following: let two of m, g and ρ be fixed, say, m and g. Then the set of particle densities ρ forms an index set of a family $\{\tau_{\infty,\lambda}\}_\rho$ of mutually inequivalent irreducible representations of CCR. The same applies to the parameters m and g. Thus masses, coupling constants and particle densities appear respectively as an index set of a family of mutually inequivalent irreducible representations of CCR.

Remark 10.11 There is a theory of interacting non-relativistic electrons with spin 1/2, which is constructed in a framework of fermion Fock space and has a structure similar to the boson theory discussed in this section in the sense that one needs only to replace the bosonic Bogoliubov transformation by a fermionic Bogoliubov transformation (see, e.g., [109, Chapter 5], [154, Chapter 4, §4.4]). In the physics literature, the theory is used to explain superconductivity and called the Bardeen–Cooper–Schrieffer (BCS) theory [42]. In quite the same way as in the theory presented above, one can construct an infinite volume theory of such a system and see that a representation of CAR inequivalent to a Fock representation of CAR appear. But, here, we omit describing the construction (cf. [69, 84]).

10.7 The Free Hermitian Klein–Gordon Quantum Field

10.7.1 Definition

We consider a concrete quantum field model describing a quantum field of free relativistic bosons with mass $m \geq 0$ in the d-dimensional position space $\mathbb{R}^d$, where $d \in \mathbb{N}$ is arbitrary. A Hilbert space of state vectors of such a system can be taken to be the boson Fock space

$$\mathscr{F}_{\mathrm{b}}(L^2(\hat{\mathbb{R}}^d)) = \oplus_{n=0}^{\infty} \otimes_{\mathrm{s}}^n L^2(\hat{\mathbb{R}}^d)$$

over $L^2(\hat{\mathbb{R}}^d)$. The space $\hat{\mathbb{R}}^d$ here denotes the momentum space of one free relativistic boson. The energy $\omega_m(\boldsymbol{k})$ of one free boson with momentum $\boldsymbol{k} \in \hat{\mathbb{R}}^d$ is given by

$$\omega_m(\boldsymbol{k}) = \sqrt{\boldsymbol{k}^2 + m^2}. \tag{10.85}$$

We denote by ω_m the multiplication operator by the function $\omega_m(\cdot)$, which is a non-negative self-adjoint operator. It is easy to see that the operator ω_m is injective.

The Hamiltonian of the model we consider is defined by

$$H_{\mathrm{b}} := \mathrm{d}\Gamma_{\mathrm{b}}(\omega_m), \tag{10.86}$$

the boson second quantization operator of ω_m.

For each $f \in L^2(\hat{\mathbb{R}}^d)$, we denote by $a(f)$ the annihilation operator on $\mathscr{F}_{\mathrm{b}}(L^2(\hat{\mathbb{R}}^d))$ with test vector f.

We define a mapping $C_d \colon L^2(\mathbb{R}^d) \to L^2(\mathbb{R}^d)$ by

$$C_d f := f^*, \quad f \in L^2(\mathbb{R}^d).$$

It is easy to see that C_d is a conjugation. The conjugation C_d is called the **complex conjugation** on $L^2(\mathbb{R}^d)$. Hence

$$L_{\mathrm{R}}(\mathbb{R}^d) := \{f \in L^2(\mathbb{R}^d) | C_d f = f\}$$

is a real Hilbert space.[9]

Let $\mathscr{F}_d\colon L^2(\mathbb{R}^d) \to L^2(\hat{\mathbb{R}}^d)$ be the d-dimensional Fourier transform (see (1.74)) and

$$\hat{C}_d := \mathscr{F}_d C_d \mathscr{F}_d^{-1}$$

Then it follows that $\hat{C}_d$ is a conjugation on $L^2(\hat{\mathbb{R}}^d)$. Concretely, one has

$$(\hat{C}_d g)(\boldsymbol{k}) = g(-\boldsymbol{k})^*, \quad g \in L^2(\hat{\mathbb{R}}^d), \quad \text{a.e.} \boldsymbol{k} \in \hat{\mathbb{R}}^d.$$

Hence

$$L^2_{\mathrm{R}}(\hat{\mathbb{R}}^d) := \{f \in L^2(\hat{\mathbb{R}}^d) | \hat{C}_d f = f\}$$

is a real Hilbert space. It follows that

$$\mathscr{F}_d(L^2_{\mathrm{R}}(\mathbb{R}^d)) = L^2_{\mathrm{R}}(\hat{\mathbb{R}}^d).$$

For a tempered distribution f, we denote by $\hat{f}$ its Fourier transform and by $\mathscr{S}'_{\mathbb{R}}(\mathbb{R}^d)$ the space of real tempered distributions on $\mathbb{R}^d$. For each $s \in \mathbb{R}$, we introduce

$$\mathscr{M}_s(\mathbb{R}^d) := \{f \in \mathscr{S}'_{\mathbb{R}}(\mathbb{R}^d) | \omega_m^s \hat{f} \in L^2(\hat{\mathbb{R}}^d)\}$$

For each $f \in \mathscr{M}_{-1/2}(\mathbb{R}^d) \cap \mathscr{M}_{1/2}(\mathbb{R}^d)$, we define

$$\phi_m(f) := \phi_{\mathrm{F}}(\omega_m^{-1/2} \hat{f}) = \frac{1}{\sqrt{2}} \overline{(a(\omega_m^{-1/2} \hat{f})^* + a(\omega_m^{-1/2} \hat{f}))}, \tag{10.87}$$

$$\pi_m(f) := \pi_{\mathrm{F}}(\omega_m^{1/2} \hat{f}) = \frac{i}{\sqrt{2}} \overline{(a(\omega_m^{1/2} \hat{f})^* - a(\omega_m^{1/2} \hat{f}))}, \tag{10.88}$$

where $\{\phi_{\mathrm{F}}(g), \pi_{\mathrm{F}}(g) | g \in L^2_{\mathrm{R}}(\hat{\mathbb{R}}^d)\}$ is the Fock representation of the Heisenberg CCR over $L^2_{\mathrm{R}}(\hat{\mathbb{R}}^d)$.

In what follows, we assume the following:

Assumption (d): In the case $m = 0$, we assume that $d \geq 2$.

Remark 10.12 If $m = 0$ and $d = 1$, then, for all $f \in \mathscr{S}_{\mathbb{R}}(\mathbb{R})$,

$$(\omega_0^{-1/2} \hat{f})(k) = \frac{1}{\sqrt{|k|}} \hat{f}(k), \quad k \in \hat{\mathbb{R}} \setminus \{0\}.$$

[9] In notation in Sect. 1.9, $L^2_{\mathrm{R}}(\mathbb{R}^d) = (L^2(\mathbb{R}^d))_{C_d}$.

Hence, if $\inf_{k\in[-\delta,\delta]}|\hat{f}(k)| > 0$ for some $\delta > 0$, then $\omega_0^{-1/2}\hat{f} \notin L^2(\hat{\mathbb{R}})$. Hence $\mathscr{S}_{\mathbb{R}}(\mathbb{R})$ is not a subset of $\mathscr{M}_{-1/2}(\mathbb{R})$. This is an example of infrared divergences which may occur in a massless quantum field theory. To avoid this type of divergence in the case $d = 1$, one can change the test function space $\mathscr{M}_{-1/2}(\mathbb{R}) \cap \mathscr{M}_{1/2}(\mathbb{R})$ to a subspace $\{f \in \mathscr{M}_{-1/2}(\mathbb{R}) \cap \mathscr{M}_{1/2}(\mathbb{R}) |\ \hat{f}$ vanishes in a neighborhood of $0 \in \hat{\mathbb{R}}\ \}$. But, in this book, we do not take such an approach for simplicity.

It is easy to see that

$$\mathscr{S}_{\mathbb{R}}(\mathbb{R}^d) \subset \mathscr{M}_{-1/2}(\mathbb{R}^d) \cap \mathscr{M}_{1/2}(\mathbb{R}^d).$$

Let

$$\pi_m := (\mathscr{F}_{\mathrm{b}}(L^2(\hat{\mathbb{R}}^d)), \mathscr{F}_{\mathrm{b},0}(L^2(\hat{\mathbb{R}}^d)), \{\phi_m(f), \pi_m(f) | f \in \mathscr{S}_{\mathbb{R}}(\mathbb{R}^d)\}). \qquad (10.89)$$

Then it is easy to see that π_m is a representation of the Heisenberg CCR over $\mathscr{S}_{\mathbb{R}}(\mathbb{R}^d)$. In fact, we have the following:

Theorem 10.12 *The representation π_m is an irreducible Weyl representation of the CCR over $\mathscr{S}_{\mathbb{R}}(\mathbb{R}^d)$.*

Proof Let

$$\hat{\pi}_m := (\mathscr{F}_{\mathrm{b}}(L^2(\hat{\mathbb{R}}^d)), \mathscr{F}_{\mathrm{b},0}(L^2(\hat{\mathbb{R}}^d)), \{\phi_{\mathrm{F}}(\omega_m^{-1/2} f), \pi_{\mathrm{F}}(\omega_m^{1/2} f) | f \in (\mathscr{S}(\hat{\mathbb{R}}^d))_{\hat{C}_d}\}),$$

where

$$(\mathscr{S}(\hat{\mathbb{R}}^d))_{\hat{C}_d} := \{f \in \mathscr{S}(\hat{\mathbb{R}}^d) | \hat{C}_d f = f\}.$$

Then $\hat{\pi}_m$ is an example of $\Pi_T(\mathscr{W})$ given by (8.120) with the following choice of $(\mathscr{H}, \mathscr{W}, T)$:

$$\mathscr{H} = L^2(\hat{\mathbb{R}}^d), \quad \mathscr{W} = (\mathscr{S}(\hat{\mathbb{R}}^d))_{\hat{C}_d}, \qquad T = \omega_m^{1/2}.$$

It is easy to see that

$$\hat{C}_d \omega_m \subset \omega_m \hat{C}_d, \quad (\mathscr{S}(\hat{\mathbb{R}}^d))_{\hat{C}_d} \subset D(\omega_m^{1/2}) \cap D(\omega_m^{-1/2}).$$

Using the du Bois–Reymond lemma, one can show that $\omega_m^{\pm 1/2}((\mathscr{S}(\hat{\mathbb{R}}^d))_{\hat{C}_d})$ are dense in $L^2_{\mathrm{R}}(\hat{\mathbb{R}}^d)$. Hence it follows from Theorem 8.15 that $\hat{\pi}_m$ is an irreducible Weyl representation of the CCR over $(\mathscr{S}_{\mathbb{R}}(\mathbb{R}^d))_{\hat{C}_d}$. We have

$$\mathscr{F}_d(\mathscr{S}_{\mathbb{R}}(\mathbb{R}^d)) = (\mathscr{S}_{\mathbb{R}}(\mathbb{R}^d))_{\hat{C}_d}.$$

Hence π_m is obtained by changing the index set $(\mathscr{S}_{\mathbb{R}}(\mathbb{R}^d))_{\hat{C}_d}$ of the representation $\hat{\pi}_m$ to $\mathscr{S}_{\mathbb{R}}(\mathbb{R}^d)$. Thus π_m is an irreducible Weyl representation of the CCR over $\mathscr{S}_{\mathbb{R}}(\mathbb{R}^d)$. ■

In conventional models of quantum scalar fields, $\phi_m(\cdot)$ and $\pi_m(\cdot)$ are taken as time-zero fields. A simple example of such models is given by

$$(\mathscr{F}_{\mathrm{b}}(L^2(\hat{\mathbb{R}}^d)), H_{\mathrm{b}}, \{\phi_m(f), \pi_m(f) | f \in \mathscr{S}_{\mathbb{R}}(\mathbb{R}^d)\}),$$

where H_{b} is defined by (10.86). In this model, the time-t fields are given by

$$\phi_m(t, f) := e^{itH_{\mathrm{b}}}\phi_m(f)e^{-itH_{\mathrm{b}}},$$
$$\pi_m(t, f) := e^{itH_{\mathrm{b}}}\pi_m(f)e^{-itH_{\mathrm{b}}}, \quad t \in \mathbb{R}, \ f \in \mathscr{S}_{\mathbb{R}}(\mathbb{R}^d).$$

Applying Corollary 6.2 with $T = \omega_m$, we have

$$\phi_m(t, f) = \phi_{\mathrm{F}}(\omega_m^{-1/2} e^{it\omega_m} \hat{f}), \tag{10.90}$$

$$\pi_m(t, f) = \pi_{\mathrm{F}}(\omega_m^{1/2} e^{it\omega_m} \hat{f}). \tag{10.91}$$

We denote by N_{b} the boson number operator on $\mathscr{F}_{\mathrm{b}}(L^2(\hat{\mathbb{R}}^d))$. It is easy to see that, for all $f \in \mathscr{S}_{\mathbb{R}}(\mathbb{R}^d)$ and $p \in \mathbb{N}$, $\omega_m^{-1/2}\hat{f} \in D(\omega_m^p)$. Hence, applying Corollary 6.8(ii) with $T = \omega_m$, we see that, for all $\Psi \in D(N_{\mathrm{b}}^{1/2})$, $\phi_m(t, f)$ is infinitely strongly differentiable and

$$\frac{d}{dt}\phi_m(t, f)\Psi = \pi_m(t, f)\Psi,$$
$$\frac{d^p}{dt^p}\phi_m(t, f)\Psi = \phi_{\mathrm{F}}(i^p \omega_m^p \omega_m^{-1/2} e^{it\omega_m} \hat{f})\Psi.$$

Let Δ be the d-dimensional generalized Laplacian:

$$\Delta := \sum_{j=1}^{d} D_j^2. \tag{10.92}$$

Then, as is well known, we have

$$(\mathscr{F}_d \Delta f)(\boldsymbol{k}) = -\boldsymbol{k}^2 \hat{f}(\boldsymbol{k}), \quad f \in D(\Delta), \ \text{a.e.} \boldsymbol{k} \in \hat{\mathbb{R}}^d.$$

Hence

$$(\mathscr{F}_d(-\Delta + m^2) f)(\boldsymbol{k}) = \omega_m(\boldsymbol{k})^2 \hat{f}(\boldsymbol{k}), \quad \text{a.e.} \boldsymbol{k} \in \hat{\mathbb{R}}^d.$$

Therefore, for all $f \in \mathscr{S}_{\mathbb{R}}(\mathbb{R}^d)$,

$$\frac{d^2}{dt^2}\phi_m(t, f)\Psi + \phi_m(t, (-\Delta + m^2)f)\Psi = 0. \tag{10.93}$$

This means that the operator-valued functional[10] $(t, f) \mapsto \phi_m(t, f)$ obeys the **free Klein–Gordon equation** with mass m. Note that $\phi_m(t, f)$ is self-adjoint. Hence the expectation value $\langle \Phi, \phi_m(t, f)\Phi \rangle$ with respect to $\Phi \in D(\phi_m(t, f))$ is a real number. Based on these facts, the operator-valued functional $\phi_m(t, f)$ is called the **free Hermitian Klein–Gordon quantum field** with mass m.

10.7.2 Boson Masses as Indices of a Family of Mutually Inequivalent Representations of CCR

As a result in the preceding subsection, we have a family $\{\pi_m\}_{m>0}$ of irreducible Weyl representations of the CCR over $\mathscr{S}_{\mathbb{R}}(\mathbb{R}^d)$, each of which gives the time-zero fields in the theory of the free Hermitian Klein–Gordon quantum field with mass m. An interesting fact in this family is the following:

Theorem 10.13 *Let $m_1, m_2 > 0$ and $m_1 \neq m_2$. Then π_{m_1} is inequivalent to π_{m_2}.*

Proof We apply Theorem 8.18 with the following case:

$$\mathscr{H} = L^2(\hat{\mathbb{R}}^d), \quad J = \hat{C}_d, \quad \mathscr{W} = \mathscr{F}_d\mathscr{S}_{\mathbb{R}}(\mathbb{R}^d),$$
$$T_1 = \omega_{m_1}^{1/2}, \quad T_2 = \omega_{m_2}^{1/2}.$$

Then it is not so difficult to show that the assumption of Theorem 8.18 is satisfied. Hence we need only to show that

$$M := \overline{\omega_{m_2}^{-1/2}\omega_{m_1}^{1/2} - \omega_{m_2}^{1/2}\omega_{m_1}^{-1/2}}$$

is not Hilbert–Schmidt. It is easy to see that M is a multiplication operator by the function

$$F(\boldsymbol{k}) = \frac{\omega_{m_1}(\boldsymbol{k}) - \omega_{m_2}(\boldsymbol{k})}{\sqrt{\omega_{m_1}(\boldsymbol{k})\omega_{m_2}(\boldsymbol{k})}} = \frac{(m_1 - m_2)(m_1 + m_2)}{(\omega_{m_1}(\boldsymbol{k}) + \omega_{m_2}(\boldsymbol{k}))\sqrt{\omega_{m_1}(\boldsymbol{k})\omega_{m_2}(\boldsymbol{k})}}, \quad \boldsymbol{k} \in \hat{\mathbb{R}}^d.$$

The function $F$is continuous. Without loss of generality, we can assume that $m_1 > m_2$. Then we have

$$\sigma(M) = \overline{F(\hat{\mathbb{R}}^d)} = [0, (m_1 - m_2)/\sqrt{m_1 m_2}].$$

[10] In fact, the correspondence $f \mapsto \phi_m(t, f)$ is an operator-valued distribution (see [28, §8.2]).

Hence, M is not Hilbert–Schmidt.[11] ■

Theorem 10.13 shows that $\{\pi_m\}_{m>0}$ is a family of mutually inequivalent irreducible Weyl representations of the CCR over $\mathscr{S}_{\mathbb{R}}(\mathbb{R}^d)$. This gives a representation theoretic meaning for boson masses. Namely, the set of positive boson masses forms an index set for a family of mutually inequivalent irreducible Weyl representations of the CCR over $\mathscr{S}_{\mathbb{R}}(\mathbb{R}^d)$. This is an interesting example of physical correspondences associated with inequivalency of representations of CCR.

Remark 10.13 Theorem 10.13 has appeared in [131, Theorem X.46]. But the proof there is completely different from the above one. Our treatment is more general in the sense that it makes clear a general abstract structure behind π_m (see Theorem 8.18). It can be shown also that π_0 is inequivalent to π_m for all $m > 0$. For the proof of this fact and related aspects, see [24].

10.8 Quantum Fields at Finite Temperatures

Let $\mathscr{H}$ be an infinite-dimensional separable Hilbert space and $h_\mathfrak{b}$ be a strictly positive self-adjoint operator on $\mathscr{H}$, physically denoting a one-boson Hamiltonian.[12] Let J be a conjugation on $\mathscr{H}$ and assume that

$$Jh_\mathfrak{b} \subset h_\mathfrak{b} J.$$

We denote by $\mathscr{H}_J$ the real part of $\mathscr{H}$ with respect to J as before.

We set

$$m_\mathfrak{b} := \inf \sigma(h_\mathfrak{b}) > 0.$$

Then it follows that, for all $\beta > 0$,

$$\|e^{-\beta h_\mathfrak{b}}\| = e^{-\beta m_\mathfrak{b}} < 1.$$

Hence, by applying a well-known theorem, we see that $1 \pm e^{-\beta h_\mathfrak{b}}$ are bijective with bounded inverses

$$(1 \pm e^{-\beta h_\mathfrak{b}})^{-1} = \sum_{n=0}^{\infty} (\mp 1)^n e^{-n\beta h_\mathfrak{b}}$$

in operator norm topology.

[11] Recall that, for any Hilbert–Schmidt operator S, $\sigma(S) \setminus \{0\}$ consists of only isolated eigenvalues with finite multiplicity.

[12] We can treat also the case where $h_\mathfrak{b}$ is non-negative with $\inf \sigma(h_\mathfrak{b}) = 0$. But, for simplicity, here we restrict ourselves to the case $\inf \sigma(h_\mathfrak{b}) > 0$.

We consider a system of free bosons described by quantum fields $\phi_{\mathrm{F}}(f)$ and $\pi_{\mathrm{F}}(f)$ $(f \in \mathscr{H}_J)$ in a canonical equilibrium state at a finite temperature. The Hamiltonian of the system is assumed to be given by $\mathrm{d}\Gamma_{\mathrm{b}}(h_{\mathrm{b}})$. We denote by $\beta > 0$ the inverse temperature of the system. Let $\mathfrak{A}_{\mathrm{F}}$ be the algebra generated by $\{\phi_{\mathrm{F}}(f), \pi_{\mathrm{F}}(f) | f \in \mathscr{H}_J\}$.

We first consider the case where $e^{-\beta h_{\mathrm{b}}}$ is trace class. Then one can show that, for all $A \in \mathfrak{A}_{\mathrm{F}}$, $Ae^{-\beta \mathrm{d}\Gamma_{\mathrm{b}}(h_{\mathrm{b}})}$ is trace class (see, e.g., [51, Proposition 5.2.28], [18, Theorem 8.18]). Hence one can define a linear functional $\langle \cdot \rangle_\beta$ on the algebra $\mathfrak{A}_{\mathrm{F}}$ by

$$\langle A \rangle_\beta := \frac{\mathrm{Tr}\,(Ae^{-\beta \mathrm{d}\Gamma_{\mathrm{b}}(h_{\mathrm{b}})})}{\mathrm{Tr}\, e^{-\beta \mathrm{d}\Gamma_{\mathrm{b}}(h_{\mathrm{b}})}}, \quad A \in \mathfrak{A}_{\mathrm{F}}.$$

Then one has

$$\langle \phi_{\mathrm{F}}(f)\phi_{\mathrm{F}}(g) \rangle_\beta = \frac{1}{2}\langle f, S_\beta g \rangle, \quad f, g \in \mathscr{H}_J, \tag{10.94}$$

where

$$S_\beta := (1 + e^{-\beta h_{\mathrm{b}}})(1 - e^{-\beta h_{\mathrm{b}}})^{-1}.$$

Note that S_β is a bijective bounded operator on $\mathscr{H}$ and S_β^{-1} is bounded. The linear functional $\langle \cdot \rangle_\beta$ is called the **canonical equilibrium state** or **the Gibbs state with the Hamiltonian** $\mathrm{d}\Gamma_{\mathrm{b}}(h_{\mathrm{b}})$ at inverse temperature β.

In the case where $e^{-\beta h_{\mathrm{b}}}$ is not trace class, $\langle \cdot \rangle_\beta$ is defined to be a quasi free state with covariance form $\langle f, S_\beta g \rangle$ $(f, g \in \mathscr{H}_J)$ (see [51, p. 40], [18, §8.12]). In this case too, (10.94) holds.

To express the right hand side of (10.94) as the expectation value of two quantum fields with respect to the Fock vacuum $\Omega_{\mathscr{H}}$, we introduce

$$\phi_\beta(f) := \phi_{\mathrm{F}}(S_\beta^{1/2} f). \tag{10.95}$$

Then we have

$$\langle \phi_{\mathrm{F}}(f)\phi_{\mathrm{F}}(g) \rangle_\beta = \langle \Omega_{\mathscr{H}}, \phi_\beta(f)\phi_\beta(g)\Omega_{\mathscr{H}} \rangle, \quad f, g, \in \mathscr{H}_J.$$

Let

$$\pi_\beta(f) := \pi_{\mathrm{F}}(S_\beta^{-1/2} f), \quad f \in \mathscr{H}_J. \tag{10.96}$$

Then, by applying Theorem 8.15 to the case where $T = S_\beta^{-1/2}$ and $\mathscr{W} = \mathscr{H}_J$, one can conclude that

$$\Pi_\beta := (\mathscr{F}_{\mathrm{b}}(\mathscr{H}), \{\phi_\beta(f), \pi_\beta(f) | f \in \mathscr{H}_J\}) \tag{10.97}$$

is an irreducible Weyl representation of the CCR over $\mathscr{H}_J$.

With regard to comparison of two representations in $\{\Pi_\beta\}_{\beta>0}$, we have the following:

Theorem 10.14 *Let $\beta_1 \neq \beta_2$ ($\beta_1, \beta_2 > 0$). Then $\Pi_{\beta_1} \cong \Pi_{\beta_2}$ if and only if $e^{-\beta_2 h_b} - e^{-\beta_1 h_b}$ is Hilbert–Schmidt.*

Proof We need only apply Theorem 8.18 to the following case: $T = S_\beta^{-1/2}$. Hence $\Pi_{\beta_1} \cong \Pi_{\beta_2}$ if and only if

$$S := S_{\beta_2}^{1/2} S_{\beta_1}^{-1/2} - S_{\beta_2}^{-1/2} S_{\beta_1}^{1/2}$$

is Hilbert–Schmidt.

By direct computations, we have

$$S = X(e^{-\beta_2 h_b} - e^{-\beta_1 h_b})$$

with

$$X := 2S_{\beta_1}^{-1/2} S_{\beta_2}^{-1/2} (1 - e^{-\beta_1 h_b})^{-1} (1 - e^{-\beta_2 h_b})^{-1}.$$

Note that X is a bijective bounded operator. Hence S is Hilbert–Schmidt if and only if so is $e^{-\beta_2 h_b} - e^{-\beta_1 h_b}$. Thus the desired result follows. ∎

Corollary 10.3 *Suppose that, for some $\beta_0 > 0$, $e^{-\beta_0 h_b}$ is Hilbert–Schmidt. Let $\beta_2 > \beta_1 \geq \beta_0$. Then $\Pi_{\beta_1} \cong \Pi_{\beta_2}$.*

Proof By the present assumption, for all $\beta > \beta_0$, $e^{-\beta h_b}$ is Hilbert–Schmidt (note that $e^{-2\beta h_b} \leq e^{-2\beta_0 h_b}$). Hence $e^{-\beta_2 h_b} - e^{-\beta_1 h_b}$ is Hilbert–Schmidt. Thus, by Theorem 10.14, $\Pi_{\beta_1} \cong \Pi_{\beta_2}$. ∎

Corollary 10.4 *Suppose that, for some $\beta_0 > 0$, $e^{-\beta_0 h_b}$ is not Hilbert–Schmidt. Let $0 < \beta_2 < \beta_1 \leq \beta_0$. Then $\Pi_{\beta_1} \ncong \Pi_{\beta_2}$.*

Proof By the present assumption, for all $\beta \in (0, \beta_0)$, $e^{-\beta h_b}$ is not Hilbert–Schmidt (note that $e^{-2\beta_0 h_b} \leq e^{-2\beta h_b}$). One has

$$e^{-\beta_2 h_b} - e^{-\beta_1 h_b} = e^{-\beta_2 h_b}(1 - e^{-(\beta_1 - \beta_2) h_b}).$$

If $e^{-\beta_2 h_b} - e^{-\beta_1 h_b}$ were Hilbert–Schmidt, then

$$e^{-\beta_2 h_b} = (e^{-\beta_2 h_b} - e^{-\beta_1 h_b})(1 - e^{-(\beta_1 - \beta_2) h_b})^{-1}$$

is Hilbert–Schmidt. But this is a contradiction. Hence $e^{-\beta_2 h_b} - e^{-\beta_1 h_b}$ is not Hilbert–Schmidt. Thus, by Theorem 10.14, $\Pi_{\beta_1} \ncong \Pi_{\beta_2}$. ∎

Example 10.4 Let $d \in \mathbb{N}$ and consider a quantum system of free relativistic bosons in the cubic box $\mathbb{I}_L^d$ (see (1.114)). Then the one-boson Hilbert space in the momentum representation is given by $\ell^2(\Gamma_L)$ (see (1.117)). In this case, $h_{\mathfrak{b}}$ is given by

$$h_{\mathfrak{b}} = \omega_m^{(L)}, \tag{10.98}$$

where $\omega_m^{(L)}$ is a function on Γ_L defined by

$$\omega_m^{(L)}(\boldsymbol{k}) = \sqrt{\boldsymbol{k}^2 + m^2}, \quad \boldsymbol{k} \in \Gamma_L.$$

Hence

$$\sigma(h_{\mathfrak{b}}) = \sigma_{\mathrm{d}}(h_{\mathfrak{b}}) = \{\omega_m^{(L)}(\boldsymbol{k}) | \boldsymbol{k} \in \Gamma_L\}.$$

Hence, for all $\beta > 0$,

$$\sum_{\boldsymbol{k} \in \Gamma_L} e^{-\beta \omega_m^{(L)}(\boldsymbol{k})} < \infty.$$

Therefore $e^{-\beta h_{\mathfrak{b}}}$ is trace class and hence Hilbert–Schmidt. Thus, by Corollary 10.3, for all $\beta_1, \beta_2 > 0$ with $\beta_1 \neq \beta_2$, $\Pi_{\beta_1} \cong \Pi_{\beta_2}$.

Example 10.5 We next consider a quantum system of free relativistic bosons in the infinite space $\mathbb{R}^d$. In this case, $\mathscr{H} = L^2(\hat{\mathbb{R}}^d)$ and $h_{\mathfrak{b}} = \omega_m$. Hence $\sigma(h_{\mathfrak{b}}) = [m, \infty)$. Therefore, for all $\beta > 0$, $e^{-\beta h_{\mathfrak{b}}}$ is not Hilbert–Schmidt. Thus, by Corollary 10.4, for all $\beta_1, \beta_2 > 0$ with $\beta_1 \neq \beta_2$, $\Pi_{\beta_1} \ncong \Pi_{\beta_2}$.

The above two examples show an essential difference between a finite volume theory and an infinite volume theory. It is noteworthy that, in the infinite volume theory in Example 10.5, the inverse temperatures appear as the indices of the family $\{\Pi_\beta\}_{\beta>0}$ of mutually inequivalent irreducible Weyl representations of the CCR over $\mathscr{H}_J$.

10.9 Van Hove Model

In quantum field theory, ground states play basic roles. It is an important issue to find criteria for existence or absence of ground states in quantum field models. For some models, it is known that there are correspondences between absence of ground states and inequivalent representations of CCR. In this section we discuss this aspect by using a simple model, called the **van Hove model** [158] or the **van Hove–Miyatake model** [112, 113] or a **fixed source model** [58, 89, 114, 117].

We first give an algebraic structure whose Hilbert space operator realizations yield representations of the van Hove model. Let T be a non-negative and injective self-adjoint operator on a Hilbert space $\mathscr{H}$ such that $T^{1/2} \in \mathfrak{S}_{\mathscr{W}}(\mathscr{H})$ (see Sect. 8.12) and $g \in \mathscr{H}$ be a vector satisfying

$$g \in D(T^{-1/2}) \setminus \{0\}. \tag{10.99}$$

The following commutation relations for algebraic elements H, $\phi(f)$ and $\pi(f)$ ($f \in D(T^2) \cap \mathscr{W}$) are called the **van Hove relations associated with** (T, g):

$$[\phi(f), \pi(f')] = i \langle f, f' \rangle, \tag{10.100}$$

$$[\phi(f), \phi(f')] = 0, \quad [\pi(f), \pi(f')] = 0, \tag{10.101}$$

$$i[H, \phi(f)] = \pi(f) - \text{Im} \left\langle T^{-1/2} g, f \right\rangle, \tag{10.102}$$

$$i[H, \pi(f)] = -\phi(T^2 f) - \text{Re} \left\langle g, T^{1/2} f \right\rangle, \; f \in D(T^2) \cap \mathscr{W}, \; f' \in \mathscr{W}. \tag{10.103}$$

Roughly speaking, the van Hove model for the pair (T, g) is defined to be a Hilbert space operator realization of $(H, \{\phi(f), \pi(f) | f \in \mathscr{W}\})$ satisfying the van Hove relations associated with (T, g).

Remark 10.14 Let $\mathscr{W} = C^\infty(T) \cap D(T^{-1/2}) \cap \mathscr{H}_J$. Then the van Hove relations associated with (T, g) show that, if $\mathscr{H}$ is infinite dimensional, then $(H, I, \{\phi(T^{2n} f), \pi(T^{2n} f) | n \in \mathbb{Z}_+, f \in \mathscr{W}\})$ forms an infinite-dimensional Lie algebra.

By Theorem 6.11, the operator H_v defined by (6.77) is self-adjoint and bounded from below. Hence the following definition is possible:

Definition 10.4 The **abstract van Hove model** or **abstract fixed source model** associated with (T, g) in the Fock representation is the triple

$$\mathsf{M}_\text{v} := (\mathscr{F}_\text{b}(\mathscr{H}), H_\text{v}, \{\phi_{T^{1/2}}(f), \pi_{T^{1/2}}(f) | f \in \mathscr{W}\})$$

(see (8.118) and (8.119)). The vector g is called the **fixed source** of the model.

One may call the model M_v the abstract van Hove–Miyatake model associated with (T, g) [112, 113].

One can easily see that $(H_\text{v}, \{\phi_{T^{1/2}}(f), \pi_{T^{1/2}}(f) | f \in \mathscr{W}\})$ is an operator realization of the van Hove relations associated with (T, g).

Example 10.6 The original concrete fixed source model is a model describing an interaction of a fixed source $\rho \in \mathscr{S}'_\mathbb{R}(\mathbb{R}^3)$ (e.g., fixed nucleons in the three-dimensional space $\mathbb{R}^3$) with a quantum neutral (Hermitian) scalar field ϕ (e.g., the quantum field of π^0 mesons), which obeys the field equation

$$\left(\frac{\partial^2}{\partial t^2} - \Delta + m^2 \right) \phi(t, \boldsymbol{x}) = \rho(\boldsymbol{x}), \quad (t, \boldsymbol{x}) \in \mathbb{R} \times \mathbb{R}^3,$$

in a suitable sense, where Δ is the three-dimensional Laplacian and $m > 0$ is the mass of one boson. Mathematically, it is a concrete realization of M_{v} with $(\mathscr{H}, T, g)$ given as follows:

$$\mathscr{H} = L^2(\hat{\mathbb{R}}^3), \quad T = \omega_m, \quad g = -\frac{\hat{\rho}}{\sqrt{\omega_m}},$$

where ω_m is given by (10.85) with $m > 0$ (the massive model) and the Fourier transform $\hat{\rho}$ of ρ is a function on $\hat{\mathbb{R}}^3$ obeying the condition that $\hat{\rho}/\sqrt{\omega_m} \in D(\omega_m^{-1/2})$ (i.e., $\hat{\rho} \in D(\omega_m^{-1})$). The distribution ρ describes a source which generates a quantum field around it. Hence the Hamiltonian of the concrete fixed source model takes the form

$$H_{\text{v}}(\rho) := H_{\text{b}} - \frac{1}{\sqrt{2}}\big(a(\omega_m^{-1/2}\hat{\rho})^* + a(\omega_m^{-1/2}\hat{\rho})\big) \tag{10.104}$$

acting in the boson Fock space $\mathscr{F}_{\text{b}}(L^2(\hat{\mathbb{R}}^3))$, where H_{b} is defined by (10.86) and $a(\cdot)$ is the annihilation operator on $\mathscr{F}_{\text{b}}(L^2(\hat{\mathbb{R}}^3))$. Detailed descriptions and analysis of the model from physics point of view are given in [89, Chapter 9]. A mathematically rigorous comprehensive study of the model M_{v} including the massless case $m = 0$ is presented in [26, Chapter 12]; the contents of this section are mainly based on it (see also [28, Chapter 13] and [58, 59, 61]).

The purpose of this section is to discuss existence and absence of ground states of the model M_{v} in relation to representations of CCR. Existence and absence of ground states of H_{v} depend on regularity of g. There are two cases to be distinguished on regularity of g:

(i) The vector $g \in \mathscr{H}$ is said to be **infrared regular with respect to** T if

$$g \in D(T^{-1}). \tag{10.105}$$

(ii) The vector $g \in \mathscr{H}$ is said to be **infrared singular with respect to** T if

$$g \notin D(T^{-1}). \tag{10.106}$$

A remark may be in order. If T^{-1} is bounded, then $D(T^{-1}) = \mathscr{H}$ and hence every $g \in \mathscr{H}$ is infrared regular. Therefore, in this case, the definition above has no substantial meaning. In other words, only in the case where T^{-1} is unbounded, the definition above is meaningful.

It is easy to see that T^{-1} is unbounded if and only if

$$m(T) := \inf \sigma(T) \tag{10.107}$$

is zero.[13]

For convenience, we introduce the following definitions: if $m(T) > 0$, then T is said to be **massive**;[14] if $m(T) = 0$, then T is said to be **massless**. Thus, if T is massive, then every $g \in \mathscr{H}$ is infrared regular with respect to T; only in the case where T is massless can g be infrared singular with respect to T.

10.9.1 The Infrared Regular Case

In this subsection, we assume (10.105). Then one can define a unitary operator by

$$U_g(T) := e^{-i\Phi_S(iT^{-1}g)}. \tag{10.108}$$

The following theorem shows that H_v can be diagonalized:

Theorem 10.15 *Assume (10.105). Then*

$$U_g(T)H_v U_g(T)^{-1} = d\Gamma_b(T) + E_g(T), \tag{10.109}$$

where

$$E_g(T) := -\frac{1}{2}\|T^{-1/2}g\|^2. \tag{10.110}$$

Proof We first consider the case where T is bounded. Then, by Example 8.9, (8.35) holds and

$$\mathscr{Q}_{C_L} = \mathscr{Q}_A = D(d\Gamma_b(T))^{1/2}) = D(\hat{H}_v^{1/2}). \tag{10.111}$$

Hence

$$\left\langle \hat{H}_v^{1/2}\Phi, \hat{H}_v^{1/2}\Phi\right\rangle = \sum_{n=1}^{\infty}\left\langle C_L(T^{1/2}e_n)\Phi, C_L(T^{1/2})\Psi\right\rangle, \quad \Phi, \Psi \in D(H_v^{1/2}),$$

where

$$L(f) := \frac{1}{\sqrt{2}}\left\langle T^{-1/2}g, T^{-1/2}f\right\rangle, \ f \in D(T^{-1/2}) \tag{10.112}$$

[13] The statement is equivalent to that T^{-1} is bounded if and only if $m(T) > 0$. This statement can be easily proved by using functional calculus.

[14] Recall that, in the case where $\mathscr{H} = L^2(\hat{\mathbb{R}}^d)$ and $T = \omega_m$ with a constant $m \geq 0$ (see (10.85)), one has $\inf \sigma(\omega_m) = m$.

and $\{e_n\}_{n=1}^{\infty}$ is a CONS of $\mathscr{H}$. For notational simplicity, we set $U := U_g(T)$. By Theorem 8.5 and its proof, we have

$$UC_L(f)U^{-1} = A(f), \quad f \in D(T^{-1/2}),$$

Hence

$$\sum_{n=1}^{\infty}\left\langle C_L(T^{1/2}e_n)\Phi, C_L(T^{1/2})\Psi\right\rangle = \sum_{n=1}^{\infty}\left\langle A(T^{1/2}e_n)U\Phi, A(T^{1/2}e_n)U\Psi\right\rangle$$
$$= \left\langle \mathrm{d}\Gamma_{\mathrm{b}}(T)^{1/2}U\Phi, \mathrm{d}\Gamma_{\mathrm{b}}(T)^{1/2}U\Psi\right\rangle.$$

Therefore

$$\left\langle \hat{H}_{\mathrm{v}}^{1/2}\Phi, \hat{H}_{\mathrm{v}}^{1/2}\Psi\right\rangle = \left\langle \mathrm{d}\Gamma_{\mathrm{b}}(T)^{1/2}U\Phi, \mathrm{d}\Gamma_{\mathrm{b}}(T)^{1/2}U\Psi\right\rangle.$$

It follows from this fact and (10.111) that operator equality

$$\hat{H}_{\mathrm{v}} = U^{-1}\mathrm{d}\Gamma_{\mathrm{b}}(T)U$$

holds. This is equivalent to (10.109).

We next consider the case where T is unbounded. For each $n \in \mathbb{N}$, the operator

$$T_n := T(n^{-1}T + 1)^{-1}$$

is a non-negative and injective bounded self-adjoint operator on $\mathscr{H}$. It is easy to see that

$$\lim_{n\to\infty} T_n f = Tf, \quad \Psi \in D(T).$$

The domain $D(T)$ is trivially a core for T. Hence, by Theorem 6.2,

$$\text{s-}\lim_{n\to\infty} e^{it\mathrm{d}\Gamma_{\mathrm{b}}(T_n)} = e^{it\mathrm{d}\Gamma_{\mathrm{b}}(T)}.$$

By the preceding result, we have

$$U_g(T_n)H_{\mathrm{v}}^{(n)}U_g(T_n)^{-1} = \mathrm{d}\Gamma_{\mathrm{b}}(T_n) + E_g(T_n),$$

where $H_{\mathrm{v}}^{(n)} := \mathrm{d}\Gamma_{\mathrm{b}}(T_n) + \frac{1}{\sqrt{2}}(A(g)^* + A(g))$. Hence, for all $t \in \mathbb{R}$,

$$U_g(T_n)e^{itH_{\mathrm{v}}^{(n)}}U_g(T_n)^{-1} = e^{it\mathrm{d}\Gamma_{\mathrm{b}}(T_n)}e^{itE_g(T_n)}. \tag{10.113}$$

It is easy to see that

$$\lim_{n\to\infty} H_{\mathrm{v}}^{(n)}\Psi = H_{\mathrm{v}}\Psi, \quad \Psi \in \mathscr{F}_{\mathrm{b,fin}}(D(T)).$$

Recall that any core for $\mathrm{d}\Gamma_{\mathrm{b}}(T)$ is a core for H_{v}. Hence $\mathscr{F}_{\mathrm{b,fin}}(D(T))$ is a core for H_{v}. Therefore, by Theorems 1.29 and 1.30,

$$\text{s-}\lim_{n\to\infty} e^{itH_{\mathrm{v}}^{(n)}} = e^{itH_{\mathrm{v}}}.$$

By functional calculus, one can show that, for all $\alpha > 0$,

$$\lim_{n\to\infty} T_n^{-1/2} f = T^{-1/2} f, \quad f \in D(T^{-1/2}),$$

$$\lim_{n\to\infty} T_n^{-1} h = T^{-1} h, \quad h \in D(T^{-1}).$$

The first formula implies that

$$\lim_{n\to\infty} E_g(T_n) = E_g(T),$$

while the second and Theorem 6.22(iii) imply that

$$\text{s-}\lim_{n\to\infty} U_g(T_n) = U_g(T), \quad \text{s-}\lim_{n\to\infty} U_g(T_n)^{-1} = U_g(T)^{-1}.$$

Now, taking the limit $n \to \infty$ in (10.113), we obtain

$$U_g(T)e^{itH_{\mathrm{v}}}U_g(T)^{-1} = e^{it\mathrm{d}\Gamma_{\mathrm{b}}(T)}e^{itE_g(T)} = e^{it(\mathrm{d}\Gamma_{\mathrm{b}}(T)+E_g(T))}.$$

By the unitary covariance of functional calculus, the left hand side is equal to $\exp(itU_g(T)H_{\mathrm{v}}U_g(T)^{-1})$. Thus (10.109) is proved. ■

Remark 10.15 There is another proof of (10.109) (see [28, Theorem 13.4(i)]).

As a consequence of Theorem 10.15 and Corollary 8.4, we obtain the following:

Corollary 10.5 *Assume (10.105). Then H_{v} has a unique ground state given by*

$$\Omega_g(T) := U_g(T)^{-1}\Omega_{\mathscr{H}}$$

up to constant multiples with the ground state energy being $E_g(T)$. Moreover, the representation $(\mathscr{F}_{\mathrm{b}}(\mathscr{H}), \mathscr{F}_{\mathrm{b},0}(\mathscr{H}), \{C_L(f)|f \in D(T^{-1/2})\})$ of the CCR over $D(T^{-1/2})$ with L given by (10.112) is equivalent to the Fock representation $\rho_{\mathrm{F}}(D(T^{-1/2}))$.

Let

$$H' := \mathrm{d}\Gamma_{\mathrm{b}}(T) + E_g(T), \tag{10.114}$$

$$\phi'(f) := U_g(T)\phi_{T^{1/2}}(f)U_g(T)^{-1}, \tag{10.115}$$

$$\pi'(f) := U_g(T)\pi_{T^{1/2}}(f)U_g(T)^{-1}, \quad f \in \mathscr{W}. \tag{10.116}$$

Then Theorem 10.15 yields the following result:

Corollary 10.6 *Assume (10.105). Then the abstract fixed source model* M_{v} *in the Fock representation is equivalent to the model* $(\mathscr{F}_{\mathrm{b}}(\mathscr{H}), H', \{\phi'(f), \pi'(f) | f \in \mathscr{V}\})$.

One can easily prove the following formulae:

$$\phi'(f) = \phi_{T^{1/2}}(f) - \mathrm{Re}\left\langle T^{-1}g, T^{-1/2}f\right\rangle, \tag{10.117}$$

$$\pi'(f) = \pi_{T^{1/2}}(f) + \mathrm{Im}\left\langle T^{-1/2}g, f\right\rangle. \tag{10.118}$$

10.9.2 The Infrared Singular Case

We next consider the case where g obeys (10.106). Let L be defined by (10.112). We fix a dense subspace $\mathscr{V} \subset D(T^{-1/2})$.

Theorem 10.16 *Assume (10.106). Let* $\mathscr{V}$ *be a dense subspace of* $\mathscr{H}$ *which is a core* $T^{-1/2}$. *Then the representation* $(\mathscr{F}_{\mathrm{b}}(\mathscr{H}), \mathscr{F}_{\mathrm{b},0}(\mathscr{H}), \{C_L(f) | f \in \mathscr{V}\})$ *is inequivalent to the Fock representation* $\rho_{\mathrm{F}}(\mathscr{V})$.

Proof Suppose that there existed a vector $h \in \mathscr{H}$ such that $L(f) = \langle h, f\rangle$, $f \in \mathscr{V}$. Then $\langle T^{-1/2}g, T^{-1/2}f\rangle = \left\langle\sqrt{2}h, f\right\rangle$. Since $\mathscr{V}$ is a core for $T^{-1/2}$, it follows that $T^{-1/2}g \in D(T^{-1/2})$ and $T^{-1/2}T^{-1/2}g = \sqrt{2}h$. The former means that $g \in D(T^{-1})$. But this contradicts (10.106). Hence such a vector h does not exist. Thus Corollary 8.4 yields the result. ■

The inequivalence stated in Theorem 10.16 is related to absence of ground states of H_{v}:

Theorem 10.17 *Assume (10.106). Suppose that* T *is bounded. Then* H_{v} *has no ground states.*

Proof Suppose that H_{v} had a ground state $\Omega \neq 0$. Then $\hat{H}_{\mathrm{v}}\Omega = 0$. Hence, by Example 8.9, $\Omega \in D(H_{\beta_L}^{1/2})$ and $H_{\beta_L}^{1/2}\Omega = 0$. Then, by Proposition 8.1(ii), $C_L(f)\Omega = 0$ for all $f \in \mathscr{W} := \mathrm{span}\{T^{1/2}e_n | n \in \mathbb{N}\}$, where $\{e_n\}_{n=1}^{\infty} \subset D(T^{1/2})$ is a CONS of $\mathscr{H}$. Since $T^{1/2}$ is bounded, it follows that $\mathscr{W}$ is dense in $\mathscr{H}$. Hence, by Lemma 8.4 with $\mathscr{D} = \mathscr{W}$, there exists a vector $h \in \mathscr{H}$ such that $L(f) = \langle h, f \rangle_{\mathscr{H}}$, $f \in \mathscr{W}$. Note that $\mathscr{W}$ is a core for $T^{-1/2}$. Hence it follows that $T^{-1/2}g \in D(T^{-1/2})$, i.e., $g \in D(T^{-1})$ and $T^{-1}g = h$. But this contradicts (10.106). Thus H_{v} has no ground states. ■

Remark 10.16 In Theorem 10.17, the boundedness of T is assumed for simplicity. But the conclusion holds in the theorem also in the case where T is unbounded and absolutely continuous. For details, see [28, Theorem 13.8].

10.9.3 Infrared Divergence

In this subsection, we point out that Theorem 10.17 is related to a characteristic phenomenon. We continue to assume (10.106) (the infrared singular case). For each $a > 0$, we define a vector $g_a \in \mathscr{H}$ by

$$g_a := E_T([a, \infty))g.$$

The number a is called an **infrared cutoff** for g. It is easy to see that $g_a \in D(T^{-1})$. Hence one can define

$$H_a := \mathrm{d}\Gamma_{\mathrm{b}}(T) + \frac{1}{\sqrt{2}}(A(g_a)^* + A(g_a)).$$

We call it the van Hove Hamiltonian with infrared cutoff a. By Corollary 10.5, H_a has a ground state

$$\Omega_a := \Omega_{g_a}(T) = e^{i\Phi_{\mathrm{S}}(iT^{-1}g_a)}\Omega_{\mathscr{H}}.$$

Theorem 10.18 *Assume (10.106). Let N_{b} be the boson number operator on $\mathscr{F}_{\mathrm{b}}(\mathscr{H})$ (see (6.11)). Then $\Omega_a \in D(N_{\mathrm{b}})$ and*

$$\lim_{a \to 0} \langle \Omega_a, N_{\mathrm{b}} \Omega_a \rangle = \infty. \tag{10.119}$$

Proof Considering the case $T = I$ in Theorem 10.15 with g replaced by $-h \in \mathscr{H}$, we have for all $h \in \mathscr{H}$

$$e^{-i\Phi_{\mathrm{S}}(ih)} N_{\mathrm{b}} e^{i\Phi_{\mathrm{S}}(ih)} = N_{\mathrm{b}} - \frac{1}{\sqrt{2}}(A(h)^* + A(h)) + \frac{1}{2}\|h\|^2.$$

Taking $h = T^{-1}g_a$, we obtain

$$e^{-i\Phi_{\mathrm{S}}(iT^{-1}g_a)} N_{\mathrm{b}} e^{i\Phi_{\mathrm{S}}(iT^{-1}g_a)} = N_{\mathrm{b}} - \frac{1}{\sqrt{2}}(A(T^{-1}g_a)^* + A(T^{-1}g_a)) + \frac{1}{2}\|T^{-1}g_a\|^2.$$

Hence, $\Omega_a \in D(N_{\mathrm{b}})$ and, using (6.39) and (6.12), we obtain

$$\langle \Omega_a, N_{\mathrm{b}} \Omega_a \rangle = \frac{1}{2} \|T^{-1} g_a\|^2.$$

Condition (10.106) implies that

$$\lim_{a \to 0} \|T^{-1} g_a\|^2 = \infty. \tag{10.120}$$

Thus (10.119) follows. ■

Formula (10.119) physically means that the mean value of bare boson numbers in the ground state Ω_a diverges as the infrared cutoff a is removed. This is one of the phenomena called **infrared divergences** peculiar to theories of *massless* quantum fields. Thus we see that an infrared divergence is associated with the absence of ground states of the Hamiltonian H_{v} with g infrared singular with respect to T.

There is another interesting phenomenon:

Theorem 10.19 *Assume (10.106). Then*

$$w\text{-}\lim_{a \to 0} \Omega_a = 0, \tag{10.121}$$

where w- lim *means weak limit.*

Proof We have by (6.106)

$$\langle \Omega_{\mathscr{H}}, \Omega_a \rangle = e^{-\|T^{-1} g_a\|^2/4}.$$

Hence, by (10.120),

$$\lim_{a \to 0} \langle \Omega_{\mathscr{H}}, \Omega_a \rangle = 0.$$

Next, let

$$\Psi := A(f_1)^* \cdots A(f_n)^* \Omega_{\mathscr{H}}, \quad f_1, \ldots, f_n \in \mathscr{H}.$$

Then, by (6.107), we have

$$\langle \Psi, \Omega_a \rangle = \frac{(-1)^n}{2^{n/2}} \left(\prod_{j=1}^n \left\langle f_j, T^{-1} g_a \right\rangle \right) e^{-\|T^{-1} g_a\|^2/4}.$$

Hence, by the Schwarz inequality,

$$|\langle \Psi, \Omega_a \rangle| \leq \frac{1}{2^{n/2}} \left(\prod_{j=1}^n \|f_j\| \right) \|T^{-1} g_a\|^n e^{-\|T^{-1} g_a\|^2/4}.$$

It is easy to see that $\lim_{t\to\infty} t^n e^{-t^2/4} = 0$. Hence $\lim_{a\to 0} \langle \Psi, \Omega_a \rangle = 0$. Thus, for all $\Phi \in \mathscr{F}_{\mathrm{b,fin}}(\mathscr{H})$,

$$\lim_{a\to\infty} \langle \Phi, \Omega_a \rangle = 0. \tag{10.122}$$

Finally, let $\Psi \in \mathscr{F}_{\mathrm{b}}(\mathscr{H})$. Since $\mathscr{F}_{\mathrm{b,fin}}(\mathscr{H})$ is dense in $\mathscr{F}_{\mathrm{b}}(\mathscr{H})$, there exists a sequence $\{\Psi_n\}_{n=1}^{\infty}$ in $\mathscr{F}_{\mathrm{b,fin}}(\mathscr{H})$ such that $\lim_{n\to\infty} \|\Psi_n - \Psi\| = 0$. Using the identity

$$\langle \Psi, \Omega_a \rangle = \langle \Psi - \Psi_n, \Omega_a \rangle + \langle \Psi_n, \Omega_a \rangle ,$$

and the Schwarz inequality with $\|\Omega_a\| = 1$, we have

$$| \langle \Psi, \Omega_a \rangle | \le \|\Psi - \Psi_n\| + | \langle \Psi_n, \Omega_a \rangle |.$$

Hence, by (10.122),

$$\limsup_{a\to 0} | \langle \Psi, \Omega_a \rangle | \le \|\Psi - \Psi_n\|.$$

Then, taking the limit $n \to \infty$, we obtain $\limsup_{a\to 0} | \langle \Psi, \Omega_a \rangle | = 0$, which is equivalent to $\lim_{a\to 0} \langle \Psi, \Omega_a \rangle = 0$. Thus (10.121) holds. ∎

Formula (10.121) may be regarded as a form of infrared divergence in the present model, suggesting the absence of ground states in the infrared singular case. We call property (10.121) the **infrared van Hove–Miyatake phenomenon** or the **infrared van Hove–Miyatake catastrophe**.[15]

10.9.4 Infrared Renormalization

Theorem 10.17 shows that, under the infrared singular condition (10.106), H_{v} with T bounded has no ground states. But, if one passes to a representation of CCR inequivalent to the Fock representation for the time-zero fields, then one can construct a fixed source model which has a ground state. We show this below. The idea is to note that the operators $\phi'(f)$ and $\pi'(f)$ given by (10.117) and (10.118) respectively can be defined also in the case where (10.106) holds if test vectors f are in a smaller space.

In this subsection, we assume (10.106). Let

$$\mathscr{V}' := D(T^{-1}) \cap D(T^{1/2}) \tag{10.123}$$

[15] Originally the term "van Hove–Miyatake catastrophe" was used in relation to ultraviolet divergence of a *massive* van Hove–Miyatake model [68]. See Sect. 10.9.6 below.

and define

$$H_{\text{ren}} := d\Gamma_b(T) + E_g(T), \tag{10.124}$$

$$\phi_{\text{ren}}(f) := \phi_{T^{1/2}}(f) - \text{Re}\left\langle T^{-1/2}g, T^{-1}f\right\rangle, \tag{10.125}$$

$$\pi_{\text{ren}}(f) := \pi_{T^{1/2}}(f) + \text{Im}\left\langle T^{-1/2}g, f\right\rangle, \quad f \in \mathscr{V}'. \tag{10.126}$$

Then it is obvious that H_{ren} is a self-adjoint operator which has a unique ground state given by

$$\Omega_{\text{ren}} := \Omega_{\mathscr{H}},$$

up to constant multiples, i.e.,

$$H_{\text{ren}}\Omega_{\text{ren}} = E_g(T)\Omega_{\text{ren}}.$$

It follows that

$$\pi_{\text{ren}}(\mathscr{V}') := (\mathscr{F}_b(\mathscr{H}), \mathscr{F}_{b,0}(\mathscr{H}), \{\phi_{\text{ren}}(f), \pi_{\text{ren}}(f) | f \in \mathscr{V}'\}) \tag{10.127}$$

is a representation of the Heisenberg CCR over $\mathscr{V}'$. Note that this representation is defined even if g is infrared singular with respect to T (i.e., $g \notin D(T^{-1})$).

Lemma 10.3 *The representation $\pi_{\text{ren}}(\mathscr{V}')$ is an irreducible Weyl representation of the CCR over $\mathscr{V}'$.*

Proof This follows from Theorem 8.15 and Proposition 8.9. ∎

One can show that $(H_{\text{ren}}, \{\phi_{\text{ren}}(f), \pi_{\text{ren}}(f) | f \in \mathscr{V}'\})$ satisfies the van Hove relations associated with (T, g). Hence

$$\mathsf{M}_{\text{v, ren}} := (\mathscr{F}_b(\mathscr{H}), H_{\text{ren}}, \{\phi_{\text{ren}}(f), \pi_{\text{ren}}(f) | f \in \mathscr{V}'\})$$

is a fixed source model in the infrared singular case, which *has a ground state.*

The following theorem shows that $\mathsf{M}_{\text{v,ren}}$ is inequivalent to M_{v} under condition (10.106).

Theorem 10.20 *Assume (10.106). Then $(\mathscr{F}_b(\mathscr{H}), \{\phi_{\text{ren}}(f), \pi_{\text{ren}}(f) | f \in \mathscr{V}'\})$ is inequivalent to $(\mathscr{F}_b(\mathscr{H}), \{\phi_{T^{1/2}}(f), \pi_{T^{1/2}}(f) | f \in \mathscr{V}'\})$.*

Proof Suppose that the two representations under consideration were equivalent. Then there exists a unitary operator U on $\mathscr{F}_b(\mathscr{H})$ such that

$$U\phi_{T^{1/2}}(f)U^{-1} = \phi_{\text{ren}}(f), \quad U\pi_{T^{1/2}}(f)U^{-1} = \pi_{\text{ren}}(f), \quad f \in \mathscr{V}'.$$

By change of test vectors f, we have

$$U\phi_{\mathrm{F}}(f)U^{-1} = \phi_{\mathrm{F}}(f) - \mathrm{Re}\left\langle T^{-1/2}g, T^{-1/2}f\right\rangle,$$
$$U\pi_{\mathrm{F}}(f)U^{-1} = \pi_{\mathrm{F}}(f) + \mathrm{Im}\left\langle T^{-1/2}g, T^{-1/2}f\right\rangle, \quad f \in D(T^{-3/2}) \cap D(T) \cap \mathscr{H}_J,$$

where we have used the fact that

$$(T^{-1/2}\mathscr{V}') \cap (T^{1/2}\mathscr{V}') = D(T^{-3/2}) \cap D(T).$$

By the functional calculus for T, $D(T^{-3/2}) \cap D(T) \cap \mathscr{H}_J$ is dense in $\mathscr{H}_J$. Hence, by Theorem 8.12, there exists a vector h such that

$$-\left\langle T^{-1/2}g, T^{-1/2}f\right\rangle = \langle h, f\rangle, \quad f \in D(T^{-3/2}) \cap D(T).$$

It follows from the functional calculus for T that $D(T^{-3/2}) \cap D(T)$ is a core for $T^{-1/2}$. Hence $T^{-1/2}g \in D(T^{-1/2})$, i.e., $g \in D(T^{-1})$ and $T^{-1}g = -h$. But this contradicts (10.106). Thus the desired result follows. ■

Generally speaking, in a quantum field model which has infrared divergences in a representation of time-zero fields (CCR), a prescription passing to another representation of the model in which there are no infrared divergences is called an **infrared renormalization** of the model.

The theory presented above describes an infrared renormalization of an infrared singular fixed source model. From this point of view, we call $\mathsf{M}_{\mathrm{v,\,ren}}$ the **infrared-renormalized van Hove model**.

A moral of the results obtained above is that, *even if a quantum field model has no ground states in a representation of time-zero fields, it may have a ground state in a representation of time-zero fields which is inequivalent to the former.*

Remark 10.17 The infrared renormalization of the van Hove model described above may be regarded as a mathematically rigorous form for the procedure given by Bloch and Nordsieck [43] which makes a simplified model in quantum electrodynamics (essentially same as the concrete massless fixed source model) free from infrared divergences (cf. also [150, Part II, Example 3.2]).

Remark 10.18 As a generalization of a fixed source model, one can consider a model of N non-relativistic quantum particles in a scalar potential interacting with a quantum scalar field. This model is called the **Nelson model** [74, 121]. It is shown in [106] that the three-dimensional massless Nelson model with $N = 1$ has no ground states in the infrared singular case. But, in this case, the same idea as above can be applied to construct a massless Nelson model which has a ground state. See [14, 139]. An infrared renormalization of the massless Nelson model using functional integral methods is made in [105, 107].

10.9.5 Representations Indexed by Sources

It may be interesting to study dependence of the representation $(\mathscr{F}_{\rm b}(\mathscr{H}), \{\phi_{\rm ren}(f), \pi_{\rm ren}(f)|f \in \mathscr{V}'\})$ on the source g. For this purpose, we write

$$\mathsf{M}_{\rm v,\,ren} = \mathsf{M}_{\rm ren}(g), \quad H_{\rm ren} = H_{\rm ren}(g),$$
$$\phi^{(g)}(f) := \phi_{\rm ren}(f), \quad \pi^{(g)}(f) := \pi_{\rm ren}(f), \quad f \in \mathscr{V}'.$$

Then

$$\pi_g := (\mathscr{F}_{\rm b}(\mathscr{H}), \{\phi^{(g)}(f), \pi^{(g)}(f)|f \in \mathscr{V}'\})$$

is an irreducible Weyl representation of the CCR over $\mathscr{V}'$. We suppose that g varies in the following subset of $\mathscr{H}$:

$$\mathscr{G} := \{g \in D(T^{-1/2})|g \notin D(T^{-1}\}. \tag{10.128}$$

Then it may be interesting to examine equivalence or inequivalence of two representations π_{g_1} and π_{g_2} for $g_1, g_2 \in \mathscr{G}$. One has the following fact:

Theorem 10.21 *Let $g_1, g_2 \in \mathscr{G}$. Then π_{g_1} is equivalent to π_{g_2} if and only if $g_1 - g_2 \in D(T^{-1})$.*

Proof Suppose that π_{g_1} is equivalent to π_{g_2}. Then there exists a unitary operator U on $\mathscr{F}_{\rm b}(\mathscr{H})$ such that, for all $f \in \mathscr{V}'$,

$$U\phi^{(g_2)}(f)U^{-1} = \phi^{(g_1)}(f), \quad U\pi^{(g_2)}(f)U^{-1} = \pi^{(g_1)}(f).$$

This is equivalent to

$$U\phi_{T^{1/2}}(f)U^{-1} = \phi_{T^{1/2}}(f) - \operatorname{Re}\left\langle T^{-1/2}(g_1 - g_2), T^{-1}f\right\rangle, \tag{10.129}$$

$$U\pi_{T^{1/2}}(f)U^{-1} = \pi_{T^{1/2}}(f) + \operatorname{Im}\left\langle T^{-1/2}(g_1 - g_2), f\right\rangle. \tag{10.130}$$

This means that $\pi_{g_1-g_2}$ is equivalent to $(\mathscr{F}_{\rm b}(\mathscr{H}), \{\phi_{T^{1/2}}(f), \pi_{T^{1/2}}(f)|f \in \mathscr{V}'\})$. Hence, by Theorem 10.20, $g_1 - g_2 \in D(T^{-1})$.

Conversely, suppose that $g_1 - g_2 \in D(T^{-1})$. Then the source $g = g_1 - g_2$ is infrared regular. Hence, by Corollary 10.6, there exists a unitary operator U on $\mathscr{F}_{\rm b}(\mathscr{H})$ such that (10.129) and (10.130) hold. Hence it follows that π_{g_1} is equivalent to π_{g_2}. ∎

Theorem 10.21 implies the following fact on inequivalence of models $\mathsf{M}_{\text{ren}}(g_1)$ and $\mathsf{M}_{\text{ren}}(g_2)$ ($g_1, g_2 \in \mathscr{G}$):

Corollary 10.7 *Let* $g_1, g_2 \in \mathscr{G}$ *and* $g_1 - g_2 \notin D(T^{-1})$. *Then* $\mathsf{M}_{\text{ren}}(g_1)$ *is inequivalent to* $\mathsf{M}_{\text{ren}}(g_2)$.

Proof Suppose that $\mathsf{M}_{\text{ren}}(g_1)$ is equivalent to $\mathsf{M}_{\text{ren}}(g_2)$. Then π_{g_1} is equivalent to π_{g_2}. Hence, by Theorem 10.21, $g_1 - g_2 \in D(T^{-1})$. ■

Remark 10.19 It is unclear if the converse of Corollary 10.7 holds.

To understand Theorem 10.21 in an intrinsic way, we define a relation $\sim$ in $\mathscr{G}$ as follows:

$$g_1, \, g_2 \in \mathscr{G}, \; g_1 \sim g_2 \overset{\text{def}}{\Longleftrightarrow} g_1 - g_2 \in D(T^{-1}). \tag{10.131}$$

It is easy to see that the relation $\sim$ is an equivalence relation. We denote by $[g]$ the equivalence class of $g \in \mathscr{G}$ with respect to $\sim$ and set $\mathscr{G}/\sim := \{[g] | g \in \mathscr{G}\}$ (the quotient set by the equivalence relation $\sim$). Theorem 10.21 is rephrased as follows: π_{g_1} is equivalent to π_{g_2} if and only if $[g_1] = [g_2]$. Therefore, for each equivalence class $[g]$, we define a representation $\pi_{[g]}$ by

$$\pi_{[g]} := \pi_g \text{ modulo equivalence.}$$

Thus we have a family $\{\pi_{[g]} | [g] \in \mathscr{G}/\sim\}$ of mutually inequivalent irreducible Weyl representations of the CCR over $\mathscr{V}'$.

10.9.6 *Ultraviolet Renormalization*

The fixed source model in the Fock representation has another type of divergences.

Example 10.7 Consider the concrete fixed source model described in Example 10.6 with $m > 0$ (the massive case). A physical case is given by the case where ρ is a finite linear combination of Dirac's delta distributions:

$$\rho(\boldsymbol{x}) = \sum_{j=1}^{n} \lambda_j \delta(\boldsymbol{x} - \boldsymbol{R}_j),$$

where $n \in \mathbb{N}$, $\delta(\cdot)$ is the three-dimensional Dirac delta distribution, $\lambda_j \in \mathbb{R} \setminus \{0\}$ is a constant and $\boldsymbol{R}_j \in \mathbb{R}^3$ denotes the position of the jth source. In this case ρ is called a **point source**. Hence

$$\hat{\rho}(\boldsymbol{k}) = \frac{1}{(2\pi)^{3/2}} \sum_{j=1}^{n} \lambda_j e^{-i\boldsymbol{k}\boldsymbol{R}_j}.$$

It is not difficult to show that

$$g_m := -\frac{\hat{\rho}}{\sqrt{\omega_m}} \tag{10.132}$$

is not in $L^2(\hat{\mathbb{R}}^3)$. Hence, in this case, the interaction term

$$-\frac{1}{\sqrt{2}}\big(a(\omega_m^{-1/2}\hat{\rho})^* + a(\omega_m^{-1/2}\hat{\rho})\big)$$

in the Hamiltonian $H_{\mathrm{v}}(\rho)$ is meaningless. Therefore the concrete fixed source model in this case cannot be defined in the Fock representation. To avoid this difficulty, one usually introduces a regularized source

$$\hat{\rho}_\kappa := \chi_\kappa \hat{\rho}$$

with $\kappa > 0$, where χ_κ is defined by (10.56). It is obvious that $\hat{\rho}_\kappa/\sqrt{\omega_m}$ is in $L^2(\hat{\mathbb{R}}^3)$. As in the case of the model H_κ discussed in Sect. 10.6.1, the parameter κ denotes an ultraviolet cutoff. Then one can define a Hamiltonian $H_{\mathrm{v},\kappa} := H(\hat{\rho}_\kappa)$ with ultraviolet cutoff κ by replacing $\hat{\rho}$ with $\hat{\rho}_\kappa$ in the Hamiltonian $H_{\mathrm{v}}(\rho)$. As we shall show in an abstract setting below, the weak limit of the ground state of $H_{\mathrm{v},\kappa}$ as $\kappa \to \infty$ (the removal of the ultraviolet cutoff) is zero. Moreover, as heuristically seen, one cannot expect that $H_{\mathrm{v},\kappa}$ converges in a suitable sense within the Fock space $\mathscr{F}_{\mathrm{b}}(L^2(\hat{\mathbb{R}}^3))$. These are examples of the so-called **ultraviolet divergences**.

We now describe ultraviolet divergence in the abstract fixed source model. For simplicity, we assume the following:

Assumption (m) The operator T is massive, i.e., $m(T) > 0$ and T is unbounded.

Then, as already remarked, $T^{-1} \in \mathfrak{B}(\mathscr{H})$. Hence conditions (10.105) and (10.106) are trivially satisfied for all $g \in \mathscr{H}$.

We want to construct a fixed source model with $g \notin \mathscr{H}$. For this purpose, we introduce a scale of Hilbert spaces. For $s > 0$, the domain $D(T^s)$ is a Hilbert space with inner product

$$\langle f, f' \rangle_s := \langle T^s f, T^s f' \rangle, \quad f, f' \in D(T^s).$$

We denote this Hilbert space by $\mathscr{H}_s$. We set $\mathscr{H}_0 := \mathscr{H}$. The sesquilinear form

$$\langle f, f' \rangle_{-s} := \langle T^{-s} f, T^{-s} f' \rangle \quad (f, f' \in \mathscr{H})$$

on $\mathscr{H}$ is an inner product of $\mathscr{H}$. We denote the completion of the inner product space $(\mathscr{H}, \langle\cdot,\cdot\rangle_{-s})$ by $\mathscr{H}_{-s}$. We have

$$\mathscr{H}_{s_1} \subset \mathscr{H}_{s_2} \subset \mathscr{H} \subset \mathscr{H}_{-s_2} \subset \mathscr{H}_{-s_1}, \quad s_1 > s_2 > 0.$$

By the Schwarz inequality, we have

$$|\langle \phi, f \rangle_{\mathscr{H}}| = |\langle T^{-s}\phi, T^{s} f \rangle| \le \|\phi\|_{-s} \|f\|_{s}, \quad \phi \in \mathscr{H}, \ f \in \mathscr{H}_s.$$

Hence, for each $f \in \mathscr{H}_s$, the mapping $u_f \colon \phi \mapsto \langle \phi, f \rangle_{\mathscr{H}}$ is a continuous anti-linear functional on $\mathscr{H}_{-s}$ with domain $D(u_f) = \mathscr{H}$. Hence it extends uniquely to a continuous anti-linear functional $\hat{u}_f$ on $\mathscr{H}_{-s}$ such that $\hat{u}_f(\phi) = u_f(\phi)$, $\phi \in \mathscr{H}$ and

$$|\hat{u}_f(\phi)| \le \|\phi\|_{-s} \|f\|_{s}, \quad \phi \in \mathscr{H}_{-s}, \ f \in \mathscr{H}_s.$$

We write

$$\hat{u}_f(\phi) = {}_{-s}\langle \phi, f \rangle_{s}, \quad \phi \in \mathscr{H}_{-s}, \ f \in \mathscr{H}_s.$$

Throughout this sesquilinear form, one can identify the dual space $\mathscr{H}_s^*$ of $\mathscr{H}_s$ with $\mathscr{H}_{-s}$.

It follows that, for each $\phi \in \mathscr{H}_{-s}$, there exists a sequence $\{\phi_n\}_{n=1}^{\infty}$ in $\mathscr{H}$ such that

$${}_{-s}\langle \phi, f \rangle_{s} = \lim_{n\to\infty} \langle \phi_n, f \rangle_{\mathscr{H}} = \lim_{n\to\infty} \langle T^{-s}\phi_n, T^{s} f \rangle_{\mathscr{H}}, \quad f \in \mathscr{H}_s.$$

It is easy to see that, for all $s \ge 0$, T is a unitary operator from $\mathscr{H}_{s+1}$ to $\mathscr{H}_s$. Hence its dual mapping (Banach space adjoint) T' from $\mathscr{H}_s^* = \mathscr{H}_{-s}$ to $\mathscr{H}_{s+1}^* = \mathscr{H}_{-(s+1)}$, defined by $(T'\phi)(f) := \phi(Tf)$ ($\phi \in \mathscr{H}_s^*$, $f \in \mathscr{H}_{s+1}^*$), is unitary.[16]

In what follows we assume the following:

Assumption (g) For some $p > 0$, $g \in \mathscr{H}_{-(p+1)}$ and $(T')^{-1} g \notin \mathscr{H}$.

Let n_p be the natural number such that $p + 1 \le n_p < p + 2$. Then $n_p \ge 2$. Note that $(T')^{-1}$ maps $\mathscr{H}_{-(s+1)}$ onto $\mathscr{H}_{-s}$. Hence, by Assumption (g), $(T')^{-n_p} g \in \mathscr{H}_{-(p+1-n_p)}$. We have $0 \le n_p - (p+1) < 1$. Hence

$$g' := (T')^{-n_p} g$$

is in $\mathscr{H}$. For each $\kappa > 0$, $E_T([0,\kappa])g'$ is in $C^{\infty}(T)$, where E_T is the spectral measure of T. Hence one can define a vector

$$g_\kappa := T^{n_p} E_T([0,\kappa]) g'.$$

[16] Apply the following theorem: *Let $\mathscr{H}$ and $\mathscr{K}$ be Hilbert spaces. Then the mapping $\mathfrak{B}(\mathscr{H}, \mathscr{K}) \ni T \mapsto T' \in \mathfrak{B}(\mathscr{K}^*, \mathscr{H}^*)$ is an isometric isomorphism. In particular,* $\|T\| = \|T'\|$. For a proof, see, e.g., [130, Theorem VI.2].

The next lemma shows that g_κ can be regarded as an ultraviolet cutoff of g:

Lemma 10.4 $\lim_{\kappa\to\infty} g_\kappa = g$ *in* $\mathscr{H}_{-n_p}$.

Proof By the unitarity of $(T')^{-1}$, we have

$$\|g_\kappa - g\|_{-n_p} = \|(T')^{-n_p} g_\kappa - (T')^{-n_p} g\|_{\mathscr{H}} = \|E_T([0,\kappa])g' - g'\|_{\mathscr{H}} = 0.$$

Hence the result follows. ∎

Lemma 10.5 $\lim_{\kappa\to\infty} \|T^{-1}g_\kappa\|_{\mathscr{H}} = \infty$.

Proof We have

$$\|T^{-1}g_\kappa\|_{\mathscr{H}} = \|T^{n_p-1}E_T([0,\kappa])g'\|.$$

The condition $(T')^{-1}g \notin \mathscr{H}$ implies that $g' \notin D(T^{n_p-1})$. Hence it follows that $\lim_{\kappa\to\infty} \|T^{n_p-1}E_T([0,\kappa])g'\| = \infty$. Thus we obtain the result. ∎

We now define a Hamiltonian of the fixed source model with an ultraviolet cutoff source g_κ:

$$H(\kappa) := \mathrm{d}\Gamma_{\mathrm{b}}(T) + \frac{1}{\sqrt{2}}(A(g_\kappa)^* + A(g_\kappa)). \tag{10.133}$$

As we already know, H_κ is self-adjoint with $D(H_\kappa) = D(\mathrm{d}\Gamma_{\mathrm{b}}(T))$ and bounded from below with

$$H(\kappa) \geq -\frac{1}{2}\|T^{-1/2}g_\kappa\|^2.$$

By Corollary 10.5, H_κ has a unique ground state

$$\Omega(\kappa) := e^{i\Phi_{\mathrm{S}}(iT^{-1}g_\kappa)}\Omega_{\mathscr{H}}.$$

Theorem 10.22 (Ultraviolet Divergences) *Under Assumption (g), the following hold:*

(i) $\lim_{\kappa\to\infty} \langle\Omega(\kappa), N_{\mathrm{b}}\Omega(\kappa)\rangle = \infty$.
(ii) $w\text{-}\lim_{\kappa\to\infty} \Omega(\kappa) = 0$.

Proof We have Lemma 10.5. Hence part (i) (resp. (ii)) can be proved in quite the same way as in the proof Theorem 10.18 (resp. Theorem 10.19). ∎

We call the fact in part (ii) of Theorem 10.22 the **ultraviolet van Hove–Miyatake phenomenon (catastrophe)** in the cutoff model $(\mathscr{F}_{\mathrm{b}}(\mathscr{H}), H(\kappa), \{\phi_{T^{1/2}}(f), \pi_{T^{1/2}}(f)\,|\,f \in \mathscr{V}\})$ (cf. [68]).

Remark 10.20 In relativistic quantum field theory, there is a model of a neutral quantum scalar field with a self-interaction on the two-dimensional space–time, called the $\boldsymbol{P(\phi)_2}$ **model**, where ϕ denotes the scalar field, P is a real polynomial of one variable bounded from below and $P(\phi)$ describes the self-interaction of the quantum field. This model is well-defined in the Fock representation for the time-zero fields without ultraviolet cutoff, but with a volume cutoff $\ell > 0$ (for the details, see [147] or [72]). In this volume cutoff model too, the van Hove–Miyatake phenomenon occurs as $\ell \to \infty$ ([147, Corollary VI.11], [72, Corollary 9.5.6]).

Using the same idea as in the infrared renormalization, one can construct a fixed source model without ultraviolet cutoff. Namely, we need only to introduce the following operators:

$$H_\infty := \mathrm{d}\Gamma_{\mathrm{b}}(T), \tag{10.134}$$

$$\phi_\infty(f) := \phi_{T^{1/2}}(f) - \mathrm{Re}\ {}_{-(p+1)}\langle g, T^{-3/2} f\rangle_{p+1}, \tag{10.135}$$

$$\pi_\infty(f) := \pi_{T^{1/2}}(f) + \mathrm{Im}\ {}_{-(p+1)}\langle g, T^{-1/2} f\rangle_{p+1}, \quad f \in \mathscr{V}_p, \tag{10.136}$$

where

$$\mathscr{V}_p := \mathscr{H}_{p+\frac{1}{2}} \cap \mathscr{H}_J.$$

It is obvious that H_∞ is a self-adjoint operator and has a unique ground state given by

$$\Omega_\infty := \Omega_{\mathscr{H}},$$

i.e.,

$$H_\infty \Omega_\infty = 0.$$

It follows from Theorem 8.15 and Proposition 8.9 that $\{\phi_\infty(f), \pi_\infty(f) | f \in \mathscr{V}_p\}$ is an irreducible Weyl representation of the CCR over $\mathscr{V}_{\mathrm{p}}$. We call this representation the **ultraviolet renormalized representation of the CCR over $\mathscr{V}_p$ with respect to** (T, g).

Moreover, $(H_\infty, \{\phi_\infty(f), \pi_\infty(f) | f \in \mathscr{V}_p\})$ satisfies the van Hove relations associated with (T, g) with the following replacements:

$$\langle T^{-1/2} g, f\rangle \to {}_{-(p+1)}\langle g, T^{-1/2} f\rangle_{p+1}, \quad \langle g, T^{1/2} f\rangle \to {}_{-(p+1)}\langle g, T^{1/2} f\rangle_{p+1},$$
$$\mathscr{V} \to \mathscr{V}_p.$$

Hence

$$\mathsf{M}_{\mathrm{v},\infty} := (\mathscr{F}_{\mathrm{b}}(\mathscr{H}), H_\infty, \{\phi_\infty(f), \pi_\infty(f) | f \in \mathscr{V}_p\})$$

is a fixed source model without ultraviolet cutoff, which has a ground state.

Remark 10.21 To define the renormalized Hamiltonian H_∞, we have dropped on the right hand side of (10.134) the ground state energy term corresponding to $E_g(T)$ in (10.124), because, in the present case, $\|T^{-1/2}g\|_{\mathscr{H}}$ is ill-defined. This is related to the following fact: Lemma 10.5 implies that $\lim_{\kappa\to\infty}\|T^{-1/2}g_\kappa\|^2=\infty$. Hence $\lim_{\kappa\to\infty}E_{g_\kappa}(T)=-\infty$. The operator H_∞ heuristically corresponds to the symbolical expression "$\lim_{\kappa\to\infty}(H(\kappa)-E_{g_\kappa}(T))$". This kind of procedure is called an **energy renormalization**.

In the same way as in the proof of Theorem 10.20, one can prove the following theorem:

Theorem 10.23 *Under Assumption (g), the representation* $(\mathscr{F}_{\mathrm{b}}(\mathscr{H}),\{\phi_\infty(f),\pi_\infty(f)\mid f\in\mathscr{V}_p\})$ *is inequivalent to* $(\mathscr{F}_{\mathrm{b}}(\mathscr{H}),\{\phi_{T^{1/2}}(f),\pi_{T^{1/2}}(f)|f\in\mathscr{V}_p\})$.

This theorem shows that the model $\mathsf{M}_{\mathrm{v},\infty}$ is inequivalent to the free field model $\mathsf{M}(T,\mathscr{V}_p)$ (see Example 10.2), although the Hamiltonian H_∞ of the model $\mathsf{M}_{\mathrm{v},\infty}$ is the same as that of the latter.

The prescription of the infrared and the ultraviolet renormalizations discussed above suggests that, *to construct a divergence free quantum field model corresponding to a model which has infrared or ultraviolet divergences in the Fock representation of CCR, one must use a CCR-representation for time-zero fields which is inequivalent to the Fock one.*

Remark 10.22 Also for the ultraviolet renormalized model

$$\mathsf{M}_\infty(g):=(\mathscr{F}_{\mathrm{b}}(\mathscr{H}),H_\infty,\{\phi_\infty(f),\pi_\infty(f)|f\in\mathscr{V}_p\}),$$

considerations similar to those in Sect. 10.9.5 can be made.

Example 10.8 In the original concrete fixed source model without ultraviolet cutoff (see Example 10.7), we have

$$\omega_m^{-s}g_m\in L^2(\hat{\mathbb{R}}^3)$$

for all $s>1$ and $\omega_m^{-1}g_m\notin L^2(\hat{\mathbb{R}}^3)$. Hence Assumption (g) with $\mathscr{H}=L^2(\hat{\mathbb{R}}^3)$, $T=\omega_m$ and $g=g_m$ is satisfied. Therefore we can apply the abstract theory given above to the concrete fixed source model without ultraviolet cutoff to obtain an ultraviolet renormalized model $\mathsf{M}_\infty(g_m)$.

In concluding descriptions on Bose field theories, we want to make a remark from a point of view of constructive quantum field theory [73, 80, 95]. In the standard method of construction of an interacting model of quantum scalar field, one usually takes as time-zero fields the Fock representation $\{\phi_m(f),\pi_m(f)|f\in\mathscr{S}_{\mathbb{R}}(\mathbb{R}^d)\}$. But, to construct a model with the full relativistic symmetry [45, 80, 85, 149, 152], this choice of time-zero fields is not adequate due to **Haag's theorem** which states essentially that a relativistic quantum field model whose time-t fields are unitarily equivalent to those of a free quantum field at a fixed time is completely equivalent to the free field model [82, 83, 108]. Haag's theorem suggests that, to construct a non-trivial relativistic quantum field model (a relativistic quantum field model describing an interaction of elementary particles), one must use as time-zero fields ones inequivalent to $\{\phi_m(f), \pi_m(f) | f \in \mathscr{S}_{\mathbb{R}}(\mathbb{R}^d)\}$.[17] In this sense too, it is important to study irreducible representations of CCR inequivalent to $\{\pi_m\}_{m\geq 0}$. A more detailed account of this aspect can be found in [150, Part II, Chapter 3]. It should be mentioned that it still is a very important open problem to construct a *non-trivial relativistic quantum field model on the four-dimensional Minkowski space–time* (see also [95]).

10.10 Free Quantum Dirac Fields

In Sect. 10.7, we have seen that the masses of scalar bosons form the index set of a family of mutually inequivalent Weyl representations of CCR, which consists of time-zero fields of free quantum scalar fields. In this section, we show that, for fermion masses also, there is a fact corresponding to it. For simplicity, we restrict ourselves to the case of Dirac particles, i.e., relativistic charged particles with spin 1/2 and consider free quantum Dirac fields on the four-dimensional space–time. In Sect. 4.5.12, we have already described the free Dirac operator H_{D} of mass $m \geq 0$. We follow the notation there.

10.10.1 Eigenvectors of $h_{\mathrm{D}}(\boldsymbol{k})$ and Some Operators

For simplicity, we work with the standard representation for the matrices $(\boldsymbol{\alpha}, \beta)$ given by (4.79). We denote by $\{\boldsymbol{e}_r\}_{r=1}^4$ the standard basis of $\mathbb{C}^4$: $\boldsymbol{e}_r = (\delta_{rr'})_{r'=1}^4$. For all $\boldsymbol{k} \in \hat{\mathbb{R}}^3$ and $s = 1, 2$, we define the following vectors in $\mathbb{C}^4$:

$$u_m(\boldsymbol{k}, s) := u_{\mathrm{D}}(\boldsymbol{k})^{-1}\boldsymbol{e}_s \in \mathbb{C}^4, \quad v_m(\boldsymbol{k}, s) := u_{\mathrm{D}}(\boldsymbol{k})^{-1}\boldsymbol{e}_{s+2} \in \mathbb{C}^4. \tag{10.137}$$

By (4.88), we have

$$h_{\mathrm{D}}(\boldsymbol{k})u_m(\boldsymbol{k}, s) = E_m(\boldsymbol{k})u_m(\boldsymbol{k}, s), \quad h_{\mathrm{D}}(\boldsymbol{k})v_m(\boldsymbol{k}, s) = -E_m(\boldsymbol{k})v_m(\boldsymbol{k}, s).$$

Namely, $u_m(\boldsymbol{k}, s)$ (resp. $v_m(\boldsymbol{k}, s)$) is an eigenvector of $h_{\mathrm{D}}(\boldsymbol{k})$ with positive (resp. negative) energy $E_m(\boldsymbol{k})$ (resp. $-E_m(\boldsymbol{k})$). Since $u_{\mathrm{D}}(\boldsymbol{k})^{-1}$ is unitary, it follows that,

[17] This is not restricted to relativistic quantum field models. In fact, we have so far seen that this kind of structure exists also in non-relativistic quantum field models discussed in the previous sections (e.g., infinite volume theories of bosons, van Hove model without infrared and/or ultraviolet cutoffs).

for each $\boldsymbol{k} \in \hat{\mathbb{R}}^d$, the set $\{u_m(\boldsymbol{k}, s), v_m(\boldsymbol{k}, s) | s = 1, 2\}$ is a complete orthonormal basis of $\mathbb{C}^4$. Hence

$$\langle u_m(\boldsymbol{k}, s), u_m(\boldsymbol{k}, s') \rangle_{\mathbb{C}^4} = \delta_{ss'}, \tag{10.138}$$

$$\langle v_m(\boldsymbol{k}, s), v_m(\boldsymbol{k}, s') \rangle_{\mathbb{C}^4} = \delta_{ss'}, \tag{10.139}$$

$$\langle u_m(\boldsymbol{k}, s), v_m(\boldsymbol{k}, s') \rangle_{\mathbb{C}^4} = 0, \quad s, s' = 1, 2. \tag{10.140}$$

and

$$\sum_{s=1}^{2} \left(u_{mr}(\boldsymbol{k}, s) u_{mr'}(\boldsymbol{k}, s)^* + v_{mr}(\boldsymbol{k}, s) v_{mr'}(\boldsymbol{k}, s)^* \right) = \delta_{rr'}, \quad r, r' = 1, 2, 3, 4, \tag{10.141}$$

where $u_{mr}(\boldsymbol{k}, s)$ (resp. $v_{mr}(\boldsymbol{k}, s)$) is the rth component of the vector $u_m(\boldsymbol{k}, s)$ (resp. $v_m(\boldsymbol{k}, s)$).

Let

$$\hat{\mathscr{H}}_{\mathrm{D}} := \mathscr{F}_3 \mathscr{H}_{\mathrm{D}} = L^2(\hat{\mathbb{R}}^3; \mathbb{C}^4), \tag{10.142}$$

where $\mathscr{F}_3$ is the three-dimensional Fourier transform. Then one has the orthogonal decomposition

$$\hat{\mathscr{H}}_{\mathrm{D}} = \hat{\mathscr{H}}_{\mathrm{D}+} \oplus \hat{\mathscr{H}}_{\mathrm{D}-} \tag{10.143}$$

with

$$\hat{\mathscr{H}}_{\mathrm{D}\pm} := L^2(\hat{\mathbb{R}}^3; \mathbb{C}^2).$$

We define linear operators $T_{m\pm} \colon \hat{\mathscr{H}}_{\mathrm{D}} \to \hat{\mathscr{H}}_{\mathrm{D}\pm}$ by

$$T_{m+} f := (u_m(\cdot, s)^* f)_{s=1}^2 \in \hat{\mathscr{H}}_{\mathrm{D}+}, \tag{10.144}$$

$$T_{m-} f := (\tilde{v}_m(\cdot, s) \tilde{f})_{s=1}^2 \in \hat{\mathscr{H}}_{\mathrm{D}-}, \quad f \in \hat{\mathscr{H}}_{\mathrm{D}}, \tag{10.145}$$

where, for $w = (w_r)_{r=1}^4 : \hat{\mathbb{R}}^3 \to \mathbb{C}^4$,

$$(wf)(\boldsymbol{k}) := \sum_{r=1}^{4} w_r(\boldsymbol{k}) f_r(\boldsymbol{k}), \quad f = (f_r)_{r=1}^4 \in \hat{\mathscr{H}}_{\mathrm{D}}, \ \boldsymbol{k} \in \hat{\mathbb{R}}^3$$

and

$$\tilde{w}(\boldsymbol{k}) := w(-\boldsymbol{k}), \quad \boldsymbol{k} \in \hat{\mathbb{R}}^3.$$

It follows that $T_{m\pm}$ are bounded with $\|T_{m\pm}\| \leq \sqrt{2}$.

It is easy to see that

$$(T_{m+}^* f_+)_r = \sum_{s=1}^{2} u_{mr}(\cdot, s) f_{+s}, \quad f_+ = (f_{+s})_{s=1}^2 \in \hat{\mathscr{H}}_{\mathrm{D}+}, \tag{10.146}$$

$$(T_{m-}^* f_-)_r = \sum_{s=1}^{2} v_{mr}(\cdot, s)^* \tilde{f}_{-s}, \quad f_- = (f_{-s})_{s=1}^2 \in \hat{\mathscr{H}}_{\mathrm{D}-}, \ r = 1, 2, 3, 4. \tag{10.147}$$

We denote by C the complex conjugation on $\hat{\mathscr{H}}_{\mathrm{D}}$:

$$Cf := (f_r^*)_{r=1}^4.$$

Throughout this section below, for notational simplicity, the conjugation of a bounded linear operator $A \in \mathfrak{B}(\hat{\mathscr{H}}_{\mathrm{D}})$ with respect to C (see (8.51)) is denoted by $\overline{A}$ (this is not the closure of A):

$$\overline{A} := A_C = CAC.$$

We regard $\hat{\mathscr{H}}_{\mathrm{D}\pm}$ as subspaces of $\hat{\mathscr{H}}_{\mathrm{D}}$ in the natural way so that C acts also on $\hat{\mathscr{H}}_{\mathrm{D}\pm}$.

Important properties of $T_{m\pm}$ are summarized in the next lemma:

Lemma 10.6

$$T_{m+}^* T_{m+} + \overline{T}_{m-}^* \overline{T}_{m-} = I, \tag{10.148}$$

$$T_{m\pm} T_{m\pm}^* = I, \tag{10.149}$$

$$T_{m+} \overline{T}_{m-}^* = 0, \quad T_{m-} \overline{T}_{m+}^* = 0. \tag{10.150}$$

Proof Throughout the proof, we write $T_\pm$ (resp. $u(\boldsymbol{k}, s)$, $v(\boldsymbol{k}, s)$) for $T_{m\pm}$ (resp. $u_m(\boldsymbol{k}, s)$, $v_m(\boldsymbol{k}, s)$). Let $f, g \in \hat{\mathscr{H}}_{\mathrm{D}}$. Then

$$\begin{aligned}
&\langle T_+ f, T_+ g\rangle + \langle \overline{T}_- f, \overline{T}_- g\rangle \\
&= \sum_{s=1}^{2} \int_{\mathbb{R}^3} \left\{ (u(\cdot, s) f^*)(\boldsymbol{k}) (u(\cdot, s)^* g)(\boldsymbol{k}) + (v(\cdot, s) f^*)(\boldsymbol{k}) (v(\cdot, s)^* g)(\boldsymbol{k}) \right\} d\boldsymbol{k} \\
&= \sum_{r,r'=1}^{4} \int_{\mathbb{R}^3} f_r(\boldsymbol{k})^* g_{r'}(\boldsymbol{k}) \left(\sum_{s=1}^{2} \left(u_r(\boldsymbol{k}, s) u_{r'}(\boldsymbol{k}, s)^* + v_r(\boldsymbol{k}, s) v_{r'}(\boldsymbol{k}, s)^* \right) \right) d\boldsymbol{k} \\
&= \sum_{r=1}^{4} \int_{\mathbb{R}^3} f_r(\boldsymbol{k})^* g_r(\boldsymbol{k}) d\boldsymbol{k},
\end{aligned}$$

where we have used (10.141) to obtain the last equality. Hence

$$\langle T_+ f, T_+ g\rangle + \langle \overline{T}_- f, \overline{T}_- g\rangle = \langle f, g\rangle .$$

This implies (10.148).

We have by (10.146)

$$\begin{aligned}\langle T_+^* f, T_+^* g\rangle &= \sum_{r=1}^{4} \sum_{s,s'=1}^{2} \langle u_r(\cdot, s) f_s, u_r(\cdot, s') g_{s'}\rangle \\ &= \int_{\hat{\mathbb{R}}^3} \langle u(\boldsymbol{k}, s), u(\boldsymbol{k}, s')\rangle f_s(\boldsymbol{k})^* g_{s'}(\boldsymbol{k}) d\boldsymbol{k} \\ &= \langle f, g\rangle ,\end{aligned}$$

where we have used (10.138). Hence $T_+ T_+^* = I$. Similarly, using (10.147), one can show that $T_- T_-^* = I$. One can see that orthogonality (10.140) implies (10.150). ■

Lemma 10.6 immediately yields the following result:

Lemma 10.7 *For all $m \geq 0$,*

$$T_m := (T_{m+}, T_{m-})$$

is an element of $\mathfrak{T}_*(\hat{\mathscr{H}}_{\mathrm{D}})$, where $\mathfrak{T}_*(\hat{\mathscr{H}}_{\mathrm{D}})$ is defined by (9.18) with $\mathscr{K}$ replaced by $\hat{\mathscr{H}}_{\mathrm{D}}$.

It follows from (10.149) that $\|T_{m\pm}^*\| = 1$ and hence

$$\|T_{m\pm}\| = 1. \tag{10.151}$$

10.10.2 Construction of a Free Quantum Dirac Field

We construct a free quantum Dirac field on the fermion Fock space $\mathscr{F}_{\mathrm{f}}(\hat{\mathscr{H}}_{\mathrm{D}})$ over $\hat{\mathscr{H}}_{\mathrm{D}}$. Note that we work with momentum representation. We denote by $b(f)$ ($f \in \hat{\mathscr{H}}_{\mathrm{D}}$) the annihilation operator on $\mathscr{F}_{\mathrm{f}}(\hat{\mathscr{H}}_{\mathrm{D}})$.

For each $f \in \hat{\mathscr{H}}_{\mathrm{D}}$, we define

$$\hat{\psi}_m(f) := b(T_{m+} f, 0) + b(0, T_{m-} f^*)^*, \quad f \in \hat{\mathscr{H}}_{\mathrm{D}} \tag{10.152}$$

and set

$$\hat{\rho}_m := (\mathscr{F}_{\mathrm{f}}(\hat{\mathscr{H}}_{\mathrm{D}}, \{\hat{\psi}_m(f), \hat{\psi}_m(f)^* | f \in \hat{\mathscr{H}}_{\mathrm{D}}\}). \tag{10.153}$$

Remark 10.23 In a standard construction of a free quantum Dirac field, the projection method as mentioned in Remark 9.3 is used (see, e.g., [124, 136] or [156, §10.1]). But we find that the projection method is somewhat inconvenient when discussing a family of quantum Dirac fields indexed by mass m, since the orthogonal decomposition of $\hat{\mathscr{H}}_D$ in the projection method depends on m. Thus we take a slightly different approach in which the orthogonal decomposition (10.143) is fixed independently of m.

By Lemma 10.7 and an application of Theorem 9.4, we obtain the following proposition:

Proposition 10.4 *For each $m \geq 0$, $\hat{\rho}_m$ is an irreducible representation of the CAR over $\hat{\mathscr{H}}_{\mathrm{D}}$.*

Let

$$\psi_m(t, f) := \hat{\psi}_m(e^{ith_{\mathrm{D}}}\hat{f}), \quad t \in \mathbb{R},\ f \in D(H_{\mathrm{D}}), \tag{10.154}$$

and

$$\rho_m(t) := (\mathscr{F}_{\mathrm{f}}(\hat{\mathscr{H}}_{\mathrm{D}}), \{\psi_m(t, f), \psi_m(t, f)^* | f \in \hat{\mathscr{H}}_{\mathrm{D}}\}). \tag{10.155}$$

Since $e^{ith_{\mathrm{D}}}$ is unitary for all $t \in \mathbb{R}$, Proposition 10.4 yields the following result:

Proposition 10.5 *For each $m \geq 0$ and $t \in \mathbb{R}$, $\rho_m(t)$ is an irreducible representation of the CAR over $\hat{\mathscr{H}}_{\mathrm{D}}$.*

Theorem 10.24 *Let $m \geq 0$ and $f \in D(H_{\mathrm{D}})$. Then the operator-valued function: $t \mapsto \psi_m(t, f) \in \mathscr{F}_{\mathrm{f}}(\hat{\mathscr{H}}_{\mathrm{D}})$ on $\mathbb{R}$ is differentiable in the operator norm topology and obeys the free functional Dirac equation*

$$i\frac{d\psi_m(t, f)}{dt} = \psi_m(t, H_{\mathrm{D}}f).$$

Proof Let $\varepsilon \in \mathbb{R} \setminus \{0\}$ and

$$\phi_\varepsilon := i\frac{\psi_m(t+\varepsilon, f) - \psi_m(t, f)}{\varepsilon} - \psi_m(t, H_{\mathrm{D}}f).$$

Then we have

$$\phi_\varepsilon = \hat{\psi}_m(e^{ith_{\mathrm{D}}}g_\varepsilon),$$

where

$$g_\varepsilon := \left(-i\frac{(e^{i\varepsilon h_{\mathrm{D}}} - 1)}{\varepsilon} - h_{\mathrm{D}}\right)\hat{f}.$$

We have for all $h \in \hat{\mathscr{H}}_{\mathrm{D}}$

$$\|\hat{\psi}_m(h)\| \leq \|T_{m+}h\| + \|T_{m-}h^*\| \leq 2\|h\|,$$

where we have used (10.151). Hence $\|\phi_\varepsilon\| \leq 2\|g_\varepsilon\|$. Since $\hat{f} \in D(h_{\mathrm{D}})$, it follows that $\lim_{\varepsilon \to 0} \|g_\varepsilon\| = 0$. Hence $\lim_{\varepsilon \to 0} \|\phi_\varepsilon\| = 0$. Thus the desired result follows. ■

Based on Theorem 10.24, we call $\psi_m(t, f)$ a **free quantum Dirac field** on the four-dimensional space–time.[18] Note that $\hat{\psi}_m(\hat{f}) = \psi_m(0, f)$ is the time-zero field of it.

Let

$$\mathbb{H}_m := \mathrm{d}\Gamma_{\mathrm{f}}(E_m), \tag{10.156}$$

the fermion second quantization operator of the multiplication operator by E_m on $\hat{\mathscr{H}}_{\mathrm{D}}$, acting in $\mathscr{F}_{\mathrm{f}}(\hat{\mathscr{H}}_{\mathrm{D}})$, where E_m is defined by (4.85).

Theorem 10.25 *For all $f \in \mathscr{H}_{\mathrm{D}}$,*

$$\psi_m(t, f) = e^{it\mathbb{H}_m}\hat{\psi}_m(\hat{f})e^{-it\mathbb{H}_m}, \quad t \in \mathbb{R}. \tag{10.157}$$

Proof By (10.152), we have

$$\hat{\psi}_m(e^{ith_{\mathrm{D}}}\hat{f}) = b(T_{m+}e^{ith_{\mathrm{D}}}\hat{f}, 0) + b(0, T_{m-}(e^{ith_{\mathrm{D}}}\hat{f})^*)^*.$$

By (10.144), we have for a.e. $\boldsymbol{k} \in \hat{\mathbb{R}}^3$

$$\begin{aligned}(T_{m+}e^{ith_{\mathrm{D}}}\hat{f})_s(\boldsymbol{k}) &= \left\langle u_m(\boldsymbol{k}, s), e^{ith_{\mathrm{D}}(\boldsymbol{k})}\hat{f}(\boldsymbol{k})\right\rangle_{\mathbb{C}^4} = \left\langle e^{-ith_{\mathrm{D}}(\boldsymbol{k})}u_m(\boldsymbol{k}, s), \hat{f}(\boldsymbol{k})\right\rangle_{\mathbb{C}^4} \\ &= e^{itE_m(\boldsymbol{k})}\left\langle u_m(\boldsymbol{k}, s), \hat{f}(\boldsymbol{k})\right\rangle_{\mathbb{C}^4},\end{aligned}$$

where we have used the fact that $e^{-ith_{\mathrm{D}}(\boldsymbol{k})}u_m(\boldsymbol{k}, s) = e^{-itE_m(\boldsymbol{k})}u_m(\boldsymbol{k}, s)$. Similarly, using the fact that $e^{-ith_{\mathrm{D}}(-\boldsymbol{k})}v_m(-\boldsymbol{k}, s) = e^{itE_m(-\boldsymbol{k})}v_m(-\boldsymbol{k}, s)$ and $E_m(-\boldsymbol{k}) = E_m(\boldsymbol{k})$, one can show that

$$(T_{m-}(e^{ith_{\mathrm{D}}}\hat{f})^*)_s(\boldsymbol{k}) = e^{itE_m(\boldsymbol{k})}\left\langle \tilde{v}_m(\boldsymbol{k}, s), \tilde{\hat{f}}(\boldsymbol{k})\right\rangle_{\mathbb{C}^4}.$$

[18] For a general definition of a free quantum Dirac field, see [28, §12.3.1].

Hence

$$\begin{aligned}\hat{\psi}_m(e^{ith_{\mathrm{D}}}\hat{f}) &= b(e^{itE_m}T_{m+}\hat{f},0)+b(0,e^{itE_m}T_{m-}\hat{f}^{\,*})^* \\ &= e^{it\mathbb{H}_m}\left(b(T_{m+}\hat{f},0)+b(0,T_{m-}\hat{f}^{\,*})^*\right)e^{-it\mathbb{H}_m} \\ &= e^{it\mathbb{H}_m}\hat{\psi}_m(\hat{f})e^{-it\mathbb{H}_m},\end{aligned}$$

where we have used Theorem 7.11. Thus (10.157) is derived. ■

Formula (10.157) shows that $\mathbb{H}_m$ is the Hamiltonian of the free quantum Dirac field $\psi_m(t,f)$ of mass m in momentum representation (up to constant additions).

10.10.3 Inequivalence of Free Quantum Dirac Fields of Different Masses

We now prove the main result in this chapter:

Theorem 10.26 *Let $m_1 \neq m_2$ ($m_1, m_2 \geq 0$). Then the representation $\hat{\rho}_{m_1}$ is inequivalent to $\hat{\rho}_{m_2}$.*

Proof By Theorem 9.6, we need only to prove that, if $m_1 \neq m_2$, then $T_{m_1+}\overline{T}^*_{m_2-}$ or $T_{m_1-}\overline{T}^*_{m_2+}$ is not Hilbert–Schmidt. It is easy to see that

$$(T_{m_1+}\overline{T}^*_{m_2-}f)_r(\boldsymbol{k}) = \sum_{s=1}^{2} K_{rs}(\boldsymbol{k})\tilde{f}_s(\boldsymbol{k}), \quad f=(f_s)_{s=1}^2 \in \hat{\mathscr{H}}_{\mathrm{D}-},\ r=1,2, \tag{10.158}$$

where

$$K_{rs}(\boldsymbol{k}) := \langle u_{m_1}(\boldsymbol{k},r), v_{m_2}(\boldsymbol{k},s)\rangle_{\mathbb{C}^4}, \quad \boldsymbol{k}\in\hat{\mathbb{R}}^3,\ r,s=1,2. \tag{10.159}$$

To make explicit the dependence of $u_{\mathrm{D}}(\boldsymbol{k})$ on m, we write

$$d_m(\boldsymbol{k}) := u_{\mathrm{D}}(\boldsymbol{k}).$$

Then, by (10.137), we have

$$K_{rs}(\boldsymbol{k}) = \langle \boldsymbol{e}_r, d_{m_1}(\boldsymbol{k})d_{m_2}(\boldsymbol{k})^*\boldsymbol{e}_{s+2}\rangle.$$

Using (4.86), (4.87), the orthogonality $\langle \boldsymbol{e}_r, \boldsymbol{e}_{s+2}\rangle = 0$ $(r, s = 1, 2)$ and the fact that $\beta \boldsymbol{e}_{s+2} = -\boldsymbol{e}_{s+2}$, we obtain

$$K_{rs}(\boldsymbol{k}) = (m_2 - m_1) F_{m_1,m_2}(\boldsymbol{k}) \langle \boldsymbol{e}_r, (\boldsymbol{\alpha} \cdot \boldsymbol{k}) \boldsymbol{e}_{s+2}\rangle, \quad r, s = 1, 2, \tag{10.160}$$

where

$$F_{m_1,m_2}(\boldsymbol{k}) := \frac{1}{2\sqrt{E_{m_1}(\boldsymbol{k}) E_{m_2}(\boldsymbol{k})(m_1 + E_{m_1}(\boldsymbol{k}))(m_2 + E_{m_2}(\boldsymbol{k}))}} \times \left(1 + \frac{m_1 + m_2}{E_{m_1}(\boldsymbol{k}) + E_{m_2}(\boldsymbol{k})}\right). \tag{10.161}$$

It follows from (9.16) that $(\boldsymbol{\alpha} \cdot \boldsymbol{k})\boldsymbol{e}_{s+2} \in \ker(\beta - 1)$, $s = 1, 2$. Hence

$$\sum_{r=1}^{2} |\langle \boldsymbol{e}_r, (\boldsymbol{\alpha} \cdot \boldsymbol{k})\boldsymbol{e}_{s+2}\rangle|^2 = \|(\boldsymbol{\alpha} \cdot \boldsymbol{k})\boldsymbol{e}_{s+2}\|^2 = \boldsymbol{k}^2,$$

where we have used (4.83). Therefore

$$\sum_{r=1}^{2} |K_{rs}(\boldsymbol{k})|^2 = (m_2 - m_1)^2 F_{m_1,m_2}(\boldsymbol{k})^2 \boldsymbol{k}^2. \tag{10.162}$$

Without loss of generality, one can assume that $m_1 > m_2 \geq 0$. Suppose that $T_{m_1+}\overline{T}^*_{m_2-}$ were Hilbert–Schmidt. We denote by $\hat{K}_{rs}$ the multiplication operator by the function $K_{rs}(\boldsymbol{k})$ on $L^2(\hat{\mathbb{R}}^3)$. Then, by (10.158), $\hat{K}_{rs}$ is Hilbert–Schmidt. Hence $L := \sum_{r=1}^{2} \hat{K}^*_{rs}\hat{K}_{rs}$ is Hilbert–Schmidt. By (10.162), L is the multiplication operator by the function

$$G(\boldsymbol{k}) := (m_2 - m_1)^2 F_{m_1,m_2}(\boldsymbol{k})^2 \boldsymbol{k}^2,$$

which is bounded, non-negative and continuous on $\hat{\mathbb{R}}^3$. Hence, by Theorem A.1(iv) in Appendix A, L is a bounded self-adjoint operator on $\hat{\mathscr{H}}_{\mathrm{D}-}$ and

$$\sigma(L) = \overline{\{G(\boldsymbol{k}) | \boldsymbol{k} \in \hat{\mathbb{R}}^3\}}.$$

It is easy to see that $\lim_{|\boldsymbol{k}|\to\infty} G(\boldsymbol{k}) = 0$. Since G is not a constant function, it follows that $M := \max_{\boldsymbol{k}\in\hat{\mathbb{R}}^3} G(\boldsymbol{k}) > 0$. Hence, by the middle value theorem, $\sigma(L) = [0, M]$. But this contradicts the Hilbert–Schmidtness of L, because the spectrum of a self-adjoint Hilbert–Schmidt operator is purely discrete in $\mathbb{R} \setminus \{0\}$. Therefore we arrive at a contradiction. Thus $T_{m_1+}\overline{T}^*_{m_2-}$ is not Hilbert–Schmidt. ■

Theorem 10.26 shows that $\{\hat{\rho}_m\}_{m\geq 0}$ is a family of mutually inequivalent irreducible representations of the CAR over $\hat{\mathscr{H}}_{\mathrm{D}}$. This gives a representation theoretical meaning to masses of Dirac particles.

Remark 10.24 The contents of this section are taken partly from the paper [27]. It is shown in fact that $\{\hat{\rho}_m\}_{m\geq 0}$ is an example in a general class of irreducible representations of CAR which are mutually inequivalent. For the details, see [27].

Appendix A
Multiplication Operators

Let (X, Σ, μ) be a measure space, i.e., X is a non-empty set, Σ is a σ-field of subsets of X and μ is a measure on the measurable space (X, Σ). It is well known that the space

$$L^2(X, d\mu) := \left\{ f : X \to \mathbb{C} \cup \{\pm\infty\},\ \Sigma\text{-measurable} \middle| \int_X |f(x)|^2 d\mu(x) < \infty \right\}$$

of square integrable functions on (X, Σ, μ) is a Hilbert space with inner product

$$\langle f, g \rangle_{L^2(X,d\mu)} := \int_X f(x)^* g(x) d\mu(x), \quad f, g \in L^2(X, d\mu),$$

where, for $f, g \in L^2(X, d\mu)$, the equality "$f = g$" is defined to be $f(x) = g(x)$ for μ-a.e.$x \in X$ ("μ-a.e.x" means "almost everywhere (a.e.) x with respect to μ").

A Σ-measurable function on X is said to be **essentially bounded** if there exists a constant $C \geq 0$ such that $|f(x)| \leq C$, μ-a.e.$x \in X$. In this case we denote the infimum of such constants C by $\|f\|_\infty$, which is called the **essential supremum** of f and written

$$\|f\|_\infty = \text{ess.sup}_{x \in X} |f(x)|. \tag{A.1}$$

For a Σ-measurable function F on X with $|F(x)| < \infty$ for μ-a.e.x, one can define a linear operator M_F on $L^2(X, d\mu)$ as follows:

$$D(M_F) := \left\{ f \in L^2(X, d\mu) \middle| \int_X |F(x) f(x)|^2 d\mu(x) < \infty \right\},$$

$$(M_F f)(x) := F(x) f(x), \quad \mu\text{-a.e.}x \in X,\ f \in D(M_F).$$

© Springer Nature Singapore Pte Ltd. 2020

A. Arai, *Inequivalent Representations of Canonical Commutation and Anti-Commutation Relations*, Mathematical Physics Studies,
https://doi.org/10.1007/978-981-15-2180-5

The **essential range** of F is defined by

$$\text{ess.ran}(F) := \{\lambda \in \mathbb{C} | \forall \varepsilon > 0, \mu(\{x \in X | \ |F(x) - \lambda| < \varepsilon\}) > 0\}.$$

It is easy to see that ess.ran(F) is a closed set.

Example A.1 For any continuous function F on $\mathbb{R}^d$, ess.ran$(F) = \overline{F(\mathbb{R}^d)}$, where $F(\mathbb{R}^d) := \{F(x) | x \in \mathbb{R}^d\}$ and, for a subset $S \subset \mathbb{R}^d$, $\overline{S}$ denotes the closure of S.

The following theorem states basic properties of the multiplication operator M_F:

Theorem A.1 *Let M_F be as above. Then:*

(i) *M_F is a closed operator.*
(ii) *If F is essentially bounded, then $D(M_F) = L^2(X, d\mu)$ and M_F is bounded with $\|M_F\| \le \|F\|_\infty$.*
(iii) *If F is not essentially bounded, then $D(M_F)$ is dense, $D(M_F) \neq L^2(X, d\mu)$ and M_F is unbounded.*
(iv) *$(M_F)^* = M_{F^*}$.*
(v) *If $F: X \to \mathbb{R} \cup \{\pm\infty\}$ with $F(x) \in \mathbb{R}$ for μ-a.e.x, then M_F is self-adjoint and*

$$\sigma(M_F) = \text{ess.ran}(F). \tag{A.2}$$

Proof

(i) Let $\{f_n\}_{n=1}^\infty$ be a sequence in $D(M_F)$ such that $f_n \to f \in L^2(X, d\mu)$ and $M_F f_n \to g \in L^2(X, d\mu)$ as $n \to \infty$. Then there exists a subsequence $\{f_{n_k}\}_{k=1}^\infty$ of $\{f_n\}_n$ such that $\lim_{k\to\infty} f_{n_k}(x) = f(x)$ and $\lim_{k\to\infty} F(x) f_{n_k}(x) = g(x)$ for μ-a.e.$x \in X$. Hence $F(x)f(x) = g(x)$ for μ-a.e.$x \in X$. Thus $f \in D(M_F)$ and $M_F f = g$.
(ii) An easy exercise.
(iii) The proof of this part is not so difficult, but takes much space (careful measure-theoretical arguments are needed). Hence we omit it. See, e.g., [23, Example 2.17] or [31, Example 1.33].
(iv) It is obvious that $D(M_F) = D(M_{F^*})$. Let $f \in D(M_{F^*})$. Then, for all $g \in D(M_F)$,

$$\langle f, M_F g\rangle = \int_X (F^* f)(x) g(x) d\mu(X) = \langle M_{F^*} f, g\rangle.$$

Hence $f \in D((M_F)^*)$ and $(M_F)^* f = M_{F^*} f$. Therefore $M_{F^*} \subset (M_F)^*$.

Conversely, let $f \in D((M_F)^*)$. For each $n \in \mathbb{N}$, we define $\chi_n : X \to \mathbb{R}$ as follows: if $0 \le |F(x)| \le n$, then $\chi_n(x) := 1$ and if $|F(x)| > n$, then $\chi_n(x) := 0$. Then, for all $h \in L^2(X, d\mu)$, $\chi_n h \in D(M_F)$ and $\chi_n F^* f$ is in $L^2(X, d\mu)$. Hence

$$\langle \chi_n (M_F)^* f, g\rangle = \langle f, M_F(\chi_n g)\rangle = \langle \chi_n F^* f, g\rangle.$$

Hence $\chi_n(M_F)^* f = \chi_n F^* f$. Note that $\lim_{n\to\infty} \chi_n(x) = 1$, μ-a.e.x. Therefore

$$((M_F)^* f)(x) = F(x)^* f(x), \quad \mu\text{-a.e.}x.$$

This implies that $F^* f \in L^2(X, d\mu)$ (hence $f \in D(M_{F^*})$) and $M_{F^*} f = (M_F)^* f$. Therefore $(M_F)^* \subset M_{F^*}$.

(v) The self-adjointness of M_F in the present case follows from (iv). For a proof of (A.2), see, e.g., [31, Theorem 2.27]. ■

Appendix B
Spectral Measures and Functional Calculus

For each $d \in \mathbb{N}$, we denote by $\boldsymbol{B}^d$ the Borel field of the d-dimensional Euclidean vector space $\mathbb{R}^d$. Let $\mathscr{H}$ be a Hilbert space and $\mathsf{P}(\mathscr{H})$ be the set of all orthogonal projections on $\mathscr{H}$.[1] A mapping E: $\boldsymbol{B}^d \to \mathsf{P}(\mathscr{H})$; $\boldsymbol{B}^d \ni B \mapsto E(B) \in \mathsf{P}(\mathscr{H})$ is called a **d-dimensional spectral measure** or a **d-dimensional resolution of identity**[2] if (E.1) and (E.2) below hold:

(E.1) $E(\emptyset) = 0, \ E(\mathbb{R}^d) = I.$
(E.2) (complete additivity) If $B_n \in \boldsymbol{B}^d$, $n \in \mathbb{N}$ and $B_n \cap B_m = \emptyset$ $(n \neq m)$, then

$$E(\cup_{n=1}^{\infty} B_n) = \text{s-}\lim_{N\to\infty} \sum_{n=1}^{N} E(B_n),$$

where s-lim means strong limit (see Sect. 1.1.2).

It is not so difficult to prove that (E.1) and (E.2) imply

(E.3) For all $B_1, B_2 \in \boldsymbol{B}^d$, $E(B_1)E(B_2) = E(B_1 \cap B_2)$.

Hence, in particular, $E(B_1)$ commutes with $E(B_2)$:

$$E(B_1)E(B_2) = E(B_2)E(B_1), \quad B_1, B_2 \in \boldsymbol{B}^d.$$

Thus the set $\{E(B)|B \in \boldsymbol{B}^d\}$ is a family of commuting orthogonal projections on $\mathscr{H}$ satisfying (E.1)–(E.3). Property (E.2) implies the following:

(E.4) (monotonicity) If $A \subset B$, $A, B \in \boldsymbol{B}^d$, then, for all $\Psi \in \mathscr{H}$,

$$\langle \Psi, E(A)\Psi \rangle \leq \langle \Psi, E(B)\Psi \rangle .$$

[1] An operator $P \in \mathfrak{B}(\mathscr{H})$ is called an **orthogonal projection** if $P = P^*$ and $P = P^2$.
[2] It is called also a **projection-valued measure**.

© Springer Nature Singapore Pte Ltd. 2020

A. Arai, *Inequivalent Representations of Canonical Commutation and Anti-Commutation Relations*, Mathematical Physics Studies,
https://doi.org/10.1007/978-981-15-2180-5

The smallest closed set $C_0 \in \boldsymbol{B}^d$ such that $E(C_0) = I$ is called the **support** of E. We write $C_0 = \text{supp}\, E$. Hence $E(\text{supp}\, E) = I$.

For each $\Psi \in \mathscr{H}$, there exists a bounded measure $\mu_\Psi \colon \boldsymbol{B}^d \to [0, \infty)$ defined by

$$\mu_\Psi(B) := \|E(B)\Psi\|^2, \quad B \in \boldsymbol{B}^d.$$

For a Borel measurable function $f \colon \mathbb{R}^d \to \mathbb{C} \cup \{\pm\infty\}$, the Lebesgue integral $\int_{\mathbb{R}^d} f d\mu_\Psi \in \mathbb{C} \cup \{\pm\infty\}$ of f with respect to μ_Ψ (if it exists) is written as $\int_{\mathbb{R}^d} f(\lambda) d \langle \Psi, E(\lambda)\Psi \rangle$ or $\int_{\mathbb{R}^d} f(\lambda) d\|E(\lambda)\Psi\|^2$:

$$\int_{\mathbb{R}^d} f(\lambda) d \langle \Psi, E(\lambda)\Psi \rangle := \int_{\mathbb{R}^d} f(\lambda) d\|E(\lambda)\Psi\|^2 := \int_{\mathbb{R}^d} f(\lambda) d\mu_\Psi(\lambda).$$

For each pair $(\Phi, \Psi) \in \mathscr{H} \times \mathscr{H}$, the mapping $\mu_{\Phi,\Psi} \colon \boldsymbol{B}^d \to \mathbb{C}$ defined by

$$\mu_{\Phi,\Psi}(B) := \langle \Phi, E(B)\Psi \rangle, \quad B \in \boldsymbol{B}^d,$$

is a complex-valued measure. Hence one can define the Lebesgue–Stieltjes integral $\int_{\mathbb{R}^d} f(\lambda) d\mu_{\Psi,\Phi}(\lambda)$ with respect to $\mu_{\Phi,\Psi}$ under the condition that

$$\int_{\mathbb{R}^d} |f(\lambda)| d|\mu_{\Psi,\Phi}|(\lambda) < \infty,$$

where $|\mu_{\Psi,\Phi}|(\cdot)$ is the total variation of $\mu_{\Psi,\Phi}(\cdot)$. We write

$$\int_{\mathbb{R}^d} f(\lambda) d\mu_{\Psi,\Phi}(\lambda) = \int_{\mathbb{R}^d} f(\lambda) d \langle \Psi, E(\lambda)\Phi \rangle.$$

For a Borel measurable function f on $\mathbb{R}^d$, the subset

$$\mathscr{D}_f := \left\{ \Psi \in \mathscr{H} | \int_{\mathbb{R}^d} |f(\lambda)|^2 d\|E(\lambda)\Psi\|^2 < \infty \right\}$$

is a subspace. One can show that, for all $\Psi \in \mathscr{D}_f$ and $\Phi \in \mathscr{H}$,

$$\left| \int_{\mathbb{R}^d} f(\lambda) d \langle \Phi, E(\lambda)\Psi \rangle \right| \leq \left(\int_{\mathbb{R}^d} |f(\lambda)|^2 d\|E(\lambda)\Psi\|^2 \right)^{1/2} \|\Phi\|.$$

By this fact and the Riesz representation theorem, there exists a unique linear operator $T_E(f)$ on $\mathscr{H}$ such that

$$D(T_E(f)) = \mathscr{D}_f,$$

$$\langle \Phi, T_E(f)\Psi \rangle = \int_{\mathbb{R}^d} f(\lambda) d \langle \Phi, E(\lambda)\Psi \rangle, \quad \Psi \in \mathscr{D}_f, \Phi \in \mathscr{H}.$$

The operator $T_E(f)$ is written *symbolically* as

$$T_E(f) = \int f(\lambda) dE(\lambda).$$

One has

$$\|T_E(f)\Psi\|^2 = \int_{\mathbb{R}^d} |f(\lambda)|^2 d\|E(\lambda)\Psi\|^2, \quad \Psi \in D(T_E(f)).$$

The next theorem summarizes some fundamental properties of $T_E(f)$.

Theorem B.1 *Let f and g be Borel measurable functions on $\mathbb{R}^d$.*

(i) *For all $B \in \mathbf{B}^d$, $E(B)T_E(f) \subset T_E(f)E(B)$.*
(ii) *The domain $D(T_E(f))$ is dense in $\mathscr{H}$ if and only if $E(\{\lambda | |f(\lambda)| = +\infty\}) = 0$. In this case, $T_E(f)$ is a densely defined closed operator and $T_E(f)^* = T_E(f^*)$.*
(iii) *If f is continuous on $\mathbb{R}^d$, then $T_E(f)$ is a densely defined closed operator and $T_E(f)^* = T_E(f^*)$.*
(iv) *If f is bounded on* supp E, *i.e., $C_f := \sup_{\lambda \in \mathrm{supp}\, E} |f(\lambda)| < \infty$, then $T_E(f) \in \mathfrak{B}(\mathscr{H})$ and $\|T_E(f)\| \leq C_f$.*
(v) *If $|f(\lambda)| = 1$ for all $\lambda \in$* supp E, *then $T_E(f)$ is unitary.*
(vi) *If f is real-valued and $E(\{\lambda \in \mathbb{R}^d | |f(\lambda)| = +\infty\}) = 0$, then $T_E(f)$ is self-adjoint. In particular, if f is a real-valued continuous function on $\mathbb{R}^d$, then $T_E(f)$ is self-adjoint.*
(vii) *$T_E(f) + T_E(g) \subset T_E(f + g)$. In particular, if g is bounded on* supp E, *then $T_E(f) + T_E(g) = T_E(f + g)$.*
(viii) *$D(T_E(fg)) \cap D(T_E(g)) = D(T_E(f)T_E(g))$ and $T_E(f)T_E(g) \subset T_E(fg)$. In particular, if g is bounded on* supp E, *then $T_E(f)T_E(g) = T_E(fg)$.*
(ix) *If $f_n \colon \mathbb{R}^d \to \mathbb{C} \cup \{\pm\infty\}$ $(n \in \mathbb{N})$ is Borel measurable with*

$$\sup_{n \in \mathbb{N}, \lambda \in \mathrm{supp}\, E} |f_n(\lambda)| < \infty$$

and

$$\lim_{n \to \infty} f_n(\lambda) = f(\lambda), \quad \lambda \in \mathrm{supp}\, E,$$

then $T_E(f_n), T_E(f) \in \mathfrak{B}(\mathscr{H})$ and $s\text{-}\lim_{n \to \infty} T_E(f_n) = T_E(f)$.
(x) *(spectral mapping theorem) If f is continuous on $\mathbb{R}^d$, then*

$$\sigma(T_E(f)) = \overline{\{f(\lambda) | \lambda \in \mathrm{supp}\, E\}},$$

where $\sigma(T_E(f))$ is the spectrum of $T_E(f)$.

Calculus of the mapping $T_E\colon f \mapsto T_E(f) \in \mathfrak{L}(\mathscr{H})$ (the set of linear operators on $\mathscr{H}$) as stated in the above theorem is called **functional calculus** with respect to the spectral measure E.

The following theorem is one of the deepest theorems in Hilbert space theory:

Theorem B.2 (Spectral Theorem) *For any self-adjoint operator A on $\mathscr{H}$, there exists a unique one-dimensional spectral measure E_A such that $A = \int_{\mathbb{R}} \lambda\, dE_A(\lambda)$.*

Example B.1 Let M_F be the multiplication operator discussed in Appendix A. Suppose that $F(x) \in \mathbb{R}$ μ-a.e.x. Then, by Theorem A.1(v), M_F is self-adjoint. For all $g \in L^2(X, d\mu)$ and $f \in D(M_F)$, we have

$$\langle g, M_F f\rangle = \int_X g(x)^* F(x) f(x) d\mu(x) = \int_X F(x) d\mu_{g,f}(x), \tag{B.1}$$

where $\mu_{g,f}(Y) := \int_Y g(x)^* f(x) d\mu(x) = \langle g, \chi_Y f\rangle_{L^2(X,d\mu)}$, $Y \in \Sigma$ (χ_Y is the characteristic function of Y). For a Borel set B of $\mathbb{R}$, we denote by $F^{-1}(B)$ the inverse image of F:

$$F^{-1}(B) := \{x \in X | F(x) \in B\}$$

and define

$$\nu^F_{g,f}(B) := \mu_{g,f}(F^{-1}(B)) = \left\langle g, \chi_{F^{-1}(B)} f\right\rangle_{L^2(X,d\mu)}.$$

Then $\nu^F_{g,f}$ is a complex measure on $(\mathbb{R}, \boldsymbol{B}^1)$. It follows from (B.1) that

$$\langle g, M_F f\rangle = \int_{\mathbb{R}} \lambda d\mu^F_{g,f}(\lambda).$$

Hence the spectral measure E_{M_F} of M_F is of the following form:

$$E_{M_F}(B) f = \chi_{F^{-1}(B)} f, \quad f \in L^2(X, d\mu),\ B \in \boldsymbol{B}^1,$$

which means that E_{M_F} is the multiplication operator by the characteristic function $\chi_{F^{-1}(B)}$ of $F^{-1}(B)$. It follows from this fact that, for all $f, g \in L^2(X, d\mu)$ and $t \in \mathbb{R}$,

$$\left\langle g, e^{itM_F} f\right\rangle = \int_{\mathbb{R}} e^{it\lambda} d\nu^F_{g,f}(\lambda) = \int_X e^{itF(x)} g(x)^* f(x) d\mu(x) = \left\langle g, e^{itF} f\right\rangle.$$

Hence

$$e^{itM_F} f = e^{itF} f, \quad f \in L^2(X, d\mu),$$

i.e., e^{itM_F} is the multiplication operator by the function e^{itF}.

Let A be a self-adjoint operator on $\mathscr{H}$. Then, by an application of the theory of $T_E(f)$ to the case where $d = 1$ and $E = E_A$, one has an operator $T_{E_A}(f)$ for each Borel measurable function f on $\mathbb{R}$. Usually this operator is written $f(A)$:

$$f(A) = T_{E_A}(f) = \int_{\mathbb{R}} f(\lambda) dE_A(\lambda).$$

The correspondence $A \mapsto f(A)$ gives an operator-valued function on the set of self-adjoint operators on $\mathscr{H}$. The following theorem is very useful in operator calculus.

Theorem B.3 (Unitary Covariance of Functional Calculus) *Let A be a self-adjoint operator on a Hilbert space $\mathscr{H}$ and U be a unitary operator from $\mathscr{H}$ to a Hilbert space $\mathscr{K}$. Then, for all Borel measurable functions f on $\mathbb{R}$,*

$$Uf(A)U^{-1} = f(UAU^{-1}).$$

In relation to spectral measures, we need a theorem which is used in the main text of this book. For this purpose, we first present a lemma.

Lemma B.1 *Let A be a densely defined closed operator on a Hilbert space $\mathscr{H}$ and S be a bounded self-adjoint operator on $\mathscr{H}$ such that*

$$SA \subset AS. \tag{B.2}$$

Then

$$(S - z)^{-1} A \subset A(S - z)^{-1}, \quad \forall z \in \rho(S). \tag{B.3}$$

Proof It follows from (B.2) that, for all $n \in \mathbb{N}$, $S^n A \subset AS^n \cdots (*)$. Let $z \in \mathbb{C}$ be such that $\|S\| < |z|$, where $\|S\|$ is the operator norm of S. Then $z \in \rho(S)$ and, putting $S_N := -\sum_{n=0}^{\infty} z^{-n-1} S^n$, we have $(S - z)^{-1} = \lim_{N\to\infty} S_N$ in the operator norm topology (i.e. $\lim_{N\to\infty} \|S_N - (S - z)^{-1}\| = 0$). By $(*)$, for all $\Psi \in D(A)$, $S_N A\Psi = AS_N\Psi$. The left hand side converges to $(S - z)^{-1} A\Psi$ as $N \to \infty$. Hence $AS_N\Psi \to (S - z)^{-1} A\Psi$ as $N \to \infty$. Since A is closed, it follows that $(S - z)^{-1}\Psi \in D(A)$ and $A(S - z)^{-1}\Psi = (S - z)^{-1} A\Psi$. Hence $(S - z)^{-1} A \subset A(S - z)^{-1}$.

Next, let $\Phi \in D(A^*)$, $\Psi \in D(A)$ and define functions f and g on $\rho(S)$ by

$$f(z) := \left\langle A^*\Phi, (S - z)^{-1}\Psi \right\rangle, \quad g(z) := \left\langle \Phi, (S - z)^{-1} A\Psi \right\rangle, \quad z \in \rho(S).$$

It is obvious that f and g are analytic on $\rho(S)$ and, by the preceding result, $f(z) = g(z)$ for all $z \in \rho(S)$ satisfying $|z| > \|S\|$. Since S is self-adjoint, $\mathbb{C} \setminus \mathbb{R} \subset \rho(S)$ and $\sigma(S)$ is a bounded closed subset of $\mathbb{R}$. Hence $\rho(S)$ is arcwise connected. Therefore we can apply the theorem of identity to conclude that $f(z) = g(z)$ for all $z \in \rho(S)$. This implies that $(S - z)^{-1}\Psi \in D(A^{**}) = D(A)$ and (B.3) holds. ■

Remark B.1 If A in Lemma B.1 is bounded, then (B.3) trivially holds. Hence Lemma B.1 is non-trivial only if A is unbounded.

Theorem B.4 *Let A be a densely defined closed operator on a Hilbert space $\mathscr{H}$ and S be a bounded self-adjoint operator on $\mathscr{H}$ such that*

$$SA \subset AS. \tag{B.4}$$

Then, for all Borel sets $B \subset \mathbb{R}$, $E_S(B)A \subset AE_S(B)$, where E_S is the spectral measure of S.

Proof By the Stone formula, we have

$$\frac{1}{2}\big(E_S([a,b]) + E_S((a,b))\big) = \text{s-}\lim_{\varepsilon\downarrow 0}\frac{1}{2\pi i}\int_a^b \Big((S-\lambda-i\varepsilon)^{-1} - (S-\lambda+i\varepsilon)^{-1}\Big)\,d\lambda,$$

where the integral on the right hand side is taken in the strong sense and s- lim means strong limit. By Lemma B.1, $(S-z)^{-1}A \subset A(S-z)^{-1}$, $z \in \rho(S)$. Hence, for all $\Psi \in D(A)$,

$$\Big((S-\lambda-i\varepsilon)^{-1} - (S-\lambda+i\varepsilon)^{-1}\Big)A\Psi = A\Big((S-\lambda-i\varepsilon)^{-1} - (S-\lambda+i\varepsilon)^{-1}\Big)\Psi.$$

This equation and the closedness of A imply that $\big(E_S([a,b]) + E_S((a,b))\big)\Psi \in D(A)$ and

$$A\big(E_S([a,b]) + E_S((a,b))\big)\Psi = \big(E_S([a,b]) + E_S((a,b))\big)A\Psi.$$

Hence it follows that

$$AE_S(\{a\})\Psi = E_S(\{a\})A\Psi, \quad AE_S((a,b])\Psi = E_S((a,b])A\Psi.$$

Let $\Phi \in D(A^*)$ and

$$\mu_1(B) := \langle A^*\Phi, E_S(B)\Psi\rangle, \quad \mu_2(B) := \langle \Phi, E_S(B)A\Psi\rangle, \quad B \in \boldsymbol{B}^1.$$

Then μ_1 and μ_2 are bounded complex measures satisfying that $\mu_1(\{a\}) = \mu_2(\{a\})$ and $\mu_1((a,b]) = \mu_2((a,b])$ for all $a, b \in \mathbb{R}$ with $a < b$. Hence, by the uniqueness of Hopf's extension theorem, $\mu_1 = \mu_2$. Thus $E_S(B)\Psi \in D(A^{**}) = D(A)$ and $AE_S(B)\Psi = E_S(B)A\Psi$, i.e. $E_S(B)A \subset AE_S(B)$. ■

Appendix C
Direct Sum Hilbert Spaces and Direct Sum Operators

C.1 Finite Direct Sums of Hilbert Spaces and Operators

Let $N \geq 2$ ($N \in \mathbb{N}$) and $\mathscr{H}_1, \ldots, \mathscr{H}_N$ be Hilbert spaces. Then the direct product

$$\mathscr{H}_1 \times \cdots \times \mathscr{H}_N = \{\Psi = (\Psi_1, \ldots, \Psi_N) | \Psi_j \in \mathscr{H}_j, j = 1, \ldots, N\}$$

becomes a complex vector space with the following operations of addition and scalar multiplication:

$$\Psi + \Phi := (\Psi_1 + \Phi_1, \ldots, \Psi_N + \Phi_N),$$
$$\alpha\Psi := (\alpha\Psi_1, \ldots, \alpha\Psi_N), \quad \Psi, \Phi \in \mathscr{H}_1 \times \cdots \times \mathscr{H}_N, \ \alpha \in \mathbb{C}.$$

This vector space becomes a Hilbert space with the following inner product:

$$\langle \Psi, \Phi \rangle := \sum_{j=1}^{N} \langle \Psi_j, \Phi_j \rangle_{\mathscr{H}_j}.$$

This Hilbert space, denoted by $\oplus_{j=1}^{N} \mathscr{H}_j$, is called the **direct sum Hilbert space** of $\mathscr{H}_1, \ldots, \mathscr{H}_N$.

Let A_j be a linear operator on $\mathscr{H}_j$. Then the **direct sum operator** $\oplus_{j=1}^{N} A_j$ of $A_1, \ldots, A_N$ on $\oplus_{j=1}^{N} \mathscr{H}_j$ is defined as follows:

$$D(\oplus_{j=1}^{N} A_j) := \hat{\oplus}_{j=1}^{N} D(A_j) = \{(\Psi_1, \ldots, \Psi_N) | \Psi_j \in D(A_j), \ j = 1, \ldots, N\}$$
$$\text{(the algebraic direct sum of } D(A_j), \ j = 1, \ldots, N),$$
$$(\oplus_{j=1}^{N} A_j)\Psi := (A_1\Psi_1, \ldots, A_N\Psi_N), \quad \Psi \in D(\oplus_{j=1}^{N} A_j).$$

© Springer Nature Singapore Pte Ltd. 2020

A. Arai, *Inequivalent Representations of Canonical Commutation and Anti-Commutation Relations*, Mathematical Physics Studies,
https://doi.org/10.1007/978-981-15-2180-5

With regard to spectral properties of $\oplus_{j=1}^N A_j$, the following theorem holds:

Theorem C.1 *Let* $A := \oplus_{j=1}^N A_j$. *Then:*

(i) *For all* $z \in \mathbb{C}$, $\ker(A - z) = \oplus_{j=1}^N \ker(A_j - z)$.
(ii) $\sigma(A) = \cup_{j=1}^N \sigma(A_j)$.
(iii) $\sigma_{\mathrm{p}}(A) = \cup_{j=1}^N \sigma_{\mathrm{p}}(A_j)$.

Proof

(i) We have

$$A - z = \oplus_{j=1}^N (A_j - z), \tag{C.1}$$

which implies the desired result.

(ii) The statement of this part is equivalent to that $\rho(A) = \cap_{j=1}^N \rho(A_j)$. Hence we prove this. Let $z \in \rho(A)$. Then $A - z$ is injective and $(A - z)^{-1}$ is bounded. Hence, by (i), each $A_j - z$ is injective. The boundedness of $(A - z)^{-1}$ implies that, for a constant $C > 0$, $\|(A - z)\Psi\| \geq C\|\Psi\|$, $\Psi \in D(A)$. Note that

$$\|(A - z)\Psi\|^2 = \sum_{j=1}^N \|(A_j - z)\Psi_j\|^2, \quad \Psi \in D(A). \tag{C.2}$$

Fixing $j = 1, \ldots, N$ and taking $\Psi \in D(A)$ as $\Psi_k = \delta_{jk}\eta$ $(k = 1, \ldots, N)$ with $\eta \in D(A_j)$ being arbitrary, we obtain $\|(A_j - z)\eta\| \geq C\|\eta\|$. Hence $(A_j - z)^{-1}$ is bounded. Moreover, Ran $(A - z)$ is dense in $\oplus_{j=1}^N \mathscr{H}_j$. By this property and (C.1), for each $j = 1, \ldots, N$, Ran $(A_j - z)$ is dense in $\mathscr{H}_j$. Hence $z \in \rho(A_j)$. Thus $\rho(A) \subset \cap_{j=1}^N \rho(A_j)$.

Conversely, let $z \in \cap_{j=1}^N \rho(A_j)$. Then each $A_j - z$ is injective with $(A_j - z)^{-1}$ being bounded and Ran $(A_j - z)$ is dense in $\mathscr{H}_j$. The former implies that $A - z$ is injective with $(A - z)^{-1}$ being bounded (use (C.2)), while the latter implies that Ran $(A - z)$ is dense in $\oplus_{j=1}^N \mathscr{H}_j$. Hence $z \in \rho(A)$. Thus $\cap_{j=1}^N \rho(A_j) \subset \rho(A)$.

(iii) This easily follows from (i). ■

Let $\mathscr{M}$ be a closed subspace of a Hilbert space $\mathscr{H}$ and $P_{\mathscr{M}}$ be the orthogonal projection onto $\mathscr{M}$. Then, by the projection theorem, for any $\Psi \in \mathscr{H}$, there exists a unique pair $(\Psi_{\mathscr{M}}, \Psi_{\mathscr{M}^\perp}) \in \mathscr{M} \oplus \mathscr{M}^\perp$ such that

$$\Psi = \Psi_{\mathscr{M}} + \Psi_{\mathscr{M}^\perp}.$$

It is shown that the correspondence $\Psi \mapsto (\Psi_{\mathscr{M}}, \Psi_{\mathscr{M}^\perp})$ defines a unitary transformation from $\mathscr{H}$ to $\mathscr{M} \oplus \mathscr{M}^\perp$. In this sense, we identify $\mathscr{H}$ with $\mathscr{M} \oplus \mathscr{M}^\perp$ and write

$$\mathscr{H} = \mathscr{M} \oplus \mathscr{M}^\perp.$$

Let $A \in \mathfrak{L}(\mathscr{H})$ be reduced by $\mathscr{M}$. Then it is easy to see that

$$A = A_{\mathscr{M}} \oplus A_{\mathscr{M}^{\perp}},$$

where $A_{\mathscr{M}}$ is the reduced part of A to $\mathscr{M}$.

C.2 Infinite Direct Sums of Hilbert Spaces and Operators

The concept of the direct sum of a finite number of Hilbert spaces is extended in natural way to the case of an infinite number of Hilbert spaces. Let $\mathscr{H}_n$, $n \in \mathbb{Z}_+$ be Hilbert spaces. Then

$$\oplus_{n=0}^{\infty}\mathscr{H}_n := \left\{ \Psi = \{\Psi^{(n)}\}_{n=0}^{\infty} |, \Psi^{(n)} \in \mathscr{H}_n,\ n \geq 0, \sum_{n=0}^{\infty} \|\Psi^{(n)}\|^2 < \infty \right\}$$

is a Hilbert space with inner product

$$\langle \Psi, \Phi \rangle := \sum_{n=0}^{\infty} \left\langle \Phi^{(n)}, \Psi^{(n)} \right\rangle_{\mathscr{H}_n}, \qquad \Psi, \Phi \in \oplus_{n=0}^{\infty}\mathscr{H}_n.$$

The Hilbert space $\oplus_{n=0}^{\infty}\mathscr{H}_n$ is called the **infinite direct sum Hilbert space** of $\{\mathscr{H}_n\}_{n=0}^{\infty}$.

For a subspace $\mathscr{D}_n$ of $\mathscr{H}_n$ ($n \in \mathbb{N}$), one can define a subspace of $\oplus_{n=0}^{\infty}\mathscr{H}_n$:

$$\hat{\oplus}_{n=0}^{\infty}\mathscr{D}_n := \{\Psi = \{\Psi^{(n)}\}_{n=0}^{\infty} | \ \Psi^{(n)} \in \mathscr{D}_n,\ n \geq 0, \exists n_0 \text{ such that } \Psi^{(n)} = 0,\ \forall n \geq n_0\}.$$

This subspace is called the **algebraic infinite direct sum** of $\{\mathscr{D}_n\}_{n=0}^{\infty}$. It is easy to see that, if $\mathscr{D}_n$ is dense in $\mathscr{H}_n$ for all $n \geq 1$, then $\hat{\oplus}_{n=0}^{\infty}\mathscr{D}_n$ is dense in $\oplus_{n=0}^{\infty}\mathscr{H}_n$.

Let A_n be a linear operator on $\mathscr{H}_n$. Then one can define a linear operator $\oplus_{n=0}^{\infty}A_n$ on $\oplus_{n=0}^{\infty}\mathscr{H}_n$ as follows:

$$D(\oplus_{n=0}^{\infty}A_n) := \left\{ \Psi \in \oplus_{n=0}^{\infty}\mathscr{H}_n \middle| \Psi^{(n)} \in D(A_n),\ n \geq 0, \sum_{n=0}^{\infty} \|A_n\Psi^{(n)}\|^2 < \infty \right\},$$

$$(\oplus_{n=0}^{\infty}A_n)\Psi := \{A_n\Psi^{(n)}\}_{n=0}^{\infty}, \quad \Psi \in D(\oplus_{n=0}^{\infty}A_n).$$

We call the operator $\oplus_{n=0}^{\infty}A_n$ the direct sum of $\{A_n\}_{n=0}^{\infty}$.

Theorem C.2 *Let each A_n be self-adjoint and $A := \oplus_{n=0}^{\infty}A_n$ as above. Then:*

(i) *The operator A is self-adjoint. Moreover, for all $t \in \mathbb{R}$,*

$$e^{itA} = \oplus_{n=0}^{\infty}e^{itA_n}. \tag{C.3}$$

(ii) *If each A_n is non-negative, then A is non-negative. Moreover, for all $\beta > 0$, $e^{-\beta A}$ is bounded and*

$$e^{-\beta A} = \oplus_{n=0}^{\infty} e^{-\beta A_n}. \tag{C.4}$$

(iii) $\sigma(A) = \overline{\cup_{n=0}^{\infty}\sigma(A_n)}$.
(iv) $\sigma_{\mathrm{p}}(A) = \cup_{n=0}^{\infty}\sigma_{\mathrm{p}}(A_n)$.

Proof We set

$$\mathscr{H} := \oplus_{n=0}^{\infty}\mathscr{H}_n. \tag{C.5}$$

(i) Let $\Phi \in \mathscr{H}$. Then, for each $n \in \mathbb{N}$, by the self-adjointness of A_n, there exists a vector $\Psi^{(n)} \in D(A_n)$ such that $(A_n + i)\Psi^{(n)} = \Phi^{(n)}$. We have $\|\Phi^{(n)}\|^2 = \|A_n\Psi^{(n)}\|^2 + \|\Psi^{(n)}\|^2$. Since $\sum_{n=0}^{\infty}\|\Phi^{(n)}\|^2 = \|\Phi\|^2 < \infty$, it follows that $\Psi := \{\Psi^{(n)}\}_{n=0}^{\infty}$ is in $D(A)$ and $(A+i)\Psi = \Phi$. Hence $\mathrm{Ran}\,(A+i) = \mathscr{H}$. Similarly, one can show that $\mathrm{Ran}\,(A-i) = \mathscr{H}$. Thus, by a general criterion, A is self-adjoint.

It is obvious that A is reduced by each $\mathscr{H}_n$ identified with $\{\Psi \in \mathscr{H} | \Psi^{(k)} = 0, k \neq n\}$ and its reduced part is A_n identified with $0 \oplus \cdots \oplus A_n \oplus 0 \cdots$. Hence, by Proposition 1.16(v), e^{itA} is reduced by each $\mathscr{H}_n$ and its reduced part is e^{itA_n}. Thus (C.3) holds.

(ii) For all $\Psi \in D(A)$,

$$\langle \Psi, A\Psi\rangle = \sum_{n=0}^{\infty}\left\langle \Psi^{(n)}, A_n\Psi^{(n)}\right\rangle.$$

Since each $\left\langle \Psi^{(n)}, A_n\Psi^{(n)}\right\rangle$ is non-negative, $\langle \Psi, A\Psi\rangle \geq 0$. Formula (C.4) can be proved in the same way as in the case of (C.3).

(iii) Let $z \in \rho(A)$. Then $A - z$ is bijective. This implies that each $A_n - z$ is bijective. Hence $z \in \rho(A_n)$. Therefore $\rho(A) \subset \cup_{n=0}^{\infty}\rho(A_n)$. Thus $\cup_{n=0}^{\infty}\sigma(A_n) \subset \sigma(A)$. Taking the closure of both sides, we obtain $\overline{\cup_{n=0}^{\infty}\sigma(A_n)} \subset \sigma(A)$.

Let $\lambda \in \sigma(A) \setminus \overline{\cup_{n=0}^{\infty}\sigma(A_n)}$. Then there exists a constant $\delta > 0$ such that

$$\inf_{n\geq 1}\inf_{\mu\in\sigma(A_n)}|\lambda - \mu| \geq \delta.$$

By functional calculus for A_n, we have $\|(A_n - \lambda)f\| \geq \delta\|f\|$, $f \in D(A_n) \cdots (*)$. Since $\lambda \in \rho(A_n)$, $A_n - \lambda$ is bijective. Hence, for each $\Phi \in \mathscr{H}$ and $n \in \mathbb{N}$, there exists a vector $\Psi^{(n)} \in D(A_n)$ such that $(A_n - \lambda)\Psi^{(n)} = \Phi^{(n)}$. By $(*)$, we have

$$\|\Phi^{(n)}\|^2 \geq \delta^2\|\Psi^{(n)}\|^2.$$

Hence

$$\sum_{n=0}^{\infty} \|\Psi^{(n)}\|^2 \le \frac{1}{\delta^2} \sum_{n=0}^{\infty} \|\Phi^{(n)}\|^2 < \infty.$$

Therefore $\Psi \in \mathscr{H}$. Moreover,

$$\sum_{n=0}^{\infty} \|A_n \Psi^{(n)}\|^2 \le 2 \sum_{n=0}^{\infty} \|(A_n - \lambda)\Psi^{(n)}\|^2 + 2\lambda^2 \sum_{n=0}^{\infty} \|\Psi^{(n)}\|^2 < \infty.$$

Hence $\Psi \in D(A)$ and $(A - \lambda)\Psi = \Phi$. Thus $\lambda \in \rho(A)$, implying that $\sigma(A) = \overline{\cup_{n=0}^{\infty} \sigma(A_n)}$.

(iv) Let $\lambda \in \sigma_{\rm p}(A)$. Then there exists a non-zero vector $\Psi \in D(A)$ such that $A\Psi = \lambda\Psi$. Hence there exists an $n_0 \in \mathbb{N}$ such that $\Psi^{(n_0)} \neq 0$ and $A_{n_0}\Psi^{(n_0)} = \lambda\Psi^{(n_0)}$. Hence $\lambda \in \sigma_{\rm p}(A_n)$. Thus $\sigma(A) \subset \cup_{n=0}^{\infty} \sigma_{\rm p}(A_n)$.

Conversely, let $\lambda \in \cup_{n=0}^{\infty} \sigma_{\rm p}(A_n)$. Then there exists an $n_0 \in \mathbb{N}$ such that $\lambda \in \sigma_{\rm p}(A_{n_0})$. Hence there exists a non-zero vector $f \in D(A_{n_0})$ and $A_{n_0} f = \lambda f$. Let $\Psi \in \mathscr{H}$ be such that $\Psi^{(n)} = 0$, $n \neq n_0$ and $\Psi^{(n)} = f$. Then $\Psi \in D(A) \setminus \{0\}$ and $A\Psi = \lambda\Psi$. Hence $\lambda \in \sigma_{\rm p}(A)$. Thus $\cup_{n=0}^{\infty} \sigma_{\rm p}(A_n) \subset \sigma_{\rm p}(A)$. ∎

Remark C.1 For more detailed spectral properties of infinite direct sum operators, see [28, Theorem 4.2].

C.3 A Theorem on Essential Self-adjointness

Let $\mathscr{H}$ be the Hilbert space given by (C.5). Then it is easy to show that the subspace

$$\mathscr{D}_0 := \{\Psi \in \mathscr{H} | \exists n_0 \text{ such that } \Psi^{(n)} = 0 \text{ for all } n \ge n_0 \}$$

is dense in $\mathscr{H}$. Let $\hat{N}$ be the direct sum of operators $nI_{\mathscr{H}_n}$ ($I_{\mathscr{X}}$ is the identity operator on $\mathscr{X}$):

$$\hat{N} := \oplus_{n=0}^{\infty} n I_{\mathscr{H}_n}.$$

Then, by Theorem C.2, $\hat{N}$ is a non-negative self-adjoint operator with

$$\sigma(\hat{N}) = \sigma_{\rm p}(\hat{N}) = \mathbb{Z}_+.$$

Let A be a self-adjoint operator on $\mathscr{H}$ which is reduced by each $\mathscr{H}_n$ ($n \in \mathbb{Z}_+$). Then $D(A) \cap \mathscr{H}_n$ is dense in $\mathscr{H}_n$ and hence

$$\mathscr{D} := D(A) \cap \mathscr{D}_0$$

is dense in $\mathscr{H}$.

Let B be a symmetric operator on $\mathscr{H}$ satisfying

$$\mathscr{D}_0 \subset D(B).$$

Then $\mathscr{D} \subset D(A) \cap D(B)$ and

$$(A + B) \restriction \mathscr{D}$$

is a symmetric operator. We assume the following:

(i) There exists a constant $c > 0$ such that

$$\|B\Psi\| \le c\|(\hat{N} + 1)^2\Psi\|, \quad \Psi \in \mathscr{D}.$$

(ii) There exists an integer $p \in \mathbb{Z}_+$ such that, for all $m, n \in \mathbb{Z}_+$ satisfying $|m - n| > p$,

$$\left\langle \Psi^{(m)}, B\Psi^{(n)} \right\rangle = 0.$$

Theorem C.3 *Let A and B be as above. Suppose that A + B is bounded from below on $\mathscr{D}$. Then A + B is essentially self-adjoint on $\mathscr{D}$.*

For proofs of this theorem, see [8, 96].

Remark C.2 In [8], a generalized version of Theorem C.3 is proved.

Appendix D
Spectra of a Self-adjoint Operator

Let A be a self-adjoint operator on a Hilbert space $\mathscr{H}$ and E_A be its spectral measure. For any vector $\Psi \in \mathscr{H}$, the mapping $\mu_\Psi \colon \boldsymbol{B}^1 \to [0, \infty)$ ($\boldsymbol{B}^1$ is the Borel field of $\mathbb{R}$) defined by

$$\mu_\Psi(B) := \|E_A(B)\Psi\|^2, \quad B \in \boldsymbol{B}^1$$

is a bounded measure on $(\mathbb{R}, \boldsymbol{B}^1)$.

A vector $\Psi \in \mathscr{H}$ is said to be **absolutely continuous** with respect to A if the measure μ_Ψ is absolutely continuous with respect to the Lebesgue measure $\mu_{\mathrm{L}}^{(1)}$ on $\mathbb{R}$.

In what follows, we write

$$|B| := \mu_L^{(1)}(B), \quad B \in \boldsymbol{B}^1. \tag{D.1}$$

If μ_Ψ is singular with respect to μ_{L}, then Ψ is said to be **singular** with respect to A.

We denote by $\mathscr{H}_{\mathrm{ac}}(A)$ (resp. $\mathscr{H}_{\mathrm{s}}(A)$) the set of vectors absolutely continuous (resp. singular) with respect to A. It is easy to see that $\mathscr{H}_{\mathrm{ac}}(A)$ and $\mathscr{H}_{\mathrm{s}}(A)$ are subspaces of $\mathscr{H}$. The subspace $\mathscr{H}_{\mathrm{ac}}(A)$ (resp. $\mathscr{H}_{\mathrm{s}}(A)$) is called the **absolutely continuous** (resp. **singular**) **subspace** of A.

Theorem D.1 *The subspaces $\mathscr{H}_{\mathrm{ac}}(A)$ and $\mathscr{H}_{\mathrm{s}}(A)$ are mutually orthogonal closed subspaces and*

$$\mathscr{H} = \mathscr{H}_{\mathrm{ac}}(A) \oplus \mathscr{H}_{\mathrm{s}}(A).$$

Moreover, A is reduced by $\mathscr{H}_{\mathrm{ac}}(A)$ and $\mathscr{H}_{\mathrm{s}}(A)$.

Proof See, e.g., [98, p. 518, Theorem 1.5]. ■

© Springer Nature Singapore Pte Ltd. 2020

A. Arai, *Inequivalent Representations of Canonical Commutation and Anti-Commutation Relations*, Mathematical Physics Studies,
https://doi.org/10.1007/978-981-15-2180-5

We define

$$\mathscr{H}_{\mathrm{p}}(A) := \mathrm{Ran}\,(E_A(\sigma_{\mathrm{p}}(A))),$$

where $\sigma_{\mathrm{p}}(A)$ is the point spectrum of A (see Sect. 1.1.3). It is obvious that $\mathscr{H}_{\mathrm{p}}(A)$ is closed. It follows that $\mathscr{H}_{\mathrm{p}}(A)$ is equal to the closure of the subspace spanned by all eigenvectors of A if $\sigma_{\mathrm{p}}(A) \neq \emptyset$.

Lemma D.1 $\mathscr{H}_{\mathrm{p}}(A) \subset \mathscr{H}_{\mathrm{s}}(A)$.

By Lemma D.1, one can define

$$\mathscr{H}_{\mathrm{sc}}(A) := \mathscr{H}_{\mathrm{p}}(A)^{\perp} \cap \mathscr{H}_{\mathrm{s}}(A) \subset \mathscr{H}_{\mathrm{s}}(A).$$

This closed subspace is called the **singular continuous subspace** of A.

By Theorem D.1, one has the orthogonal decomposition

$$\mathscr{H} = \mathscr{H}_{\mathrm{ac}}(A) \oplus \mathscr{H}_{\mathrm{sc}}(A) \oplus \mathscr{H}_{\mathrm{p}}(A) \tag{D.2}$$

and each closed subspace $\mathscr{H}_{\#}(A)$ ($\#$ =ac, sc, p) reduces A. We denote by $A_{\#}$ the reduced part of A to $\mathscr{H}_{\#}(A)$. Hence A is written as the direct sum of them:

$$A = A_{\mathrm{ac}} \oplus A_{\mathrm{sc}} \oplus A_{\mathrm{p}}. \tag{D.3}$$

The operator A_{ac} (resp. A_{sc}, A_{p}) is called the **absolutely continuous part** (resp. **singular continuous part, point-spectral part**) of A. Thus, introducing

$$\sigma_{\mathrm{ac}}(A) := \sigma(A_{\mathrm{ac}}), \quad \sigma_{\mathrm{sc}}(A) := \sigma(A_{\mathrm{sc}}), \quad \sigma_{\mathrm{pp}}(A) := \sigma(A_{\mathrm{p}}),$$

we have by Theorem C.1

$$\sigma(A) = \sigma_{\mathrm{ac}}(A) \cup \sigma_{\mathrm{sc}}(A) \cup \sigma_{\mathrm{pp}}(A).$$

The set $\sigma_{\mathrm{ac}}(A)$ (resp. $\sigma_{\mathrm{sc}}(A)$, $\sigma_{\mathrm{pp}}(A)$) is called the **absolutely continuous spectrum** (resp. **singular continuous spectrum**, **pure point spectrum**) of A. We remark that $\sigma_{\mathrm{ac}}(A)$, $\sigma_{\mathrm{sc}}(A)$ and $\sigma_{\mathrm{pp}}(A)$ are not necessarily mutually disjoint.

We remark also that

$$\sigma_{\mathrm{pp}}(A) = \overline{\sigma_{\mathrm{p}}(A)}.$$

Note that $\sigma_{\mathrm{p}}(A)$ is not necessarily closed, but $\sigma_{\mathrm{pp}}(A)$ is closed.

If $\mathscr{H} = \mathscr{H}_{\mathrm{ac}}(A)$, then A is said to be **purely absolutely continuous** or simply **absolutely continuous**; in this case, A is said to have a purely absolutely continuous spectrum.

If $\mathscr{H} = \mathscr{H}_{\mathrm{sc}}(A)$, then A is said to be **purely singular continuous** or simply **singular continuous**; in this case, A is said to have a purely singular continuous spectrum.

If $\mathscr{H} = \mathscr{H}_{\mathrm{p}}(A)$, then A is said to be **purely point spectral** or to have a purely point spectrum.

As is well known, in the context of quantum theory where $-A$ is a Hamiltonian, the quantity $\langle \Phi, e^{itA}\Psi \rangle$ with unit vectors $\Phi, \Psi \in \mathscr{H}$ represents the probability amplitude of a state Ψ at time 0 to a state Φ at time $t \in \mathbb{R}$.[1]

The absolute continuity of A implies the decay in time t of this quantity:

Theorem D.2 *Suppose that A is absolutely continuous. Then, for all $\Psi, \Phi \in \mathscr{H}$,*

$$\lim_{t \to \pm\infty} \left\langle \Phi, e^{itA}\Psi \right\rangle = 0. \tag{D.4}$$

Proof By the polarization identity, it is sufficient to prove (D.4) in the case where $\Phi = \Psi$. By functional calculus for A, we have

$$\left\langle \Psi, e^{itA}\Psi \right\rangle = \int_{\mathbb{R}} e^{it\lambda} d\|E_A(\lambda)\Psi\|^2.$$

By the absolute continuity of A, there exists a non-negative function $f \in L^1(\mathbb{R})$ such that

$$\|E_A(B)\Psi\|^2 = \int_B f(\lambda)d\lambda, \quad B \in \boldsymbol{B}^1.$$

Hence

$$\left\langle \Psi, e^{itA}\Psi \right\rangle = \int_{\mathbb{R}} e^{it\lambda} f(\lambda)d\lambda.$$

Then, by the Riemann–Lebesgue lemma, the right hand side converges to 0 as $t \to \pm\infty$. ■

The following proposition is useful for showing that A is absolutely continuous.

Proposition D.1 *Suppose that there exists a dense subspace $\mathscr{D}$ in $\mathscr{H}$ such that, for all $\Psi \in \mathscr{D}$, the measure $\langle \Psi, E_A(\cdot)\Psi \rangle$ is absolutely continuous. Then A is absolutely continuous.*

Proof Let $\Psi \in \mathscr{H}$. Then there exists a sequence $\{\Psi_n\}_{n=0}^{\infty}$ in $\mathscr{D}$ (i.e., $\Psi_n \in \mathscr{D}$, $\forall n \in \mathbb{N}$) such that $\Psi_n \to \Psi$ $(n \to \infty)$. Let $B \in \boldsymbol{B}^1$ be such that $|B| = 0$. Then, by

[1] The quantity $|\langle \Phi, e^{itA}\Psi \rangle|^2$ represents the transition probability of a state Ψ at time 0 to a state Φ at time t, i.e., the probability that Φ is found at time t under the condition that the initial state (time-zero state) is Ψ.

the present assumption, $\langle \Psi_n, E_A(B)\Psi_n \rangle = 0$, $n \in \mathbb{N}$. Taking the limit $n \to \infty$, we obtain $\langle \Psi, E_A(B)\Psi \rangle = 0$. Hence $\langle \Psi, E_A(\cdot)\Psi \rangle$ is absolutely continuous. ∎

Proposition D.2 *Suppose that A is absolutely continuous. Then, for any unitary operator $U: \mathcal{H} \to \mathcal{K}$ (a Hilbert space), UAU^{-1} is absolutely continuous.*

Proof The spectral measure $E_{UAU^{-1}}$ of UAU^{-1} is given by

$$E_{UAU^{-1}}(B) = UE_A(B)U^{-1}, \; B \in \boldsymbol{B}^1.$$

Let $|B| = 0$. Then, by the absolute continuity of A, $E_A(B) = 0$. Hence $E_{UAU^{-1}}(B) = 0$. Thus UAU^{-1} is absolutely continuous. ∎

By abuse of notation, for each $n \in \mathbb{N}$, the n-dimensional Lebesgue measure of $B \in \boldsymbol{B}^n$ is denoted by $|B|$. For a subset $M \in \boldsymbol{B}^d$ with $|M| > 0$, we define

$$\boldsymbol{B}_M := \{B \cap M | B \in \boldsymbol{B}^d\}.$$

The following proposition gives a sufficient condition for a multiplication operator on $L^2(M)$ to be absolutely continuous.

Proposition D.3 *Let $M \in \boldsymbol{B}^d$ with $|M| > 0$ and F be a real-valued function on M a.e. finite. Suppose that, for all $B \subset \boldsymbol{B}_M$ with $|B| = 0$ and $f \in L^2(M)$, $\int_{F^{-1}(B)} |f(\boldsymbol{x})|^2 d\boldsymbol{x} = 0$. Then the multiplication operator M_F on $L^2(M)$ (which is self-adjoint by Theorem A.1(v)) is absolutely continuous and*

$$\sigma(M_F) = \sigma_{\text{ac}}(M_F) = \text{ess.ran}(F), \quad \sigma_{\text{sc}}(M_F) = \emptyset, \quad \sigma_{\text{p}}(M_F) = \emptyset. \tag{D.5}$$

Proof As we have seen in Example B.1, the spectral measure of M_F is given by $E(B) = \chi_{F^{-1}(B)}$, $B \in \boldsymbol{B}^1$ (the multiplication operator by $\chi_{F^{-1}(B)}$). Hence, for all $f \in L^2(M)$,

$$\|E(B)f\|^2 = \int_{F^{-1}(B)} |f(\boldsymbol{x})|^2 d\boldsymbol{x}.$$

If $|B| = 0$, then the right hand side vanishes by the present assumption and hence $\|E(B)f\|^2 = 0$. Thus M_F is absolutely continuous and hence (D.5) holds. ∎

Example D.1 Let $\hat{q}_j$ be the jth position operator in Sect. 2.6.1. Consider the function $F_j: \mathbb{R}^d \to \mathbb{R}$, $F_j(\boldsymbol{x}) := x_j$, $\boldsymbol{x} \in \mathbb{R}^d$. Then $\hat{q}_j = M_{F_j}$. It is obvious that F_j is continuous on $\mathbb{R}^d$ and $F_j^{-1}(B) = \{\boldsymbol{x} \in \mathbb{R}^d | x_j \in B\} = \mathbb{R}^{j-1} \times B \times \mathbb{R}^{d-j}$. Hence, for all $f \in L^2(\mathbb{R}^d)$ and $B \in \boldsymbol{B}^1$ with $|B| = 0$,

$$\int_{F_j^{-1}(B)} |f(\boldsymbol{x})|^2 d\boldsymbol{x} = \int_B \left(\int_{\mathbb{R}^{j-1} \times \mathbb{R}^{d-j}} |f(\boldsymbol{x})|^2 dx_1 \cdots dx_{j-1} dx_{j+1} \cdots dx_d \right) dx_j = 0,$$

where we have used Fubini's theorem to rewrite the integral $\int_{F_j^{-1}(B)} |f(\boldsymbol{x})|^2 d\boldsymbol{x}$ as an iterated integral. Hence, by Proposition D.3, M_{F_j}, i.e., $\hat{q}_j$ is absolutely continuous and

$$\sigma(\hat{q}_j) = \sigma_{\mathrm{ac}}(\hat{q}_j) = \overline{F_j(\mathbb{R}^d)} = \mathbb{R}.$$
$$\sigma_{\mathrm{sc}}(\hat{q}_j) = \emptyset, \quad \sigma_{\mathrm{p}}(\hat{q}_j) = \emptyset.$$

Example D.2 Let $m > 0$. Then the multiplication operator $\hat{T}_m := \boldsymbol{k}^2/2m$ on $L^2(\hat{\mathbb{R}}^d)$ is absolutely continuous (consider the case where $F(\boldsymbol{k}) = \boldsymbol{k}^2/2m$, $\boldsymbol{k} \in \hat{\mathbb{R}}^d$) and

$$\sigma(\hat{T}_m) = \sigma_{\mathrm{ac}}(\hat{T}_m) = [0, \infty), \quad \sigma_{\mathrm{sc}}(\hat{T}_m) = \emptyset, \quad \sigma_{\mathrm{p}}(\hat{T}_m) = \emptyset.$$

Let $\mathscr{F}_d \colon L^2(\mathbb{R}^d) \to L^2(\hat{\mathbb{R}}^d)$ be the Fourier transform. Then

$$\mathscr{F}_d^{-1} \hat{T}_m \mathscr{F}_d = T_m := -\frac{1}{2m} \Delta,$$

where $\Delta := \sum_{j=1}^d D_j^2$ is the generalized Laplacian. Hence, by Proposition D.2, T_m is absolutely continuous and

$$\sigma(T_m) = \sigma_{\mathrm{ac}}(T_m) = [0, \infty), \quad \sigma_{\mathrm{sc}}(T_m) = \emptyset, \quad \sigma_{\mathrm{p}}(T_m) = \emptyset.$$

Example D.3 Let $m \geq 0$. Then the multiplication operator $R_m := \sqrt{\boldsymbol{k}^2 + m^2}$ on $L^2(\mathbb{R}_k^d)$ is absolutely continuous (consider the case where $F(\boldsymbol{k}) = \sqrt{\boldsymbol{k}^2 + m^2}$, $\boldsymbol{k} \in \mathbb{R}^d$) and

$$\sigma(R_m) = \sigma_{\mathrm{ac}}(R_m) = [m, \infty), \quad \sigma_{\mathrm{sc}}(R_m) = \emptyset, \quad \sigma_{\mathrm{p}}(R_m) = \emptyset.$$

We have

$$\sqrt{-\Delta + m^2} = \mathscr{F}_d^{-1} R_m \mathscr{F}_d.$$

Hence $\sqrt{-\Delta + m^2}$ is absolutely continuous and

$$\sigma(\sqrt{-\Delta + m^2}) = \sigma_{\mathrm{ac}}(\sqrt{-\Delta + m^2}) = [m, \infty),$$
$$\sigma_{\mathrm{sc}}(\sqrt{-\Delta + m^2}) = \emptyset, \quad \sigma_{\mathrm{p}}(\sqrt{-\Delta + m^2}) = \emptyset.$$

Bibliography

1. Acerbi, F. Morchio, G., Strocchi, F.: Theta vacua, charge confinement and charged sectors from nonregular representations of CCR algebras. Lett. Math. Phys. **27**, 1–11 (1993)
2. Acerbi, F. Morchio, G., Strocchi, F.: Infrared singular fields and nonregular representations of canonical commutation relation algebras. J. Math. Phys. **34**, 899–914 (1993)
3. Aharonov, Y., Bohm, D.: Significance of electromagnetic potentials in the quantum theory. Phys. Rev. **115**, 485–491 (1959)
4. Aharonov,Y., Bohm, D.: Time in the quantum theory and the uncertainty relation for time and energy. Phys. Rev. **122**, 1649–1658 (1961)
5. Amrein, W., Jauch, J.M., Sinha, K.B.: Scattering Theory in Quantum Mechanics. Benjamin Inc., Reading (1977)
6. Antoine, J.-P., Inoue, A., Trapani, C.: Partial$*$-Algebras and Their Operator Realizations. Kluwer Academic, Dordrecht (2002)
7. Arai, A.: Noninvertible Bogoliubov transformations and instability of embedded eigenvalues. J. Math. Phys. **32**, 1838–1846 (1991)
8. Arai, A.: A theorem on essential self-adjointness with application to Hamiltonians in nonrelativistic quantum field theory. J. Math. Phys. **32**, 2082–2088 (1991)
9. Arai, A.: Momentum operators with gauge potentials, local quantization of magnetic flux, and representation of canonical commutation relations. J. Math. Phys. **33**, 3374–3378 (1992)
10. Arai, A.: Gauge theory on a non-simply connected domain and representations of canonical commutation relations. J. Math. Phys. **36**, 2569–2580 (1995)
11. Arai, A.: Representation of canonical commutation relations in a gauge theory, the Aharonov-Bohm effect, and the Dirac-Weyl operator. J. Nonlinear Math. Phys. **2**, 247–262 (1995)
12. Arai, A.: Canonical commutation relations in a gauge theory, the Weierstrass Zeta-function, and infinite dimensional Hilbert space representations of the quantum group $U_q(sl_2)$. J. Math. Phys. **37**, 4203–4218 (1996)
13. Arai, A.: Representation-theoretic aspects of two-dimensional quantum systems in singular vector potentials: canonical commutation relations, quantum algebras, and reduction to lattice quantum systems. J. Math. Phys. **39**, 2476–2498 (1998)
14. Arai, A.: Ground state of the massless Nelson model without infrared cutoff in a non-Fock representation. Rev. Math. Phys. **13**, 1075–1094 (2001)
15. Arai, A.: Generalized weak Weyl relation and decay of quantum dynamics. Rev. Math. Phys. **17**, 1071–1109 (2005)
16. Arai, A.: Mathematical Aspects of Quantum Phenomena. Asakura-Shoten, Tokyo (2006, in Japanese)

© Springer Nature Singapore Pte Ltd. 2020

A. Arai, *Inequivalent Representations of Canonical Commutation and Anti-Commutation Relations*, Mathematical Physics Studies,
https://doi.org/10.1007/978-981-15-2180-5

17. Arai, A.: Spectrum of time operators. Lett. Math. Phys. **80**, 211–221 (2007)
18. Arai, A.: Mathematical Principles of Quantum Statistical Mechanics. Kyoritsu-Shuppan, Tokyo (2008, in Japanese)
19. Arai, A.: On the uniqueness of weak Weyl representations of the canonical commutation relation. Lett. Math. Phys. **85**, 15–25 (2008). Erratum: Lett. Math. Phys **89**, 287 (2009)
20. Arai, A.: Necessary and sufficient conditions for a Hamiltonian with discrete eigenvalues to have time operators. Lett. Math. Phys. **87**, 67–80 (2009)
21. Arai, A.: Functional Integral Methods in Quantum Mathematical Physics. Kyoritsu-Shuppan, Tokyo (2010, in Japanese)
22. Arai, A.: Hilbert space representations of generalized canonical commutation relations. J. Math. **2013**, 7 (2013). Article ID 308392
23. Arai, A.: Hilbert Space and Quantum Mechanics. Kyoritsu-Shuppan, Tokyo (1997, in Japanese). Revised and enlarged edition (2014)
24. Arai, A.: A family of inequivalent Weyl representations of canonical commutation relations with applications to quantum field theory. Rev. Math. Phys. **28**, 26, 1650007 (2016)
25. Arai, A.: Fock Spaces and Quantum Fields I. Nippon-Hyoron-Sha, Tokyo (2000, in Japanese). Revised and enlarged edition (2017)
26. Arai, A.: Fock Spaces and Quantum Fields II. Nippon-Hyoron-Sha, Tokyo (2000, in Japanese). Revised and enlarged edition (2017)
27. Arai, A.: Inequivalence of quantum Dirac fields of different masses and the underlying general structures involved. In: Functional Analysis and Operator Theory for Quantum Physics, EMS Series Congr. Report. European Mathematical Society, Zürich, pp. 31–53 (2017)
28. Arai, A.: Analysis on Fock Spaces and Mathematical Theory of Quantum Fields. World Scientific, Singapore (2018)
29. Arai, A.: Inequivalent representation of canonical commutation relations in relation to Casimir effect. Hokkaido University Preprint Series in Mathematics, vol. 1118 (2018). https://doi.org/10.14943/85920
30. Arai, A.: Singular Bogoliubov transformations and inequivalent representations of canonical commutation relations. Rev. Math. Phys. **31**, 20, 1950026 (2019)
31. Arai, A., Ezawa, H.: Mathematical Structures of Quantum Mechanics I. Asakura-Shoten, Tokyo (1999, in Japanese)
32. Arai, A., Ezawa, H.: Mathematical Structures of Quantum Mechanics II. Asakura-Shoten, Tokyo (1999, in Japanese)
33. Arai, A., Hiroshima, F.: Ultra-weak time operators of Schrödinger operators. Ann. Henri Poincaré **18**, 2995–3033 (2017)
34. Arai, A., Matsuzawa, Y.: Construction of a Weyl representation from a weak Weyl representation of the canonical commutation relation. Lett. Math, Phys. **83**, 201–211 (2008)
35. Arai, A., Matsuzawa, Y.: Time operators of a Hamiltonian with purely discrete spectrum. Rev. Math. Phys. **20**, 951–978 (2008)
36. Araki, H.: Hamiltonian formalism and the canonical commutation relations in quantum field theory. J. Math. Phys. **1**, 492–504 (1960)
37. Araki, H.: On representations of the canonical commutation relations. Commun. Math. Phys. **20**, 9–25 (1971)
38. Araki, H.: On quasifree states of CAR and Bogoliubov automorphisms. Publ. Res. Inst. Math. Sci. Kyoto Univ. **6**, 385–442 (1970/1971)
39. Araki, H.: Mathematical Theory of Quantum Fields. Iwanami-Shoten, Tokyo (1993, in Japanese). English edition: Oxford University Press, Oxford (1999)
40. Araki, H., Woods, E.J.: Representations of the canonical commutation relations describing a nonrelativistic infinite free Bose gas. J. Math. Phys. **4**, 637–662 (1963)
41. Araki, H., Wyss, W.: Representations of canonical anticommutation relations. Helv. Phys. Acta **37**, 136–159 (1964)
42. Bardeen, J., Cooper, L.N., Schrieffer, J.R.: Theory of superconductivity. Phys. Rev. **108**, 1175–1204 (1957)
43. Bloch, F., Nordsieck, A.: Note on the radiation field of the electron. Phys. Rev. **52**, 54–59 (1937)

44. Bogoliubov, N.: On the theory of superfluidity. J. Phys. (USSR) **11**, 23–32 (1947)
45. Bogoliubov, N.N., Logonov, A.A., Todorov, I.T.: Introduction to Axiomatic Quantum Field Theory. Benjamin, Reading (1975)
46. Born, M., Jordan, P.: Zur Quantenmechanik. Zeitschrift für Physik **34**, 858–888 (1925)
47. Born, M., Heisenberg, W., Jordan, P.: Zur Quantenmechanik II. Zeitschrift für Physik **35**, 557–615 (1926)
48. Bozejko, M., Speicher, R.: An example of a generalized Brownian motion. Commun. Math. Phys. **137**, 519–531 (1991)
49. Bracchi, L., Picasso, L.E.: Nonequivalent representations of canonical commutation relations in quantum mechanics: the case of the Aharonov-Bohm effect. Am. J. Phys. **75**, 268–271 (2007)
50. Bracchi, L., Picasso, L.E.: On the reducibility of the Weyl algebra for a semibounded space. Lett. Math. Phys. **89**, 277–285 (2009)
51. Bratteli, O., Robinson, D.W.: Operator Algebras and Quantum Statistical Mechanics, 2nd edn., vol. 2. Springer, Berlin (1997)
52. Bruneaua, L., Dereziński, J.: Bogoliubov Hamiltonians and one-parameter groups of Bogoliubov transformations. J. Math. Phys. **48**, 022101 (2007)
53. Brunetti, R., Fredenhagen, K.: Remarks on time–energy uncertainty relations. Rev. Math. Phys. **14**, 897–903 (2002)
54. Brunetti, R., Fredenhagen, K.: Time of occurrence observable in quantum mechanics. Phys. Rev. A **66**, 044101 (2002)
55. Brunetti, R., Fredenhagen, K., Hoge, M.: Time in quantum physics: from an external parameter to an intrinsic observable. Found. Phys. **40**, 1368–1378 (2010)
56. Busch, P.: The time–energy uncertainty relation. In: Muga, G., Mayato, R.S., Egusquiza, I. (eds.) Time in Quantum Mechanics, Chapter 3, 2nd edn., vol. 1. Springer, Berlin (2008)
57. Busch, P., Grabowski, M., Lahti, P.: Operational Quantum Physics. Springer, Berlin (1995)
58. Cook, J.M.: The mathematics of second quantization. Trans. Am. Math. Soc. **74**, 222–245 (1953)
59. Dereziński, J.: Van Hove Hamiltonians—exactly solvable models of the infrared and ultraviolet problem. Ann. Henri Poincaré **4**, 713–738 (2003)
60. Dereziński, J.: Quadratic Hamiltonians and their renormalization. In: Methods of Spectral Analysis in Mathematical Physics. Operator Theory: Advances and Applications, vol. 186, pp. 89–102. Birkhauser, Basel (2009)
61. Dereziński, J.: Bosonic quadratic Hamiltonians. J. Math. Phys. **58**, 121101 (2017)
62. Dereziński, J., Gérard, C.: Mathematics of Quantization and Quantum Fields. Cambridge University Press, Cambridge (2013)
63. Dimock, J.: Quantum Mechanics and Quantum Field Theory. Cambridge University Press, Cambridge (2011)
64. Dirac, P.A.M.: The quantum theory of the emission and absorption of radiation. Proc. R. Soc. London, Ser. A **114**, 243–265 (1927)
65. Dorfmeister, G., Dorfmeister, J.: Classification of certain pairs of operators (P, Q) satisfying $[P, Q] = -i\mathrm{Id}$. J. Funct. Anal. **57**, 301–328 (1984)
66. Dykema, K., Nica, A.: On the Fock representation of the q-commutation relations. J. Reine Angew. Math. **440**, 201–212 (1993)
67. Emch, G.: Algebraic Methods in Statistical Mechanics and Quantum Field Theory. Wiley, New York (1971)
68. Ezawa, H.: A note on the van Hove–Miyatake catastrophe. Prog. Theor. Phys. **30**, 545–549 (1963)
69. Ezawa, H.: The representation of canonical variables as the limit of infinite space volume: the case of the BCS model. J. Math. Phys. **5**, 1078–1090 (1964)
70. Ezawa, H.: Quantum mechanics of a many-boson system and the representation of canonical variables. J. Math. Phys. **6**, 380–404 (1965)
71. Ezawa, H.: Structures of quantum mechanics. In: Yukawa, H., Toyoda, T. (eds.) Japanese. Quantum Mechanics II, Part VI. Iwanami-Shoten, Tokyo (1978)

72. Ezawa, H., Arai, A.: Quantum Field Theory and Statistical Mechanics. Nippon-Hyoron-Sha, Tokyo (1988, in Japanese)
73. Fernández, R., Fröhlich, J., Socal, A.: Random Walks, Critical Phenomena, and Triviality in Quantum Field Theory. Springer, Berlin (1992)
74. Fröhlich, J.: On the infrared problem in a model of scalar electrons and massless scalar bosons. Ann. Inst. Henri Poincaré Sec. A **XIX**, 1–103 (1973)
75. Fuglede, B: On the relation $PQ - QP = -i\mathrm{I_d}$. Math. Scand. **20**, 79–88 (1967)
76. Galapon, E.A.: Self-adjoint time operator is the rule for discrete semi-bounded Hamiltonians. Proc. R. Soc. Lond. **A 458**, 2671–2689 (2002)
77. Gel'fand, I.M., Shilov, G.E.: Generalized Functions, vol. 1. Academic, New York (1964)
78. Gel'fand, I.M., Shilov, G.E.: Generalized Functions, vol. 2. Academic, New York (1968)
79. Gérard, C., Tiedra de Aldecoa, R.: Generalized definition of time delay in scattering theory. J. Math. Phys. **48**, 122101-1–122101-15 (2007)
80. Glimm, J., Jaffe, A.: Quantum Physics. Springer, New York (1981/1987)
81. Goldin, G.A., Menikoff, R., Sharp, D.H.: Representations of a local current algebra in nonsimply connected space and the Aharonov-Bohm effect. J. Math. Phys. **22**, 1664–1668 (1981)
82. Greenberg, O.W.: Haag's theorem and clothed operators. Phys. Rev. **115** 706–710 (1959)
83. Haag, R.: On quantum field theories. Danske Vid. Selsk. Mat.-Fys. Medd. **29**(12), 1–37 (1955)
84. Haag, R.: The mathematical structure of the Bardeen–Cooper–Schrieffer model. Nuovo Cimento **25**, 287–299 (1962)
85. Haag, R.: Local quantum physics. In: Second Revised and Enlarged Edition. Springer, Berlin (1996)
86. Heisenberg, W.: Über quantentheoretische Umdeutung kinematischer und mechanischer Beziehungen. Z. Phys. **33**, 879–893 (1925)
87. Heisenberg, W.: Über den anschaulichen Inhalt der quantentheoretischen Kinematik und Mechanik. Z. Phys. **43**, 172–198 (1927)
88. Hellström, L., Silvestrov, S.D.: Commuting Elements in q-Deformed Heisenberg Algebras. World Scientific, Singapore (2000)
89. Henley, E.M., Thirring, W.: Elementary Quantum Field Theory. McGraw–Hill Book Company, Inc., New York (1962)
90. Hirokawa, M.: Weyl's relation on a doubly connected space and Aharonov Bohm effect. Publ. Res. Inst. Math. Sci. **982**, 240–257 (1997)
91. Hirokawa, M.: Canonical quantization on a doubly connected space and the Aharonov-Bohm phase. J. Funct. Anal. **174**, 322–363 (2000)
92. Hiroshima, F., Kuribayashi, S., Matsuzawa, Y.: Strong time operator of generalized Hamiltonians. Lett. Math. Phys. **87**, 115–123 (2009)
93. Hiroshima, F., Sasaki, I., Spohn, H., Suzuki, A.: Enhanced binding in quantum field theory. COE Lecture Note, vol. 38. Institute of Mathematics for Industry, Kyushu University, Kyushu (2012)
94. Inoue, A.: Tomita–Takesaki Theory in Algebras of Unbounded Operators. Lecture Notes in Mathematics, vol. 1699. Springer, Berlin (1988)
95. Jaffe, A., Witten, E.: Quantum Yang-Mills Theory. http://www.claymath.org/sites/default/files/yangmills.pdf
96. Jørgensen, P.E.T: Essential self-adjointness of semibounded operators. Math. Ann. **237**, 187–192 (1978)
97. Jørgensen, P.E.T.: Operators and Representation Theory. Dover, New York (1988/2008)
98. Kato, T.: Perturbation Theory for Linear Operators, 2nd edn. Springer, Berlin (1976)
99. Kevrekidis, P.G., Frantzeskakis, D.J., Carretero-González, R. (eds.) Emergent Nonlinear Phenomena in Bose-Einstein Condensates. Springer, Berlin (2008)
100. Kuroda, S.-T.: Spectral Theory II. Iwanami-Shoten, Tokyo (1979, in Japanese)
101. Kuroda, S.-T.: Functional Analysis. Kyoritsu-Shuppan, Tokyo (1980, in Japanese)
102. Kurose, H., Nakazato, H.: Geometric construction of $*$-representation of the Weyl algebra with degree 2. Publ. RIMS Kyoto Univ. **32**, 555–579 (1996)

103. Landau, L.D., Lifshitz, E.M.: Quantum Mechanics: Non-relativistic Theory, 3rd edn. revised and enlarged, Repr. with corrections. Butterworth-Heinemann, Oxford (1991)
104. Lax, P.D.: Functional Analysis. Wiley-Interscience, New York (2002)
105. Lörenczi, J., Hiroshima, F., Betz, V.: Feynman–Kac-Type Theorems and Gibbs Measures on Path Space. De Gruyter, Berlin (2011)
106. Lörinczi, J., Minlos, R.A., Spohn, H.: The infrared behaviour in Nelson's model of a quantum particle coupled to a massless scalar field. Ann. Henri Poincaré **3**, 269–295 (2002)
107. Lörinzi, J., Minlos, R.A., Spohn, H.: Infrared regular representation of the three-dimensional massless Nelson model. Lett. Math. Phys. **59**, 189–198 (2002)
108. Lupher, T.: Who proved Haag's theorem? Int. J. Theor. Phys. **44** 1995–2005 (2005)
109. Martin, Ph. A, Rothen, F.: Many-Body Problems and Quantum Field Theory. 2nd edn. Springer, Berlin (2002/2004)
110. Miyamoto, M.: A generalized Weyl relation approach to the time operator and its connection to the survival probability. J. Math. Phys. **42**, 1038–1052 (2001)
111. Miyamoto, M.: The various power decays of the survival probability at long times for free quantum particle. J. Phys. A **35**, 7159–7171 (2002)
112. Miyatake, O.: On the non-existence of solution of field equations in quantum mechanics. J. Inst. Polytech. Osaka City Univ. Ser. A Math. **2**, 89–99 (1952)
113. Miyatake, O.: On the singularity of the perturbation-term in the field quantum mechanics. J. Inst. Polytech. Osaka City Univ. Ser. A Math. **3**, 145–155 (1952)
114. Miyatake, O., Kato, Y.: Eigenvalue Problems. Maki-Shoten, Tokyo (1971, in Japanese)
115. Muga, G., Mayato, R.S., Egusquiza, I. (eds.): Time in Quantum Mechanics, 2nd edn., vol. 1. Springer, Berlin (2008)
116. Muga, G., Mayato, R.S., Egusquiza, I. (eds.): Time in Quantum Mechanics, vol. 2. Springer, Berlin (2009)
117. Mugibayashi, N., Kato, Y.: Regular perturbation and asymptotic limits of operators in fixed-source theory. Prog. Theor. Phys. **3**, 300–310 (1964)
118. Nakatsugawa, K., Fujii, T., Saxena, A., Tanda, S.: Time operators and time crystals: self-adjointness by topology change. arXiv:1711.10179v2 [quant-ph] 24 Jun 2019. J. Phys. A: Math. Theor. 29 August 2019. https://iopscience.iop.org/article/10.1088/1751-8121/ab3f46
119. Nam, P.T., Napiórkowski, M., Solovej, J.P.: Diagonalization of bosonic quadratic Hamiltonians by Bogoliubov transformations. J. Funct. Anal. **270**, 4340–4368 (2016)
120. Nelson, E.: Analytic vectors. Ann. Math. **70**, 572–615 (1959)
121. Nelson, E.: Interaction of nonrelativistic particles with a quantized scalar field. J. Math. Phys. **5**, 1190–1197 (1964)
122. Obata, N.: White Noise Calculus and Fock Spaces. Lecture Notes in Mathematics, vol. 1577. Springer, Berlin (1994)
123. Ottesen, J.T.: Infinite Dimensional Groups and Algebras in Quantum Physics. Springer, Berlin (1995)
124. Palmer, J.: Scattering automorphisms of the Dirac field. J. Math. Anal. Appl. **64**, 189–215 (1978)
125. Pauli, W.: General Principles of Quantum Mechanics. Springer, Berlin (1980)
126. Petz, D.: An invitation to the algebra of canonical commutation relations. In: Leuven Notes in Mathematical and Theoretical Physics. Leuven University, Leuven (1990)
127. Pfeifer, P., Fröhlich, J.: Generalized time–energy uncertainty relations and bounds on lifetimes of resonances. Rev. Mod. Phys. **67**, 759–779 (1995)
128. Powers, R.T.: Self-adjoint algebras of unbounded operators. Commun. Math. Phys. **21**, 85–124 (1971)
129. Putnam, C.R.: Commutation Properties of Hilbert Space Operators and Related Topics. Springer, Berlin (1967)
130. Reed, M., Simon, B: Methods of Modern Mathematical Physics I: Functional Analysis. Academic, New York (1972). Revised and enlarged edition (1980)
131. Reed, M., Simon, B: Methods of Modern Mathematical Physics II: Fourier Analysis and Self-adjointness. Academic, New York (1975)

132. Reed, M., Simon, B.: Methods of Modern Mathematical Physics III: Scattering Theory. Academic, New York (1979)
133. Reeh, H.: A remark concerning canonical commutation relations. J. Math. Phys. **29**, 1535–1536 (1988)
134. Richard, S., Tiedra de Aldecoa, R.: On a new formula relating localisation operators to time operators. In: Spectral Analysis of Quantum Hamiltonians. Operator Theory: Advances and Applications, vol. 224, pp. 301–338. Birkhäuser/Springer Basel AG, Basel (2012)
135. Ruijsenaars, S.N.M.: On Bogoliubov transformations for systems of relativistic charged particles. J. Math. Phys. **18**, 517–526 (1977)
136. Ruijsenaars, S.N.M.: Charged particles in external fields. II. The quantized Dirac and Klein-Gordon theories. Commun. Math. Phys. **52**, 267–294 (1977)
137. Ruijsenaars, S.N.M.: On Bogoliubov transformations. II. The general case. Ann. Phys. **116**, 105–134 (1978)
138. Ruijsenaars, S.N.M.: The Aharonov-Bohm effect and scattering theory. Ann. Phys. **146**, 1–34 (1983)
139. Sasaki, I.: Ground state of the massless Nelson model in a non-Fock representation. J. Math. Phys. **46**, 102107, 12 (2005)
140. Schmüdgen, K.: On the Heisenberg commutation relation. I. J. Funct. Anal. **50**, 8–49 (1983)
141. Schmüdgen, K.: On the Heisenberg commutation relation. II. Publ. RIMS, Kyoto Univ. **19**, 601–671 (1983)
142. Schmüdgen, K.: Unbounded operator algebras and representation theory. In: Operator Theory: Advances and Applications, vol. 37. Birkhauser, Basel (1990)
143. Schmüdgen, K.: Unbounded Self-adjoint Operators on Hilbert Space. Springer, Dordrecht (2012)
144. Schrödinger, E.: Collected Papers on Wave Mechanics. 2nd English edn. Chelsea, New York (1978)
145. Schwartz, L.: Méthode Mathématiques Pour Les Sciences Physique. Hermann, Paris (1961). Japanese Edition: Methods of Physical Mathematics. Iwanami-Shoten, Tokyo (1966)
146. Shale, D.: Linear symmetries of free boson fields. Trans. Am. Math. Soc. **103**, 149–167 (1962)
147. Simon, B.: The $P(\phi)_2$ Euclidean Quantum Field Theory. Princeton University Press, Princeton (1974)
148. Simon, B.: Functional Integration and Quantum Physics. Academic, New York (1979)
149. Streater, R.F., Wightman, A.S.: PCT, Spin and Statistics, and All That. Benjamin, Reading (1964)
150. Strocchi, F.: Symmetry Breaking. Lecture Notes in Physics, vol. 643. Springer, Berlin (2005)
151. Strocchi, F.: An introduction to the mathematical structure of quantum mechanics. In: A Short Course for Mathematicians. Advanced Series in Mathematical Physics, 2nd edn., vol. 28. World Scientific, Singapore (2008)
152. Strocchi, F.: An Introduction to the Non-Perturbative Foundations of Quantum Field Theory. Oxford University, Oxford (2013)
153. Takahashi, Y.: Quantum Field Theory I. Baifukan, Tokyo (1974, in Japanese)
154. Takano, F.: Many-Body Problems. Baifukan, Tokyo (1975, in Japanese)
155. Teranishi, N: A note on time operators. Lett. Math. Phys. **106**, 1259–1263 (2016)
156. Thaller, B.: The Dirac Equation. Springer, Berlin (1992)
157. Tonomura, A.: The Quantum World Unveiled by Electron Waves. World Scientific, Singapore (1997)
158. van Hove, L.: Les difficulté de divergence pour un modéle particulier de champ quantifié. Physica **18**, 145–159 (1952)
159. Vogels, J.M., Xu, K., Raman, C., Abo-Shaeer, J.R., Ketterle, W.: Experimental observation of the Bogoliubov transformation for a Bose–Einstein condensed gas. Phys. Rev. Lett. **88**, 060402-1–060402-4 (2002)
160. von Neumann, J.: Die Eindeutigkeit der Schrödingerschen Operatoren. Math. Ann. **104**, 570–578 (1931)
161. von Neumann, J.: Mathematische Grundlagen der Quantenmechanik. Springer, Berlin (1932)

162. Wada, K.: Spectrum of N-body Stark Hamiltonians—Application of generalized Weyl relations (2017). Preprint
163. Wigner, E.P.: On the time–energy uncertainty relation. In: Salam, A., Wigner, E.P. (eds.) Aspects of Quantum Theory, pp. 237–247. Cambridge University, Cambridge (1972)
164. Yamanouchi, T., Sugiura, M.: Introduction to Continuous Groups. Baifukan, Tokyo (1960, in Japanese)
165. Yosida, K.: Functional Analysis. Springer, Berlin (1965). Reprint of 1980 edition is available (1995)
166. Yosida, K., Ito, S. (eds.) Functional Analysis and Differential Equations. Iwanami-Shoten, Tokyo (1976, Japanese)

Index

© Springer Nature Singapore Pte Ltd. 2020
A. Arai, *Inequivalent Representations of Canonical Commutation and Anti-Commutation Relations*, Mathematical Physics Studies, https://doi.org/10.1007/978-981-15-2180-5

GPSR Compliance
The European Union's (EU) General Product Safety Regulation (GPSR) is a set of rules that requires consumer products to be safe and our obligations to ensure this.

If you have any concerns about our products, you can contact us on

ProductSafety@springernature.com

In case Publisher is established outside the EU, the EU authorized representative is:

Springer Nature Customer Service Center GmbH
Europaplatz 3
69115 Heidelberg, Germany

www.ingramcontent.com/pod-product-compliance
Ingram Content Group UK Ltd.
Pitfield, Milton Keynes, MK11 3LW, UK
UKHW021900190726
13853UKWH00003B/1359

* 9 7 8 9 8 1 1 5 2 1 8 2 9 *